Lecture Notes in Computer Science 3832

Commenced Publication in 1973
Founding and Former Series Editors:
Gerhard Goos, Juris Hartmanis, and Jan van Leeuwen

AF790090

David Zhang Anil K. Jain (Eds.)

Advances in Biometrics

International Conference, ICB 2006
Hong Kong, China, January 5-7, 2006
Proceedings

Springer

Volume Editors

David Zhang
The Hong Kong Polytechnic University, Department of Computing
Hung Hom, Kowloon, Hong Kong, China
E-mail: csdzhang@comp.polyu.edu.hk

Anil K. Jain
Michigan State University, Department of Computer Science and Engineering
3115 Engineering Building, East Lansing, MI 48824-1226, USA
E-mail: jain@cse.msu.edu

Library of Congress Control Number: 2005937781

CR Subject Classification (1998): I.5, I.4, K.4.1, K.4.4, K.6.5, J.1

ISSN 0302-9743
ISBN-10 3-540-31111-4 Springer Berlin Heidelberg New York
ISBN-13 978-3-540-31111-9 Springer Berlin Heidelberg New York

Springer is a part of Springer Science+Business Media

springer.com

Printed in Germany

Typesetting: Camera-ready by author, data conversion by Scientific Publishing Services, Chennai, India
Printed on acid-free paper SPIN: 11608288 06/3142 5 4 3 2 1 0

Preface

Biometrics has emerged as a reliable person identification method that can overcome some of the limitations of the traditional automatic personal identification methods. With significant advances in biometric technology and a corresponding increase in the number of applications incorporating biometrics, it is essential that we bring together researchers from academia and industry as well as practitioners to share ideas, problems and solutions for the development and successful deployment of state-of-the-art biometric systems.

The International Conference on Biometrics (ICB 2006) followed the successful International Conference on Biometric Authentication (ICBA 2004) to facilitate this interaction. ICB 2006 received a large number of high-quality research papers. After a careful review of 192 submissions, 104 papers were accepted for presentation. In addition to these technical presentations, the results of the Face Authentication Competition (FAC 2006) were also announced. This conference provided a forum for practitioners to discuss their experiences in applying the state-of-the-art biometric technologies that will further stimulate research in biometrics.

We are grateful to Vijayakumar Bhagavatula, Norihiro Hagita, and Behnam Bavarian for accepting our invitation to give keynote talks at ICB 2006. In addition, we would like to express our gratitude to all the contributors, reviewers, Program Committee members and Organizing Committee members whose efforts made ICB 2006 a very successful conference. We also wish to acknowledge the International Association of Pattern Recognition (IAPR), the Hong Kong Polytechnic University, Motorola, Omron, NSFC and Springer for sponsoring this conference. Special thanks are due to Josef Kittler, Tieniu Tan, Jane You, Michael Wong, Jian Yang and Zhenhua Guo for their support, advice and hard work in various aspects of conference organization.

We hope that the fruitful technical interactions made possible by this conference benefited research and development efforts in biometrics.

October 2005

David Zhang
Anil K. Jain

Organization

General Chairs

Anil K. Jain (Michigan State University, USA)
Roland Chin (Hong Kong University of Science and Technology, Hong Kong, China)

Program Chairs

David Zhang (Hong Kong Polytechnic University, Hong Kong, China)
Jim Wayman (San Jose State University, USA)
Tieniu Tan (Chinese Academy of Sciences, China)
Joseph P. Campbell (MIT Lincoln Lab., USA)

Competition Coordinators

Josef Kittler (University of Surrey, UK)
James Liu (Hong Kong Polytechnic University, Hong Kong, China)

Exhibition Coordinators

Stan Li (Chinese Academy of Sciences, China)
Kenneth K.M. Lam (Hong Kong Polytechnic University, Hong Kong, China)

Local Arrangements Chairs

Jane You (Hong Kong Polytechnic University, Hong Kong, China)
Yiu Sang Moon (Chinese University of Hong Kong, Hong Kong, China)

Tutorial Chair

George Baciu (Hong Kong Polytechnic University, Hong Kong, China)

Publicity Chairs

Arun Ross (West Virginia University, USA)
Davide Maltoni (University of Bologna, Italy)
Yunhong Wang (Beihang University, China)

Program Committee

Mohamed Abdel-Mottaleb (University of Miami, USA)
Simon Baker (Carnegie Mellon University, USA)
Samy Bengio (IDIAP, Switzerland)
Bir Bhanu (University of California, USA)
Prabir Bhattacharya (Concordia University, Canada)
Josef Bigun (Halmstad University and Chalmers University of Technology, Sweden)
Horst Bunke (Institute of Computer Science and Applied Mathematics, Switzerland)
Raffaele Cappelli (University of Bologna, Italy)
Keith Chan (Hong Kong Polytechnic University, Hong Kong, China)
Ke Chen (University of Manchester, UK)
Xilin Chen (Harbin Institute of Technology, China)
Gerard Chollet (ENST, France)
Sarat Dass (Michigan State University, USA)
John Daugman (Cambridge University, UK)
Bernadette Dorizzi (INT, France)
Patrick Flynn (Notre Dame University, USA)
Sadaoki Furui (Tokyo Institute of Technology, Japan)
Wen Gao (Chinese Academy of Sciences, China)
Patrick Grother (NIST, USA)
Larry Heck (Nuance, USA)
Javier Hernando (UPC, Spain)
Lawrence A. Hornak (West Virginia University, USA)
Wen Hsing Hsu (National Tsing Hua University, Taiwan)
Behrooz Kamgar-Parsi (Naval Research Lab., USA)
Jaihie Kim (Yonsei University, Korea)
Alex Kot (Nanyang Technological University, Singapore)
Ajay Kumar (IIT Delhi, India)
Kin Man Lam (Hong Kong Polytechnic University, Hong Kong, China)
Shihong Lao (Omron Corporation, Japan)
Seong-Whan Lee (Korea University, Korea)
Lee Luan Ling (State University of Campinas, Brazil)
Zhiqiang Liu (City University of Hong Kong, Hong Kong, China)
John S. Mason (Swansea University, UK)
Tsutomu Matsumoto (Yokohama National University, Japan)
Jiri Navratil (IBM, USA)
Mark Nixon (University of Southampton, UK)
Sharath Pankanti (IBM, USA)
Jonathon Philips (NIST, USA)
Ioannis Pitas (Thessaloniki University, Greece)
Salil Prabhakar (DigitalPersona Inc., USA)
Nalini Ratha (IBM, USA)
James Reisman (Siemens Corporate Research, USA)

Table of Contents

Face Verification Contest 2006

Face

Fingerprint

Iris

Speech and Signature

Biometric Fusion and Performance Evaluation

Gait and Keystroke

Others

Performance Characterisation of Face Recognition Algorithms and Their Sensitivity to Severe Illumination Changes

Kieron Messer[1], Josef Kittler[1], James Short[1], G. Heusch[2], Fabien Cardinaux[2], Sebastien Marcel[2], Yann Rodriguez[2], Shiguang Shan[3], Y. Su[3], Wen Gao[3], and X. Chen[3]

[1] University of Surrey, Guildford, Surrey, GU2 7XH, UK
[2] Dalle Molle Institute for Perceptual Artificial Intelligence, CP 592, rue du Simplon 4, 1920 Martigny, Switzerland
[3] Institute of Computing Technology, Chinese Academy of Sciences, China

Abstract. This paper details the results of a face verification competition [2] held in conjunction with the Second International Conference on Biometric Authentication. The contest was held on the publically available XM2VTS database [4] according to a defined protocol [15]. The aim of the competition was to assess the advances made in face recognition since 2003 and to measure the sensitivity of the tested algorithms to severe changes in illumination conditions. In total, more than 10 algorithms submitted by three groups were compared[1]. The results show that the relative performance of some algorithms is dependent on training conditions (data, protocol) as well as environmental changes.

1 Introduction

Over the last decade the development of biometric technologies has been greatly promoted by an important research technique instrument, namely comparative algorithm performance characterisation via competitions. Typical examples are the NIST evaluation campaign in voice based speaker recognition from telephone speech recordings, finger print competition, and face recognition and verification competitions. The main benefit of such competitions is that they allow different algorithms to be evaluated on the same data, using the same protocol. This makes the results comparable to much greater extent than in the case of an unorchestrated algorithm evaluation designed by individual researchers, using their own protocols and data, where direct comparison of the reported methods can be difficult because tests are performed on different data with large variations in test and model database sizes, sensors, viewing conditions, illumination and background. Typically, it is unclear which methods are the best and for which scenarios they should be used. The use of common datasets along with evaluation protocols can help alleviate this problem.

[1] This project was supported by EU Network of Excellence Biosecure.

D. Zhang and A.K. Jain (Eds.): ICB 2006, LNCS 3832, pp. 1–11, 2005.

In face recognition, the two main series of competitions has been run by NIST and the University of Surrey [13, 8, 14] respectively. For the purpose of the excersise, NIST collected a face database, known as FERET. A protocol for face identification and face verification [17] has been defined for the FERET database. However, only a development set of images from the database are released to researchers. The remaining are sequestered by the organisers to allow independent testing of the algorithms. To date three evaluations have taken place, the last one in the year 2000, and an account of these, together with the main findings can be found in [16].

More recently, two Face Recognition Vendor Tests [3] have been carried out, the first in 2000 and the second in 2002. The tests are done under supervision and have time restrictions placed on how quickly the algorithms should compute the results. They are aimed more at independently testing the performance of commercially available systems, however academic institutions are also able to take part. In the more recent test 10 commercial systems were evaluated. The FERET and FRVT have recently evolved in a new initiative known as Face Recognition Grand Challenge which is promoting research activities both in 2D and 3D face recognition.

The series of competitions organised by the University of Surrey commenced in the year 2000. It was initiated by the European Project M2VTS which focused on the development of multimodal biometric personal identity authentication systems. As part of the project a large database of talking faces was recorded. For a subset of the data, referred to as the XM2VTS database, two experimental protocols, known as Lausanne Protocol I and II, were defined to enable a cooperative development of face and speaker verification algorithms by the consortium of research teams involved in the project. The idea was to open this joint development and evaluation of biometric algorithms to wider participation.

In the year 2000 a competition on the XM2VTS database using the Lausanne protocol [15] was organised [13]. As part of AVBPA 2003 a second competition on exactly the same data and testing protocol was organised [8]. All the data from the XM2VTS database is available from [4]. We believe that this open approach increases, in the long term, the number of algorithms that will be tested on the XM2VTS database. Each research institution is able to assess their algorithmic performance at any time.

The competition was subsequently extended to a new database, known as the BANCA database [5] which was recorded as part of a follow up EU project, BANCA. The database was captured under 3 different realistic and challenging operating scenarios. Several protocols have also been defined which specify which data should be used for training and testing. Again this database is being made available to the research community through [1]. The first competition on the BANCA database was held in 2004 and the results reported in [14].

In this paper, the competition focuses once again on XM2VTS data with two objectives. First of all it is of interest to measure the progress made in face verification since 2003. The other was to gauge the sensitivity of face verification algorithms to severe changes to illumination conditions. This test was carried out on a section of the XM2VTS database containing face images acquired in side lighting. As with the previous competition, the current event was held under the auspices of EU Project Biosecure.

The rest of this paper is organised as follows. In the next section the competition rules and performance criterion are described. Section 3 gives an overview of each algorithm which entered the competition and in the following section the results are detailed. Finally, some conclusions are drawn in Section 4.

2 The Competition

All experiments were carried out using images acquired from the XM2VTS database on the standard and darkened image sets. The XM2VTS database can be aquired through the web-page given by [4].

There were two separate parts to the competition.

Part I: Standard Test. The XM2VTS database contains images of 295 subjects, captured over 4 sessions in a controlled environment. The database uses a standard protocol. The Lausanne protocol splits the database randomly into training, evaluation and test groups [15]. The training group contains 200 subjects as clients, the evaluation group additional 25 subjects as impostors and the testing group another 70 subjects as impostors.
There are two testing configurations of the XM2VTS database. In the first configuration, the client images for training and evaluation, were collected from each of the first three sessions. In the second configuration, the client images for training were collected from the first two sessions and the client images for evaluation from the third.

Part II: Darkened Images. In addition to the controlled images, the XM2VTS database contains a set of images with varying illumination. Each subject has four more images with lighting predominantly from one side; two have been lit from the left and two from the right.

To assess the algorithmic performance the *False Rejection Rate* P_{FR} and *False Acceptance Rate* P_{FA} are typically used. These two measures are directly related, i.e. decreasing the false rejection rate will increase the number of false acceptances. The point at which $P_{FR} = P_{FA}$ is known as the EER (Equal Error Rate).

Fig. 1. Example images from XM2VTS database

Fig. 2. Example images from dark set of XM2VTS database

3 Overview of Algorithms

In this section the algorithms that participated in the contest are summarised.

3.1 Dalle Molle Institute for Perceptual Artificial Intelligence (IDIAP)

IDIAP proposed three different classifiers, used with two distinct preprocessing steps, resulting in a total of six complete face authentication systems. The preprocessing steps aims to enhance the image or to reduce the effect of illumination changes. The first preprocessing step we used is simple histogram equalization whereas the second one is the illumination normalization model first proposed by Gross & Brajovic [10] and described in details in [9].

The first two classification systems (called GMM and HMM) are based on local features and statistical models while the third one (called PCA-LDA) uses discriminant holistic features with a distance metric.

IDIAP-GMM. The GMM based system uses DCT-mod2 features [18] and models faces using Gaussian Mixture Models (GMMs) [6]. In DCTmod2 feature extraction, each given face is analyzed on a block by block basis: from each block a subset of Discrete Cosine Transform (DCT) coefficients is obtained; coefficients which are most affected by illumination direction changes are replaced with their respective horizontal and vertical deltas, computed as differences between coefficients from neighboring blocks. A GMM is trained for each client in the database. To circumvent the problem of small amount of client training data, parameters are obtained via Maximum a Posteriori (MAP) adaptation of a generic face GMM: the generic face GMM is trained using Maximum Likelihood training with faces from all clients. A score for a given face is found by taking the difference between the log-likelihood of the face belonging to the claimed identity (estimated with the client specific GMM) and log-likelihood of the face belonging to an impostor (estimated with the generic face GMM). A global threshold is used in making the final verification decision.

IDIAP-HMM. The HMM based system uses DCT features and models faces using Hidden Markov Models (HMMs). Here, we use a simple DCT Feature extraction: each given face is analyzed on a block by block basis; from each block, DCT coefficients are obtained; the first fifeteen coefficients compose the feature vector corresponding to the block. A special topology of HMM is used to model the client faces which allows the use of local features. The HMM represents a face as a sequence of horizontal strips from the forehead to the chin. The emission probabilities of the HMM are estimated with mixture of gaussians modeling the set of blocks that composes a strip. A further description of this model is given in [7]. A HMM is trained for each client in the database using MAP adaptation. A score for a given face is found by taking the difference between the log-likelihood of the face belonging to the claimed identity (estimated with the client specific GMM) and log-likelihood of the face belonging to an impostor (estimated with the generic face GMM). A global threshold is used in making the final verification decision.

IDIAP-PCA/LDA. Principal component analysis (PCA) is first applied on the data so as to achieve decorrelation and dimensionality reduction. The projected face images into the coordinate system of eigenvectors (Eigenfaces) are then used as features to derive the optimal projection in the Fisher's linear discriminant sense (LDA) [12]. Considering a set of N images $\{x_1, x_2, ..., x_N\}$, an image x_k is linearly projected to obtain the feature vector y_k:

$$y_k = W^T x_k \qquad k = 1, 2, ..., N$$

where $W^T = W_{lda}^T W_{pca}^T$. Finally, classification is performed using a metric: considering two feature vectors, a template y_t and a sample y_s, their correlation is computed according to:

$$1 - \frac{< y_t, y_s >}{||y_t|| \cdot ||y_s||}$$

3.2 Chinese Academy of Sciences

The adopted method, Gabor Feature based Multiple Classifier Combination (CAS-GFMCC), is an ensemble learning classifier based on the manipulation of Gabor features with multiple scales and orientations. The basic procedure of CAS-GFMCC is described as follows: First, face images are aligned geometrically and normalized photometrically by using region-based histogram equalization. Then, Gabor filters with 5 scales and 8 orientations are convolved with the normalized image and the magnitude of the transform results are kept for further processing. These high dimensional Gabor features, with a dimension of 40 times of the original normalized face images, are then adaptively divided into multiple groups. For each feature group, one classifier is learnt through Fisher discriminant analysis, which will result in an ensemble of classifier. These classifiers are then combined using a fusion strategy. In addition, face image re-lighting techniques are exploited to adapt the method for more robustness to the face images with complex illumination (named by CAS-GFMCC-L). For automatic evaluation case, AdaBoost-based methods are exploited for both the localization of the face and facial landmarks (the two eyes). Please refer to http://www.jdl.ac.cn/project/faceId/index_en.htm for more details of our methods.

3.3 University of Surrey (UniS)

Two algorithms have been tested using the competition protocol. The first algorithm (Unis-Components) applies client-specific linear discriminant analysis to a number of components of the face image. Firstly, twelve sub-images are obtained. The images are found relative to the eye positions, so that no further landmarking is necessary. These images are of the face, both eyes, nose and mouth and of the left and right halves of each, respectively. All twelve images have the same number of pixels, so that the images of smaller components will effectively be of higher resolution. These components are then normalised using histogram equalisation. Client-specific linear discriminant analysis [11] is applied to these sub-images separately. The resulting scores for each of the components are fused using the sum rule.

The second algorithm (UniS-Lda) is based on the standard LDA. Each image is first photometrically normalised using filtering and histogram equalisation. The corrected images are then projected into an LDA space which has been designed by first reducing the dimensionality of the image representation space using PCA. The similarity of probe and template images is measured in the LDA space using normalised correlation. In contrast to the results reported in the AVBPA2003 competition, here the decision threshold is globally optimal rather than client specific. For the automatic registration of the probe images, an SVM based face detection and eye localisation algorithm was used. Exactly the same system was used in Part II of the competition, without any adjustmenst of the system parameters, including the decision threshold.

4 Results and Discussion

4.1 Part I

Most of the algorithm entries provide results for Part I of the competition with manually registered images, which is aimed at establishing a bench mark for the other parts. As there were so few entrants, the competition was used as a framework for comparative evaluation of different algorithms from two of the groups, rather than just the best performing entry. This offered an interesting insight into the effectiveness of different decision making schemes under the same photometric normalisation conditions, and the dependence of each decision making scheme on different photometric normalisation methods. Interestingly, the best combination of preprocessing and decision making methods investigated by IDIAP differed from one evaluation protocol to another.

In general the performance of the algorithms achieved under Protocol II was better. This is probably the consequence of more data being available for training and the evaluation data available for setting the operational thresholds being more representative, as it was recorded in a completely different session. The best performing algorithm was CAS, which also achieved the best results on the BANCA database in the previous competition. The CAS algorithm outperformed the winning algorithm on the XM2VTS database at the AVBPA03 competition [8].

Table 1. Error rates according to Lausanne protocol for configuration I with manual registration

	Evaluation Set			Test Set		
Method	FA	FR	TER	FA	FR	TER
ICPR2000-Best	-	-	5.00	2.30	2.50	4.80
AVBPA03-Best	1.16	1.05	2.21	0.97	0.50	1.47
IDIAP-HE/GMM	2.16	2.16	4.32	2.00	1.50	3.50
IDIAP-HE/HMM	2.48	2.50	4.98	2.57	1.50	4.07
IDIAP-HE/PCA/LDA	3.16	3.33	6.49	3.72	2.00	5.72
IDIAP-GROSS/GMM	2.20	2.17	4.37	2.32	2.00	4.32
IDIAP-GROSS/HMM	6.00	6.00	12.0	6.31	4.75	11.06
IDIAP-GROSS/PCA/LDA	5.96	6.00	11.96	7.04	4.50	11.54
UNIS-Components	5.50	5.50	11.00	4.44	3.50	7.94
UNIS-Lda	1.66	1.67	3.33	1.66	1.25	2.91
CAS	0.80	0.80	1.63	0.96	0.00	0.96

Table 2. Error rates according to Lausanne protocol for configuration II with manual registration

	Evaluation Set			Test Set		
Method	FA	FR	TER	FA	FR	TER
AVBPA03-Best	0.33	0.75	1.08	0.25	0.50	0.75
IDIAP-HE/GMM	1.00	1.00	2.00	0.04	4.75	4.79
IDIAP-HE/HMM	1.75	1.75	3.50	1.80	1.25	3.05
IDIAP-HE/PCA/LDA	1.64	1.75	3.39	1.86	3.25	5.11
IDIAP-GROSS/GMM	1.00	1.00	2.00	1.15	1.00	2.15
IDIAP-GROSS/HMM	5.25	5.25	10.50	5.13	3.25	8.38
IDIAP-GROSS/PCA/LDA	3.25	3.25	6.50	4.01	5.75	9.76
UNIS-Components	2.64	2.75	5.39	1.99	1.75	3.74
UNIS-Lda	1.00	1.00	2.00	1.26	0.00	1.26
CAS	0.24	0.25	0.49	0.26	0.25	0.51

Table 3. Error rates according to Lausanne protocol for configuration I with automatic registration in test phase

	Evaluation Set			Test Set		
Method	FA	FR	TER	FA	FR	TER
ICPR2000-Best	-	-	14.0	5.80	7.30	13.10
AVBPA03-Best	0.82	4.16	4.98	1.36	2.50	3.86
CAS	1.00	1.00	2.00	0.57	1.57	1.57

Only one of the algorithms, CAS, was also subjected to the test on automatically registered images. The automatic registration was accomplished with a CAS in house face detection and localisation method. By default, CAS is the winning entry. However, the achievement of the CAS method should not be

Table 4. Error rates according to Lausanne protocol for configuration II with auto registration in test phase

	Evaluation Set			Test Set		
Method	FA	FR	TER	FA	FR	TER
AVBPA03-Best	0.63	2.25	2.88	1.36	2.00	3.36
CAS	0.49	0.50	0.99	0.28	0.50	0.78

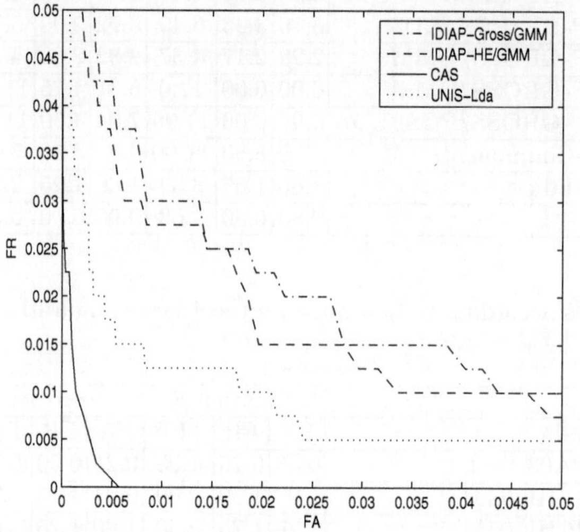

Fig. 3. ROC curves for configuration I with manual registration

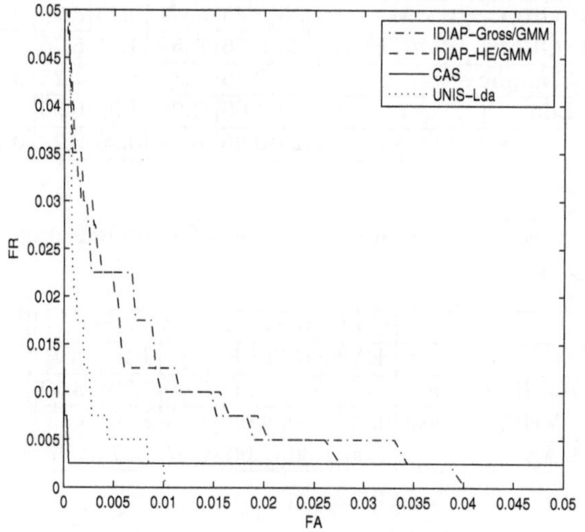

Fig. 4. ROC curves for configuration II with manual registration

underrated, as the overall performance shown in Table 3 and Table 4 is very impressive. The results show only a slight degradation, in comparison with the

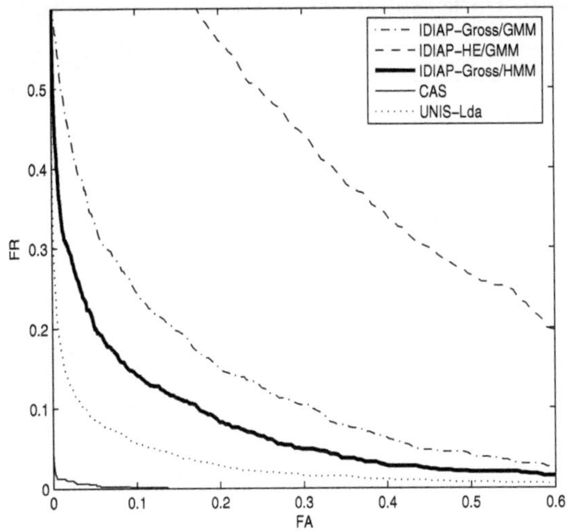

Fig. 5. ROC curves for the dark test set with manual registration

manually registered figures. More over, the results are a significant improvement over the previously best reported results.

Figures 3, 4, 5 provide the ROC curves for the better performing methods. It is interesting to note that if the operating points were selected aposteriori, then the performance of the algorithms would be even better. This suggests that if the evaluation data set was more extensive and therefore fully representative, the error rates could be reduced even further.

4.2 Part II

This part of the competition provided a useful insight into the sensitivity of the tested algorithms to severe changes in subject illumination. In some cases the performance degraded by an order of magnitude. Surprisingly, the error rates of some of the lower ranking methods, such as the Unis-Components and IDIAP LDA based procedures, deteriorated only by a factor of two. Again, the CAS approach achieved the best performance, which was an order of magnitude better than the second best algorithm. The comparability of the results was somewhat affected by the interesting idea of CAS to relight the training and evaluation set data to simulate the illumination conditions of the test set. This has no doubt limited the degree of degradation from good conditions to side lighting. However, it would have been interesting to see how well the system would perform on the original frontal lighting data sets. This would better indicate the algorithm sensitivity to changes in lighting conditions.

The CAS algorithm was the only entry in Part II, automatic registration category. Again the reported results are consistently excellent, demonstrating a high degree of robustness of the CAS system and the overall high level of performance.

Table 5. Darkened image set with manual registration

	Evaluation Set			Test Set		
Method	FA	FR	TER	FA	FR	TER
IDIAP-HE/GMM				6.20	77.37	88.68
IDIAP-HE/HMM				12.78	60.75	73.53
IDIAP-HE/PCA/LDA				2.41	29.50	31.91
IDIAP-GROSS/GMM				10.54	23.75	34.29
IDIAP-GROSS/HMM				8.14	15.86	24.00
IDIAP-GROSS/PCA/LDA				6.49	18.75	25.24
UNIS-Components				4.01	17.38	21.39
UNIS-Lda				17.88	0.98	18.86
CAS	1.18	1.17	2.35	0.77	1.25	2.02

Table 6. Darkened image set with automatic registration

	Evaluation Set			Test Set		
Method	FA	FR	TER	FA	FR	TER
CAS	1.18	1.17	2.35	1.25	1.63	2.88

5 Conclusions

The results of a face verification competition [2] held in conjunction with the Second International Conference on Biometric Authentication have been presented. The contest was held on the publically available XM2VTS database [4] according to a defined protocol [15]. The aim of the competition was to assess the advances made in face recognition since 2003 and to measure the sensitivity of the tested algorithms to severe changes in illumination conditions. In total, more than 10 algorithms submitted by three groups were compared. The results showed that the relative performance of some algorithms is dependent on training conditions(data, protocol). All algorithms were affected by environmental changes. The performance degraded by a factor of two or more.

References

1. *BANCA; http://www.ee.surrey.ac.uk/banca/.*
2. *BANCA; http://www.ee.surrey.ac.uk/banca/icba2004.*
3. *Face Recognition Vendor Tests; http://www.frvt.org.*
4. *The XM2VTSDB; http://www.ee.surrey.ac.uk/ Research/VSSP/xm2vtsdb/.*
5. E. Bailly-Bailliere, S. Bengio, F. Bimbot, M. Hamouz, J. Kittler, J. Mariethoz, J. Matas, K. Messer, V. Popovici, F. Poree, B. Ruiz, and J. P. Thiran. The BANCA database and evaluation protocol. In *Audio- and Video-Based Biometric Person Authentication: Proceedings of the 4th International Conference, AVBPA 2003*, volume 2688 of *Lecture Notes in Computer Science*, pages 625–638, Berlin, Germany, June 2003. Springer-Verlag.

6. F. Cardinaux, C. Sanderson, and S. Bengio. User authentication via adapted statistical models of face images. *To appear in IEEE Transactions on Signal Processing*, 2005.
7. Fabien Cardinaux. Local features and 1D-HMMs for fast and robust face authentication. Technical report, 2005.
8. Kieron Messer et al. Face verification competition on the xm2vts database. In *4th International Conference on Audio and Video Based Biometric Person Authentication*, pages 964–974, June 2003.
9. F. Cardinaux G. Heusch and S. Marcel. Efficient diffusion-based illumination normalization for face verification. Technical report, 2005.
10. R. Gross and V. Brajovic. An Image Preprocessing Algorithm for Illumination Invariant Face Recognition. In *International Conference on Audio- and Video-Based Biometric Person Authentication*, 2003.
11. J. Kittler, Y. P. Li, and J. Matas. Face verification using client specific fisher faces. *The Statistics of Directions, Shapes and Images (2000) 63–66.*
12. S. Marcel. A symmetric transformation for lda-based face verification. In *Proc. Int. Conf. Automatic Face and Gesture Recognition (AFGR), Seoul, Korea*, 2004.
13. J Matas, M Hamouz, K Jonsson, J Kittler, Y P Li, C Kotropoulos, A Tefas, I Pitas, T Tan, H Yan, F Smeraldi, J Bigun, N Capdevielle, W Gerstner, S Ben-Yacoub, Y Abdeljaoued, and E Mayoraz. Comparison and face verification results on the xm2vts database. In A Sanfeliu, J J Villanueva, M Vanrell, R Alquezar, J Crowley, and Y Shirai, editors, *Proceedings of International Conference on Pattern Recognition, Volume 4*, pages 858–863, 2000.
14. K Messer, J Kittler, M Sadeghi, M Hamouz, A Kostin, and et al. Face authentication test on the banca database. In J.Kittler, M Petrou, and M Nixon, editors, *Proc. 17th Intern. Conf. on Pattern Recognition*, volume IV, pages 523–529, Los Alamitos, CA, USA, August 2004. IEEE Computer Society Press.
15. K Messer, J Matas, J Kittler, J Luettin, and G Maitre. XM2VTSDB: The Extended M2VTS Database. In *Second International Conference on Audio and Video-based Biometric Person Authentication*, March 1999.
16. P. J. Phillips, H. Moon, P. Rauss, and S. A. Rizvi. The feret evaluation methodology for face-recognition algorithms. volume 22, pages 1090–1104, October 2000.
17. P.J. Phillips, H. Wechsler, J.Huang, and P.J. Rauss. The FERET database and evaluation procedure for face-recognition algorithm. *Image and Vision Computing*, 16:295–306, 1998.
18. C. Sanderson and K.K. Paliwal. Fast features for face authentication under illumination direction changes. *Pattern Recognition Letters*, 24(14):2409–2419, 2003.

Assessment of Blurring and Facial Expression Effects on Facial Image Recognition*

Mohamed Abdel-Mottaleb** and Mohammad H. Mahoor

Department of ECE, University of Miami,
1251 Memorial Drive, Coral Gables, FL 33146
mottaleb@miami.edu, mmahoor@umsis.miami.edu

Abstract. In this paper we present methods for assessing the quality of facial images, degraded by blurring and facial expressions, for recognition. To assess the blurring effect, we measure the level of blurriness in the facial images by statistical analysis in the Fourier domain. Based on this analysis, a function is proposed to predict the performance of face recognition on blurred images. To assess facial images with expressions, we use Gaussian Mixture Models (GMMs) to represent images that can be recognized with the Eigenface method, we refer to these images as "Good Quality", and images that cannot be recognized, we refer to these images as "Poor Quality". During testing, we classify a given image into one of the two classes. We use the FERET and Cohn-Kanade facial image databases to evaluate our algorithms for image quality assessment. The experimental results demonstrate that the prediction function for assessing the quality of blurred facial images is successful. In addition, our experiments show that our approach for assessing facial images with expressions is successful in predicting whether an image has a good quality or poor quality for recognition. Although the experiments in this paper are based on the Eigenface technique, the assessment methods can be extended to other face recognition algorithms.

Keywords: Face recognition, Image Quality Assessment, Facial expressions, Blurring Effect, Gaussian Mixture Model.

1 Introduction

Face recognition has become one of the most important applications of image analysis and computer vision in recent years. Nowadays, the use of face recognition systems for biometrics is considered by many governments for security in important buildings such as airports and military bases. The performance of biometric systems such as *fingerprint*, *face*, and *iris* recognition highly rely on the quality of the captured images. Thus, the demand for a preprocessing

* This work is supported in part through an award from the NSF Center for Identification Technology Research (CITeR).

** Corresponding author.

D. Zhang and A.K. Jain (Eds.): ICB 2006, LNCS 3832, pp. 12–18, 2005.

module to assess the quality of input images for the biometric systems is obvious. The quality measures of a captured image can then determine whether the image is acceptable for further processing by the biometric system, or another image needs to be captured. The importance of the facial image quality and its effects on the performance of the face recognition systems was also considered by Face Recognition Vendor Test (FRVT) protocols [1]. For example, FRVT2002 [2] consists of two tests: the High Computational Intensity (HCInt) test and the Medium Computational Intensity (MCInt) test. The HCInt test examines the effect of changing the size of the database on system performance. On the other hand, the MCInt measures the performances on different categories of images that include images with different effects such as changes in illumination, and pose variations.

In the literature, few researchers have addressed the performance of face recognition systems with lower quality images [3]. In [4], Draper *et al.* built two statistical models to examine how features of the human face could influence the performance of three different face recognition algorithms: principle components analysis (PCA), an interpersonal image difference classifier (IIDC), and an elastic bunch graph matching (EBGM) algorithm. They examined 11 features: race, gender, age, glasses uses, facial hair, bangs, mouth state, complexion, state of eyes, make up use, and facial expressions. Their study, based on two statistical models, showed that images with certain features are easier to recognize by certain methods. For example, subjects who close their eyes are easier to recognize using PCA than EBGM. Considering the results in their paper, it is obvious that there is a need for systems to assess the quality of facial images for face recognition.

In this paper, we develop novel algorithms for assessing the quality of facial images with respect to the effects of blurring and facial expressions. These algorithms can be used in developing a facial image quality assessment system (FIQAS) that works as a preprocessing module for any face recognition method. The idea of FIQAS is to assess the quality of facial images and either reject or accept them for the recognition step. We focus on assessing the effect of blurring and facial expressions on facial images. In order to develop the algorithms for assessing the quality of facial images, the challenge is to measure the level or the intensity[1] of the factors that affect the quality of the facial images. For example, a facial image could have an expression with intensity in a range starting from neutral to maximum. Obviously, the recognition of a facial image with exaggerated expressions is more difficult than the recognition of a facial image with a light expression. For blurring effect, measuring the level of blurriness is possible. On the other hand, measuring the intensity of face expression is difficult because of the absence of the reference neutral face image.

Considering the issues discussed above, we take two different strategies to assess the quality of facial images: one strategy for blurring effect and another strategy for facial expressions. For blurring effect, we develop a function for predicting the performance rate of the Eigenface recognition method on images

[1] In this paper, the word intensity is synonymous with the word level.

with different levels of blurriness. In case of facial expressions, where measuring the intensity of an expression is difficult, we classify the images into two different classes: "Good Quality" images, and "Poor Quality" images; and then model the images based on Gaussian Mixture Models (GMMs). The GMMs are trained using the Cohen-Kande face database, where the class assignment of the training images is based on whether the Eigenface method succeeds or fails in recognizing the face. The results are encouraging and can be easily extended to assess quality for other face recognition methods.

The rest of this paper is organized as follows: Section 2 introduces the algorithms for assessing the quality of facial images affected by blurring and facial expressions. Section 3 presents experimental results. Conclusions and the future works are discussed in Section 4.

2 Algorithms for Quality Assessment of Facial Images

We assume that the facial images do not have illumination problem. In fact, illumination is one of the important factors that could affect the performance of a face recognition system, but in this paper we assume that the images are only affected either by blurring or by facial expressions. Following, we will present our algorithms for assessing the facial images with respect to blurring and expressions.

2.1 Blurring Effect Assessment

To assess the quality of facial images with respect to blurring, we measure the intensity of blurriness. Based on this measure, we define a function to predict the recognition rate of the Eigenface method. An image with sharp edges and without blurring effects has more energy at the higher spatial frequencies of its Fourier transform than the lower spatial frequencies. In other words, an image with fine details and edges has flatter 2-D spatial frequency response than a blurred image.

There are different techniques to measure the energy of the high frequency content of an image. One technique is to analyze the image in the Fourier domain and calculate the energy of the high frequency content of the image by statistical analysis. One statistical measure that can be used for this purpose is the Kurtosis. In the following subsection, we review this measure and discuss the advantages and disadvantages of it. Then in the last subsection, we introduce the function that predicts the performance rate of face recognition on a given image based on the blurriness of the image.

Image Sharpness Measurement Using the Kurtosis. An elegant approach for image sharpness measurement is used in electron microscope [5]. This approach is based on the statistical analysis of the image using Fourier transform. Kurtosis is a measure of the departure of a probability distribution from Gaussian (normal) distribution. For a one dimensional random variable x with mean μ_x and statistical moments up to the fourth degree, the Kurtosis is defined by Kotz and Johnson [6]:

$$\kappa = m_4/m_2^2 \quad (1)$$

where m_4 and m_2 are the fourth and second moments respectively. For a normal distribution, the value of the $\kappa = 3$. Therefore, the value of κ can be compared with 3 to determine whether the distribution is "flat-topped" or "peaked" relative to a Gaussian. In other words, the smaller the value of the Kurtosis, the flatter the distribution. For a multi-dimensional random variable, Y, the Kurtosis is defined as:

$$\kappa = E[(Y - \mu_Y)^t \Sigma^{-1} (Y - \mu_Y)]^2 \quad (2)$$

where Σ is the covariance matrix and μ_Y is the mean vector.

In this work, we use the value of Kurtosis (Eq. 2) for predicting the face recognition rate. Our experiments show that this measure has a linear response within a wide range of blurring. In our experiments the facial images were blurred using a Gaussian mask with different values of the σ. The average value of the Kurtosis for facial images without blurring is 10 and it increases with larger values of σ.

Face Recognition Performance Prediction. Figure 1(a) shows the recognition rate of the Eigenface method versus the Kurtosis measure. The figure shows that the recognition rate decreases with larger values of the Kurtosis measure (higher blurriness). To assess the quality of an unknown face image degraded by blurring, we define a function that predicts the recognition rate of the Eigenface from the Kurtosis measure. This function is obtained by linear regression of the data in Figure 1(a):

$$R(\kappa) = R_{max} + a_1 * (\kappa - 10) + a_2 * (\kappa - 10)^2 \quad (3)$$

where R_{max} is the maximum recognition rate of the specific face recognition system (e.g. Eigenface in our work), and the parameters a_1 and a_2 can be determined by linear least mean square error regression. As shown in the experiments Section, this function is capable of predicting the recognition rate of the Eigenface method on images affected by blurring. The same procedure can be used to develop quality measures and prediction functions for other face recognition methods.

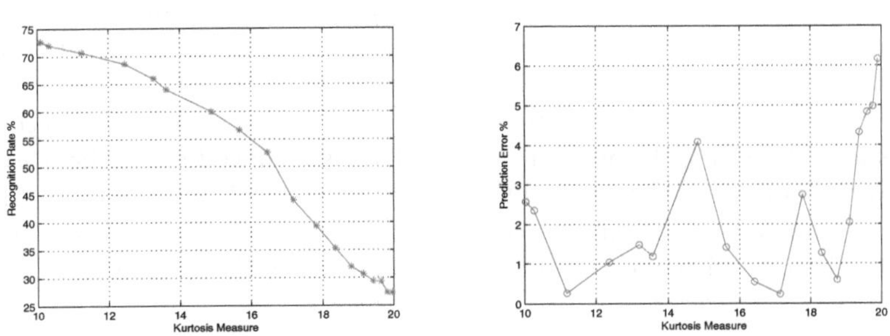

(a) (b)

Fig. 1. (a) Recognition rate of the Eigenface method versus Kurtosis measure. (b)Prediction error versus Kurtosis measure.

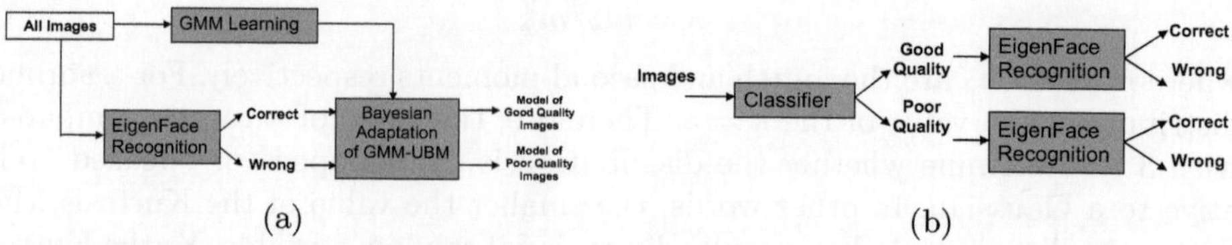

Fig. 2. System diagram for assessing the quality of facial images with expressions: (a) Training the GMM-UBM models, (b) Testing the models for classification

2.2 Facial Expression Effect Assessment

In facial expression analysis, the temporal dynamics and intensity of facial expressions can be measured by determining either the geometric deformation or the density of wrinkles that appear in certain regions of the face [7]. For example the degree of smiling is proportional to the magnitude of the cheek movement and the rise of the corners of the mouth. Since there are interpersonal variations with regard to the amplitudes of the facial actions, it is difficult to determine the absolute facial expression intensity for a given subject without referring to an image of the neutral face of the subject. In this work, we assume that we do not have the image of the neutral face of the subject during the operation of the system, as a result, we follow a different approach from the one we use in the blurring effect. Figure 2(a) shows a block diagram of our algorithm. In order to train the system, we use a database of facial images that contains for each subject an image with neutral face and images with different expressions with varying intensities.

During training, we use the Eigenface recognition method, for recognizing these facial images. The result of this step would be two subsets of facial images: one set that could be recognized correctly, called "Good Quality" images, and the other set that could not be recognized correctly, called "Poor Quality" images. Next, we adapt the Gaussian Mixture Model (GMM) based on Universal Background Model (UBM) to model these two classes of facial images. During the image assessment phase, for a given test image, we use the GMM-UBM models to classify the facial image into one of the two classes, i.e., good quality or poor quality image for face recognition. For a review of the GMM-UBM models, we refer the readers to the work in [8] that has been successfully applied in speaker verification. During testing, as shown in Figure 2(b), given a test image, we test if the image belongs to the class of images with good quality or poor quality. This is achieved using the *Maximum Likelihood* decision rule. We applied this approach to the Cohn-Kanade database [9]. Our experiments show that the accuracy of the system is 75% in discriminating between the images with good quality and the images with poor quality.

3 Experiments and Results

We use the images in the FERET gallery [1] to evaluate our algorithm for predicting the recognition rate of the Eigenface method on images with blurring

Table 1. Classifier performance: (a) Different expressions. (b) Total performance.

		Correct Classification (%)	Incorrect Classification (%)
Good Quality	Joy	73.66	26.34
	Anger	67.68	32.32
	Fear	81.25	18.75
	Disgust	67.05	32.95
	Surprise	33.58	66.41
	Sadness	61.46	38.54
Poor Quality	Joy	25.00	75.00
	Anger	33.33	66.67
	Fear	0.00	100.00
	Disgust	37.50	62.50
	Surprise	6.45	93.55
	Sadness	0.00	0.00

(a)

	Classifier performance%
True Positive	75.67
False Positive	29.03
True Negative	70.97
False Negative	24.33

(b)

effect. The FERET gallery includes 600 images for 150 different subjects. Each subject has four images, one is frontal with no expression, one is frontal with joy expression, and two are near frontal. In our experiments we only use the frontal images. To apply the Kurtosis measure to a facial image, we first detect the face and normalize the illumination in the images. For face detection, we use boosted face detector [10] which is implemented by OpenCV library [11]. Then, we normalize the size of the detected face area to 128×128 pixels. To test this measure, we use a Gaussian filter to blur the neutral face images in the FERET gallery and the Kurtosis to measure the intensity of blurring effect. We split the gallery into two separate sets of equal sizes for the training and the testing phases. We experiment with different values for σ, of the Gaussian filter, to obtain images with different levels of blurriness. We estimate the coefficients of Equation 3 by applying regression to the data in Figure 1(a). Figure 1(b) shows the error in predicting the recognition rate of the Eigenface method for the images in the test set.

To evaluate our approach for assessing the quality of facial images with facial expressions, we use the Cohn-Kanade face database which includes 97 subjects with different facial expressions captured in video sequences. Each sequence starts with a neutral face expression and the expression's intensity increases toward the end of the sequence. We split the database into two separate sets of equal sizes for the training and the testing. For training the classifiers, we need two sets of facial images. The first set includes images that are correctly recognized by the Eigenface recognition method. The second set includes images that the face recognition system fails to recognize. The two sets are obtained by applying the face recognition to all the images in the training set.

To train the GMM-UBM model, we select the frames of the neutral faces and the frames with high intensity expression for both training and testing the GMMs. Table 1(a) shows the performance of the classification for assessing the quality of facial images with different expressions. Table 1(b) shows the total performance of the system. The surprise expression is the expression that highly degrades the performance of the face recognition system. This is due to the fact that for the surprise expression the muscles in the upper and the lower parts of the face are deformed. In other words, the change in face appearance with surprise expression is more than the change for the other expressions.

4 Conclusion

In this paper, we presented methods for assessing the quality of facial images affected by blurring and facial expressions. Our experiments show that our methods are capable of predicting the performance of the Eigenface method on the images. In the future, we will work on finding a measure for assessing the quality of facial images with respect to illumination. We will also integrate the different measures of image quality to produce a single measure that indicates the overall quality of a face image.

References

1. Phillips, P.J., Rizvi, H.M.S., Pauss, P.: The feret evaluation methodology for face-recognition algorithms. IEEE Trans. on Pattern Analysis and Machine Intelligence **22** (2000)
2. Phillips, P.J., Grother, P.J., Michaels, R.J., Blackburn, D.M., Tabassi, E., Bone, J.M.: Face recognition vendor test 2002: Evaluation report. Technical report, NISTIR 6965, Available online at http://www.frvt.org (2003)
3. Zhao, W., R. Chellappa, P.P., Rosenfeld, A.: Face recognition: A literature survey. ACM Computing Surveys **35**(4) (2003) 399–458
4. Givens, G., Beveridge, R., Draper, B., Grother, P., Phillips, J.: Statistical models for assessing how features of the human face affect recognition. In: Proceedings of the 17th International Conference on Pattern Recognition. (2004)
5. Zhang, N.F., Postek, M.T., Larrabee, R.D., Vladar, A.E., Kerry, W.J., Josnes, S.N.: Image sharpness measurement in the scanning electron microscope part iii. Scanning **21**(4) (1999) 246–252
6. Kotz, S., Johnson, N.: Encyclopedia of statistical sciences. In: Wiely. (1982) 415–426
7. Fasel, B., Luettin, J.: Automatic facial expression analysis: A survey. Pattern Recognition **36** (2003) 259–275
8. Reynolds, D.A., Quatieri, T.F., Dunn, R.B.: Speaker verification using adapted gaussian mixture models. Digital Signal Processing **10** (2000) 19–41
9. Kandade, T., Cohn, J., Tian, Y.: Comprehensive database for face expression analysis. In: In Proceedings of the 4th IEEE International Conference of Automatic Face and Gesture Recognition (FG00). (2000) 46–53
10. Viola, P., Jones, M.J.: Rapid object detection using a boosted cascade of simple features. In: IEEE CVPR. (2001) 511–518
11. OpenCV: Open source computer vision library. Technical report, Intel Corp, Available at http://www.intel.com/research/mrl/research/opencv/ (2000)

Ambient Illumination Variation Removal by Active Near-IR Imaging

Xuan Zou, Josef Kittler, and Kieron Messer

Centre for Vision, Speech and Signal Processing,
University of Surrey, United Kingdom
{x.zou, j.kittler, k.messer}@surrey.ac.uk

Abstract. We investigate an active illumination method to overcome the effect of illumination variation in face recognition. Active Near-Infrared (Near-IR) illumination projected by a Light Emitting Diode (LED) light source is used to provide a constant illumination. The difference between two face images captured when the LED light is on and off respectively, is the image of a face under just the LED illumination, and is independent of ambient illumination. In preliminary experiments across different illuminations, across time, and their combinations, significantly better results are achieved in both automatic and semi-automatic face recognition experiments on LED illuminated faces than on face images under ambient illuminations.

1 Introduction

Face has been widely adopted as a useful biometric trait for personal identification for long time. However, for practical face recognition systems, several major problems remain to be solved. The effect of variation in the illumination conditions is one of those challenging problems [10].

Existing approaches addressing this problem fall into two main categories. The first category includes methods attempting to model the behaviours of the face appearance change as a function of illumination. However, the modelling of the image formation generally requires the assumption that the surface of the object is Lambertian, which is violated for real human face. In the other category, the goal is to remove the influence of illumination changes from face images or to extract face features that are invariant to illumination. Various photometric normalization techniques have been introduced to pre-process face images, and a comparison of five photometric normalisation algorithms used in a pre-processing stage for face verification on the Yale B database, the BANCA database and the XM2VTS database can be found in [7]. Face shape (depth map or surface normal) [1] or face images in multiple spectra [5] are used in face recognition as illumination invariant features. However, face shape acquisition always requires additional devices and is usually computationally expensive. The problem with using multi-spectral images is that although invisible spectral images can be invariant to visible illumination change, there can be variation in the invisible spectra of ambient illumination.

D. Zhang and A.K. Jain (Eds.): ICB 2006, LNCS 3832, pp. 19–25, 2005.

In this paper we present a completely different approach to address the illumination variation problem. Rather than studying passively the variation of illumination itself or attempting to extract illumination invariant feature, we actively create an active and invariant illumination condition for both gallery images and probe images. Two face images are captured for every subject. The first capture is done when the LED lamp is on, and the other capture is done when LED is off. The difference of these two images is an image of the face illuminated only by the Near-IR illumination provided by the LED lamp, and is independent of environmental illumination. Meanwhile, the invisibility of Near-IR illumination ensures that the capture is non-intrusive.

The rest of the paper is organized as follows: A brief review of the previous applications of active Near-IR illumination in computer vision is presented in Section 2. Section 3 describes the hardware of capture system and the acquisition of a face database. We give the details and results of the recognition experiments performed on this face database in Section 4, and conclusions in Section 5.

2 Active Near-IR Illumination

Active vision is not new in the computer vision area. In structure/coded light approaches, light patterns are projected onto object surfaces to facilitate 3D surface reconstruction. Active illumination is often used for shadow removal.

The Near-IR band falls into the reflective portion of the infrared spectrum, between the visible light band (0.3μm-0.6μm) and the thermal infrared band (2.4μm-100μm). Thus it has advantages over both visible light and thermal infrared. Firstly, since it can be reflected by objects, it can serve as active illumination source, in contrast to thermal infrared. Secondly, it is invisible, making active Near-IR illumination unobtrusive. In [9] IR patterns are projected to the human face to solve the correspondence problem in multi-camera 3D face reconstruction. Dowdall and et. al performed face detection on Near-IR face images [2]. Skin region is detected based on the fact that skin has different responses to the upper band and the lower band of Near-IR illumination. Morimoto and Flickner [6] proposed a multiple face detector which deployed a robust eye detector, exploiting the retro-reflectivity of the eyes. One Near-IR light set is used to provide bright pupil image, whilst another setting is used to generate dark pupil image, while keeping similar brightness in the rest of the scene. The pupils are very prominent and easy to detect in the difference image. Similar eye detectors using active illumination are used in [4] for 3D face pose estimation and tracking.

Although active Near-IR illumination has been widely used in face processing as detailed above, the novel idea advocated in this paper is to use it to provide constant and non-intrusive illumination for face recognition.

3 Face Database Acquisition

A database of face images of 40 subjects has been captured indoor. This database contains two subsets: ambient faces (faces under only ambient illumination) and

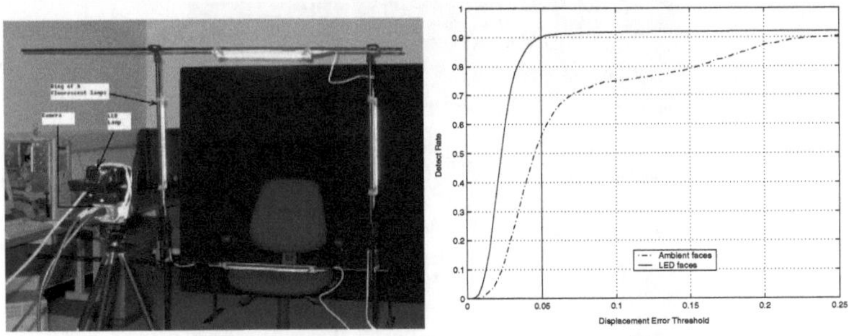

(a) (b)

Fig. 1. (a) A picture of face capture system. (b) The automatic eye center detection results for LED faces and ambient faces.

LED faces (faces under only LED illumination). Two capture sessions have been conducted with a time interval of several weeks. For each session, 4 different illumination configurations are used with light sources directed individually from left, bottom, right and top. 6 recordings were acquired for each illumination configuration. LED illumination is provided by a LED lamp with peak out wavelength at 850nm. This lamp is attached close to the Near-IR sensor so that the reflective component of the Near-IR light from the eyes will be projected straight into the camera. See fig1(a). This allows us to obtain face images with prominent bright pupils. For each recording a face image under ambient illumination only and one image under combined ambient and LED illumination are captured. A LED face image is obtained by taking the difference of these two images. Therefore, we have 40*2*4*6 =1920 ambient faces and the same amount of LED faces. See [11] for more details about face capture and system setup.

4 Experiments and Results

4.1 Face Localisation

For all face images, we manually marked the eye centres as ground truth positions, and also used two different automatic localization algorithms for ambient faces and LED faces respectively. For ambient faces, we used the algorithm based on Gaussian Mixture Model (GMM) face feature detector and enhanced appearance model [3], which has been trained on 1000 images from BANCA face database. For LED faces we used a simple correlation-based localization algorithm has been applied to LED faces. We used a different approach for LED faces because usually bright pupils can be found in LED faces and they can serve as strong features for eye localization. General face detectors which have not been trainined on faces with bright pupils do not work on LED faces. From the localisation errors shown on fig 1(b), it is evident that the illumination variations directly lead to the poor performance on ambient faces. With the help of the bright pupils and the consistency in LED illumination, the simple correlation-based approach gives much better results on LED faces.

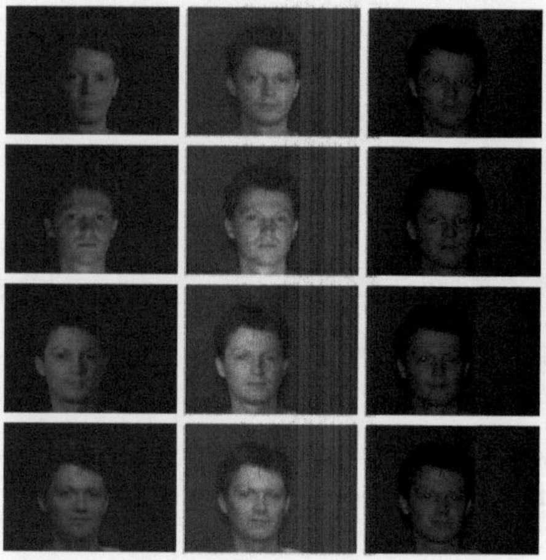

Fig. 2. Ambient faces (the left column), combined illumination faces (the middle column) and LED illuminated faces (the right column) under 4 different illumination configurations. The ambient illumination change caused significant differences in the appearance of the whole face. All important facial features look very different in different illumination conditions. Ambient faces and LED faces are relatively dark because the aperture for the camera is adjusted to avoid the saturation of the combined illuminated faces.

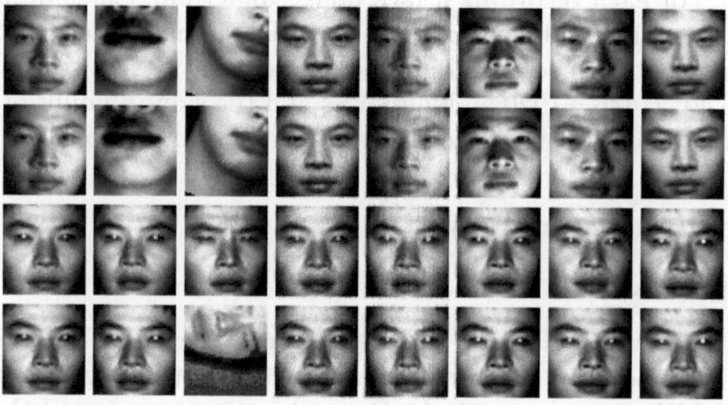

Fig. 3. Resulting images after the histogram equalization is performed for manually and automatically registered ambient faces (top 2 rows) and for corresponding LED faces (bottom 2 rows). It is obvious that data from LED faces exhibits much less variation as compared to the data from ambient faces. Bright pupils are prominent in LED faces. There are localisation errors in some automatically registered faces.

Face images are registered according to the manually marked or automatically detected eye centre positions, then cropped and sampled to the same size (55*50). Histogram equalization is applied subsequently. Fig 3 shows some samples of faces after the histogram equalization has been performed. The resulting images are then projected to an LDA subspace obtained from XM2VTS face database.

This LDA subspace is constructed from the PCA projections of all the 2360 face images of 295 subjects in XM2VTS face database and is supposed to be a subspace focusing on discrimitive information among subjects.

4.2 Recognition Experiments and Results

In the above LDA subspace, several different face recognition tests have been carried out on manually registered and automatically registered subsets of LED faces and ambient faces. A machine learning toolbox named WEKA [8] developed by University of Waikato has been used to perform experiments on the above data set. We applied Support Vector Machine (SVM) as the classifier, because it performed well in our previous experiments [11].

The whole dataset was divided into different subsets to serve as training sets and test sets in different test protocols. The rules of naming a subset are listed as below: 1)S_i for data in Session i, $i = 1, 2$; 2)C_i for data in Illumination Condition i, $i = 1..4$; 3)X_i for data in all illumination conditions except condition i, $i = 1..4$; 4) M for manually registered data, A for automatically registered data. For instance, MC_2S_1 stands for the manually registered data in Session 1 with Illumination Condition 2, AX_1S_2 stands for the automatically registered data in Session 2 with Illumination Conditions 2,3 and 4.

In the first experiment we measured the face recognition error across different sessions, and/or across different illumination conditions within each manually marked subset and within each automatically registered subset, respectively. Table 1 shows the error rates obtained under each test protocol. Each row corresponds to one test protocol. In a *Cross Session* test training set and test set are from different sessions. In a *Cross Illum.* test training set contains data with one illumination and test set contains data with the other illumination conditions. The error rate show in the table under a specific test protocol is the average error among all tests under this protocol. For example, the test error under the first protocol is the average of the errors of 2 subtests. In one of these two subtests, data from Session 1 is used for training and Session 2 data for testing, while in the other, Session 2 data is used for training and Session 1 data for testing. It is shown for all tests that the test results on LED faces are consistently much better than on ambient faces, regardless of the way the faces were registered. The advantage that LED faces offer over ambient faces is significant. The tests on manually registered data of LED faces achieved error rates close to zero.

Table 1. Error in face recognition experiment 1 (in percentage)

Ambient Faces		LED Faces		Test Protocol		
Manu. Reg.	Auto. Reg.	Manu. Reg.	Auto. Reg.	Training Set	Test Set	Description
1.61	13.70	0.05	5.16	S_i	$S_{(3-i)}$	Cross Session
42.57	67.22	0.07	3.26	C_i	X_i	Cross Illum.
52.95	72.74	1.75	8.87	C_iS_j	$X_iS_{(3-j)}$	Cross Both

Table 2. Error in face recognition experiment 2 (in percentage)

Ambient Faces	LED Faces	Test Protocol		
		Training Set	Test Set	Description
24.95	7.92	MS_i	$AS_{(3-j)}$	Cross Session
60.07	7.81	MC_i	AX_i	Cross Illum.
68.14	9.53	C_iS_j	$X_iS_{(3-j)}$	Cross Both

It can also be shown that cross-illumination tests on ambient faces gave very poor results. Among the tests on manually registered ambient faces (see the first column), if the training data contains data with all illumination conditions, the error rate is as low as 1.61%. However if training data does not contain any illumination condition appearing in test data, the test error rate inreased to 42.57%. If the training data and test data are from two different sessions, the result is even worse with an error rate of 52.95%. In sharp contrast, the test results on LED faces are consistently good for cross-session tests, cross-illumination tests and the tests invloving their combination. Even in the combination test, which is the most difficult one, the test error rate for manually registered LED faces is as low as 1.75%.

Due to errors in automatic eye localization, each test on automatically registered data obtained poorer results than the same test on manually registered data. However, the increases of errors on ambient faces are much larger than those on LED faces. This is the outcome of the relatively good performance of automatic eye localization on LED faces.

The second experiment reports the results of face recognition tests across manually registered data and automatically registered data. The test protocols are the same as those in the first experiment except that manually registered data serves as training set and automatically registered data as test set. Table 2 shows the test errors. Again, test errors on LED faces are much smaller than on ambient faces. Moreover, compared to corresponding tests in the previous experiment, the test errors are similarly poor on ambient faces, and slightly worse on LED faces. The combined cross-session and cross-illumination test in this experiment represents a practical application scenario of automatic face recognition. Usually the gallery images are manually registered, while the probe images are captured at a different time, under a different illumination condition and faces are automatically registered. The error on this test for LED face is 9.53%, but for ambient faces it is 68.14%, which is extremely poor.

5 Conclusion and Future Work

We proposed in this paper a novel way to overcome the illumination problem in face recognition by using active Near-IR illumination. Active Near-IR illumination provides a constant invisible illumination condition and faciliates the automatic eye detection by introducing bright pupils. Significantly better results have been obtained on LED faces than on ambient faces in cross-illumination test, cross-session and combined tests. The proposed active Near-IR illumination approach to face recognition is promising for face recognition. Further work will be the development of a more specific eye detection algorithm for Near-IR illuminated faces to improve the performance of automatic system.

References

1. K. W. Bowyer, K. Chang, and P. Flynn. A survey of approaches to three-dimensional face recognition. In *Proceedings of International Conference on Pattern Recognition*, 2004.
2. J. Dowdall, I. Pavlidis, and G. Bebis. Face detection in the near-ir spectrum. *Image Vis. Comput.*, 21:565–578, 2003.
3. M. Hamouz, J. Kittler, J. K. Kamarainen, P. Paalanen, and H. Kalaviainen. Affine-invariant face detection and localization using gmm-based feature detector and enhanced appearance model. In *Proceedings of the Sixth International Conference on Automatic Face and Gesture Recognition*, pages 67–72, May 2004.
4. Qiang Ji. 3d face pose estimation and tracking from a monocular camera. *Image and Vision Computing*, 20:499–511, 2002.
5. S.G. Kong, J. Heo, B. Abidi, J. Paik, and M. Abidi. Recent advances in visual and infrared face recognition - a review. *Computer Vision and Image Understanding*, 2004.
6. C. H. Morimoto and M. Flickner. Real-time multiple face detection using active illumination. In *Proceedings of the Fourth International Conference on Automatic Face and Gesture Recognition*, 2000.
7. J. Short, J. Kittler, and K. Messer. A comparison of photometric normalisation algorithm for face verification. In *Proceedings of the Sixth International Conference on Automatic Face and Gesture Recognition*, pages 254–259, May 2004.
8. I. H. Witten and E. Frank. *Data Mining: Practical Machine Learning Tools and Techniques with JAVA Implementations.* Morgan Kaufmann, 1999.
9. I. A. Ypsilos, A. Hilton, and S. Rowe. Video-rate capture of dynamic face shape and appearance. In *Proceedings of the Sixth International Conference on Automatic Face and Gesture Recognition*, pages 117–122, May 2004.
10. W. Zhao, R. Chellappa, and A. Rosenfeld. Face recognition: A literature survey. *ACM Computing Surveys*, 35:399–458, December 2003.
11. X. Zou, J. Kittler, and K. Messer. Face recognition using active near-ir illumination. In *Proceedings of British Machine Vision Conference*, 2005.

Rapid 3D Face Data Acquisition Using a Color-Coded Pattern and a Stereo Camera System

Byoungwoo Kim, Sunjin Yu, Sangyoun Lee, and Jaihie Kim

Biometrics Engineering Research Center,
Dept. of Electrical and Electronics Engineering, Yonsei University,
134 Shinchon-dong, Seodaemun-gu, Seoul 120-749, Korea
{bwkim, biomerics, syleee, jhkim}@yonsei.ac.kr

Abstract. This paper presents a rapid 3D face data acquisition method that uses a color-coded pattern and a stereo camera system. The technique works by projecting a color coded pattern on an object and capturing two images with two cameras. The proposed color encoding strategy not only increased the speed of feature matching but also increased the accuracy of the process. We then solved the correspondence problem between the two images by using epipolar constraint, disparity compensation based searching range reduction, and hue correlation. The proposed method was applied to 3D data acquisition and time efficiency was compared with previous methods. The time efficiency of the suggested method was improved by about 40% and reasonable accuracy was achieved.

1 Introduction

Although current 2D face recognition systems have reached a certain level of maturity, their performance has been limited by external conditions such as head pose and lighting. To alleviate these conditions, 3D face recognition methods have recently received significant attention, and the appropriate 3D sensing techniques have also been highlighted [1][2].

Previous approaches in the field of 3D shape reconstruction in computer vision can be broadly classified into two categories; active and passive sensing. Although the stereo camera, a kind of passive sensing technique, infers 3D information from multiple images, the human face has an unlimited number of features. Because of this, it is difficult to use dense reconstruction with human faces. Therefore, passive sensing is not an adequate choice for 3D face data acquisition.

On the other hand, active sensing projects a special pattern onto the subject and reconstructs shapes from reflected pattern imaging with a CCD camera. Because active sensing is better at matching ambiguity and also provides dense feature points, it can act as an appropriate 3D face-sensing device.

The most simple approach in structured lighting is to use a single-line stripe pattern, which greatly simplifies the matching process, although only a single line of 3D data points can be obtained with each image shot. To speed up the acquisition of 3D range data, it is necessary to adopt a multiple-line stripe pattern instead. However, the matching process then becomes much more difficult. One possibility is to use color information to simplify this difficulty [2][3].

D. Zhang and A.K. Jain (Eds.): ICB 2006, LNCS 3832, pp. 26 – 32, 2005.

Furthermore, in the single-camera approach, it is necessary to find the correspondence between the color stripes projected by the light source and the color stripes observed in the image. In general, due to the different reflection properties (or surface albedos) of object surfaces, the color of the stripes recorded by the camera is usually different from that of the stripes projected by the light source (even when the objects are perfectly Lambertian.) It is difficult to solve these problems in many practical applications [4].

On the other hand, this does not affect our color-lighting/stereo system if the object is Lambertian, because the color observed by the two cameras will be the same, even though this observed color may not be exactly the same as the color projected by the light source. Therefore, by adding one more camera, the more difficult problem of lighting-to-image correspondence is replaced by an easier problem of image-to-image stereo correspondence. Here, the stereo correspondence problem is also easier to solve than traditional stereo correspondence problems because an effective color pattern has been projected onto the object [4].

In this paper, we show how we have developed and implemented a new method for 3D range data acquisition that combines color structured lighting and stereo vision. In the proposed system, we developed a new coded color pattern and a corresponding point matching algorithm. Once the correspondence problem was solved, the 3D range data was computed by the triangulation technique. Triangulation is a well-established technique for acquiring range data with corresponding point information [5].

This paper is organized as follows; in section 2, we address system calibration, and section 3 discusses generating a new color-coded pattern. Stereo matching methods are dealt with in section 4. In section 5, experimental results are presented. Finally, section 6 concludes the paper.

2 Camera Calibration

Calibration is the process of estimating the parameters that determine a projective transformation from the 3D space of the world onto the 2D space of image planes. A set of 3D-2D point pairs for calibration was obtained with a calibration rig. If we know 6 point pairs, calibration matrix is uniquely determined. However, in many cases, since there exists errors, more than 6 point pairs are recommended, and it results in over-determined problem. Then the stereo camera system was calibrated with the DLT (Direct Linear Transform) algorithm [5][6].

3 Color-Coded Pattern Generation

The color-coded pattern generates an effective color sequence that can solve the corresponding problem and provide strong line edge segments. For pattern design, line segments have been effectively used in many 3D data acquisition systems, so we have exploited these line segments in our pattern design [7].

Previous research has shown that the HSI model is an effective color model for stereo matching [3][8]. Using the line features and the HSI color model, a set of unique color encoded vertical stripes was generated.

Each color-coded stripe was obtained as follows. Stripe color was denoted as $stripe(\rho,\theta) = \rho e^{j\theta}$, where ρ is the saturation value and θ is the hue value in the HS polar coordinate system shown in Fig. 1. To obtain a distinctive color sequence, we defined four sets of color. Each set contained three colors whose hue was separated by 120° within the set. We used only one saturation value (saturation=1) because hue information was enough to distinguish each stripe for matching process. Finally, the stripe color equation was denoted as (1).

$$color(m,n) = e^{j(mH_{jmp} + \varepsilon n)} \tag{1}$$

(a) Hue-Saturation polar coordinates (b) Generated color-coded pattern

Fig. 1. Generation of color-coded pattern

Next, the color-coded sequence was obtained as follows. First, we chose one out of the four, and 3 elements from this set were used. The next set elements were then used sequentially. After a 12-color sequence was generated, the next 12 color stripes were generated in the same manner. Fig. 1(b) shows the generated color-coded pattern.

4 Stereo Matching

In this section, rectification, and the corresponding points matching method are introduced. The color stripes to be projected onto the face were captured by both the left and right cameras. The captured images were then processed and were represented by thinned color lines. Then, the preprocessed image pairs were rectified using calibration information. Finally we found the corresponding point pairs quickly using the proposed method.

4.1 Epipolar Rectification

After thinning, the obtained image pairs were rectified using the camera calibration information. This step transforms the images so that the epipolar lines are aligned horizontally. In this case, the stereo matching was able to take advantage of the epipolar constraint and the search space was reduced to one dimension. Rectification is important when finding the corresponding points of the left image (i_l, j_l). We only needed to look along the scanline $j_r = j_l$ in the right image [5][9].

4.2 Disparity Compensation

To minimize computational complexity, we needed to restrict the searching ranges. After rectification, the difference between the pair of stereo images was small and was caused by horizontal shifts, it was necessary to compensate for the disparity of the stereo images. We used the SAD (Sum of Absolute Difference) to get the disparity value. Because it would take too much time to compensate for every image row, we only did so at multiples of 100 rows. We compensated at the K_{th} row using the following equation:

$$SAD_K = \sum_{i}^{N_x} \sum_{j}^{N_y} \left| Hue_L(i,j) - Hue_R(i+k,j) \right| \tag{2}$$

where N_x and N_y are 3 by 3 block size, and $Hue_L(i,j)$ is the hue value of the i, j pixel positions in the left image. At the equal row line, we found the minimum SAD:

$$SAD^p{}_{MIN} = MIN(\sum SAD_K) \tag{3}$$

Finally we found the background disparity of the whole image by maximizing equation (4):

$$SAD_{MAX} = MAX(\sum SAD^p{}_{MIN}) \tag{4}$$

By this process, we found K, which is the background disparity of the stereo images:

$$Right^{compensated} = Right_{t-K} \tag{5}$$

$$Left^{compensated} = Left_{t+K} \tag{6}$$

4.3 Stereo Matching

At the stereo matching step, we obtained the corresponding pairs of the two captured images. We found the hue distribution of two images very similar. However, the hue distribution of the captured left image and that of the captured right image are more similar than the hue distribution of the pattern image. Matching between the two captured images is more robust and accurate than between one of the captured images and the pattern image. This result confirms one of the major benefits of our new proposed system.

Up to the thinning step, we obtained two images that contained thinned color lines. With the epipolar constraint, the corresponding point pair fell on the epipolar line. With this constraint, the searching range was reduced to a line. Furthermore, we needed to limit the searching range of the epipolar line. Because the same color stripes were used twice in the designed color sequence, one point of the left image was matched twice on the epipolar line. To solve this problem, we used the disparity compensation method to restrict the searching range. So we never considered matching pixels with a disparity of more than (K+40), or less than –(K+40). We only compared the hue values of about 4 points on the epipolar line. In this case, there was no chance of getting two corresponding pairs. Three constraints including the epipolar constraint, disparity compensation-based searching range reduction, and hue information allowed us to find the corresponding points very rapidly. This is another major benefit of the proposed method. Fig. 2. shows the matching process.

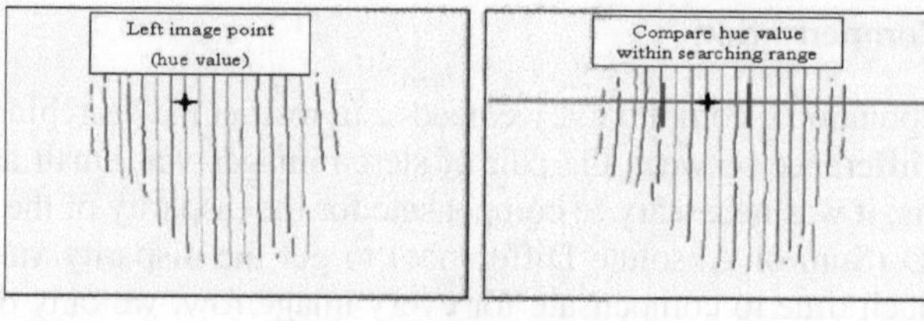

Fig. 2. Matching process

4.4 3D Reconstruction

Triangulation is the process of obtaining a real 3D position from two intersecting lines [5]. These lines are defined by the corresponding pairs and information from each calibration. After camera calibration, the triangulation method was used to obtain the 3D depth information. The triangulation method was solved with the SVD(Singular Value Decomposition) algorithm and 3D points were reconstructed [6].

5 Experiments

The system underwent an initializing step prior to inferring the 3D coordinates. After initializing the step, the color-coded pattern illuminated the subject. Corresponding point matching was then followed.

5.1 Accuracy Test

To test accuracy, we used a skin-colored box. We estimated the width, height and degree of the box. The metric RMS error between the real value and the reconstructed value was used as the accuracy measure. Table 1 shows the obtained results. From table 1, we can see that our system produced the maximum 2.39% RMS(Root Mean Squared) error when compared to the real values.

Table 1. The Accuracy test results

	Width	Length	Height	Degree A	Degree B	Degree C
Real value	14.5	12.5	9.5	90	88	92
Reconstruction result	13.89	11.32	9.28	88.32	86.12	89.48
RMS error	0.6211	1.2135	0.2641	1.86	2.35	2.39

unit: cm, degrees

Table 2. Time efficiency test results

Process	Dataset1			Dataset2		
	Previous Method1	Previous Method2	Proposed Method	Previous Method1	Previous Method2	Proposed Method
Preprocessing	3904	2942	1206	3889	3124	1284
Matching	736	720	946	814	749	856
Triangulation	242	237	244	287	264	255
Total Time	4888	3899	2396	4990	4137	2395
Total Points	5620	5644	5723	6920	6425	6324
Time / Point	0.8690	0.6908	0.4187	0.7210	0.6439	0.3787

unit: ms

Table 3. Computation time of the proposed matching method versus the DP matching method

	Proposed method	DP matching
Corresponding pairs	6947	7012
Time	945	1344

unit: ms

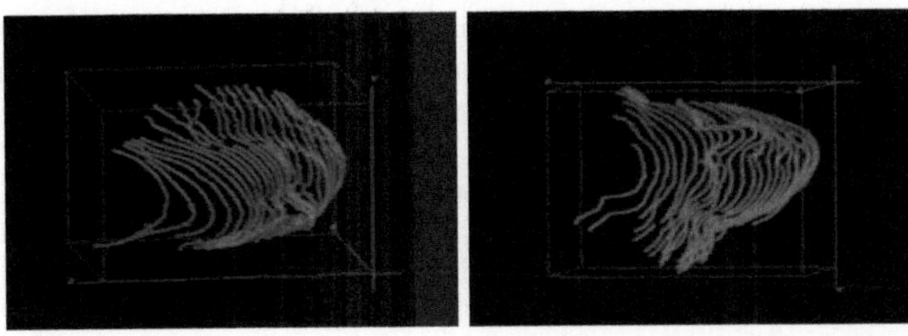

Fig. 3. 3D reconstruction results: Facial range data from two different viewing points

5.2 Time Efficiency

To test time efficiency, we estimated one 3D point reconstruction time. This is because, even for the same object, the number of reconstructed data points were different for each acquisition system. This made it impossible to estimate time efficiency by reconstruction time per total data points. We compared our system with a previous method [10][11]. The results are shown in table 2. We found that the time efficiency of our system improved by about 40% compared to a previous method. Table 3 also shows the comparison results between the proposed matching algorithm and the DP matching algorithm [7][12]. Performance of the proposed matching algorithm improved by about 30% compared to the DP matching algorithm. Fig. 3 shows the results of 3D face data reconstruction.

6 Conclusions

One significant advantage of our approach is that there is no need to find the correspondence between the color stripes projected by the light source and the color stripes observed in the image. In general, it is quite difficult to solve the above matching problem because the surface albedos are usually unknown. By not having to deal with this, we were able to focus on the easier image-to-image stereo correspondence problem. This process was also easier than traditional stereo correspondence because a good color pattern was projected onto the object.

Experimental results show a value of about 2% of depth error for the polyhedral object, but its performance decreased a little around the curved object. Also, the time efficiency of the proposed system is better than previous color structured lighting methods and the DP matching method. A drawback of this system is that color-coded stripes are usually sensitive to ambient light effects. Also, for dense reconstruction, the number of lines needs to be increased. Therefore, future works will include developing a more robust color pattern for ambient illumination and dense reconstruction.

Acknowledgement. This work was supported by the Korea Science and Engineering Foundation (KOSEF) through the Biometrics Engineering Research Center (BERC) at Yonsei University.

References

1. H.S. Yang, K.L. Boyer and A.C. Kak.: Range data extraction and interpretation by structured light. Proc. 1st IEEE Conference on Artificial Intelligence Applications, Denver, CO, (1984) 199-205.
2. K.L. Boyer and AC. Kak.: Color-encoded structured light for rapid active ranging. IEEE Trans. Pattern Analysis and Machine Intelligence, (1987) 14-28.
3. C.H. Hsieh, C.J. Tsai, Y.P. Hung and SC. Hsu.: Use of chromatic information in region-based stereo. Proc. IPPR Conference on Computer Vision, Graphics, and Image Processing, Nantou, Taiwan, (1993) 236-243
4. C. Chen, Y. Hung, C. Chiang, and J. Wu.: Range data acquisition using color structured lighting and stereo vision. Image and Vision Computing, Mar. (1997) 445-456
5. Emanuele Trucco and Alessandro Verri.: Introductory Techniques for 3-D Computer Vision, Prentice Hall (1998)
6. R. Hartley and A. Zisserman.: Multiple view Geometry in computer vision, Cambridge University Press (2000)
7. Y. Ohta, and T. Kanade.: Stereo by intra and inter scan line search using dynamic programming. IEEE Transactions on Pattern Analysis and Machine Intelligence, Vol. 7, No. 2, Mar. (1985) 139-154
8. R.C. Gonzales and R.E. Woods.: Digital Image Processing, Addison-Wesley, Reading, MA, (1992).
9. H. Jahn.: Parallel Epipolar Stereo Matching. IEEE int. Conf. on Pattern Recognition, ICPR2000, (2000) 402-405
10. Dongjoe Shin.: The hard calibration of structured light for the Euclidian reconstruction of face data. Master's Thesis. Dept. of Electrical and Electronic Engineering. Yonsei University. (2004)
11. Sungwoo Yang, Sangyoun Lee and Jaihie Kim.: Rapid Shape Acquisition for Recognition Using Absolutely Coded Pattern. Comm. Int. Symp. Intell. Signal Process., Comm. Systems (ISPACS). Seoul, Korea. Nov. (2004) 620-624
12. L. Zhang, and B. Curless, and S. M. Seitz.: Rapid Shape Acquisition Using Color Structured Light and Multi-pass Dynamic Programming," Proc. of First International Symposium on 3D Data Processing Visualization and Transmission, Jun. (2002) 24-36

Face Recognition Issues in a Border Control Environment

Marijana Kosmerlj, Tom Fladsrud, Erik Hjelmås, and Einar Snekkenes

NISlab, Department of Computer Science and Media Technology,
Gjøvik University College,
P. O. Box 191, N-2802 Gjøvik, Norway
`marijana.kosmerlj@ergo.no`, `tom.fladsrud@edb.com`,
`{erikh, einars}@hig.no`

Abstract. Face recognition has greatly matured since the earliest forms, but still improvements must be made before it can be applied in high security or large scale applications. We conducted an experiment in order to estimate percentage of Norwegian people having one or more look-alikes in Norwegian population. The results indicate that the face recognition technology may not be adequate for identity verification in large scale applications. To survey the additional value of a human supervisor, we conducted an experiment where we investigated whether a human guard would detect false acceptances made by a computerized system, and the role of hair in human recognition of faces. The study showed that the human guard was able to detect almost 80 % of the errors made by the computerized system. More over, the study showed that the ability of human guard to recognize a human face is a function of hair: false acceptance rate was significantly higher for the images where the hair was removed compared to where it was present.

1 Introduction

After September 11, 2001, the interest in use of physiological and behavioural characteristics to identify and verify identity of an individual has increased rapidly worldwide. These physiological and behavioural characteristics are believed to be distinct to each individual and can therefore be used to increase the binding between the travel document and the person who holds it.

In May, 2003, the International Civil Aviation Organization (ICAO) adopted a global, harmonized blueprint for the integration of biometric identification information into passports [1, 2]. The blueprint requires that a high-capacity contact-less integrated circuit containing a raw image file of the holder's face in addition to other identity information be included in the machine readable passports. Inclusion of the additional biometric technologies, fingerprint and iris, is optional.

The purpose of biometric passports is to prevent the illegal entry of travellers into a specific country, limit the use of fraudulent documents and make the border control more efficient [2].

D. Zhang and A.K. Jain (Eds.): ICB 2006, LNCS 3832, pp. 33–39, 2005.

In this paper we focus on the ability of the biometric authentication and the face technology to prevent identity theft in a border control setting with an assumed adversary environment. We claim that the face recognition technology alone is not adequate for identity verification in large scale applications, such as border control, unless it is combined with additional security measures.

2 Face as a Biometric in Border Controls

As a biometric identifier, the face has the advantage that it is socially acceptable and easily collectable. However, the face has large intra-person variability causing face recognition systems to have problems dealing with pose, illumination, facial expression and aging. The current state of the art in face recognition is 90 % verification at 1 % false accept rate under the assumption of the controlled indoor lighting [3].

2.1 Adversary Model in a Border Control Context

In "best practices" standard for testing and reporting on biometric system performance [4], the calculation of the false acceptance rate is based on the "zero effort" impostors. These impostors submit their biometric identifier as if they were attempting successful verification against their own template. In environments where it is realistic to assume that impostors will actively try to fool a biometric system, the false acceptance rate computed in the traditional way will not be representative for the actual percentage of impostors falsely accepted by the biometric system. An example of such an environment is a border control.

In order to propose a new way of calculating false acceptance rate in a border control context, we have modelled a possible adversary in this environment. In this model the adversary is a world wide organization that sells travel documents to people who for some reason need a new identity. The organization does not have the knowledge and the skills about the reproduction and alteration techniques for travel documents. Instead it cooperates with people who are willing to sell or lend their own travel documents, and with people who are willing to steal travel documents. Since the ICAO has recommended use of face as mandatory biometric identifier, they have been preparing for these new biometric based passports. They have obtained access to several face databases of people in different countries and they have purchased several face recognition systems which are used to found look-alikes for their customers. In a border control scenario where the identity of passport holders is verified by use of a face recognition system, there is a high probability that an impostor holding the passport of his "look-alike", will pass the identity verification.

In such adversary environment, a more adequate measure for the true false acceptance rate would be the proportion of the impostors who will be falsely accepted as their look-alikes in the target population.

2.2 Experimental Results and Discussions

We conducted an experiment in order to estimate the percentage of Norwegian people having one or more look-alikes in the Norwegian population. Subjects in the experiment were selected from several face databases: Ljubljani CVL Face Database [5], XM2VTS Database [6], AR Face Database [7], photos of Norwegian students at Gjøvik University College (HIG face database) [8] and several thousands of Norwegian passport photos [8]. In order to limit the effect of side views, lighting conditions and occlusions on the verification performance, frontal and approximately frontal facial images without occlusions and with varying but controlled lighting conditions were selected for the experiment.

We used the CSU Face Identification Evaluation System 5.0 [9] to generate similarity scores between our facial images. We determined the eye coordinates of the HIG photos manually. The eye coordinates of the passport photos were automatically determined with help of a Matlab script with an error rate of 16 %. The images were randomly assigned to four disjoint data sets: one training data set and three test data sets. The training data set was created by random selection of 1336 subjects from the HIG photo database, 50 subjects from the CVL database, 100 subjects from the XM2VTS database and 50 subjects from the AR database. The test data set I was created by random selection of two images of each subject from the XM2VTS database, the CVL database and the AR database. The test data set II contained the rest of the HIG photos whereas the data set III was created by random selection of 10 000 images from several thousands of passport photos.

The images with the eye coordinates were processed by the preprocessing script of the CSU software that removed unwanted image variations. In this process the hair is removed from the images such that only the face from forehead to chin and cheek to cheek is visible. After the training of the face recognition

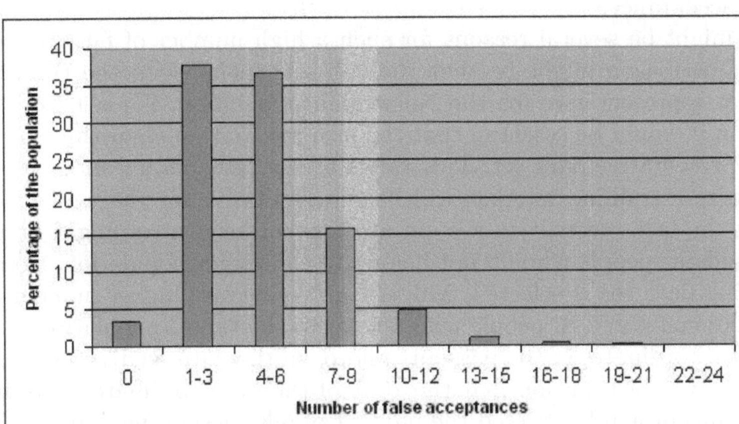

Fig. 1. The frequency distribution for the number of false acceptances in the test set II (1 % FAR, 14 % FRR)

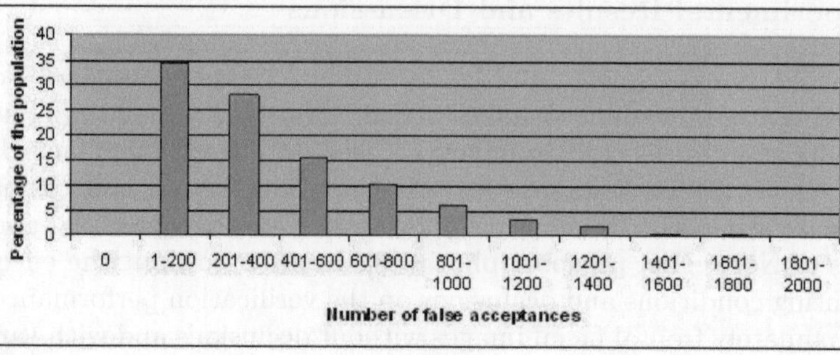

Fig. 2. The frequency distribution for the number of false acceptances in the test set III (1 % FAR, 14 % FRR)

algorithms and calculation of distance scores for data set I, we calculated the verification performance of the face recognition algorithms at several operating points. The face recognition algorithm with best performance was selected for the last part of the experiment where we calculated frequency distributions for the number of false acceptances in the data set II and III at selected operating points.

Figure 1 and Figure 2 show respectively the relative frequency distribution for the number of false acceptances in the test set II and III for the threshold value that corresponds to 1 % FAR. At the operating point of 1 % FAR, 97 % the subjects in the data set II generated one or more false acceptances while 99.99 % of the subjects in the data set III generated more than one false acceptance. We repeated the experiment at the operating point of 0.1 % FAR. The results showed that majority of the subjects in the data set II did not generate any false acceptances while 92 % of the subjects in the data set III generated more than one false acceptance.

There might be several reasons for such a high number of false acceptances in data. One reason might be that the subjects included in the training data set are not representative for the Norwegian population. For a border control application it would be essential that the face recognition algorithms be trained with a representative data set. This raises a new research question: is it possible to create a training set that will be representative for the whole world? If not, then the face recognition system used in border control might be population dependent: people who do not belong to the target population, from which the training data set is selected, will probably generate higher number of false acceptances compared to people who belong to the target population.

The eye coordinates of the passport photos in the data set III were generated automatically, which means that the 16 % of the eye coordinates were not correct. This has probably affected the number of false acceptances in the passport data set.

Additional information about the experiment can be found in MSc thesis of M. Kosmerlj [8].

3 The Effect of Additional Human Inspection

Based on our discoveries of look-alikes that might be able to pass a computerized face recognition environment, a natural next step would be to investigate whether an additional human guard would detect these false acceptances by the computerized system. In the previous experiment the computerized face recognition system compared normalized images without hair while in a real-world situation the people passing a control post will have hair. Therefore it is natural to investigate how good a human guard will be at recognizing human faces, both faces with hair, and faces without hair. This way we could see whether humans' face recognition process is affected by the presence of hair or not. If an impostor is able to find someone he or she resembles, this person may alter his hair style, colour etc., to amplify the similarities between her and the target person.

3.1 Experimental Results and Discussions

The data set was a subset of the data set used in the experiment in Sect. 2. From this data set we chose the images of the persons that generated high number of false acceptances and the images of their look-alikes. Only subjects from the CVL-, XM2VTS- and the AR Face face databases are used since the two other databases did not include more than one image of each subject.

A control group of 61 persons were divided into two groups. The division was made simple by having every other participant evaluate images of faces with hair, while the other evaluated faces where the hair was removed. Half of each group was presented the images in reverse order to eliminate variance due to difficult images instead of variance due to mental weariness. Group 1 consisted of 31 participants that were presented with image-pairs where an oval was used to remove the hair and background from the pictures. Group 2 consisted of 30 participants that were presented image-pairs where the depicted persons' hair was visible.

The participants were presented with several image-pairs that could be composed by two images of the same individual taken at different times, or of an image-pair composed by one image of one individual and the other image consisting of an image of his or her look-alike. Each participant had to mark the image-pairs as either being of the same individual or of someone else.

The analysis of the experimental results reveals, as shown in Fig. 3, that the false acceptance rate on the image-pairs where the hair is removed are significantly higher than for those where the hair is present. When looking at false rejections, there seem to be no significant difference in this error rate.

The hair is a feature that can be easily manipulated, indicating that there is in fact a great opportunity for an impostor to circumvent both the system and the human guard using simple and cheap methods. When combining this with facial make-up and the influence eyebrows, the colour of the eyes and beard have on human face recognition performance we see that using a human supervisor to increase the security may be insufficient. A better solution to achieve higher security would then be to employ multi-modal biometric systems [10–12].

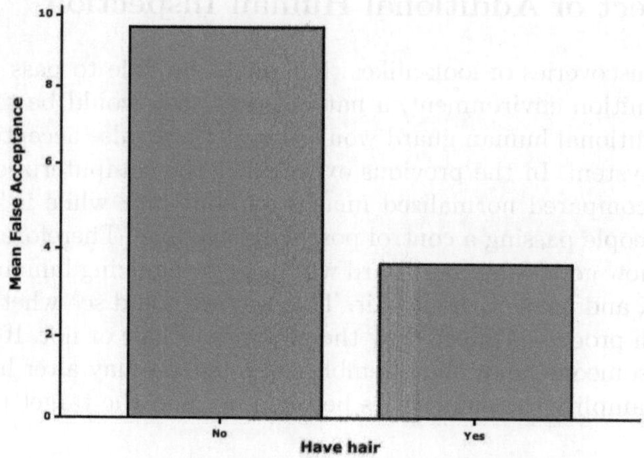

Fig. 3. The histogram shows a graphical overview of the false acceptances of the two groups with and without hair

There were only 3 image-pairs that have not been guessed wrong from one or more participant in the experiment where the hair was removed, while when hair was present there were 18. This may indicate that the hair is a feature that plays a major role in distinguishing several of the faces. It may also indicate that the face-images are very much alike. This makes it even more likely that they may be falsely considered as the same person also in a border control environment. In such environment the human supervisor may also relay more on the decision of the computer based system and this could affect his decision.

It should be noted that only 45 of the 60 image-pairs in the experiment where the hair and background was removed were actually composed of face images of different persons, while 15 image-pairs were composed of images of the same person to control the results. This produces a average false acceptance rate of 21,36 %. Combining this with the observation that most of the face image-pairs where evaluated wrong by more than one individual, we have an indication that the human supervision does not provide sufficient additional security.

Additional information about the experiment are provided in the Master's thesis of Tom Fladsrud [13].

4 Concluding Remarks

Automatic identity verification of a passport holder by use of a face recognition system may not give significant additional security against identity theft in a border control setting unless additional security measures are used. Using a human supervisor to increase the security may be insufficient, especially because hair, which is a feature that is easy to manipulate, plays such a significant role in human evaluation of faces.

The false acceptance rate as measured in the face recognition community does not give the correct picture of the true false acceptance rate that can be expected in a border control application with non zero-effort impostors. The more representative measure for the true acceptance rate would, for example, be the percentage of the impostors who have at least 20 look-alikes in the target population.

Acknowledgments

The face images used in this work have been provided, among others, by the Computer Vision Laboratory, University of Ljubljana, Slovenia [5], Computer Vision Center (CVC) at the U.A.B. [7], Centre for Vision, Speech and Signal Processing at the University of Surrey [6] and the Gjøvik University College [14].

References

1. ICAO: Biometrics Deployment of Machine Readable Travel Documents. ICAO TAG MRTD/NTWG. Technical Report, Version 1.9. Montreal. (May 2003)
2. United States General Accounting Office: Technology Assessment: Using Biometrics for Border Security. (November 14, 2002)
3. P.J. Phillips, P. Grother, R.J Micheals, D.M. Blackburn, E Tabassi and J.M. Bone: FRVT 2002: Evaluation Report. (March 2003)
4. A. J. Mansfield and J.L. Wayman: Best Practices in Testing and Reporting Performance of Biometric Devices. Version 2.01. (August 2002)
5. Faculty of Computer and Information Science, University of Ljubljana, Slovenia: (CVL FACE DATABASE)
6. K. Messer, J. Matas, J. Kittler, J. Luettin and G. Maitre: XM2VTSDB: The Extended M2VTS Database. In Second International Conference on Audio and Video-based Biometric Person Authentication (March 1999)
7. A.M. Martinez and R. Benavente: The AR face database, CVC Tech. Report #24. (1998)
8. M. Kosmerlj: Passport of the Future: Biometrics against Identity Theft? MSc thesis. Gjøvik University College, NISlab. Master's thesis (June 30, 2004)
9. Ross Beveridge, David Bolme, Marcio Teixeira and Bruce Draper: The CSU Face Identification Evaluation System User's Guide: Version 5.0. Computer Science Department Colorado State University. (May 1, 2003)
10. Anil K. Jain, Arun Ross and Salil Prabhakar: An Introduction to Biometric Recognition. IEEE Transactions on circuits and systems for video technology **14** (January 2004)
11. Arun Ross and Anil Jain: Information Fusion in Biometrics. Pattern Recognition Letters **24** (2003) 2115–2125
12. Anil K. Jain and Arun Ross: Multibiometric Systems. Communications of the ACM **47** (January 2004)
13. T. Fladsrud: Face Recognition Software in a border control environment: Non-zero-effort-attacks' effect on False Acceptance Rate. MSc thesis. Gjøvik University College, NISlab. Master's thesis (June 30, 2005)
14. (Gjøvik University College http://www.hig.no, http://www.nislab.no)

Face Recognition Using Ordinal Features*

ShengCai Liao, Zhen Lei, XiangXin Zhu, Zhenan Sun, Stan Z. Li, and Tieniu Tan

Center for Biometrics and Security Research & National Laboratory of Pattern Recognition, Institute of Automation, Chinese Academy of Sciences, 95 Zhongguancun Donglu Beijing 100080, China
http://www.cbsr.ia.ac.cn

Abstract. In this paper, we present an ordinal feature based method for face recognition. Ordinal features are used to represent faces. Hamming distance of many local sub-windows is computed to evaluate differences of two ordinal faces. AdaBoost learning is finally applied to select most effective hamming distance based weak classifiers and build a powerful classifier. Experiments demonstrate good results for face recognition on the FERET database, and the power of learning ordinal features for face recognition.

1 Introduction

It is believed that the human vision system uses a series of levels of representation, with increasing complexity. A recent study on local appearance or fragment (or local region) based face recognition [7] shows that features of intermediate complexity are optimal for basic visual task of classification, and mutual information for classification is maximized in a middle range of fragment size. Existing approaches suggest a trade-off between the complexity of features and the complexity of the classification scheme. Using fragment features is advantageous [8] in that they reduce the number of features used for classification for richer information content of the individual features, and that a linear classifier may suffice when proper fragment features are selected; on the other hand, with simple generic features, the classifier has to use higher-order properties of their distributions.

However, whether to use fragment or generic features remain a question. While using fragment features may be advantages for classification between apparently different classes, such as between a car and a face, the conclusion may not apply for object classes in which the differences in their appearances are not so obvious, eg faces of different individuals. For the latter case, more elementary and generic feature should provide better discriminative power. This in general requires a nonlinear classifier in which higher order constraints are incorporated.

In this regard, we consider a class of simple features: the ordinal relationship. Ordinal features are defined based on the qualitative relationship between two image regions and are robust against various intra-class variations [3, 5, 6]. For example, they invariant to monotonic transformations on images and is flexible enough to represent different local

* This work was supported by Chinese National 863 Projects 2004AA1Z2290 & 2004AA119050.

D. Zhang and A.K. Jain (Eds.): ICB 2006, LNCS 3832, pp. 40–46, 2005.

structures of different complexity. Sinha [5] shows that several ordinal measures on facial images, such as those between eye and forehead and between mouth and cheek, are invariant with different persons and imaging conditions, and thereby develops a ratio-template for face detection. Schneiderman [4] uses an ordinal representation for face detection.

Face recognition is a more difficult problem than face detection. While ordinal features have shown excellent separability between the face class and the rest of the world, it remains a question whether it is powerful enough for face recognition [6]. Thoresz [6] believes ordinal features are only suited for face detection but too weak for fine discrimination tasks, such as personal identification.

In this paper, we present an ordinal feature based method for face recognition. Ordinal features are generated using ordinal filters and are used to represent faces. Hamming distance of many local sub-windows is computed to evaluate differences of two ordinal faces. AdaBoost learning is finally applied to select most effective ordinal features and build a powerful classifier. Experiments demonstrate good results for face recognition on the FERET database.

The contributions of this work are summarized as follows: While ordinal features have been used for face detection, its application in face recognition is for the first time. We will show that ordinal features when properly selected using statistical learning method can do well for face based personal identification. The second contribution is that unlike manual feature selection as in [5], we propose to use a statistical learning method for selecting effective ordinal features and thereby constructing a strong classifier for face recognition.

The rest of this paper is organized as follows. In Section 2, we introduce ordinal features. In Section 3, AdaBoost learning is applied to select most discriminative feature, while removing large redundance in the feature set, and learn boosted classifiers. Section 4 describes weak classifiers for ordinal features learning. Experimental results are presented in Section 5.

2 Ordinal Features

Ordinal features come from a simple and straightforward concept that we often use. For example, we could easily rank or order the heights or weights of two persons, but it is hard to answer their precise differences. For computer vision, the absolute intensity information associated with an face can vary because it can changes under various illumination settings. However, ordinal relationships among neighborhood image pixels or regions present some stability with such changes and reflect the intrinsic natures of the face.

An ordinal feature encodes an ordinal relationship between two concept. Figure 1 gives an example in which the average intensities between regions A and B are compared to give the ordinal code of 1 or 0. Ordinal features are efficient to compute. Moreover, the information entropy of the measure is maximized because the ordinal code has nearly equal probability of being 1 or 0 for arbitrary patterns.

While differential filters, such as Gabor filters, are sufficient for comparison of neighboring regions', Balas and Sinha [1] extend those filters to "dissociated dipoles" for

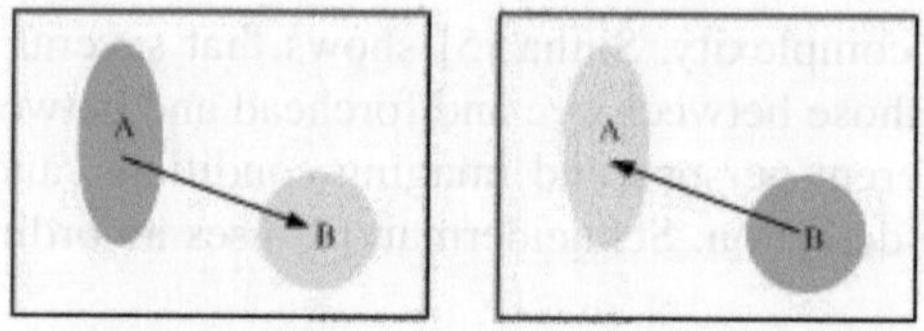

Fig. 1. Ordinal measure of relationship between two regions. An arrow points from the darker region to the brighter one. Left: Region A is darker than B, i.e. $A \prec B$. Right: Region A is brighter than B, i.e. $A \succ B$.

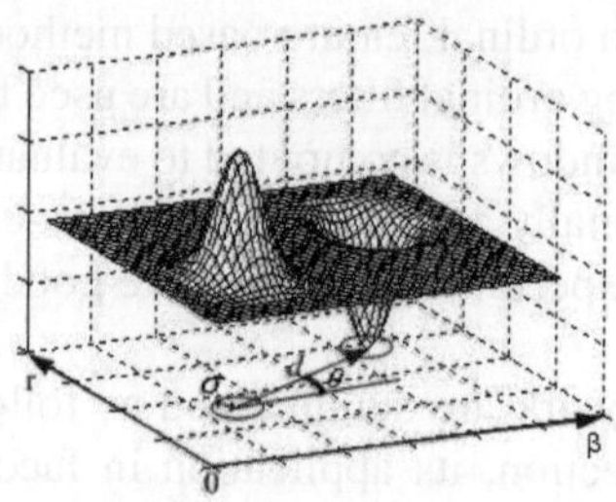

Fig. 2. Dissociated dipole operator

non-local comparison, shown in Figure 2. Like differential filters, a dissociated dipole also consists an excitatory and an inhibitory lobe, but the limitation on the relative position between the two lobes is removed. There are three parameters in dissociated dipoles:

- The scale parameter σ: For dipoles with a Gaussian filter, the standard deviation σ is an indicator of the scale.
- The inter-lobe distance d: This is defined as the distance between the centers of the two lobes.
- The orientation θ: This is the angle between the line joining the centers of the two lobes and the horizontal line. It is in the range from 0 to 2π.

We extend dissociated dipoles to dissociated multi-poles, as shown Figure 3. While a dipole tells us the orientation of a slope edge, a multi-pole can represent more complex image micro-structures. A multi-pole filter can be designed for a specific macro-structure, by using appropriate lobe shape configuration. This gives much flexibility for filter design.

To be effective for face recognition or image representation, there are three rules in development of dissociated multi-poles (DMPs):

- Each lobe of a DMP should be a low-pass filter. On one hand, the intensity information within the region of the lobe should be statistically estimated; on the other hand, the image noise should attenuated by low-pass filtering.
- To obtain the locality of the operator, the coefficients of each lobe should be arranged in such a way that the weight of a pixel is inverse proportional to its distance from the lobe center. Gaussian mask satisfies this; there are other choices as well.

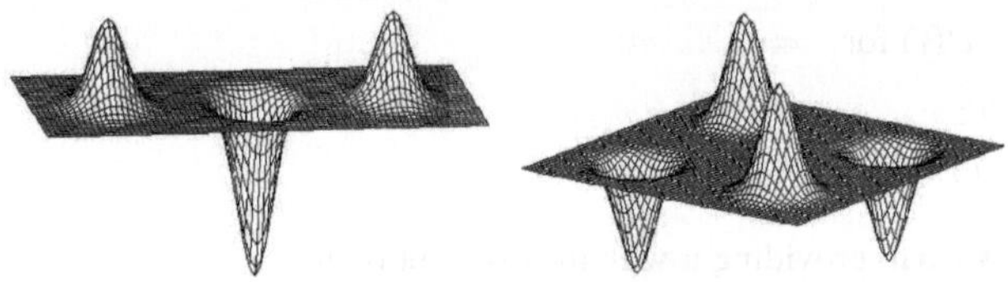

Fig. 3. Dissociated multi-pole: tri- and quad-pole filters

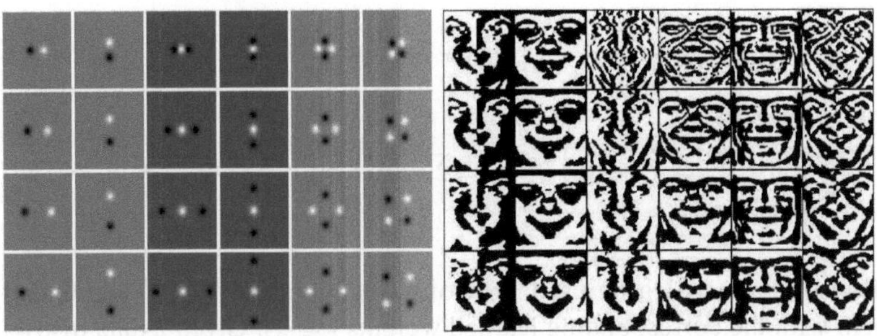

Fig. 4. The 24 ordinal filters used in the experiments, and the corresponding filtered images of a face

- The sum of all lobes' coefficients should be zero, so that the ordinal code of a non-local comparison has equal probability being 1 or 0. Thus the entropy of a single ordinal code is maximized. In the examples shown in Figure 3, the sum of two excitatory lobes' weights is equal to the inhibitory lobes' total absolute weights.

In this paper, we use 24 disassociated multi-pole ordinal filters as shown in Fig.4. The filter sizes are all 41x41 pixels. The Gaussian parameter is uniformly $\sigma = \pi/2$. The inter-pole distances are $d = 8, 12, 16, 20$ for the 2-poles and 4-poles, and $d = 4, 8, 12, 16$ for the 3-poles. For 2-poles and 3-poles, the directions are 0 and $\pi/2$; for the 4-poles, the directions are 0 and $\pi/4$. A more complete set would include a much larger number of filters with varying parameters. Optimization of parameters would take into consideration of the final performance as well as costs in memory and training speed.

3 AdaBoost Learning

Because the large Ordinal feature set contains much redundant information, A further processing is need to remove the redundancy and build effective classifiers. This is done in this work by using the following AdaBoost algorithm [2]:

Input: Sequence of N weighted examples

$\{(x_1, y_1, w_1), (x_2, y_2, w_2), \ldots, (x_n, y_n, w_n)\}$;

Initial distribution P over the n examples;

Weak learning algorithm **WeakLearn**;

Integer T specifying number of iterations;

Initialize $w_i^1 = P(i)$ for $i = 1, \dots, n$;

For $t = 1, \dots, T$:

1. Set $p_i^t = w_i^t / \sum_i w_i^t$;
2. Call **WeakLearn**, providing it with the distribution p;

   get back hypothesis $h_t(x_i) \in \{0, 1\}$ for each x_i;
3. Calculate the error of h_t : $\epsilon_t = \sum_{i=1}^N p_i^t |h_t(x_i) - y_i|$;
4. Set $\beta_t = \frac{\epsilon_i}{(1-\epsilon_t)}$;
5. Set the new weights to $w_i^{t+1} = \beta_i^{1-|h_t(x_i)-y_i|}$;

Output the hypothesis

$$H(x) = \begin{cases} 1 \text{ if } \sum_{t=1}^T \left(\log \frac{1}{\beta_t}\right) h_t(x) \geq \sum_{t=1}^T \left(\log \frac{1}{\beta_t}\right) \\ 0 \text{ otherwise} \end{cases}$$

AdaBoost iteratively learns a sequence of weak hypotheses $h_t(x)$ and linearly combines them with the corresponding learned weights $\log \frac{1}{\beta_t}$. Given a data distribution p, AdaBoost assumes that a **WeakLearn** procedure is available for learning a sequence of most effective weak classifiers $h_t(x)$. This will be discussed in the next section.

4 Weak Classifiers

The simplest weak classifier can be constructed for each pixel and each filter type, which we call single bit weak classifier (SBWC). We can concatenate all the filtered images into a complete filtered image. Consider every pixel in the complete image as a bit. An SBWC outputs 0 or 1 according to the bit value. At each iteration, the AdaBoost learning selects the bit by which the performance is the best, eg causing the lowest weighted error over the training set.

A more involved weak classifier can be designed based on a spatially local subwindow instead of a single bit. The advantage is that some statistic over a local subwindow can be more stable than that at a bit. In this scheme, the Hamming distance can be calculated between the ordinal values in the two corresponding subwindows. The Hamming distance as a weak classifier can be used to make a weak decision for the classification. The use of subwindows gives one more dimension of freedom, the subwindow size. A different size leads to a different weak classifier. The two types of week classifiers will be evaluated in the experiment.

5 Experiments

The proposed method is tested on the FERET face database. The training set contains 540 images from 270 subjects. The test set contains 1196 galleries and 1195 probes from 1196 subjects. All images are cropped to 142 pixels high by 120 pixels wide, according to the eyes positions. The 24 ordinal filters are applied to all the images.

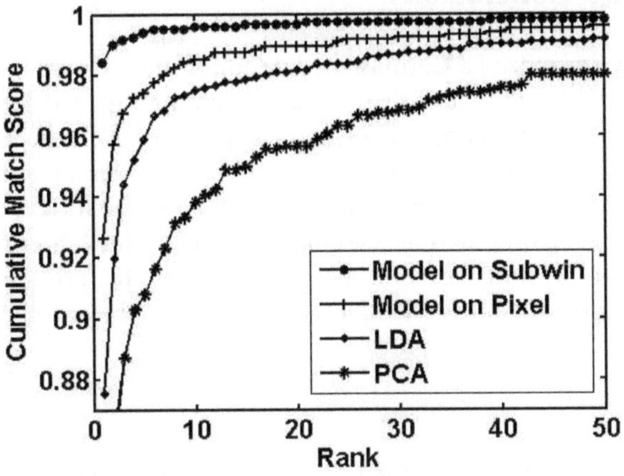

Fig. 5. Cumulative match curves of 4 compared methods

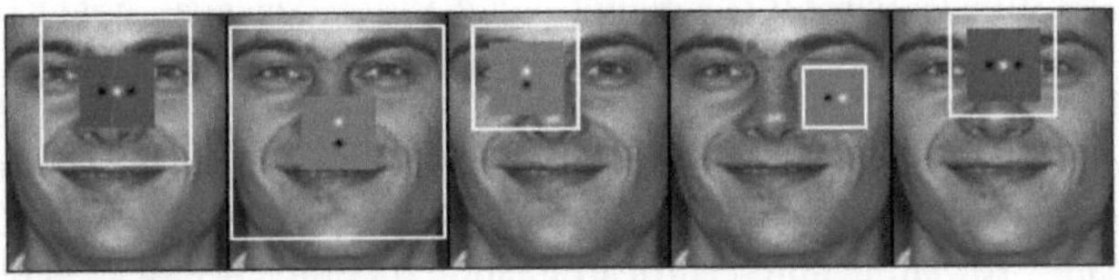

Fig. 6. The first 5 features and associated subwindow sizes selected by AdaBoost learning

The experiments evaluate the two AdaBoost learning based methods. The first is with the SBWC for feature selection and classifier learning. The second uses the local subwindow of ordinal features to construct Hamming distance based weak classifiers for AdaBoost learning. These two methods are be compared with the standard PCA and LDA methods (derived using the intensity images).

For the first method, a total of 173 weak classifiers, are trained to reach the training error rate of zero on the training set.

For the second method, 20 subwindow sizes are used: 6x6, 12x12, ..., 120x120 where the length of the side is incremented by 6. A single strong classifiers, consisting of 34 weak classifiers, is trained to reach the error rate of zero on the training set. The first 5 learned weak classifiers are shown in Fig.6. In the figure, the type of the filter and the subwindow size indicates the corresponding weak classifier.

Figs.5 describes the performances of the tested methods, in terms of the accumulative match curves, where the first method is named "Model on Pixel" and the second named "Model on Subwin". "Model on Subwin" performs the best, "Model on Pixel" the second, followed by LDA and PCA. The rank one recognition rates for the four methods are 98.4%, 92.5%, 87.5%, and 80.0%, respectively. This shows that the methods based ordinal features with statistical learning give good face recognition performances. Of the two proposed methods, "Model on Subwin" method is obviously advantageous: it needs fewer weak classifiers yet achieves a very good result.

6 Summary and Conclusions

In this paper, we have proposed a learning method for ordinal feature based face recognition. While it was believed that ordinal features were only suited for face detection and too weak for fine discrimination tasks, such as personal identification that[6], our preliminary results show that ordinal features with statistical learning can be powerful enough for complex tasks such as personal identification. In the future, we will investigate the effects of varying ordinal filter parameters, and find how intermediate features such as fragments can be built based on the simple ordinal features, and how to construct higher order ordinal features effectively using statistical learning.

References

1. B. Balas and P. Sinha. "Toward dissociated dipoles: Image representation via non-local comparisons". CBCL Paper #229/AI Memo #2003-018, MIT Computer Science and Artificial Intelligence Laboratory, Cambridge, MA, USA, August 2003.
2. Y. Freund and R. Schapire. "A decision-theoretic generalization of on-line learning and an application to boosting". *Journal of Computer and System Sciences*, 55(1):119–139, August 1997.
3. J. Sadr, S. Mukherjee, K. Thoresz, , and P. Sinha. "Toward the fidelity of local ordinal encoding". In *Proceedings of the Fifteenth Annual Conference on Neural Information Processing Systems*, Vancouver, British Columbia, Canada, December 3-8 2001.
4. H. Schneiderman. "Toward feature-centric evaluation for efficient cascaded object detection". In *Proceedings of IEEE Computer Society Conference on Computer Vision and Pattern Recognition*, pages 1007–1013, Washington, DC, USA, June 27 - July 2 2004.
5. P. Sinha. "Toward qualitative representations for recognition". In *Proceedings of the Second International Workshop on Biologically Motivated Computer Vision*, pages 249–262, Tubingen, Germany, November 22-24 2002.
6. K. J. Thoresz. On qualitative representations for recognition. Master's thesis, MIT, July 2002.
7. S. Ullman, M. Vidal-Naquet, and E. Sali. "Visual features of intermediate complexity and their use in classification". *Nature Neuroscience*, 5(7), 2002.
8. M. Vidal-Naquet and S. Ullman. "Object recognition with informative features and linear classification". In *Proceedings of IEEE International Conference on Computer Vision*, Nice, France, 2003.

Specific Sensors for Face Recognition

Walid Hizem, Emine Krichen, Yang Ni, Bernadette Dorizzi,
and Sonia Garcia-Salicetti

Département Electronique et Physique, Institut National des Télécommunications,
9 Rue Charles Fourier, 91011 Evry France
Tel: (33-1) 60.76.44.30 , (33-1) 60.76.46.73
Fax: (33-1) 60.76.42.84
{Walid.Hizem, Emine.Krichen, Yang.Ni, Sonia.Salicetti,
Bernadette.Dorizzi}@int-evry.fr

Abstract. This paper describes an association of original hardware solutions associated to adequate software software for human face recognition. A differential CMOS imaging system [1] and a Synchronized flash camera [2] have been developed to provide ambient light invariant images and facilitate segmentation of the face from the background. This invariance of face image demonstrated by our prototype camera systems can result in a significant software/hardware simplification in such biometrics applications especially on a mobile platform where the computation power and memory capacity are both limited. In order to evaluate our prototypes we have build a face database of 25 persons with 4 different illumination conditions. These solutions with appropriate cameras give a significant improvement in performance (on the normal CCD cameras) using a simple correlation based algorithm associated with an adequate preprocessing. Finally, we have obtained a promising results using fusion between different sensors.

1 Introduction

The face recognition systems are composed of a normal video camera for image capturing and a high speed computer for the associated image data processing. But this structure is not well suited for mobile device such as PDA or mobile phone configuration where both computation power and memory capacity are limited. The use of biometrics in mobile devices is becoming an interesting choice to replace the traditional PIN code and password due to its commodity and higher security.

The high complexity of face recognition in a cooperative context comes largely from the face image variability due to illumination changes. Indeed, a same human face can have very different visual aspects under different illumination source configurations. Research on face recognition offers numerous possible solutions. First, geometric feature-based methods [3] are insensitive to a certain extent to variations in illumination since they are based on relations between facial features (eyes, nose, mouth); the problem of these methods is the quality of the detection of such features, which is far from being straightforward, particularly in bad illumination conditions. Also, statistical methods like Principal Components Analysis [4], Fisherfaces [5], and Independent Components Analysis [6] emerged as an alternative

D. Zhang and A.K. Jain (Eds.): ICB 2006, LNCS 3832, pp. 47 – 54, 2005.

to a certain variability of facial appearance. Such methods, despite success in certain conditions, have the drawback of being reliable only when the face references used by the system and the face test images present similar illumination conditions, which is why some studies have proposed to model illumination effects [7]. So large computation power and memory capacities have to be dedicated to compensate this variability. Consequently reducing this image variability at the face image capturing stage can result in a significant both hardware and software simplification.

In this paper, we present an association of hardware and software solutions to minimize the effect of ambient illumination on face recognition. We have used two dedicated cameras and an appropriate pre-processing to suppress the ambient light. We have also built a database under different illumination conditions and with different cameras. Then a pixel correlation algorithm has been used for testing purpose.

In the following sections, we will present the two cameras. Then, we show the influence of illumination on face recognition. And finally, we describe our protocols and the results of our method.

2 Camera Presentation

2.1 Active Differential Imaging Camera - DiffCam

In a normal scene there is not a big illumination variation between two successive frames: the illumination remains static. So to eliminate it, a differentiation operation can be used in this case. We have applied inside a specially design CMOS image sensor with an analog memory in-situ in each pixel (Fig. 1). The integration of this in-situ analog memory permits a parallel image capture and further an on-chip differentiation computation.

The working sequence is the following: 1) the first image is captured by illuminating the subject's face with an infrared light source and 2) the second is captured by turning this light source off. The two captured images will be subtracted from each other during the image readout phase by using on-chip analog computation circuits on the sensor chip as shown Fig. 2. We have designed and fabricated a prototype CMOS sensor with 160*120 pixels by using a standard 0.5µm single poly CMOS technology. The pixel size is 12µm. A 8-bit ADC converter has been integrated equally on the sensor chip, which reduces considerably the system design complexity[1].

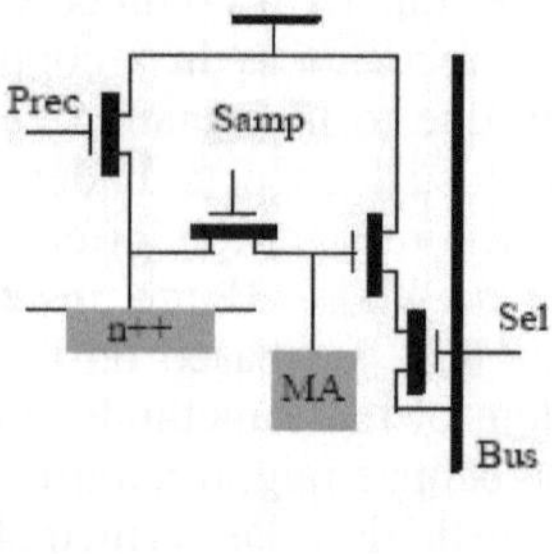

Fig. 1. Structure of the pixel

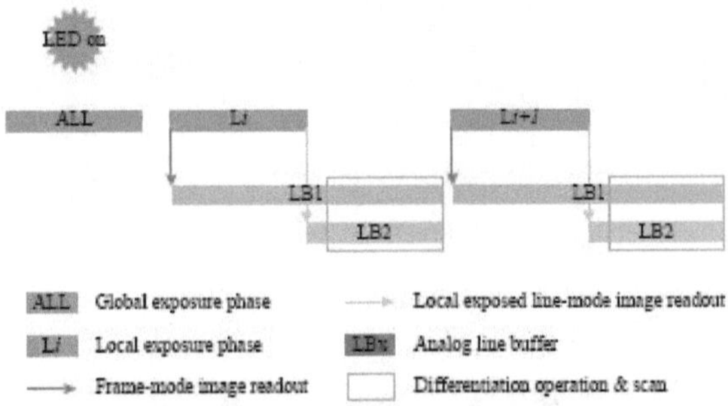

Fig. 2. The function principle and sequence of the active differential imaging system[1]

Compared to other analog/digital implementations such [8] [9], our solution requires not only single analog memory in each pixel, which gives an important pixel size reduction, but also neither off-chip computation nor image frame buffer memory.

A prototype camera with parallel port interface has been built by using two micro controllers. The infrared flash has been built with 48 IR LEDs switched by a MOSFET. A synchronization signal is generated from the microcontroller controlling the sensor. The pulse length is equal to the exposure time (50µs, the frame time is 10ms). The peak current in the LEDs is about 1A but due to the small duty cycle (1/200), the average current is low.

2.2 Synchronized Flash Infrared Camera – FlashCam

Another possible way to attenuate the ambient light contribution in an image is to use a synchronized flash infrared illumination. As shown in (Fig. 3), in classic integration-mode image sensor, the output image results from a photoelectric charge accumulation in pixels.

As has been indicated, the stationarity of ambient light makes its contribution proportional to its exposure time. So the idea here is to diminish the ambient light contribution by reducing the exposure time and at the same time using a powerful infrared flash, synchronized with this short exposure time. The images obtained by this imaging mode result mostly from the synchronized flash infrared light. This imaging mode has the advantage to work with a standard CCD sensor.

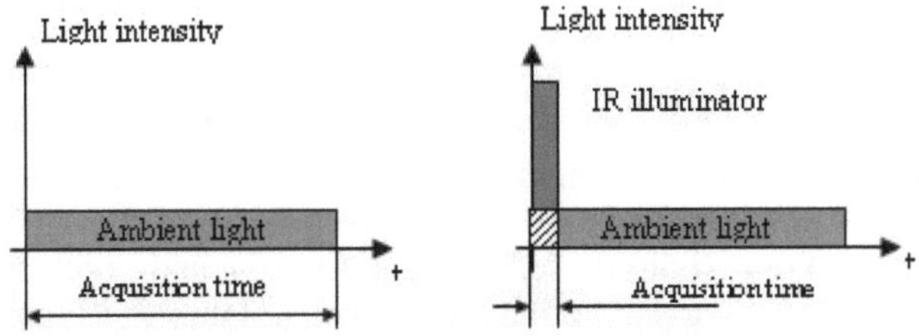

Fig. 3. Principle of the synchronized pulsed illumination camera[2]

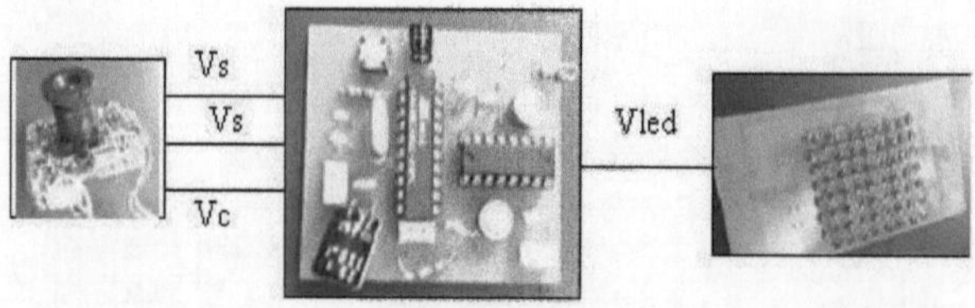

Fig. 4. The functional architecture of the prototype

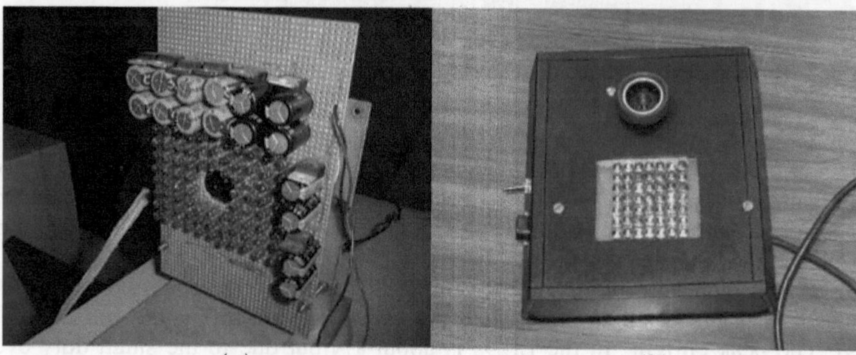

(a) (b)

Fig. 5. (a) The active Differential Imaging System (b) The Synchronized pulsed flash camera

An experimental camera has been built by modifying a PC camera with a CCD sensor. CMOS sensor based PC cameras cannot be used here, because the line sequential imaging mode used in APS CMOS image sensors is not compatible with a short flash-like illumination. The electronic shutter and synchronization information has been extracted from the CCD vertical driver. This information is fed into a micro controller which generates a set of control signals for infrared illuminator switching operation as shown in Fig. 4. The same LED based infrared illuminator has been used for this prototype camera. Fig. 5 shows the two prototype cameras.

3 Database

3.1 Description

To compare the influence of the illumination on faces, a database with 25 persons has been constructed by using three cameras: DiffCam, FlashCam and also a normal CCD camera. There are 4 sessions in this database with different illumination conditions: Normal light (base1), no light (base2), facial (base3) and right side illumination (base4). In the last two sessions, we have used a desk lamp to illuminate the face. In each session we have taken 10 images per person per camera. So we have 40 images per person per camera. The resolution of the images from the DiffCam is 160×120, the resolution obtained from the FlashCam and the normal CCD Camera images are

Fig. 6. Samples of the face database

Fig. 7. Samples of the face expression

320×280. The captured images are frontal faces; the subject was about 50cm from the device. There are small rotations of the faces on the three axes and also expression on faces. Indeed, anyone could wear glasses, regardless of whether spot reflections obscured the eyes. Face detection is done manually using the eyes location. Samples of this database are shown in Fig.6. (for the same person and different illumination conditions). Samples of different face expressions are shown in the Fig.7.

3.2 Protocol

For the experimentation, we have chosen 5 images for each person as test images and 5 as reference ones. We have two scenarios: The first consists in comparing images from the same camera and the same illumination condition. The second compares images from the same camera but from different session (illumination conditions change): there are six comparisons in this scenario: Normal light versus no light (base 1 vs base 2), Normal light versus facial illumination (base 1 vs base 3), Normal light versus right side illumination (base 1 vs base 4), No light session versus facial light (base 2 vs base 3), no light versus right side illumination (base 2 vs base 4) and facial illumination versus right side illumination (base 3 vs base 4).

4 Preprocessing and Recognition Algorithm

First the faces are detected and normalized. We have performed a series of preprocessing to attenuate the effect of illumination on face images. The best result has been found with a local histogram equalization associated with a Gaussian filter. In order to take benefits from the face symmetry and to reduce the effect of lateral illumination, we have added a second preprocessing calculating a new image.

$$I_s = \frac{I(x,y) + I(-x,y)}{2} .$$

We have applied this preprocessing to the images acquired with the normal CCD cam as they will be more perturbed by illumination effects. For the other images we've applied only an histogram equalization.

The verification process is done by computing the Euclidian distance between a reference image (template) and a test image.

4.1 Experimental Result

We have splitten our database into 2 sets, the templates set and the test set. As 10 images are available for each client and each session, we consider 5 images as client's templates, and the remaining 5 images as test images. Each test image of a client is compared to a template of the same client using the preprocessing and recognition algorithms above described, and the minimum distance between each test image and the 5 templates is kept. We obtain this way 125 intra class distances. Each test image is also compared to the other sets of 5 templates of the other persons of the database in order to compute the impostor distances. So we have 3000 inter class distances. The following tables (Tab.1 and Tab.2) compare the performance (in terms of EER) in function of the type of the camera. For the first Camera, we have two results: the first corresponds to preprocessed images. The second one uses images without preprocessing.

The first scenario (images from the same session) shows a general good and equivalent performance for each camera for the different illumination conditions.

In the second scenario, the reference images are taken from one session and the test images are taken from another session (different illumination conditions). Using the images, taken from the first camera without preprocessing, gives 50% of EER in nearly all the tests. Using a preprocessing improves significantly the results, which proves its usefulness for the Normal CCD Camera in order to attenuate the illumination effects. Comparing the normal camera and the FlashCam, we notice that

Table 1. Scenario 1

EER	Base 1		Base 2		Base 3		Base 4	
Normal CCD	3.4%	6%	6%	6%	3.2%	5,5%	4.5%	4,7%
FlashCam	5%		4.2%		3.2%		2%	
DiffCam	5,6%		2%		3.5%		4.1%	

Table 2. Scenario 2

EER	Base 1vs2		Base 1vs3		Base 1vs4		Base 2vs3		Base 2vs4		Base 3vs4	
Normal CCD	20%	39%	38%	53%	24.5%	54%	40%	56%	30%	50,7%	25%	37,6%
FlashCam	26%		27%		22%		28%		22%		23%	
DiffCam	15,7%		14%		21%		9.5%		13%		15%	

flashCam gives an improvement of the EER especially in the tests: Base1vs3, Base2vs3 and Base2vs4. In all these tests we observe a stable EER for the flashCam: this suggests a stronger similarity between the images acquired under different illumination conditions than the ones from the normal CCD. The relative high EER of the FlashCam is due to the quality of some images for which the flash did not give a sufficient light due to battery weakness. The correlation algorithm might be not suitable for the flashCam.

We have tried the eigenfaces algorithms but it gives worse results. We have to investigate other methods.

Comparing the FlashCam and the DiffCam, we observe that the second camera gives better results in all tests. The noticeable improvement is on tests: Base 2vs3, Base 1vs2 and Base 3vs4. This indicates the existence of residual influence of ambient light on the output images from FlashCam. On the contrary, we confirm real suppression of the ambient light by the differentiation operation.

4.2 Fusion Results

We have done other tests to know if the three cameras can be associated to give better results. For this purpose, we have done a simple mean between the scores given by the three cameras, (after some normalization) Table 3 shows the results of this fusion scheme and compares them to the best single camera performance.

We notice that in most cases, the fusion improves the best single camera results: this is due to the complementarities between the infrared images that eliminate the ambient light and the details in faces and the normal camera that compensates this lack of details.

Table 3. Fusion result of the three cameras

EER	Base 1vs2	Base 1vs3	Base 1vs4	Base 2vs3	Base 2vs4	Base 3vs4
3 cameras fusion	11.2%	10.7%	18%	13.9%	14.3%	9.6%
Best single camera	15,7%	14%	21%	9.5%	13%	15%

5 Conclusion

In this paper, we have presented two specialized hardware developed in our laboratory dedicated to face recognition biometric applications. The first one is based on temporal differential imaging and the second is based on synchronized flash light. Both cameras have demonstrated a desired ambient light suppression effect. After a specific preprocessing, we have used a simple pixel-level correlation based recognition method on a database constructed with varying illumination effects. The obtained performance is very encouraging and our research direction in the future is focused on a SoC integration of both sensing and recognition functions on a same smart CMOS sensor targeted for mobile applications.

References

1. Y. Ni, X.L. Yan, "CMOS Active Differential Imaging Device with Single in-pixel Analog Memory", Proceedings of IEEE European Solid-State Circuits Conference (ESSCIRC'02), pp. 359-362, Florence, Italy, Sept. 2002.
2. W. Hizem, Y. NI and E. Krichen, "Ambient light suppression camera for human face recognition" CSIST Pekin 2005
3. R. Brunelli, T. Poggio, "Face Recognition: Features vs. Templates", IEEE Transactions on Pattern Analysis and Machine Intelligence, Vol. 15, N° 10, pp. 1042-1053, October 1993.
4. M. A. Turk and A. P. Pentland. Face Recognition Using Eigenfaces. In Proc. of IEEE Conference on Computer Vision and Pattern Recognition, pages 586 – 591, June 1991.
5. Jian Li, Shaohua Zhou, Shekhar, C., "A comparison of subspace analysis for face recognition", Proceedings of ICASSP'2003 (IEEE Intl. Conf. on Acoustics, Speech, and Signal Processing), 2003.
6. M.S. Bartlett, J.R. Movellan, T.J. Sejnowski, "Face recognition by Independent Component Analysis", IEEE Transactions on Neural Networks, Vol. 13, N°6, pp. 1450-1464, Nov. 2002.
7. Athinodors S Georghiades, Peter N.Belhumeur, David J.Kriegman, "From Few to Many: Illumination Cone Models for Face Recognition under Variable Lighting and Pose", IEEE Transactions on Pattern Analysis and Machine Intelligence, 2001, pp. 643 – 660.
8. Hiroki Miura & al. "A 100Frame/s CMOS Active Pixel Sensor for 3D-Gesture Recognition System", Proceeding of ISSCC98, pp. 142-143
9. A. Teuner & al. "A survey of surveillance sensor systems using CMOS imagers", in 10th International Conference on Image Analysis and Processing, Venice, Spet. 1999.

Fusion of Infrared and Range Data: Multi-modal Face Images

Xin Chen, Patrick J. Flynn, and Kevin W. Bowyer

Dept. of Computer Science and Engineering, University of Notre Dame,
Notre Dame, IN 46556 USA
{xchen2, flynn, kwb}@nd.edu

Abstract. Infrared and range imagery are intriguing sensing modalities for face recognition systems. They may offer better performance than other modalities due to their robustness to environmental effects and deliberate attempts to obscure identity. Furthermore, a combination of these modalities may offer additional discrimination power. Toward this end, we present a semi-automatic system that captures range and infrared data of a human subject's face, registers and integrates multiple 3D views into one model, and applies the infrared measurements as a registered texture map.

1 Introduction

Although current face recognition systems employing intensity imagery have achieved very good results for faces that are taken in a controlled environment, they perform poorly in less uncontrolled situations. This motivates the use of non-intensity image modalities to supplement (or replace) intensity images [1]. Two major environmental problems in face recognition are illumination and pose variations [2]. Representations of the image and the stored model that are relatively insensitive to changes in illumination and viewpoint are therefore desirable. Examples of such representations include edge maps, image intensity derivatives and directional filter responses. It has been claimed [3] that no single one of these representations is sufficient by itself to withstand lighting, pose, and expression changes. Within-class variability introduced by changes in illumination is larger than the between-class variability in the data, which is why the influence of varying ambient illumination severely affects classification performance [4]. Thermal imagery of faces is nearly invariant to changes in ambient illumination [5], and may therefore yield lower within-class variability than intensity, while maintaining sufficient between-class variability to ensure uniqueness [1].

Well-known face recognition techniques, (for example, PCA), not only successfully applies to infrared images [6], they also perform better on infrared imagery than on visible imagery in most conditions [7] [8]. Calibrated 3D (range) images of the face are also minimally affected by photometric or scale variations. Therefore, they are receiving increasing attention in face recognition applications. Gordon [9] developed a curvature-based system employing Cyberware cylindrical scans. Beumier and Acheroy showed that recognition using surface matching from parallel profiles possesses high discrimination power, and also highlighted system sensitivity to absolute

D. Zhang and A.K. Jain (Eds.): ICB 2006, LNCS 3832, pp. 55 – 63, 2005.

gray level when range and intensity are considered jointly [10]. Yacoob and Davis [11] solved the related problem of face component labeling. Lapreste et al. [12] proposed a primal approach to face characterization from 3D images based on a structural analysis. Chua and Jarvis [13] proposed point-based features for free-form object recognition that could be used to match faces. Achermann et al. [14] also presented a system for face recognition using range images as input data, the results of their experiments show clearly that face recognition with range images is a challenging and promising alternative to techniques based on intensity.

Multimodal analyses seem to show promise in this domain. Recognition rates are improved by the combination of 3D and grey data, as reported by Beumier and Acheroy [10]. Wang et al. [15] proposes a face recognition algorithm based on both of the range and gray-level facial images. Chang et al. [16] designed a vector phase-only filter to implement a face recognition between range face (stored in the database) and intensity face (taken as the input), which is insensitive to illumination, but not scale and orientation invariant.

Since both infrared and range data are insensitive to variations caused by illumination, viewpoint, facial expressions and facial surface material changes, it is hoped that a combination of these two modalities may offer additional performance improvements for face recognition. Yet little multimodal experimental data of this sort exists. This paper presents a system that can semi-automatically produce a large dataset of integrated 3D model texture-mapped with IR data. As such, it offers a significant database building capability that can be used to good effect for large-scale face recognition trials from a limited database of experimental imagery.

2 Processing Method

The system described here takes as input multiple range and infrared images of the face, and produces a single 3D model with overlaid thermal sample values. The technical challenges in this task include interpolation of low-resolution IR values onto a high-resolution 3D mesh, registration of range views, and accommodation of some facial shape change between acquisitions. Our discussion focuses on two novel stages: mapping infrared data onto range data and view integration. The mapping stage assigns each range pixel an IR value and the integration stage combines two different view range images into one model.

2.1 Data Acquisition

Structured light acquisition systems use the projection of a known pattern of light (in our case, a laser stripe) to recover 3D coordinates [17]. Our acquisition proceeds as follows. A human subject is imaged in two poses corresponding to views offset 45 degrees (vertical rotation) on either side of frontal. Images are acquired roughly simultaneously from a Minolta Vivid 900 range camera and an Indigo Systems Merlin-Uncooled microbolometer array that senses long-wave infrared (LWIR) imagery. The cameras are placed side-by-side and standoff to the human subject is approximately two meters. Our LWIR camera is radiometrically calibrated but (other than

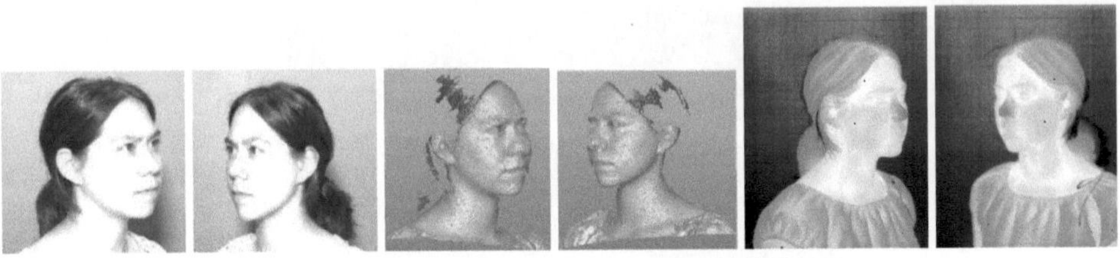

Fig. 1. Example images (color, range and infrared)

maintaining calibration during acquisition) we do not currently exploit the thermal calibration. After some trivial file manipulation, we have two 640x480 arrays of range and registered color intensity data, and two 320x240 arrays of infrared measurements.

2.2 Mapping Infrared Data onto Range Image

A. Spatial Transformation

A spatial transformation defines a geometric relationship between each point in the range/color and IR images. This is a 2-D image projection, since the cameras are assumed to be nearly coincident relative to the standoff. The goal is to obtain a mapping (X(u, v), Y(u, v)) between range image raster coordinates (u, v) and the corresponding position (x, y) in the IR image raster. X(,.,) and Y(,.,) are obtained through manual feature selection. Since the mapping will not in general take integer coordinates to integer coordinates, an interpolation stage is used to fill in destination raster values [20]. The form of the mappings X(,.,) and Y(,.,) is the affine transformation, with coefficients estimated from corresponding points (an assumption of affinity is appropriate given the standoff assumption above).

Rather than estimate a single affine coordinate map, we estimate maps independently within corresponding triangles identified in the image. The six coefficients a_{ij} are estimated from point triplet correspondences chosen manually. The more triangles into which the face is divided, the more precise the mapping will be. To infer the affine transformation, we need to provide at least three corresponding point pairs with the constraint on the point set in color image to consist of non-collinear points. When more than three correspondence points are available and when these points are known to contain errors, it is common practice to approximate the coefficients by solving an over-determined system of equations. However, it is unnecessary to use more than three point pairs to infer an affine transformation in our case since we can easily identify corresponding pairs with tolerable data errors.

Therefore, our method is to manually select approximately ten feature points and obtain a Delaunay triangulation of the convex hull of the point set in both images. The feature points chosen include anatomically robust locations such as the pupil, eye corner, brows, nose tip, and mouth. Normally, the features are more difficult to obtain in the IR image. Only coordinates within the convex hull of the points chosen can be mapped to the range image coordinate system. Figure 2 shows a typical triangular decomposition of IR and range (depicted with the registered color) images for one subject.

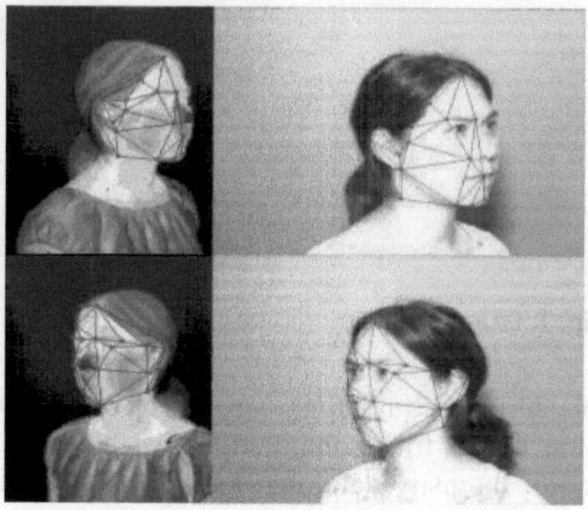

Fig. 2. Triangulation of color image of range data and grayscale image of IR data

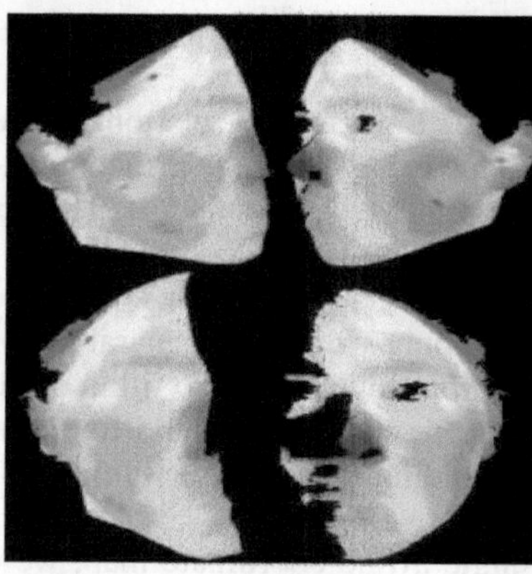

Fig. 3. Range face mapped with IR data

B. Temperature Interpolation

Once a mapping between range raster coordinates and corresponding IR pixels has been established, the IR measurements are used to "fill in" the range image mesh, causing the range image to gain another measurement (IR) at each pixel location. This requires an interpolation step. Figure 3 is a mapped result of the left-side face pose, rotated to see different views.

C. Initial Registration

We can estimate a rotation and translation that aligns the two objects roughly. Three non-collinear points are enough to compute the transformation since we can fix the 6 degrees of freedom in 3D space.

We manually select 3 points in the left and right pose range image respectively. The selection is not perfectly precise and need not be. We always select easily identified human face feature points to reduce the data error (eye corners, chin points, nose tip). Experience has shown that some guidelines should be followed when selecting points. Tangent edge points (jump edges) should not be picked since their positions are not reliably estimated. The triplet of points should not be nearly collinear because the transformation estimate may be ill-conditioned.

Before registration, we arbitrarily set the left-turn pose face surface fixed in 3D coordinate system; the right-turn pose face surface is moved to be in best alignment. We call former surface model shape, the latter data shape. As a result, let p_i be a selected point set on a data shape P to be aligned with a selected point x_i from a model point set X.

D. Modified ICP Registration

With the corresponding point set selected, we implement the quaternion-based algorithm for registration. It makes data shape P move to be in best alignment with model shape X.

Let $q_R = [q_0\ q_1\ q_2\ q_3]^T$ be a unit quaternion, where $q_0 \geq 0$ and $q_0^2+q_1^2+q_2^2+q_3^2=1$.The corresponding 3×3 rotation matrix is given by

$$R(\boldsymbol{q}) = \begin{matrix} q_0^2+q_1^2-q_2^2-q_3^2 & 2(q_1q_2-q_0q_3) & 2(q_1q_3-q_0q_2) \\ 2(q_1q_2+q_0q_3) & q_0^2+q_2^2-q_1^2-q_3^2 & 2(q_2q_3-q_0q_1) \\ 2(q_1q_3-q_0q_2) & 2(q_2q_3+q_0q_1) & q_0^2+q_3^2-q_1^2-q_2^2 \end{matrix}.$$

The translation component of the registration transform is denoted $q_T = [q_4\ q_5\ q_6]^T$. The complete registration state vector q is denoted $[q_R\ q_T]^T$. The mean square registration error (to be minimized) is

$$f(\boldsymbol{q}) = \frac{1}{N_p} \sum_{i=1}^{N_p} \|x_i - R(\boldsymbol{q}_r)p_i - \boldsymbol{q}_t\|^2$$

Our goal is to minimize f(q) subject to the constraint that the number of corresponding points is as large as possible. Besl and McKay[19] proposed an automatic surface registration algorithm called ICP which registers two surfaces starting from an initial coarse transformation estimate. This algorithm has been shown to converge fast but not necessarily towards the global optimal solution.

ICP is not useful if only a subset of the data point shape P corresponds to the model shape X or a subset of the model shape X, In our situation, assuming that one view is to be treated as the model shape and one as the data shape, these two shapes have only a narrow strip of overlapping area. ICP requires modification in our application. Another restriction of ICP is that the two surfaces are from rigid objects. However, the human face deforms non-rigidly continuously in our situation due to respiration and the body's unconscious balance control (subjects are standing when imaged). Again, this shows that ICP cannot be directly applied in our application.

"Closest points" that are "too far apart" are not considered to be corresponding points and marked as invalid so they have no influence during the error minimization. This is accomplished through an "outlier detection" phase. We define a threshold dynamically. In each ICP step, we "trust" the previous step's result and make use of the mean square distance calculated from that step as the threshold for the current step. This method can prevent introduction of unlikely corresponding point pairs while giving a good registration quickly.

In ICP, a good starting configuration for the two surfaces P and X is essential to a successful convergence. However, the range of successful starting configurations is quite large which does not impose difficult constraints to the operator when entering a pose estimate for P and X. Fortunately, it is fairly easy to manually select three corresponding points in each view to obtain a tolerable data error. The initial registration not only gives a useful approximation of registration but also provides an approximate average distance between the corresponding point pairs in the two images. Specifically, we can use the mean square distance calculated from the mean square objective minimization as our first threshold for modified ICP algorithm.

The modified ICP algorithm is defined as follows:

- Input: Two face surfaces P and X containing respectively N_P and N_X vertices, an initial transformation $q_0 = (R_0, t_0)$ which registers P and x approximately; and a mean square distance computed in initial registration using the default threshold T.
- Output: A transformation q=(R, t) which registers P and X.
- Initial Configuration: Apply the transformation (R_0, t0) to P.
- Iteration: Build the set of closest point pairs (p, x). If their distance exceeds T, discard the pair. Find the rigid transformation (R, t) that minimizes the mean square objective function. Update R and t. Set T = f (q). Repeat until convergence of f (q).

We use a kd-tree data structure to facilitate nearest-neighbor searches in the ICP update step. In order to verify the registration quality and terminate the iterative algorithm, the mean-square distance is generally used. We can also use the number of corresponding point pairs as a sign to stop the iterations. Figure 4 (left) shows the result of the registered range images of a human face scanned in two different poses. Figure 4 (right) shows the registration result of the two face surfaces mapped with IR data.

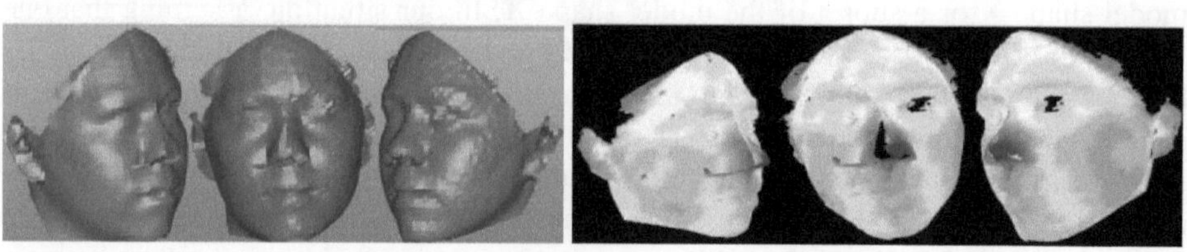

Fig. 4. Registered face surfaces (left) and registered face surfaces mapped with IR data (right)

E. Mesh Fusion

There are several methods to integrate registered surfaces acquired from different views. We propose a new mesh fusion algorithm that is particularly simple and useful in our human face applications. It erodes the overlapping surface of the data face shape until the overlap disappears, then constructs a frontier mesh region to connect them.

Due to the complexity of the human face surface, we expect that there will be multiple disjoint regions of overlap between the data and model meshes. Schutz et al. [18] proposed a mesh fusion algorithm which can deal with such problems. Our approach is simpler and relies on the distinctive nature of the registered meshes arising from our sensing set-up. We preserve the model mesh as a continuous region without holes while maximizing the face area it can cover by carefully selecting feature points which construct the convex hull (mentioned in Section 2.2A). The model mesh remains intact while the data mesh is eroded in the overlapping region. Vertices in the data mesh that are within a threshold distance of a vertex in the model mesh are removed; this process continues until no vertices are removed. The threshold value is determined empirically, and in our case a 5 to 10mm value works well. The result, as depicted in Figure 5, is a pair of faces with a curvilinear frontier between them.

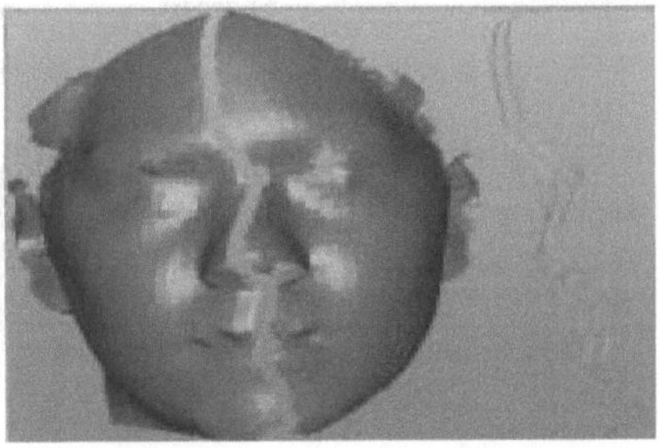

Fig. 5. Gap and frontiers

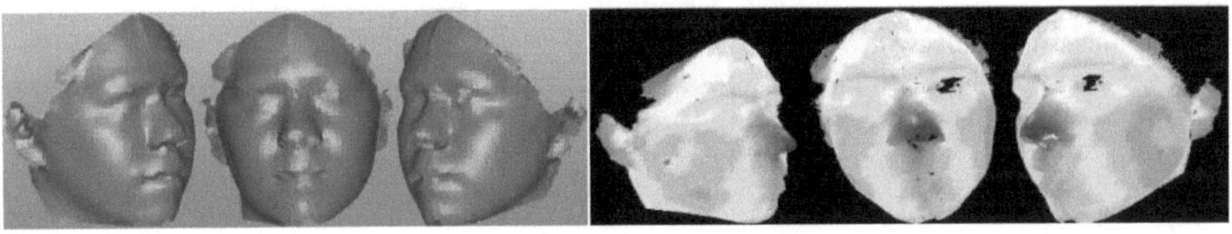

Fig. 6. Mesh integration results

The frontier is a distinguished set of vertices. Any point inside the convex hull of either mesh whose left/right adjacent pixel is eroded is labeled as a frontier point. Holes in the image due to missing data are not considered. These vertices are placed in a linked list. The gap enclosed by the two frontiers is filled with triangles. The frontier list of the two face surfaces is sorted in incremental y coordinate order. Figure 6 illustrates the mesh fusion result as shaded 3D mesh data and seamlessly integrated IR overlays.

3 Summary and Conclusions

The system described in this paper has been used to process several sets of multimodal imagery of experimental subjects acquired in a data collection protocol. Inspection of these results suggests that IR detail and overall 3D shape of the face are well preserved, and that the range image integration step works reasonably well. However, spurious range points are not always eliminated by the filtering procedure, missing data due to the lack of range points on the eyeball yields a model with holes, and radiometrically calibrated IR data is currently not incorporated into the model. This is the focus of current research. Results to date suggest that this approach to creation of synthetic head models with IR attributes, which can then be rendered to produce IR images from any viewpoint, offers a potentially valuable source of data to multimodal face recognition systems.

References

1. Jain, A., Bolle, R. and Pankanti, S., Biometrics: Personal Identification in Networked Society, Kluwer Academic Publishers, 1999.
2. Zhao, W., Chellappa, R., Rosenfeld, A. and Phillips, J. "Face Recognition: A Literature Survey", Univ. of MD Tech. Rep. CFAR-TR00-948, 2000.
3. Adini, Y., Moses, Y. and Ullman, S. "Face Recognition: The Problem of Compensating for Changes in Illumination Direction", Proc. ECCV, A:286-296, 1994.
4. Wilder, J., Phillips, P.J., Jiang, C. and Wiener, S. "Comparison of Visible and Infrared Imagery for face Recognition", Proc. Int. Conf. Autom. Face and Gesture Recog., 192-187, 1996.
5. Wolff, L., Socolinsky, D. and Eveland, C. "Quantitative Measurement of Illumination Invariance for Face Recognition Using Thermal Infrared Imagery", Proc. Workshop Computer Vision Beyond the Visible Spectrum, Kauai, December 2001.
6. Cutler, R. "Face Recognition Using Infrared Images and Eigenfaces", website http://cs.umd.edu/rgc/face/face.htm, 1996.
7. Socolinsky, S. and Selinger, A., "A Comparative Analysis of face Recognition Performance with Visible and Thermal Infrared Imagery", Tech Rep., Equinox Corp., 2001.
8. Selinger, A. and Socolinsky, D. "Appearance-Based Facial Recognition Using Visible and Thermal Imagery: A Comparative Study", Proc. Int. Conf. Pattern Recognition, Quebec City, 2002.
9. Gordon, G. "Face Recognition based on Depth Maps and Surface Curvature", Proc. SPIE 1570, 234-247, 1991.
10. Beumier, C. and Acheroy, M., "Automatic Face Verification from 3D and Grey Level Clues", Proc. 11th Portuguese Conference on Pattern Recognition (RECPAD 2000), Sept. 2000.

11. Yacoob, Y. and Davis, L. "Labeling of Human Face Components from Range Data," CVGIP 60(2):168-178, Sept. 1994.
12. Lapreste, J., Cartoux, J. and Richetin, M. "Face Recognition from Range Data by Structral Analysis", NATO ASI Series v. F45 (Syntactic and Structural Pattern Recognition), Springer, 1988.
13. Chua, C. and Jarvis, R. "Point Signatures: A New Representation for 3D Object Recognition", Int. J. Comp. Vision 25(1):63-85, 1997.
14. Achermann, B. and Jiang, X. and Bunke, H., "Face Recognition using Range Images", Proc. International Conference on Virtual Systems and MultiMedia '97 (Geneva,Switzerland), Sept. 1997, pp. 129-136.
15. Wang, Y., Chua, C. and Ho, Y. "Facial Feature Detection and Face Recognition from 2D and 3D Images", Pattern Recognition Letters 23(10):1191-1202, August 1991.
16. Chang, S., Rioux, M. and Domey, J. "Recognition with Range Images and Intensity Images", Optical Engineering 36(4):1106-1112, April 1997.
17. Beumier, C. and Acheroy, M., "Automatic Face Authentification from 3D Surface", Proc. 1998 British Machine Vision Conference, Sept. 1998, pp. 449-458.
18. Schutz, C., Jost, T. and Hugli, H. "Semi-Automatic 3D Object Digitizing System Using Range Images", Proc. ACCV, 1998.
19. Besl, P.J. and McKay, N.D., "A Method for Registration of 3-D Shapes", IEEE Trans. on PAMI 14(2):239-256, February 1992.
20. Wolberg, G., Digital Image Warping, Wiley-IEEE Press, 1990.

Recognize Color Face Images Using Complex Eigenfaces

Jian Yang[1], David Zhang[1], Yong Xu[2], and Jing-yu Yang[3]

[1] Department of Computing, Hong Kong Polytechnic University, Kowloon, Hong Kong
{csjyang, csdzhang}@comp.polyu.edu.hk
http://www4.comp.polyu.edu.hk/~biometrics/
[2] Bio-Computing Research Center and Shenzhen graduate school,
Harbin Institute of Technology, Shenzhen, China
laterfall2@yahoo.com.cn
[3] Department of Computer Science, Nanjing University of Science and Technology,
Nanjing 210094, P.R. China
yangjy@public1.ptt.js.cn

Abstract. A strategy of color image based human face representation is first proposed. Then, based on this representation, complex Eigenfaces technique is developed for facial feature extraction. Finally, we test our idea using the AR face database. The experimental result demonstrates that the proposed color image based complex Eigenfaces method is more robust to illumination variations than the traditional grayscale image based Eigenfaces.

1 Introduction

In recent years, face recognition has become a very active research area. Up to now, numerous techniques for face representation and recognition have been developed [1]. However, almost all of these methods are based on grayscale (intensity) face images. Even if the color images are available, the usual way is to convert them into grayscale images and then base on them to recognize. Obviously, in the process of image conversion, some useful discriminatory information contained in the face color itself is lost. More specifically, if we characterize a color image using color model such as HSV (or HSI), there are three basic color attributes, i.e., hue, saturation and intensity (value). Converting color images into grayscale ones means that the intensity component is merely employed while the two other components are discarded. Does there exist some discriminatory information in hue and saturation components? If so, how to make use of these discriminatory information for recognition? And, as we know, the intensity component is sensitive to illumination conditions, which leads to the difficulty of recognition based on grayscale images. Now, another issue is: can we combine the color components of image effectively to reduce the disadvantageous effect resulting from different illumination conditions as far as possible?

In this paper, we try to answer these questions. We make use of two color components, saturation and intensity (rather than the single intensity component), and combine them together by a complex matrix to represent face. Then, the classical Eigenfaces [2,3] is generalized for recognition. The experimental result on AR face database demonstrates that the suggested face representation and recognition method outperforms the usual grayscale image based Eigenfaces.

D. Zhang and A.K. Jain (Eds.): ICB 2006, LNCS 3832, pp. 64–68, 2005.

2 Face Representation in HSV Color Space

Since it is generally considered that the HSV model is more similar to human perception of color, this color model is adopted in this paper. The common RGB model can be converted into HSV by the formulations provided in [4]. Fig. 1 shows the three HSV components, i.e., hue, saturation and (intensity) value corresponding to image (a), (b) and (c), respectively. From Fig. 1, it is easy to see that the illumination conditions of image (a), (b) and (c) are different and the component hue is most sensitive to lighting variation. So, we decide to use the saturation and value components to represent face. These two components can be combined together by a complex matrix

$$\text{Complex-matrix} = \mu_1 S + i\mu_2 V \tag{1}$$

where i is imaginary unit, μ_1 and μ_2 are called combination parameters.

Note that the parameters μ_1 and μ_2 are introduced to reduce the effect of illumination variations. Here, we select $\mu_1 = 1/m_1$, $\mu_2 = 1/m_2$, where, m_1 is the mean of all elements of component S, and m_2 is the mean of all elements of component V.

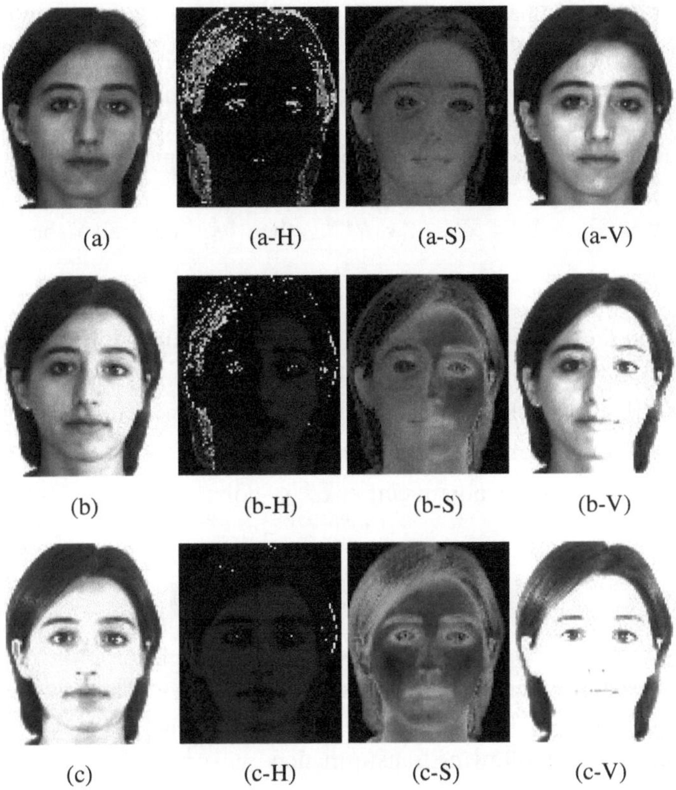

Fig. 1. Three images under different illumination conditions and their corresponding hue (H), saturation (S) and value (V) component images

The complex-matrix is used to represent color face. It can be converted into same dimensional complex vector X, which is called the image vector.

3 Complex Eigenfaces Technique

In documents [5, 6], principle component analysis (PCA) is generalized to suit for feature extraction in complex feature space. In a similar way, the Eigenfaces technique can be generalized.

The total covariance matrix S_t in complex image vector space is defined by

$$S_t = \frac{1}{M}\sum_{i=1}^{m}(X_i - \overline{X})(X_i - \overline{X})^H \tag{2}$$

where H is the denotation of conjugate transpose; M is the total number of training samples; $\overline{X}$ denotes the mean vector of training samples.

It is easy to know S_t is a non-negative definite Hermite matrix.

Since n-dimensional image vectors will result in an $n \times n$ covariance matrix S_t, if the dimension of image vector is very high, it is very difficult to calculate S_t's eigenvectors directly. As we know, in face recognition problems, the total number of training samples m is always much smaller than the dimension of image vector n, so, for computational efficiency, we can adopt the following technique to get the S_t's eigenvectors.

Let $Y = (X_1 - \overline{X}, \cdots, X_m - \overline{X})$, $Y \in R^{n \times m}$, then S_t can also be denoted by $S_t = \frac{1}{M} YY^H$.

Form matrix $R = Y^H Y$, which is a $m \times m$ non-negative definite Hermite matrix. Since R's size is much smaller than that of S_t, it is much easier to get its eigenvectors. If we work out R's orthonormal eigenvectors $v_1, v_2, \cdots, v_m$, and suppose the associated eigenvalues satisfy $\lambda_1 \geq \lambda_2 \geq \cdots \geq \lambda_m$, then, it is easy to prove that the orthonormal eigenvectors of S_t corresponding to nonzero eigenvalues are

$$u_i = \frac{1}{\sqrt{\lambda_i}} Y v_i \text{ , } i = 1, \cdots, r \; (r \leq m-1) \tag{3}$$

And, the associated eigenvalues are exactly $\lambda_i, i = 1, \cdots, r$. The first d eigenvectors (eigenfaces) are selected as projection axes, and the resulting feature vector of sample X can be obtained by the following transformation

$$Y = \Phi^H X \text{ , where, } \Phi = (u_1, \cdots, u_d) \tag{4}$$

4 Experiment

We intend to test our idea on AR face database, which was created by Aleix Martinez and Robert Benavente in CVC at the U.A.B [7]. This database contains over 4,000 color images corresponding to 126 people's faces (70 men and 56 women). Images feature frontal view faces with different facial expressions, illumination conditions, and occlusions (sun glasses and scarf). The pictures were taken at the CVC under strictly controlled conditions. No restrictions on wear (clothes, glasses, etc.), make-up, hair style, etc. were imposed to participants. Each person participated in two sessions, separated by two weeks (14 days) time. The same pictures were taken in both sessions. Each section contains 13 color images. Some examples are shown in web page (http://rvl1.ecn.purdue.edu/~aleix/aleix_face_DB.html).

1-1 1-5 1-6 1-7 1-14 1-18 1-19 1-20

Fig. 2. The training and testing samples of the first man in the database, where, (1-1) and (1-14) are training samples, the remaining are testing samples

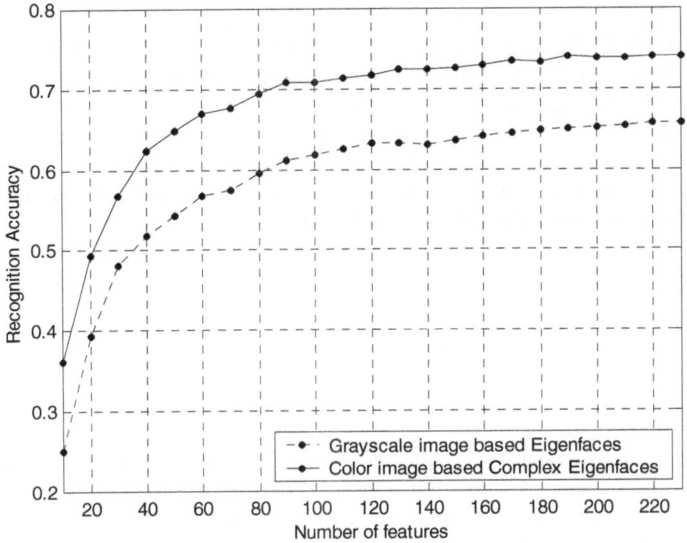

Fig. 3. Comparison of the proposed color image based Complex Eigenfaces and the traditional grayscale image based Eigenfaces under a nearest neighbor classifier (NN)

In this experiment, 120 different individuals (65 men and 55 women) are randomly selected from this database. We manually cut the face portion from the original image and resize it to be 50×40 pixels. Since the main objective of this experiment is to compare the robustness of face representation approaches in variable illumination conditions, we use the first image of each session (No. 1 and No. 14) for training, and the other images (No. 5, 6, 7 and No. 18, 19, 20), which are taken under different illumination conditions and without occlusions, are used for testing. The training and testing samples of the first man in the database are shown in Fig. 2.

The images are first converted from RGB space to HSV space. Then, the saturation (S) and value (V) components of each image are combined together by Eq. (1) to represent face. In the resulting complex image vector space, the developed complex Eigenfaces technique is used for feature extraction. In the final feature space, a nearest neighbor classifier is employed. When the number of selected features varies from 10 to 230 with an interval of ten, the corresponding recognition accuracy is illustrated in Fig. 3.

For comparison, another experiment is performed using the common method. The color images are first converted to gray-level ones by adding the three color channels, i.e., $I = \frac{1}{3}(R+G+B)$. Then based on these grayscale images, classical Eigenfaces [2,3] technique is used for feature extraction and a nearest neighbor classifier is employed for classification. The recognition accuracy is also illustrated in Fig. 3.

From Fig. 3, it is obvious that the proposed color image based complex Eigenfaces is superior to the traditional grayscale image based Eigenfaces. The top recognition accuracy of the complex Eigenfaces reaches 74.0%, which is an increase of 8.3% compared to the Eigenfaces (65.7%). This experimental result also demonstrates that color image based face representation and recognition is more robust to illumination variations.

5 Conclusion

In this paper, we first propose a new strategy for representing color face images, that is, to combine the two color attributes, saturation and value, together by a complex matrix. Then, a technique called complex Eigenfaces is developed for feature extraction. The experimental results indicate that the proposed color image based complex Eigenfaces outperforms the traditional grayscale image based Eigenfaces and also demonstrate that the developed color image based face representation and recognition method is more robust to illumination variations.

References

1. W. Zhao, R. Chellappa, A. Rosenfeld, and P. Phillips, Face recognition: A literature survey. Technical Report CAR-TR-948, UMD CS-TR-4167R, August (2002)
2. M. Turk and A. Pentland, Eigenfaces for recognition. J. Cognitive Neuroscience, 3(1) (1991) 71-86
3. M. Turk and A. Pentland, Face recognition using Eigenfaces. Proc. IEEE Conf. On Computer Vision and Pattern Recognition, (1991) 586-591
4. Y. Wang and B. Yuan, A novel approach for human face detection from color images under complex background. Pattern Recognition, 34 (10) (2001) 1983-1992
5. J. Yang, J.-y. Yang, Generalized K-L transformed based combined feature extraction. Pattern Recognition, 35 (1) (2002) 295-297
6. J. Yang, J.-y. Yang, D. Zhang, J. F. Lu, Feature fusion: parallel strategy vs. serial strategy. Pattern Recognition, 36 (6) (2003) 1369-1381
7. A.M. Martinez and R. Benavente, The AR Face Database. CVC Technical .Report #24, June (1998)

Face Verification Based on Bagging RBF Networks

Yunhong Wang[1], Yiding Wang[2], Anil K. Jain[3], and Tieniu Tan[4]

[1] School of Computer Science and Engineering, Beihang University, Beijing, 100083, China
yhwang@buaa.edu.cn
[2] Graduate School, Chinese Academy of Sciences, Beijing, 100049, China
ydwang@gscas.ac.cn
[3] Department of Computer Science & Engineering,
Michigan State University, East Lansing, MI 48824
jain@cse.msu.edu
[4] National Laboratory of Pattern Recognition, Institute of Automation, Chinese Academy of Sciences, P.O. Box 2728, Beijing 100080, P.R. China
tnt@nlpr.ia.ac.cn

Abstract. Face verification is useful in a variety of applications. A face verification system is vulnerable not only to variations in ambient lighting, facial expression and facial pose, but also to the effect of small sample size during the training phase. In this paper, we propose an approach to face verification based on Radial Basis Function (RBF) networks and bagging. The technique seeks to offset the effect of using a small sample size during the training phase. The RBF networks are trained using all available positive samples of a subject and a few randomly selected negative samples. Bagging is then applied to the outputs of these RBF-based classifiers. Theoretical analysis and experimental results show the validity of the proposed approach.

1 Introduction

Systems based on face recognition and verification play an important role in applications such as access control, credit card authentication, video surveillance, etc., where the identity of a user has to be either determined or validated. Although face recognition and face verification use similar algorithms [1], they are two different problems with different inherent complexities [2]. Recognition is an N-class problem, where the input face image is mapped to one of the N possible identities, whereas verification is a 2-class problem, where the input image is mapped to one of two classes, genuine or impostor. In other words, recognition necessitates a one-to-many matching, while verification requires a one-to-one matching. In designing a classifier for face verification, both positive and negative learning samples are needed. Usually, a very small number of positive (genuine) samples and a very large number of negative (impostor) samples are available during training. Thus, the classifier will be over-fitting the impostor samples while it is learning using only a few positive samples. Simply put, the generalization ability of the classifier during the training stage is very low. This could be one reason why face

D. Zhang and A.K. Jain (Eds.): ICB 2006, LNCS 3832, pp. 69 – 77, 2005.

verification systems do not achieve high matching accuracy. In this paper, we will introduce a technique to decrease this effect by non-equilibrium training. A radial basis function (RBF) network is a good classifier for face recognition because it has the ability to reduce misclassifications among the neighboring classes [6]. Another advantage of RBF network is that it can learn using both positive and negative samples [10]. This property motivates the choice of RBF network for face verification. We train several RBF networks for verification, and we boost the performance by bagging the results of these various networks. There are many methods for face verification described in the literature [3][4][5][10]. Most of them operate by training a classifier that is unique for each subject, although the structure of the classifier is the same for all subjects. Theoretically, the number of possible impostor samples for a subject should be much larger than the number of genuine samples. In practice, only a subset of impostor samples is used for training and, hence, the impostor space cannot be established very well. However, we cannot collect samples of all possible impostors. This makes it difficult to arrive at a reasonable estimation of the probability space of impostors. Therefore, we will not attempt to estimate the probability of impostor space by using all possible impostor samples. Rather, we use some of the samples selected randomly from the impostor database (along with all available genuine samples) in the training stage of each RBF classifier. The number of training samples for each classifier is small compared to the dimensionality of the data (number of features). Usually, a classifier that is constructed using a small training set is biased and has a large variance since the classifier parameters are poorly estimated. Consequently, such a classifier may be weak, having a poor performance [7]. Bagging is a good technique to combine weak classifiers resulting in a powerful decision rule.

In this paper, we use bagging to combine the RBF classifiers in order to improve the accuracy of face verification. The rest of the paper is organized as follows: Section 2 introduces face feature extraction and a classifier based on RBF networks and bagging; experimental results are given in Section 3; Section 4 presents a discussion and summary of the work.

2 Face Verification

2.1 The Problem of Face Verification

The verification problem can be formulated as follows. Classify a test sample S (a face image) into one of the following two classes: ω_0 (genuine) or ω_1 (impostor). Let Y be a feature vector extracted from S, then

Assign $S \rightarrow \omega_j$ if

$$P(\omega_j|Y) = \max_{k=0}^{1} P(\omega_k|Y) \quad j=0, 1 \tag{1}$$

Where $P(w_k|Y)$ denotes the posteriori probability of w_k given Y.

2.2 Feature Representation Using Eigenface

We use the eigenface technique to represent a face image [9]. Let the ith sample face image be represented as an N-dimensional vector X_i, i=1,2…n. The scatter matrix S of all the n samples is computed as

$$S = \sum_i (X_i - \mu)(X_i - \mu)^T \tag{2}$$

Where μ is the mean vector. Here, only a portion of the available database is used to create the eigenspace. For each image X, we obtain a feature vector Y by projecting X onto the subspace generated by the M principal directions, according to the following equation:

$$Y = W^T X \tag{3}$$

Images are compared by finding the distance between their corresponding feature vectors. In our face verification problem, we represent each face sample as a 40-dimensional (M=40) and a 10-dimensional (M=10) vector, respectively. Since the first 3 eigenvectors are related to the variation in illumination (see, Pentland [9]), we eliminate the first 3 eigenvectors for every face sample. Each subject is trained using a different classifier.

2.3 RBF Neural Network

The output of the jth hidden node in a RBF network can be expressed as [11]:

$$O_{hk} = \Phi(\|Y_k - \mathbf{C}_j\|) \quad \text{j=1,2,...} \; N_0 \tag{4}$$

where Y_k is a M-dimensional input vector, $\mathbf{C}_j$ is the center of the jth RBF network, N_0 is the number of hidden units, and $\Phi(\cdot)$ is a nonlinear, radial symmetric function whose center is C_j. We use the Gaussian function as the basis function. The output of the hidden layer can be written as:

$$O_{hk} = \Phi(\|Y_\mathbf{k} - \mathbf{C}_j\|) = \exp\left[-\sum_{i=1}^{M} \frac{(y_i - C_{ij})^2}{2\rho_j^2}\right] \tag{5}$$

The output of the ith output unit of the RBF network is:

$$z_{ki} = \sum_h w_{ih} \Phi(\|Y_k - \mathbf{C_j}\| + w_{k0} \tag{6}$$

We use the training samples to compute the center, $\mathbf{C}_j$. The ρ_j s are selected based on the method in [6], namely, that ρ_j is computed based on inter-class and intra-class distances. One of the advantages of RBF network is that it can be trained using both positive and negative samples. Note that since we are dealing with a verification problem, we can build an individual network for each subject in the database. This is because, for verification, an unknown individual must claim his identity first, and therefore we would know which network to use.

2.4 Bagging

Bagging has proven to be a useful method to improve the classifier performance by combining individual classifiers. Such a combination often gives better results than any of the individual classifiers [8]. As mentioned above, RBF classifiers we have used are weak classifiers. It is necessary to boost their performance by using a bagging technique. Bagging is implemented in the following way [8]:

1. Let b = 1,2, …, B be the training samples available for a subject. The following two steps are done for each b:
   (a) Take a bootstrap replicate Z^b of the training data set Z.
   (b) Construct a classifier $C^b(z)$ (with a decision boundary $C^b(z) = 0$) on Z^b.
2. Combine classifiers $C^b(z)$, b = 1,2,…B, by a simple majority voting (the most often predicted label) to obtain the final decision rule as follows:

$$\beta(z) = \arg\max_{y\in\{-1,1\}} \sum_b \delta_{\mathrm{sgn}(C^b(z)),y} \quad \text{Where } \delta_{i,j} = \begin{cases} 1, i = j \\ 0, i \neq j \end{cases}$$

is the Kronecker symbol, $y \in \{-1,1\}$ is a decision (class label) of the classifier.

Note that there are multiple RBF networks trained for one subject so the decision is made by majority voting. The network classifiers are combined using the bagging rule. We have used different number of classifiers (1, 5, 10, 15, and 20, respectively) in our experiments. To evaluate the performance of bagging, we conduct another experiment in which only a single RBF network is used for every subject. The features are once again extracted using the eigenface technique. The negative (impostor) samples correspond to the genuine samples of all the other subjects. All the negative samples are used during training along with all the available positive samples. This method is represented as PCA+RBF in Table 2 and Table 4.

2.5 Universal and Individual Eigenface Method

We compare the proposed method to two existing approaches to face verification: the universal eigenface and individual eigenface methods [2]. The universal eigenface method constructs an eigenspace using all the training data available for all the subjects. The templates are the coefficients of the projected vectors in the above eigenspace. The distance between the coefficients of a test image and the template is used as a matching score. If the matching score exceeds a threshold, it is declared to be an impostor. Here we use different thresholds for each subject. The thresholds are proportional to the variability of interclass and intraclass. The basic idea of the individual eigenface method [2] is to capture the intra-class variations of each subject, such as changes in expression, illumination, age, etc. In the individual PCA approach, one eigenspace is constructed for each training subject. The residue of a test vector to that vector's individual eigenspace (i.e., the squared norm of the difference between a test vector and its representation in the eigenspace) is used to define the matching score. The thresholds are computed from the training set. We set different thresholds for each subject.

3 Experimental Results

3.1 Database

We use the ORL [8], Yale [12] and NLPR databases for face verification. While the first two are well known public domain face databases, NLPR consists of face images taken in national lab of pattern recognition (NLPR) at two different time instances. Examples of typical face images from NLPR databases are shown in Figure 1. The ORL database contains 40 subjects and 400 images. The Yale database contains 15 subjects and 165 images. The NLPR database contains 19 subjects and 266 images. The above databases are composed of faces with reasonable variations in expression, pose and lighting. We use 6 samples per subject as the positive (genuine) training data. All the images are preprocessed to decrease the effect of variations in illumination. To test our proposed approach on a larger database, we combine all the three databases. The integrated database includes the ORL, NLPR, Yale and MIT databases. The first three databases are introduced above. The MIT database contains 16 subjects and 432 images. There are 90 subjects in the integrated database.

Fig. 1. NLPR face Database

3.2 Experimental Results

The experiments were conducted in the following way: Firstly, all the images in the training set are mapped to the eigenspace to generate the projected feature vectors. For each subject, we select 6 samples (randomly) as positive training samples and the images of the remaining subjects are regarded as impostors. Half of the samples coming from the other subjects are used as negative training data. Secondly, 6 samples are randomly selected from the negative training data and combined with the positive training data. This data is used to train several RBF classifiers. Finally, the output of these RBF classifiers are bagged using the methods described in 2.4. The results are shown in Tables 1 and 2. Table 2 gives the verification results using the universal and individual eigenface methods. The PCA+RBF method is used to refer to the technique where we first extract features of each subject via the eigenface method, and then use all the training samples to construct a RBF classifier for verification. This method has been mentioned in Section 2.4. We use 40-dimensional and 10-dimensional eigenfeatures to realize the proposed methods.

Table 1. Verification error of bagging RBF classifiers (all the impostor subjects are used in training) on the ORL database

Number of classifiers	FRR (%) Dimension of features		FAR (%) Dimension of features	
	40	10	40	10
1	17.75	12.50	24.48	21.32
5	2.50	2.30	3.87	3.79
10	0	0	3.06	1.18
15	0	0	4.51	2.56
20	0	0	6.00	4.53

Table 2. The Verification Error Rates on Yale and NLPR databases (10 dimensional feature space; all the impostor subjects are used in training)

Number classifiers	Yale Database		NLPR database	
	FRR	FAR	FRR	FAR
1	20.32	28.58	25.33	32.38
5	10.26	12.63	16.42	21.84
10	3.72	4.19	4.85	6.62
15	4.27	6.35	5.62	8.39
20	4.86	7.07	6.16	8.85

It has to be noted that in an operational system, some of the impostor samples, with respect to a single subject, cannot be acquired during the training phase. We refrain from using all the impostor subjects for training and testing the performance of the verification system. Instead, we use a subset of the impostors during training. Thus, the testing database contains some subjects that are not presented in the training database. The results of bagging the RBF classifiers are shown in Tables 4 and 5. Table 6 gives the verification results of the universal and individual eigenface methods. We can see that the results with the NLPR and Yale databases are not as good as that of the ORL database. The reason is that there are more variations in the NLPR and Yale databases which affect the accuracy of face verification. This can not be compensated for by learning. The error rates of bagging on the integrated database is even lower than that

Table 3. The verification error rates of Universal eigenface (UEigenface), individual eigenface(IEigenface), PCA+RBF, and bagging , (10 dimensional feature space)

Verification System	ORL Database		Yale Database		NLPR Database		Integrated Database	
	FRR	FAR	FRR	FAR	FRR	FAR	FRR	FAR
UEigenface	4.51	4.20	6.74	6.79	8.19	8.86	12.57	15.72
IEigenface	3.30	2.81	5.41	4.92	6.63	6.06	8.15	7.82
PCA+ RBF	5.00	8.74	14.57	12.59	18.71	16.62	23.71	21.80
Bagging RBF	0	1.18	3.72	4.19	4.85	6.62	5.73	6.88

Table 4. Verification Error Rates of bagging RBF classifiers on ORL database

Number of classifiers	FRR (%) Dimension		FAR (%) Dimension	
	40	10	40	10
1	19.78	13.55	24.48	25.65
5	4.50	2.60	8.73	4.82
10	0	0	3.12	1.24
15	0	0	4.64	2.69
20	0	0	6.25	4.62

Table 5. The Verification Error Rates on Yale and NLPR databases (10 dimensional feature space, part of the impostor database is used in training)

Number of classifiers	Yale Database		NLPR database	
	FRR	FAR	FRR	FAR
1	21.57	27.65	28.39	34.63
5	11.48	12.93	17.16	20.96
10	3.74	4.31	5.02	6.52
15	4.47	6.81	5.78	8.85
20	4.92	7.85	6.42	9.10

Table 6. The verification rates of Universal Eigenface (UEigenface), Individual Eigenface(IEigenface), PCA+RBF, and bagging, (10 dimensional feature space, part of the impostor subjects are used in training)

Verification System	ORL Database		Yale Database		NLPR Database		Integrated Database	
	FRR	FAR	FRR	FAR	FRR	FAR	FRR	FAR
UEigenface	4.59	5.3	8.84	9.19	11.10	12.21	12.91	14.64
IEigenface	3.60	3.48	6.72	5.53	8.83	8.80	9.71	9.10
PCA+ RBF	5.00	9.89	18.16	19.37	24.31	28.89	25.92	26.61
Bagging RBF	0	1.24	3.74	4.31	5.02	6.52	5.62	6.61

on the NLPR. The reason is that the eigenface created by the different databases emphasizes different variations (illumination, pose etc.). The differences among the subjects are not emphasized in this way. That means that the eigenfeature of every subject is not as 'significant' as in the individual database.

3.3 Discussions

We have applied bagging on the outputs of multiple RBF classifiers to improve the performance of a face verification system. It has been shown that the proposed method has better matching performance than the universal eigenface and individual eigenface methods. One of the advantages of the proposed approach is the use of a subset of the subjects as impostors during training without compromising the verification performance. Another advantage of the proposed approach is that its verification accuracy is not proportional to the number of classifiers. Using a large number of classifiers does not result in a higher accuracy of verification. In our proposed system, 10 classifiers are sufficient for bagging. The reason for this could be that the feature vector is 10-dimensional while the number of training samples for every classifier is only 12.Experimental results show that the 10-dimensional feature vector gives better verification results than the 40-dimensional feature vector. The error rate of bagging RBF does not decrease dramatically when only a subset of the impostor samples is employed in training, while other face verification methods do not have this advantage. The error rates on Yale database and NLPR database are high because there are many variations in illumination and pose in these two databases. Considering that only 6 randomly selected samples are used in the training phase, these results are reasonably good. This is typical in real systems since often we can only get a small number of positive samples that may not be typical ones for a person.

4 Conclusions

In summary, the proposed approach not only has a good accuracy but also has a good generalization capability. The accuracy may be attributed to the following: (i) The RBF classifier can learn not only from positive samples but also from negative samples. (ii) We have selected the negative samples randomly and combined them with an equal number of positive samples. This will decrease the over-fitting of negative subjects. (iii) The random choice of negative samples enhances the generalization ability that is useful when all the impostor samples are not available.

Acknowledgements

We would like to thank Arun Ross for a careful reading of this paper. This research was supported by a grant from the Chinese NSFC (No.60332010).

References

1. C. L. Kotropoulos, C. L. Wilson, S. Sir Ohey, Human and Machine Recognition of Faces: a Survey, Proc. IEEE 83(5), (1995), 705-741.
2. Xiaoming Liu, Tsuhan Chen, and B. V. K. Vijaya Kumar, Face authentication for multiple subjects using eigenflow, Pattern Recognition, 36(2), (2003), 313-328,
3. Gian luca marcialis and Fabio Roli, Fusion of LDA and PCA for Face Verification, Biometric Authentication, LNCS 2359, Proc. of ECCV 2002 Workshop, (2002), 30-37.
4. http://www.ece.cmu.edu/~marlene/kumar/Biometrics_AutoID.pdf
5. Yunhong Wang, Tieniu Tan and Yong Zhu, Face Verification Based on Singular Value Decomposition and Radial Basis Function Neural Network, Proceedings of Asian Conference on Computer Vision, ACCV, (2002), 432-436.
6. Meng Joo Er, Shiqian Wu, Juwei Lu, and Hock Lye Toh, Face Recognition With Radial Basis Function (RBF) Neural Networks, IEEE Trans. on NN, 13(3), (2002), 697-710.

7. Marina Skurichina and Robert P. W. Duin, Bagging, Boosting and the Random Subspace Method for Linear Classifiers, Pattern Analysis & Applications, 5 (2002) 121-135,
8. Ferdinando Samaria and Andy Harter, Parameterization of a Stochastic Model for Human Face Identification, in Proc. 2nd IEEE workshop on Applications of Computer Vision, Sarasota, FL, 1994.
9. M. Turk and A. Pentland, Eigenfaces for Recognition, Journal of Cognitive Neuroscience, 3(1), (1991), 71-86.
10. Simon Haykin, Neural Networks: A Comprehensive Foundation, MacMillan Publishing Company, 1994.
11. http://cvc.yale.edu/projects/yalefaces/yalefaces.html.

Improvement on Null Space LDA for Face Recognition: A Symmetry Consideration

Wangmeng Zuo[1], Kuanquan Wang[1], and David Zhang[2]

[1] School of Computer Science and Technology, Harbin Institute of Technology, Harbin, 150001, China

[2] Biometrics Research Centre, Department of Computing, The Hong Kong Polytechnic University, Hung Hom, Kowloon, Hong Kong

Abstract. The approximate bilateral symmetry of human face has been explored to improve the recognition performance of some face recognition algorithms such as Linear Discriminant Analysis (LDA) and Direct-LDA (D-LDA). In this paper we summary the ways to generate virtual sample using facial symmetry, and investigate the three strategies of using facial symmetric information in the Null Space LDA (NLDA) framework. The results of our experiments indicate that, the use of facial symmetric information can further improve the recognition accuracy of conventional NLDA.

1 Introduction

It is well known that face has an approximate bilateral symmetry, which has been investigated in psychology and anthropology to study the relation of facial symmetry and facial attractiveness [1]. As to face recognition, Zhao et al. have utilized the facial symmetry to generate virtual mirrored training images [2]. More recently, the mirrored images are used as both training and gallery images [3]. Rather than the mirrored image, Marcel proposed another symmetric transform to generate virtual images [4].

Facial asymmetry also contains very important discriminative information for person identification. In [5], psychologists found the potential role of facial asymmetry in face recognition by humans. Recently, Liu revealed the efficacy of facial asymmetry in face recognition over expression variation [6]. Soon after they find that facial asymmetry can also be used to facial expression recognition.

While comparing with facial asymmetry, facial symmetry still has its advantageous properties. The measurement of facial asymmetry is based on the normalization of facial image according to the inner canthus (C_1, C_2) of each eye and the philtrum (C_3). The accurate location of these three points, however, is practically very difficult due to the complexity of lighting and facial variation. Besides, the asymmetric discriminative information greatly suffers from the variation of lighting and pose. In [6], Liu investigate only the frontal face recognition problem. But for facial symmetry, it is natural to believe that face image has the symmetric illumination and pose variations.

Most current work on facial symmetry concentrates on two aspects, *how to generate virtual images* and *how to use virtual images*. For the first problem, Zhao pro-

D. Zhang and A.K. Jain (Eds.): ICB 2006, LNCS 3832, pp. 78 – 84, 2005.

posed to generate the mirrored images [2] and Marcel proposed a symmetric transform to generate virtual samples [4]. For the second problem, most researchers use the LDA and Direct-LDA (D-LDA) frameworks [2, 3, 4].

In this paper, we extend to use facial symmetry in the Null Space LDA (NLDA) framework. Because of the particularity of NLDA, the common strategy to use symmetric information may be ineffective. We thus investigate some novel strategies, and comparatively evaluate them on two FERET face subsets.

2 Null Space LDA: A Brief Review

Null Space LDA (NLDA) is a natural extension of conventional LDA when the within-class scatter matrix $\mathbf{S}_w$ is singular [7, 8]. LDA intends to obtain the discriminant vector by maximizing the Fisher's linear discriminant criterion. When the within-class scatter matrix $\mathbf{S}_w$ is singular, we could find a subspace spanned by $\mathbf{U} = [\varphi_1, \varphi_2, \cdots, \varphi_d]$ (hereafter this subspace is named as the null space of $\mathbf{S}_w$) that satisfies

$$\mathbf{U}^T \mathbf{S}_b \mathbf{U} > 0 \text{ and } \mathbf{U}^T \mathbf{S}_w \mathbf{U} = 0 . \tag{1}$$

where $\mathbf{S}_b$ is the between-class scatter matrix. In this subspace, the Fisher's discriminant criterion degenerates to

$$J'_{FLD}(\mathbf{w}) = \left|\mathbf{w}^T \mathbf{U}^T \mathbf{S}_b \mathbf{U} \mathbf{w}\right| = \left|\mathbf{w}^T \tilde{\mathbf{S}}_b \mathbf{w}\right| . \tag{2}$$

Another way to construct $\tilde{\mathbf{S}}_b$ is to find an orthonormal basis for the range of $\mathbf{S}_w$, $\mathbf{Q} = [\mathbf{u}_1, \mathbf{u}_2, \cdots, \mathbf{u}_{d_w}]$. Then $\mathbf{S}_b$ can be projected into the null space of $\mathbf{S}_w$ by $\tilde{\mathbf{S}}_b = \mathbf{S}_b - \mathbf{Q}(\mathbf{Q}^T \mathbf{S}_b \mathbf{Q})\mathbf{Q}^T$. The discriminant vectors of NLDA are obtained by calculating the eigenvectors of $\mathbf{U}^T\mathbf{S}_b\mathbf{U}$. By choosing the eigenvectors $\mathbf{W} = [\mathbf{w}_1, \mathbf{w}_2, \cdots, \mathbf{w}_{d_{\mathrm{NLDA}}}]$ corresponding to the first d_{NLDA} largest eigenvalues, we obtain the NLDA projector

$$\mathbf{T}_{\mathrm{NLDA}} = \mathbf{U}\mathbf{W} . \tag{3}$$

From the previous discussion, the NLDA projector is easy to be calculated once we find the null or the range space of $\mathbf{S}_w$. Next, we review two methods in addressing this issue: by solving eigen-problems [7, 8] and by Gram-Schmidt orthogonalization [8].

2.1 Constructing $\tilde{\mathbf{S}}_b$ by Solving Eigen-Problems

To obtain the null space of $\mathbf{S}_w$, Yang proposed to first calculate all the eigenvectors $\mathbf{\Phi} = [\phi_1, \phi_2, \cdots, \phi_{d_{\mathrm{PCA}}}]$ corresponding to positive eigenvalues of the total scatter matrix $\mathbf{S}_t$. With the PCA projector $\mathbf{\Phi}$, we can construct a $d_{\mathrm{PCA}} \times d_{\mathrm{PCA}}$ matrix $\tilde{\mathbf{S}}_w$

$$\tilde{\mathbf{S}}_w = \mathbf{\Phi}^T \mathbf{S}_w \mathbf{\Phi} . \tag{4}$$

Then, we calculate the eigenvectors corresponding to the zero eigenvalues of $\tilde{\mathbf{S}}_w$. Yang has proved that the subspace spanned by $\mathbf{V}=[\mathbf{v}_{d_w+1},\mathbf{v}_{d_w+2},\cdots,\mathbf{v}_{d_{PCA}}]$ is the null space of $\tilde{\mathbf{S}}_w$ [7]. So we can obtain $\tilde{\mathbf{S}}_b=\mathbf{U}^T\mathbf{S}_b\mathbf{U}$, where $\mathbf{U}$ is defined as

$$\mathbf{U}=\mathbf{\Phi V}=[\mathbf{\Phi v}_{d_w+1},\mathbf{\Phi v}_{d_w+2},\cdots,\mathbf{\Phi v}_{d_{PCA}}]. \tag{5}$$

Actually, we can obtain $\tilde{\mathbf{S}}_b$ without calculating the eigenvectors of $\mathbf{S}_t$. In [8], Cevikalp proposed to compute the eigenvectors $\mathbf{Q}=[\mathbf{u}_1,\mathbf{u}_2,\cdots,\mathbf{u}_{d_w}]$ corresponding to the positive eigenvalues of $\mathbf{S}_w$. Then $\mathbf{S}_b$ can be projected into the null space of $\mathbf{S}_w$ by

$$\tilde{\mathbf{S}}_b=\mathbf{S}_b-\mathbf{Q}(\mathbf{Q}^T\mathbf{S}_b\mathbf{Q})\mathbf{Q}^T. \tag{6}$$

2.2 Constructing $\tilde{\mathbf{S}}_b$ by Gram-Schmidt Orthogonalization

Gram-Schmidt orthogonalization is introduced to speed up the computation of $\tilde{\mathbf{S}}_b$. Both the two methods in Section 2.1 require $O(N^3)$ floating point multiplications. Actually all orthogonal basis for the range of $\mathbf{S}_w$ are equivalent. From this aspect, Cevikalp proposed a fast method with $O(N^2)$ multiplications to constructing $\tilde{\mathbf{S}}_b$ [8].

Give a training set $\mathbf{X}=\{\mathbf{x}_1^{(1)},\cdots,\mathbf{x}_{N_1}^{(1)},\mathbf{x}_1^{(2)},\cdots,\mathbf{x}_j^{(i)},\cdots,\mathbf{x}_{N_C}^{(C)}\}$, we should first find the independent difference vector which spanned the difference subspace $\mathbf{B}$. In [8], Cevikalp had proved the equivalence of the difference subspace and the range space of $\mathbf{S}_w$. Then Gram-Schmidt orthogonalization procedure is used to find an orthonormal basis $\mathbf{Q}=[\beta_1,\beta_2,\cdots,\beta_{N-C}]$ of $\mathbf{B}$. Next we can project the between-class scatter matrix $\mathbf{S}_b$ into the null space of $\mathbf{S}_w$ by $\tilde{\mathbf{S}}_b=\mathbf{S}_b-\mathbf{Q}(\mathbf{Q}^T\mathbf{S}_b\mathbf{Q})\mathbf{Q}^T$.

3 Strategies of Using Facial Symmetry in the NLDA Framework

In this Section, we investigate the ways to utilize facial symmetry in the NLDA framework from two aspects. First, we summarize the ways to generate virtual images. Second, we investigate three possible methods to use facial symmetric transform.

3.1 Two Ways to Generate Virtual Images Using Facial Symmetry

We name the way to generate virtual images using facial symmetry as facial symmetric transform. So far, there are mainly two kinds of facial symmetric transform, *SymmTrans-I* and *SymmTrans-II*, defined as follows:

Definition 1. Given a facial image $A=(a_{i,j})_{m\times n}$, *SymmTrans-I* is defined to transform $\mathbf{A}$ to a new image $\mathbf{A}'=(a'_{i,j})_{m\times n}$ by $a'_{i,j}=a_{i,(n-j+1)}$.

Definition 2. Given a facial image $A=(a_{i,j})_{m\times n}$, *SymmTrans-II* is defined to transform $\mathbf{A}$ to a new image $\mathbf{A}''=(a''_{i,j})_{m\times n}$ by $a''_{i,j}=(a_{i,j}+a_{i,(n-j+1)})/2$.

(a) (b) (c)

Fig. 1. Illustration of the results of facial symmetric transform: (a) original image; and the virtual images generated by (b) SymmTrans-I and (c) SymmTrans-II

These two facial symmetric transform had been reported in some literature. The virtual image generated by SymmTrans-I is usually called as mirrored image and has been applied in [2, 3]. The virtual image generated by SymmTrans-II has been used in [4] and Marcel finds that SymmTrans-II can alleviate the effect of small pose variation. As an example, Fig. 1 illustrates the results of these two facial symmetric transform.

In NLDA, the image **A** should always be mapped to an image vector **a** in advance. Thus the virtual images generated by SymmTrans-I and SymmTrans-II should also be mapped to their corresponding image vectors, $\mathbf{a}'$ and $\mathbf{a}''$.

3.2 Three Methods to Use the Facial Symmetric Transform

Facial symmetric transform can be used to generate virtual training images, virtual NLDA projector, or virtual gallery images. In this section, we investigate these three ways of utilizing facial symmetric information in the NLDA framework.

Generally, the NLDA-based face recognition system involves two stages, training and testing. In the training stage, the NLDA projector is obtained by learning from the training set and the gallery images are then projected into gallery feature vectors. In the testing stage, an image from the probe set is first projected into probe feature vector and then a nearest neighbor classifier is used to recognize the probe feature vector.

Fig. 2~4 illustrates the framework of using virtual training images, using virtual projector and virtual gallery images in NLDA-based face recognition. In Fig. 2, we use facial symmetric transform to obtain a virtual training set. Then both the training set and the virtual training set are used in the NLDA learning to obtain the projector (*SymmNLDA-I*). This is the most popular strategy of using facial symmetric transform and has been adopted in [2, 4]. But for NLDA, this strategy may be ineffective because the addition of virtual training set may decrease the discriminative information in the null space of $\mathbf{S}_w$, and further degrade the recognition accuracy of NLDA.

In Fig. 3, facial symmetric transform is used to obtain a virtual projector. We use the NLDA projector and the virtual projector to extract two feature vectors, and then we combine the classification results based on these two feature vectors (*SymmNLDA-II*). For details of the combination rule, see [9]. In Fig. 4, facial symmetric transform is used to obtain a virtual gallery set. Then both the gallery set and the virtual gallery set are used to construct the generalized gallery feature sets (*SymmNLDA-III*).

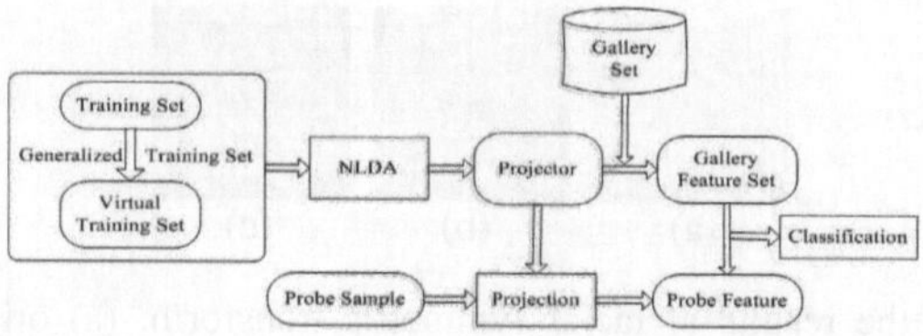

Fig. 2. An illustration of using virtual training set in the NLDA framework (SymmNLDA-I)

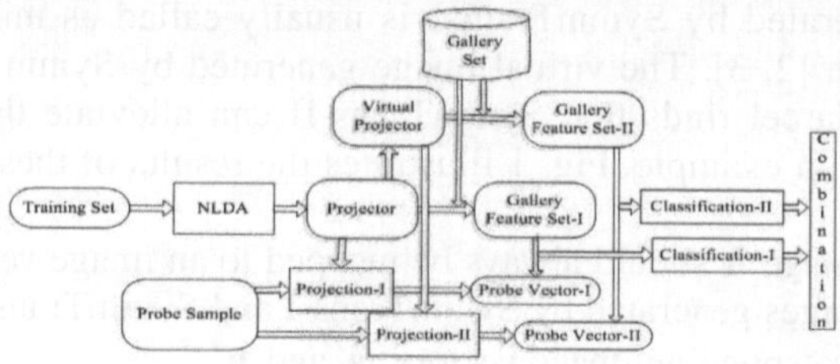

Fig. 3. An illustration of using virtual projector in the NLDA framework (SymmNLDA-II)

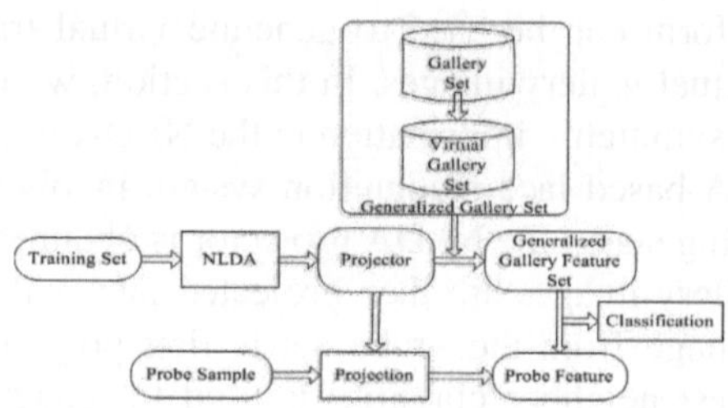

Fig. 4. An illustration of using virtual gallery set in the NLDA framework (SymmNLDA-III)

4 Experimental Results and Discussions

In this section, we use two face subsets from the FERET database (FERET-1 and FERET-2) to evaluate the facial symmetry in NLDA. To simplify the problem, we just compare the recognition rate of the three methods using SymmTrans-II.

4.1 Experimental Results on FERET-1 Database

In this section, we chose a subset from the FERET database (FERET-1) which includes 1,400 images of 200 individuals (each individual has seven images). The seven images of each individual consist of three front images and four profile images. The facial portion of each original image was cropped to a size of 80×80 and pre-processed using histogram equalization. Fig. 5 presents 7 cropped images of a person.

Fig. 5. Seven images of one person from the FERET-1 database

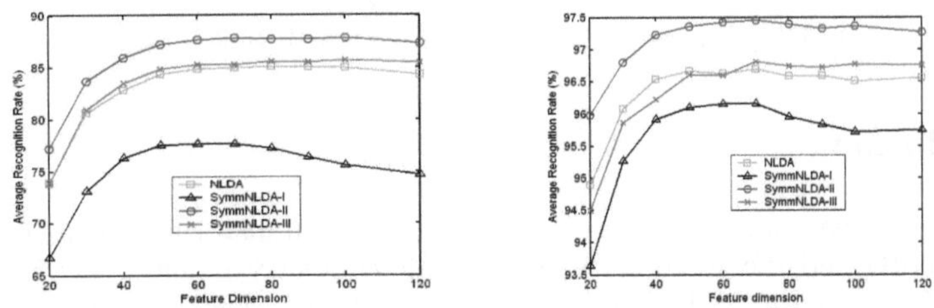

(a) (b)

Fig. 6. Plots of the ARRs of NLDA, SymmNLDA-I, SymmNLDA-II, SymmNLDA-III: (a) FERET-1, and (b) FERET-2

The experimental setup is summarized as follows: First all the images of 100 persons are randomly selected for training. We use the 100 neutral frontal images of the other 100 persons as gallery images, and the remaining images as probe images. We run the recognition method 10 times to calculate the average recognition rate (ARR).

Fig. 6(a) depicts the ARRs obtained using NLDA, SymmNLDA-I, SymmNLDA-II, and SymmNLDA-III. SymmNLDA-II and SymmNLDA-III achieve higher ARRs than NLDA and the highest ARR is obtained using SymmNLDA-II. But the ARR of SymmNLDA-I is much lower than that of NLDA, though the addition of virtual training samples has been reported to improve the recognition performance for subspace LDA and D-LDA [2, 4, 3]. NLDA extracts the discriminative information in the null space of $\mathbf{S}_w$. The addition of virtual training samples, however, enriches facial information in the range space of $\mathbf{S}_w$, and may degrade the recognition performance of NLDA.

4.2 Experimental Results on FERET-2 Database

We use a FERET subset consisted of 1195 people with two images (fa/fb) for each person (FERET-2). The facial portion of each image was cropped to a size of 80×80 and pre-processed by histogram equalization. Fig. 7 shows the ten pre-processed images of five persons. In our experiment, we randomly select 495 persons to construct the training set. Then, the 700 regular frontal images (fa) of the other 700 persons are used as gallery set, and the remained 700 images (fb) are used as probe set. We run the face recognition method 10 times and calculate the average recognition rate.

Fig. 6(b) illustrates the ARRs obtained using NLDA, SymmNLDA-I, SymmNLDA-II, and SymmNLDA-III. SymmNLDA-II and SymmNLDA-III also achieve higher maximum ARR than conventional NLDA and the highest ARR is obtained using SymmNLDA-II. But the ARR of SymmNLDA-I is lower than that of NLDA.

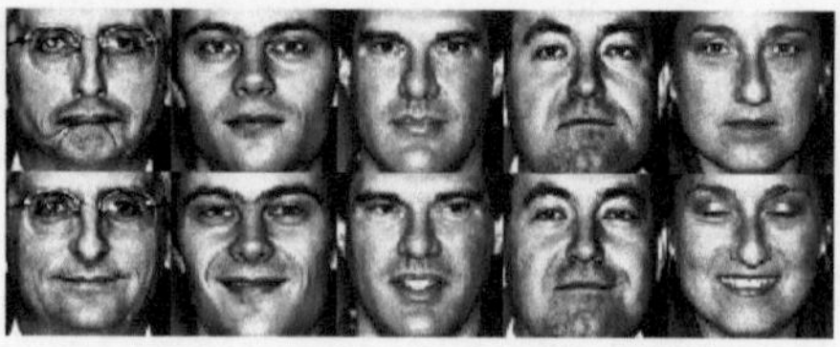

Fig. 7. Ten images of five persons from the FERET-2 database

5 Conclusion

In this paper we summary the facial symmetric transform (SymmTrans-I and SymmTrans-II) and the methods to use facial symmetry in the NLDA framework (SymmNLDA-I, SymmNLDA-II and SymmNLDA-III). Two face subsets from the FERET database are used to evaluate these methods. Experimental results show that SymmNLDA can further improve the recognition performance of NLDA. For a database of 1195 persons with expression variation, SymmNLDA-II achieves an average recognition rate of 97.46% with 495 persons for training and 700 persons for testing.

Acknowledgements

The work is partially supported by the NSFC fund under the contract No. 60332010 and No. 90209020.

References

1. Grammer, K., and Thornhill, R.: Human (Homo sapiens) facial attractiveness and selection: The role of symmetry and averageness. Journal of Comparative Psychology, 108 (1994) 233-242.
2. Zhao W, Chellappa R, Phillips P.J.: Subspace Linear Discriminant Analysis for Face Recognition. Tech Report CAR-TR-914, Center for Automation Research, University of Maryland (1999).
3. Lu, J., Plataniotis, K.N., and Venetsanopoulos, A.N.: Regularization studies of linear discriminant analysis in small sample size scenarios with application to face recognition. Pattern Recognition Letters, 26 (2005) 181-191.
4. Marcel, S., A symmetric transformation for LDA-based face verification. Proc. 6th IEEE Int'l Conf. Automatic Face and Gesture Recognition (2004) 207-212.
5. Troje, N.F., and Buelthoff, H.H.: How is bilateral symmetry of human faces used for recognition of novel views?. Vision Research, 38 (1998) 79-89.
6. Liu, Y., Schmidt, K.L., Cohn, J.F., and Mitra, S.: Facial asymmetry quantification for expression invariant human identification. CVIU, 91 (2003) 138-159.
7. Yang, J., Zhang, D., and Yang, J.Y.: A generalized K-L expansion method which can deal with Small Smaple Size and high-dimensional problems. PAA, 6(2003), 47-54.
8. Cevikalp, H., Neamtu, M., Wilkes, M., and Barkana, A.: Discriminative common vectors for face recognition. IEEE Trans. PAMI, 27(2005), 4-13.
9. Marcialis, G.L., Roli, F.: Fusion of appearance-based face recognition algorithms. Pattern Analysis and Applications, 7(2004), 151-163.

Automatic 3D Face Recognition Using Discriminant Common Vectors

Cheng Zhong, Tieniu Tan, Chenghua Xu, and Jiangwei Li

National Laboratory of Pattern Recognition, Institute of Automation, Chinese Academy of Sciences, Beijing, 100080, P.R. China
{czhong, tnt, chxu, jwli}@nlpr.ia.ac.cn

Abstract. In this paper we propose a fully automatic scheme for 3D face recognition. In our scheme, the original 3D data is automatically converted into the normalized 3D data, then the discriminant common vector (DCV) is introduced for 3D face recognition. We also compare DCV with two common methods, i.e., principal component analysis (PCA) and linear discriminant analysis (LDA). Our experiments are based on the CASIA 3D Face Database, a challenging database with complex variations. The experimental results show that DCV is superior to the other two methods.

1 Introduction

Automatic identification of human faces is a very challenging research topic, which has gained much attention during the last few years. Most of this work, however, is focused on intensity or color images of faces [1]. There is a commonly accepted claim that face recognition in 3D is superior to 2D because of the invariance of 3D sensors to illumination and pose variation. Recently with the development of 3D acquisition system, 3D face recognition has attracted more and more interest and a great deal of research effort has been devoted to this topic.

Many methods have been proposed for 3D face recognition over the last two decades. Some earlier research on curvatures analysis has been proposed for face recognition based on the high-quality 3D data, which can characterize delicate features [2] [3]; In [4], a 3D morphable model is described with a linear combination of the shape and texture of multiple exemplars. This model can be fitted to a single image to obtain the individual parameters, which are used to characterize the personal features; Chua et al. [5] treat face recognition as a 3D non-rigid surface matching problem and divided the human face into rigid and non-rigid regions. The rigid parts are represented by point signatures to identify the individual. Beumier et al. [6] develop a 3D acquisition prototype based on structure light and built a 3D face database. They also propose two methods of surface matching and central/lateral profiles to compare two instances. Chang et al. [7] use PCA on both 2D intensity images and 3D depth images, and fuse 2D and 3D results to obtain the final performance. Their results show that the combination

D. Zhang and A.K. Jain (Eds.): ICB 2006, LNCS 3832, pp. 85–91, 2005.

of 2D and 3D features is very effective for characterizing a person. However, it should be noted that the existing methods usually have a high computational cost [4] [6], involve small databases [3] [5] or depend on manual labeled points [7].

In this paper, we introduce a fully automatic 3D face recognition scheme. The flowchart is shown in Fig. 1. First, we preprocess the input 3D data. Second, we use DCV to project the normalized 3D data from the original high dimensional space to low dimensional subspace spanned by DCV. Third, we use nearest neighbor (NN) classifier to classify the 3D face images. We also make a detailed comparison between DCV, LDA and PCA to test their performance for 3D face recognition.

The main contributions of this paper are as follows: (1) We introduce the DCV method into 3D face recognition; (2) We make a detailed comparison between PCA, LDA and DCV.

The rest of this paper is organized as follows. In Section 2, we describe the 3D face data preprocessing. A detailed description on DCV is illustrated in Section 3. Section 4 shows the experimental results, and finally we conclude this paper in Section 5.

2 3D Face Data Preprocessing

Fig. 2 shows some examples from the CASIA 3D Face Database. The original images have many problems, such as different poses and much noise, so data preprocessing is necessary before recognition. Data preprocessing includes the following three steps:

The first step is nose location. In this step, we use local features to obtain some nose tip candidate points, and a trained SVM classifier is used to find the nose tip point [8].

The second step is the registration. In this step, we construct a mesh model corresponding to each 3D face and the ICP algorithm is applied to the mesh models to complete the registration [9].

The third step is data normalization. In this step, we follow the method as stated in [7], but here we use a double mask scheme. Because the margin region contains more noise than the region of interest, we first adopt a large mask. After we fill holes and smooth the data, we adopt a small mask to obtain the region of interest, which is the final output depth image. Fig. 3 shows some normalized 3D images after the data preprocessing.

3 3D Face Representation Using DCV

In this section, we mainly describe how to represent 3D face images using DCV. The main procedures can be summarized as follows: first we need to calculate common vector (CV) images from the given training set; second we calculate the DCV based on the obtained CV images; finally we represent the original 3D faces using DCV. Next we will describe these procedures in detail.

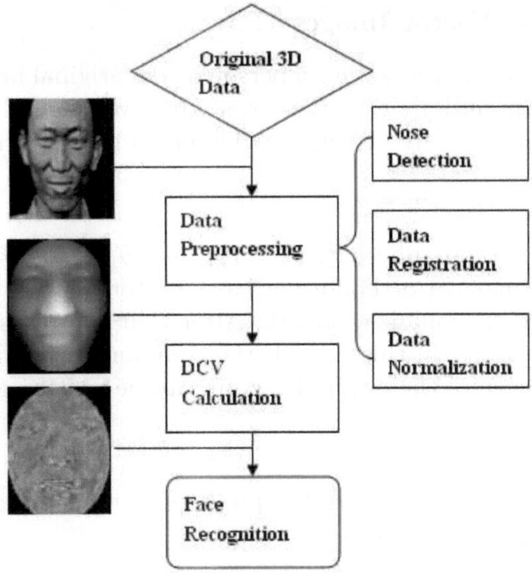

Fig. 1. The flowchart of our automatic 3D face recognition

Fig. 2. Original 3D images

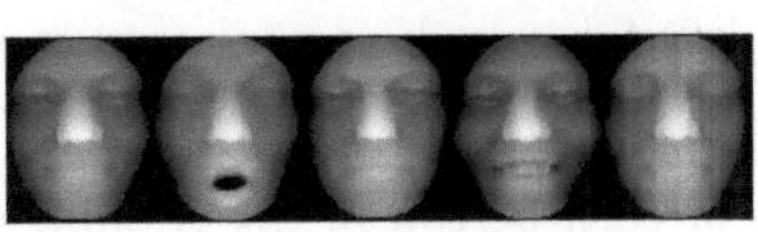

Fig. 3. Preprocessed 3D images

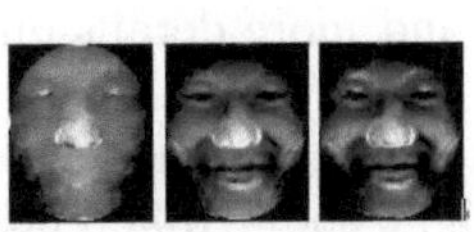

Fig. 4. Comparison of different common vector images, the first one is common vector image of five images with neural expression, the second one is common vector image of five images with different expressions and the third one is common vector image of the above ten images

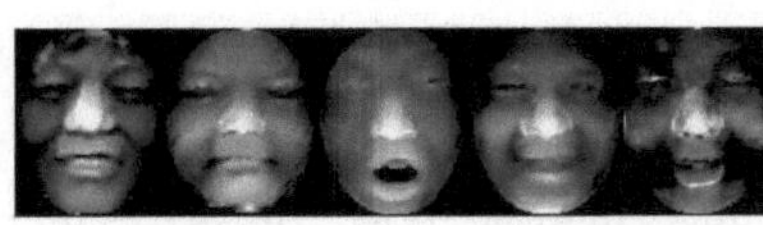

Fig. 5. Common vector images

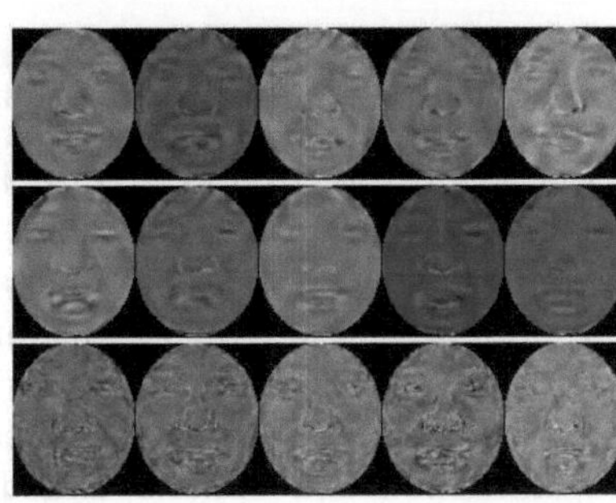

Fig. 6. Comparison of eigenfaces, fisherfaces and discriminant common vector images. The first row shows the eigenfaces, the second row shows the fisherfaces, and the third row shows discriminant common vector images

3.1 Common Vector Images

Supposed that in training set each person has m original images $\{a_1, a_2, \cdots, a_m\}$. We convert them into m original vectors , then we define the $(m-1)$ dimensional difference subspace B by taking the differences between the vectors, i.e.

$$b_i = a_{i+1} - a_1 (i = 1, 2, \cdots, m-1) \tag{1}$$

B is spanned by these difference vectors. Since $b_1, b_2, \cdots, b_{m-1}$ are not expected to be orthonormal, an orthonormal basis vector set can be obtained by using Gram-Schmit orthogonalization [10]. After that the basis vector set for B will be $\{z_1, z_2, \cdots, z_{m-1}\}$ in this case. If the common vector of one person is called as a_{com}, then each of the original vectors can be written as

$$a_i = a_{i,dif} + a_{com} \tag{2}$$

the difference vectors $a_{i,dif}$ are the projections of the original vectors onto the difference subspace B, that is

$$a_{i,dif} =< a_i, z_1 > z_1 + < a_i, z_2 > z_2 + \cdots + < a_i, z_{m-1} > z_{m-1} \tag{3}$$

We can obtain m difference vectors from m original vectors. The common vector a_{com} is chosen as

$$a_{com} = a_i - a_{i,dif} \qquad \forall i = 1, 2, \cdots, m \tag{4}$$

It can be seen as the projection of the original vectors onto the indifference subspace. As $a_{com} = a_1 - a_{1,dif} = a_2 - a_{2,dif} = \cdots = a_m - a_{m,dif}$, we can obtain only one common vector from the m original vectors of one person and more details on this may be found in [11]. Fig. 5 shows some common vector images.

3.2 Discriminant Common Vectors

After we obtain the common vector images of each person in the training set, we attempt to compute discriminant common vectors. DCV is the projection that maximizes the total scatter across the common vector images. We can use PCA to obtain the discriminant common vectors and more details on this may be found in [12].

After we obtain DCV, we can project the original high dimensional space into the low dimensional subspace spanned by DCV. Fig. 6 shows different eigenface, fisherface and the discriminant common vector images, respectively. From this figure, we can find that the discriminant common vector images contain more detailed information than eigenfaces or fisherfaces.

4 Experimental Results and Discussion

To make a comparison between PCA, LDA and DCV methods, we have done many experiments on CASIA 3D Face Database. There are 123 persons in the database, and each person has 37 or 38 images. In our experiment we only use 5 images with neural expression and 5 images with different expressions (smile, laugh, anger, surprise, eye closed) for each person. First we construct a small 3D face database (DB1), which including 5 images with neural expression and 2 images with common expressions (smile and eye closed). Second we use the whole set of images to construct a larger 3D face database (DB2), which includes 5 images with neural expression and 5 images with different expressions. The comparisons of the three methods are all based on the same training sets and testing sets. In all experiments, we use the NN classifier with Mahalanobis cosine distance.

4.1 Experiments on DB1

We list the recognition rate in two cases. First, we use the first three images with neural expression as the training set (Experiment 1), and the remained images as the testing set. Second, we use one image with neural expression, one image with smile expression, one image with eye closed expression as the training set (Experiment 2), and the remaining images as the testing set. The results are shown in Table 1.

4.2 Experiments on DB2

We list the recognition rate in four cases. First, we use the first three images with neural expression as the training set (Experiment 3), and the remained images as the testing set. Second, we use five images with neural expression as the training set (Experiment 4), and the remained images as the testing set. Third, we use one image with neural expression, one image with laugh expression, one image with surprising expression as the training set (Experiment 5), and the remained images as the testing set. Fourth, we use the five images with different expressions as the training set (Experiment 6), and the remained images as the testing set. The results are shown in Table 2.

4.3 Experimental Results Analysis

From Table 1 and Table 2, we can make the following observations:

1) When the intra-class variation is large, we obtain a better performance;
2) In most cases, DCV obtains the best performance;
3) Although the size of training set 4 is larger than training set 3, its performance is worse.

Because the DCV performance mainly depends on the common vectors obtained, here we explain the reasons using common vectors. Fig. 4 shows the common vector images in different situations. We find when in training set one person contains much intra-class variation, the common vector image is almost the same as that of the whole set of images, which means the training set is a very

Table 1. Rank one recognition rates on DB1

Methods	Experiment1	Experiment2
DCV	99%	99.2%
LDA	97.4%	98.4%
PCA	92.9%	94.7%

Table 2. Rank one recognition rates on DB2

Methods	Experiment3	Experiment4	Experiment5	Experiment6
DCV	90.7%	84.6%	96.1%	98.5%
LDA	86.9%	87.4%	93.1%	97.4%
PCA	83.5%	80.8%	92.6%	94.0%

Table 3. Recognition rates on different size of training set

Size	2	3	4	5
Verification rate	90.2%	90.7%	87.4%	84.6%

good representation of the whole set of images, so all methods obtain better performance in this case. From the Section 3 we can find that if training set is a good representation of the whole set of images, DCV is a better choice. Not only it exploits the structures of the original high dimensional space, but also it is the best optimization of the Fisher linear discriminant rule. So in most cases, it performs better than the other two methods. But because DCV exploits more information than other methods from the training set, its recognition performance also depends more on training set. Table. 3 shows the recognition rates with different size of training sets only containing images with neural expression. Although the size of training set is increased, the recognition rate drops. We encounter the over fitting problem here. Because training set is not a representation of the whole set of images, the result we obtained is lack of the generalization ability. In this case, the projection cannot get a good performance in the testing set.

4.4 Discussion

As to computation cost, we only consider the eigen-analysis which is the most time-consuming procedure. Suppose we have N images in the training set, which can be divided into c classes ($N > c$). Then eigen-analysis is performed on one matrix in DCV ($c \times c$), one matrix in PCA ($N \times N$) and two matrices in LDA (one is ($N \times N$) , the other is ($(N - c) \times (N - c)$)).The comparison illustrates DCV is the most efficient method of the three.

There are also some drawbacks of our experiments. Because of the limitation of the CASIA 3D Face Database, we only have the 3D face data in one session and we cannot test the influence on the DCV method due to session variations.There are also some other public 3D face databases, such as FRGC, but it is a manual labeled database and its 3D face data does not suit to our preprocessing algorithm. Using the given points, our experimental results show that DCV also performs better than LDA and PCA on FRGC1.0.

5 Conclusions

In this paper, we have presented a fully automatic system integrating efficient DCV representation for 3D face recognition. We have also compared our proposed method with two other commonly used methods, i.e., PCA and LDA on a large 3D face database. All the experiments are performed in a fully automatic way. From the experimental results, we find that DCV obtains a better performance than LDA and PCA.

Acknowledgement

This work is funded by research grants from the National Basic Research Program (Grant No. 2004CB318110).

References

1. R. Chellapa, C. L. Wilson, and S. Sirohey. Human and machine recognition of faces: A survey. In *Proceedings of the IEEE*, pages 705–740, May 1995.
2. G. Gordon. Face recognition based on depth and curvature features. In *Proc. CVPR*, pages 108–110, June 1992.
3. J. C. Lee and E. Milios. Matching range images of human faces. In *Proc. ICCV*, pages 722–726, 1990.
4. V. Blanz and T. Vetter. Face identification based on fitting a 3d morphable model. *IEEE Trans. PAMI*, (9):1063–1074, 2003.
5. C. S. Chua, F. Han, and Y. K. Ho. 3d human face recognition using point signature. In *Proc. FG*, pages 233–239, 2000.
6. C. Beumier and M. Acheroy. Automatic 3d face authentication. *Image and Vision Computing*, (4):315–321, 2000.
7. K. I. Chang, K. W. Bowyer, and P. J. Flynn. An evaluation of multi-model 2d+3d face biometrics. *IEEE Trans. PAMI*, (4):619–624, 2005.
8. Chenghua Xu, Yunhong Wang, Tieniu Tan, and Long Quan. Robust nose detection in 3d facial data using local characteristics. In *Proceedings of the IEEE, International Conference of Image Processing*, pages 1995–1998, 2004.
9. Chenghua Xu, Yunhong Wang, Tieniu Tan, and Long Quan. Automatic 3d face recognition combining global geometric features with local shape variation information. In *Proceedings of the IEEE, International Conference Automatic Face and Gesture Recognition*, pages 308–313, 2004.
10. M. Keskin. Orthogonalization process of vector space in work station and matlab medium. In *Elect. Electron. Eng. Dept., Osmangazi Univ., Eskisehir, Turkey*, July 1994.
11. M. B. Gulmezoglu, V. Dzhafarov, and A. Barkana. The common vector approach and its relation to principal component analysis. *IEEE Transactions on Speech and Audio Processing*, (6):655–662, 2001.
12. H. Cevikalp, M. Neamtu, M. Wilkes, and A. Barkana. Discriminative common vectors for face recognition. *IEEE Trans. PAMI*, (1):4–13, 2005.

Face Recognition by Inverse Fisher Discriminant Features

Xiao-Sheng Zhuang[1], Dao-Qing Dai[1,*], and P.C. Yuen[2]

[1] Center for Computer Vision and Department of Mathematics,
Sun Yat-Sen (Zhongshan) University, Guangzhou 510275 China
Tel: (86)(20)8411 3190; Fax: (86)(20)8403 7978
stsddq@mail.sysu.edu.cn

[2] Department of Computer Science, Hong Kong Baptist University, Hong Kong
pcyuen@comp.hkbu.edu.hk

Abstract. For face recognition task the PCA plus LDA technique is a famous two-phrase framework to deal with high dimensional space and singular cases. In this paper, we examine the theory of this framework: (1) LDA can still fail even after PCA procedure. (2) Some small principal components that might be essential for classification are thrown away after PCA step. (3) The null space of the within-class scatter matrix S_w contains discriminative information for classification. To eliminate these deficiencies of the PCA plus LDA method we thus develop a new framework by introducing an inverse Fisher criterion and adding a constrain in PCA procedure so that the singularity phenomenon will not occur. Experiment results suggest that this new approach works well.

1 Introduction

Face recognition [8, 18] technique has wide applications. Numerous algorithms have been proposed. Among various solutions, the most successful are those appearance-based approaches. Principle component analysis (PCA) and linear discriminant analysis (LDA) are two classic tools widely used in the appearance-based approaches for data reduction and feature extraction. Many state-of-the-art methods, such as Eigenfaces and Fisherfaces [2], are built on these two techniques or their variants. Although successful in many cases, in real-world applications, many LDA-based algorithms suffer from the so-called "small sample size problem"(SSS) [12]. Since SSS problem is common, it is necessary to develop new and more effective algorithms to deal with them. A number of regularization techniques that might alleviate this problem have been suggested [4-7]. Many researchers have been dedicated to searching for more effective discriminant subspaces [15-17].

A well-known approach, called Fisher discriminant analysis (FDA), to avoid the SSS problem was proposed by Belhumeur, Hespanha and Kriegman [2]. This method consists of two steps: PCA plus LDA. The first step is the use of principal

* Corresponding author.

D. Zhang and A.K. Jain (Eds.): ICB 2006, LNCS 3832, pp. 92–98, 2005.

component analysis for dimensionality reduction. The second step is the application of LDA for the transformed data. The basic idea is that after the PCA step the within-class scatter matrix for the transformed data is not singular. Although the effectiveness of this framework in face recognition are obvious, see [2, 9, 13, 18] and the theoretical foundation for this framework has been also laid [16] yet in this paper we find out that (1) LDA can still fail even after the PCA procedure. (2) Some small principal components that might be essential for classification are thrown away after PCA step. (3) The null space of the within-class scatter matrix S_w contains discriminative information for classification.

In this paper, motivated by the success and power of the PCA plus LDA in pattern classification tasks, considering the importance of the information in the null space of the within-class scatter matrix, and in view of the limitation of the PCA step, we propose a new framework for face recognition.

This paper is organized as follows. In Section 2, we start the analysis by briefly reviewing the two latter methods. We point out the deficiency of the PCA plus LDA method. Following that, our new method is introduced and analyzed in Section 3. In section 4, experiments are presented to demonstrate the effectiveness of the new method. Conclusions are summarized in Section 5.

2 The PCA Plus LDA Approach and Its Deficiency

Suppose that there are K classes, labelled as $G_1, G_2, ..., G_K$. We randomly select n_j samples $X_j^{(i)} (i = 1, 2, ..., n_j)$ from each class $G_j, j = 1, 2, ..., K$ for training. Set $N = \sum_{j=1}^{K} n_j$, $\mu_j = \frac{1}{n_j} \sum_{i=1}^{n_j} X_i^{(j)}, j = 1, 2, \cdots, K$ and $\mu = \frac{1}{N} \sum_{j=1}^{K} \sum_{i=1}^{n_j} X_i^{(j)}$. Let the between-class scatter matrix and the within-class scatter matrix be defined by $S_b = \frac{1}{N} \sum_{j=1}^{K} n_j (\mu_j - \mu)(\mu_j - \mu)^T$, $S_w = \sum_{j=1}^{K} \sum_{i=1}^{n_j} (X_i^{(j)} - \mu_j)(X_i^{(j)} - \mu_j)^T$, $S_t = S_b + S_w$ is the total scatter matrix.

2.1 The PCA Procedure

PCA is a technique now commonly used for dimensionality reduction in face recognition. The goal of PCA is to find out a linear transformation or projection matrix $W_{PCA} \in \mathbb{R}^{d \times d'}$ that maps the original $d-$dimensional image space into an $d'-$dimensional feature space $(d' < d)$ and maximize the determinant of the total scatter of the projected samples, i.e.,

$$W_{PCA} = \underset{W \in \mathbb{R}^{d \times d'}}{arg\, max} |W^T S_t W|. \tag{1}$$

2.2 The LDA Procedure

The aim of LDA is also to find a projection matrix as in PCA that maximizes the so-called **Fisher criterion**:

$$W_{LDA} = \underset{W \in \mathbb{R}^{d \times d'}}{arg\, max} \frac{|W^T S_b W|}{|W^T S_w W|}. \tag{2}$$

2.3 The Deficiency of PCA Plus LDA Approach

When applying the PCA plus LDA approach the following remarks should be considered.

- LDA can still fail even after PCA procedure. For the PCA projected data we get the matrix S'_w, S'_b and S'_t. Then there might exist a direction α such that $\alpha^T S'_t \alpha = \alpha^T S'_b \alpha$ so that $\alpha^T S'_w \alpha = 0$. Hence the matrix S'_w is still singular.
- Some small principal components that might be essential for classification are thrown away after PCA step. Since in PCA step, it just chooses d' eigenvectors corresponding to the first d' largest eigenvalues of S_t. It is very likely that the remainder contains some potential and valuable discriminatory information for the next LDA step.
- The null space of the within-class scatter matrix S_w contains discriminative information for classification. For a projection direct β, if $\beta^T S_w \beta = 0$ and $\beta^T S_b \beta \neq 0$, obviously, the optimization problem (2) is maximized.

3 Inverse Fisher Discriminant Analysis

In this section, we shall develop a new Fisher discriminant analysis algorithm based on **the inverse Fisher criterion**

$$W_{IFDA} = \underset{W \in \mathbb{R}^{d \times d'}}{arg\,min} \frac{|W^T S_w W|}{|W^T S_b W|}. \tag{3}$$

In contrast with LDA or FDA, we name the procedure using the above criterion as **the inverse Fisher discriminant analysis (IFDA)**. Obviously, the Fisher criterion (2) and inverse Fisher criterion (3) are equivalent, provided that the within-class scatter matrix S_w and the between-class scatter matrix S_b are not singular. However, we notice that the rank of the between-class scatter matrix $S_b \in \mathbb{R}^{d \times d}$ satisfies $rank(S_b) \leq K - 1$. Thus, the difficulty of SSS problem still exists for this new criterion.

On the other hand, let us come back to exploit the principle component analysis. For the optimization problem (1), it gives optimal projection vectors that have the largest variance and PCA just selects d' eigenvectors corresponding to the first d' largest eigenvalues of S_t but ignores the smaller ones. If we want to take those eigenvectors into account, we should abandon or modify such criterion for vector selection. Here we present a new criterion by modifying the equation (1) as follow:

$$\begin{aligned} W_{PCA\_S} &= \underset{W \in \mathbb{R}^{d \times d'}}{arg\,max} |W^T S_t W| \\ &= [w_1 \; w_2 \; \cdots \; w_{d'}] \\ s.t. \;\; & w_i^T S_b w_i > w_i^T S_w w_i, \;\; ||w_i|| = 1, \;\; i = 1, 2, \cdots, d' \end{aligned} \tag{4}$$

We name it as **PCA with selection (PCA_S)**. The reduced matrix $S'_b = W^T_{PCA\_S} S_b W_{PCA\_S}$ might still be singular. It is obvious that we should not work in the null space of the reduced within-covariance matrix S'_b. We further project S'_b onto its range space and denote this operation as $W_{proj} \in \mathbb{R}^{d' \times d''}$ $(d'' \leq d')$.

We now introduce our new framework. Firstly, we apply our modified PCA procedure to lower the dimension from d to d' and get a projection matrix $W_{PCA\_S} \in \mathbb{R}^{d \times d'}$. Moreover we project onto the range space of the matrix S'_b and get a projection matrix $W_{proj} \in \mathbb{R}^{d' \times d''}$. Finally, we use IFDA to find out the feature representation in the lower dimensionality feature space $\mathbb{R}^{d''}$ and obtain a transformation matrix W_{IFDA}. Consequently, we have the transformation matrix W_{opt} of our new approach as follow

$$W_{opt}^T = W_{IFDA}^T \cdot W_{proj}^T \cdot W_{PCA\_S}^T,$$

where W_{PCA_S} is the result of the optimization problem (4) and

$$\begin{aligned} W_{IFDA} &= \arg\min_W \frac{|W^T W_{proj}^T W_{PCA\_S}^T S_w W_{PCA\_S} W_{proj} W|}{|W^T W_{proj}^T W_{PCA\_S}^T S_b W_{PCA\_S} W_{proj} W|} \\ &= \arg\min_W \frac{|W^T W_{proj}^T S'_w W_{proj} W|}{|W^T W_{proj}^T S'_b W_{proj} W|} \\ &= \arg\min_W \frac{|W^T S''_w W|}{|W^T S''_b W|} \end{aligned} \tag{5}$$

We call the columns of the transform W_{opt} **the inverse Fisher face (IFFace)** and this new approach as **IFFace** method. Before we go to the end of this part, we make some comments on our new framework.

- Those eigenvectors with respect to the smaller eigenvalues of S_t are taken into account in our modified PCA step.
- Our inverse Fisher criterion can extract discriminant vectors in the null space of S_w rather than just throw them away.

4 Experiment Results

In this section, experiments are designed to evaluate the performance of our new approach: IFFace. Experiment for comparing the performance between FisherFace and IFFace is also done.

Two standard databases from the Olivetti Research Laboratory(ORL) and the FERET are selected for evaluation. These databases could be utilized to test moderate variations in pose, illumination and facial expression. The Olivetti set contains 400 images of 40 persons. Each one has 10 images of size 92×112 with variations in pose, illumination and facial expression. For the FERET set we use 432 images of 72 persons. Each person has 6 images whose resolution after cropping is also 92×112 (See Figure 1). Moreover we combine the two to get a new larger set, the ORLFERET, which has 832 images of 112 persons. We implement our IFFace algorithm and test its performance on the above three databases. On 'Decision Step', We use the l^2 metric as the distance measure. For

Fig. 1. Example images of two subjects(the first row) and the cropped images(the second row) with the FERET database

the classifier we use the nearest neighbor rule. The recognition rate is calculated as the ratio of number of successful recognition and the total number of test samples. The experiments are repeated 50 times on each database and average recognition rates are reported.

4.1 Performance of the IFFace Method

We run our algorithm for the ORL database and the FERET database separately. Figure 2 shows the recognition rates from Rank 1 to Rank 10 for different training sample size with ORL in left and FERET in right. From Figure 2, we can see that, when the training sample size is 5, the recognition rates of Rank 5 for both databases are nearly 99%. These results indicate the effectiveness of our new IFFace method in real-world applications.

4.2 Comparison Between IFFace Method and FisherFace Method

As we know, LDA is based on an assumption that all classes are multivariate Gaussian with a common covariance matrix. For ORL database or FERET database, the assumption is reasonable since a great deal of experiments on these

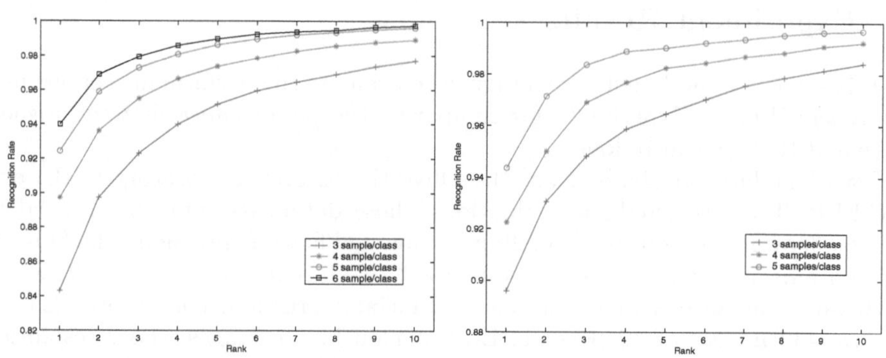

Fig. 2. Recognition rates from Rank 1 to Rank 10 for different training sample per class with ORL database (left) and FERET database (right)

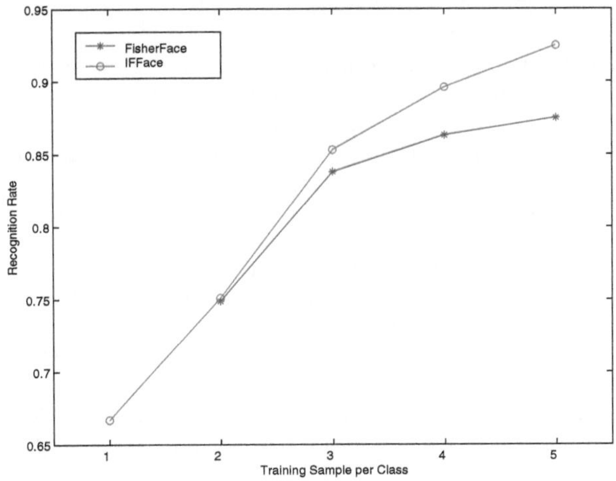

Fig. 3. Comparison between FisherFace and IFFace on the ORLFERET database

two database using FisherFace algorithm have substantiated the efficiency of this two-phrase algorithm. However, when each class has different covariance matrix, this algorithm might not work very well. Therefore, the combination of the two databases would result in a bigger database having different covariance matrix for different classes.

From Figure 3 we can see that IFFace outperforms FisherFace for every number of training sample for each class, take 5, for example, the average recognition rates are 92.5% for IFFace, while for FisherFace it is only 87.6%. This experiment suggests that our IFFace method can work well even in the case that the covariance matrices for different classes are not all the same.

5 Conclusion

In this paper, we proposed a new Fisher discriminant analysis framework: PCA with selection plus IFDA to eliminate deficiencies of the PCA plus LDA method. Based on this framework, we present a new algorithm for face recognition named IFFace method. The algorithm is implemented and experiments are also carried out to evaluate this method. Comparison is made with the PCA plus LDA approach. Further work will be on feature selections and kernel versions.

Acknowledgments

This project is supported in part by grants from NSF of China(Grant No: 60175031, 10231040, 60575004), the Ministry of Education of China, NSF of GuangDong and Sun Yat-Sen University.

References

1. G. Baudat and F. Anouar, Generalized discriminant analysis using a kernel approach, *Neural Computation*, Vol. 12, no. 10(2000), 2385-2404.
2. P. N. Belhumeur, J. P. Hespanha, and D. J. Kriegman, Eigenfaces vs. Fisherfaces: Recognition using class specific linear projection, *IEEE Trans. Pattern Analysis and Machine Intelligence.*, Vol. 19(1997), 711-720.
3. L. F. Chen, H. Y. M. Liao, J. C. Lin, M. D. Kao, and G. J. Yu, A new LDA-based face recognition system which can solve the small sample size problem, *Pattern Recognition*, Vol. 33, no. 10(2000), 1713-1726.
4. W. S. Chen, P. C. Yuen, J. Huang and D. Q. Dai, Kernel machine-based one-parameter regularized Fisher discriminant method for face recognition, *IEEE Trans. on Systems, Man and Cybernetics-part B: Cybernetics*, Vol. 35, no. 4(2005), 657-669.
5. D. Q. Dai and P. C. Yuen, Regularized discriminant analysis and its applications to face recognition, *Pattern Recognition*, Vol. 36, no.3(2003), 845-847.
6. D. Q. Dai and P. C. Yuen, A wavelet-based 2-parameter regularization discriminant analysis for face recognition, *Lecture Notes in Computer Science*, Vol. 2688(2003), 137-144.
7. D. Q. Dai and P. C. Yuen, Wavelet based discriminant analysis for face recognition, *Applied Math. and Computation*, 2005(in press), doi: 10.1016/j.amc.2005.07.044
8. A. K. Jain, A. Ross and S. Prabhakar, An introduction to biometric recognition, *IEEE Transactions on Circuits and Systems for Video Technology*, Vol. 14, No. 1(2004), 4-20.
9. C. J. Liu and H. Wechsler, A shape- and texture-based enhanced fisher classifier for face recognition, *IEEE Trans. Image Processing*, Vol. 10, no. 4(2001), 598-608.
10. S. Mika, G. Rätsch, J Weston, B. Schölkopf, A. Smola, and K.-R. Müller, Constructing descriptive and discriminative nonlinear features: rayleigh coefficients in kernel feature spaces, *IEEE Trans. Pattern Analysis and Machine Intelligence*, Vol. 25, no. 5(2003), 623-628.
11. I. Pima and M. Aladjem, Regularizedd discriminant analysis for face recognition, *Pattern Recognition*, Vol. 37(2004), 1945-1948.
12. S. J. Raudys and A. K. Jain, Small sample size effects in statistical pattern recognition: Recommendations for practitioners, *IEEE Trans. Pattern Anal. Machine Intell.*, Vol. 13(1991), 252-264.
13. D. L. Swets and J. Weng, Using discriminant eigenfeatures for image retrieval, *IEEE Trans. Pattern Analysis and Machine Intelligence*, Vol. 18, no. 8(1996), 831-836.
14. J. Yang, A. F. Frangi, J. Y. Yang, D. Zhang, and Z. Jin, KPCA plus LDA: A complete kernel fisher discriminant framework for feature extraction and recognition, *IEEE Trans. Pattern Analysis and Machine Intelligence*, Vol. 27, no. 2(2005), 230-244.
15. J. P. Ye, R. Janardan, C. H. Park, H. Park, An optimization criterion for generalized discriminant analysis on undersampled problems, *IEEE Transactions on Pattern Analysis and Machine Intelligence*, Vol. 26 (8)(2004) 982-994.
16. H. Yu and J. Yang, A direct LDA algorithm for high-dimensional data-with application to face recognition, *Pattern Recognition*, Vol. 34, no. 10(2001), 2067-2070.
17. B. Zhang, H. Zhang, and S. Sam Ge, Face recognition by applying wavelet subband representation and kernel associative memory, *IEEE Transactions on Neural Networks*, Vol. 15, No. 1(2004), 166-177.
18. W. Zhao, R. Chellappa, P. J. Phillips and A. Rosenfeld, Face recognition: A literature survey, *ACM Comput. Surv.*, Vol. 35 (4)(2003), 399-459.

3D Face Recognition Based on Facial Shape Indexes with Dynamic Programming

Hwanjong Song, Ukil Yang, Sangyoun Lee, and Kwanghoon Sohn*

Biometrics Engineering Research Center,
Dept. of Electrical & Electronic Eng., Yonsei University,
134 Shinchon-dong, Seodaemun-gu, Seoul, 120-749, Korea
{ultrarex, starb612}@diml.yonsei.ac.kr,
{syleee, khsohn}@yonsei.ac.kr

Abstract. This paper describes a 3D face recognition method using facial shape indexes. Given an unknown range image, we extract invariant facial features based on the facial geometry. We estimate the 3D head pose using the proposed error compensated SVD method. For face recognition method, we define and extract facial shape indexes based on facial curvature characteristics and perform dynamic programming. Experimental results show that the proposed method is capable of determining the angle of faces accurately over a wide range of poses. In addition, 96.8% face recognition rate has been achieved based on the proposed method with 300 individuals with seven different poses.

1 Introduction

Over the past few decades, face recognition technologies have made great progress with 2D images, which have played an important role in many applications such as identification, crowd surveillance and access control [1-2]. Although most of the face recognition researches have shown reasonable performance, there are still many unsolved problems in applications with variable environments such as those involving pose, illumination and expression changes. With the development of 3D acquisition system, face recognition based on 3D information is attracting in order to solve problems of using 2D images. A few 3D face recognition approaches have been reported on face recognition using 3D data acquired by 3D sensors [3-5] and stereo-based systems [6]. Especially, most works mentioned above exploited a range image. The advantages of range images are the explicit representation of 3D shape, invariance under change of illumination.

In this paper, we concentrate on the face recognition system using two different 3D sensors. For our system, we utilize the structured light approach for acquiring range data as a probe image and 3D full laser scanned faces for stored images. Fig. 1 briefly presents the whole process of the proposed method. The remainder of this paper is organized as follows: Section 2 describes the representation of 3D faces for the probe

* Corresponding author.

D. Zhang and A.K. Jain (Eds.): ICB 2006, LNCS 3832, pp. 99 – 105, 2005.

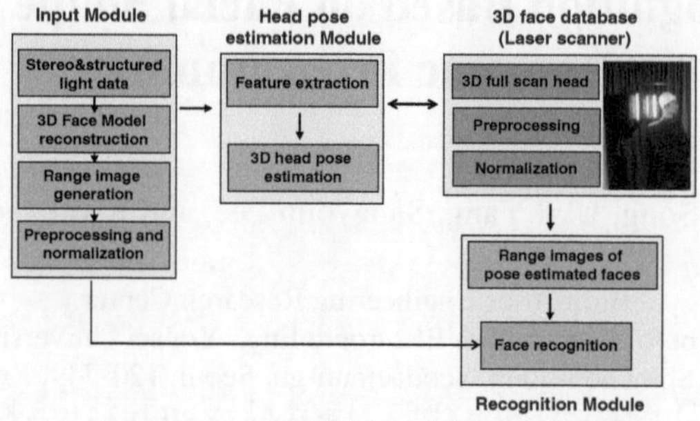

Fig. 1. The block diagram of the proposed method

and store images and describes the extraction of 3D facial feature points. Section 3 introduces an EC-SVD. In section 4, face recognition method is described. In section 5, test performance is analyzed to explain the efficiency of the proposed algorithm. Finally, section 6 concludes by suggesting future directions.

2 Representation of 3D Faces

We acquire a 3D face model from the Genex 3D FaceCam® which is a structured light system in a controlled background. Noise filtering is performed for eliminating the background by some toolkit and we have used the same filter on all images. The orthogonal projection, the range mapping, and projecting uniformly to pixel locations in the image plane are performed with a 3D face model and we generate the range image of the acquired face model. Since the generated range image has some holes to fill due to overlapped or missing the discrete mesh, we use the bilinear interpolation technique.

3D face data is recorded with the *Cyberware*™ Model 3030PS/RGB highly realistic laser scanner with both shape and texture data. For each 3D face, the scans represent face shapes in cylindrical coordinates relative to a vertical axis centered with respect to the head. In angular steps, angle covers 230°, which means that we scan from the left ear to the right ear. All the faces that we consider are in normalized face space and they are located based on the original face data in the limited range of $[-\sigma,\sigma], [-\varepsilon,\varepsilon], [0,Z]$ for the X, Y, and Z axis.

We extract feature points using 3D geometric information. To find the nose peak point (NPP), we select the region from the maximal depth to the depth value lower by three which is empirically found. We calculate the center of gravity of that selected region and treat as an initial NPP. Then we calculate the variances of the horizontal and vertical profiles. We find the points where the minimal variance of the horizontal profiles and the maximal variance of the vertical profiles. We can vertically and almost symmetrically divide the face using the YZ plane which includes the NPP and Y axis, and obtains the face dividing curvature. On the face center curve, we extract facial feature points using curvature characteristics. We finally select six points, which are a minimum point of the nose ridge, the left and right inner eye corner points, a NPP and two nose base points.

3 3D Head Pose Estimation

We describe a 3D head pose estimation algorithm by using 3D facial features. We use them for calculating the initial head pose of the input face based on the Singular Value Decomposition (SVD) method [7]. We utilized EC-SVD to compensate for the rest of the errors which had not yet been recovered from the SVD method [8].

We establish a complete rotation matrix with an assumption that there still exist some errors to compensate for as,

$$R = R_X R_Y R_Z = R_{SVD_x} R_{\theta_x} R_{SVD_y} R_{\theta_y} R_{SVD_z} R_{\theta_z} \qquad (1)$$

Where R : 3×3 rotation matrix , $R_X = R_{SVD_x} R_{\theta_x}$, R_{SVD_x} , R_{SVD_y} , R_{SVD_z} : Rotation matrix obtained from the SVD, R_{θ_x} , R_{θ_Y} , R_{θ_Z} : Error rotation matrix. Since the inverse of the complete rotation matrix must be an input rotated face of frontal view,

$$\mathbf{p_i} = R^{-1}\mathbf{p'_i} = R_Z^{-1} R_Y^{-1} R_X^{-1} \mathbf{p'_i} = R_{\theta_z}^{-1} R_{SVD_z}^{-1} R_{\theta_y}^{-1} R_{SVD_y}^{-1} R_{\theta_x}^{-1} R_{SVD_x}^{-1} \mathbf{p'_i} \qquad (2)$$

where $\mathbf{p'_i}$, $\mathbf{p_i}$ are feature vectors before and after rotation.

After rotating the estimated angle obtained from the SVD method about the X axis, the error θ_x is supposed to be computed for compensating. To estimate θ_x, we exploit the X axis rotation matrix for evaluation. The key feature point is the NPP because all the NPPs of the 3D face model and the input are normalized to the fixed point $p(0,0,z)$ when the face is frontal. We can estimate θ_x from the follow equation.

$$\mathbf{p'} = R_X^{-1}\mathbf{n} = R_{\theta_x}^{-1} R_{SVD_x}^{-1} \mathbf{n} \qquad (3)$$

$$\therefore \theta_x = \arctan\left(\frac{y\cos\theta_{SVD_x} - z\sin\theta_{SVD_x}}{y\sin\theta_{SVD_x} + z\cos\theta_{SVD_x}} \right) \qquad (4)$$

The similar refinement procedure is applied to estimate the error θ_y.

$$\therefore \theta_y = \arctan\left(\frac{x\cos\theta_{SVD_y} - z'\sin\theta_{SVD_y}}{x\sin\theta_{SVD_y} + z'\cos\theta_{SVD_y}} \right) \qquad (5)$$

The error angle for θ_z can be obtained from the method in [8]. When the face vector is denoted as $\vec{F}(a,b,c)$, which is a vertical vector connected from the minimum point of the nose ridge to the center point of the left and right eyebrow.

$$\theta_z = \arcsin\left(\frac{-a}{\sqrt{a^2 + b^2 + c^2}} \right) \qquad (6)$$

4 Face Recognition

In this section, we present a novel face recognition method using the face curvature shape indexes with dynamic programming. Fig. 2 describes the proposed procedure for face recognition. We extract feature points which are defined as areas with large shape variation measured by shape index calculated from principal curvatures [9].

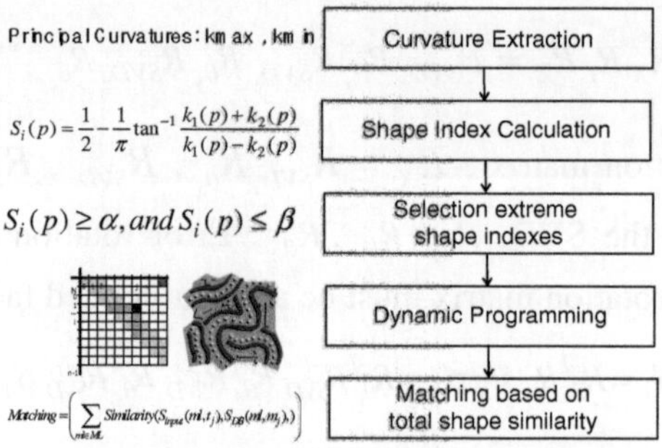

Fig. 2. The proposed face recognition procedure

Shape index $S_i(p)$, a quantitative measure of the shape of a surface point p, is

$$S_i(p) = \frac{1}{2} - \frac{1}{\pi}\tan^{-1}\frac{k_1(p)+k_2(p)}{k_1(p)-k_2(p)} \quad (7)$$

Where $k_1(p)$ and $k_2(p)$ are maximum and minimum principal curvatures. These shape indexes are in the range of [0, 1]. As we can see from [10], there are nine well-known shape categories and their locations on the shape index scale. Among those shape indexes, we select the extreme concave and convex points of curvatures as feature points. These feature points are distinctive for recognizing faces. Therefore, we select those shape indexes as feature points, $feature_i(p)$, if a shape index $S_i(p)$ satisfies the following condition.

$$feature_i(p) = \begin{cases} \partial \le S_i(p) < 1, & concavity \\ 0 < S_i(p) \le \beta, & convexity \end{cases} \quad (8)$$

where $0 < \partial, \beta < 1$. With these selected facial shape indexes, we perform a dynamic programming in order to recognize the faces in the database [11].

We define a similarity measure and Total Shape Similarity Score (TSSS) as follow.

$$Similarity(S_{input}, S_{DB}) = 1 - \left| feature_{input} - feature_{DB} \right| \quad (9)$$

$$TSSS = \sum_{n} Similarity(S_{input}(i, c_j), S_{DB}(i, c_j, n),) \quad (10)$$

where S is denoted as facial shape index. C_j is a face curvature and n is the number faces in the database. Score is a summation of the individual similarity score for each pair of matching descriptors.

5 Experimental Results

We test the 3D head pose estimation based on the proposed EC-SVD and face recognition rate under pose varying environments by using two different 3D sensors.

To evaluate the proposed EC-SVD algorithm, we first extract six facial feature points based on the geometrical configuration, Fig. 3 shows range images of the selected facial feature points of frontal (top row), left (middle row) and right (bottom row) pose variations. To estimate the head pose of an input data, we test range data on various rotation angles. The results are tabulated in Table 1. We obtain various head poses of individuals and we acquire 7 head poses per person such as frontal, ±15 and ±30 for the Y axis, ±15 for the X axis as probe images.

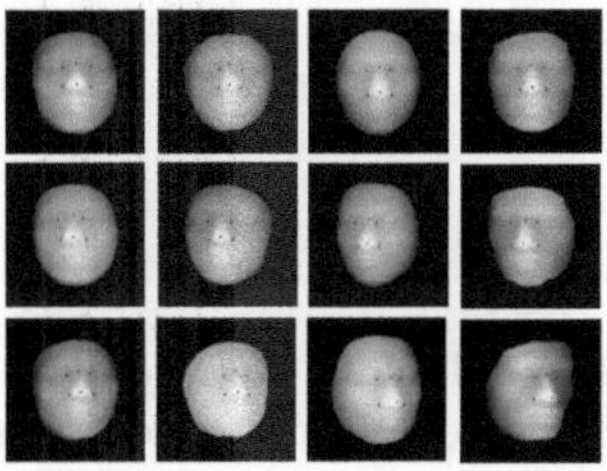

Fig. 3. 3D facial feature extraction for head pose estimation: Top row(frontal), second row(right pose) and third row(left pose)

Table 1. Mean absolute rotation error (degree) and translational error for each axis

Test Images	Mean Absolute Error using SVD(Degree)			Mean Absolute Error using EC-SVD (Degree)			Average Translation errors (RMSE)
	X axis	Y axis	Z axis	X axis	Y axis	Z axis	
Face01	3.0215	4.3265	5.0124	0.8738	1.0125	1.5923	2.756
Face14	3.0214	3.5216	5.1579	1.8991	0.9236	2.0080	2.457
Face23	2.3549	3.6546	3.0646	0.8680	1.1532	1.3783	3.175
Average for all faces	**2.8765**	**3.6563**	**3.8565**	**0.8565**	**0.9654**	**1.5212**	**2.614**

From the results shown in Table 1, we can confirm that the EC-SVD algorithm provides an estimated head pose for a different range of head poses. The error angle for each axis is compensated for any head poses when we normalize the NPP to the fixed point on the Z axis. Less than 1.6 degree error is resulted from our test results for each X, Y and Z axis and it is highly acceptable for pose invariant face recognition. The proposed EC-SVD algorithm recovers the error angle remained by the SVD method, and it can be efficiently applied to pose invariant face.

For the identification of a pose estimated range image, we compare the proposed method with the correlation matching and 3D Principal Component Analysis (PCA) [12]. For the proposed method, we first perform surface reconstruction of the faces from range images. We acquire very smooth facial surfaces from the 3D faces in the database, but discrete lines are appeared on the input face due to structured light patterns. Therefore, we extract curvatures which should be distinctive features for individuals, and adopt this feature which can be utilized for face recognition.

We extract 20 curvatures from the nose peak point which is in the center curvature. We select them based on sampling by two pixels towards the horizontal direction. Among them, we select facial shape indexes based on the threshold as mentioned in

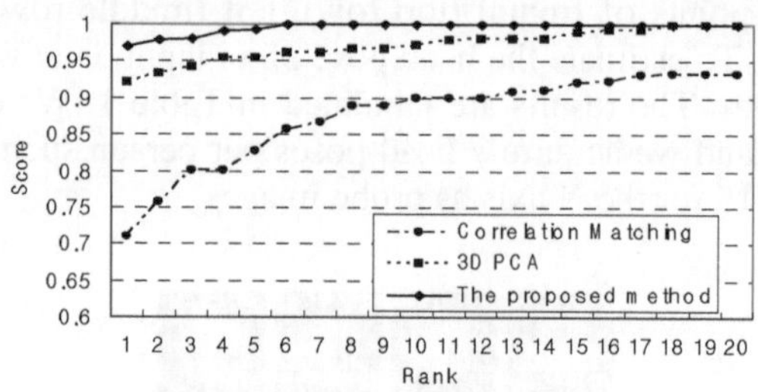

Fig. 4. Comparison of the face recognition rates under different poses

section 4. The determined threshold value α for concave points is 0.75, and β is 0.25 for convex points. These values are selected based on the nine well-known shapes. We compare facial curvatures based on facial shape indexes based on dynamic programming for various head poses.

To describe the face matching, we tabulated matching results based on DP with facial shape indexes. When an input face is selected, we compare all the faces in the database based on the sum of facial shape indexes with DP, finally get a Total Shape Similarity Score (TSSS) for matching. From the experimental results, even though we get the less number of shape indexes than some faces, the TSSS of the identical face in the database is the highest among them. That is, facial shape indexes are distinctive features for face recognition. As we can see from Fig. 4, we have higher recognition rate according to the proposed method. We have 72% recognition rate for the correlation matching and 92% at first rank by the 3D PCA. However, we obtain 96.8% based on the proposed method at first rank under seven different poses. From the simulation results, we have effectively utilized facial shape indexes for pose invariant face recognition and achieved satisfactory recognition rate based on the proposed method.

6 Conclusion

In this paper, we proposed the face recognition method based on facial shape indexes by using two different 3D sensors under pose varying environments. We utilized the advantages of each 3D sensor such as real time 3D data acquisition system for the input and high quality images of 3D heads for the database.

As we can see from the results, we obtained accurate 3D head pose estimation results using the EC-SVD procedure, and the final estimation errors of the 3D head pose estimation in our proposed method were less than 1.6 degree on average for each axis. In addition, our 3D facial feature extraction is automatically performed and assured that geometrically extracted feature points were efficient to estimate the head pose. For face recognition, we used facial shape indexes for recognizing faces with dynamic programming. We obtained 96.8% face recognition rate at first rank based on the proposed method which is highly acceptable results for pose invariant face recognition. We are now researching expression invariant face recognition with more 3D faces.

Acknowledgments

This work was supported by the Korea Science and Engineering Foundation (KOSEF) through the Biometrics Engineering Research Center (BERC) at Yonsei University.

References

1. R. Chellappa, C. L Wilson, and S. Sirohey, "Human and machine recognition of faces : A survey," Proceedings of the IEEE, vol. 83, pp. 705-740, May 1995.
2. W. Zhao, R. Chellappa, P. J. Phillips, and A. Rosenfeld, "Face recognition: A literature survey," ACM, Computing Surveys, Vol. 35, No.4, Dec. 2003.
3. H. T. Tanaka, M. Ikeda and H. Chiaki, "Curvature-based face surface recognition using spherical correlation," Proceedings of the Third International Conference on Automatic Face and Gesture Recognition, pp.372-377, 1998.
4. C. S. Chua, F. Han, and Y. K. Ho, "3D human face recognition using point signature," Proc. of the Fourth International Conference on Automatic Face and Gesture Recognition, pp.233-238, 2000.
5. C. Hesher, A. Srivastava, and G. Erlebacher, "A novel technique for face recognition using range images," Proceedings of the Seventh Int'l Symp. on Signal Processing and Its Applications, 2003.
6. G. Medioni and R. Waupotitsch, "Face recognition and modeling in 3D," Proceedings of the IEEE Int'l Workshop on Analysis and Modeling of Faces and Gestures (AMFG 2003), pp. 232-233, 2003.
7. T.S. Huang, A.N. Netravali, "Motion and structure from feature correspondences: A Review," Proceedings of the IEEE, vol. 82, no. 2, pp. 252-268, 1994.
8. H. Song, J. Kim, S. Lee and K. Sohn, "3D sensor based face recognition," Applied Optics, vol. 44, No. 5, pp.677-687, Feb. 2005.
9. G. G. Gordon, "Face recognition based on depth maps and surface curvature," SPIE Proceedings : Geometric Methods in Computer Vision, San Diego, CA, Proc. SPIE **1570**, 1991.
10. C. Dorai and A. K. Jain, "COSMOS-A Representation Scheme for 3D Free-Form Objects," IEEE Trans. on Pattern Anal. and Machine Intell., vol. 19, no. 10, pp. 1115-1130, Oct. 1997.
11. D. P. Bertsekas, Dynamic Programming and Optimal Control : 2nd Edition, ISBNs : 1-886529-09-4, Nov. 2000.
12. K. Chang, K. Bowyer, and P. Flynn, "Face **recognition** using 2D and 3D facial data," Proceeding of the Multimodal User Authentication Workshop, pp 25–32, 2003.

Revealing the Secret of FaceHashing

King-Hong Cheung[1], Adams Kong[1,2], David Zhang[1],
Mohamed Kamel[2], and Jane You[1]

[1] Biometrics Research Centre, Department of Computing,
The Hong Kong Polytechnic University, Hung Hom, Kowloon, Hong Kong
{cskhc, cswkkong, csdzhang, csyjia}@comp.polyu.edu.hk
[2] Pattern Analysis and Machine Intelligence Lab, University of Waterloo,
200 University Avenue West, Ontario, Canada
mkamel@uwaterloo.ca

Abstract. Biometric authentication has attracted substantial attention over the past few years. It has been reported recently that a new technique called FaceHashing, which is proposed for personal authentication using face images, has achieved perfect accuracy and zero equal error rates (EER). In this paper, we are going to reveal that the secret of FaceHashing in achieving zero EER is based on a false assumption. This is done through simulating the claimants' experiments. Thus, we would like to alert the use of "safe" token.

1 Introduction

Biometric systems for personal authentication have been proposed for various applications based on single or a combination of biometrics, such as face [1], fingerprint [2], [3], iris [4] and palmprint [5] over the past few decades. Although biometric authentication poses several advantages over the classical authentication technologies, all biometric verification systems make two types of errors [6]: 1) misrecognizing measurements from two different persons to be from the same person, called false acceptance and 2) misrecognizing measurements from the same person to be from two different persons, called false rejection. [6]-[7]

The performance of a biometric system is usually assessed by two indexes: false acceptance rate (FAR) and false rejection rate (FRR). These two performance indexes are controlled by adjusting a threshold but it is impossible to reduce FAR and FRR simultaneously. Another important performance index of a biometric system is equal error rate (EER), which is at the point where FAR and FRR are equal. The EER of a system with perfect accuracy is zero.

Recently, a group of researchers proposed a new personal authentication approach called FaceHashing [8]-[11]. It is based on BioHashing [12], which has been widely applied in other biometrics [12]-[15], that combines facial features and tokenized (pseudo-) random number (TRN). The authors reported zero EERs for faces that does not rely on advanced feature representations or complex classifiers. Even with Fisher Discrimination Analysis (FDA), face recognition can still achieve perfect accuracy [8]. Those impressive results and claims of perfection aroused our interest and motivated our study on FaceHashing described below.

D. Zhang and A.K. Jain (Eds.): ICB 2006, LNCS 3832, pp. 106 – 112, 2005.

This paper is organized as follows. Section 2 presents the foundation for our study by giving a general review of biometric verification systems and FaceHashing. Section 3 gives the details of the simulation of FaceHashing. Section 4 reveals the secret and the true performance of FaceHashing and Section 5 offers our conclusions.

2 Review of Biometric Verification System and FaceHashing

In this paper, we concerned with biometric verification systems to which FaceHashing belongs. In this section, we will set the foundation for our study by reviewing some major characteristics of biometric verification systems of our interests and summarizing the processes in FaceHashing.

2.1 Biometric Verification System

Biometric verification systems conduct one-to-one matching in personal authentication using two pieces of information: a claimed identity and biometric data. [7] The input biometric data is compared with biometric templates associated with the claimed identity in a given database. Fig. 1 illustrates the operation flow of a typical biometric verification system.

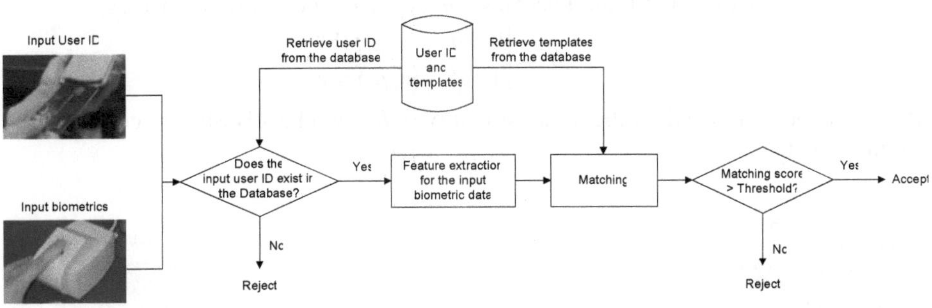

Fig. 1. Operation flow of a biometric verification system

User identities should be unique to each person, as to a primary key in a database. They can be stored in smart card or in the form of keyboard/pad input. It is worth to pointed out that user identities may, therefore, be shared, lost, forgotten and duplicated like token/knowledge in traditional authentication technologies. Nonetheless, for biometric authentication, in order to pass through the verification system, user must possess a valid user identity and valid biometric features, which is verified by the biometric verification system. We would like to point out that a biometric verification system will not perform any comparison of biometrics template/data if the user identity is not valid. We have to make clear, moreover, that a biometric verification system should not depend solely on user identity or its equivalent. Therefore, it can accept user identities that are not secrets, such as personal names.

If "token" or "knowledge" representing the user identity in verification would not be forgotten, lost or stolen, it made the introduction of biometric system less meaningful except for guarding against multiple users using the same identity through sharing or duplicating "token" or "knowledge". If, further, "token" or "knowledge" would not be shared or duplicated, introducing biometrics became meaningless.

2.2 Summary of FaceHashing

We recapitulate the mostly used method [9]-[11] (also in [12]-[15]), while another method has been reported [8] which differs by thresholding and selection of basis forming TRN [8]. Two major processes in FaceHashing [8]-[11]: facial feature extraction and discretization are illustrated in Fig. 2. Different techniques may be employed to extract features and our analysis is of more interests in discretization, the secret of FaceHashing, which is conducted in four steps:

1) Employ the input token to generate a set of pseudo-random vectors, $\{r_i \in \Re^M \mid i = 1,.....,m\}$ based on a seed.
2) Apply the Gram-Schmidt process to $\{r_i \in \Re^M \mid i = 1,.....,m\}$ and thus obtain TRN, a set of orthonormal vectors $\{p_i \in \Re^M \mid i = 1,.....,m\}$.
3) Calculate the dot product of ***v***, the feature vector obtained from first step and each orthnonormal vector in TRN, $\boldsymbol{p}_i$, such that $\langle v, p_i \rangle$.
4) Use a threshold τ to obtain FaceHash, ***b*** whose elements are defined as

$$b_i = \begin{cases} 0 & if \quad \langle v, p_i \rangle \le \tau \\ 1 & if \quad \langle v, p_i \rangle > \tau \end{cases},$$

where i is between 0 and m, the dimensionality of $\boldsymbol{b}$. Two FaceHashs are compared by hamming distance.

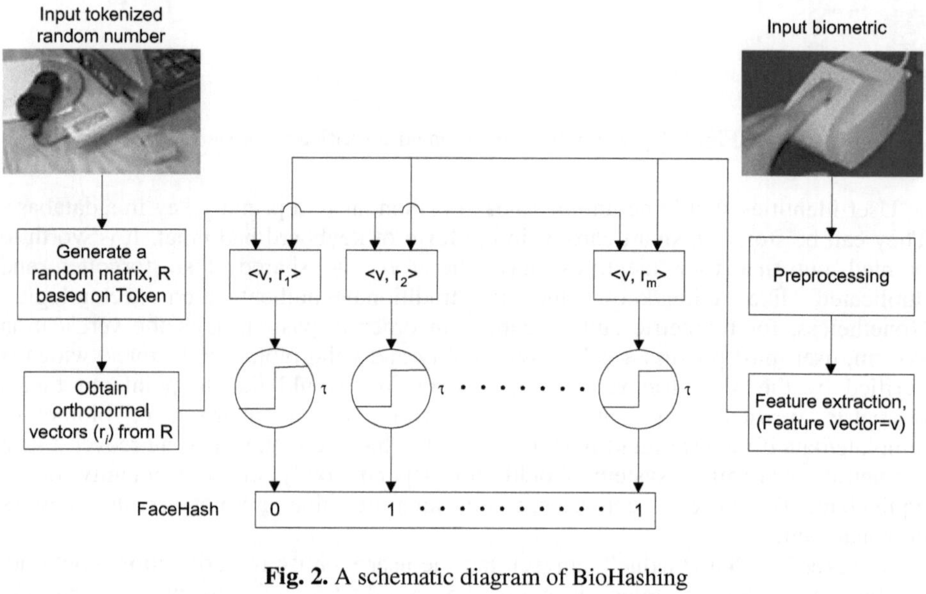

Fig. 2. A schematic diagram of BioHashing

3 FaceHashing Simulated: Experiments and Results

In this section, we will lay down the details of simulating the FaceHashing experiments for our study. A publicly available face database, the ORL face database[16], which is also used in [9]-[11], and a well known feature extraction technique, Principal Component Analysis (PCA), also termed Eigenface for face recognition [17]-[18] are chosen for this simulation so that all the results reported in this paper are reproducible.

3.1 Experimental Setup

The ORL face database contains 10 different images for each of 40 distinct subjects. For some of the subjects, the images were taken at different times, varying lighting slightly, facial expressions (open/closed eyes, smiling/non-smiling) and facial details (glasses/no-glasses). All the images are taken against a dark homogeneous background and the subjects are in up-right, frontal position (with tolerance for some side movement). The size of each image is 92×112 of 8-bit grey levels. Samples of a subject in ORL database is shown in Fig. 3.

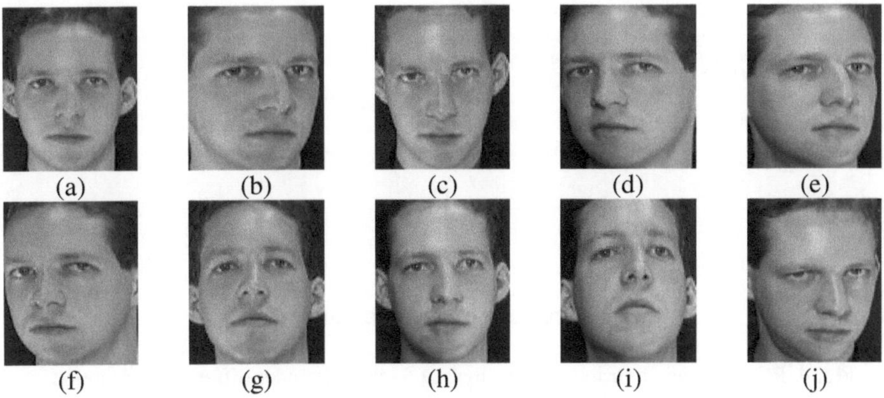

Fig. 3. Sample face images used in the ORL database

Principal components are obtained from all images of the ORL database. Each subject is assigned a unique token and the same token is used for different dimensions of the FaceHash under consideration. Table 1 lists the dimensions of the FaceHash and the corresponding thresholds (τ).

Table 1. Thresholds used for various dimensions of FaceHash

FaceHash dimension	Threshold for FaceHash (τ)
10	0
25	0
50	0
75	0
100	0

3.2 Experimental Results

We simulated FaceHashing [8]-[11] with different dimensions of FaceHash and their performances are reported in the form of Receiver Operating Characteristic (ROC) curves as a plot of the genuine acceptance rates (GAR) against the false acceptance rates (FAR) for all possible operating points in Fig. 4 using dotted lines with markers. It can be seen that as the FaceHashs increase in dimensionality, the EERs gradually decrease to zero. The results of our simulation are inline with the reported results [8]-[11]. Providing FaceHash is large enough, it was possible to achieve zero EER.

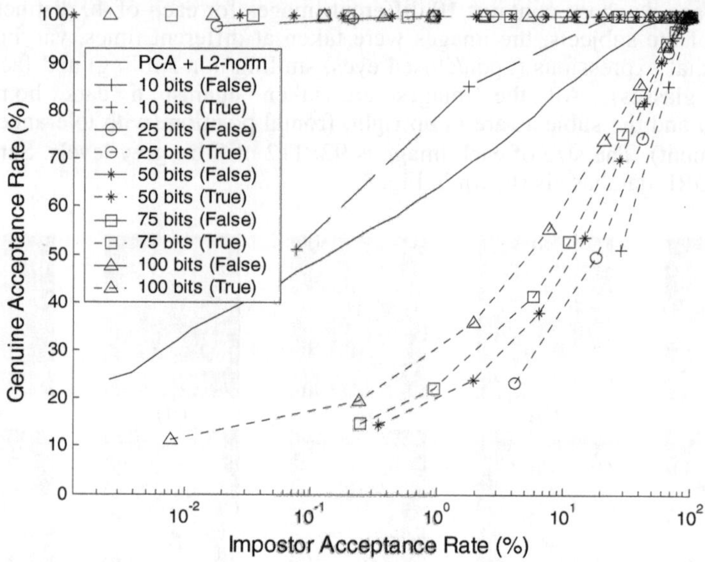

Fig. 4. ROC curves of various dimensions of FaceHash under different assumptions

4 The Secret of FaceHashing

In Section 3, we simulated FaceHashing in achieving zero EER, as in [8]-[11]. Obviously, the high performance of BioHashing is not resulted from the biometric features. In our simulation above, we are able to obtain zero EER by applying only a simple feature extraction method, PCA, but in general, even with advanced classifiers, such as support vector machines, PCA is impossible to yield 100% accuracy along with zero EER. We are going to reveal the secret of FaceHashing in this section.

4.1 The Secret of FaceHashing in Achieving Zero EER

The TRN is generated from a token (seed) which is unique among different persons and applications [8]-[11]. The token and thus the TRN for each user used in enrollment and verification is the same; different users (and applications), moreover, have different tokens and thus different TRNs. It is trivial that the token and TRN are unique across users as well as applications. Contrasting a token in FaceHashing with a user identity of a biometric verification system, as described in Section 2, it is obvious that the token, and thus the TRN serve as a user identity.

The outstanding performance reported in FaceHashing [8]-[11] is based on the use of TRN. They assume that no impostor has a valid token/TRN. That is, they assume that the token, an user identity equivalent, will not be lost, stolen, shared and duplicated. If their assumption is true, introducing any biometric becomes meaningless since the system can rely solely on the tokens without a flaw. Undoubtedly, their assumption does not hold in general. In their experiments, as simulated above in Section 3, they determine the genuine distribution correctly using the same token/user identity and different biometrics template/data of the same person. They determine the impostor distribution incorrectly, nevertheless, using different token/user identity and biometrics template/data of different person. As explained in Section 2, matching of biometrics template/data should not be performed because of the mismatch of the user identity equivalent, the token/TRN. Although FaceHashing does not explicitly verify the token as what is done on user identity, their determination of impostor distribution should not assume the token will not be lost, stolen, shared and duplicated. This also helps explaining why the performance of FaceHashing is better when the number of bits in FashHashs increases. It is because the effect of TRN becomes more significant as FashHashs' dimension (bits) increases.

4.2 The True Performance of FaceHashing

As discussed in Section 4.1, the impostor distribution should be determined under the assumption that impostors have valid TRNs, just as the general practice of evaluating a biometric verification system. The true performance of FaceHashing, in the form of ROC curves, for each dimension of FaceHash tested in Section 3 is shown in Fig. 4. The solid line without marker is the ROC curve when using PCA and Euclidean distance. The dashed lines with markers are the ROC curves assuming token, stolen, shared and duplicated. The dotted lines with markers are the ROC curves when using the general assumption for evaluating a biometric verification system, i.e. the true performance. It is easily observed that the true performance of FaceHash is even worse than that of using PCA and Euclidean distance. In opposite to results reported in [9]-[11], the performance of FaceHashing is far from perfect.

5 Conclusion

We, first, have reviewed the key concepts and components of a biometric verification system and FaceHashing. We, then, have revealed that the outstanding achievements of FaceHashing, zero EER, is achieved based on a false assumption that the token/TRN would never be lost, stolen, shared or duplicated. We also point out that it would be meaningless to combine the TRN with biometric features for verification if the assumption held. We used a public face database and PCA to simulate FaceHashing in achieving zero EER based on the false assumption. Afterwards, we uncover the true performance of FaceHashing, which is not as good as using PCA with Euclidean distance, with a valid assumption that is generally accepted by the research community. We would like to raise this issue to alert the use of "safe" token.

References

1. Chellappa, R., Wilson, C.L., Sirohey, A.: Human and machine recognition of faces: A survey. Proceedings of the IEEE **83** (1995) 705-740
2. Jain, A., Hong, L., Bolle, R.: On-line fingerprint verification. IEEE Transactions on Pattern Analysis and Machine Intelligence **19** (1997) 302-314
3. Bhanu, B., Tan, X.: Fingerprint indexing based on novel features of minutiae triplets. IEEE Transactions on Pattern Analysis and Machine Intelligence **25** (2003) 616-622
4. Daugman, J.: High confidence visual recognition of persons by a test of statistical independence. IEEE Transactions on Pattern Analysis and Machine Intelligence **15** (1993) 1148-1161
5. Zhang, D., Kong, W.K., You J., Wong, M.: On-line palmprint identification. IEEE Transactions on Pattern Analysis and Machine Intelligence **25** (2003) 1041-1050
6. Jain, A.K., Ross, A., Prabhakar, S.: An Introduction to Biometric Recognition. IEEE Transactions on Circuits and Systems for Video Technology **14** (2004) 4-20
7. Jain, A., Bolle, R., Pankanti, S. (eds.): Biometrics: Personal Identification in Networked Society. Kluwer Academic Publishers, Boston Mass (1999)
8. Teoh, A.B.J., Ngo, D.C.L, Goh, A.: An integrated dual factor authenticator based on the face data and tokenised random number. In: Zhang, D., Jain, A.K. (eds.): Biometric Authentication. Lecture Notes in Computer Science, Vol. 3072. Springer-Verlag, Berlin Heidelberg NewYork (ICBA 2004) 117-123
9. Ngo, D.C.L, Teoh, A.B.J., Goh, A.: Eigenspace-based face hashing. In: Zhang, D., Jain, A.K. (eds.): Biometric Authentication. Lecture Notes in Computer Science, Vol. 3072. Springer-Verlag, Berlin Heidelberg NewYork (ICBA 2004) 195-199
10. Teoh, A.B.J., Ngo, D.C.L, Goh, A.: Personalised cryptographic key generation based on FaceHashing. Computers and Security Journal **7** (2004) 606-614
11. Teoh, A.B.J., Ngo, D.C.L.: Cancellable biometerics featuring with tokenised random number. Pattern Recognition Letters **26** (2005) 1454-1460
12. Teoh, A.B.J., Ngo, D.C.L, Goh, A.: BioHashing: two factor authentication featuring fingerprint data and tokenised random number. Pattern Recognition **37** (2004) 2245-2255
13. Connie, T., Teoh, A., Goh, M., Ngo, D: PalmHashing: A Novel Approach for Dual-Factor Authentication. Pattern Analysis and Application **7** 255-268
14. Pang, Y.H., Teoh, A.B.J., Ngo, D.C.L.: Palmprint based cancelable biometric authentication system. International Journal of Signal Processing **1** (2004) 98-104
15. Connie, T., Teoh, A., Goh, M., Ngo, D: PalmHashing: a novel approach to cancelable biometrics. Information Processing Letter **93** (2005) 1-5
16. Samaria, F., Harter, A.: Parameterisation of a stochastic model for human face identification. Proceedings of the 2nd IEEE Workshop on Applications of Computer Vision, Sarasota (Florida), (1994) 138-142 (paper and ORL face database both available online at http://www.uk.research.att.com/facedatabase.html)
17. Martinez, A.M., Kak, A.C.: PCA versus LDA. IEEE Transactions on Pattern Analysis and Machine Intelligence **23** (2001) 228-233
18. Turk, M., Pentland, A.: Eigenfaces for recognition. Journal of Cognitive Neuroscience **3** (1991) 71-86

Person Authentication from Video of Faces: A Behavioral and Physiological Approach Using Pseudo Hierarchical Hidden Markov Models

Manuele Bicego[1], Enrico Grosso[1], and Massimo Tistarelli[2]

[1] DEIR - University of Sassari, via Torre Tonda 34 - 07100 Sassari - Italy
[2] DAP - University of Sassari, piazza Duomo 6 - 07041 Alghero (SS) - Italy

Abstract. In this paper a novel approach to identity verification, based on the analysis of face video streams, is proposed, which makes use of both physiological and behavioral features. While physical features are obtained from the subject's face appearance, behavioral features are obtained by asking the subject to vocalize a given sentence. The recorded video sequence is modelled using a Pseudo-Hierarchical Hidden Markov Model, a new type of HMM in which the emission probability of each state is represented by another HMM. The number of states are automatically determined from the data by unsupervised clustering of expressions of faces in the video. Preliminary results on real image data show the feasibility of the proposed approach.

1 Introduction

In the recent years biometrics research has grown in interest. Because of its natural interpretation (human visual recognition is mostly based on face analysis) and the low intrusiveness, face-based recognition, among others, is one of the most important biometric trait. Face analysis is a fecund research area, with a long history, but typically based on analysis of still images [15]. Recently, the analysis of video streams of face images has received an increasing attention [16, 8, 6, 3]. A first advantage in using video is the possibility of employing redundancy present in the video sequence to improve still images recognition systems, for example using voting schemes, or choosing the faces best suited for the recognition process, or also to build a 3D representation or super-resolution images. Besides these motivations, recent psychophysical and neural studies [5, 10] have shown that dynamic information is very crucial in human face recognition process. These findings inspired the development of true spatio-temporal video-based face recognition systems [16, 8, 6, 3].

All video-based approaches presented in the literature are mainly devoted to the recognition task, and to the best of our knowledge, a video-based authentication system has never been proposed. Moreover, in all video-based systems, only physiological visual cues are used: the process of recognition is based on the face appearance. When the subject is cooperative, as for authentication, also a behavioral cue can be effectively employed. For example, the subject may be

D. Zhang and A.K. Jain (Eds.): ICB 2006, LNCS 3832, pp. 113–120, 2005.

asked to vocalize a predefined sentence, such as counting from 1 to 10 or to pronounce his/her name. Each individual has its own characteristic way of vocalizing a given sentence, which could change both the appearance of the face and the temporal evolution of the visual patterns. These differences are mainly due to typical accents, pronounce, velocity of speaking, and so on. By including these behavioral features, i.e. by asking the subject to vocalize a predefined sentence, the characteristic dynamic features in the video stream are enhanced.

The system presented in this paper makes use of physiological and behavioral visual cues for person authentication, based on pseudo hierarchical Hidden Markov Models (HMM). HMMs are sequential tools largely applied in Pattern Recognition applications, and recently also employed in video-based face analysis [8, 3]. HMMs are quite appropriate for the representation of dynamic data; nonetheless, the emission probability function of a standard continuous HMM (Gaussians or Mixture of Gaussians [8, 3]) is not sufficient to fully represent the variability in the appearance of the face. In this case, it is more appropriate to apply a more complex model, such as another HMM [13, 1]. In summary, the proposed method is based on the modelling of the entire video sequence with an HMM in which the emission probability function of each state consists in another HMM itself (see Fig. 1), resulting in a pseudo-hierarchical HMM.

Determining the number of states (namely the model selection problem) is a key issue when using HMMs, and is typically selected a priori. In the method adopted, a model selection analysis has been carried out by assigning to each state of the PH-HMM a different facial expression. The problem of finding the number of states is then casted into the problem of finding all different facial expressions in the video stream. The facial expressions have been identified using an unsupervised clustering approach, where the number of clusters has been automatically determined with the Bayesian Inference Criterion [14].

2 Hidden Markov Models and Pseudo Hierarchical Hidden Markov Models

A discrete-time Hidden Markov Model $\boldsymbol{\lambda}$ can be viewed as a Markov model whose states cannot be explicitly observed: a probability distribution function is associated to each state, modelling the probability of emitting symbols from that state. More formally, a HMM is defined by the following entities [12]:

- $H = \{H_1, H_2, \cdots, H_K\}$ the finite set of the possible hidden states;
- the transition matrix $\mathbf{A} = \{a_{ij},\ 1 \leq j \leq K\}$ representing the probability to go from state H_i to state H_j;
- the emission matrix $\mathbf{B} = \{b(o|H_j)\}$, indicating the probability of the emission of the symbol o when system state is H_j (continuous or discrete)
- $\boldsymbol{\pi} = \{\pi_i\}$, the initial state probability distribution;

Given a set of sequences $\{S^k\}$, the training of the model is usually performed using the standard Baum-Welch re-estimation [12].

The evaluation step (*i.e.* the computation of the probability $P(S|\boldsymbol{\lambda})$, given a model $\boldsymbol{\lambda}$ and a sequence S to be evaluated) is performed using the *forward-backward procedure* [12].

2.1 Pseudo Hierarchical-HMM

The emission probability of a standard HMM is typically modelled using simple probability distributions, like Gaussians or Mixture of Gaussians. Nevertheless, in the case of sequences of face images, each symbol of the sequence is a face image, and a simple Gaussian could be not sufficiently accurate to properly and effectively model the probability of emission. In the PH-HMM, the emission probability is modelled using another HMM, which has been proven to be very accurate in describing faces [13, 9, 1]. The differences between standard HMMs and PH-HMM are briefly sketched in Fig. 1(a).

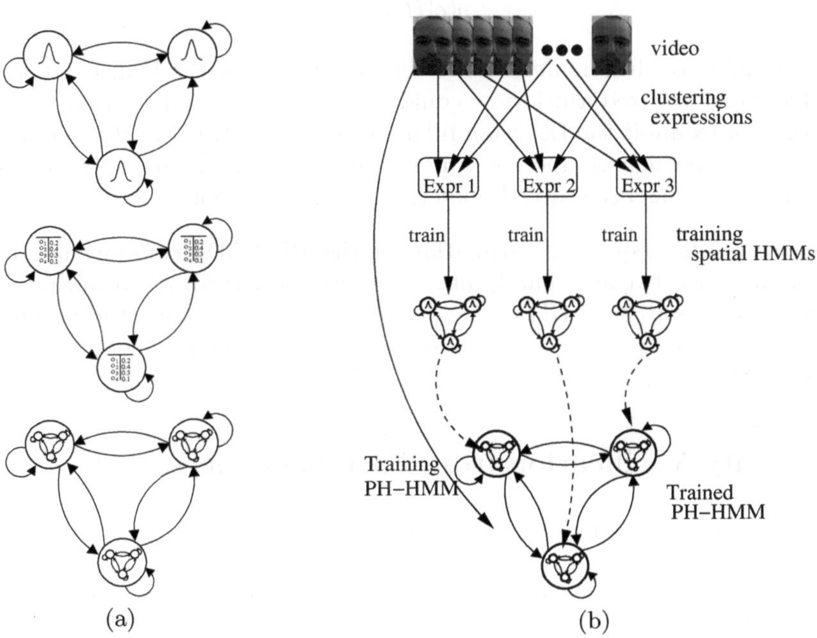

Fig. 1. (a) Differences between standard HMMs and PH-HMM, where emission probabilities are displayed into the state: (top) standard Gaussian emission; (center) standard discrete emission; (bottom) Pseudo Hierarchical HMM: in the PH-HMM the emissions are HMMs. (b) Sketch of the enrollment phase of the proposed approach.

The PH-HMM can be useful when the data have a double sequential profile. This is when the data is composed of a set of sequences of symbols $\{S^k\}$, $S^k = s_1^k, s_2^k, \cdots, s_T^k$, where each symbol s_i^k is a sequence itself: $s_i^k = o_{i1}^k, o_{i2}^k, \cdots, o_{iT_i}^k$. Let us call S^k the first-level sequences, whereas s_i^k denotes second-level sequences.

Fixed the number of states K of the PH-HMM, for each class C the training is performed in two sequential steps:

1. *Training of emission.* The first level sequence $S^k = s_1^k, s_2^k, \cdots, s_T^k$ is "unrolled", i.e. the $\{s_i^k\}$ are considered to form an unordered set U (no matter the order in which they appear in the first level sequence). This set is subsequently split in K clusters, grouping together similar $\{s_i^k\}$. For each cluster j, a standard HMM $\boldsymbol{\lambda}_j$ is trained, using the second-level sequences contained in that cluster. These HMMs $\boldsymbol{\lambda}_j$ represents the emission HMMs.
2. *Training of transition and initial states matrices.* Considering that the emission probability functions are determined by the emission HMMs, the transition and the initial states probability matrices of the PH-HMM are estimated using the first level sequences. In other words, the standard Baum Welch procedure is used, recalling that

$$b(o|H_j) = \boldsymbol{\lambda}_j$$

The number of clusters determines the number of the PH-HMM states. This value could be fixed a priori or could be directly determined from the data (using for example the Bayesian Inference Criterion [14]). In this phase, only the transition matrix and the initial state probability are estimated, since the emission has been already determined in the previous step.

Because of the sequential estimation of the PH-HMM components (firstly emission and then transition and initial state probabilities), the resulting HMM is a "pseudo" hierarchical HMM. In a truly hierarchical model, the parameters $\mathbf{A}$, $\boldsymbol{\pi}$ and $\mathbf{B}$ should be jointly estimated, because they could influence each other (see for example [2]).

3 Identity Verification from Face Sequences

Any identity verification system is based on two steps: off-line enrollment and on-line authentication.

The enrollment consists of the following sequential steps (for simplicity we assume only one video sequence $S = s_1, s_2, \cdots, s_T$, the generalization to more than one sequence is straightforward):

1. The video sequences S is analyzed to detect all faces sharing similar expression, i.e. to find clusters of expressions. Firstly, each face image s_i of the video sequence is processed, with a standard raster scan procedure, to obtain a sequence used to train a standard spatial HMM [1]. The resulting HMM models, one for each face of the video sequence, are then clustered in different groups based on their similarities [11]. Faces in the sequence with similar expression are grouped together independently from their appearance in time. The number of different expressions are automatically determined from the data using the Bayesian Inference Criterion [14].

2. For each expression cluster, a spatial face HMM is trained. In this phase all the sequences of the cluster are used to train the HMM, while in the first step one HMM for sequence has been built. At the end of the process, K HMMs are trained. We refer to these HMMs as "spatial" HMMs, because they are related to the spatial appearance of the face. In particular, each spatial HMM models a particular expression of the face in the video sequence. These models represents the emission probabilities functions of the PH-HMM.
3. The transition matrix and the initial state probability of the PH-HMM are estimated from the sequence $S = s_1, s_2, \cdots, s_T$, using the Baum-Welch procedure and the emission probabilities found in the previous step (see Sect. 2). This process aims at determining the temporal evolution of facial expressions in the video sequence. The number of states is fixed to the number of discovered clusters, this representing a sort of model selection criterion.

In summary, the main idea is to determine the facial expressions in the video sequence, modelling each of them with a spatial HMM. The expressions change during time is then modelled by the transition matrix of the PH-HMM, the "temporal" model (see Fig. 1(b))).

3.1 Spatial HMM Modelling

The process to build spatial HMMs is used in two stages of the proposed algorithm: in clustering expressions, where one HMM is trained for each face, and in the PH-HMM emission probabilities estimation, where one HMM is trained for each cluster of faces. Apart from the number of sequences used, in both cases the method consists of two steps. The former is the extraction of a sequence of sub images of fixed dimension from the original face image. This is obtained by sliding a fixed sized square window over the face image, in a raster scan fashion and keeping a constant overlap during the image scan.

For each of these sub-images, a set of low complexity features have been extracted, such as first and higher order statistics: the gray level mean, variance, Kurtosis and skewness (which are the third and the fourth moment of the data).

After the image scanning and feature extraction process, a sequence of $D \times R$ features is obtained, where D is the number of features extracted from each sub image (4), and R is the number of image patches. The learning phase is then performed using standard Baum-Welch re-estimation algorithm [12]. In this case the emission probabilities are all Gaussians, and the number of states is set to be equal to four. The learning procedure is initialized using a Gaussian clustering process, and stopped after likelihood convergence.

3.2 Clustering Facial Expressions

The goal of this step is to group together all face images in the video sequence with the same appearance, namely the same facial expression. The result is rather to label each face of the sequence corresponding to its facial expression, independently from their position in the sequence. In fact, it is possible that two not contiguous faces share the same expression, in this sense, the sequence of faces is unrolled before the clustering process.

Since each face is described with an HMM sequence, the expression clustering process is casted into the problem of clustering sequences represented by HMMs [11, 7]. Considering the unrolled set of faces $s_1, s_2, \cdots, s_T$, where each face s_i is a sequence $s_i = o_{i1}, o_{i2}, \cdots, o_{iT_i}$, the clustering algorithm is based on the following steps:

1. Train one standard HMM $\boldsymbol{\lambda}_i$ for each sequence s_i.
2. Compute the distance matrix $D = \{D(s_i, s_j)\}$, where $D(s_i, s_j)$ is defined as:
$$D(s_i, s_j) = \frac{P(s_j|\boldsymbol{\lambda}_i) + P(s_i|\boldsymbol{\lambda}_j)}{2}$$
This is a natural way for devising a measure of similarity between stochastic sequences. The validity of this measure in the clustering context has been already demonstrated [11].
3. Given the similarity matrix D, a pairwise distance-matrix-based method (the agglomerative complete link approach [4], in this case) is applied to perform the clustering.

In typical clustering applications the number of clusters is defined a priori. As it is impossible to arbitrarily establish the number of facial expressions in a sequence of facial images, the number of clusters has been estimated from the data, using the standard Bayesian Inference Criterion (BIC) [14], a penalized likelihood criterion.

3.3 PH-HMM Modelling

From the extracted set of facial expressions, the PH-HMM is trained. The different PH-HMM emission probability functions (spatial HMMs) model the facial expressions, while the temporal evolution of the facial expressions in the video sequence is modelled by the PH-HMM transition matrix. In particular, for each facial expression cluster, one spatial HMM is trained, using all faces belonging to the cluster (see section 3.1). The transition and the initial state matrices are estimated using the procedure described in section 2. One of the most important issues when training a HMM is model selection: in the presented approach, the number of states of the PH-HMM directly derives from the previous stage (number of clusters), representing a direct smart approach to the model selection issue.

3.4 Face Authentication

After building the PH-HMM the face authentication process, for identity verification, is straightforward. Given an unknown sequence and a claimed identity, the sequence is fed to the corresponding PH-HMM, which returns a probability value. If this value is over a predetermined threshold, the claimed identity is confirmed, otherwise it is denied.

4 Experimental Results

The system has been preliminary tested using a database composed of 5 subjects. Each subject is requested to vocalize ten digits, from one to ten. A minimum of five sequences for each subject have been acquired, in two different sessions.

The proposed approach has been tested against three other HMM-based methods, which do not fully exploit the spatio-temporal information. The first method, called "1 HMM for all", applies one spatial HMM (as described in section 3.1) to model all images in the video sequence. In the authentication phase, given an unknown video sequence, all the composing images are fed into the HMM, and the sum of their likelihoods represents the matching score. In the second method, called "1 HMM for cluster", one spatial HMM is trained for each expression cluster, using all the sequences belonging to that cluster. Given an unknown video, all images are fed into the different HMMs (and summed as before): the final matching score is the maximum among the different HMMs' scores. The last method, called "1 HMM for image", is based on training one HMM for each image in the video sequence. As in the "1 HMM for cluster" method, the matching score is computed as the maximum between the different HMMs' scores.

In all experiments only one video sequence for each subject has been used for the enrollment phase. Testing and training sets were always disjoint:in table 1 the Equal Error Rates for the four methods are reported.

Table 1. Authentication results for different methods

Method	EER
Still Image: 1 HMM for all	10.00%
Still Image: 1 HMM for cluster	11.55%
Still Image: 1 HMM for image	13.27%
Video: PH-HMM	8.275%

It is worth noting that when incorporating temporal information into the analysis a remarkable advantage is obtained, thus confirming the importance of dynamic face analysis. The applied test database is very limited and clearly too small to give a statistically reliable estimate of the performances of the method. On the other hand, the results obtained on this limited data set already show the applicability and the potential of the method in a real application scenario. The results obtained will be further verified performing a more extensive test.

5 Conclusions

In this paper a novel approach to video based face authentication is proposed, using both physiological and behavioral features. The video sequence is modelled using Pseudo Hierarchical HMM, in which the emission probability of each state is represented by another HMM. The number of states has been determined from the data by unsupervised clustering of facial expressions in the video. The system has been preliminary tested on real image streams, showing promising results. On the other hand, more tests are required, also in comparison with other techniques, to fully evaluate the real potential of the proposed method.

References

1. M. Bicego, U. Castellani, and V. Murino. Using Hidden Markov Models and wavelets for face recognition. In *IEEE. Proc. of Int. Conf on Image Analysis and Processing*, pages 52–56, 2003.
2. S. Fine, Y. Singer, and N. Tishby. The hierarchical hidden markov model: Analysis and applications. *Machine Learning*, 32:41–62, 1998.
3. A. Hadid and M. Pietikäinen. An experimental investigation about the integration of facial dynamics in video-based face recognition. *Electronic Letters on Computer Vision and Image Analysis*, 5(1):1–13, 2005.
4. A.K. Jain and R. Dubes. *Algorithms for clustering data.* Prentice Hall, 1988.
5. B. Knight and A. Johnston. The role of movement in face recognition. *Visual Cognition*, 4:265–274, 1997.
6. K.C. Lee, J. Ho, M.H. Yang, and D. Kriegman. Video-based face recognition using probabilistic appearance manifolds. In *Proc. Int. Conf. on Computer Vision and Pattern Recognition*, 2003.
7. C. Li. *A Bayesian Approach to Temporal Data Clustering using Hidden Markov Model Methodology.* PhD thesis, Vanderbilt University, 2000.
8. X. Liu and T. Chen. Video-based face recognition using adaptive hidden markov models. In *Proc. Int. Conf. on Computer Vision and Pattern Recognition*, 2003.
9. A.V. Nefian and M.H. Hayes. Hidden Markov models for face recognition. In *Proc. Int. Conf. on Acoustics, Speech and Signal Processing (ICASSP)*, pages 2721–2724, Seattle, 1998.
10. A.J. OToole, D.A. Roark, and H. Abdi. Recognizing moving faces: A psychological and neural synthesis. *Trends in Cognitive Science*, 6:261–266, 2002.
11. A. Panuccio, M. Bicego, and V. Murino. A Hidden Markov model-based approach to sequential data clustering. In *Structural, Syntactic and Statistical Pattern Recognition*, volume LNCS 2396, pages 734–742. Springer, 2002.
12. L. Rabiner. A tutorial on Hidden Markov Models and selected applications in speech recognition. *Proc. of IEEE*, 77(2):257–286, 1989.
13. F. Samaria. *Face recognition using Hidden Markov Models.* PhD thesis, Engineering Department, Cambridge University, October 1994.
14. G. Schwarz. Estimating the dimension of a model. *The Annals of Statistics*, 6(2):461–464, 1978.
15. W. Zhao, R. Chellappa, P.J. Phillips, and A. Rosenfeld. Face recognition: A literature survey. *ACM Computing Surveys*, 35:399 – 458, 2003.
16. S. Zhou, V. Krueger, and R. Chellappa. Probabilistic recognition of human faces from video. *Computer Vision and Image Understanding*, 91:214–245, 2003.

Cascade AdaBoost Classifiers with Stage Optimization for Face Detection

Zongying Ou, Xusheng Tang, Tieming Su, and Pengfei Zhao

Key Laboratory for Precision and Non-traditional Machining Technology of Ministry of Education, Dalian University of Technology, Dalian 116024, P.R. China
ouzyg@dlut.edu.cn

Abstract. In this paper, we propose a novel feature optimization method to build a cascade Adaboost face detector for real-time applications, such as teleconferencing, user interfaces, and security access control. AdaBoost algorithm selects a set of weak classifiers and combines them into a final strong classifier. However, conventional AdaBoost is a sequential forward search procedure using the greedy selection strategy, the weights of weak classifiers may not be optimized. To address this issue, we proposed a novel Genetic Algorithm post optimization procedure for a given boosted classifier, which yields better generalization performance.

1 Introduction

Many commercial applications demand a fast face detector, such as teleconferencing, user interfaces, and security access control [1]. Several face detection techniques have been developed in recent years [2], [3], [4], [5]. Due to the variation of poses, facial expressions, occlusion, environment lighting conditions etc., fast and robust face detection is still a challenging task.

Recently, Viola [3] introduced an boosted cascade of simple classifiers using Haar-like features capable of detecting faces in real-time with both high detection rate and very low false positive rates, which is considered to be one of the fastest systems. Central part of this method is a feature selection and combination algorithm based on AdaBoost [6]. Some of the recent works on face detection following Viola-Jones approach also explore alternative-boosting algorithms such as Float-Boost [7], Gentle-Boost [8], and Asymmetric AdaBoost [8]. In essence, Adaboost is a sequential learning approach based on one-step greedy strategy. It is reasonably expected that a post global optimization processing will further upgrade the performance of Adaboost. This paper investigates performance improvement of cascade Adaboost classifier by post stage optimization using Genetic Algorithm.

The remainder of this paper is organized as follows. In section 2 the Adaboost learning procedure proposed in [3] is introduced. The stage Optimization procedure based on Genetic Algorithms is presented in section 3. Section 4 provides the experimental results and conclusion is drawn in section 5.

D. Zhang and A.K. Jain (Eds.): ICB 2006, LNCS 3832, pp. 121 – 128, 2005.

2 Cascade of AdaBoost Classifiers and Performance Evaluation

There are three elements in the Viola-Jones framework: the cascade architecture, a set of Haar-like features, and AdaBoost algorithm for constructing classifier.

A cascade of face classifiers is a decision tree where at each stage a classifier is trained and formed to detect almost all frontal faces while rejecting a certain fraction of non-face patterns. Those image-windows that are not rejected by a stage classifier in the cascade sequence will be processed by the succeed stage classifiers. The cascade architecture can dramatically increases the speed of the detector by focusing attention on promising regions of the images.

Each stage classifier was trained using the Adaboost algorithm [6]. The idea of boosting is selecting and ensemble a set of weak learners to form a strong classifier by repeatedly learning processing over the training examples. In i stage, T numbers of weak classifiers h_{ij} and ensemble weights α_{ij} are yielded by learning. Then a stage strong classifier H_i (x) is:

$$H_i(x) = \begin{cases} 1 & \sum_{j=1}^{T} \alpha_{ij} h_{ij}(x) \geq \theta_i \\ -1 & otherwise \end{cases} . \quad (1)$$

The stage threshold θ_i is adjusted to meet the detection rate goal.

As conventional AdaBoost is a sequential forward search procedure based on the greedy selection strategy, the coefficients may not be optimal globally. Ideally, given $\{h_1,...h_T\}$, one solves the optimization problem for all weak classifier coefficients $\{\alpha_1,...\alpha_T\}$. The task becomes to construct a learning function that minimizes misclassification error.

3 Genetic Algorithms for Stage Optimization

To achieve high detection performance, the false rejection rate (FRR) and the false acceptance rate (FAR) should be both as low as possible. We take the minimum FAR as optimal object function, and take the FRR within an allowance magnitude as constraint condition. The weight α_{ij} and threshold θ. are the optimal parameters in optimization processing. For a given sets of positive and negative samples $\{(x_1,y_1)...(x_k,y_k)\}$ where $y_i=\pm 1$, given the FRR f, the optimization model can be written as:

$$\begin{aligned} &\arg\min_{\alpha_i,\theta}(num(y_i^n \neq H(x_i^n|\alpha_i,\theta)) / num(x_i^n) \\ &s.t.\ \ \mathrm{num}(y_i^p \neq H(x_i^n|\alpha_i,\theta)) / num(x_i^n|) \leq f \end{aligned} . \quad (2)$$

The function num (·) means the numbers of samples and the superscript p and n denote the positive and negative samples respectively. A true gradient decent cannot be implemented since the H (x) is not continuous. To address this issue, we use the Genetic algorithms to optimize the parameter.

3.1 Individual Representation and Fitness Function

In order to apply genetic search a mapping must be established between concept descriptions and individual in the search population. Assume that the stage classifier contains T weak classifiers (h_i) with T weight values α_i and threshold b. This information is encoded in a string as Fig.1.

α_1	α_2	...	α_T	b

Fig. 1. The representational structure of individual

The fitness function concerns accuracy measures-high hit rate (*hit*) and low false acceptance rate (*f*), and is defined as follow:

$$F = \begin{cases} 1 - n^- / N^- + m^+ / M^+ & if \quad m^+ / M^+ \geq hit \\ m^+ / M^+ & if \quad m^+ / M^+ < hit \end{cases} . \tag{3}$$

where: m^+ is the number of labeled positive samples correctly predicted,
M^+ is the total number of labeled positives samples in the training set,
n^- is the number of labeled negative samples wrongly predicted,
N^- is the total number of labeled negative samples in the training set,
hit is the hit rate of the original stage classifier in the training set.

3.2 Cascade Face Classifiers GA Post Optimization Learning Framework

We adapted "bootstrap" method [10] to reduce the size of the training set needed. The negative images are collected during training, in the following manner, instead of collecting the images before training is started.

1. Create an initial set of nonface images by collecting m numbers of random images. Create an initial set of face images by selecting l numbers of representative face images. Given a total stage number T_S, the final cumulatively false acceptance rate f.
2. Set stage number $S = 1$
3. Train a stage face classifier using these $m+l$ numbers of samples by Discrete Adaboost [3].
4. Using **GA algorithm [11]** to optimize the stage classifier.
5. Add this stage face classifier to ensemble a cascade face classifier system. Run the system on an image of scenery that contains no faces and filter out m numbers of negative images that the system incorrectly identifies as face to update the negative samples.
6. $S=S+1$;
7. If ($S < T_S$ and (m/the numbers of detected image)>f) Go to step 3.
8. Else Exit.

4 Experimental Results

The training face image set is provided by P.Carbonetto [12], which contains 4916 face images of size of 24×24. The non-faces samples are collected from various sources using the "bootstrap" method as mentioned above. Each stage 9000 non-face samples are used.

Two cascade face detection systems consisting of 30 stages were trained: One is with conventional AdaBoost [3] and the other is with our novel post-optimization procedure for each stage classifier. A Harr-like candidate feature set as used in [3] is adopted for Adaboost processing, and the selected weak classifiers is combined to form a stage classifier.

Parameters used for evolution were: 70% of all individuals undergo crossover, 0.5% of all individuals were mutated. The GA terminated if the population was converged to a good solution so that no better individual was found within the next 2000 generations. If convergence did not occur within 10000 generations, the GA was stopped as well.

We tested our systems on the CMU dataset [2] and the non-faces test set of CBCL face database [13]. The CMU dataset has been widely used for comparison of face detectors [2,3,7,8]. It consists of 130 images with 507-labeled frontal faces. The non-faces test set of CBCL face database contains 23,573 non-faces images, which resize to 24×24 pixel.

The criterion [8] is used to evaluate the precision of face localization. A hit was declared if and only if

- The Euclidian distance between the center of a detected and actual face was less than 30% of the width of the actual face as well as
- The width of the detected face was within ±50% of the actual face width.

During detection, a sliding window was moved pixel by pixel over the picture at each scale. Starting with the original scale, the features were enlarged by 20% until exceeding the size of the picture in at least one dimension.

Often multiple faces are detected at near by location and scale at an actual face location. Therefore, multiple nearby detection results were merged. Receiver Operating Curves (ROC) was constructed by varying the required number of detected faces per actual face before merging into a single detection result.

Fig.2 shows changes of weights of composed weak classifiers in the first stage in the process of GA optimization. There are total 14 weak classifiers in this stage. In training process, two methods are used to generate initial individual. One initializes the weight individual near the original weight yielded by conventional Adaboost. The other randomly initializes the weight individual. As can be seen in Fig.3 the first method reach the optimizations object (FAR=0.394) very quickly with about 66 iterations. Both methods can reach same optimization level, though the randomly initialized weights method takes much more iteration before convergence after GA post-optimization. The false acceptance rate on training set was about 15% lower than before, while keeping the hit rate constant at 99.95% as shown in Fig.3. In Fig.2 we can see the weight of the 12th weak classifier of the first stage is close to zero. The small weight implies the less important in discrimination the weak classifier will be. With this heuristic, the weak classifier whose weight closes to zero can be removed. This will lead to fewer weak classifiers and consequently decrease the total processing work in classifying. Just as shown in Fig.3 after deleting the 12th weak classifier and re-post optimization, the false acceptance rate will be change to 0.41,which is about 3.9% higher than without post optimal processing.

Table 1. A comparison of the false acceptance rate of total 16 stages in a cascade Adaboost processing with and without post GA optimization on the non-face test set of CBCL Database

Stage NO.	False acceptance rate		Stage NO.	False acceptance rate	
	conventional AdaBoost	With GA-post-optimization		conventional AdaBoost	With GA-post-optimization
1	0.7572	0.6440	9	0.1243	0.1118
2	0.6637	0.5500	10	0.1614	0.1453
3	0.4817	0.4045	11	0.0706	0.0607
4	0.4221	0.3413	12	0.1240	0.1066
5	0.6774	0.5758	13	0.2027	0.1724
6	0.3157	0.2715	14	0.2257	0.1918
7	0. 3560	0.3100	15	0.2468	0.2087
8	0.3349	0.2947	16	0.3052	0.2503
Final Cascade system FAR				0.0013	0.00067

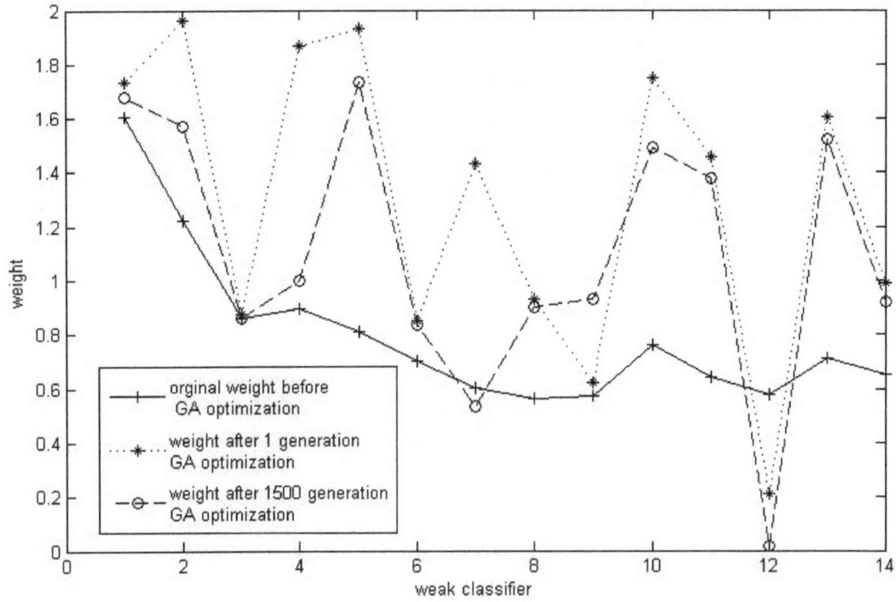

Fig. 2. The weight values of weak classifier in stage 1 with and without GA post optimization

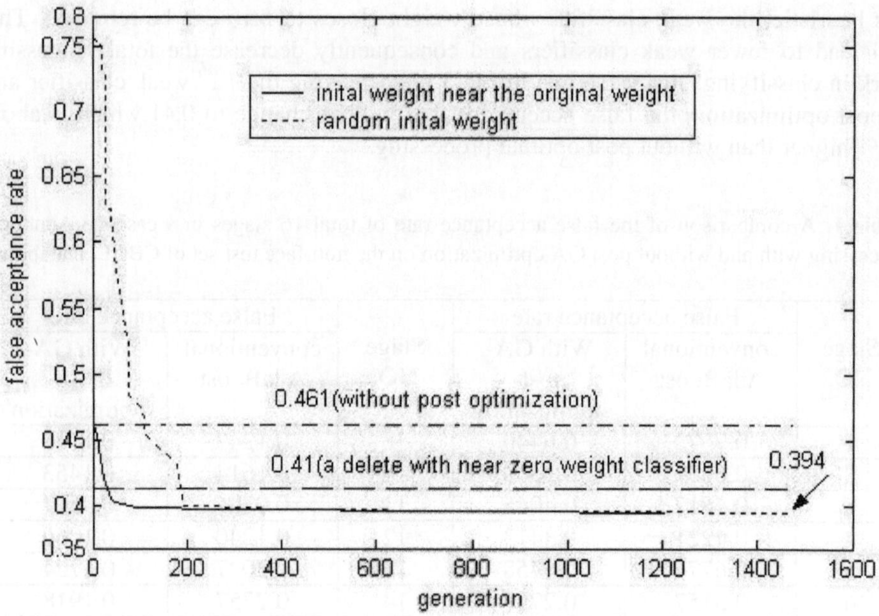

Fig. 3. The changes of false acceptance rate of stage 1 in cascade Adaboost with post GA optimization on training set (keeping hit rate constant)

Table 2. A Comparison of detection rate for various face detectors on the MIT+CMU test set

Detector \ False Acceptance number	10	31	50	95	167
With GA-post optimization(our)	81.3%	89.9%	92.4%	93.5%	94.1%
Without GA-post optimization(our)	80.9%	89.3%	91.5%	92.9%	93.5%
Viola-Jones(voting Adaboost) [3]	81.1%	89.7%	92.1%	93.2%	93.7%
Viola-Jones(Discrete Adaboost) [3]	79.1%	88.4%	91.4%	92.9%	93.9%
Rowley-Baluja-Kanade [3]	83.2%	86.0%	-	-	90.1%

We tested two face detection systems on the non-faces test set of CBCL face database. As cascade structure adopt the more non-face sub-window discard in early stage, the quicker detection speed will be achieved. From Table 1 we also can see that the face detector by GA post optimization discards more non-face image with same number of stage. This means GA post optimization can upgrade effectively the detection speed and accuracy. The average decrease of false acceptance rate is about 14.5%. Table 1 also shows that the final FAR of the classifier with post optimization was about 50% (0.00067 vs. 0.0013) lower than the classifier without post optimization.

Table 2 lists the detection rates corresponding to specified false acceptance numbers for our two systems (with and without post optimization) as well as other published systems (the data is adopted from Ref.[3]). The test database is MIT+CMU test set. As shown from Table.2, GA-post-optimization boosting outperformed the conventional Adaboost.

5 Conclusion

Adaboost is an excellent machine-learning algorithm, which provides an effective approach in selecting the discriminating features and combining them to form a strong discriminating classifier. Based on above framework, many face detection algorithms have got much success in practice. However, in essence Adaboost is a sequentially one-step forward greedy algorithm. It is expected that a global optimization will further improve the performance of Adaboost. A stage post GA optimization schema for cascade Adaboost face detector is presented in this paper. The experiment example shows that the false acceptance rate can be decrease 15% (from 0.461% to 0.39%) in one stage while the hit rate of stage keeps the same level on train set. The decrease rates of false acceptance rate in different stage on test set are about the similar value as shown in table 1, which means the classifier with GA post optimization achieves higher detection rate than the conventional Adaboost classifier. A total average decrease rate of false acceptance rate is about 50%, which implies that the cascade detector will decrease a similar percentage of processing work in repeating treating the non-face image regions, which will lead to increase the detection speed. The experiment also shows that the hit rate and the false acceptance rate can be both simultaneously upgrading with stage post optimization.

Reference

1. Yang, M. H., Kriegman, D. J., and Ahuja, N.: Detecting Faces in Images: A Survey. IEEE Transactions on Pattern Analysis and Machine Intelligence, Vol. 24 (2002) 34-58
2. Rowly, H., Baluja, S., and Kanade, T.: Neural network-based face detection. PAMI, Vol. 20 (1998) 23-38
3. Viola, P., Jones, M.: Rapid object detection using a boosted cascade of simple features. IEEE CVPR, (2001) 511~518
4. Romdhani, S., Torr, P., Schoelkopf, B., and Blake, A.: Computationally efficient face detection. In Proc. Intl. Conf. Computer Vision, (2001) 695–700
5. Henry, S., Takeo, K.: A statistical model for 3d object detection applied to faces and cars. In IEEE Conference on Computer Vision and Pattern Recognition. (2000)
6. Freund, Y., Schapire, R.: A diction-theoretic generalization of on-line learning and an application to boosting. Journal of Computer and System Sciences, Vol. 55 (1997) 119-139
7. Li, S.Z., Zhang, Z.Q., Harry, S., and Zhang, H.J.: FloatBoost learning for classification. In Proc.CVPR, (2001) 511-518
8. Lienhart, R., Kuranov, A., and Pisarevsky, V.: Empirical analysis of detection cascades of boosted classifiers for rapid object detection. Technical report, MRL, Intel Labs, (2002)
9. Viola, P., Jones, M.: Fast and robust classification using asymmetric AdaBoost and a detector cascade. In NIPS 14, (2002)

10. Sung, K.K.: Learning and Example Selection for Object and Pattern Detection. PhD thesis, MIT AI Lab, January (1996)
11. Goldberg, D.E.: Genetic algorithms in search, optimization, and machine learning, Addison-Wesley, Reading, A (1989)
12. Carbonetto, P.: Viola training data (Database). URL http://www.cs.ubc.ca/~pcarbo
13. http://cbcl.mit.edu/projects/cbcl/software-datasets/FaceData1Readme.html

Facial Image Reconstruction by SVDD-Based Pattern De-noising

Jooyoung Park[1,*], Daesung Kang[1], James T. Kwok[2], Sang-Woong Lee[3], Bon-Woo Hwang[3], and Seong-Whan Lee[3]

[1] Department of Control and Instrumentation Engineering, Korea University Jochiwon, Chungnam, 339-700, Korea
[2] Department of Computer Science, Hong Kong University of Science and Technology, Clear Water Bay, Hong Kong
[3] Department of Computer Science and Engineering, Korea University, Anam-dong, Seongbuk-ku, Seoul 136-713, Korea

Abstract. The SVDD (support vector data description) is one of the most well-known one-class support vector learning methods, in which one tries the strategy of utilizing balls defined on the feature space in order to distinguish a set of normal data from all other possible abnormal objects. In this paper, we consider the problem of reconstructing facial images from the partially damaged ones, and propose to use the SVDD-based de-noising for the reconstruction. In the proposed method, we deal with the shape and texture information separately. We first solve the SVDD problem for the data belonging to the given prototype facial images, and model the data region for the normal faces as the ball resulting from the SVDD problem. Next, for each damaged input facial image, we project its feature vector onto the decision boundary of the SVDD ball so that it can be tailored enough to belong to the normal region. Finally, we obtain the image of the reconstructed face by obtaining the pre-image of the projection, and then further processing with its shape and texture information. The applicability of the proposed method is illustrated via some experiments dealing with damaged facial images.

1 Introduction

Recently, the support vector learning method has grown up as a viable tool in the area of intelligent systems. Among the important application areas for the support vector learning, we have the one-class classification problems [1, 2]. In the problems of one-class classification, we are in general given only the training data for the normal class, and after the training phase is finished, we are required to decide whether each test vector belongs to normal class or abnormal class. One of the most well-known support vector learning methods for the one-class problems is the SVDD (support vector data description) [1, 2]. In the SVDD, balls are used for expressing the region for the normal class. Since balls on the input domain can express only limited class of regions, the SVDD in general enhances its

* Corresponding author.

D. Zhang and A.K. Jain (Eds.): ICB 2006, LNCS 3832, pp. 129–135, 2005.

expressing power by utilizing balls on the feature space instead of the balls on the input domain. In this paper, we extend the main idea of the SVDD for the reconstruction of partially damaged facial images [3]. Utilizing the morphable face model [4, 5, 6], the projection onto the spherical decision boundary of the SVDD, and a solver for the pre-image problem, we propose a new method for the problem of reconstructing facial images. The proposed method deals with the shape and texture information separately, and its main idea consists of the following steps: First, we solve the SVDD problem for the data belonging to the given prototype facial images, and model the data region for the normal faces as the ball resulting from the SVDD problem. Next, for each damaged input facial image, we perform de-noising by projecting its feature vector onto the spherical decision boundary on the feature space. Finally, we obtain the image of the reconstructed face by obtaining the pre-image of the projection with the strategy of [7], and further processing with its shape and texture information.

The remaining parts of this paper are organized as follows: In Section 2, preliminaries are provided regarding the SVDD, morphable face model, forward warping, and backward warping. Our main results on the facial image reconstruction by the SVDD-based learning are presented in Section 3. In Section 4, the applicability of the proposed method is illustrated via some experiments. Finally, in Section 5, concluding remarks are given.

2 Preliminaries

2.1 Support Vector Data Description

The SVDD method, which approximates the support of objects belonging to normal class, is derived as follows [1, 2]: Consider a ball B with the center $a \in \Re^d$ and the radius R, and the training data set D consisting of objects $x_i \in \Re^d$, $i = 1, \cdots, N$. Since the training data may be prone to noise, some part of the training data could be abnormal objects. The main idea of the SVDD is to find a ball that can achieve two conflicting goals simultaneously. First, it should be as small as possible, and with equal importance, it should contain as many training data as possible. Obviously, satisfactory balls satisfying these objectives can be obtained by solving the following optimization problem:

$$\begin{array}{ll} \min & L_0(R^2, a, \xi) = R^2 + C \sum_{i=1}^{N} \xi_i \\ \text{s. t.} & \|x_i - a\|^2 \le R^2 + \xi_i,\ \xi_i \ge 0, \quad i = 1, \cdots, N. \end{array} \tag{1}$$

Here, the slack variable ξ_i represents the penalty associated with the deviation of the i-th training pattern outside the ball. The objective function of (1) consists of the two conflicting terms, *i.e.*, the square of radius, R^2, and the total penalty $\sum_{i=1}^{N} \xi_i$. The constant C controls relative importance of each term; thus called the trade-off constant. Note that the dual problem of (1) is:

$$\begin{array}{ll} \max_\alpha & \sum_{i=1}^{N} \alpha_i \langle x_i, x_i \rangle - \sum_{i=1}^{N} \sum_{j=1}^{N} \alpha_i \alpha_j \langle x_i, x_j \rangle \\ \text{s. t.} & \sum_{i=1}^{N} \alpha_i = 1,\ \alpha_i \in [0, C], \quad \forall i. \end{array} \tag{2}$$

From the Kuhn-Tucker condition one can express the center of the SVDD ball as $a = \sum_{i=1}^{N} \alpha_i x_i$, and can compute the radius R utilizing the distance between a and any support vector x_i on the ball boundary. After the training phase is over, one may decide whether a given test point $x \in \Re^d$ belongs to the normal class utilizing the following criterion: $f(x) \triangleq R^2 - \|x - a\|^2 \geq 0$. In order to express more complex decision regions in $\Re^d$, one can use the so-called feature map $\phi : \Re^d \to F$ and balls defined on the feature space F. Proceeding similarly as the above and utilizing the kernel trick $\langle \phi(x), \phi(z) \rangle = k(x, z)$, one can find the corresponding feature-space SVDD ball B_F in F, whose center and radius are a_F and R_F, respectively. If the Gaussian function $K(x, z) = \exp(-\|x - z\|^2/\sigma^2)$ is chosen for the kernel K, one has $K(x, x) = 1$ for each $x \in \Re^d$, which is assumed throughout this paper. Finally, note that in this case, the SVDD formulation is equivalent to

$$\begin{array}{ll} \min_\alpha & \sum_{i=1}^{N} \sum_{j=1}^{N} \alpha_i \alpha_j K(x_i, x_j) \\ \text{s. t.} & \sum_{i=1}^{N} \alpha_i = 1, \ \alpha_i \in [0, C], \quad \forall i, \end{array} \tag{3}$$

and the resulting criterion for the normality is represented by

$$\begin{aligned} & f_F(x) \triangleq R_F^2 - \|\phi(x) - a_F\|^2 \\ & = R_F^2 - 1 + 2\sum_{i=1}^{N} \alpha_i k(x_i, x) - \sum_{i=1}^{N} \sum_{j=1}^{N} \alpha_i \alpha_j k(x_i, x_j) \geq 0. \end{aligned} \tag{4}$$

2.2 Morphable Face Model, Forward Warping and Backward Warping

Our reconstruction method is based on the morphable face model introduced by Beymer and Poggio [4], and developed further by Vetter *et al.* [5, 6]. Assuming that the pixelwise correspondence between facial images has already been established, a given facial image can be separated into the shape information and texture information. The two-dimensional shape information is coded as the displacement fields from a reference face, which plays the role of the origin in further information processing. On the other hand, the texture information is coded as an intensity map of the image which results from mapping the face onto the reference face. The shape of a facial image is represented by a vector $S = (d_1^x, d_1^y, \cdots, d_N^x, d_N^y)^T \in \Re^{2N}$, where N is the number of pixels in facial image, (d_k^x, d_k^y) the x, y displacement of a pixel that corresponds to a pixel x_k in the reference face and can be denoted by $S(x_k)$. The texture is represented as a vector $T = (i_1, \cdots, i_N)^T \in \Re^N$, where i_k is the intensity of a pixel that corresponds to a pixel x_k among N pixels in the reference face and can be denoted by $T(x_k)$. Before explaining our reconstruction procedure, we specify two types of warping processes: forward warping and backward warping. Forward warping warps a texture expressed in the reference face onto each input face by using its shape information. This process results in an input facial image. Backward warping warps an input facial image onto the reference face by using its shape information. This process yields a texture information expressed in reference shape. More details on the forward and backward warping can be found in reference [5].

3 Facial Image Reconstruction by SVDD-Based Learning

In the SVDD, the objective is to find the support of the normal objects, and anything outside the support is viewed as abnormal. On the feature space, the support is expressed by a reasonably small ball containing a reasonably large portion of the $\phi(x_i)$. A central idea of this paper is to utilize the ball-shaped support on the feature space for the purpose of correcting input facial images distorted by noises. More precisely, with the trade-off constant C set appropriately[1], we can find a region where the shape (or texture) data belonging to the normal facial images without noise generally reside. When a facial image (which was originally normal) is given as a test input x in a distorted form, the network resulting from the SVDD is supposed to judge that the distorted x does not belong to the normal class. The role of the SVDD has been conventionally up to this point, and the problem of curing the noise might be thought beyond the scope of the SVDD. However, here we observe that since the decision region of the SVDD is a simple ball B_F on the feature space F, it is quite easy to let the feature vector $\phi(x)$ of the distorted test input x move toward the center a_F of the ball B_F until it reaches the decision boundary so that it can be tailored enough to be counted normal. Of course, since the movement starts from the distorted feature $\phi(x)$, there are plenty of reasons to believe that the tailored feature $P\phi(x)$ still contain essential information about the original facial image. Thus, we claim that the tailored feature $P\phi(x)$ is the de-noised version of the feature vector $\phi(x)$. The above arguments together with additional step for finding the pre-image of $P\phi(x)$ comprise the essence of our method for facial image recovery. More precisely, our reconstruction procedure consists of the following steps:

1. Find the shape vectors $S_1, \cdots, S_N$ and texture vectors $T_1, \cdots, T_N$ for the given N prototype facial images.
2. Solve the SVDD problems for the shape and texture data belonging to the given prototype facial images, respectively, and model the data region for the shape and texture vectors of the normal faces as the balls resulting from the SVDD solutions, respectively.
3. For each damaged input facial image, perform the following:
   (a) Find the shape vector S of the damaged input facial image.
   (b) Perform de-noising for S by projecting its feature vector, $\phi_s(S)$, onto the spherical decision boundary of the SVDD ball on the feature space.
   (c) Estimate the shape of the recovered face, $\hat{S}$, by obtaining the pre-image of the projection $P\phi_s(S)$.
   (d) Find the texture vector T of the damaged input facial image.
   (e) Perform de-noising for T by projecting its feature vector, $\phi_t(T)$, onto the spherical decision boundary of the SVDD ball on the feature space.
   (f) Estimate the texture of the recovered face, $\hat{T}$, by obtaining the pre-image of the projection $P\phi_t(T)$.

---

[1] In our experiments, $C = 1/(N \times 0.2)$ was used for the purpose of de-noising.

(g) Synthesize a facial image for the reconstructed one by forward warping the estimated texture $\hat{T}$ with the estimated shape $\hat{S}$.

Steps 1, 3(a), and 3(d) are well explained in the previous studies of morphable face models [5, 8], and step 2 can be performed by the standard SVDD procedure. Steps 3(b)-(c) and 3(e)-(f) are carried out by the same mathematical procedure except that the shape about a pixel is a two-dimensional vector while the texture is one-dimensional. Therefore in the following description for steps 3(b)-(c) and 3(e)-(f), a universal notation is used for both S and T, *i.e.*, we will denote the object under consideration by $x \in \Re^d$, which can be interpreted as S or T according to which steps we are dealing with. Similarly, the feature maps for $\phi_s(\cdot)$ and $\phi_t(\cdot)$ are both denoted by $\phi(\cdot)$. As mentioned before, in step 2 of the proposed method, we solve the SVDD (3) for the shape (or texture) vectors of the prototype facial images $D \triangleq \{x_i \in \Re^d | i = 1, \cdots, N\}$. As a result, we find the optimal α_i along with a_F and R_F^2. In steps 3(b) and 3(e), we consider each damaged test pattern x. When the decision function f_F of (4) yields a nonnegative value for x, the test input is accepted normal as it is, and the de-noising process is bypassed. Otherwise, the test input x is considered to be abnormal and distorted by noise. To recover the de-noised pattern, an SVDD-based projection approach recently proposed by us [9] is used, in which we move the feature vector $\phi(x)$ toward the center a_F up to the point where it touches the ball B_F. Thus, the outcome of this movement is the following:

$$P\phi(x) = a_F + \frac{R_F}{\|\phi(x) - a_F\|}(\phi(x) - a_F). \tag{5}$$

Obviously, this movement is a kind of the projection, and can be interpreted as performing de-noising in the feature space. Note that as a result of the projection, we have the obvious result $\|P\phi(x) - a_F\| = R_F$. Also, note that with $\lambda \triangleq R_F / \|\phi(x) - a_F\|$, the equation (5) can be further simplified into

$$P\phi(x) = \lambda\phi(x) + (1 - \lambda)a_F, \tag{6}$$

where λ can be computed from

$$\lambda^2 = \frac{R_F^2}{\|\phi(x) - a_F\|^2} = \frac{R_F^2}{(1 - 2\sum_i \alpha_i K(x_i, x) + \sum_i \sum_j \alpha_i \alpha_j K(x_i, x_j))}. \tag{7}$$

In step 3(c) and 3(f), we try to find the pre-image of the de-noised feature $P\phi(x)$. If the inverse map $\phi^{-1} : F \to \Re^d$ is well-defined and available, this final step attempting to get the de-noised pattern via $\hat{x} = \phi^{-1}(P\phi(x))$ will be trivial. However, the exact pre-image typically does not exist [10]. Thus, we need to seek an approximate solution $\hat{x}$ instead. For this, we follow the strategy of [7], which uses a simple relationship between feature-space distance and input-space distance [11] together with the MDS (multi-dimensional scaling) [12]. After obtaining the de-noised vectors $\hat{S}$ and $\hat{T}$ from the above steps, we synthesize a facial image by forward warping the texture information $\hat{T}$ onto the input face by using the shape information $\hat{S}$. This final synthesis step is well explained in [5, 8].

4 Experiments

For illustration of the proposed method, we used two-dimensional images of Caucasian faces that were rendered from a database of three-dimensional head models recorded with a laser scanner ($Cyberware^{TM}$) [5, 6]. The resolution of the images was 256 by 256 pixels, and the color images were converted to 8-bit gray level images. Out of the 200 facial images, 100 images were randomly chosen as the prototypes for the SVDD training (step 2), and the other images were used for testing our method. For the test data set, some part of each test image was damaged with random noises. When extracting the S and T information from the damaged test input images, manual intervention based on the method of [13] was additionally employed. The first row of Fig. 1 shows the examples of the damaged facial images. The second and third row of Fig. 1 show the facial images reconstructed by the proposed method and the original facial images, respectively. From the figure we see that most of the reconstructed images are similar to the original ones.

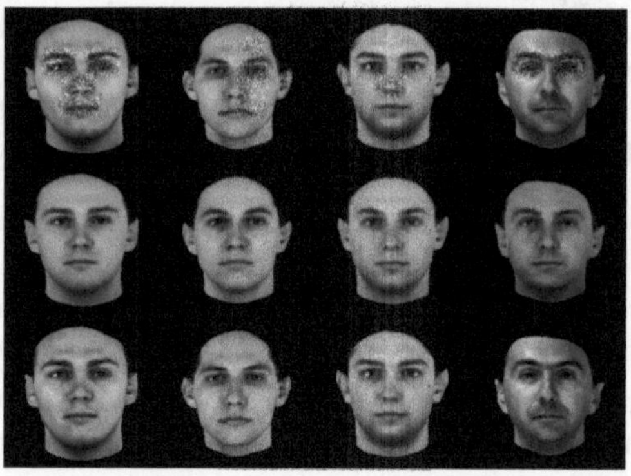

Fig. 1. Examples of facial images reconstructed from the partially damaged ones. The images on the top row are the damaged facial images, and those on the middle row are the facial images reconstructed by the proposed method. Those on the bottom row are the original face images.

5 Concluding Remarks

In this paper, we addressed the problem of reconstructing facial images from partially damaged ones. Our reconstruction method depends on the separation of facial images into the shape vectors S and texture vectors T, the SVDD-based de-noising for each of S and T, and finally the synthesis of facial images from the de-noised shape and texture information. In the SVDD-based de-noising, we utilized the SVDD learning, the projection onto the SVDD balls in the feature space, and a method for finding the pre-image of the projection. Experimental results show that reconstructed facial images are natural and plausible like original facial images. Works yet to be done include extensive comparative studies, which will reveal the strength and weakness of the proposed method, and further use of the proposed reconstruction method to improve the performance of face recognition systems.

Acknowledgments

We would like to thank the Max-Planck-Institute for providing the MPI Face Database.

References

1. D. Tax and R. Duin, "Support Vector Domain Description," *Pattern Recognition Letters*, vol. 20, pp. 1191–1199, 1999.
2. D. Tax, *One-Class Classification, Ph.D. Thesis*, Delft University of Technology, 2001.
3. B.-W. Hwang and S.-W. Lee, "Reconstruction of partially damaged face images based on a morphable face model," *IEEE Transactions on Pattern Analysis and Machine Intelligence*, vol. 25, pp. 365-372, 2003.
4. D. Beymer and T. Poggio, "Image representation for visual learning," *Science*, vol. 272, pp. 1905-1909, 1996.
5. T. Vetter and N. E. Troje, "Separation of texture and shape in images of faces for image coding and synthesis," *Journal of the Optical Society of America A*, vol. 14, pp. 2152-2161, 1997.
6. V. Blanz, S. Romdhani, and T. Vetter, "Face identification across different poses and illuminations with a 3d morphable model," *Proceedings of the 5th International Conference on Automatic Face and Gesture Recognition*, Washington, D.C., pp. 202-207, 2002.
7. J. T. Kwok and I. W. Tsang, "The pre-image problem in kernel methods," *IEEE Transactions on Neural Networks*, vol. 15, pp. 1517–1525, 2004.
8. M. J. Jones, P. Sinha, T. Vetter, and T. Poggio, "Top-down learning of low-level vision task[Brief Communication]," *Current Biology*, vol. 7, pp. 991-994, 1997.
9. J. Park, D. Kang, J. Kim, I. W. Tsang, and J. T. Kwok, "Pattern de-noising based on support vector data description," To appear in Proceedings of International Joint Conference on Neural Networks, 2005.
10. S. Mika, B. Schölkopf, A. Smola, K. R. Müller, M. Scholz, and G. Rätsch, "Kernel PCA and de-noising in feature space," *Advances in Neural Information Processing Systems*, vol. 11, pp. 536–542, Cambridge, MA: MIT Press, 1999.
11. C. K. I. Williams, "On a connection between kernel PCA and metric multidimensional scaling," *Machine Learning*, vol. 46, pp. 11–19, 2002.
12. T. F. Cox and M. A. A. Cox, "Multidimensional Scaling," *Monographs on Statistics and Applied Probability*, vol. 88, *2nd Ed.*, London, U.K.: Chapman & Hall, 2001.
13. B.-W. Hwang, V. Blanz, T. Vetter, H.-H. Song and S.-W. Lee, "Face Reconstruction Using a Small Set of Feature Points," *Lecture Notes in Computer Science*, vol. 1811, pp. 308-315, 2000.

Pose Estimation Based on Gaussian Error Models

Xiujuan Chai[1], Shiguang Shan[2], Laiyun Qing[2], and Wen Gao[1,2]

[1] School of Computer Science and Technology, Harbin Institute of Technology, 150001 Harbin, China
[2] ICT-ISVISION Joint R&D Lab for Face Recognition, ICT, CAS, 100080 Beijing, China
{xjchai, sgshan, lyqing, wgao}@jdl.ac.cn

Abstract. In this paper, a new method is presented to estimate the 3D pose of facial image based on statistical Gaussian error models. The basic idea is that the pose angle can be computed by the orthogonal projection computation if the specific 3D shape vector of the given person is known. In our algorithm, Gaussian probability density function is used to model the distributions of the 3D shape vector as well as the errors between the orthogonal projection computation and the weak perspective projection. By using the prior knowledge of the errors distribution, the most likely 3D shape vector can be referred by the labeled 2D landmarks in the given facial image according to the maximum posterior probability theory. Refining the error term, thus the pose parameters can be estimated by the transformed orthogonal projection formula. Experimental results on real images are presented to give the objective evaluation.

1 Introduction

Human head pose estimation is the key step towards the multi-view face recognition[1] and other multimedia applications, such as the passive navigation, industry inspection and human-computer interface and so on [2]. With these applications more and more techniques are investigated to realize the robust pose estimation.

Existing pose estimation algorithms can be classified into two main categories, one is the model-based algorithm, and the other is the appearance-based method. Model-based methods first assume a 3D face model to depict face. Then erect the relation of the features between 2D and 3D, finally the conventional pose estimation techniques are used to recover the pose information. Appearance-based algorithms suppose that there is one and only correlation between the 3D pose and the characteristics of 2D facial image. So the aim is to find this mapping relation from lots of training images with the known 3D poses. Here, the characteristics of the facial image conclude not only the intensities, color but also the intensity gradient and all kinds of image transformations etc.

Many appearance-based approaches have been reported on pose estimation. Hogg proposed a method to construct the mapping relation between 2D facial image and the 3D face pose by using artificial Neural Networks [3]. Later, Darrell performed face

D. Zhang and A.K. Jain (Eds.): ICB 2006, LNCS 3832, pp. 136 – 143, 2005.

detection and pose estimation by eigen-space method [4]. A separate eigen-space was erected for every pose of each training face. Given an image, projecting it to each eigen-space, the face and its pose were determined by the eigen-space which has the minimum error term. The similar idea was also appeared in paper [5]. Exclusive correlation between the 3D pose and its projection to the eigen-space is the potential assumption of this kind of eigen-space methods. A skin color model based pose estimation algorithm was proposed in [6], where the head was modeled by the combination of the skin/hair regions. In summary, the appearance based methods usually need lots of facial images under many poses for different persons to do training. They are simple in computing, however, not very accurate since many of them require interpolation.

Many model-based approaches have also been reported in the literature. Most of them model a face with some feature, for example the cylinder, the eclipse, or some key feature points. Then the 2D features are matched to the corresponding 3D feature to get the face pose. Nikolaidis determined the face pose by the equilateral triangle composed by the eyes and mouth [7]. Similarly, Gee used a facial model based on the ratios of four world lengths to depict the head [8,9]. Under the assumption of the weak perspective projection, the ratio of the 2D/3D lengths and the plane skew-symmetry are investigated to compute the normal and estimate the head pose finally. Except these methods, more complicated models were also proposed to tackle the pose estimation problem. Lee used a general 3D face model to synthesis eight different poses facial images [10]. The correlation between the input image and the modeled images were calculated to give the pose estimation results. More complicated, Ji and Hu assumed that the shape of a 3D face could be approximated by an ellipse and the aspect ratio of 3D face ellipse was given in advance [11]. So the ratio of the detected major axis and minor axis was used to calculate the face pose. To sum up, these model-based methods are more reliable and robust if the features can be detected accurately.

Our pose estimation method is also a model-based algorithm. In this paper, face is modeled by five landmarks. Using the MAP theory, the specific 3D shape vector corresponding to the given face is inferred and then used to get the accurate 3D pose.

The remaining parts of the paper are organized as follows: In Section 2, a simple pose estimation idea based on orthogonal projection is introduced. Then against the two problem existed in the above method, we propose a novel pose estimation algorithm based on the Gaussian error models in Section 3. Some pose estimation results of our algorithm are presented in Section 4, followed by short conclusion and discussion in the last section.

2 Pose Estimation Based on Simple Orthogonal Projection

We know that the head can be approximated as a 3D rigid body within the 3D coordinate system, hence the pose variation also satisfied the regular pattern of rigid motion. The face images under different poses can be regarded as the different projections in 2D image plan for different rotations around the head center.

In this paper, the pose variation is denoted by the tilt-yaw-pitch rotation matrix. The definition of the rotation angles is illustrated as Fig.1.

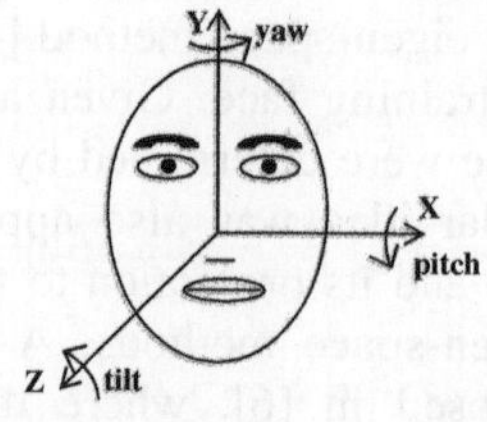

Fig. 1. The definition of the three rotation angles

Thus the rotation matrix **R** can be represented by:

$$\mathbf{R} = \mathbf{R_Z}(\gamma)\mathbf{R_Y}(\beta)\mathbf{R_X}(\alpha) = \begin{pmatrix} \cos\gamma & -\sin\gamma & 0 \\ \sin\gamma & \cos\gamma & 0 \\ 0 & 0 & 1 \end{pmatrix} \begin{pmatrix} \cos\beta & 0 & \sin\beta \\ 0 & 1 & 0 \\ -\sin\beta & 0 & \cos\beta \end{pmatrix} \begin{pmatrix} 1 & 0 & 0 \\ 0 & \cos\alpha & -\sin\alpha \\ 0 & \sin\alpha & \cos\alpha \end{pmatrix}. \quad (1)$$

In our method, five landmarks are used to model the head, which are the left and right iris centers, nose tip, left and right mouth corners respectively. The five points of 2D facial image can be written as a 2×5 matrix $\mathbf{S}_f$, where $\mathbf{S}_f = \begin{pmatrix} x_1 & x_2 & x_3 & x_4 & x_5 \\ y_1 & y_2 & y_3 & y_4 & y_5 \end{pmatrix}$. In a similar way the corresponding 3D points can be reshaped into a 3×5 matrix $\mathbf{S}$. Basing on the orthogonal projection theory, the following equation holds:

$$\mathbf{S}_f = c\mathbf{PRS} + \mathbf{T}, \quad (2)$$

where c is the scale factor, $\mathbf{T}$ is the 2D translation vector on x and y orientation. And $\mathbf{P} = \begin{pmatrix} 1 & 0 & 0 \\ 0 & 1 & 0 \end{pmatrix}$ is a transformation matrix to throw away the z information. We can obtain the pose parameters from equation (2) if the 3D head model $\mathbf{S}$ is known. Because the $\mathbf{S}$ is unknown for a specific given face, the average 3D face model can be used to substitute the specific $\mathbf{S}$ to get the approximate pose angles.

3 Gaussian Error Models Based Pose Estimation Algorithm

The above method will get good solutions for the faces whose 3D structures are similar with the average 3D face model. While, it will lead to large errors for those faces having remarkable different 3D structures compared with the general face. We think that there are two major factors introducing the deviations:

- ✧ The 3D shape vector $\mathbf{S}$ is different from each other. The using of the average shape $\mathbf{S}$ inevitably imports the deviation more or less.
- ✧ The facial images we estimate are almost generated by weak perspective projection. The orthogonal projection computation with the feature landmarks of real facial image will also generates indeterminate deviations.

Considering these factors, we modify the equation (2) as:

$$\mathbf{S}_f = \mathbf{PRS} + \mathbf{e}. \quad (3)$$

In this equation, the 2D shape vector $\mathbf{S}_f$ and the 3D shape vector $\mathbf{S}$ is aligned to the same standard position and scale to statistic the error distribution. Error term $\mathbf{e}$ is a 2×5 matrix. The distribution of the error terms can be modeled by a Gaussian probability density function.

Our pose estimation based on Gaussian error models algorithm consists of 2 steps: statistical error models computation and the pose estimation of facial image. In the following paragraphs, the two steps will be described in turn.

3.1 Learning the Gaussian Error Models

Our training set includes 100 laser-scanned 3D faces selected from the USF Human ID 3-D database [12]. The 3D shape vectors can be denoted as $\{\mathbf{S}_1 \;\; \mathbf{S}_2 \;\; \cdots \;\; \mathbf{S}_n\}$, where $n = 100$. $\mathbf{S}_i$ is a 3x5 matrix. The mean vector and the covariance matrix of these vectors can be computed by : $\boldsymbol{\mu}_S = \frac{1}{n}\sum_{i=1}^{n}\mathbf{S}_i$ and $C_s = \frac{1}{n}\sum(\mathbf{S}_i - \boldsymbol{\mu}_s)(\mathbf{S}_i - \boldsymbol{\mu}_s)^T$.

To simplify the statistical procedure, the error term $\mathbf{e}_\mathbf{R}^n$ under each sampling pose for a face is computed by the imaging formula directly. Computing the orthogonal projection and the weak perspective projection for the five points respectively, we get the two vectors: orthogonal projection vector $\mathbf{V}_{orth}$ and perspective projection vector $\mathbf{V}_{per}$. In order to normalize these two vectors, we align them in scale and make them have the same barycenter. Then we have error term $\mathbf{e}_\mathbf{R}^n$ by $\mathbf{e}_\mathbf{R}^n = \mathbf{V}_{per}^n - \mathbf{V}_{orth}^n$, where n is the index of different training shape. Under each sampling pose, we statistic the error mean vector $\boldsymbol{\mu}_\mathbf{e}^\mathbf{R}$ and the covariance matrix $\mathbf{C}_\mathbf{e}^\mathbf{R}$. Having these statistical Gaussian error models, the concrete pose estimation algorithm is described later.

3.2 Pose Estimation Based on the Gaussian Error Models

When given a facial image, first, let average 3D shape $\overline{\mathbf{S}}$ be $\mathbf{S}$ and the error term $\mathbf{e}$ be zero. So the approximate pose $\mathbf{R}_0$ can be computed by equation (3): $\mathbf{S}_f = \mathbf{PRS} + \mathbf{e}$. Set the $\mathbf{R} = \mathbf{R}_0$, the specific 3D shape of the given face is computed by the maximum posterior probability and the error term $\mathbf{e}$ can be calculated subsequently. In the first place, the mean vector $\boldsymbol{\mu}_\mathbf{e}^\mathbf{R}$ and the covariance matrix $\mathbf{C}_\mathbf{e}^\mathbf{R}$ of the error are refined by the simple neighborhood weighted strategy.

After refining the mean and covariance of error, we can recover the specific 3D shape $\mathbf{S}$ for the given face. As we all know that $\mathbf{S}_{MAP} = \arg\max_S(P(\mathbf{S} \mid \mathbf{S}_f))$. It is difficult to compute the $\arg\max_S(P(\mathbf{S}) \mid \mathbf{S}_f)$ directly, so we use the Bayes' rule $P(\mathbf{S} \mid \mathbf{S}_f)P(\mathbf{S}_f) = P(\mathbf{S}_f \mid \mathbf{S})P(\mathbf{S})$ to simplify $\mathbf{S}_{MAP}$. As the $\mathbf{S}_f$ is definite, the $P(\mathbf{S}_f)$ is a constant, thus we have:

$$\mathbf{S}_{MAP} = \arg\max P(\mathbf{S}_f \mid \mathbf{S})P(\mathbf{S}) \tag{4}$$

where $P(\mathbf{S})$ is the Gaussian probability density function we have learned in advance. From equation (3), if the $\mathbf{S}$ is fixed, and then $P(\mathbf{S}_f)$ is also a Gaussian probability density function with mean $(\mathbf{PRS} + \boldsymbol{\mu}_\mathbf{e}^\mathbf{R})$ and covariance matrix $\mathbf{C}_\mathbf{e}^\mathbf{R}$. So, we have:

$$\mathbf{S}_{MAP} = \arg\max_\mathbf{S} \left(Gauss\left(\mathbf{PRS} + \boldsymbol{\mu}_\mathbf{e}^\mathbf{R}, \mathbf{C}_\mathbf{e}^\mathbf{R}\right) \times Gauss\left(\boldsymbol{\mu}_\mathbf{S}, \mathbf{C}_\mathbf{S}\right)\right) \quad (5)$$

Using log probability for the right segment of equation (5), then set the first derivative with respect S to 0 to get the maximum probability, we get:

$$-(\mathbf{PR})^T \cdot (\mathbf{C}_\mathbf{e}^\mathbf{R})^{-1} \cdot \left(\mathbf{S}_f - \mathbf{PRS} - \boldsymbol{\mu}_\mathbf{e}^\mathbf{R}\right) + (\mathbf{C}_\mathbf{S})^{-1}(\mathbf{S} - \boldsymbol{\mu}_\mathbf{S}) = 0 \quad (6)$$

Rearranging the equation (6), we can obtain the following linear equation: $\mathbf{A} * \mathbf{S} = \mathbf{T}$, where, $\mathbf{A} = (\mathbf{PR})^T \left(\mathbf{C}_\mathbf{e}^\mathbf{R}\right)^{-1} (\mathbf{PR})$, $\mathbf{T} = (\mathbf{PR})^T \left(\mathbf{C}_\mathbf{e}^\mathbf{R}\right)^{-1} \left(\mathbf{S}_f - \boldsymbol{\mu}_\mathbf{e}^\mathbf{R}\right) + (\mathbf{C}_\mathbf{S})^{-1} \boldsymbol{\mu}_\mathbf{S}$.

Thus, the specific 3D shape vector $\mathbf{S}$ for the given face is recovered. And finally, the accurate pose angle can be calculated according to the equation (3).

4 Experiments and Results

Pose estimation is an opened problem so far. It is difficult to estimate the accurate angles for only given one facial image. Through many experiments, we think that the orthogonal projection computation (OPC) is a reasonable solution to this problem. So in our experiments, we compare our results with those of the orthogonal projection computation using average 3D shape vector.

4.1 Experiments with Single Image

First, we carry our experiment on some images in FERET database [13] and the results examples are given by Fig.2. To present the visualized evaluation, a 3D face

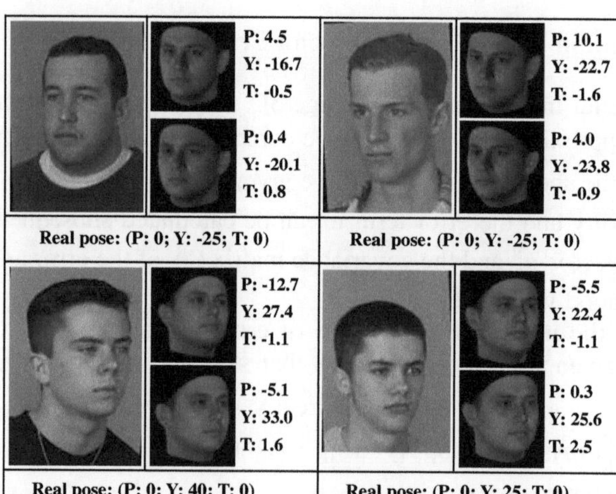

Fig. 2. The pose estimation results for the images in FERET database

model is rendered according to the pose estimated by our Gaussian error models (GEMs) algorithm and the orthogonal projection computation (OPC) algorithm respectively. The estimated pose angles are listed right to the rendered faces. For each test image, the upper rendered pose face is for the OPC results and the lower one is for the result of our algorithm. The real image poses are also given below the input images to be the references. From these results, we can see that our Gaussian error models based pose estimation improves the two major problems in orthogonal projection computation and achieves good performance.

4.2 Experiments with Image Series

We also take our experiment on an image series that captured the variation of the face turning from left to right. The image series is recorded by a real-time image captured system frame by frame. At the same time, the real pose angles can be provided by the special sensor equipment.The example images of pose variations are shown in Fig.3.

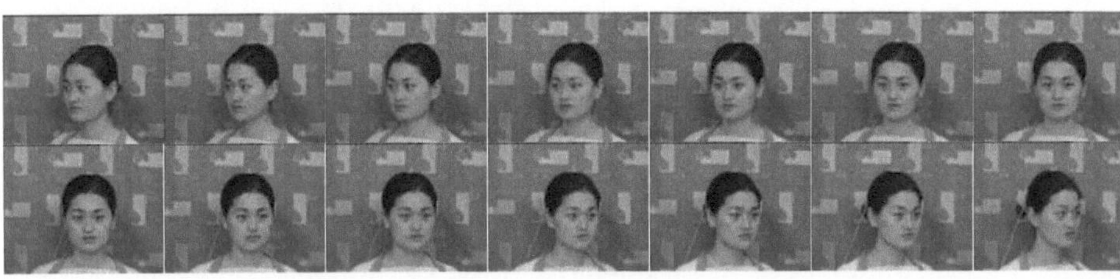

Fig. 3. The examples of the pose image series

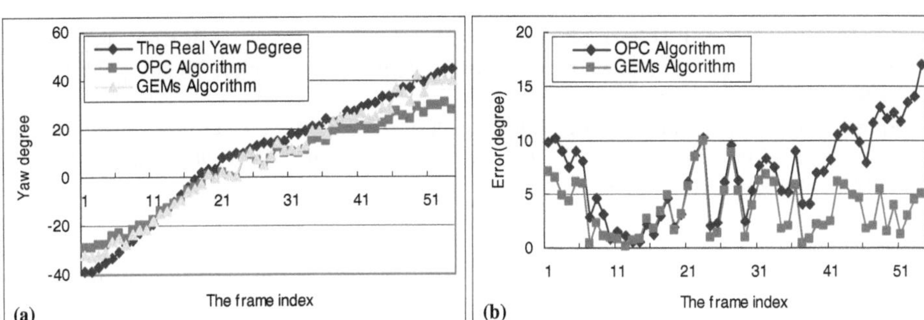

Fig. 4. (a) is the pose estimation results and (b) is the estimation deviation

In our test series, there are 54 images. The pose changes from left 39 degree to right 45 degree. The pitch is maintained nearly horizontal so only the yaw angle is statistic here. In this experiment, we also compare the results between the orthogonal projection computation (OPC) and the Gaussian error models algorithm (GEMs). The pose estimation results are given by Fig.4 (a) and the deviations to real yaw angles are presented by Fig.4 (b). The quantitative deviations of this image series for OPC algorithm and GEMs algorithm are 6.9 degree and 3.6 degree respectively. From these experimental results, we can see that the estimation pose angles by our Gaussian error models method are close to the real degrees and the deviations are small enough for many related applications.

5 Conclusion

In this paper a novel Gaussian error models based algorithm is proposed to perform pose estimation. Five key points are used to model the face. Assuming the 2D landmarks of given facial image have been located, orthogonal projection computation can be used to compute a coarse pose by using a general average 3D model. For considering the difference of specific face and the error term between the orthogonal projection and weak perspective projection, we use Gaussian probability density function to model the distribution of the two variables respectively. Based on the prior knowledge, the specific 3D shape vector corresponding to the given face can be inferred by MAP theory. Finally, the more accurate pose angles can be calculated easily using the transformation of the orthogonal projection formula. The experimental results show that our pose estimation algorithm is robust and reliable for estimating the pose of real facial images.

We should note that the locations of five landmarks in 2D images are necessary for pose estimation, hence the many efforts in the future, for example, the more accurate feature alignment, will make our algorithm more practicable in daily applications.

References

1. S.Y.Lee, Y.K. Ham, R.H.Park, Recognition of Hman Front Faces using Knowledge-based Feature Extraction and Neuro-Fuzzy Algorithm. Pattern Recognition 29(11), (1996) 1863-1876.
2. Shinn-Ying Ho, H.L.Huang, "An Analytic Solution for the Pose Determination of Human Faces from a Monocular Image", Pattern Recognition Letters, 19, (1998) 1045-1054.
3. T. Hogg, D. Rees, H. Talhami. Three-dimensional Pose from Two-dimensional images: a Novel Approach using Synergetic Networks. IEEE International Conference on Neural Networks. 2(11), (1995) 1140-1144.
4. T. Darrell, B. Moghaddam, A. P. Pentland. Active Face Tracking and Pose Estimation in an Interactive Room. IEEE Computer Society Conference on Computer Vision and Pattern Recognition. (1996) 67-72.
5. H. Murase, S. Nayar. Visual Learning and Recognition of 3-d Objects from Appearance. International Journal of Computer Vision, 14, (1995) 5-24
6. Q. Chen, H.Wu, T. Shioyama, T. Shimada. A Robust Algorithm for 3D Head Pose Estimation. IEEE International Conference on Multimedia Computing and Systems. (1999) 697-702.
7. A. Nikolaidis, I. Pitas. Facial Feature Extraction and Determination of Pose. Pattern Recognition, 33, (2000) 1783-1791.
8. A. Gee, R. Cipolla, "Determining the Gaze of Faces in Images", Image and Vision Computing 12, (1994) 639-647.

9. A. Gee, R. Cipolla. Fast Visual Tracking by Temporal Consensus. Image and Vision Computing. 14, (1996) 105-114.
10. C. W. Lee, A. Tsukamato. A Visual Interaction System using Real-time Face Tracking. The 28th Asilomar Conference on Signals, Systems and Computers. (1994) 1282-1286.
11. Q. Ji, R. Hu. 3D Face Pose Estimation and Tracking from a Monocular Camera. Image and Vision Computing. (2002) 1-13.
12. V. Blanz and T. Vetter, "A Morphable Model for the Synthesis of 3D Faces", In Proceedings, SIGGRAPH'99, (1999) 187-194.
13. P. Phillipse, H. Moon, S. Rizvi and P. Rauss, "The FERET Evaluation for Face-Recognition Algorithms", IEEE Trans. on PAMI, 22, (2000) 1090-1103.

A Novel PCA-Based Bayes Classifier and Face Analysis

Zhong Jin[1,2], Franck Davoine[3], Zhen Lou[2], and Jingyu Yang[2]

[1] Centre de Visió per Computador,
Universitat Autònoma de Barcelona, Barcelona, Spain
zhong.jin@cvc.uab.es
[2] Department of Computer Science,
Nanjing University of Science and Technology,
Nanjing, People's Republic of China
jyyang@mail.njust.edu.cn
[3] HEUDIASYC - CNRS Mixed Research Unit,
Compiègne University of Technology,
60205 Compiègne cedex, France
franck.davoine@hds.utc.fr

Abstract. The classical Bayes classifier plays an important role in the field of pattern recognition. Usually, it is not easy to use a Bayes classifier for pattern recognition problems in high dimensional spaces. This paper proposes a novel PCA-based Bayes classifier for pattern recognition problems in high dimensional spaces. Experiments for face analysis have been performed on CMU facial expression image database. It is shown that the PCA-based Bayes classifier can perform much better than the minimum distance classifier. And, with the PCA-based Bayes classifier, we can obtain a better understanding of data.

1 Introduction

In recent years, many approaches have been brought to bear on pattern recognition problems in high dimensional space. Such high-dimensional problems occur frequently in many applications, including face recognition, facial expression analysis, handwritten numeral recognition, information retrieval, and content-based image retrieval. The main approach applies an intermediate dimension reduction method, such as principal component analysis (PCA), to extract important components for linear discriminant analysis (LDA) [1, 2].

PCA is a classical, effective and efficient data representation technique. It involves a mathematical procedure that transforms a number of (possibly) correlated variables into a (smaller) number of uncorrelated variables called principal components.

The classical Bayes classifier plays an important role in statistical pattern recognition. Usually, it is not easy to use a Bayes classifier for pattern recognition problems in high dimensional space. The difficulty is in solving the singularity of covariance matrices since pattern recognition problems in high dimensional spaces are usually so-called undersampled problems.

D. Zhang and A.K. Jain (Eds.): ICB 2006, LNCS 3832, pp. 144–150, 2005.

In this paper, we seek a PCA-based Bayes classifier by combining PCA technique and Bayesian decision theory. It is organized as follows. Section 2 gives an introduction to Bayesian decision theory. A PCA-based Bayes classifier is proposed in Section 3. Experiments for face analysis are performed in Section 4. Finally, conclusions are given in Section 5.

2 Bayesian Decision Theory

Bayesian decision theory is fundamental in statistical pattern recognition.

2.1 Minimum-Error-Rate Rule

Let $\{\omega_1, \cdots, \omega_c\}$ be the finite set of c states of nature ("categories"). Let the feature vector x be a d-dimensional vector-values random variable and let $p(x|\omega_j)$ be the state-conditional probability density function for x, with the probability density function for x conditioned on ω_j being the true state of nature. Let $P(\omega_j)$ describe the prior probability that nature is in state ω_j. The target is to make a decision for the true state of nature.

It is natural to seek a decision rule that minimizes the probability of error, that is, the error rate. The Bayes decision rule to minimize the average probability error calls for making a decision that maximizes the posterior probability $P(\omega_i|x)$. It can formally be written as the argument i that maximizes the posterior probability $P(\omega_i|x)$, that is,

$$x \rightarrow \omega_i \quad \text{with} \quad i = \arg\max_j P(\omega_j|x). \tag{1}$$

The structure of a Bayes classifier is determined by the conditional densities $p(x|\omega_j)$ as well as by the prior probabilities $P(\omega_j)$. Under the assumption of the same prior probabilities $P(\omega_j)$ $(j = 1, \cdots, c)$ for all the c classes, the minimum-error-rate rule of Eq. (1) can be achieved by use of the state-conditional probability density function $p(x|\omega_j)$ as follows

$$x \rightarrow \omega_i \quad \text{with} \quad i = \arg\max_j p(x|\omega_j). \tag{2}$$

Of the various density functions that have been investigated, none has received more attention than the multivariate normal or Gaussian density. In this paper, it is assumed that $p(x|\omega_j)$ is a multivariate normal density in d dimensions as follows

$$p(x|\omega_j) = \frac{1}{(2\pi)^{d/2}|\Sigma_j|^{1/2}} \exp\left[-\frac{1}{2}(x-\mu_j)^t \Sigma_j^{-1}(x-\mu_j)\right], \tag{3}$$

where μ_j is the d-component mean vector and Σ_j is the $d \times d$ covariance matrix.

2.2 Minimum Distance Classifier

The simplest case occurs when the features are statistically independent and when each feature has the same variance, σ^2. In this case, the covariance matrix is diagonal, being merely σ^2 times the identity matrix I, that is

$$\Sigma_j = \sigma^2 \mathrm{I} \quad (j = 1, \cdots, c). \tag{4}$$

Thus, the minimum-error-rate rule of Eqs. (2),(3) and (4) can be expressed as follows

$$x \to \omega_i \quad \text{with} \quad i = \arg\min_j ||x - \mu_j||^2, \tag{5}$$

where $||\cdot||$ denotes the Euclidean norm. This is the commonly used minimum distance classifier.

2.3 Limitation of Bayes Classifier

In a high dimensional space, some classes may lie on or near a low dimensional manifold. In other words, for some classes, the covariance matrices Σ_j may be singular in high dimensional space. Such a limitation exists even in 2-dimensional spaces. A two-class problem is shown in Fig. 1. In the example, one class de-

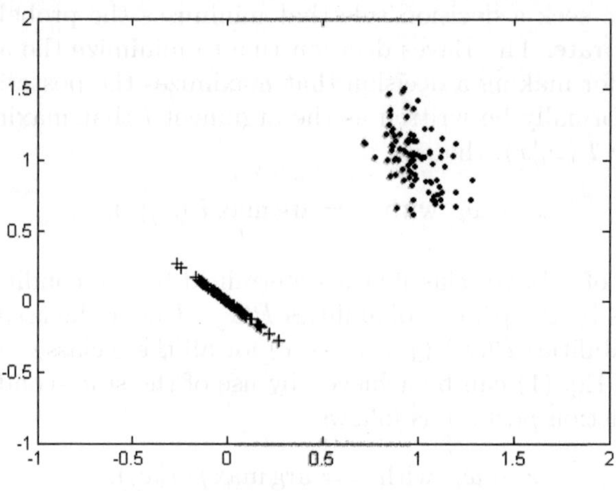

Fig. 1. A two-class problem

generates in a 1-dimensional line so that the Bayes classifier can not directly be used to perform classification. Anyway, the minimum distance classifier can be used to perform classifications. However, we fail to have a correct understanding of data since the constraint of Eq. (4) is not satisfied in the two-class problem.

3 PCA-Based Bayes Classifier

One solution to the above limitation of Bayes classifier is to describe Gaussian density of Eq. (3) by using principal component analysis (PCA). We are going to propose a novel PCA-based Bayes classifier in this section.

3.1 PCA Model

Let $\Psi_j = (\psi_{j1}, \cdots, \psi_{jd})$ be the matrix whose columns are the unit-norm eigenvectors of the covariance matrix Σ_j of Eq. (3). Let $\Lambda_j = diag(\lambda_{j1}, \cdots, \lambda_{jd})$ be the diagonal matrix of the eigenvalues of Σ_j, where λ_{ji} are the eigenvalues corresponding to the eigenvectors ψ_{ji} $(i = 1, \cdots, d)$. We have

$$\Psi_j^t \Sigma_j \Psi_j = \Lambda_j. \tag{6}$$

If the covariance matrix Σ_j is non-singular, all the corresponding eigenvalues are positive. Otherwise, some eigenvalues may be zero. In general, assume that λ_{ji} $(i = 1, \cdots, d)$ are ranked in order from larger to smaller as follows:

$$\lambda_{j1} \geq \cdots \geq \lambda_{jd_j} > \lambda_{j(d_j+1)} = \cdots = \lambda_{jd} = 0, \tag{7}$$

where d_j is the number of non-zero eigenvalues of the covariance matrix Σ_j.

Recently, a perturbation approach has been proposed [3]. However, for practical application problems, the dimension d may be too high to obtain all the d eigen-vectors.

3.2 Novel Perturbation Approach

Assume that all the eigen-vectors corresponding to non-zero eigenvalues are available. Let

$$z = (\psi_{11}, \cdots, \psi_{1d_1}, \cdots\cdots, \psi_{c1}, \cdots, \psi_{cd_c})^t x. \tag{8}$$

This is a linear transformation from the original d-dimensional x space to a new $\bar{d}$-dimensional z space, where

$$\bar{d} = \sum_{j=1}^{c} d_j. \tag{9}$$

Suppose

$$\bar{d} < d. \tag{10}$$

Thus, the new z space can be regarded as a "compact" space of the original x space. Instead of the Bayes classifier of Eq. (2) in x space, a Bayes classifier can be introduced in the z space

$$x \rightarrow \omega_i \quad \text{with} \quad i = \arg\max_j p(z|\omega_j). \tag{11}$$

Obviously, $p(z|\omega_1)$ has formally a Gaussian distribution since the transformation of Eq. (8) is linear. We are going to propose a novel perturbation approach to determine $p(z|\omega_1)$ in the rest of this section.

Conditional Distribution $p(z|\omega_1)$. We know that $(\psi_{11}, \cdots, \psi_{1d_1})$ are eigenvectors corresponding to the non-zero eigenvalues of the covariance matrix Σ_1. In general, the $\bar{d} - d_1$ eigen-vectors $(\psi_{21}, \cdots, \psi_{2d_2}, \cdots\cdots, \psi_{c1}, \cdots, \psi_{cd_c})$ are not the eigen-vectors corresponding to zero eigenvalues of the covariance matrix Σ_1.

Firstly, let

$$(\xi_1, \cdots, \xi_{\bar{d}}) \Leftarrow (\psi_{11}, \cdots, \psi_{1d_1}, \psi_{21}, \cdots, \psi_{2d_2}, \cdots\cdots, \psi_{c1}, \cdots, \psi_{cd_c}). \quad (12)$$

Then, perform the Gram-Schmit orthogonalization for each $j (j = 2, \cdots, \bar{d})$ as follows.

$$\xi_j \Leftarrow \xi_j - \sum_{i=1}^{j-1} (\xi_j^t \xi_i) \xi_i, \quad (13)$$

$$\xi_j \Leftarrow \xi_j / ||\xi_j||. \quad (14)$$

Giving that $(\psi_{11}, \cdots, \psi_{1d_1}, \psi_{21}, \cdots, \psi_{2d_2}, \cdots\cdots, \psi_{c1}, \cdots, \psi_{cd_c})$ has a rank of $\bar{d}$, that is, these eigen-vectors are linearly independent, the Gram-Schmit orthogonalization of Eqs. (12-14) is a linear transformation

$$(\xi_1, \cdots, \xi_{\bar{d}}) = A(\psi_{11}, \cdots, \psi_{1d_1}, \cdots\cdots, \psi_{c1}, \cdots, \psi_{cd_c}), \quad (15)$$

where A is a non-singular upper triangle $\bar{d} \times \bar{d}$ matrix.

Theorem 1. Let

$$y = (\xi_1, \cdots, \xi_{\bar{d}})^t x. \quad (16)$$

The covariance matrix of $p(y|\omega_1)$ is a diagonal matrix

$$diag(\lambda_{11}, \cdots, \lambda_{1d_1}, 0, \cdots\cdots, 0). \quad (17)$$

The proof of Theorem 1 is omitted here.

Denote the diagonal elements of the covariance matrix in Eq. (17) as $\bar{\lambda}_i$ $(i = 1, \cdots, \bar{d})$. By changing the zero-diagonal elements of the covariance matrix in Eq. (17) with a perturbation factor ε, that is

$$\bar{\lambda}_{d_1+1} = \cdots = \bar{\lambda}_{\bar{d}} = \varepsilon, \quad (18)$$

we can determine $p(y|\omega_1)$ as follows

$$p(y|\omega_1) = \prod_{i=1}^{\bar{d}} \frac{1}{(2\pi\bar{\lambda}_i)^{\frac{1}{2}}} exp\left[-\frac{(y_i - \bar{\mu}_i)^2}{2\bar{\lambda}_i}\right], \quad (19)$$

where

$$\bar{\mu}_i = \xi_i^t \mu_1 \ (i = 1, \cdots, \bar{d}). \quad (20)$$

From Eqs. (8), (15) and (16), we have

$$z = A^{-1} y. \quad (21)$$

Then, a novel perturbation approach to determine $p(z|\omega_1)$ can be proposed

$$p(z|\omega_1) = p(y|\omega_1)|A^{-1}|, \quad (22)$$

where $|A^{-1}|$ is the determinant of the inverse matrix of A.

Conditional Distribution $p(z|\omega_j)$. It is now ready to propose an algorithm to determine the conditional distribution $p(z|\omega_j)$ $(j = 2, \cdots, c)$.

Step 1. Initialize $(\xi_1, \cdots, \xi_{\bar{d}})$ firstly by assigning d_j eigen-vectors of the covariance matrix Σ_j and then by assigning all the other $\bar{d} - d_j$ eigen-vectors of the covariance matrix Σ_i $(i \neq j)$, that is,

$$(\xi_1, \cdots, \xi_{\bar{d}}) \Leftarrow (\psi_{j1}, \cdots, \psi_{jd_j}, \psi_{11}, \cdots, \psi_{1d_1}, \cdots\cdots, \psi_{c1}, \cdots, \psi_{cd_c}). \quad (23)$$

Step 2. Perform the Gram-Schmit orthogonalization according to Eqs. (13) and (14). Thus, we obtain the matrix A in Eq. (15).
Step 3. Substitute $(\lambda_{j1}, \cdots, \lambda_{jd_j})$ for $(\lambda_{11}, \cdots, \lambda_{1d_1})$ in Eq. (17). Substitute d_j for d_1 in Eq. (18). Substitute μ_j for μ_1 in Eq. (20). Thus, we can obtain the conditional distribution $p(y|\omega_j)$ by performing the transformation of Eq. (16) and substituting ω_j for ω_1 in Eq. (19).
Step 4. Obtain the conditional distribution $p(z|\omega_j)$ by substituting ω_j for ω_1 in Eq. (22).

4 Experiments

In this section, experiments for face analysis have been performed on CMU facial expression image database to test the effectiveness of the proposed PCA-based Bayes classifier.

From CMU-Pittsburgh AU-Coded Facial Expression Database [4], 312 facial expression mask images can be obtained by using a spatial adaptive triangulation technique based on local Gabor filters [5]. Six facial expressions are concerned as follows: anger, disgust, fear, joy, unhappy, and surprise. For each expression, there are 52 images with a resolution of 55 × 59, the first 26 images of which have moderate expressions while the last 26 images of which have intensive expressions.

In experiments, for each expression, the first $k (k = 5, 10, 15, 20, 25)$ images are for training and all the other images are for test. Experiments have been performed by using the proposed PCA-based Bayes classifier and the minimum distance classifier, respectively. Experimental results with different k are listed in Table 1.

From Table 1, we can see that the proposed PCA-based Bayes classifier performs obviously better than the minimum distance classifier. As the number of

Table 1. Classification rates on CMU facial expression image database

Images	k	Minimum distance	PCA-based Bayes
55 × 59	5	25.53%	27.30%
	10	29.76%	63.89%
	15	50.45%	73.87%
	20	59.38%	88.02%
	25	64.02%	95.68%

training samples k increases, the classification rate by using the proposed classifier increases much faster than that by using the minimum distance classifier. It means that the proposed classifier can perform much more efficient than the minimum distance classifier.

5 Conclusions

In this paper, we have proposed a novel PCA-based Bayes classifier in high dimensional spaces. Experiments for face analysis have been performed on CMU facial expression image database. It is shown that the proposed classifier performs much better than the minimum distance classifier. With the proposed classifier, we can not only improve the classification rate, but also obtain a better understanding of data.

Acknowledgements

This work was supported by Ramón y Cajal research fellowship from the Ministry of Science and Technology, Spain and the National Natural Science Foundation of China under Grant No. 60473039.

References

1. K. Fukunaga. *Introduction to Statistical Pattern Recognition.* Academic Press, 1990.
2. Zhong Jin, Jingyu Yang, Zhongshan Hu, and Zhen Lou. Face recognition based on the uncorrelated discriminant transformation. *Pattern Recognition*, 34(7):1405-1416, 2001.
3. Z. Jin, F. Davoine, and Z. Lou. An effective EM algorithm for PCA mixture model. In *Structural, Syntactic and Statistical Pattern Recongnition*, volume 3138 of *Lecture Notes in Computer Science*, pp. 626-634, Lisbon, Portugal, Aug. 18-20 2004. Springer.
4. Takeo Kanade, Jeffrey F. Cohn, and Yingli Tian. Comprehensive database for facial expression analysis. In *Proceedings of the Fourth International Conference of Face and Gesture Recognition*, pages 46-53, Grenoble, France, 2000.
5. S. Dubuisson, F. Davoine, and M. Masson. A solution for facial expression representation and recognition. *Signal Processing: Image Communication*, 17(9):657-673, 2002.

Highly Accurate and Fast Face Recognition Using Near Infrared Images*

Stan Z. Li, RuFeng Chu, Meng Ao, Lun Zhang, and Ran He

Center for Biometrics and Security Research & National Laboratory of Pattern Recognition,
Institute of Automation, Chinese Academy of Sciences,
95 Zhongguancun Donglu Beijing 100080, China
http://www.cbsr.ia.ac.cn

Abstract. In this paper, we present a highly accurate, realtime face recognition system for cooperative user applications. The novelties are: (1) a novel design of camera hardware, and (2) a learning based procedure for effective face and eye detection and recognition with the resulting imagery. The hardware minimizes environmental lighting and delivers face images with frontal lighting. This avoids many problems in subsequent face processing to a great extent. The face detection and recognition algorithms are based on a local feature representation. Statistical learning is applied to learn most effective features and classifiers for building face detection and recognition engines. The novel imaging system and the detection and recognition engines are integrated into a powerful face recognition system. Evaluated in real-world user scenario, a condition that is harder than a technology evaluation such as Face Recognition Vendor Tests (FRVT), the system has demonstrated excellent accuracy, speed and usability.

1 Introduction

Face recognition has a wide range of applications such as face-based video indexing and browsing engines, multimedia management, human-computer interaction, biometric identity authentication, and surveillance. Interest and research activities in face recognition have increased significantly in the past years [16, 17, 5, 20].

In *cooperative user* scenarios, a user is required cooperate with the face camera to have his/her face image captured properly, in order to be grated for the access; this is in contrast to more general scenarios, such as face recognition under surveillance. There are many cooperative user applications, such as access control, machine readable traveling documents (MRTD), ATM, computer login, e-commerce and e-government. In fact, many face recognition systems have been developed for such applications.

However, even in such a favorable condition, most existing face recognition systems, academic and commercial, are confounded by even moderate illumination changes. When the lighting differs from that for the enrollment, the system would either fail to recognize (false rejection) or make mistaken matches (false acceptance).

* This work was supported by Chinese National 863 Projects 2004AA1Z2290 & 2004AA119050.

D. Zhang and A.K. Jain (Eds.): ICB 2006, LNCS 3832, pp. 151–158, 2005.

To avoid the problem caused by illumination changes (and other changes), several solutions have been investigated into. One technique is to use 3D (in many case, 2.5D) data obtained from a laser scanner or 3D vision method (*cf.* papers [3, 21]). Because 3D data captures geometric shapes of face, such systems are affected less by environmental lighting and it can cope with rotated faces because of the availability of 3D (2.5D) information for visible points. The disadvantages are the increased cost and slowed speed as well as the artifacts due to speculation. Recognition performances obtained using a single 2D image or a single 3D image are similar. [4].

Invisible imagery has recently received increased attention in the computer vision community, as seen from the IEEE workshop series [6, 13]. Thermal or far infrared imagery has been used for face recognition (*cf.* and a survey paper [10]). While thermal based face recognition systems are advantages for detecting disguised faces or when there is no control over illumination, they are subject to environmental temperature, emotional and health conditions, and generally do not perform as well as 2D based systems for the cooperative scenario.

The use of near infrared (NIR) imagery brings a new dimension for applications of invisible lights for face detection and recognition [7, 11, 14]. In [7], face detection is performed by analyzing horizontal projections of the face area using the fact that eyes and eyebrows regions have different responses in the lower and upper bands of NIR. In [11], a method of homomorphic-filtering is used as a pre-processing before extracting facial features. In [14], face recognition is done using hyperspectral images captured in 31 bands over an NIR range of 0.7μm-1.0μm; invariant features are extracted from such images.

In this paper, we present a highly accurate, real-time system for face recognition in cooperative user applications. The contributions are the following: First, we present a novel design of camera hardware. The camera delivers filtered NIR images containing mostly relevant, intrinsic information for face detection and recognition, with extrinsic factors minimized. This alleviates much difficulty in subsequent processing. Second, we present learning based algorithms, using a local feature representation, for effective face/eye detection and face recognition in filtered NIR images. The algorithms can achieve high accuracies with high speed. The most important contribution is the methodology learned from the building of this successful system for how to make face recognition really work.

The present system has been tested for a real application of access control and time attendance. This is a scenario evaluation[15], an evaluation condition that is harder than a technology evaluation such as FRVT tests. The working conditions are under varying indoor locations and illumination conditions, with cooperative users. After a period of one month, the system has demonstrated excellent accuracy, speed, usability and stability under varying indoor illumination, even in the complete darkness. It has achieved an equal error rate below 0.3%.

The rest of the paper is organized as follows: Section 2 describes the design of the imaging hardware and presents an analysis of characteristics and amicability of resulting images for subsequent face processing. Section 3 describes the software part, including the feature representation, and the learning based methods for face/eye detection and face recognition. Section describes the system evaluation (Section 4).

2 Imaging Hardware

The goal of making the special-purpose hardware is to avoid the problems arising from environmental lighting, towards producing nearly idealized face images for face recognition. By the word "idealized", we mean that the lighting is frontal and of suitable strength. Environmental lighting is generally existing but from un-controlled directions and it is difficult to normalize it well by using an illumination normalization method. This is in fact a major obstacle in traditional face recognition.

To overcome the problem, we decide to use some active lights mounted on the camera to provide frontal lighting and to use further means to reduce environmental lighting to minimum. We propose two principles for the active lighting: (1) the lights should be strong enough to produce clear frontal-lighted face image but not cause disturbance to human eyes, and (2) the resulting face image should be affected as little as possible after minimizing the environmental lighting.

Our solution for (1) is to mount near infrared (NIR) light-emitting diodes (LEDs) on the hardware device to provide active lighting. When mounted on the camera, the LEDs provide the best possible straight frontal lighting, better than mounted anywhere else.

For (2), we use a long pass optical filter on the camera lens to cut off visible light while allowing NIR light to pass. The long pass filter is such that the wavelength points for 0%, 50%, 88%, and 99% passing rates are 720, 800, 850, and 880nm, respectively. The filter cuts off visible environmental lights ($<$ 700nm) while allowing the NIR light (850nm) to pass.

As a result, this imaging hardware device not only provides appropriate active frontal lighting but also minimizes lightings from other sources. Figure 1 shows example images of a face illuminated by both frontal NIR and a side environmental light.We can see that the lighting conditions are likely to cause problems for face recognition with the conventional color (and black and white) images, the NIR images are mostly frontal-lighted by the NIR lights only, with minimum influence from the environmental light, and are very suitable for face recognition. The effect of remaining NIR component of environmental lights in the NIR image (such as due to the lamp light for making the example images) is much weak than that of the NIR LED lights.

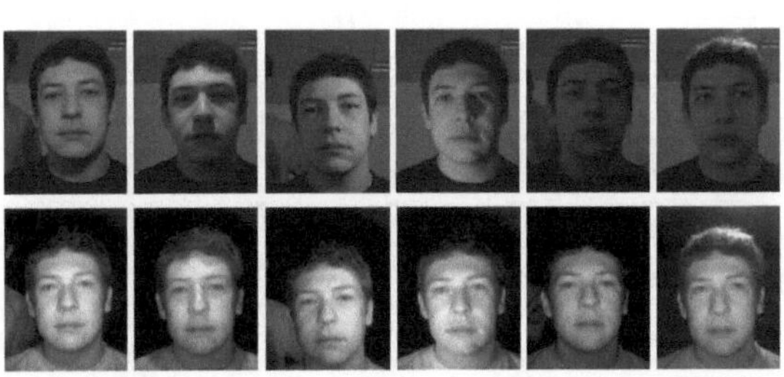

Fig. 1. Upper-row: 5 color images of a face. Lower-row: The corresponding NIR-filtered images.

3 Learning-Based Algorithms

Both detection and matching are posed as a two-class problem of classifying the input into the positive or negative class. The central problem in face/eye detection is to classify each scanned sub-window into either face/eye or non-face/eye; the positive sub-windows are post-processed by merging multiple detects in nearby locations. For face matching, the central problem is to develop a matching engine or a similarity/distance function for the comparison of two cropped face images. In this regard, we adopt the intrapersonal and extrapersonal dichotomy proposed in [12], and train a classifier for the two-class classification. The trained classifier outputs a similarity value, based on which the classification can be done with a confidence threshold.

3.1 Learning for Face/Eye Detection

A cascade of classifiers are learned from face/eye and non-face/eye training data. For face detection, an example is a 21x21 image, containing a face or nonface pattern. For eye detection, an example is a 21x15 image, containing an eye or noneye pattern. Sub-regions of varying sizes from 5×5 to 11×11 with step size 3 in both directions are used for computing the LBP histogram features for the local regions, which generates all possible features composed of all the 59 scalar features at all the locations.

Figure 2 show statistics on the training results. On the left shows the face and nonface distributions as functions of number of weak classifiers. We can see that the two classes are well separated, and a large number (more than 95% in the data) of nonface examples are rejected at the first two stages. The ROC indicates that the overall detection rate is 96.8% given the false alarm rate of 10^{-7}. On the right compares the ROC curves with that of the baseline algorithm of [18].

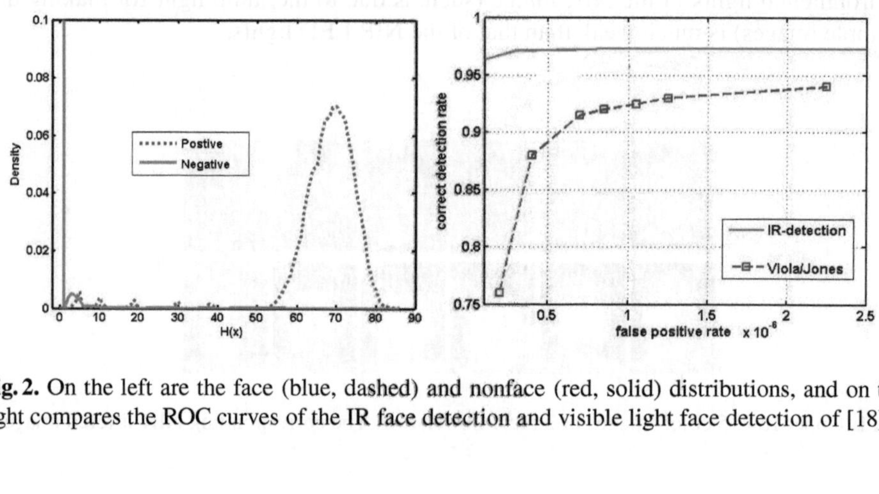

Fig. 2. On the left are the face (blue, dashed) and nonface (red, solid) distributions, and on the right compares the ROC curves of the IR face detection and visible light face detection of [18]

3.2 Learning for Face Recognition

Recently, the LPB representation has been used for face detection and recognition. In [1, 9], an input face image is divided into 42 blocks of size w by h pixels. Instead of using the LBP patterns for individual pixels, the histogram of 59 bins over each block in the image is computed to make a more stable representation of the block. The Chi-square distance is used for the comparison of the two histograms (feature vectors)

$$\chi^2(S, M) = \sum_{b=1}^{B} \frac{(S_b - M_b)^2}{(S_b + M_b)} \tag{1}$$

where S_b and M_b are to the probabilities of bin b for the corresponding histograms in the gallery and probe images and B is the number of bins in the distributions. The final matching is based on the weighted chi-square distance over all blocks.

We believe that the above scheme lacks optimality. First, a partition into blocks is not optimized in any sense and ideally all possible pixel locations should be considered. Second, manually assigning a weight to a block is not optimized. Third, there should be better matching schemes than using the block comparison with the Chi-distance.

Therefore, we adopt a statistical learning approach [19], instead of using a Chi-square distance [1, 9] and weighted sum of block matches for matching between two faces. The need for a learning is also due to the complexity of the classification. The classification here is inherently a nonlinear problem. An AdaBoost learning procedure [8] is used for these purposes, where we adopt the intrapersonal and extrapersonal dichotomy [12] to convert the multi-class problem into one of two-class. See [19] for more details of the methods.

Figure 3 shows the ROC curve for the present method obtained on a test data set, which shows a verification rate (VR) of 90% at FAR=0.001 and 95% at FAR=0.01. In comparison, the corresponding VR's for the PCA (with Mahalanobis distance) and LDA on the same data set are 42% and 31%, respectively, FAR=0.001; and 62% and 59% at FAR=0.01. (Note that it is not unusual that LDA performs worse than PCA [2].)

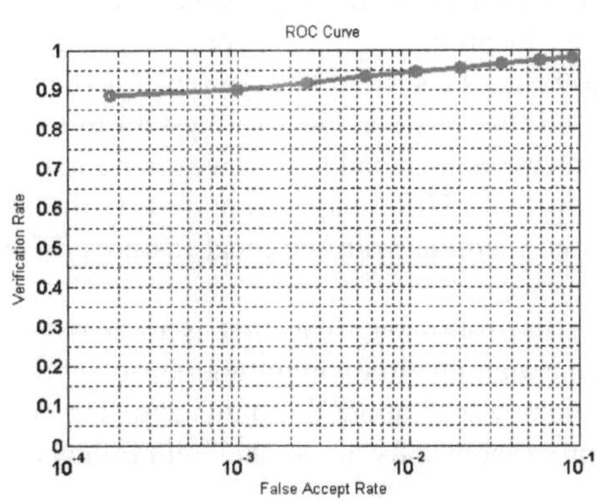

Fig. 3. ROC Curves for verification on a test data set

4 System Evaluation

Our tests are in the form of scenario evaluation [15], for 1-N identification in an access control and time attendance application in an office building. The participation protocol was the following: 1470 persons were enrolled under environmental conditions different from those of the client sites, with 5 templates per enrolled person recorded. Of these persons, 100 were workers in the building and most others were collected from other sources unrelated to the building environment. The 100 workers were used as the genuine clients while the others were used as the background individuals. On the other hand, additional 10 workers were used as the regular imposters, and some visitors were required to participate as irregular imposters. This provided statistics for calculating correct rejection rate and false acceptance rate.

The 100 clients and 10 imposters were required to report to the system 4 times a day to take time attendance, twice in the morning and twice in the evening when they started working and left the office for lunch and for home. Not all workers followed this rule strictly. Some did more than 4 times a day. Some clients deliberately challenged the system by doing strange face or occluding the face with a hand, so that the system did not recognize them. We counted these as visitor imposter sessions. Only those client sessions which were reported having problems getting recognized were counted as false rejections. On the other hand, the imposters were encouraged to challenge the system to get false acceptances. The results show that the system achieved an equal error rate below 0.3%. Hence, we conclude that the system has achieved high performance for cooperative face recognition.

5 Summary and Conclusions

We have presented a highly accurate and fast face recognition system for cooperative user applications. The novel design of the imaging hardware delivers face images amicable for face processing. The statistical learning procedures with local features give to highly accurate and fast classifiers for face/eye detection and face recognition. These, together with engineering inspirations, have made a successful system. Evaluated in real-world user scenario tests, the system has demonstrated excellent accuracy, speed and usability. We believed that this was the best system in the world for cooperative face recognition. The success is ascribed to two reasons: First, the classification tasks herein are made very easy with NIR images captured by the novel hardware device. Second, the learning based methods with the local features by their own are powerful classification engines.

Future work includes the following: The first is to study the performance of the matching engine for face matching after a long time-lapse, while the system has had no problem with faces previously seen about 8 months ago. The second is to improve the imaging hardware and processing software to deal with influence of NIR component in outdoor sunlight.

- *Two patents have been filed for the technology described in this paper.*

References

1. T. Ahonen, A. Hadid, and M.Pietikainen. "Face recognition with local binary patterns". In *Proceedings of the European Conference on Computer Vision*, pages 469–481, Prague, Czech, 2004.
2. J. R. Beveridge, K. She, B. A. Draper, and G. H. Givens. "A nonparametric statistical comparison of principal component and linear discriminant subspaces for face recognition". In *Proceedings of IEEE Computer Society Conference on Computer Vision and Pattern Recognition*, pages I:535–542, 2001.
3. K. W. Bowyer, Chang, and P. J. Flynn. "A survey of 3D and multi-modal 3d+2d face recognition". In *Proceedings of International Conference Pattern Recognition*, pages 358–361, August 2004.
4. K. I. Chang, K. W. Bowyer, and P. J. Flynn. "An evaluation of multi-modal 2D+3D face biometrics". *IEEE Transactions on Pattern Analysis and Machine Intelligence*, page to appear, 2005.
5. R. Chellappa, C. Wilson, and S. Sirohey. "Human and machine recognition of faces: A survey". *Proceedings of the IEEE*, 83:705–740, 1995.
6. CVBVS. In *IEEE Workshop on Computer Vision Beyond the Visible Spectrum: Methods and Applications*, 1999-2003.
7. J. Dowdall, I. Pavlidis, and G. Bebis. "Face detection in the near-IR spectrum". *Image and Vision Computing*, 21:565–578, July 2003.
8. Y. Freund and R. Schapire. "A decision-theoretic generalization of on-line learning and an application to boosting". *Journal of Computer and System Sciences*, 55(1):119–139, August 1997.
9. A. Hadid, M. Pietikinen, and T. Ahonen. "A discriminative feature space for detecting and recognizing faces". In *Proceedings of IEEE Computer Society Conference on Computer Vision and Pattern Recognition*, volume 2, pages 797–804, 2004.
10. S. G. Kong, J. Heo, B. Abidi, J. Paik, and M. Abidi. "Recent advances in visual and infrared face recognition - A review". *Computer Vision and Image Understanding*, 97(1):103–135, January 2005.
11. D.-Y. Li and W.-H. Liao. "Facial feature detection in near-infrared images". In *Proc. of 5th International Conference on Computer Vision, Pattern Recognition and Image Processing*, pages 26–30, Cary, NC, September 2003.
12. B. Moghaddam, C. Nastar, and A. Pentland. "A Bayesain similarity measure for direct image matching". *Media Lab Tech Report* No.393, MIT, August 1996.
13. OTCBVS. In *IEEE International Workshop on Object Tracking and Classification in and Beyond the Visible Spectrum*, 2004-2005.
14. Z. Pan, G. Healey, M. Prasad, and B. Tromberg. "Face recognition in hyperspectral images". *IEEE Transactions on Pattern Analysis and Machine Intelligence*, 25(12):1552–1560, December 2003.
15. P. J. Phillips, A. Martin, C. L. Wilson, and M. Przybocki. "An introduction to evaluating biometric system". *IEEE Computer (Special issue on biometrics)*, pages 56–63, February 2000.
16. A. Samal and P. A.Iyengar. "Automatic recognition and analysis of human faces and facial expressions: A survey". *Pattern Recognition*, 25:65–77, 1992.
17. D. Valentin, H. Abdi, A. J. O'Toole, and G. W. Cottrell. "Connectionist models of face processing: A survey". *Pattern Recognition*, 27(9):1209–1230, 1994.
18. P. Viola and M. Jones. "Robust real time object detection". In *IEEE ICCV Workshop on Statistical and Computational Theories of Vision*, Vancouver, Canada, July 13 2001.

19. G. Zhang, X. Huang, S. Z. Li, Y. Wang, and X. Wu. "Boosting local binary pattern (LBP)-based face recognition". In S. Z. Li, J. Lai, T. Tan, G. Feng, and Y. Wang, editors, *Advances in Biometric Personal Authentication*, volume LNCS-3338, pages 180–187. Springer, December 2004.
20. W. Zhao and R. Chellappa. "Image based face recognition, issues and methods". In B. Javidi, editor, *Image Recognition and Classification*, pages 375–402. Mercel Dekker, 2002.
21. W. Zhao, R. Chellappa, P. Phillips, and A. Rosenfeld. Face recognition: A literature survey. *ACM Computing Surveys*, pages 399–458, 2003.

Background Robust Face Tracking Using Active Contour Technique Combined Active Appearance Model*

Jaewon Sung and Daijin Kim

Biometrics Engineering Research Center (BERC),
Pohang University of Science and Technology
{jwsung, dkim}@postech.ac.kr

Abstract. This paper proposes a two stage AAM fitting algorithm that is robust to the cluttered background and a large motion. The proposed AAM fitting algorithm consists of two alternative procedures: the active contour fitting to find the contour sample that best fits the face image and then the active appearance model fitting over the best selected contour. Experimental results show that the proposed active contour based AAM provides better accuracy and convergence characteristics in terms of RMS error and convergence rate, respectively, than the existing robust AAM.

1 Introduction

Active Appearance Models (AAMs) [1] are generative, parametric models of certain visual phenomena that show both shape and appearance variations. These variations are represented by linear models such as Principal Component Analysis (PCA), which finds a subspace reserving maximum variance of given data. The most common application of AAMs has been face modeling [1], [2], [3], [4].

Although the structure of the AAM is simple, fitting an AAM to an target image is a complex task that requires a non-linear optimization technique that requires a huge amount of computation when the standard non-linear optimization techniques such as the gradient descent method are used. Recently, a gradient based efficient AAM fitting algorithm, which is extended from an inverse compositional LK image matching algorithm [5], has been introduced by Matthews et. al. [4]. The AAM fitting problem is treated as an image matching problem that includes both shape and appearance variations with a piece-wise affine warping function. Other AAM fitting algorithms can be found in [6].

We propose a novel AAM fitting method that pre-estimates the change of the shape (motion) of an object using the active contour technique and then begins

* This work was supported by the Korea Science and Engineering Foundation (KOSEF) through the Biometrics Engineering Research Center (BERC) at Yonsei University.

D. Zhang and A.K. Jain (Eds.): ICB 2006, LNCS 3832, pp. 159–165, 2005.

existing AAM fitting algorithm using the motion compensated parameters. In this work, a CONDENSATION-like [7] active contour technique has been used to estimate the object contour effectively, thus accurately estimating the motion of the object in the image sequence.

The remainder of this paper is organized as follows. In section 2, we briefly review the original AAM fitting algorithm and active contour technique. In Section 3, we explain how the active contour technique can be incorporated into the AAM fitting algorithm to make it robust to the large motion. In section 4, experimental results are presented. Finally, we draw a conclusion.

2 Theoretical Backgrounds

2.1 Active Appearance Models

In 2D AAMs [1], [4], the 2D shape $\mathbf{s}$ of an object is represented by a triangulated 2D mesh and it is assumed that the varying shape can be approximated by a linear combination of a mean shape $\mathbf{s}_0$ and orthogonal shape bases $\mathbf{s}_i$ as

$$\mathbf{s} = \mathbf{s}_0 + \sum_{i=1}^{n} p_i \mathbf{s}_i, \tag{1}$$

where p_i are the shape parameters and $\mathbf{s} = (x_i, y_1, ..., x_l, y_l)^T$. The appearance is defined in the mean shape $\mathbf{s}_0$ and the appearance variation is modeled by a linear combination of a mean appearance A_0 and orthogonal appearance bases A_i as

$$A = A_0 + \sum_{i=1}^{m} \alpha_i A_i, \tag{2}$$

where α_i are the appearance parameters and A_i represents the vectorized appearance. To build an AAM, we need a set of landmarked training images. The shape and appearance bases are computed by applying PCA to the shape and appearance data that are collected and normalized appropriately.

Using an 2D AAM, the shape-variable appearance of an object in the image can be represented by

$$M(W(\mathbf{x}; \mathbf{p}')) = \sum_{i=0}^{m} \alpha_i A_i(\mathbf{x}), \tag{3}$$

where W is a coordinate transformation function from the coordinate $\mathbf{x}$ in the template image frame to the coordinate of the synthesized image frame. The parameters of the warping function are represented by $\mathbf{p}'^T = (\mathbf{p}^T, \mathbf{q}^T) = (p_1, \ldots, p_n, q_1, \ldots, q_4)$, where $\mathbf{p}$ and $\mathbf{q}$ determine the varying 2D shape of the object and its similar transformation, respectively. Four similar transformation parameters q_1, q_2, q_3, and q_4 describe the scale, rotation, horizontal and vertical translation of the shape, respectively.

2.2 The AAM Fitting Algorithm

The problem of fitting a 2D AAM to a given image can be formulated as finding the appearance and shape parameters of an AAM that minimizes the following error

$$E = \sum_{\mathbf{x} \in \mathbf{s}_0} \left[\sum_{i=0}^{m} \alpha_i A_i(\mathbf{x}) - I(W(\mathbf{x}; \mathbf{p}')) \right]^2 . \tag{4}$$

Among various gradient based fitting algorithms, we will briefly review the Inverse Compositional Simultaneous Update algorithm (SI), which is known to have the best convergence performance and the Inverse Compositional Normalization algorithm (NO), which is more efficient than the SI algorithm. The SI algorithm is derived by applying the Taylor expansion with respect to the both shape and appearance parameters. The update of model parameters $\Delta\boldsymbol{\theta}^T = \{\Delta\mathbf{p}'^T, \Delta\boldsymbol{\alpha}^T\}$ are computed as

$$\Delta\boldsymbol{\theta} = \left\{ \sum_{\mathbf{x} \in \mathbf{s}_0} SD^T(\mathbf{x}) SD(\mathbf{x}) \right\}^{-1} \sum_{\mathbf{x} \in \mathbf{s}_0} SD^T(\mathbf{x}) E(\mathbf{x}) \tag{5}$$

$$SD(\mathbf{x}) = \left[\nabla A(\mathbf{x}; \boldsymbol{\alpha})^T \frac{\partial W}{\partial \mathbf{p}'}, A_1(\mathbf{x}), \ldots, A_m(\mathbf{x}) \right], \tag{6}$$

where $SD(\mathbf{x})$ represents the steepest descent vector of the model parameters θ. The warping parameters and appearance parameters are updated as $W(\mathbf{x}; \mathbf{p}') \leftarrow W(\mathbf{x}; \mathbf{p}') \circ W(\mathbf{x}; \Delta\mathbf{p}')^{-1}$, and $\boldsymbol{\alpha} \leftarrow \boldsymbol{\alpha} + \Delta\boldsymbol{\alpha}$, respectively. The SI algorithm is inefficient because $SD(\mathbf{x})$ in (5) depends on varying parameters and must be recomputed at every iteration. The inverse compositional normalization algorithm (NO) makes use of the orthogonal property of appearance bases. This orthogonal property enables the error term in (4) to be decomposed into sums of two squared error terms:

$$\left\| A_0 + \sum_{i=1}^{m} \alpha_i A_i - I^W(\mathbf{p}') \right\|^2_{span(A_i)} + \left\| A_0 + \sum_{i=1}^{m} \alpha_i A_i - I^W(\mathbf{p}') \right\|^2_{span(A_i)^\perp}, \tag{7}$$

where $I^W(\mathbf{p}')$ means vector representation of backward warped image. The first term is defined in the subspace $span(A_i)$ that is spanned by the orthogonal appearance bases and the second term is defined in the subspace $span(A_i)^\perp$, orthogonal complement subspace. For any warping parameter $\mathbf{p}'$, the minimum value of the first term is always exactly 0. Since the norm in the second term only considers the component of the vector in the orthogonal complement of $span(A_i)$, any component in $span(Ai)$ can be dropped. As a result, the second error term can be optimized efficiently with respect to $\mathbf{p}'$ using an image matching algorithm such as the inverse compositional algorithm [6].

Robust fitting algorithms use the weighted least squares formulation that includes a weighing function into its error function. Weighted least squares formulation can be applied to NO algorithm to make it robust. Detailed derivations and explanations can be found in the [4].

2.3 Active Contour Techniques

In this paper, we locate the foreground object using a CONDENSATION-like contour-tracking technique which is based on probabilistic sampling.

A contour $\mathbf{c}$ of an object is represented by a set of boundary points $\mathbf{c} = (x_1, y_1, ..., x_v, y_v)^T$. We can represent all the possible contours within a specified contour space by a linear equation as

$$\mathbf{c} = \mathbf{c_0} + S\mathbf{y}, \tag{8}$$

where $\mathbf{c_0}$ is the mean contour, S is a shape matrix that is dependent on the selected contour space and $\mathbf{y}$ is a contour parameter vector [8].

The CONDENSATION method [8] aims to estimate the posterior probability distribution $p(\mathbf{y}|\mathbf{z})$ of the parameter vector $\mathbf{y}$ in the contour space $S_{\mathbf{y}}$ using a factored sampling, where $\mathbf{z}$ denotes the observations from a sample set. The output of a factored sampling step in the CONDENSATION method is a set of samples with weights denoted as $\{(\mathbf{s}_1, \pi_1), (\mathbf{s}_2, \pi_2), ..., (\mathbf{s}_N, \pi_N)\}$, which approximates the conditional observation density $p(\mathbf{y}|\mathbf{z})$. In the factored sampling, a sample set $\{\mathbf{s}_1, \mathbf{s}_2, ..., \mathbf{s}_N\}$ are randomly generated from the prior density $p(\mathbf{y})$ and then the weights π_i of the N generated samples are computed by

$$\pi_i = \frac{p_z(\mathbf{s}_i)}{\sum_{j=1}^{N} p_z(\mathbf{s}_j)}, \tag{9}$$

where $p_z(\mathbf{s}) = p(\mathbf{z}|\mathbf{y} = \mathbf{s})$ is the conditional observation density.

In this work, we measured $p(\mathbf{z}|\mathbf{y})$ using a fitness evaluation function that consider the quality of the image edge features found in the image and the distance between the contour sample and the image edge features as

$$p(\mathbf{z}|\mathbf{y}) \propto n_f \frac{\bar{s}_f}{\sigma_s \bar{d}_f}, \tag{10}$$

where n_f is the number of edge features that have found within a given search range along the normal direction of the contour, $\bar{s}_f$ and $\bar{d}_f$ are the mean magnitude of edge gradient and the mean distance of the n_f image edge features, and σ_s is used to compensate the different scales of the edge gradient and the distance.

3 Active Contour Based AAM

We apply the following two stages alternatively in order to track the face image. During stage I, we perform the active contour technique to find the contour sample that best fits the face image as follows,

1. Make the base shape c_0 and the shape matrix S in (8) using the fitted shape of the AAM at (t-1)-th image frame.
2. Generate N random samples $\{\mathbf{s}_1 \dots \mathbf{s}_N\}$ that are located near the computed contour $\mathbf{c}$

3. Evaluate the fitness of all generated samples using the conditional observation density function $p(\mathbf{z}|\mathbf{y})$ explained in section 2.3.
4. Choose the best sample $\mathbf{s}_{best}$ with the highest fitness value among N samples. We estimate the motion parameter $\hat{\mathbf{q}}^t$ at the next image frame t by composing two similar transformations $\mathbf{q}^{t-1}$ and $\Delta\hat{\mathbf{q}}^t$, where $\Delta\hat{\mathbf{q}}^t = \mathbf{s}_{best}$.

During stage II, we perform the active appearance model fitting algorithm over the best selected contour $\mathbf{s}_{best}$ as follows,

1. Run the AAM fitting algorithm using the shape parameter $\mathbf{p}^{t-1}$, the appearance parameter $\boldsymbol{\alpha}^{t-1}$, and the estimated motion parameter, $\hat{\mathbf{q}}^t$.
2. Obtain the optimal AAM model parameters $\mathbf{p}^t$, $\mathbf{q}^t$, and $\boldsymbol{\alpha}^t$.
3. Set the image frame index $t = t - 1$, and return to stage I until reaching the final frame.

4 Experimental Results

4.1 Comparison of Fitting Performances of Different AAM Methods

We compared the accuracy of three different AAM fitting methods such as the existing robust AAM (R-AAM), the proposed active contour based AAM (AC-AAM), and a combination of the two methods. For each methods, we measured the performances using two different types of parameter updates [6] such as the normalization method (NO-update) and the simultaneous update method (SI-update). The left and right figures of Fig 1 show the results from NO-update and SI-update, respectively. The top row of Fig. 1 shows the decreasing RMS error as the fitting algorithm is iterated, where the RMS error is defined as the mean distance between the ground truth shape points and the corresponding points of the current fitted shape. In each picture, the horizontal and vertical axis denotes the iteration index and the RMS error, respectively. Two curves are represented for each AAM method, corresponding to two differently perturbed AAM shapes, respectively. Each point over the curve is the average value of the RMS errors of 100 independent trials. Figure 1 shows that 1) the contour combined AAM fitting is converged within 5 iterations in most cases, 2) the fitting of the R-AAM method is not effective when the initial displacement is great, and 3) the proposed AC-AAM has a good convergence accuracy even if there is a great initial displacement. We also compared the convergence rate of the three different AAM fitting methods, where the convergence rate is defined by the ratio of convergence cases to all trials. The bottom row of Fig. 1 show the convergence rate where each point in the figure is the average convergence rate of 100 trials. Figure 1 shows that the difference of convergence rate between R-AAM and AC-AAM becomes larger as the initial displacement error increases, which implies that the proposed AC-AAM is more effective when the AAM shape is placed far from the target face. In the above experiments the combined AC-R-AAM shows the best convergence performance.

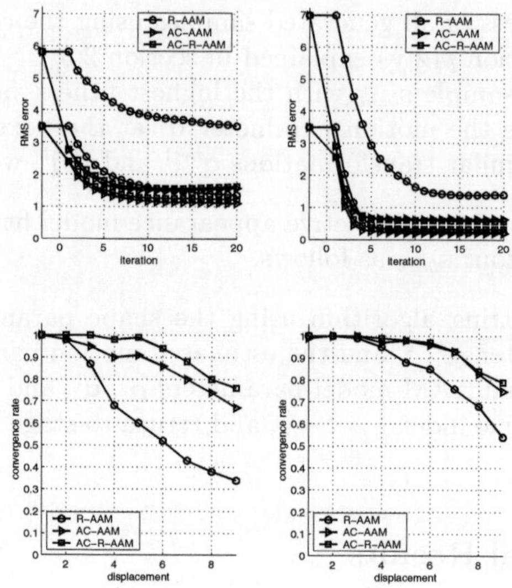

Fig. 1. Convergence characteristics of two different updates

4.2 Comparison of Execution Times Between Difference AAM Methods

Figure 2 shows the average number of iterations of the different methods, where the horizontal and the vertical axes denote the average number of iterations and the displacement σ, respectively. Each point represents the average number of iterations of independent successfully converged trails when the same stop condition if applied. From the Fig. 2, we note that the average number of iterations of AC-AAM and AC-R-AAM are almost constant even the displacement σ increases, while those of R-AAM increased rapidly as the displacement σ increases.

We measured the execution time of the different methods in our C implementation. It took about 5 msec for the active contour fitting when 50 samples, 51 contour points, and 10 pixels of search range were considered. Also, it took

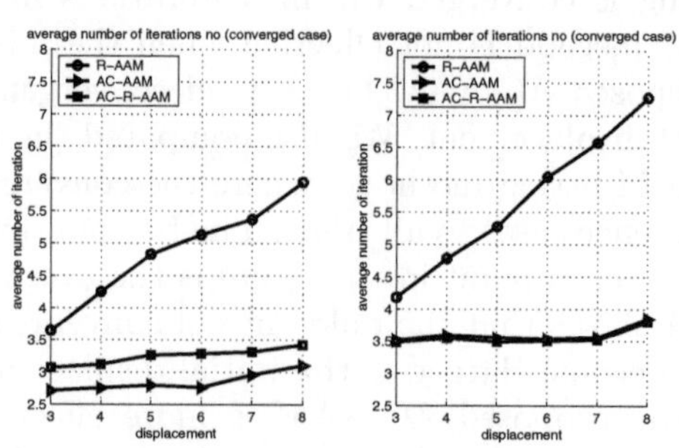

Fig. 2. Comparison of the number of iterations of three different AAM methods

about 8 msec and 26 msec for the NO-update and SI-update, respectively, in the robust AAM and it takes about 4 msec and 23 msec for the the NO-update and SI-update, respectively, in the the proposed AC-AAM.

5 Conclusion

In this paper, we proposed an active contour combined AAM fitting algorithm that is robust to a large motion of an object. Although the existing robust AAM can cope with the mismatch between currently estimated AAM instance and an input image, it does not converge well when the motion of the face is large. This comes from the fact that only the small part of the backward warped image may be used to estimate the update of parameters and it is not sufficient for correct estimation. The proposed AAM fitting method was robust to a large motion of the face because it rapidly locates the AAM instance to an area close to the correct face position. The proposed AAM fitting method was also fast because the active contour technique can estimate the large motion of the face more chiefly than the AAM fitting algorithm.

We performed many experiments to evaluate the accuracy and convergence characteristics in terms of RMS error and convergence rate, respectively. The combination of the existing robust AAM and the proposed active contour based AAM (AC-R-AAM) showed the best accuracy and convergence performance.

References

1. T.F. Cootes, G.J. Edwards, and C.J. Taylor, "Active Appearance Models," *IEEE Trans. Pattern Analysis and Machine Intelligence*, vol. 23, issue 6, pp. 681–685, 2001.
2. G.J. Edwards, C.J. Taylor, T.F. Cootes, "Interpreting Face Images Using Active Appearance Models," *Proc. of IEEE 3rd International Conference on Automatic Face and Gesture Recognition*, vol.0, pp. 300, 1998.
3. G.J. Edwards, T.F. Cootes, C.J. Taylor, "Face Recognition Using Active Appearance Models,", *Proc. of 5th European Conference on Computer Vision*, vol. 2, pp. 581, June 1998.
4. S. Baker and I. Matthews, "Active Appearance Models Revisited," CMU-RI-TR-03-01, CMU, Apr 2003.
5. B.D. Lucas T. Kanade, "An iterative image registration technique with an application to stereo vision," *Proc. of International Joint Conference on Artificial Intelligence*, 1981, pp. 674–679.
6. I. Matthews, R. Gross, and S. Baker, "Lucas-Kanade 20 Years on: A Unifying Framework: Part 3," CMU-RI-TR-03-05, CMU, Nov 2003.
7. M. Isard and A. Blake, "CONDENSATION-Conditional Density Propagation for Visual Tracking," *International Journal of Computer Vision*, vol. 29, pp. 5–28, 1998.
8. M. Isard and A. Blake, *Active Contours*, Springer, 1998.

Ensemble LDA for Face Recognition

Hui Kong[1], Xuchun Li[1], Jian-Gang Wang[2], and Chandra Kambhamettu[3]

[1] School of Electrical and Electronic Engineering,
Nanyang Technological University,
50 Nanyang Ave., Singapore 639798
[2] Institute for Infocomm Research, 21 Heng Mui Keng Terrace, Singapore 119613
[3] Department of Computer and Information Science, University of Delaware,
Newark, DE 19716-2712

Abstract. Linear Discriminant Analysis (LDA) is a popular feature extraction technique for face image recognition and retrieval. However, It often suffers from the small sample size problem when dealing with the high dimensional face data. Two-step LDA (PCA+LDA) [1,2,3] is a class of conventional approaches to address this problem. But in many cases, these LDA classifiers are overfitted to the training set and discard some useful discriminative information. In this paper, by analyzing the overfitting problem for the two-step LDA approach, a framework of **En**semble **L**inear **D**iscriminant **A**nalysis (E_nLDA) is proposed for face recognition with small number of training samples. In E_nLDA, a Boosting-LDA (B-LDA) and a Random Sub-feature LDA (RS-LDA) schemes are incorporated together to construct the total weak-LDA classifier ensemble. By combining these weak-LDA classifiers using majority voting method, recognition accuracy can be significantly improved. Extensive experiments on two public face databases verify the superiority of the proposed E_nLDA over the state-of-the-art algorithms in recognition accuracy.

1 Introduction

Linear Discriminant Analysis [4] is a well-known scheme for feature extraction and dimension reduction. It has been used widely in many applications such as face recognition [1], image retrieval [2], etc. Classical LDA projects the data onto a lower-dimensional vector space such that the ratio of the between-class scatter to the within-class scatter is maximized, thus achieving maximum discrimination. The optimal projection (transformation) can be readily computed by solving a generalized eigenvalue problem. However, the intrinsic limitation of classical LDA is that its objective function requires the within-class covariance matrix to be nonsingular. For many applications, such as face recognition, all scatter matrices in question can be singular since the data vectors lie in a very high-dimensional space, and in general, the feature dimension far exceeds the number of data samples. This is known as the *Small Sample Size* or singularity problem [4].

D. Zhang and A.K. Jain (Eds.): ICB 2006, LNCS 3832, pp. 166–172, 2005.

In recent years, many approaches have been proposed to deal with this problem. Among these LDA extensions, the two-stage LDA (PCA+LDA) has received a lot of attention, especially for face recognition [1, 2]. Direct-LDA (D-LDA) [5], Null-space based LDA (N-LDA) [3, 6] and Discriminant Common Vector based LDA (DCV) [7] have also been proposed. However, they all discard some useful subspaces for such-and-such reasons that prevent themselves from achieving higher recognition rate. Recently, Wang and Tang [8] presented a random sampling LDA for face recognition with small number of training samples. This paper concludes that both Fisherface and N-LDA encounter respective overfitting problem for different reasons. A random subspace method and a random bagging approach are proposed to solve them. A fusion rule is adopted to combine these random sampling based classifiers. A dual-space LDA approach [9] for face recognition was proposed to simultaneously apply discriminant analysis in the principal and null subspaces of the within-class covariance matrix. The two sets of discriminative features are then combined for recognition.

One common property of the above LDA techniques is that the image matrices must be transformed into the image vectors before feature extraction. More recently, a straightforward strategy was proposed for face recognition and representation, i.e., Two-Dimensional Fisher Discriminant Analysis (2DFDA) [10]. Different from conventional LDA where data are represented as vectors, 2DFDA adopts the matrix-based data representation model. That is, the image matrix does not need to be transformed into a vector beforehand. Instead, the covariance matrix is evaluated directly using the 2D image matrices. In contrast to the $\mathbf{S}_b$ and $\mathbf{S}_w$ of conventional LDA, the covariance matrices obtained by 2DFDA are generally not singular. Therefore, 2DFDA has achieved more promising results than the conventional LDA-based methods.

In this paper, by analyzing the overfitting problem for the two-step LDA approach, a framework of **En**semble **L**inear **D**iscriminant **A**nalysis (E_nLDA) is proposed for face recognition with small number of training samples. In E_nLDA, two different schemes are proposed and coupled together to construct the component weak-LDA classifier ensemble, i.e., a Boosting-LDA (B-LDA) algorithm and a Random Sub-feature LDA (RS-LDA) scheme. In B-LDA, multiple weighted-LDA classifiers are built where the weights of the component weak-LDA classifiers and those of the training samples are updated online based on AdaBoost algorithm. In RS-LDA, the component weak-LDA classifiers are created based on randomly selected PCA sub-features. Thus, the LDA ensemble comprises all the component weak-LDA classifiers created by B-LDA and RS-LDA. By combining these weak-LDA classifiers using majority voting method, recognition accuracy can be significantly improved.

It is well known that, in the two-step LDA methods (e.g., Fisherface), an intermediate PCA step is implemented before the LDA step and then LDA is performed in the PCA subspace. It can easily be seen that there are several drawbacks in the two-step LDA. Firstly, the obtained optimal transformation is a global and single projection matrix. Secondly, the overfitting problem is usually inevitable when the training set is relatively small compared to the high dimensionality of the feature vector. In addition, the constructed classifier is numerically unstable, and much discriminative information has to be discarded to construct a stable classifier. There are two major reasons that arouse the overfitting problem in the two-step LDA. The first one is the existence of the non-representative training samples (or noise/unimportant data). The second is that although $\mathbf{S}_w$ is nonsingular, $N - c$ dimensionality is still too high for the training set in many cases. When the training set is small (e.g., only two/three training samples available for each subject), $\mathbf{S}_w$ is not well estimated. A slight disturbance of noise on the training set will greatly change the inverse of $\mathbf{S}_w$. Therefore, the LDA classifier is often biased and unstable. In fact, the proper PCA subspace dimension depends on the training set.

2 Ensemble LDA

Ensemble method is one of the major developments in machine learning in the past decade, which finds a highly accurate classifier by combining many moderately accurate component classifiers. Bagging [11], Boosting [12] and Random Subspace [13] methods are the most successful techniques for constructing ensemble classifiers.

To reduce the effect of the overfitting problem in the two-step LDA, we use Ensemble LDA (E_nLDA) to improve LDA based face recognition. Two different schemes are proposed to overcome the two problems that arouse the overfittings. To erase the effect brought by the existence of the nonrepresentative training samples, a boosting-LDA (B-LDA) is proposed to dynamically update the weights of training samples so that more important (more representative) training samples have larger weights and less important (less representative) training samples have smaller weights. With iteration of updated weights for the training samples, a series of weighted component weak-LDA classifiers are constructed. To remove the effect brought by the discrepancy between the size of training set and the length of feature vectors, a random sub-feature LDA (RS-LDA) is proposed to reduce such a discrepancy.

2.1 Boosting-LDA

In this section, the AdaBoost algorithm is incorporated into the B-LDA scheme (Table 1), where the component classifier is the standard Fisherface method. A set of trained weak-LDA classifiers can be obtained via B-LDA algorithm, and the majority voting method is used to combine these weak-LDA classifiers. One point deserving attention is that a so-called nearest class-center classifier instead of nearest neighborhood classifier is used in computing the training and test error. The nearest class-center classifier is similar to the nearest neighborhood classifier except that the metric used is the distance between the test data and the centers of the training data of each class not the one between the test sample and each training sample.

Table 1. Boosting-LDA algorithm

Algorithm: Boosting-LDA
1. Input: a set of training samples with labels $\{(\mathbf{x}_1, y_1),$
$..., (\mathbf{x}_N, y_N)\}$, Fisherface algorithm, the number of cycles T.
2. Initialize: the weight of samples: $w_i^1 = 1/N$, for all $i = 1, ..., N$.
3. Do for $t = 1, ..., T$
(1)Use Fisherface algorithm to train the weak-LDA classifier h_t on the weighted training sample set.
(2)Calculate the training error of $h_t : \epsilon_t = \sum_{i=1}^{N} w_i^t,\ y_i \neq h_t(\mathbf{x}_i)$.
(3)Set weight of weak learner $h_t : \alpha_t = \frac{1}{2}\ln(\frac{1-\epsilon_t}{\epsilon_t})$.
(4)Update training samples' weights: $w_i^{t+1} = \frac{w_i^t \exp\{-\alpha_t y_i h_t(\mathbf{x}_i)\}}{C_t}$ where C_t is a normalization constant, and $\sum_{i=1}^{N} w_i^{t+1} = 1$.
4. Output: a series of component weak-LDA classifiers.

2.2 Random Sub-feature LDA

Although the dimension of image space is very high, only part of the full space contains the discriminative information. This subspace is spanned by all the eigenvectors of the total covariance matrix with nonzero eigenvalues. For the covariance matrix computed from N training samples, there are at most $N-1$ eigenvectors with nonzero eigenvalues. On the remaining eigenvectors with zero eigenvalues, all the training samples have zero projections and no discriminative information can be obtained. Therefore, for Random Sub-feature LDA, we first project the high dimension image data to the $N-1$ dimension PCA subspace before random sampling.

In Fisherface, the PCA subspace dimension should be $(N-C)$, however, Fig.1 (a) reports that the optimal result does not appear at the *120th* $(40 \times 4 - 40)$ dimension of PCA subspace when there are 4 training samples for each subject

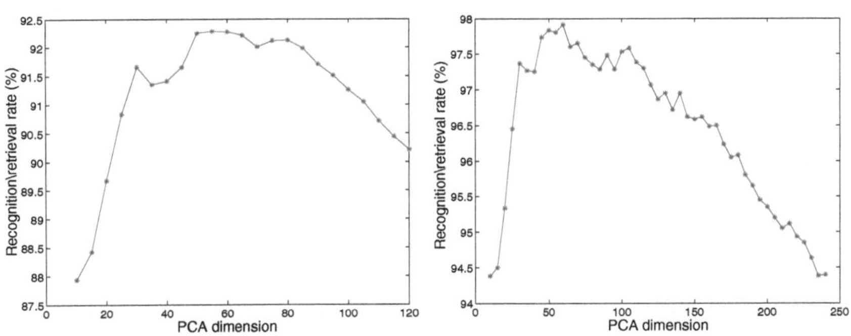

(a) (b)

Fig. 1. Recognition/retrieval accuracy of Fisherface classifier with different dimension of PCA subspace

Table 2. E_nLDA algorithm

Algorithm: E_nLDA
1. Input: a set of training samples with labels $\{(\mathbf{x}_1, y_1), ..., (\mathbf{x}_N, y_N)\}$, Fisherface algorithm, the number of cycles R.
2. Do: Apply PCA to the face training set. All the eigenfaces with zero eigenvalues are removed, and $N-1$ eigenfaces $\mathbf{U}_t = [u_1, u_2, ..., u_{N-1}]$ are retained as candidates to construct the random subspaces.
3. Do for $k = 1, ..., K$: Generate K random subspaces $\{\mathbf{S}_i\}_{i=1}^{K}$. Each random subspace $\mathbf{S}_i$ is spanned by $N_0 + N_1$ dimension. The first N_0 dimensions are fixed as the first N_0 largest eigenfaces in $\mathbf{U}_t$. The remaining N_1 dimensions are randomly selected from the other $N - 1 - N_0$ eigenfaces in $\mathbf{U}_t$
4. Do: Perform B-LDA to produce T weak-LDA classifiers in each iteration of RS-LDA.
5. Output: a set of $K \times T$ component weak-LDA classifiers.

in ORL database. A similar case appears in Fig.1 (b) where the optimal PCA dimension is about *60th* instead of *240th* ($40 \times 7 - 40$) when there are 7 training samples for each subject.

Therefore, in order to construct a stable LDA classifier, we sample a small subset of features to reduce discrepancy between the size of the training set and the length of the feature vector. Using such a random sampling method, we construct a multiple number of stable LDA classifiers. A more powerful classifier can be constructed by combining these component classifiers. A detailed description of RS-LDA is listed in Table 2.

2.3 Ensemble LDA: Combination of B-LDA and RS-LDA

Ensemble LDA (E_nLDA) can be constructed by combining B-LDA and RS-LDA. This is because that the dimension of the PCA subspace is fixed in B-LDA while the dimension of the PCA subspace is random in RS-LDA. As long as we first perform the random selection of different dimension of PCA subspace, B-LDA can be performed based on the selected PCA subspace to construct T weak-LDA classifiers. That means, if we perform K iterations of random selection (RS-LDA), $K \times T$ weak-LDA classifiers can be constructed. E_nLDA algorithm is listed in Table 2. Similarly, all the obtained component LDA classifiers can be combined via majority voting method for final classification.

3 Experiment Results

The proposed E_nLDA method is used for face image recognition/retrieval and tested on two well-known face image databases (ORL and Yale face database B). ORL database is used to evaluate the performance of E_nLDA under conditions where the pose, face expression, face scale vary. Yale face database B is used to examine the performance when illumination varies extremely.

3.1 Experiments on the ORL Database

The ORL database (http://www.cam-orl.co.uk) contains images from 40 individuals, each providing 10 different images. All images are grayscale and normalized to a resolution of 46×56 pixels. We test the recognition performance with different training numbers. k ($2 \leq k \leq 9$) images of each subject are randomly selected for training and the remaining 10-k images of each subject for testing. For each number k, 50 runs are performed with different random partition between training set and testing set. For each run, E_nLDA method is performed by training the selected fixed samples and testing on the left images. The dimension, $\{N_0, N_1\}$, for the RS-LDA is $\{15, 15\}$, $\{20, 40\}$, $\{20, 60\}$, $\{20, 80\}$, $\{20, 120\}$, $\{20, 150\}$, $\{20, 180\}$ and $\{20, 210\}$ respectively with the number of training samples for each subject changes from 2 to 9. Fig.2(a) shows the average recognition rate. From Fig.2(a), it can be seen that the performance of E_nLDA is much better than other linear subspace methods, no matter the size of training set.

3.2 Experiments on Yale Face Database B

In our experiment, altogether 640 images for 10 subjects from the Yale face database B are used (64 illumination conditions under the same frontal pose). The image size is 50×60. The recognition performance is tested with different training numbers. k ($2 \leq k \leq 12$) images of each subject are randomly selected for training and the remaining 64-k images of each subject for testing. For each number k, 100 runs are performed with different random partition between training set and testing set. For each run, E_nLDA method is performed by training the selected fixed samples and testing on the left images. The dimension, $\{N_0, N_1\}$, for the RS-LDA is $\{5, 5\}$, $\{5, 15\}$, $\{10, 20\}$, $\{10, 25\}$, $\{10, 30\}$, $\{15, 35\}$, $\{20, 40\}$, $\{30, 40\}$, $\{40, 40\}$ and $\{40, 50\}$ respectively with the number of training samples for each subject changes from 2 to 11. Fig.2(b) shows the average recognition rate. Similarly, From Fig.2(b), it can be seen that E_nLDA is the best of all the algorithms.

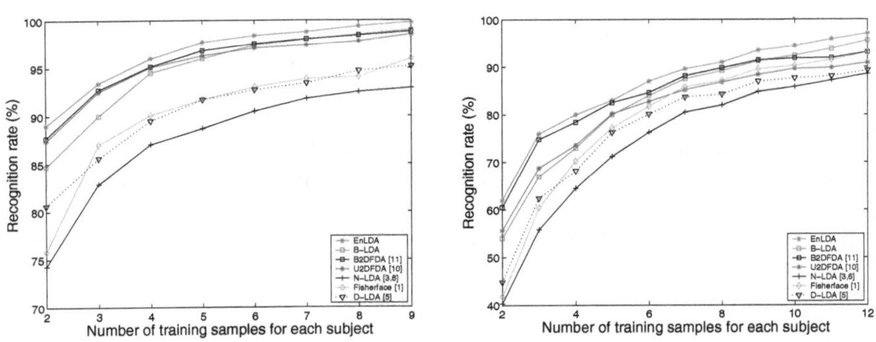

(a)Performance on the ORL database (b)Performance on the Yale face database B

Fig. 2. Recognition rate on the ORL database and the Yale face database B

4 Conclusions

In this paper, a framework of **En**semble **L**inear **D**iscriminant **A**nalysis (E_nLDA) is proposed for face recognition with small number of training samples. In E_nLDA, a Boosting-LDA (B-LDA) and a Random Sub-feature LDA (RS-LDA) schemes are coupled together to construct the total weak-LDA classifier ensemble. By combining these weak-LDA classifiers using majority voting method, recognition accuracy can be significantly improved. Extensive experiments on two public face databases verify the superiority of the proposed E_nLDA over the state-of-the-art algorithms in recognition accuracy.

References

1. Belhumeur, P., Hespanha, J., Kriegman, D.: Eigenfaces vs. fisherfaces: Recognition using class specific linear projection. IEEE Trans. on PAMI **19** (1997) 711–720
2. Swets, D., Weng, J.: Using discriminant eigenfeatures for image retrieval. IEEE Transactions on Pattern Analysis and Machine Intelligence **18** (1996) 831–836
3. Chen, L., Liao, H., Ko, M., Lin, J., Yu, G.: A new lda-based face recognition system which can solve the samll sample size problem. Pattern Recognition (2000)
4. Fukunnaga: Introduction to Statistical Pattern Recognition. Academic Press, New York (1991)
5. Yu, H., Yang, J.: A direct lda algorithm for high-dimensional data with application to face recognition. Pattern Recognition **34** (2001) 2067–2070
6. Huang, R., Liu, Q., Lu, H., Ma, S.: Solving the small sample size problem of lda. In: Proceedings of International Conference on Pattern Recognition. (2002)
7. Cevikalp, H., Neamtu, M., Wilkes, M., Barkana, A.: Discriminative common vectors for face recognition. IEEE Transactions on Pattern Analysis and Machine Intelligence **27** (2005) 4–13
8. Wang, X., Tang, X.: Random sampling lda for face recognition. In: IEEE International Conference on Computer Vision and Pattern Recognition. (2004)
9. Wang, X., Tang, X.: Dual-space linear discriminant analysis for face recognition. In: IEEE International Conference on Computer Vision and Pattern Recognition. (2004)
10. Kong, H., Wang, L., Teoh, E., Wang, J., Venkateswarlu, R.: A framework of 2d fisher discriminant analysis: Application to face recognition with small number of training samples. In: to appear in the IEEE International Conference on Computer Vision and Pattern Recognition 2005. (2005)
11. Breima, L.: Bagging predictors. Machine Learning **10** (1996) 123–140
12. Schapire, R., Singer, Y.: Improved boosting algorithms using confidence-rated predictions. Machine Learning **37** (1999) 297–336
13. Ho, T.: The random subspace method for constructing decision forests. IEEE Transactions on Pattern Analysis and Machine Intelligence (1998)

Information Fusion for Local Gabor Features Based Frontal Face Verification

Enrique Argones Rúa[1], Josef Kittler[2], Jose Luis Alba Castro[1], and Daniel González Jiménez[1]

[1] Signal Theory Group, Signal Theory and Communications Dep., University of Vigo, 36310, Spain
[2] Centre for Vision, Speech and Signal Processing, University of Surrey, Guildford GU2 7XH, UK

Abstract. We address the problem of fusion in a facial component approach to face verification. In our study the facial components are local image windows defined on a regular grid covering the face image. Gabor jets computed in each window provide face representation. A fusion architecture is proposed to combine the face verification evidence conveyed by each facial component. A novel modification of the linear discriminant analysis method is presented that improves fusion performance as well as providing a basis for feature selection. The potential of the method is demonstrated in experiments on the XM2VTS data base.

1 Introduction

Several studies in face recognition and verification reported in the literature suggest that the methods based on the analysis of facial components exhibit better performance than those using the full face image. There are a number of reasons that could explain this general behaviour. First of all when one deals with facial components, it should be easier to compensate for changes in illumination between gallery and probe images. Second, any pose changes can also be more readily corrected for small face patches, rather than for the whole image. Third, faces are not rigid objects and they undergo local deformations. Such deformations can seriously degrade a full image representation, but will affect only a small number of facial components. The unaffected facial components may still provide sufficient evidence about person's identity.

Although it has many advantages, the component based approach to face recognition poses a new problem. The evidence that is gathered by analysing and matching individual facial components has to be fused to a single decision. In this paper this fusion problem is addressed in the context of face verification. We propose a multistage fusion architecture and investigate several fusion methods that can be deployed at its respective stages. These include linear discriminant analysis (LDA) and multilayer perceptron (MLP). Most importantly, we propose a novel modification of the LDA fusion technique that brings two significant benefits: improved performance and considerable speed up of the face verification process. This is achieved by discarding those facial components that

D. Zhang and A.K. Jain (Eds.): ICB 2006, LNCS 3832, pp. 173–181, 2005.

are associated with negative coefficients of the LDA projection vector. We provide some theoretical argument in support of the proposed method. Its superior performance is demonstrated by experiments on the XM2VTS database using the standard protocols.

The paper is organised as follows. In the next Section we describe the component based face representation method used in our study. Section 3 introduces the proposed fusion architecture. The novel LDA method with feature selection capabilities is presented in Section 3.2. The experiments conducted on the XM2VTS database are described and the results discussed in Section 4. Finally, the paper is drawn to conclusion in Section 5.

2 Local Gabor Features for Frontal Face Verification: Local Texture Similarities

Gabor filters are biologically motivated convolution kernels that capture the texture information and are quite invariant to the local mean brightness, so a good face encoding approach could be to extract the texture from some equally spaced windows. The local Gabor features are basically the response of several Gabor filters, with different frequencies and orientations. In this case we use 5 different frequencies and 8 different orientations, so every Gabor jet is a vector with 40 components. These Gabor jets are located in small windows which are centered following the rectangular grid pattern that we can see in the figure 1. The face images have been normalized to align the center of the eyes and the mouth to the same windows for all the images. This grid has 13 rows and 10 columns, so we have $N = 130$ Gabor jets with 40 coefficients each encoding every frontal face image.

Let $\mathcal{P} = \{\boldsymbol{p_1}, \boldsymbol{p_2}, \ldots, \boldsymbol{p_N}\}$ denote the set of points we use to extract the texture information, and $\mathcal{J} = \{\mathcal{J}_{\boldsymbol{p_1}}, \mathcal{J}_{\boldsymbol{p_2}}, \ldots, \mathcal{J}_{\boldsymbol{p_N}}\}$ be the set of jets calculated for one face. The similarity function between two Gabor jets taken from two different images $\mathcal{I}^1$ and $\mathcal{I}^2$ results in:

$$\mathcal{S}\left(\mathcal{J}^1_{\boldsymbol{p_i}}, \mathcal{J}^2_{\boldsymbol{p_i}}\right) = < \mathcal{J}^1_{\boldsymbol{p_i}}, \mathcal{J}^2_{\boldsymbol{p_i}} > , \tag{1}$$

where $< \mathcal{J}^1_{\boldsymbol{p_i}}, \mathcal{J}^2_{\boldsymbol{p_i}} >$ represents the normalized dot product between the i-th component from $\mathcal{J}^1$ and the corresponding component from $\mathcal{J}^2$, but taking into account that only the moduli of jet coefficients are used.

Fig. 1. Rectangular grid used to take the local features

So, if we want to compare two frontal face images, we will get, using the equation 1, the following similarity set:

$$\mathcal{S}_{\mathcal{I}^1,\mathcal{I}^2} = \{\mathcal{S}\left(\mathcal{J}^1_{\boldsymbol{p_1}}, \mathcal{J}^2_{\boldsymbol{p_1}}\right), \ldots, \mathcal{S}\left(\mathcal{J}^1_{\boldsymbol{p_N}}, \mathcal{J}^2_{\boldsymbol{p_N}}\right)\} \quad (2)$$

These similarity scores then have to be combined to a single decision score output by an appropriate fusion rule.

When we have T training images for the client training we have several choices. One of them is to make a decision based on the similarity set that we can get comparing a single *user template* with the probe image. On the other hand we could use the Gabor jets of every training image as a template, and then obtain T different decision scores. This approach, which is the information fusion approach adopted in this paper and is referred as *multiple template* method, then requires the fusion of decision scores corresponding to the individual templates.

3 Information Fusion

Let us suppose that we have T different training images for every client. We can then build a set of T decision functions for the user k, and we can write them as:

$$\mathcal{D}^k_i\left(\mathcal{J}\right) = f\left(\mathcal{J}, \mathcal{J}^{k,i}\right), i \in \{1, \ldots, T\}\ , \quad (3)$$

where $\mathcal{J}^{k,i}$ denotes the ith training image for user k, and assuming that the decision functions $f\left(\cdot\right)$ computed for the respective training images are identical.

As indicated in the previous Section, the decision function $\mathcal{D}^k_i\left(\mathcal{J}\right)$ is realised as a two step operation where by in the first step we obtain similarity scores for the individual local jets and in the second stage we fuse these scores by a fusion rule, $g(\cdot)$, i.e.

$$f\left(\mathcal{J}, \mathcal{J}^{k,i}\right) = g\{\mathcal{S}\left(\mathcal{J}_{\boldsymbol{p_1}}, \mathcal{J}^{k,i}_{\boldsymbol{p_1}}\right), \ldots, \mathcal{S}\left(\mathcal{J}_{\boldsymbol{p_N}}, \mathcal{J}^{k,i}_{\boldsymbol{p_N}}\right)\} \quad (4)$$

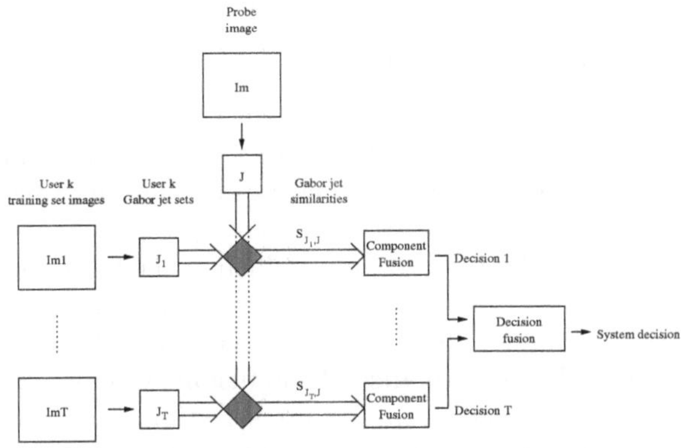

Fig. 2. Decision-fusion scheme

The decision scores obtained for the multiple templates then have to be fused. The decision fusion function can be defined as $\mathcal{D}^k(\mathcal{D}_1^k, \ldots, \mathcal{D}_T^k)$, and can be performed by any suitable fusion function such as those described in the next Section 3.1. This decision fusion function must take the final decision about the identity claim as

$$\mathcal{D}^k = h\left(D_1^k, \ldots, D_T^k\right) \tag{5}$$

An overview of the scheme is shown in figure 2.

3.1 Fusion Methods

The fusion of image component similarity scores defined in equation 4 as well as the decision score fusion in equation 5 can be implemented using one of several trainable or non trainable functions or rules for this task, as MLP, SVM, LDA, AdaBoost or the sum rule. For this experiment we will compare the performance of MLP and LDA. In figure 3 we can see an overview of the training and evaluation processes with these methods. Both LDA and MLP outputs are not thresholded in the decision score level because it could cause a loss of information in this stage.

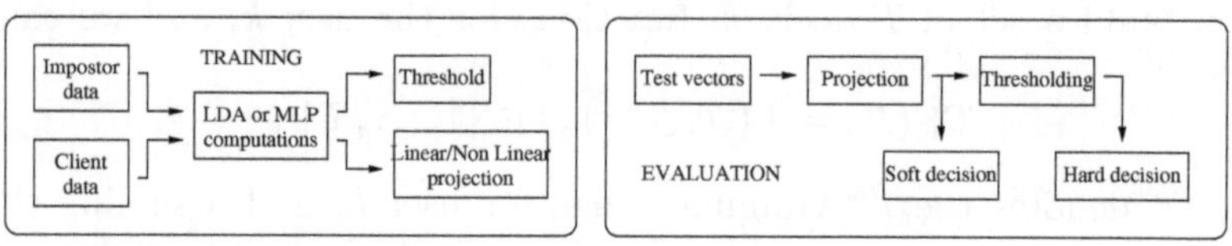

Fig. 3. LDA or MLP based fusion

The MLP that we use in this experiment is a fully connected and one hidden layer network. Based on some previous work we decided to use 3 neurons in the hidden layer to get the decision scores and 2 neurons in the hidden layer for the decision score fusion. We have trained the MLPs using the standard backpropagation algorithm.

3.2 LDA-Based Feature Selection

In a two class problem, LDA yields just one direction vector. Each component v_i of the LDA vector $\mathbf{v}$ represents the weight of the contribution of the ith component to the separability of the two classes as measured by the eigenvalue of the LDA eigenanalysis problem. At this point it is pertinent to ask whether the coefficient values could be used to judge which of the features are least useful from the point of view of class separation. If there was a basis for identifying irrelevant features, we could reduce the dimensionality of the problem and at the same time improve the performance of the fusion system. This is the normal positive outcome one can expect from feature selection.

To answer this question, let us look at the LDA solution in more detail. Let $\mathcal{X} = [x_1, \ldots, x_N]$ denote our Gabor jet similarities vector. Clearly, x_i are not independent, as ideally, all similarity values should be high for the true identity claim and vice-versa for an imposter claim. However, it is not unreasonable to assume that x_i is class conditional independent of x_j $\forall i, j | i \neq j$ and $i, j \in \{1, \ldots, N\}$. This is a relatively strong assumption, but for the sake of simplicity, we shall adopt it.

Let the mean of the ith component be denoted $\mu_{i,0} = E\{x_i | C = 0\}$ and $\mu_{i,1} = E\{x_i | C = 1\}$, where $C = 1$ when $\mathcal{X}$ comes from a true identity claim and $C = 0$ when $\mathcal{X}$ comes from a false identity claim. Let $\mu_i = \frac{1}{2}(\mu_{i,0} + \mu_{i,1})$. Further, let $\sigma_{i,0}^2 = \{(x_i - \mu_{i,0})^2 | C = 0\}$ and $\sigma_{i,1}^2 = \{(x_i - \mu_{i,1})^2 | C = 1\}$ denote the variances of the similarity scores. Let $c_i = \frac{1}{2}(\sigma_{i,0}^2 + \sigma_{i,1}^2)$.

As x_i represents similarity and the greater the similarity the higher the value of x_i, we can assume $\mu_{i,1} > \mu_{i,0}, \forall i \in \{1, \ldots, N\}$.

LDA finds a one dimensional subspace in which the separability of true clients and impostors is maximised. The solution is defined in terms of the within class and between class scatter matrices S_w and S_b respectively, i.e.

$$S_w = \begin{pmatrix} c_1 & 0 & \ldots & 0 \\ 0 & c_2 & \ldots & 0 \\ \vdots & \vdots & \ddots & \vdots \\ 0 & \ldots & 0 & c_N \end{pmatrix} \tag{6}$$

$$S_b = (\mu_1 - \mu_0)(\mu_1 - \mu_0)^T \tag{7}$$

where μ_C is the mean vector of class C composed of the above components.

Now the LDA subspace is defined by the solution to the eigenvalue problem

$$S_w^{-1} S_b \mathbf{v} - \lambda \mathbf{v} = 0 \tag{8}$$

In our face verification case equation 8 has only one non zero eigenvalue λ and the corresponding eigenvector defines the LDA subspace. It is easy to show that the eigenvector $\mathbf{v}$ is defined as

$$\mathbf{v} = S_w^{-1}(\mu_1 - \mu_0) \tag{9}$$

Recall that all the components of the difference of the two mean vectors are non negative. Then from equations 9 and 6 it follows that the components of the LDA vector $\mathbf{v}$ should also be non negative. If a component is non positive, it means that the actual training data is such that

- the observations do not satisfy the axiomatic properties of similarities
- the component has a strong negative correlations with some other components in the feature vector, so it is most likely encoding random redundant information emerging from the sampling problems, rather than genuine discriminatory information. Reflecting this information in the learned solution does help to get a better performance on the evaluation set where it is used as a disimilarity. However, this does not extend to the test set.

When LDA projection vector components have all the same sign, the similarity scores are re-enforcing each other and compensating for within class variations.

But for a negative component in the projection vector a positive similarity information in that dimension is not helping to get a general solution, and it is very likely that it is being used to overfit the LDA training data.

LDA is not an obvious choice for feature selection, but in the two class case of combining similarity evidence it appears that the method offers an instrument for identifying dimensions which have an undesirable effect on fusion. By eliminating every feature with a negative projection coefficient, we obtain a lower dimensional LDA projection vector with all projection coefficients positive. This projection vector is not using many of the original similarity features, and therefore performs the role of an LDA-based feature selection algorithm.

4 Experimental Results

Our experiments were conducted using the XM2VTS database [1], according to the Lausanne protocol [2] in both configurations.

For verification experiments this database was divided in three different sets: training set, evaluation set (used to tune the algorithms) and test set. We have 3 different images for every client training in Configuration I of the Lausanne protocol and 4 images for every client training in Configuration II.

An important consideration about the two different configurations is that Configuration I is using the same sessions to train and tune the algorithms, so the client attempts are more correlated than in Configuration II, where the sessions used to train the algorithms are different than those used to tune the algorithms. This means that Configuration I is likely to lead to an intrinsically poorer general solution.

In tables 1 and 2 we show the single decision stage performance with and without the LDA-based feature selection. If we compare the results in both tables we can clearly draw two main conclusions:

- The TER is lower using the LDA-based feature selection for both MLP and LDA decision fusion functions in both configurations in the test set but higher in the evaluation set.
- The difference between the FAR and FRR in the test set performance is lower for both configurations and decision fusion functions.

These two suggest that the LDA-based feature selection has enabled us to construct a solution exhibiting better generalisation properties than the one obtained when using all the features together. The stability of the operating point is also better.

On the other hand, in tables 3, 4 and 5 we have the overall system performance with and without the LDA-based feature selection algorithm. If we compare the results in tables 3 and 4, where the decision fusion function is LDA (without and with the feature selection respectively) we obtain a degradation of 5.42% in TER when using the feature selection in Configuration I and an improvement of 6.71% in TER when using feature selection in Configuration II.

Table 1. Single template performance with global thresholding and without feature selection

		Configuration I		Configuration II	
		FAR(%)	FRR(%)	FAR(%)	FRR(%)
LDA	Ev. Set	3.83	3.83	3.20	3.19
	Ts. Set	7.13	4.42	5.79	5.63
MLP	Ev. Set	0.90	0.94	0.76	0.75
	Ts. Set	2.21	7.42	2.50	9.50

Table 2. Single template performance with LDA-based feature selection and global thresholding

		Configuration I		Configuration II	
		FAR(%)	FRR(%)	FAR(%)	FRR(%)
LDA	Ev. Set	4.39	4.39	3.87	3.87
	Ts. Set	6.79	4.67	5.44	5.44
MLP	Ev. Set	2.89	2.89	2.15	2.19
	Ts. Set	4.24	5.00	3.18	6.63

However, if we use the MLP as the decision fusion function trained with the LDA-based feature selection features, as we can see in table 5, the results in Configuration I are much better. If we do not use feature selection prior to the MLP based similarity score fusion, the results (not listed in this paper) are much worse than those listed in table 5 for both configurations, as could be expected from the highly unbalanced results shown in table 1 for the MLP fusion method.

The overall results in Configuration I should not be considered as a reflection of the generalization power of our fusion algorithms, as the poor generalization behaviour is intrinsically imposed by the test protocol. Therefore it is reasonable to argue that the LDA-based feature selection allow us to improve the overall system performance.

Finally, the LDA-based selected features for both configurations can be seen super imposed over the face of one of the subjects of the database (for illustration purposes) in figure 4. Note that the number and location of the selected features (40 in the configuration I and 44 in the configuration II) are very simi-

Table 3. Multiple template performance using LDA without feature selection for similarity score fusion, LDA and MLP as decision fusion functions and client specific thresholding

		Configuration I		Configuration II	
		FAR(%)	FRR(%)	FAR(%)	FRR(%)
LDA	Ev. Set	1.48	1.43	0.75	0.75
	Ts. Set	3.39	3.25	1.92	2.25
MLP	Ev. Set	1.36	1.33	0.50	0.50
	Ts. Set	3.30	2.75	1.26	3.25

Table 4. Multiple template performance using LDA with feature selection for similarity score fusion, LDA and MLP as decision fusion functions, and client specific thresholding

		Configuration I		Configuration II	
		FAR(%)	FRR(%)	FAR(%)	FRR(%)
LDA	Ev. Set	1.66	1.67	0.75	0.75
	Ts. Set	3.75	3.25	1.89	2.00
MLP	Ev. Set	1.83	1.83	0.50	0.50
	Ts. Set	4.65	3.00	1.05	2.75

Table 5. Multiple template performance using LDA based feature selection, MLP as similarity score fusion function, LDA and MLP as decision fusion functions and client specific thresholding

		Configuration I		Configuration II	
		FAR(%)	FRR(%)	FAR(%)	FRR(%)
LDA	Ev. Set	1.22	1.17	0.61	0.50
	Ts. Set	2.37	2.25	1.07	5.00
MLP	Ev. Set	1.11	1.00	0.52	0.50
	Ts. Set	2.20	2.25	0.93	8.00

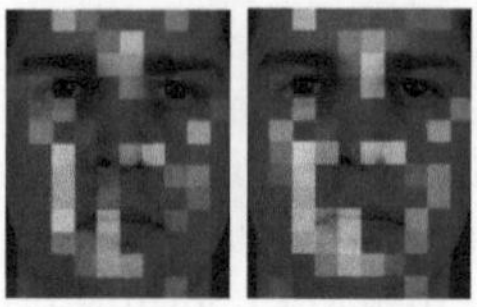

Fig. 4. LDA-based selected features for configuration I (left) and configuration II (right). The brightness is proportional to the LDA projection vector coefficient.

lar in both configurations, and even the values (represented in the figure by the window brightness) of the coefficients are also very similar. The stability and consistency of the features identified by the proposed algorithm is very encouraging. Moreover, the number of selected features is small enough to allow a high reduction in the computational complexity in the verification phase, and hence an important reduction (nearly a 60%) in the verification time.

5 Conclusions

We addressed the problem of information fusion in component based face verification where similarity scores computed for individual facial components have to be combined to reach a final decision. We proposed a multistage fusion architecture and investigated several fusion methods that could be deployed at its respective stages. These included LDA and MLP. Most importantly, we proposed a novel modification of the LDA fusion technique that brings two significant benefits: improved performance and considerable speed up of the face verification process. This was achieved by discarding those facial components that were associated with negative coefficients of the LDA projection vector. We provided some theoretical argument in support of the proposed method. Its superior performance was demonstrated by experiments on the XM2VTS database using the standard protocols. Performance improvements, on the more realistic Configuration II, varying between 7-20% were achieved with the proposed method.

References

1. K. Messer, J. Matas, J. Kittler, J. Luettin and G. Maître: XM2VTSDB: The extended M2VTSDB. International Conference on Audio and Video-based Biometric Person Authentication, 1999
2. J. Luettin and G. Maître: Evaluation protocol for the XM2FDB (Lausanne protocol). IDIAP Communication, 1998.
3. Wiskott, L., Fellous, J.M., Kruger, N. and von der Malsburg, C.: Face recognition by Elastic Bunch Graph Matching. IEEE Transactions on Pattern Analysis and Machine Intelligence, 19(7), 775–779, 1997
4. L. I. Kuncheva: "Fuzzy" versus "Nonfuzzy" in combining classifiers designed by boosting. IEEE Transactions on Fuzzy Systems, 11(6), 729–741, 2003
5. P. Silapachote, Deepak R. Karuppiah, and Allen R. Hanson: Feature selection using AdaBoost for face expression recognition. Proceedings of the Fourth IASTED International Conference on Visualization, Imaging, and Image Processing, 84-89, 2004
6. P. Viola and M. Jones: Robust Real-Time Face Detection. International Conference on Computer Vision, 2001
7. B. Heisele, P. Ho and T. Poggio: Face Recognition with Support Vector Machines: Global versus Component-based Approach. International Conference on Computer Vision, 2001
8. A. Tefas, C. Kotropoulos and Ioannis Pitas: Face verification using elastic graph matching based on morphological signal decomposition. Signal Processing 82(6), 833–851, 2002
9. R. Brunelli and T. Poggio. Face Recognition: Features versus Templates. IEEE Transactions on Pattern Analysis and Machine Intelligence, 15(10), 1042–1052, O1993
10. K. Jonsson, J. Kittler, Y. P. Li and J. Matas: Learning Support Vectors for Face Verification and Recognition. Proceedings of the Fourth IEEE International Conference on Automatic Face and Gesture Recognition, 2000
11. C. Sanderson and K.K. Paliwal: Fast feature extraction method for robust face verification. Electronics Letters Online No: 20021186, 2002
12. M. Saban and C. Sanderson: On Local Features for Face Verification. IDIAP-RR, 36, 2004
13. C. Havran, L. Hupet, J. Czyz, J. Lee, L. Vandendorpe, M. Verleysen: Independent Component Analysis for face authentication. Knowledge-Based Intelligent Information and Engineering Systems, 1207–1211, 2002
14. K. Messer, J. Kittler, M. Sadeghi, M. Hamouz, A. Kostyn, S. Marcel, S. Bengio, F. Cardinaux, C. Sanderson, N. Poh, Y. Rodriguez, K. Kryszczuk, J. Czyz, L. Vandendorpe, J. Ng, H. Cheung, and B. Tang: Face Authentication Competition on the BANCA Database.

Using Genetic Algorithms to Find Person-Specific Gabor Feature Detectors for Face Indexing and Recognition

Sreekar Krishna, John Black, and Sethuraman Panchanathan

Center for Cognitive Ubiquitous Computing (CUbiC),
Arizona State University, Tempe AZ- 85281
Tel: 480 326 6334, Fax Number: 480 965 1885
Sreekar.Krishna@asu.edu

Abstract. In this paper, we propose a novel methodology for face recognition, using person-specific Gabor wavelet representations of the human face. For each person in a face database a genetic algorithm selects a set of *Gabor features* (each feature consisting of a particular Gabor wavelet and a corresponding *(x, y)* face location) that extract facial features that are unique to that person. This set of Gabor features can then be applied to any normalized face image, to determine the presence or absence of those characteristic facial features. Because a unique set of Gabor features is used for each person in the database, this method effectively employs *multiple feature spaces* to recognize faces, unlike other face recognition algorithms in which all of the face images are mapped into a single feature space. Face recognition is then accomplished by a sequence of *face verification* steps, in which the query face image is mapped into the feature space of each person in the database, and compared to the cluster of points in that space that represents that person. The space in which the query face image most closely matches the cluster is used to identify the query face image. To evaluate the performance of this method, it is compared to the most widely used subspace method for face recognition: Principle Component Analysis (PCA). For the set of 30 people used in this experiment, the face recognition rate of the proposed method is shown to be substantially higher than PCA.

1 Introduction

Faces are an important biometric, and many computer algorithms have been proposed to identify face images. However, existing face recognition algorithms are not very robust with respect to pose angle or illumination angle variations. Humans are much better at recognizing faces when faced with these types of variations. This has prompted researchers to more closely study the ways in which humans recognize faces, and face recognition has become a proving ground for artificial intelligence researchers who are attempting to simulate human pattern recognition with computer algorithms.

Face recognition algorithms can be broadly classified into *holistic methods* and *feature-based methods*. Holistic methods attempt to recognize a face without

D. Zhang and A.K. Jain (Eds.): ICB 2006, LNCS 3832, pp. 182 – 191, 2005.

subdividing it into component parts, while feature-based methods subdivide the face into components (i.e. features) and analyze each feature, as well as its spatial location with respect to other features. The performance of holistic face recognition algorithms has been shown to be highly variable with respect to variations in pose angle, illumination angle, and facial expressions. Failures to achieve more robust face recognition using the holistic methods have motivated many researchers to study feature-based methods. This paper describes our own attempt to develop a feature-based method of face recognition that provides a higher level of performance than that of the existing holistic methods.

The rest of the paper is organized as follows: Section 2 discusses past research in the use of Gabor filters and Genetic Algorithms (GAs) in face recognition. Section 3 discusses the theoretical basis for our research. Section 4 describes the methodology we have used, including the implementation details of (1) the Gabor wavelets that we used to extract facial features, (2) the genetic algorithm that we used to select the Gabor feature detectors, and (3) the experiments that we used to evaluate the performance of the proposed algorithm. Section 5 presents the results of our experiments, and Section 6 discusses those results. Section 7 concludes the paper, and includes a discussion of future work.

2 Related Work

Classical methods of face recognition have employed statistical analysis techniques such as the Principle Component Analysis (PCA) [2] and Linear Discriminant Analysis (LDA) [3] which are logical extensions of the data analysis methods developed to investigate large datasets. These methods treat each face image as a point in a high-dimensional space, and try to associate multiple views of a person's face with a distinct cluster in that space. The problem with using these statistical methods is that small variations in capture conditions tend to scatter face images of each person across a wide expanse of this space, making it difficult to discern a distinct cluster for each person.

Faced with this problem, many researchers have attempted to extract localized facial features. Among the many available feature extractors, Gabor wavelets have been popular – possibly due to the fact that Gabor wavelets model the receptive fields of the simple cells [4]. Shen *et al.* [5] used Gabor filters in combination with a Kernel Direct Discriminant Analysis (KDDA) subspace as a classifier, and Liu *et al* proposed using Gabor filters in an Enhanced Fisher Linear Discriminant Model [7] and with Independent Component Analysis (ICA) [6]. However, none of these methods specifically select feature detectors (or the locations of their application) based on the salient features of faces. There exists some face recognition research that does take into account the localities of salient facial features [8] [9]. However, these methods rely on a human to select facial feature locations manually, leaving open the question of how much this human contribution influences the results.

Genetic Algorithms, (GAs) have been used in face recognition to search for optimal sets of features from a pool of potentially useful features that have been extracted from the face images. Liu *et al.* [10] used a GA to search for optimal components from a pool of independent components, while Xu *et al.* [11] used a GA to search for the optimal components in a pool of Kernel Principle Components.

In each of the cases described above, all of the faces in a database were indexed with a single feature set. We believe that this approach imposes a fundamental and unnecessary constraint on the recognition of faces. We suspect that people first learn to recognize faces based on person-specific features. This suggests that better recognition performance might be achieved by indexing each person's face based on a person-specific feature space. As a guide to further exploration of this approach, we propose the following research question: How does the performance of a face recognition algorithm based on person-specific features compare to the performance of a face recognition algorithm that indexes all faces with a common set of features?

3 Theory

3.1 Gabor Filters

Gabor wavelets are a family of filters derived from a mother Gabor function by altering the parameters of that function. The response of a particular Gabor filter is tuned to the spatial frequency, and the spatial orientation content of the region within its spatial extent. By employing Gabor filters with a variety of spatial extents, it is possible to index faces based on both large and small facial features. Because Gabor filter responses are similar to those of many primate cortical simple cells, and because they are able to index features based on their locality in both space and frequency, they have become one of the most widely chosen filters for image decomposition and representation. Gabor filters are defined as follows:

$$\psi_{\omega,\theta}(x,y) = \frac{1}{2\pi\sigma_x\sigma_y}.G_\theta(x,y).S_{\omega,\theta}(x,y) \qquad (1)$$

$$G_\theta(x,y) = e^{-\left(\frac{(x\cos\theta + y\sin\theta)^2}{2\sigma_x^2} + \frac{(-x\sin\theta + y\cos\theta)^2}{2\sigma_y^2}\right)} \qquad (2)$$

$$S_{\omega,\theta}(x,y) = e^{i(\omega x\cos\theta + \omega y\sin\theta)} - e^{-\frac{\omega^2\sigma^2}{2}} \qquad (3)$$

where, *(x,y)* is the 2D spatial location where the filter is centered, ω is the spatial frequency parameter of its 2D sinusoidal signal, σ_{dir}^2 represents the variance of the Gaussian mask along the specified direction – which can be either *x* or *y*. This variance determines the spatial extent of the Gabor filter, where its output is readily influenced. From the definition of Gabor wavelets, as given in Equation (1), it can be seen that Gabor filters are generated by multiplying two components: (1) the Gaussian mask $G_\theta(x,y)$ shown in Equation (2) and (2) the complex sinusoid $S_{\omega,\theta}(x,y)$ shown in Equation (3).

3.1.1 The Gaussian Mask

The 2D Gaussian mask determines the spatial extent of the Gabor filter. This spatial extent is controlled by the variance parameters (along the x and y directions) together with the orientation parameter θ. Typically, $\sigma_x = \sigma_y = \sigma$. Under such conditions the orientation parameter, θ, does not play any role, and the spatial extent of the Gabor filter will be circular.

3.1.2 The Complex Sinusoid

The 2D complex sinusoid provides the sinusoidal component of the Gabor filter. This complex sinusoid has two components (the real and the imaginary parts) which are two 2D sinusoids, phase shifted from each other by ($\pi/2$) radians. When combined with a Gaussian mask, the resulting Gabor filter kernel can be applied to a 2D array of pixel values (such as a region within a face image) to generate a complex coefficient value whose amplitude is proportional to the spatial frequency content of the array that lies within the extent of the Gaussian mask. If $\sigma_x = \sigma_y = \sigma$, then the real and imaginary parts of the Gabor coefficient produced by Equation (1) can be computed as follows.

$$\Re\left\{\psi_{\omega,\theta}(x,y)\right\} = \frac{1}{2\pi\sigma^2} G_\theta(x,y)\,\Re\left\{S_{w,\theta}(x,y)\right\} \qquad \Im\left\{\psi_{\omega,\theta}(x,y)\right\} = \frac{1}{2\pi\sigma^2} G_\theta(x,y)\,\Im\left\{S_{w,\theta}(x,y)\right\} \tag{4}$$

3.1.3 The Gabor feature (Coefficient)

In order to extract a real number Gabor coefficient at a location *(x,y)* of an image *I*, the real and imaginary parts of the filter are applied separately to the image, and the real-valued *magnitude* of the resulting complex number is used as the coefficient. Thus, the convolution coefficient C_ψ at a location *(x,y)*on an image *I* with a Gabor filter $\psi_{w,\theta}(x,y)$ is given by

$$C_\psi(x,y) = \sqrt{\left(I(x,y) * \Re\left\{\psi_{\omega,\theta}(x,y)\right\}\right)^2 + \left(I(x,y) * \Im\left\{\psi_{\omega,\theta}(x,y)\right\}\right)^2} \tag{5}$$

4 Methodology

4.1 Overview

In general, feature based face recognition methods use feature detectors that are not tailored specifically for face recognition, and they make no attempt to selectively choose feature detectors based specifically on their usefulness for face recognition. The method described in this paper uses Gabor wavelets as feature detectors, but evaluates the usefulness of each particular feature detector for distinguishing between the faces within our face database. Given the very large number of possible Gabor feature detectors, we use a Genetic Algorithm (GA) to explore the space of possibilities, with a fitness function that propagates parents with a higher ability to distinguish between the faces in the database. By selecting Gabor feature detectors that are most useful for distinguishing each person from all the other people in the database, we can define a unique (i.e. person-specific) feature space for each person.

4.2 The Image Set

All experiments were conducted with face images from the FacePix (30) database [12]. This database has face images of 30 people at various pose and illumination angles. For each person in the database, there are three sets of images. (1) The *pose angle set* contains face images of each person at pose angles from +90º to –90 º (2) The *no-ambient-light set* contains frontal face images with a spotlight placed at angles ranging from +90 º to -90 º with no ambient light, and (3) The *ambient-light set* contains frontal face images with a spot light placed at angles placed at angels from +90 º to -90 º in the presence of ambient light. Thus, for each person, there are three face images available for every angle, over a range of 180 degrees.

We selected at random two images out of each set of three frontal (0º) images for training, and used the remaining image for testing. The genetic algorithms used the training images to find a set of Gabor feature detectors that were able to distinguish each person's face from all of the other people in the training set. These feature detectors were then used to recognize the test images. The same set of training and testing images were used with PCA-based face recognition, to allow a comparison with our proposed method. Figure (1) shows some example images used in our experiments.

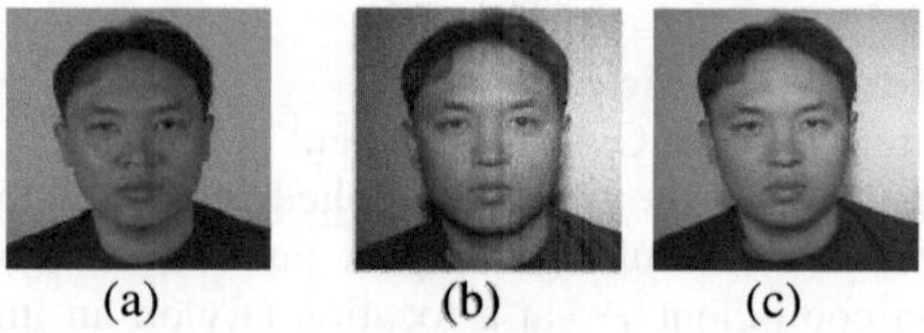

Fig. 1. (a) and (b) are the training samples of the person, while (c) is the testing sample

Fig. 2. A face image marked with 5 locations where unique Gabor features will be extracted4.3 Our Gabor features

4.3 Our Gabor Features

Each *Gabor feature* corresponds to a particular Gabor wavelet (i.e. a particular spatial frequency, a particular orientation, and a particular Gaussian-defined spatial extent) applied to a particular (*x*, *y*) location within a normalized face image. (Given that 125 different Gabor filters were generated, by varying ω, σ and θ in *5* steps each, and given that each face image contained 128*128 = 16,384 pixels, there was a pool of

125*16384 = 2,048,000 Gabor features to choose from.) We used an N-dimensional vector to represent each person's face in the database, where N represents the predetermined number of Gabor features that the Genetic Algorithm selected from this pool. Fig.2 shows an example face image, marked with 5 locations where Gabor features will be extracted (i.e. $N = 5$). Given any normalized face image, real-valued Gabor features are extracted at these locations using Equation (5). This process can be envisioned as a projection of a 16,384-dimensional face image onto an N dimensional subspace, where each dimension is represented by a single Gabor feature detector.

Thus, the objective of the proposed methodology is to extract an N dimensional real-valued person-specific feature vector to characterize each person in the database. The N (x, y) locations (and the spatial frequency and spatial extent parameters of the N Gabor wavelets used at these locations) are chosen by a GA, with a fitness function that takes into account the ability of each Gabor feature detector to distinguish one face from all the other faces in the database.

4.4 Our Genetic Algorithm

Every GA is controlled in its progress through generations with a few control parameters namely, (1) the number of generations of evolution (n_g), (2) the number of parents per generation (n_p), (3) the number of parents cloned per generation (n_c), (4) the number of parents generated through cross over (n_{co}) and (5) the number of mutations in every generation (n_m). In our experiments, the GA used the following empirically-chosen GA parameters: $n_g = 50$, $n_p = 100$, $n_c = 6$, $n_{co} = 35$ and $n_m = 5$.

4.4.1 Our Fitness Function

The fitness function of a genetic algorithm determines the nature, and the efficiency, of the search conducted within the parameter space. Our fitness function F consists of an equation with two independent terms. The term D is a distance measure that represents the ability of a parent (i.e. the ability of its Gabor feature detectors) to distinguish one person's face images from those of all the other people in the database. The other term C represents the degree of correlation between the textural qualities of the spatial locations of the N Gabor feature detectors within each parent, which are determined by applying all 125 Gabor filters to that location. These two terms are assigned weighting factors, as follows:

$$F = w_D D - w_C C \tag{6}$$

where, w_D is the weighting factor for the Distance measure D, and w_C is the weighting factor for the Correlation measure C.

The Distance Measure D

Let M_i represent a set of Gabor features extracted for person i, where $i = 1 \ldots J$ and where J is the total number of people in the database. For each person i, let all the images of person i be marked as *positives*, and all the other images be marked as *negatives*. If there are N Gabor features detectors, then $M_{n,i} = \{m_{1,i}, m_{2,i}, \ldots, m_{N,i}\}$

represents the N Gabor feature detectors, $P_{l,i} = \{p_{1,i}, p_{2,i}, \ldots, p_{L,i}\}$ represents the L positive images, and $N_{k,i} = \{n_{1,i}, n_{2,i}, \ldots, n_{K,i}\}$ represents the K negative images of person i.

The distance measure D is then defined as:

$$D = \min_{l,k}\left[\delta_N\left(\phi_N\left(p_{l,i}\right), \phi_N\left(n_{k,i}\right)\right)\right] \tag{7}$$

Where, $\phi_N(X)$ is the projection of the 16,384-dimensional face image onto an N-dimensional subspace, where the N dimensions are represented by $M_{n,i} = \{m_{1,i}, m_{2,i}, \ldots, m_{N,i}\}$, and $\delta_N(A,B)$ is the N-dimensional Euclidean distance between A and B.

The Correlation Measure C

C is a penalty on the fitness of a parent that is levied if there is a correlation between the textural qualities at the N spatial locations of the Gabor feature detectors of that parent. (The textural qualities of a location are determined by applying all 125 Gabor filters at that location.) This penalty is needed to suppress the GA's tendency to select multiple feature detectors within a single distinctive facial feature, such as a mustache. Application of the 125 Gabor filters to each of the N locations produces the following 125-column, N row matrix:

$$A = \begin{bmatrix} g_{1,1} & g_{1,2} & \cdots & g_{1,N} \\ g_{2,1} & g_{2,2} & \cdots & g_{2,N} \\ \vdots & \vdots & \vdots & \vdots \\ g_{125,1} & g_{125,2} & \cdots & g_{125,N} \end{bmatrix} \tag{8}$$

Where, $g_{x,y}$ is the real-number Gabor coefficient obtained by applying the xth Gabor filter of the 125-filter pool at the location of the yth Gabor feature detector. C can now be defined as follows:

$$C = \log(\det(diag(B))) - \log(\det(B)) \tag{9}$$

Where, $B = \frac{1}{124} A^T A$ is the correlation matrix.

Normalization of D *and* C

Since D and C are two independent measures, before they can be used in Equation (6), they need to be normalized to a common scale. For each generation, before the fitness values are computed to rank the parents, parameters D and C are each normalized to range between 0 and 1. This is done as follows

$$D_{norm} = \frac{D - D_{Min}}{D_{Max} - D_{Min}} \qquad C_{norm} = \frac{C - C_{Min}}{C_{Max} - C_{Min}} \tag{10}$$

5 Results

To evaluate the relative importance of the two terms (*D* and *C*) in the fitness function, we ran the proposed algorithm on the training set several times with 5 feature detectors per chromosome, while changing the weighting factors in the fitness function for each run, setting w_D to 0, .25, .50, .75, and 1.00, and computing $w_C = (1 - w_D)$. Figure 3 shows the recognition rate achieved in each case.

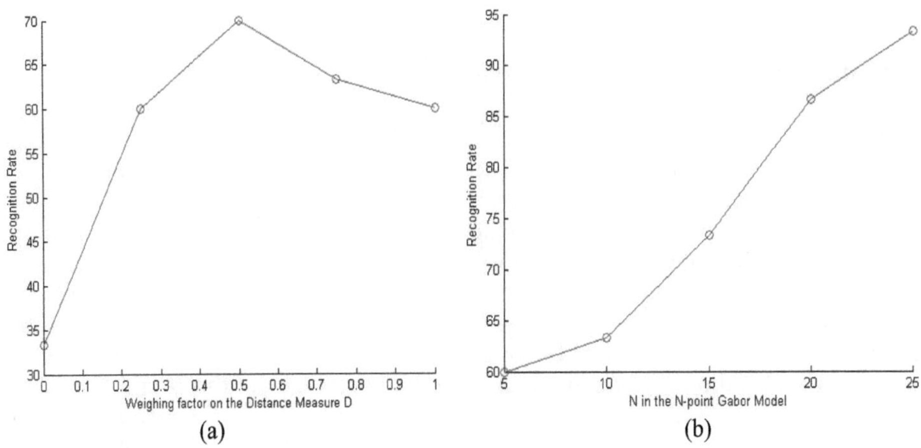

Fig. 3. (a) Recognition rate with varying weighing factor for the distance measure *D* (b) The recognition rate versus the number Gabor feature detectors

We also ran the proposed algorithm on the training set 5 times, while changing the number of Gabor feature detectors per parent chromosome for each run to 5, 10, 15, 20, and 25. In all the trials, w_D=0.5. Figure 4 shows the recognition rate achieved in each case.

6 Discussion of the Results

Fig. 3(b) shows that the recognition rate of the proposed algorithm when trained with 5, 10, 15, 20, and 25 Gabor feature detectors increases monotonically, as the number of Gabor feature detectors (*N*) is increased. This can be attributed to the fact that increasing the number of Gabor features essentially increases the number of dimensions for the Gabor feature detector space, allowing for greater spacing between the positive and the negative clusters.

Fig. 3(a) shows that for $N = 5$ the recognition rate was optimal when the distance measure *D* and the correlation measure *C* were weighted equally, in computing the fitness function *F*. The dip in the recognition rate for w_D=1.0 indicates the significance of using the correlation factor *C* in the fitness function. The penalty introduced by *C* ensures that the GA searches for Gabor features with different textural patterns. If no such penalty were to be imposed, the GA might select Gabor features that are clustered on one salient feature on an individual, such as a mole.

The best recognition results for the proposed algorithm (93.3%) were obtained with 25 Gabor feature detectors. The best recognition performance for the PCA algorithm was reached at about 15 components, and flattened out beyond that point, providing a recognition rate for the same set of faces that was less than 83.3%. This indicates that, for the face images used in this experiment (which included substantial illumination variations) the proposed method performed substantially better than the PCS algorithm.

7 Conclusions and Future Work

For the set of 30 face images used in these experiments (which included a wide range of illumination variations) person-specific indexing (as implemented by our proposed algorithm) provided better recognition rates than Principal Component Analysis (PCA). Furthermore (unlike PCA which flattened out after 15 components) the recognition rates for the proposed algorithm increase monotonically with increasing numbers of Gabor features. Based on Fig 4, it seems reasonable to expect that recognition rates for the proposed algorithm will continue to increase as more Gabor features detectors are added, and this will be further explored in future work. Future research will also thoroughly explore the relative importance of the *D* and *C* terms in the fitness function *F* as the number of Gabor feature detectors is increased, and will evaluate the performance of the proposed method on a much larger face database.

References

[1] *Holland, J. H.*, **Adaptation in natural and artificial systems**, The University of Michigan Press, 1975.

[2] *Turk, M. and Pentland, A.*, **Face Recognition Using Eigenfaces**, Proceedings of IEEE Conference on Computer Vision and Pattern Recognition, 1991, pp 586-591.

[3] *Etemad, K. and Chellappa, R.*, **Discriminant analysis for recognition of human face images**, Journal of Optical Society of America, 1997, pp 1724-1733.

[4] *Lee, T. S.*, **Image representation using 2D Gabor wavelets,** IEEE Transactions on Pattern Analysis and Machine Intelligence, Vol. 18(10), Oct. 1996, pp 959 – 971.

[5] *Shen, L. and Bai L.*, **Gabor wavelets and kernel direct discriminant analysis for face recognition**, Proceedings of the 17th International Conference on Pattern Recognition, 2004, ICPR 2004, Vol. 1(23-26), Aug. 2004, pp 284 – 287.

[6] *Liu, C. and Wechsler, H.*, **Independent component analysis of Gabor features for face recognition,** IEEE Transactions on Neural Networks, Vol. 14(4), July 2003, pp 919 – 928.

[7] *Liu, C. and Wechsler, H.*, **Gabor feature based classification using the enhanced fisher linear discriminant model for face recognition,** IEEE Transactions on Image Processing, Vol. 11(4), April 2002, pp 467 – 476.

[8] *Duc, B.; Fischer, S.; Bigun, J.*, **Face authentication with sparse grid Gabor information,** IEEE International Conference on Acoustics, Speech, and Signal Processing, 1997. ICASSP-97, Vol. 4(21-24), April 1997, pp 3053 – 3056.

[9] *Kalocsai, P.; Neven, H.; Steffens, J.*, **Statistical analysis of Gabor-filter representation** Third IEEE International Conference on Automatic Face and Gesture Recognition, 1998. Proceedings, 14-16, April 1998, pp 360 – 365.

[10] *Liu, Y. and Chongqing*, **Face recognition using kernel principal component analysis and genetic algorithms,** Proceedings of the 12th IEEE Workshop on Neural Networks for Signal Processing, Sept. 2002, pp 337 – 343.

[11] *Xu, Y., Li, B., Wang, B.*, **Face recognition by fast independent component analysis and genetic algorithm,** The Fourth International Conference on Computer and Information Technology, 2004, CIT '04, 14-16, Sept. 2004, pp 194 – 198.

[12] *Black, J., Gargesha, M., Kahol, K., Kuchi, P., Panchanathan, S.*, **A Framework for Performance Evaluation of Face Recognition Algorithms**, ITCOM, Internet Multimedia Systems II, Boston, July 2002.

The Application of Extended Geodesic Distance in Head Poses Estimation

Bingpeng Ma[1,3], Fei Yang[1,3], Wen Gao[1,2,3], and Baochang Zhang[2]

[1] Institute of Computing Technology,
Chinese Academy of Sciences, Beijing 100080, China
[2] Department of Computer Science and Engineering,
Harbin Institute of Technology, Harbin, China, 150001
[3] Graduate School of the Chinese Academy of Sciences, Beijing 100039, China

Abstract. This paper we proposes an extended geodesic distance for head pose estimation. In ISOMAP, two approaches are applied for neighborhood construction, called k-neighbor and ϵ-neighbor. For the k-neighbor, the number of the neighbors is a const k. For the other one, all the distances between the neighbors is less than ϵ. Either the k-neighbor or the ϵ-neighbor neglects the difference of each point. This paper proposes an new method called the kc-neighbor, in which the neighbors are defined based on c time distance of the k nearest neighbor, which can avoid the neighborhood graph unconnected and improve the accuracy in computing neighbors. In this paper, SVM rather than MDS is applied to classify head poses after the geodesic distances are computed. The experiments show the effectiveness of the proposed method.

1 Introduction

Dimension reduction techniques are widely used for the analysis of complex sets of data, such as face images. For face images, classical dimensionality reduction methods include Eigenface[1], Linear Discriminant Analysis(LDA)[2], Independent Component Analysis(ICA)[3], etc, all of which are linear methods. The linear methods have their limitations. On one hand, they cannot reveal the intrinsic distribution of a given data set. On the other hand, if there are changes in poses, facial expression and illumination, the projections may not be appropriate and the corresponding reconstruction error may be much higher. For a pair of points on the manifold, their Euclidean distance may not accurately reflect their intrinsic similarity and, consequently, is not suitable for determining intrinsic embedding or pattern classification. For example, Fig.1 is the data points sampled of Swissroll[4]. The Euclidean distance between point $\mathbf{x}$ and point $\mathbf{y}$ is deceptively small in the three-dimensional input space though their geodesic distance on a intrinsic two-dimensional manifold is large.

The recently proposed ISOMAP[5], LLE[6] and Laplacian Eigenmaps[7] algorithms are popular non-linear dimensionality reduction methods. The ISOMAP method computes pair-wise distances in the geodesic space of the manifold, and then performs classical Multidimensional Scaling(MDS)[8] to map data points

D. Zhang and A.K. Jain (Eds.): ICB 2006, LNCS 3832, pp. 192–198, 2005.

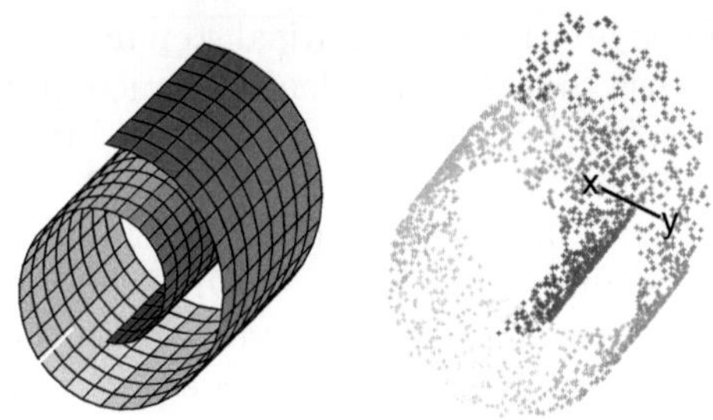

Fig. 1. The data points of Swissroll

from their high-dimensional input space to low-dimension coordinates of a non-linear manifold. In ISOMAP, the geodesic distances can reflect the intrinsic low-dimensional geometry of the manifold, but it can't reduce dimension when the number of samples is very large. And, MDS is applied for visualization in low-dimension, which can't deal with non-linear data.

In this paper, kc-neighbor is applied to compute the geodesic distances for the head-pose estimation, which is necessary in a variety of applications such as face recognition. The problem is difficult because it is an estimation for an inherently three dimensional quantity from two dimensional image data.

In this paper, each face image with a certain pose is considered as a point in the high-dimension manifold. First the neighborhood is constructed using the kc-neighbor method. Then the geodesic distance are computed for all the pairwise points. Finally, SVM is applied to classify each point into pose classes using the geodesic distances from other points. Compared with k-neighbor and ϵ-neighbor of ISOMAP, kc-neighbor can correctly reflect the relation between each point and its neighbors, and SVM classifiers can improve the accuracy of the pose estimation. Experimental results on data sets show that kc-ISOMAP improves estimation accuracy.

The remaining part of this paper is organized as follows. In section 2, we describe kc-neighbor. In section 3, we introduce the SVM classifiers. Then, two databases are used to evaluate the performance of kc-neighbor in Section 4. Finally we conclude this work in section 5.

2 The Extended Geodesic Distance

ISOMAP's global coordinates provide a simple way to analyze and manipulate the high-dimensional observation in terms of their intrinsic nonlinear degrees of freedom. In ISOMAP, nonlinear features are extracted based on estimating geodesic distances and embedded by MDS. The basic idea is that for the neighbor points on a manifold, the Euclidean distances provide a fair approximation of geodesic distances, whereas for faraway points the geodesic distances are estimated by the shortest pathes through neighboring points.

The construction of neighborhood is a critical step in the ISOMAP. Neighbors should be local in the sense that the Euclidean distances are fair approximation of the geodesic distances. Tenenbaun et. al.[5] proposed two methods for neighborhood construction, called k-ISOMAP and ϵ-ISOMAP. The k-ISOMAP means defines the graph G over all data points by connecting points $\mathbf{x}_i$ and $\mathbf{x}_j$ if $\mathbf{x}_i$ is one of the k nearest neighbors of $\mathbf{x}_j$. In ϵ-ISOMAP method, the graph G is defined by connecting each point to all the points within the fixed radius ϵ. The neighborhood relation is symmetric by definition and the numbers of neighbors is different for each point. The choice of an appropriate ϵ is a difficult task. If ϵ is too small the resulting graph becomes sparse and the unconnected subgraphs often exist, while if ϵ is too large the idea of connecting local patches gets lost. In both cases the approximation error increases.

Due to the inhomogeneous density of the samples it seems more data-sensitive to define the k nearest points of x_i as its neighbors. The k-neighbor method will not generate any isolated point.But if more than k points cluster, they will form an unconnected subgraph. Furthermore, the rule is not symmetric in the sense that x_j is a neighbor point of x_i does not necessarily imply that x_i is also a neighbor point of x_j, so that G has to be symmetrized afterwards.

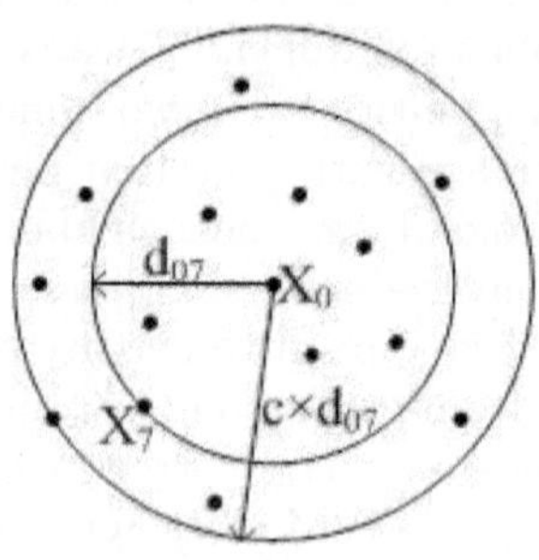

Fig. 2. kc-neighbor. $\mathbf{x}_7$ is the 7th nearest neighbor and the radius is d_{07}. In kc-neighbor method, all the points whose radius less than c times the d_{07} are $\mathbf{x}_0$'s neighbors.

To consider changes in the sample density, kc-neighbor method is presented in this paper. In this method the neighbors of a point $\mathbf{x}_i$ include all the points that lie inside the ϵ-ball with the radius ϵ equal to c times distance between $\mathbf{x}_i$ and its k-th nearest neighbor. If a point $\mathbf{x}_i$ have k neighbors in k-ISOMAP, and d_{07} is the distance of the k-th neighbor, we define all the points which are closer than c times d_{07} as the neighbors of $\mathbf{x}_i$. Three reasons lead us to present this idea. First, the sample density varies, so a fixed rule will not apply effectively to all points. When using k-neighbor, we think all the points have the same number of neighbors. As to ϵ-neighbor, all the neighbor points are within the same distance. Second, compared with k-neighbor, kc-neighbor can avoid unconnected subgraph because we specify the different numbers of neighbors for different points. Compared with ϵ-neighbor, kc-neighbor uses a dynamic ϵ for different points and makes all the points have roughly the same number of neighbors. At last, kc-neighbor does not increase the computing complexity because we use the sort result when finding neighbors.

Based on the kc-neighbor, we present the kc-ISOMAP method. Compared with the k-ISOMAP, the main difference of the kc-ISOMAP is using kc-neighbor to replace the k-neighbor. Given a training set $\{\mathbf{x}_i, i = 1, \ldots, m\}$, the first step of kc-ISOMAP determines the nearest neighbors for point $\mathbf{x}_i$ based on the Euclidean distances $d_X(\mathbf{x}_i, \mathbf{x}_j)$ in the input space $\mathbb{X}$. These neighborhood relations are represented as a weighted graph G in which

$$d_G(\mathbf{x}_i, \mathbf{x}_j) = \begin{cases} d_X(\mathbf{x}_i, \mathbf{x}_j), & \text{if } \mathbf{x}_i \text{ and } \mathbf{x}_j \text{ are neighbors} \\ \infty, & \text{otherwise} \end{cases} \quad (1)$$

In the second step, kc-ISOMAP estimates the geodesic distances $d_M(\mathbf{x}_i, \mathbf{x}_j)$ between all pairs of points on the manifold $\mathbb{M}$ by computing their shortest path distance $d_G(\mathbf{x}_i, \mathbf{x}_j)$ in the graph G. In generally, Floyd-Warshall algorithm is used to compute the geodesic distances $d_M(\mathbf{x}_i, \mathbf{x}_j)$:

$$d_M(\mathbf{x}_i, \mathbf{x}_j) = min\{d_G(\mathbf{x}_i, \mathbf{x}_j), d_G(\mathbf{x}_i, \mathbf{x}_p)) + d_G(\mathbf{x}_p, \mathbf{x}_j)\} \quad (2)$$

3 SVM Classification

In ISOMAP, after computing geodesic distances, MDS is applied at the aim of visualization in low-dimension. From a non-technical point of view, the purpose of MDS is to provide a visual representation of the pattern of proximities (i.e., similarities or distances) among a set of objects, which does not contribute to the improvement of the classification accuracy. In this paper, SVM classifers are used to replace MDS after computing the geodesic distances.

SVM is a quadratic optimization problem in order to maximize the margin between examples of two classes either in the original input space or in an implicitly mapped higher dimensional space by using kernel functions. Though new kernels are being proposed by researchers, we still use the basic RBF(radial basis function) kernel.

Generally, SVM are used for 2-class problems. In this paper, we use "one against one" approach to slove the k-class problem. In this approach each classifier is constructed to seperate a two classes, and totally $k(k-1)/2$ classifiers are constructed and a voting strategy is used to get the "winner" class.

4 Experiment

We test kc-ISOMAP method using the public FERET[11] database and the CAS-PEAL[12] database. Fig.3 and Fig.4 show some subjects in the FERET and CAS-PEAL database. The FERET database contains 1400 images of 200 persons, varying in facial expression(open/closed eyes, smiling/non smiling), and each person has seven horizontally poses $\{-40°, -25°, -15°, 0°, +15°, +25°, +40°\}$. The persons of FERET database come from Asia, Europe, and Africa.

The CAS-PEAL database contains seven poses $\{-45°, -30°, -15°, 0°, +15°, +30°, +45°\}$ of 1400 persons. In order to compare the results with FERET database, we use a subset of the CAS-PEAL database including 1400 images of 200 persons with the subject ID ranging from 401 to 600. Unlike FERET, the persons of CAS-PEAL are all from Asian.

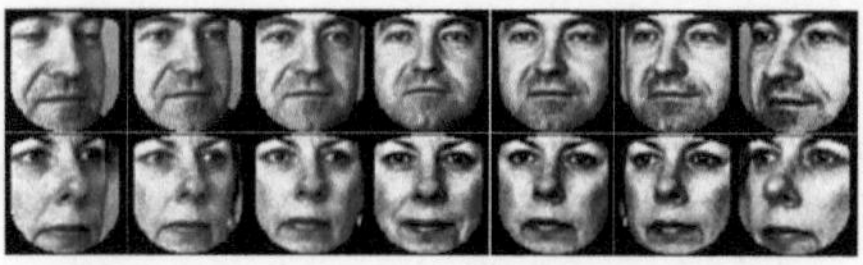

Fig. 3. Face images in the FERET database

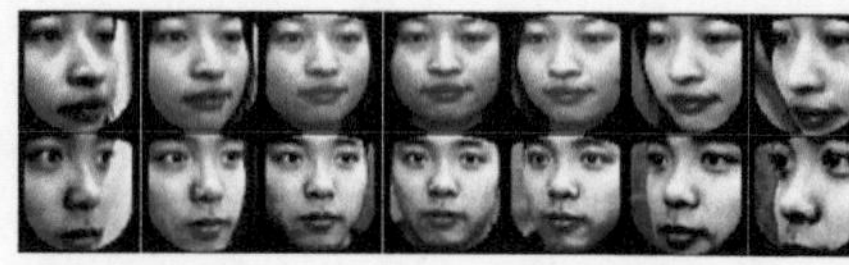

Fig. 4. Face images in the CAS-PEAL database

We first label the positions of eyes by hand, and then crop the images to 32×32 pixels. We use histogram equalization to reduce the influence of lighting, represent each image by a raster scan vector of the intensity values, and finally normalize each vector to be zero-mean unit-variance vectors.

In experiment, we use cross-validation in order to avoid over-training. We first sort the images in the database by file name, and divide them into 3 parts. One part is taken as testing set and the other two as training set. Repeat three times so that each part has been taken as testing set. All the testing results are the mean results of all testing sets.

In computing geodesic distance, it is difficult to compute geodesic distance of new samples. In our experiment, we use the general ways which are used in ISOMAP and LLE. We first compute the geodesic distances of all samples, which donot consider the difference between the training samples and the testing samples. And then the geodesic distances can be divided two parts: the training set and the testing set. In actual application, the geodesic distances of the testing samples can be computed by executing the Floyd-Warshall algorithm.

We compare the following three methods: the P-k-ISOMAP, the k-ISOMAP and kc-ISOMAP. In the P-k-ISOMAP, we first use PCA to reduce the dimension of the images from 1024 to 245, which reserves 99.9 percent of the total energy of eigenvalues, and then use the new samples to compute the geodesic distances. In the the k-ISOMAP, we directly use the images to compute the geodesic distances. In all the three methods, SVM classifiers are performed for pose estimation and different ks are used to find the influence caused by k. And three different values of $c(1.05, 1.1, 1.2)$ are selected to discover the influence of c in kc-ISOMAP. The experiment results are shown in Tab.1, Fig.5 and Fig.6.

Form the table and the figures, we can know kc-ISOMAP have an improvement in pose estimation on both the FERET database and the CAS-PEAL database. The accuracy rate of both the ISOMAP and kc-ISOMAP improve remarkably with the increase of the number of neighbors when the number of neighbors is ranging from 3 to 9, but it tends to stabilize with the increase of the number of the neighbors. It means that the selection of k is very important

Table 1. Error Rate Comparison of Different Pose Estimation

method	CAS-PEAL	FERET
P-k-ISOMAP $k = 7$	13.79	24.21
P-k-ISOMAP $k = 14$	12.64	23.52
P-k-ISOMAP $k = 21$	11.21	22.78
k-ISOMAP $k = 7$	14.29	25.35
k-ISOMAP $k = 14$	11.86	23.35
k-ISOMAP $k = 21$	11.07	22.78
kc-ISOMAP(c=1.1) k=7	11.14	22.93
kc-ISOMAP(c=1.1) k=14	10.86	21.36
kc-ISOMAP(c=1.1) k=21	9.29	21.21

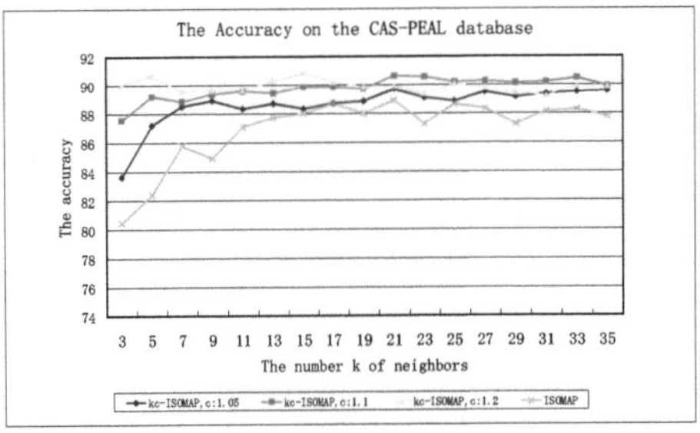

Fig. 5. The results of pose estimation on the CAS-PEAL database

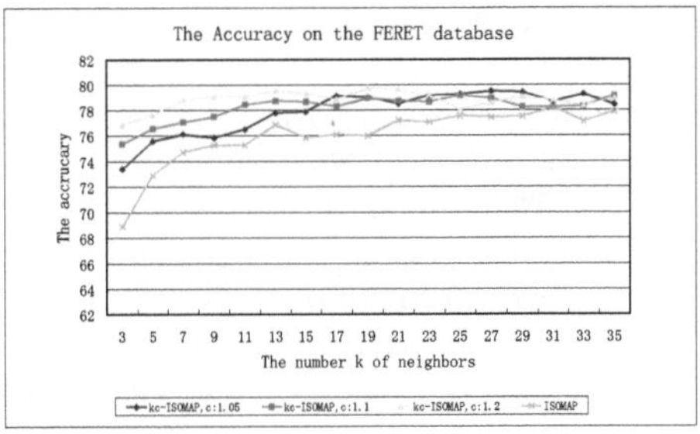

Fig. 6. The results of pose estimation on the FERET database

for the preservability of the pose manifold. If the number of neighbors is much smaller, the structure of the manifold can not be maintained, at this instance, the improvement of kc-ISOMAP is more apparent. With the increase of k, the predominance of kc-ISOMAP is decreased, but kc-ISOMAP always can obtain better accuracy than k-ISOMAP, which means kc-ISOMAP can maintain the structure of manifold better because kc-ISOMAP more care about the neighborhood relation of the each samples. And from Fig.5 and Fig.6, we can know the results of the difference c are nearly equal, which means that the value of c is not important, but it can cause the dynamic number of the neighbors and then improve the accuracy.

5 Conclusion and Future Work

This paper proposes a novel method to extend the geodesic distance in ISOMAP. Compared with the traditional geodesic distance, this method considers the dynamic number of neighbors for each point, which makes the relation of neighbors more correctly. And after computing the geodesic distances, it apply SVM classifiers replace MDS because the MDS is the method to preserve the feature of samples, which can not improve the accuracy rate of classify and SVM is the best classifier when the number of the training samples is enough to find the correct support vectors. The experiment shows *kc*-ISOMAP can improve the accuracy of the poses estimation.

Acknowledgements

This research is partially sponsored by Natural Science Foundation of China under contract No.60332010, and No.60473043, "100 Talents Program" of CAS, ShangHai Municipal Sciences and Technology Committee(No.03DZ15013), and ISVISION Technologies Co., Ltd.

References

1. M.Turk, and A. Pentland, "Eigenfaces for Recognition", Journal of Cognitive Neuroscience, (3) 71-86,1991.
2. P. N. Belhumeur, J. P. Hespanha and D. J. Kriegman, "Eigenfaces vs. Fisherfaces: Recognition using class specfic linear projection", IEEE Trans. on PAMI, Vol. 19, No. 7, 711-720, 1997.
3. Marian Stewart Bartlett, Terrence J. Sejnowski, "Independent components of face images: A representation for face recognition", Proceedings of the 4th Annual Jount Symposium on Neural Computation, Pasadena, CA, May 17, 1997.
4. Ming-HsuanYang, "Extended Isomap for Classification", ICPR (3) 2002: 615-618.
5. J.B.Tenenbaum, V.de Silva, and J.C.Langford, "A global geometric framework for nonlinear dimensionality reduction", Science 290: 2319-2323, 2000.
6. Roweis.S. and Saul.L., "Nonlinear dimensionality reduction by locally linear embedding". Science 290:2323-2326, 2000
7. M. Belkin and P. Niyogi, "Laplacian eigenmaps and spetral techniques for embedding and clustering", Advances in Neural Information Processing Systems, vol. 15, 2001.
8. Trevor F. Cox and Michael A. A. Cox, "Multidimensional Scaling", CRC Press, 2000.
9. Cortes,C. and Vapnik, V, "Support vector network", Machine Learning, 20:273:297, 1995.
10. Chih-Chung Chang and Chih-Jen Lin, LIBSVM : a library for support vector machines, 2001. Software available at http://www.csie.ntu.edu.tw/ cjlin/libsvm
11. Phillips P.J., Moon H., etc. The FERET evaluation methodology for face - recognition algorithms, IEEE Transactions on Pattern Analysis and Machine Intelligence, 2000, 22(10):1090-1104.
12. Wen Gao, Bo Cao, Shiguang Shan, Xiaohua Zhang, Delong Zhou, The CAS-PEAL Large-Scale Chinese Face Database and Baseline Evaluations, technical report of JDL, 2004, http://www.jdl.ac.cn/ peal/peal_tr.pdf.

Improved Parameters Estimating Scheme for E-HMM with Application to Face Recognition

Bindang Xue[1], Wenfang Xue[2], and Zhiguo Jiang[1]

[1] Image processing center, Beihang University, Beijng 100083, China
{xuebd, jiangzg}@buaa.edu.cn
[2] Institute of Automation, Chinese Academy of Sciences,
100088, Beijing, China
wfxue@nlpr.ia.ac.cn

Abstract. This paper presents a new scheme to initialize and re-estimate Embedded Hidden Markov Models(E-HMM) parameters for face recognition. Firstly, the current samples were assumed to be a subset of the whole training samples, after the training process, the E-HMM parameters and the necessary temporary parameters in the parameter re-estimating process were saved for the possible retraining use. When new training samples were added to the training samples, the saved E-HMM parameters were chosen as the initial model parameter. Then the E-HMM was retrained based on the new samples and the new temporary parameters were obtained. Finally, these temporary parameters were combined with saved temporary parameters to form the final E-HMM parameters for representing one person face. Experiments on ORL databases show the improved method is effective.

1 Introduction

Face recognition has been an active research topic recently and remains largely unsolved [1, 2]. Based on the recognition principle, diverse existing face recognition approaches can be briefly classified as three catalogues: geometric feature-based, principle component analysis (PCA)-like based and model based. Due to the ability to "learn" model parameters, several face recognition systems were based on E-HMM and this method appears having more promising potential [3-6].The key problem using E-HMM for face recognition is how to train the model parameters for discovering intrinsic relations between face images and human face, and further building appropriate models based on these relations. However, the problem of choosing the initial model parameters for the training process and the problem of retraining model parameters were still left as open problems. In earlier work, Davis and Lovell had studied the problem of learning from multiple observation sequences [7] and the problem of ensemble learning [8] with multiple observation sequences being provided at one time. But how to deal with multiple observation sequences being provided at different time has not been addressed. While the retraining problem of E-HMM for face recognition is just like this problem. Under new environment, in order to improve the recognition accuracy, news training samples sets are added to the training samples sets. So it is needed to re-estimate the model parameters based on the newly formed sample sets. In this paper, a segmental scheme is presented to solve this problem.

D. Zhang and A.K. Jain (Eds.): ICB 2006, LNCS 3832, pp. 199 – 205, 2005.

2 E-HMM for face

A human face can be sequentially divided from top to bottom as forehead, eyes, nose, mouth and chin. Hence a human face can be viewed as a region chain. In such a way a human face can be defined as 1-D HMM. In essence, a human face image is a two dimensional object which should process as a 2-D HMM. To simplify the model processing, a specified pseudo 2-D HMM scheme is proposed. This model extends all top-down sub-regions in 1-D HMM as sub-sequences from left-hand side to right-hand side separately and uses extended sub-1-D HMM defining these sub-sequences hierarchically. This pseudo face 2-D HMM is also called E-HMM[3]. The face 2-D HMM scheme shown as fig. 1, composed of five super states (forehead, eyes, nose, mouth and chin) vertically, and the super states are extended as {3, 6, 6, 6, 3} sub states (embedded states) horizontally.

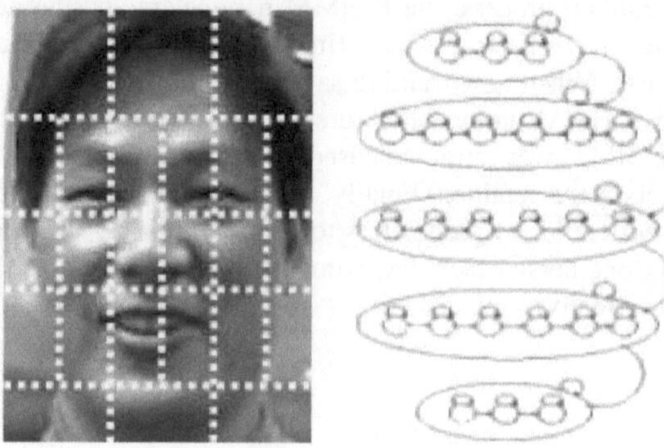

Fig. 1. E-HMM for face

An E-HMM structure can be defined by the following elements:

Super states parameters:

· N : the number of super states.

· Π : the initial super state probability distribution.

· A : the super state transition matrix. $A = \{a_{ij}, 1 \le i, j, \le N\}$.

· Λ : embedded 1D-HMMs, named super state. $\Lambda = \{\Lambda^i, 1 \le i \le N\}$.

Sub-states parameters:

· N^i : the number of sub states embedded in super state Λ^i . $S^i = \{s_k^i, 1 \le k \le N^i\}$.

· Π^i : the initial sub state probability distribution embedded in super state Λ^i , $\Pi^i = \{\Pi_k^i, 1 \le k \le N^i\}$.

. A^i : sub-states transition matrix in super state Λ^i . $A^i = \{a_{kl}^i, 1 \le k, l \le N^i\}$.

. B^i : the sub states output probability function in super state Λ^i , $B^i = \{b_k^i(o_{xy})\}$,where o_{xy} represent the observation vector at row x and column y $(x = 1, \cdots, X, y = 1, \cdots, Y)$, the sub-states output probability function that is typically used is a finite mixture of Gaussian probability density function(P.D.F.)

$$b_k^i(o_{xy}) = \sum_{f=1}^{F} C_{kf}^i N(o_{xy}, \mu_{kf}^i, U_{kf}^i), \quad (1 \le k \le N^i) \tag{1}$$

Where $N(o_{xy}, \mu_{kf}^i, U_{kf}^i)$ represents f^{th} Gaussian P.D.F. with the mean vector μ_{kf}^i and the covariance matrix U_{kf}^i, C_{kf}^i is the mixture coefficient for the fth mixture of the output probability function of sub state k in super state Λ^i.

So an E-HMM can be defined as $\lambda = (N, A, \Pi, \Lambda)$, where N is the number of the super states. $\Lambda = \{\Lambda^1, \cdots, \Lambda^N\}$, $\Lambda^i = \{N^i, \Pi^i, A^i, B^i\}$, Λ^i represents the super state i, N^i is the number of embedded sub states in super state Λ^i.

3 Training of the E-HMM

Given a set face images taken from the same person, model training is estimating the corresponding model parameters and saving them in a face database. The strategy for generating observation vector sequence and the training method are similar to the methods as described in article [3].

For describing the algorithm simply, it is useful to define the following variables:

· $StartSuperstate(i)$: represents the expected number of super state Λ^i at column $y = 1$ given R observation sequences;

· $StartState(i,k)$: represents the expected number of sub state s_k^i at row $x = 1$ in super state Λ^i given R observation sequences;

· $SuperTransition(i, j)$: represents the expected transition number from super state Λ^i to super state Λ^j;

· $StateTransition(i,k,l)$: represents the expected transition number from sub state s_k^i to sub state s_l^i in super state Λ^i;

· $SuperTransform(i)$: represents the expected transition number from super state Λ^i;

· $StateTransform(i,k)$: represents the expected transition number from sub state s_k^i in super state Λ^i;

· $Component(i,k,f)$: represents the expected number of fth mixture element of the output probability function of the sub state s_k^i.

Based on the above variables, part parameters of the EHMM can be further re-estimated using the following formulas:

$$\overline{\Pi}_i = \frac{StartSuperstate(i)}{\sum_{j=1}^{N} StartSuperstate(j)} \tag{2}$$

$$\overline{a}_{ij} = \frac{SuperTransition(i, j)}{SuperTransform(i)} \tag{3}$$

$$\overline{\Pi}_k^i = \frac{StartState(i,k)}{\sum_{k=1}^{N^i} StartState(i,k)} \tag{4}$$

$$\overline{a}_{kl}^i = \frac{StateTransition(i,k,l)}{StateTransform(i,k)} \tag{5}$$

$$\overline{C}_{kf}^i = \frac{Component(i,k,f)}{\sum_{f=1}^{F} Component(i,k,f)} \tag{6}$$

4 Improved Parameters Estimating Scheme for E-HMM

In this paper, current training sample sets is referred as R_1 ,and the model parameters can be iteratively estimated based on R_1 using formulas (2)-(6).During the estimating procedure, the variables defined above are labeled as $StartSuperstate^{R_1}(i),\cdots$ .When the training procedure is finished, the model parameters $\overline{\lambda}_1$ are saved, at the same time, the temporary variables $StartSuperstate^{R_1}(i),\cdots,Component^{R_1}(i,k,f)$ are also saved. Once new sample sets R_2 is obtained, the whole sample sets include R_1 and R_2 .The segmental retraining scheme is that only the temporary variables $StartSuperstate^{R_2}(i)$ based on R_2 are needed to be re-estimated, then the last model parameter will be formed by combining $StartSuperstate^{R_2}(i),\ \cdots,Component^{R_2}(i,k,f)$ with recoded $StartSuperstate^{R_1}(i),\ \mathrm{t}\ \cdots,Component^{R_1}(i,k,f)$.

Another problem is how to choose a set of initial model parameters. The initial model parameters have great effect on the training procedure of the model. For example, choosing different initial model parameters will affect the convergence of the iterative training algorithm and the face recognition right rate. But there is no method to choose ideal initial model parameters now .One scheme to solve this problem is that we can divide the training sample sets into two parts R_1 and R_2 , the initial model parameters $\overline{\lambda}_1 = (\overline{\Pi}_1, \overline{A}_1, \overline{\Lambda}_1)$ are estimated based on sample sets R_1 ,then we can estimate parameters $\overline{\lambda}_2 = (\overline{\Pi}_2, \overline{A}_2, \overline{\Lambda}_2)$ referring $\overline{\lambda}_1 = (\overline{\Pi}_1, \overline{A}_1, \overline{\Lambda}_1)$ as initial model parameters. In the end, it is easy to combine $\overline{\lambda}_1 = (\overline{\Pi}_1, \overline{A}_1, \overline{\Lambda}_1)$ with $\overline{\lambda}_2 = (\overline{\Pi}_2, \overline{A}_2, \overline{\Lambda}_2)$ to form the final model parameters $\overline{\lambda} = (\overline{\Pi}, \overline{A}, \overline{\Lambda})$ .The initial model parameter comes from part training sample sets, so that it is better than other methods such as random initializing or choosing experiential values.

The formulas of the improved parameter estimating scheme for E-HMM are described as below:

$$\overline{\Pi}_i = \frac{StartSuperstate^{R_1}(i) + StartSuperstate^{R_2}(i)}{\sum_{j=1}^{N} StartSuperstate^{R_1}(j) + \sum_{j=1}^{N} StartSuperstate^{R_2}(j)} \tag{7}$$

$$\overline{a}_{ij} = \frac{SuperTransition^{R_1}(i,j) + SuperTransition^{R_2}(i,j)}{SuperTransform^{R_1}(i) + SuperTransform^{R_2}(i)} \tag{8}$$

$$\overline{\Pi}_k^i = \frac{StartState^{R_1}(i,k) + StartState^{R_2}(i,k)}{\sum_{k=1}^{N^i} StartState^{R_1}(i,k) + \sum_{k=1}^{N^i} StartState^{R_2}(i,k)} \quad (9)$$

$$\overline{a}_{kl}^i = \frac{StateTransition^{R_1}(i,k,l) + StateTransition^{R_2}(i,k,l)}{StateTransform^{R_1}(i,k) + StateTransform^{R_2}(i,k)} \quad (10)$$

$$\overline{C}_{kf}^i = \frac{Component^{R_1}(i,k,f) + Component^{R_2}(i,k,f)}{\sum_{f=1}^{F} Component^{R_1}(i,k,f) + \sum_{f=1}^{F} Component^{R_2}(i,k,f)} \quad (11)$$

5 Face Recognition Experiments and Results

The goal of the experiment on face recognition is just to evaluate the proposed segmental parameter re-estimating scheme, so a small database ORL face database [10] is chosen as the test datasets. ORL database contains 400 images of 40 individuals, with 10 images per individual at the resolution of 92×112 pixels. The images of the same person are taken at different times, under slightly varying lighting conditions, and with different facial expressions. Some people are captured with or without glasses. The head of the people in the images are slightly titled or rotated. Images of one person from ORL database show as Fig.2.

Firstly, the first six face images of one person are used to train the E-HMM, and the remaining four images are used to test the system. In order to evaluate the improved parameter estimating scheme, we divide the first six training images into two equal parts R_1 and R_2. At first step, R_1 is used to train the model to get the initial model parameters $\overline{\lambda}_1 = (\overline{\Pi}_1, \overline{A}_1, \overline{\Lambda}_1)$, then R_2 is used to train the model parameters $\overline{\lambda}_2 = (\overline{\Pi}_2, \overline{A}_2, \overline{\Lambda}_2)$. At last, the final model parameters $\overline{\lambda} = (\overline{\Pi}, \overline{A}, \overline{\Lambda})$ are obtained quickly based on the improved parameter estimating scheme presented in this paper.

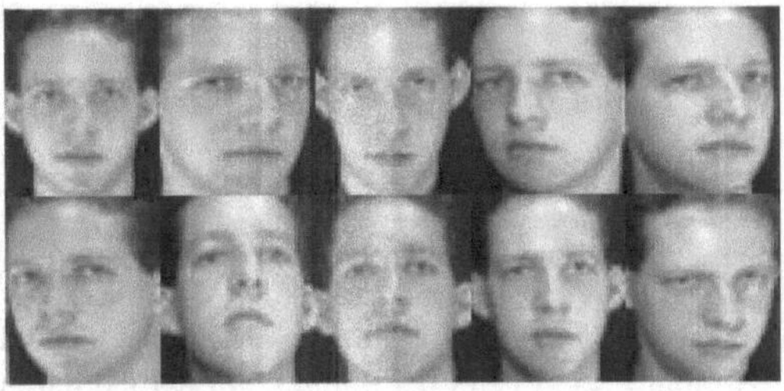

Fig. 2. Images of one person from ORL database

Given a test face image, recognition is to find the best matching E-HMM model within a given face model database and predicting the matching probability. Usually the model corresponding to the maximum likelihood is assumed to be the right choice revealing the identity among the given face model database. Let there are P individuals in the database, given a face image t, the matching maximum likelihood probability rule is prescribed as:

$$P(O^t \mid \lambda_k) = Max(P(O^t \mid \lambda_p)) \quad (1 \le k, p \le P) \tag{12}$$

So the recognition result is that the face image t is corresponding to k^{th} person in the database.

Table1.simply compares the recognition results of HMM trained using different parameter estimating method. The improved scheme achieves 99.5% correct recognition rate on ORL face database.

Table 1. Recognition results of different methods

Methods	Right recognition rate (%)
Pseudo-HMM[11]	90-95
E-HMM[3]	98.5
Segmental scheme	99.5

6 Conclusions

This paper describes an improved segmental scheme to initialize and re-estimate E-HMM parameter. The advantage of the improved parameter estimating scheme is that the E-HMM parameters re-estimating process has good ability of adaptation: when new sample set was added to the training sample, the information of the new sample set could be conveniently combined into the E-HMM, and the calculation complex was reduced. Besides, the improved parameter estimating scheme provides an answer for the problem of choice initial E-HMM parameters.

Future work will focus on sequential learning algorithm for E-HMM with application to face recognition.

References

1. 1. Chellappa R., Wilson C.L., Sirohey S. Human and machine recognition of face: A survey. Proc. IEEE, 1995,83(5):705-740.
2. Zhao W., Face recognition: A Literature Survey. CS-TR-4167, University of Maryland, Oct. 2000
3. 3 A.V. Nefian, M.H. Hayes, Maximum likelihood training of the embedded HMM for face detection and recognition, Proc. of the IEEE International Conference on Image Processing, ICIP 2000, Vol. 1, 10-13 September 2000, Vancouver, BC, Canada, pp. 33-36.
4. S. Eickeler, S. Muller, etc. Recognition of JPEG compressed face images based on statistical methods.Image and Vision Computing , 2000 (18):279–287.
5. F. Wallhoff,S. Eickeler,etc. A comparison of discrete and continuous output modeling techniques for a pseudo-2D hidden Markov model face recognition system.. Proceedings of International Conference on Image Processing, 2001(2):685 –688.

6. H, Othman ,T. Aboulnasr. A simplified second-order HMM with application to face recognition, in the IEEE International Symposium on Circuits and Systems, 2001(2): 161 – 164.
7. Davis, Richard I. A. and Lovell, Brian C. and Caelli, Terry. Improved Estimation of Hidden Markov Model Parameters from Multiple Observation Sequences. In *International Conference on Pattern Recognition*, Quebec City, Canada, August 11-14 II, 2002,:168-171.
8. Davis, Richard I. A. and Lovell, Brian C. Comparing and Evaluating HMM Ensemble Training Algorithms Using Train and Test and Condition Number Criteria. Pattern Analysis and Applications .2003 (6):327-336.
9. 9 Rabiner L., A tutorial on HMM and selected applications in speech recognition , Proc. IEEE ,1989, 77(2):257-286.
10. ORL Face database , Cambridge ,AT&T Laboratories Cambridge. (http://www.uk.research.att.com/facedatabase.html)
11. Samaria F., Face Recognition Using Hidden Markov Models, PhD thesis, University of Cambridge,1994.

Component-Based Active Appearance Models for Face Modelling

Cuiping Zhang and Fernand S. Cohen

Eletrical and Computer Engineering Department,
Drexel University, Philadelphia PA 19104, USA
{zcp, fscohen}@cbis.ece.drexel.edu

Abstract. The Active Appearance Model (AAM) is a powerful tool for modelling a class of objects such as faces. However, it is common to see a far from optimal local alignment when attempting to model a face that is quite different from training faces. In this paper, we present a novel component-based AAM algorithm. By modelling three components inside the face area, then combining them with a global AAM, face alignment achieves both local as well as global optimality. We also utilize local projection models to locate face contour points. Compared to the original AAM, our experiment shows that this new algorithm is more accurate in shape localization as the decoupling allows more flexibility. Its insensitivity to different face background patterns is also clearly manifested.

1 Introduction

Face recognition has received a lot of attention in the past decades. Detecting a face and aligning its facial features are usually the first step, therefore crucial for most face applications. Among numerous approaches, the Active Appearance Model (AAM) [1] and the Active Shape Model (ASM) [2] are 2 popular generative models that share a lot in common. As a successor of the ASM, the AAM is computationally efficient and has been intensively studied by many researchers.

The AAM has several inherent drawbacks as a global appearance based model. First, It has a simple linear update rule stemming from a first order Taylor series approximation of an otherwise complex relationship between the model parameters and the global texture difference. Clearly, any factor that is part of the global texture will affect the AAM's performance (examples are the global illumination, partial occlusions, etc.). In a converged AAM, the local alignment results may need further refinement to meet the accuracy requirement of many applications. Secondly, the gradient descent information near the face contour seeps the background pattern in the training set. Hence, the AAM can't perform well for test face images with unseen backgrounds.

With all these problems associated with the AAM in mind, in this paper, we propose a component-based AAM that groups landmark points inside the face area into three natural components in addition to a globally defined AAM. The independence of the sub-AAMs leads to a more accurate local alignment

D. Zhang and A.K. Jain (Eds.): ICB 2006, LNCS 3832, pp. 206–212, 2005.

result. For the model points on a face contour, a strategy similar to the ASM is adopted. The ASM iteratively adjusts any model point along its normal direction so that an associated texture pattern is in accordance with a typical distribution. Our new method makes full use of what we already have during the AAM procedure and local projection models are built on a standard shape frame. The revised projection models, together with the component-based analysis improve the overall modelling performance, especially on the test set.

The paper is organized as follows: In section 2, the original AAM is briefly introduced. Section 3 presents the idea of the component-based AAM. In section 4, details about our local projection models are given. Section 5 presents our experiment results and discussions. Last section is the conclusion.

2 AAM Basic Idea

In the AAM, a face's shape is defined as a sequence of the coordinates of all landmark points. Let $\mathbf{S}_0$ be the mean shape of all training images. A shapeless texture vector is generated after warping the face patch inside the convex hull of all landmark points to the mean shape. Fig. 1(a) shows a face image overlapped with landmark points and the resulted shapeless texture is in Fig. 1(c).

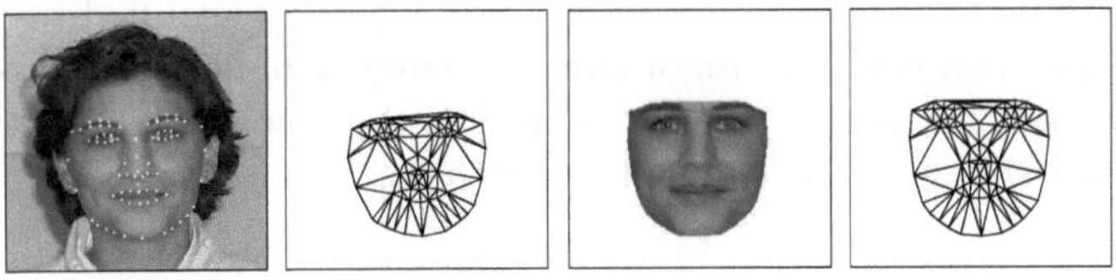

(a) (b) (c) (d)

Fig. 1. (a)landmark points. (b)Face mesh. (c)Shapeless texture. (d)Base face mesh.

All raw shape vectors need to be aligned to a common coordinate system. This normalization introduces a similarity transformation between a original face vector on the image frame and its normalized one on the model frame. Similarly, All raw texture vectors also undergo offsetting and scaling operations for normalization purpose.

PCA is used to model the shape and texture variations. A normalized shape $\mathbf{x}$ and texture $\mathbf{g}$ can be formulated as $\mathbf{x} = \overline{\mathbf{x}} + \mathbf{P}_s \cdot \mathbf{b}_s$ and $\mathbf{g} = \overline{\mathbf{g}} + \mathbf{P}_g \cdot \mathbf{b}_g$. The column vectors in the matrices $\mathbf{P}_s$ and $\mathbf{P}_g$ are the principal modes for the shape and texture variations of the training set. They span the shape and texture subspaces, respectively. The vectors $\mathbf{b}_s$ and $\mathbf{b}_g$, as the projected coefficients in the subspaces, are named shape and texture parameters. They can be concatenated for further de-correlation in a mixed eigen-subspace and the projected coefficients $\mathbf{c}$ encode both shape and texture information. The reconstruction of the shape vector $\mathbf{x}$ and the texture vector $\mathbf{g}$ from $\mathbf{c}$ is straightforward and easy.

The complete appearance parameter set includes 4 similarity pose parameters Ψ due to the coordinate normalization and the mixed parameter vector $\mathbf{c}$, i.e., $\mathbf{p} = \{\Psi, \mathbf{c}\}$. Modelling an unknown face in a test image is a process searching for the optimal appearance parameter set that best describes the face. In an iterative search, let the texture residual $\mathbf{r}(\mathbf{p})$ (also referred as difference image) be the difference of the reconstructed model texture $\mathbf{g}_m$ and the texture $\mathbf{g}_s$ extracted from the test image, $\mathbf{r}(\mathbf{p}) = \mathbf{g}_s - \mathbf{g}_m$. The matching error is measured as the RMS of the texture residual $\mathbf{r}(\mathbf{p})$. The AAM assumes a linear relationship between $\mathbf{r}(\mathbf{p})$ and the update for the model parameters $\delta\mathbf{p}$: $\delta\mathbf{p} = -\mathbf{R} \cdot \mathbf{r}(\mathbf{p})$, where $\mathbf{R}$ is a constant gradient descent matrix estimated from the training images [1]. As face image backgrounds are encoded, it is suggested to use a random background so that $\mathbf{R}$ is irrelevant of the background patterns in the training set [1]. However, useful heuristic information for face contour is also lost as a result.

3 Component-Based AAM

Based on the fact that local shape depends only on local appearance pattern, we propose a Component-based AAM in an effort to gain better feature localization. The basic idea is to group landmark points to components and train the local models independently. To avoid possible confusion, we refer to the original AAM as the global AAM. Three components on the mean shape frame are highlighted in Fig. 2(a). Landmark points are naturally grouped to balance the added computational cost and algorithm efficiency. Columns 2(b) to 2(d) show the components of the person in Fig. 1(a). The top row shows local shapes and the bottom row shows warped shapeless textures.

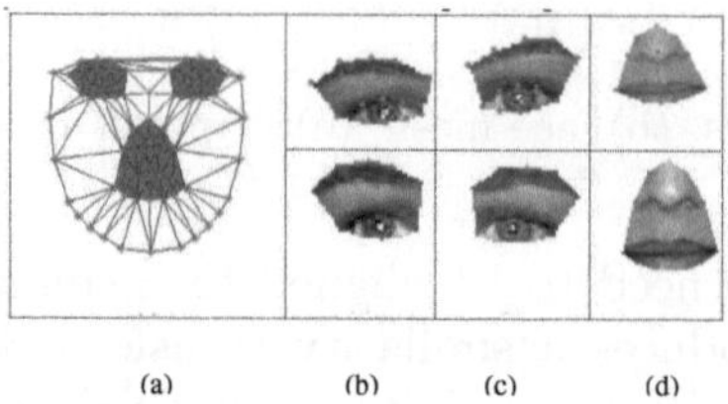

Fig. 2. (a) Definition for local components. (b) Left eyebrow and eye. (c) Right eyebrow and eye. (d) Nose and mouth.

Our component-based AAM is a combination of one global AAM and three sub-models. As part of the global face patch, all components are normalized to the same common coordinate system as that for the global face. This establishes clear correspondence between the global model and the sub-models. Not only all sub-models share the same 2D pose parameters as the whole face, but the component shapes, textures and texture residuals are just fixed entries in their counterparts of the global model. Sub-models are trained separately.

During the modelling process, the component-based AAM algorithm switches between the global model and the sub-models alternatively. After one iteration of the global AAM, we have current estimates of: the global shape $\mathbf{x}$, the texture $\mathbf{g}$, the texture residual $\mathbf{r}(\mathbf{p})$ and the global matching error is e_0. The various steps to model local components are detailed as follows(for ith sub-model, i = 1 to 3):

- **Global to local mapping:** Generate the sub-model shape $\mathbf{x}_i$, texture $\mathbf{g}_i$ and texture residual $\mathbf{r}_i(\mathbf{p}_i)$ by looking up the fixed entries in $\mathbf{x}, \mathbf{g}$, and $\mathbf{r}(\mathbf{p})$. Project $\{\mathbf{x}_i, \mathbf{g}_i\}$ onto local subspaces, $p_i = \{\Psi, \mathbf{c}_i\}$.
- **Local AAM prediction:** Apply local AAM to obtain new sub-model shape vector $\mathbf{x}_i'$, texture vector $\mathbf{g}_i'$ and local 2D pose Ψ_i.
- **Local to global mapping:** Use $\{\mathbf{x}_i', \mathbf{g}_i'\}$ to update the corresponding entries of the global texture vector $\mathbf{g}$ and component points on the image frame.
- **Decision making:** If the new global parameters lead to a smaller matching error, accept the update.

In summary, sub-models update component points independently. Meanwhile, they are united and confined within a global AAM. In this way, error propagation between local components is reduced and modelling ability is enhanced locally. In [3], sub-models are constructed to model vertebra. However, they basically repeat the same sub-model for a sequence of triplet vertebrae and propagate their results, therefore different from our approach.

4 Local Projection Models

When a test face is presented in a background unseen in the training set, the AAM often fails, especially for face contour points. Since landmark points on a face contour are usually the strongest local edge points, we developed a method similar to the ASM to complete our component-based AAM. The ASM moves a landmark point along local normal direction so that its profile conforms to a typical distribution. Instead of using the edge strength along the profile directly, we believe that edge information would be more prominent and stable after taking the local average. Further, we associate our local projection models with the triangulation result of the landmark points. Fig. 3(a) is the mesh of landmark points for the person in Fig. 1(a). Fig. 3(b) shows the mean shape.

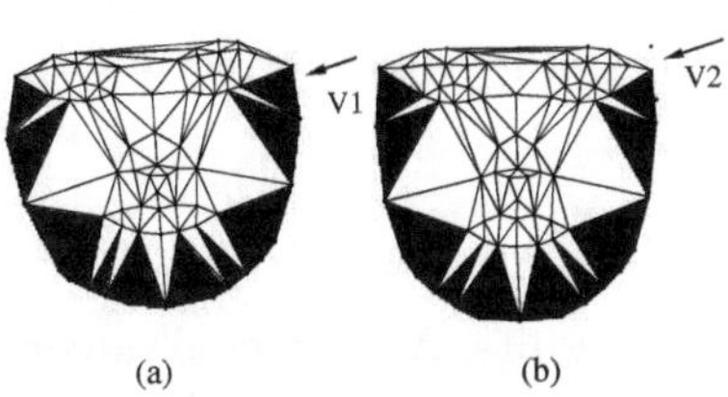

Fig. 3. Mesh definition. (a)Shape of the person in Fig. 1(a). (b)Base shape mesh.

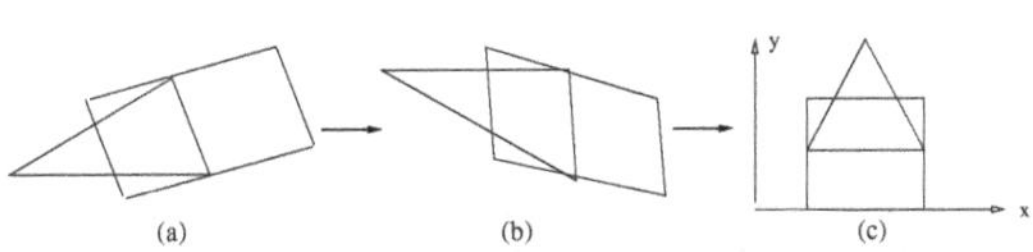

Fig. 4. Triangle-parallelogram pairs. a) Original image frame. b) Mean shape frame. c)Standard pair.

Triangles sitting on the face boundary form a special "face component". They are filled with black color. Their bottom sides form the face contour. Assume each black triangle is associated with a parallelogram with the bottom side of the triangle being the parallelogram's middle line. Our local projection models are built based on the analysis of the edge map inside these parallelograms. Fig. 4 illustrates how a triangle V_1 on the face image is transformed to V_2 on the base frame and subsequently to V_0 on a standard triangle which is an isosceles triangle. After these transformations, any projection along the face contour direction in the face image is now simplified to a summation along the x (or y) axis.

The piece-wise affine transform parameters between V_1 and V_2 are available in the basic AAM model fitting process. The transforms between V_0 and all the triangles of the base shape could be computed in advance. Clearly, with the help of the base shape and a standard triangle-parallelogram pair, local projection models can lock face contour points to the locally strongest edge points. It is much easier and robust compared to the ASM. The regions of interest for the local projection models are proportional to current face landmark points. Therefore there is no scaling problem at all.

5 Experiment Results and Discussion

Our face database includes 138 nearly frontal images from various different face databases[4][5][6]. All images were roughly resized and cropped to 256 by 256. We believe a blend face database is the best way to robust test. We sequentially picked 80 images to train the face shape subspace and rest of them as the test set.

We also tested on the Japanese Female Facial Expression database (JAFFE)[7], which contains 213 images of 7 facial expressions of 10 female models. The only pre-procession we conducted is to scale original 200 by 200 images to standard size of 256 by 256.

In an iterative realization, the global AAM is run first and when it fails to converge, local sub-models are launched, followed by the local projection models to lock the real face boundary. Search will stop when the stopping criterions are met.

To evaluate the fitting quality, we manually labelled all landmark points and created distance maps for all images. The model fitting quality is then measured by the average point to edge distance. Within the same framework, we tested and compared three different algorithms: the AAM search; the AAM with the component analysis (AAM_CA); the AAM with the component analysis and the local projection models (AAM_CA_LPM).

5.1 Component-Based AAM Search

Fig. 5 compares AAM and AAM_CA model fitting results. As expected, a converged global AAM usually couldn't achieve optimal local alignment. Better localization of the component feature points could be seen on the bottom row. Table 1 shows average point to edge errors for algorithms with and w/o the component analysis. Only face component points are considered.

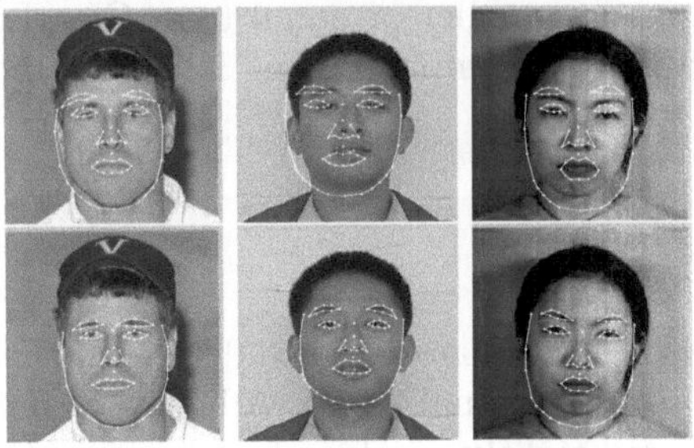

(a) (b) (c)

Fig. 5. AAM (top row) versa AAM_CA (bottom row). (a) Training set. (b) Test set. (c) JAFFE database.

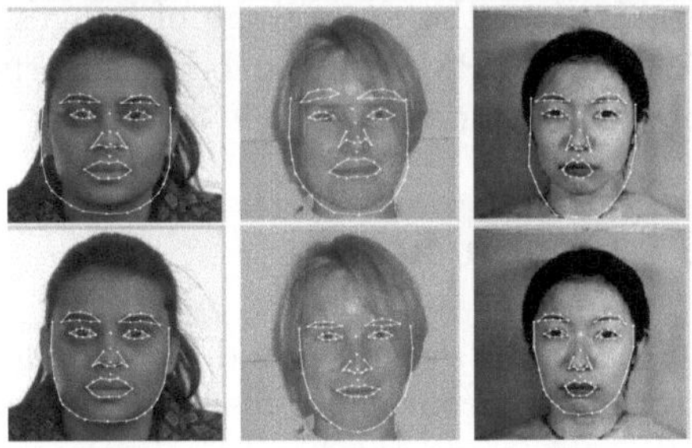

(a) (b) (c)

Fig. 6. AAM_CA (top row) versa AAM_CA_LPM (bottom row). (a) Training set. (b) Test set. (c) JAFFE database.

Table 1. Average error(contour excluded)

Algorithms	Training	Test	JAFFE
AAM	2.0661	3.5513	3.1696
AAM_CA	1.8988	3.2429	2.9377

Table 2. Average error(contour only)

Algorithms	Training	Test	JAFFE
AAM_CA	3.5298	4.7153	7.4741
AAM_CA_LPM	3.2909	3.8430	4.1356

5.2 Face Contour Detection with Local Projection Models

We compared AAM_CA and AAM_CA_LPM model fitting results to show how the integration of local projection models can help to solve the boundary problem. Fig. 6 shows some examples. Table 2 compares the average point to edge errors. It is interesting to see in Fig. 5(b), boundary points are correctly aligned due to the component analysis. Also Fig. 6(b) has correct component points. Apparently the integration of the local AAM analysis and local projection models makes our fitting algorithm more accurate and robust.

Convergent rate curves are compared for different algorithms in Fig. 7. A good approximation of an error density function can be obtained from the histogram of the resulted point errors for all images. Given a number ε in x-axis,

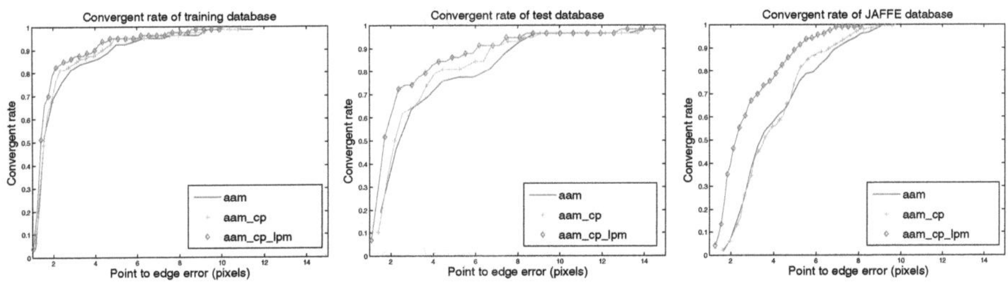

(a) (b) (c)

Fig. 7. Curves of convergent rate versa error threshold. (a) Training set. (b) Test set. (c) JAFFE database.

y-axis gives the percentage of images with errors smaller or equal to ε. Clearly AAM_CA_LPM has the best performance and the improvement is especially prominent for the JAFFE database.

6 Conclusion

In this paper, we proposed a component-based AAM algorithm to deal with the lack of the accuracy of feature localization in the original AAM. All component sub-models and the local projection models are tightly combined and smoothly interact with the global AAM model by sharing intermediate results. Robust and accurate face alignment makes it possible to extend the research to face recognition, 3D modelling etc. Extending our algorithm to images taken from different viewpoints is straightforward.

References

1. Cootes, T., Edwards, G., Taylor, C.: Active appearance models. IEEE Trans. PAMI **23** (2001) 681–685
2. Cootes, T., Taylor, C., Cooper, D., Graham, J.: Active shape models: Their training and application. CVGIP: Imaging Understanding **61** (1995) 38–59
3. Roberts, M., Cootes, T., Adams, J.: Linking sequences of active appearance sub-models via constraints: an application in automated vertebral morphometry. In: 14th British Machine Vision Conference. Volume 1. (2003) 349–358
4. Zhang, C., Cohen, F.: Face shape extraction and recognition using 3d morphing and distance mapping. In: 4th IEEE International Conference on Automatic Face and Gesture Recognition, Grenoble, France (2000)
5. Phillips, P., Moon, H., Rauss, P., Rizvi, S.: The feret evaluation methodology for face recognition algorithms. In: Proceedings of IEEE Computer Vision and Pattern Recognition. (1997) 137–143
6. Unknown: The phychological image collection at stirling. (http://pics.psych.stir.ac.uk/)
7. Lyons, M.J., Akamatsu, S., Kamachi, M., Gyoba, J.: Coding facial expressions with gabor wavelets. In: Third IEEE International Conference on Automatic Face and Gesture Recognition, Nara, Japan (1998) 200–205

Incorporating Image Quality in Multi-algorithm Fingerprint Verification

Julian Fierrez-Aguilar[1,*], Yi Chen[2], Javier Ortega-Garcia[1], and Anil K. Jain[2]

[1] ATVS, Escuela Politecnica Superior, Universidad Autonoma de Madrid, Avda. Francisco Tomas y Valiente, 11 Campus de Cantoblanco 28049 Madrid, Spain
{julian.fierrez, javier.ortega}@uam.es

[2] Department of Computer Science and Engineering, Michigan State University, East Lansing, MI 48823, USA
{chenyi1, jain}@cse.msu.edu

Abstract. The effect of image quality on the performance of fingerprint verification is studied. In particular, we investigate the performance of two fingerprint matchers based on minutiae and ridge information as well as their score-level combination under varying fingerprint image quality. The ridge-based system is found to be more robust to image quality degradation than the minutiae-based system. We exploit this fact by introducing an adaptive score fusion scheme based on automatic quality estimation in the spatial frequency domain. The proposed scheme leads to enhanced performance over a wide range of fingerprint image quality.

1 Introduction

The increasing need for reliable automated personal identification in the current networked society, and the recent advances in pattern recognition, have resulted in the current interest in *biometric systems* [1]. In particular, automatic fingerprint recognition [2] has received great attention because of the commonly accepted distinctiveness of the fingerprint pattern, the widespread deployment of electronic acquisition devices, and the wide variety of practical applications ranging from access control to forensic identification.

Our first objective in this work is to investigate the effects of varying image quality [3] on the performance of automatic fingerprint recognition systems. This is motivated by the results of the Fingerprint Verification Competition (FVC 2004) [4]. In this competition fingerprint images with lower image quality than those in FVC 2002 were used. As a result, the error rates of the best matching systems in FVC 2004 were found to be an order magnitude worse than those reported in earlier competitions (FVC 2000, FVC 2002). Similar effects have also been noticed in other recent comparative benchmark studies [5].

We also investigate the effects of varying image quality on a multi-algorithm approach [6] based on minutiae- and ridge-based matchers. These two matchers

* This work was carried out while J. F.-A. was a visiting researcher at Michigan State University.

D. Zhang and A.K. Jain (Eds.): ICB 2006, LNCS 3832, pp. 213–220, 2005.

provide complementary information commonly exploited by score-level fusion [7, 8]. Finally, we incorporate the idea of quality-based score fusion [9] into this multiple algorithm approach. In particular, an adaptive score-level fusion technique based on quality indices computed in the spatial frequency domain is presented and evaluated.

The paper is structured as follows. In Sect. 2 we summarize related work on the characterization of fingerprint image quality, and describe the fingerprint image quality measure used in this work. In Sect. 3 we summarize the individual fingerprint matching systems used here. The proposed quality-based score fusion scheme is introduced in Sect. 4. Database, experimental protocol, and results obtained are given in Sect. 5. Finally, conclusions are drawn in Sect. 6.

2 Assessment of Fingerprint Image Quality

Local image quality estimates have been traditionally used in the segmentation and enhancement steps of fingerprint recognition [10]. On the other hand, global quality measures have been traditionally used as indicators to identify invalid images. These indicators may result in failure to enroll or failure to acquire events that are handled either manually or automatically [2].

More recently, there is increasing interest in assessing the fingerprint image quality for a wide variety of applications. Some examples include: study of the effects of image quality on verification performance [3], comparison of different sensors based on the quality of the images generated [11], and comparison of commercial systems with respect to robustness to noisy images [5].

A number of fingerprint quality measures have been proposed in the literature. Most of them are based on operational procedures for computing local orientation coherence measures [12]. Some examples include: local Gabor-based filtering [10, 13], local and global spatial features [14], directional measures [15], classification-based approaches [16], and local measures based on intensity gradient [17]. In the present work we use the global quality index computed in the spatial frequency domain detailed in [17], which is summarized below.

2.1 Fingerprint Image Quality Index

Good quality fingerprint images bear a strong ring pattern in the power spectrum, indicating a dominant frequency band associated with the period of the ridges. Conversely, in poor quality images the ridges become unclear and non-uniformly spaced, resulting in a more diffused power spectrum. We thus assess the global quality of a fingerprint image by evaluating the energy distribution in the power spectrum.

A region of interest (ROI) in the power spectrum is defined to be a ring-shaped band with radius ranging from the minimum to the maximum observed frequency of ridges [17]. Fig. 1 shows three fingerprint images with increasing quality from left to right. Their corresponding power spectrums are shown in the second row. Note that the fingerprint image with good quality presents a strong

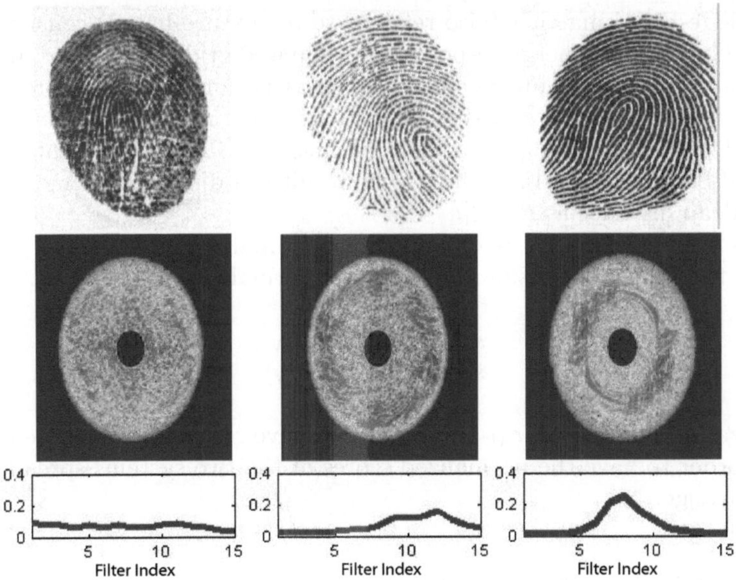

Fig. 1. Three sample fingerprint images with increasing image quality from left to right (top row), their corresponding power spectrum (middle row), and their energy distribution across concentric rings in the spatial frequency domain. It can be observed that the better the fingerprint quality, the more peaked is its energy distribution, indicating a more distinct dominant frequency band. The resulting quality measure for each fingerprint image from left to right is 0.05, 0.36, and 0.92, respectively.

ring pattern in the power spectrum (Fig. 1(c)), while a poor quality fingerprint presents a more diffused power spectrum (Fig. 1(a)). Multiple bandpass filters are designed to extract the energy in a number of ring-shaped concentric sectors in the power spectrum. The global quality index is defined in terms of the energy concentration across these sectors within the ROI.

In particular, bandpass filters are constructed by taking differences of two consecutive Butterworth functions [17]. In the third row of Fig. 1, we plot the distribution of the normalized energy across the bandpass filters. The energy distribution is more peaked as the image quality improves from (a) to (c). The resulting quality measure Q is based on the entropy of this distribution, which is normalized linearly to the range $[0, 1]$.

3 Fingerprint Matchers

We use both the minutia-based and the ridge-based fingerprint matchers developed at the Spanish ATVS/Biometrics Research Lab.

The minutiae-based matcher follows the approach presented in [18] with the modifications detailed in [3] and the references therein, resulting in a similarity measure based on dynamic programming.

The ridge-based matcher (also referred to as texture-based) consist of correlation of Gabor-filter energy responses in a squared grid as proposed in [19] with some modifications. No image enhancement is performed in the present work. Also, once the horizontal and vertical displacements maximizing the correlation are found, the original images are aligned and the Gabor-based features are re-computed before the final matching. The result is a dissimilarity measure based on Euclidean distance as in [19].

Scores from both matchers s'_M and s'_R are normalized into similarity matching scores in the range $[0, 1]$ using the following normalization functions:

$$\begin{aligned} s_M &= \tanh(s'_M / c_M) \\ s_R &= \exp(-s'_R / c_R) \end{aligned} \tag{1}$$

Normalization parameters c_M and c_R are positive real numbers chosen heuristically in order to have the normalized scores of the two systems spread out over the $[0, 1]$ range.

4 Quality-Based Score Fusion

The proposed quality-based multi-algorithm approach for fingerprint verification follows the system model depicted in Fig. 2.

The proposed method is based on the sum rule fusion approach. This basic fusion method consists of averaging the matching scores provided by the different matchers. Under some mild statistical assumptions [20, 21] and with the proper matching score normalization [22], this simple method is demonstrated to give good results for the biometric authentication problem. This fact is corroborated in a number of studies [21, 23]. Let the similarity scores s_M and s_R provided by the two matchers be already normalized to be comparable. The fused result using the sum rule is $s = (s_M + s_R)/2$.

Our basic assumption for the adaptive quality-based fusion approach is that verification performance of one of the algorithms drops significantly as compared to the other one under image quality degradation. This behavior is observed in

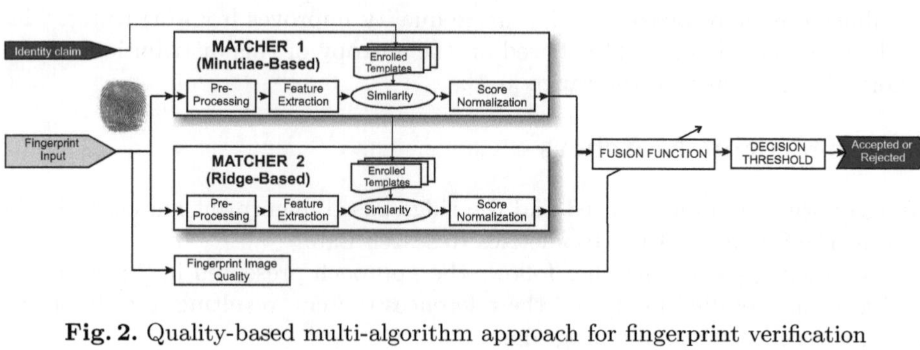

Fig. 2. Quality-based multi-algorithm approach for fingerprint verification

our minutia-based M with respect to our ridge-based R matcher. The proposed adaptive quality-based fusion strategy is as follows:

$$s_Q = \frac{Q}{2} s_M + (1 - \frac{Q}{2}) s_R, \qquad (2)$$

where Q is the input fingerprint image quality. As the image quality worsens, more importance is given to the matching score of the more robust system.

5 Experiments

5.1 Database and Experimental Protocol

We use a subcorpus of the MCYT Bimodal Biometric Database [24] for our study. Data consist of 7500 fingerprint images from all the 10 fingers of 75 subjects acquired with an optical sensor. We consider the different fingers as different users enrolled in the system, resulting in 750 users with 10 impressions per user. Some example images are shown in Fig. 1.

We use one impression per finger as template (with low control during the acquisition, see [24]). Genuine matchings are obtained comparing the template to the other 9 impressions available. Impostor matchings are obtained by comparing the template to one impression of all the other fingers. The total number of genuine and impostor matchings are therefore 750×9 and 750×749, respectively.

We further classify all the fingers in the database into five equal-sized quality groups, from I (low quality), to V (high quality), based on the quality measure Q described in Sect. 2, resulting in 150 fingers per group. Each quality group contains 150×9 genuine and 150×749 impostor matching scores.

Distribution of fingerprint quality indices and matching scores for the two systems considered are given in Fig. 3.

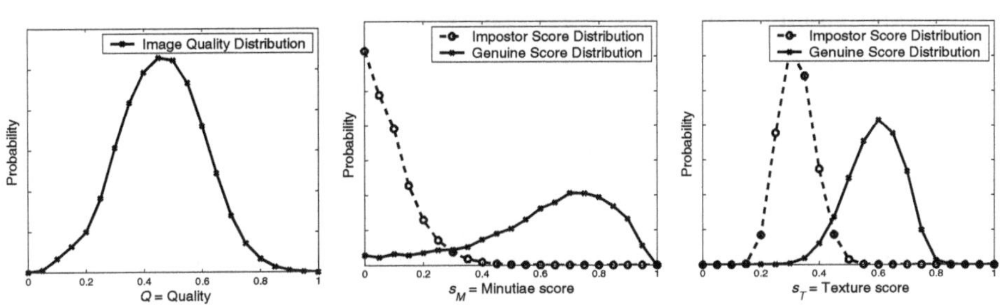

Fig. 3. Image quality distribution in the database (left) and matching score distributions for the minutiae (center) and texture matchers (right).

5.2 Results

Verification performance results are given in Fig. 4 for the individual matchers (minutiae- and texture-based), their combination through the sum fusion rule,

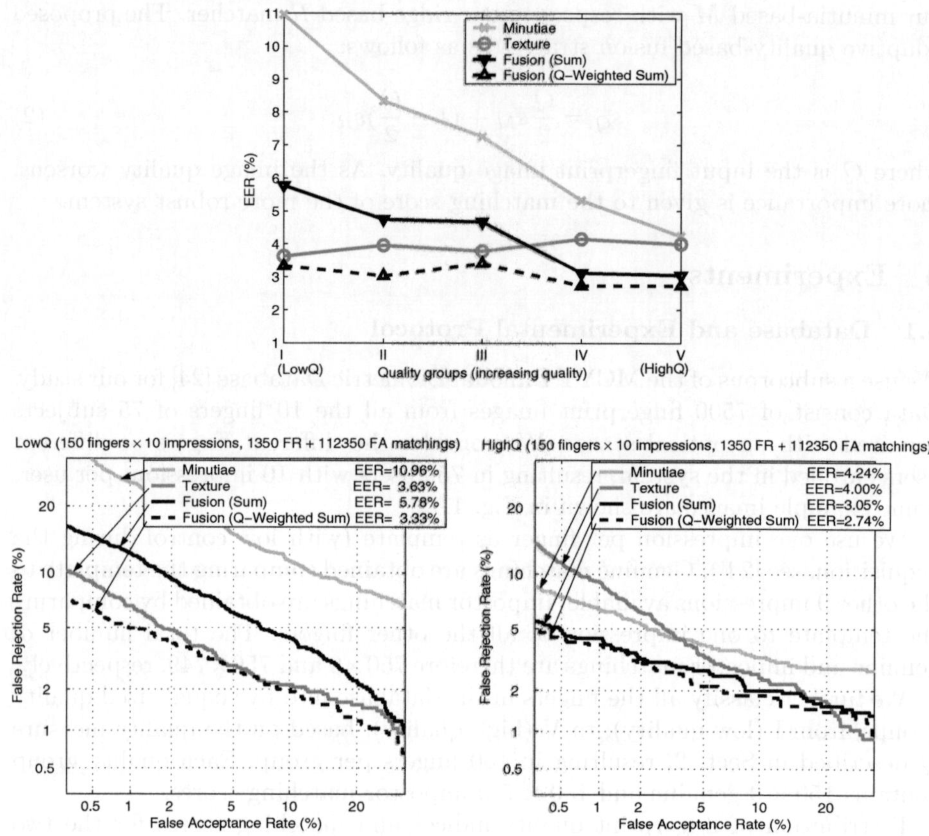

Fig. 4. Verification performance of the individual matchers (minutiae- and texture-based), their combination through the sum fusion fusion rule, and the proposed quality-based weighted sum for increasing image quality.

and the proposed quality-based weighted sum for different quality groups. We observe that the texture-based matcher is quite robust to image quality degradation. Conversely, the minutia-based matcher degrades rapidly with low quality images. As a result, the fixed fusion strategy based on the sum rule only leads to improved performance over the best individual system in medium to good quality images.

The proposed adaptive fusion approach results in improved performance for all the image quality groups, outperforming the standard sum rule approach, especially in low image quality conditions where the performance of individual matchers becomes more different.

Finally, in Fig. 5 we plot the verification performance for the whole database. Relative verification performance improvement of about 20% is obtained by the proposed adaptive fusion approach for a wide range of verification operating points as compared to the standard sum rule.

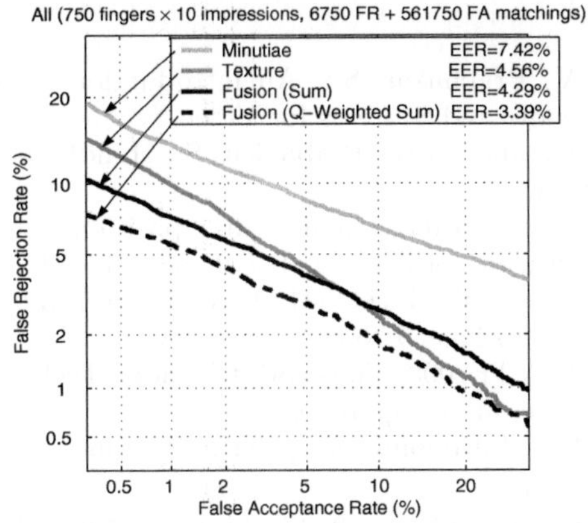

Fig. 5. Verification performance for the whole database

6 Discussion and Conclusions

The effects of image quality on the performance of two common approaches for fingerprint verification have been studied. It has been found that the approach based on ridge information outperforms the minutiae-based approach in low image quality conditions. Comparable performance is obtained on good quality images.

It must be emphasized that this evidence is based on particular implementations of well known algorithms, and should not be taken as a general statement. Other implementations may lead to improved performance of any approach over the other in varying image quality conditions. On the other hand, the robustness observed of the ridge-based approach as compared to the minutiae-based system has been observed in other studies. One example is the Fingerprint Verification Competition in 2004 [4], where low quality images where used and leading systems used some kind of ridge information [8].

This difference in robustness against varying image quality has been exploited by an adaptive score-level fusion approach using quality measures estimated in the spatial frequency domain. The proposed scheme leads to enhanced performance over the best matcher and the standard sum fusion rule over a wide range of fingerprint image quality.

Acknowledgements

This work has been supported by Spanish MCYT TIC2003-08382-C05-01 and by European Commission IST-2002-507634 Biosecure NoE projects. Authors also thank Luis-Miguel Muñoz-Serrano and Fernando Alonso-Fernandez for their valuable development work. J. F.-A. is supported by a FPI scholarship from Comunidad de Madrid.

References

1. Jain, A.K., Ross, A., Prabhakar, S.: An introduction to biometric recognition. IEEE Trans. on Circuits and Systems for Video Technology **14** (2004) 4–20
2. Maltoni, D., Maio, D., Jain, A.K., Prabhakar, S.: Handbook of Fingerprint Recognition. Springer (2003)
3. Simon-Zorita, D., et al.: Image quality and position variability assessment in minutiae-based fingerprint verification. IEE Proc. VISP **150** (2003) 402–408
4. Maio, D., Maltoni, D., et al.: FVC2004: Third Fingerprint Verification Competition. In: Proc. ICBA, Springer LNCS-3072 (2004) 1–7
5. Wilson, C., et al.: FpVTE2003: Fingerprint Vendor Technology Evaluation 2003 (NISTIR 7123) website: http://fpvte.nist.gov/.
6. Jain, A.K., Ross, A.: Multibiometric systems. Communications of the ACM **47** (2004) 34–40
7. Ross, A., Jain, A.K., Reisman, J.: A hybrid fingerprint matcher. Pattern Recognition **36** (2003) 1661–1673
8. Fierrez-Aguilar, J., et al.: Combining multiple matchers for fingerprint verification: A case study in FVC2004. In: Proc. ICIAP, Springer LNCS-3617 (2005) 1035–1042
9. Fierrez-Aguilar, J., Ortega-Garcia, J., et al.: Discriminative multimodal biometric authentication based on quality measures. Pattern Recognition **38** (2005) 777–779
10. Hong, L., Wang, Y., Jain, A.K.: Fingerprint image enhancement: algorithm and performance evaluation. IEEE Trans. on PAMI **20** (1998) 777–789
11. Yau, W.Y., Chen, T.P., Morguet, P.: Benchmarking of fingerprint sensors. In: Proc. BIOAW, Springer LNCS-3087 (2004) 89–99
12. Bigun, J., et al.: Multidimensional orientation estimation with applications to texture analysis and optical flow. IEEE Trans. on PAMI **13** (1991) 775–790
13. Shen, L., Kot, A., Koo, W.: Quality measures for fingerprint images. In: Proc. AVBPA, Springer LNCS-2091 (2001) 266–271
14. Lim, E., Jiang, X., Yau, W.: Fingerprint quality and validity analysis. In: Proc. ICIP (2002) 469–472
15. Ratha, N., Bolle, R., eds.: Automatic Fingerprint Recognition Systems. Springer (2004)
16. Tabassi, E., Wilson, C., Watson, C.: Fingerprint image quality (NIST Research Report NISTIR 7151, August 2004)
17. Chen, Y., Dass, S., Jain, A.: Fingerprint quality indices for predicting authentication performance. In: Proc. AVBPA, Springer LNCS-3546 (2005) 160-170
18. Jain, A.K., Hong, L., Pankanti, S., Bolle, R.: An identity authentication system using fingerprints. Proceedings of the IEEE **85** (1997) 1365–1388
19. Ross, A., Reisman, J., Jain, A.K.: Fingerprint matching using feature space correlation. In: Proc. BIOAW, Springer LNCS-2359 (2002) 48–57
20. Bigun, E.S., et al.: Expert conciliation for multimodal person authentication systems by Bayesian statistics. In: Proc. AVBPA, Springer LNCS-1206 (1997) 291–300
21. Kittler, J., Hatef, M., Duin, R., Matas, J.: On combining classifiers. IEEE Trans. on PAMI **20** (1998) 226–239
22. Jain, A., Nandakumar, K., Ross, A.: Score normalization in multimodal biometric systems. Pattern Recognition (2005) (to appear).
23. Ross, A., Jain, A.: Information fusion in biometrics. Pattern Recognition Letters **24** (2003) 2115–2125
24. Ortega-Garcia, J., Fierrez-Aguilar, J., et al.: MCYT baseline corpus: A bimodal biometric database. IEE Proc. VISP **150** (2003) 395–401

A New Approach to Fake Finger Detection Based on Skin Distortion*,**

A. Antonelli, R. Cappelli, Dario Maio, and Davide Maltoni

Biometric System Laboratory - DEIS,
University of Bologna, via Sacchi 3, 47023 Cesena - Italy
{athos, cappelli, maio, maltoni}@csr.unibo.it

Abstract. This work introduces a new approach for discriminating real fingers from fakes, based on the analysis of human skin elasticity. The user is required to move the finger once it touches the scanner surface, thus deliberately producing skin distortion. A multi-stage feature- extraction technique captures and processes the significant information from a sequence of frames acquired during the finger movement; this information is encoded as a sequence of *DistortionCodes* and further analyzed to determine the nature of the finger. The experimentation carried out on a database of real and fake fingers shows that the performance of the new approach is very promising.

1 Introduction

Thanks to the largely-accepted uniqueness of fingerprints and the availability of low-cost acquisition devices, fingerprint-based authentication systems are becoming more and more popular and are being deployed in several applications: from logon to PC, electronic commerce, ATMs, to physical access control for airports and border control [7]. On the other hand, as any other security system, fingerprint recognition is not totally spoof-proof; the main potential attacks can be classified as follows [1][4]: 1) attacking the communication channels, including replay attacks on the channel between the sensor and the rest of the system and other types of attacks; 2) attacking specific software modules (e.g. replacing the feature extractor or the matcher with a Trojan horse); 3) attacking the database of enrolled templates; 4) presenting fake fingers to the sensor. The feasibility of the last type of attack has been recently proved by some researchers [2][3]: current fingerprint recognition systems can be fooled with well-made fake fingers, created with the collaboration of the fingerprint owner or from latent fingerprints (in that case the procedure is more difficult but still possible). Some approaches recently proposed in the literature to address this problem can be found in [5] [6]. This work introduces a novel method for discriminating fake fingers from real ones based on the analysis of a peculiar characteristic of the human skin: the elasticity. Some preliminary studies showed that when a real finger moves on a scanner surface, it produces a significant amount of distortion, which is quite different from that produced by fake fingers. Usually fake fingers are more rigid than skin and

* This work was partially supported by European Commission (BioSec - FP6 IST-2002-001766).
** Patent pending (IT #BO2005A000399).

D. Zhang and A.K. Jain (Eds.): ICB 2006, LNCS 3832, pp. 221 – 228, 2005.

the deformation is lower and, even if made of highly elastic materials, it seems very difficult to precisely emulate the specific way a real finger is distorted, because is related to how the external skin is anchored to the underlying derma and influenced by the position and shape of the finger bone. The rest of this work is organized as follows: section 2 describes the proposed approach, section 3 reports the experimentation carried out to validate the new technique and section 4 draws some conclusions.

2 The Fake Finger Detection Approach

The user is required to place a finger onto the scanner surface and, once in touch with it, to apply some pressure while rotating the finger in a counter-clockwise direction (this particular movement has been chosen after some initial tests, as it seems comfortable for user and it produces the right amount of deformation). A sequence of frames is acquired at high frame rate (at least 20 fps) during the movement and analyzed to extract relevant features related to skin distortion. At the beginning of the sequence, the finger is assumed relaxed (i.e. non-distorted), without any superficial tension. A pre-processing stage is performed to simplify the subsequent steps; in particular:

- any frame such that the amount of rotation with respect to the previous one (inter-frame rotation) is less than θ_{min} ($\theta_{min} = 0.25°$ in our experimentation) is discarded (the inter-frame rotation angle is calculated as described in section 2.2);
- only frames acquired when the (accumulated) finger rotation is less than ϕ_{max} ($\phi_{max} = 15°$ in our experimentation) are retained: when angle ϕ_{max} is reached, the sequence is truncated (the rotation angle of the finger is calculated as described in section 2.5).

Let $\{F_1, F_2, ..., F_n\}$ be a sequence of *n* images that satisfies the above constraints; the following steps are performed on each frame F_i (figure 1):

- isolation of the fingerprint area from the background;
- computation of the optical flow between the current frame and the next one;
- computation of the distortion map;
- temporal integration of the distortion map;
- computation of the DistortionCode from the integrated distortion map.

For each image F_i, the isolation of the fingerprint area from the background is performed by computing the gradient of the image block-wise: let $\mathbf{p} = [x, y]^T$ be a generic pixel in the image and $F_i(\mathbf{p})$ a square image block (with side 12 in our tests) centred in $\mathbf{p}$: each $F_i(\mathbf{p})$ whose gradient module exceeds a given threshold is associated to the foreground. Only the foreground blocks are considered in the rest of the algorithm.

2.1 Computation of the Optical Flow

Block-wise correlation is computed to detect the new position $\mathbf{p}'$ of each block $F_i(\mathbf{p})$ in frame F_{i+1}. The vector $\Delta\mathbf{p}_i = \mathbf{p}' - \mathbf{p}$ denotes, for each block $F_i(\mathbf{p})$, the

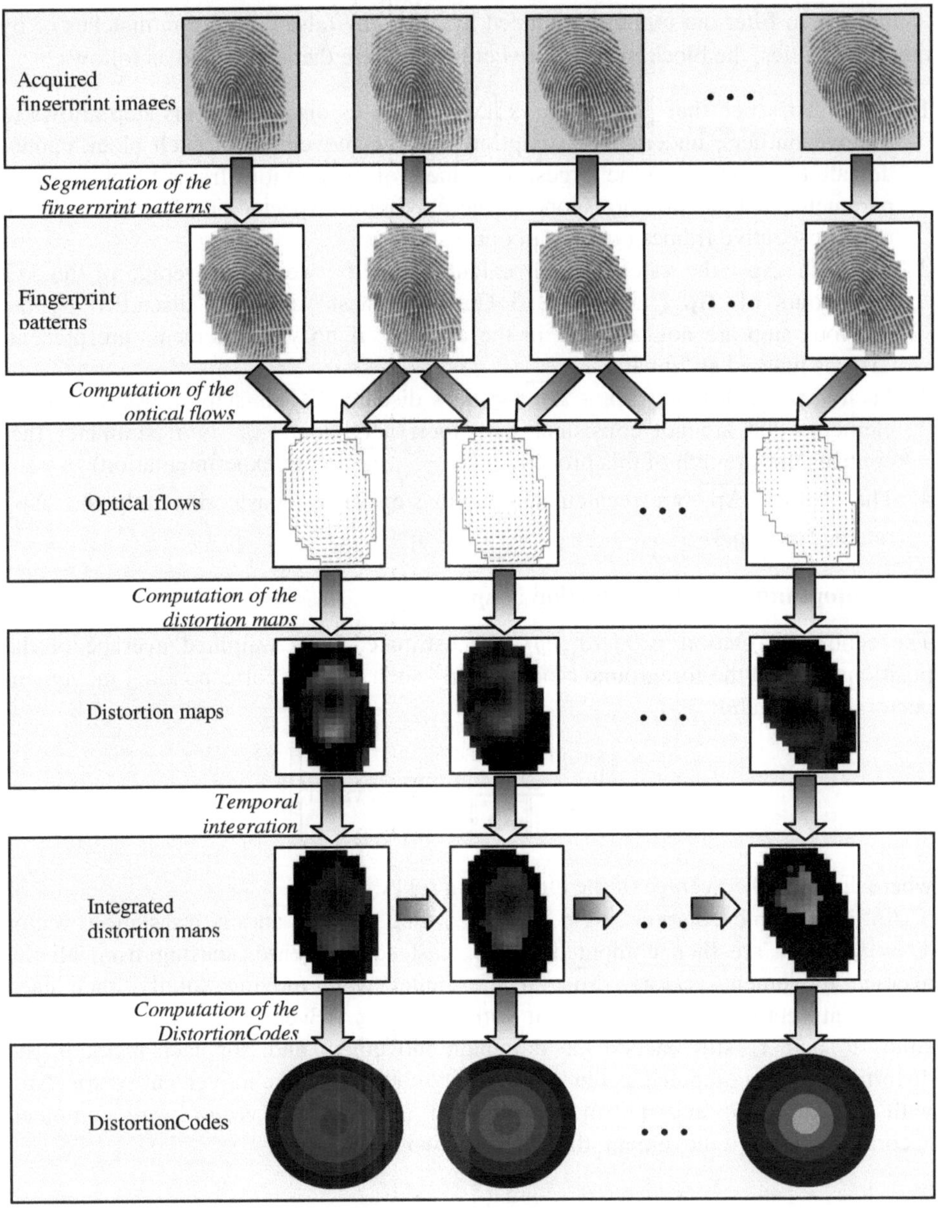

Fig. 1. The main steps of the feature extraction approach: a sequence of acquired fingerprint images is processed to obtain a sequence of DistortionCodes

estimated horizontal and vertical movements ($\Delta \mathbf{p}_i = [\Delta x, \Delta y]^T$); these movement vectors are known in the literature as *optical flow*. This method is in theory only translation-invariant but, since the images are taken at a fast frame rate, for small blocks it is possible to assume a certain rotation- and deformation-invariance.

In order to filter out outliers produced by noise, by false correlation matches or by other anomalies, the block movement vectors $\Delta\mathbf{p}_i$ are then processed as follows.

1. Each $\Delta\mathbf{p}_i$ such that $\|\Delta\mathbf{p}_i\| \geq \max_{\Delta\mathbf{p}_{i-1}} \|\Delta\mathbf{p}_{i-1}\| + \alpha$ is discarded. This step allows to remove outliers, under the assumption that the movement of each block cannot deviate too much from the largest movement of the previous frame blocks; α is a parameter that should correspond to the maximum expected acceleration between two consecutive frames ($\alpha = 3$ in our tests).
2. For each $\Delta\mathbf{p}_i$, the value $\overline{\Delta\mathbf{p}_i}$ is calculated as the weighted average of the 3x3 neighbours of $\Delta\mathbf{p}_i$, using a 3x3 Gaussian mask; elements discarded by the previous step are not included in the average: if no valid elements are present, $\overline{\Delta\mathbf{p}_i}$ is marked as "invalid".
3. Each $\Delta\mathbf{p}_i$ such that $\left\|\Delta\mathbf{p}_i - \overline{\Delta\mathbf{p}_i}\right\| \geq \beta$ is discarded. This step allows to remove elements that are not consistent with their neighbours; β is a parameter that controls the strength of this procedure ($\beta = \sqrt{3/2}$ in our experimentation).
4. The values $\overline{\Delta\mathbf{p}_i}$ are recalculated (as in step 2) by considering only the $\Delta\mathbf{p}_i$ retained at step 3.

2.2 Computation of the Distortion Map

The centre of rotation $\mathbf{c}_i = [cx_i, cy_i]^T$ is estimated as a weighted average of the positions $\mathbf{p}$ of all the foreground blocks $F_i(\mathbf{p})$ such that the corresponding movement vector $\overline{\Delta\mathbf{p}_i}$ is valid:

$$\mathbf{c}_i = \mathrm{E}\left[\left\{\mathbf{p}\frac{1}{1+\left\|\overline{\Delta\mathbf{p}_i}\right\|} \mid \overline{\Delta\mathbf{p}_i} \text{ is valid}\right\}\right], \tag{1}$$

where $\mathrm{E}[\mathrm{A}]$ is the average of the elements in set A.

An inter-frame rotation angle θ_i (according to $\mathbf{c}_i$) and a translation vector $\mathbf{t}_i = [tx_i, ty_i]^T$ are then computed in the least square sense, starting from all the average movement vectors $\overline{\Delta\mathbf{p}_i}$. If the finger were moving solidly, then each movement vector would be coherent with θ_i and $\mathbf{t}_i$. Even if the movement is not solid, θ_i and $\mathbf{t}_i$ still encode the dominant movement and, for each block $\mathbf{p}$, the distortion can be computed as the incoherence of each average movement vector $\overline{\Delta\mathbf{p}_i}$ with respect to θ_i and $\mathbf{t}_i$. In particular, if a movement vector were computed according to a solid movement, then its value would be:

$$\widetilde{\Delta\mathbf{p}_i} = \begin{bmatrix} \cos\theta_i & \sin\theta_i \\ -\sin\theta_i & \cos\theta_i \end{bmatrix} (\mathbf{p}_i - \mathbf{c}_i) + \mathbf{c}_i + \mathbf{t}_i - \mathbf{p}_i \tag{2}$$

and therefore the distortion can be defined as the residual:

$$D_i(\mathbf{p}) = \begin{cases} \left\|\widetilde{\Delta\mathbf{p}_i} - \overline{\Delta\mathbf{p}_i}\right\| & \text{if } \overline{\Delta\mathbf{p}_i} \text{ is valid} \\ \mathit{undefined} & \text{otherwise} \end{cases} \tag{3}$$

A *distortion map* is defined as a block-wise image whose blocks encode the distortion values $D_i(\mathbf{p})$.

2.3 Temporal Integration of the Distortion Map

The computation of the distortion map, made on just two consecutive frames, is affected by the following problems:

- the movement vectors are discrete (because of the discrete nature of the images) and in case of small movement the loss of accuracy might be significant;
- errors in seeking the new position of blocks could lead to a wrong distortion estimation;
- the measured distortion is proportional to the amount of movement between the two frames (and therefore depend on the finger speed), without considering previous tension accumulated/released. This makes difficult to compare a distortion map against the distortion map in another sequence.

An effective solution to the above problems is to perform a temporal-integration of the distortion map, resulting into an *integrated distortion map*. The temporal integration is simply obtained by block-wise summing the current distortion map to the distortion map "accumulated" in the previous frames. Each integrated distortion element is defined as:

$$TID_i(\mathbf{p}) = \begin{cases} TID_{i-1}(\mathbf{p}) + D_i(\mathbf{p}) & \text{if } \left\|\widetilde{\Delta\mathbf{p}_i}\right\| > \left\|\overline{\Delta\mathbf{p}_i}\right\| \text{ and } \overline{\Delta\mathbf{p}_i} \text{ is valid} \\ TID_{i-1}(\mathbf{p}) & \text{if } \overline{\Delta\mathbf{p}_i} \text{ is invalid} \\ 0 & \text{if } \left\|\widetilde{\Delta\mathbf{p}_i}\right\| \leq \left\|\overline{\Delta\mathbf{p}_i}\right\| \end{cases} \quad (4)$$

with $TID_0(\mathbf{p}) = 0$.

The rationale behind the above definition is that if the norm of the average movement vector $\overline{\Delta\mathbf{p}_i}$ is smaller than the norm of the estimated solid movement $\widetilde{\Delta\mathbf{p}_i}$, then the block is moving slower than expected and this means it is accumulating tension. Otherwise, if the norm of $\overline{\Delta\mathbf{p}_i}$ is larger than the norm of $\widetilde{\Delta\mathbf{p}_i}$, the block is moving faster than expected, thus it is slipping on the sensor surface, releasing the tension accumulated.

The integrated distortion map solves most of the previously listed problems: i) discretization and local estimation errors are no longer serious problems because the integration tends to produce smoothed values; ii) for a given movement trajectory, the integrated distortion map is quite invariant with respect to the finger speed.

2.4 The Distortion Code

Comparing two sequences of integrated distortion maps, both acquired under the same movement trajectory, is the basis of our fake finger detection approach. On the other hand, directly comparing two sequences of integrated distortion maps would be computationally very demanding and it would be quite difficult to deal with the unavoidable local changes between the sequences.

To simplify handling the sequences, a feature vector (called *DistortionCode* for the analogy with the FingerCode introduced in [9]) is extracted from each integrated distortion map: *m* circular annuli of increasing radius ($r \cdot j$, $j = 1..m$, where *r* is the radius of the smaller annulus) are centred in $\mathbf{c}_i$ and superimposed to the map (*r*=20 and *m*=5 in our experimentation). For each annulus, a feature d_{ij} is computed as the average of the integrated distortion elements of the blocks falling inside it:

$$d_{ij} = \mathrm{E}\left[\{TID_i(\mathbf{p}) \mid \mathbf{p} \text{ belongs to annulus } j\}\right] \tag{5}$$

A DistortionCode $\mathbf{d}_i$ is obtained from each frame F_i, *i*=1..*n*-1:

$$\mathbf{d}_i = \left[d_{i1}, d_{i2}, \ldots, d_{im}\right]^T$$

A *DistortionCode sequence V* is then defined as:

$$V = \{\mathbf{v}_1, \mathbf{v}_2, \ldots, \mathbf{v}_{n-1}\}, \text{ where } \mathbf{v}_k = \mathbf{d}_k \Big/ \sqrt{\sum_{i=1..n-1} \|\mathbf{d}_i\|^2} \tag{6}$$

The obtained DistortionCode sequence characterizes the deformation of a particular finger under a specific movement. Further sequences from the same finger do not necessarily lead to the same DistortionCode sequence: the overall length might be different, because the user could produce the same trajectory (or a similar trajectory) faster or slower. While a minor rotation accumulates less tension, during a major rotation the finger could slip and the tension be released in the middle of the sequence.

2.5 The Distortion Match Function

In order to discriminate a real from a fake finger, the DistortionCode sequence acquired during the enrolment and associated to a given user is compared with the DistortionCode sequence acquired at verification/identification time. Let $V_T = \{\mathbf{v}_{T,1}, \mathbf{v}_{T,2}, \ldots \mathbf{v}_{T,n_T}\}$ and $V_C = \{\mathbf{v}_{C,1}, \mathbf{v}_{C,2}, \ldots \mathbf{v}_{C,n_C}\}$ be the sequence acquired during the enrolment (template sequence) and the new one (current sequence), respectively; a *Distortion Match Function* $DMF(V_T, V_C)$ compares the template and current sequence and returns a score in the range [0..1], indicating how much the current sequence is similar to the template (1 means maximum similarity).

A Distortion Match Function must define how to: 1) calculate the similarity between two DistortionCodes, 2) align the DistortionCodes by establishing a correspondence between the DistortionCodes in the two sequences V_T and V_C, and finally 3) measure the similarity between the two aligned sequences.

A simple Euclidean distance between two DistortionCodes has been adopted as to comparison metric (step 1). As to step 2), DistortionCodes are aligned according to the accumulated rotation angles ϕ_i ($\phi_i = \sum_{k=1..i} \theta_k$, where θ_i is the inter-frame rotation angle between the frames *i* and *i*+1); re-sampling through interpolation is performed to deal with discretization; the result of step 2) is a new DistortionCode sequence $\tilde{V}_T = \{\tilde{\mathbf{v}}_{T,1}, \tilde{\mathbf{v}}_{T,2}, \ldots, \tilde{\mathbf{v}}_{T,n_C}\}$, obtained from V_T after the alignment with V_C; $\tilde{V}_T$ has the same cardinality of V_C. The final similarity can be simply computed (step 3) as the average Euclidean distance of corresponding DistortionCodes in $\tilde{V}_T$ and V_C:

$$DMF\left(V_T,V_C\right)=1-\sum_{i=1..n_C}\left\|\mathbf{v}_{C,i}-\tilde{\mathbf{v}}_{T,i}\right\| \Big/ \sqrt{m}\cdot n_C \qquad (7)$$

The normalization coefficient ($\sqrt{m}\cdot n_C$) ensures that the score is always in the range [0..1].

3 Experimental Results

A fingerprint scanner that embeds a fake-finger detection mechanism has to decide, for each transaction, if the current sample comes from a real finger or from a fake one. This decision will be unavoidably affected by errors: a scanner could reject real fingers and/or accept fake fingers.

Let FAR_{fd} be the proportion of transactions with a fake finger that are incorrectly accepted and let FRR_{fd} be the proportion of transactions with a real finger that are incorrectly rejected. In the following, the EER_{fd} (that is the value such that FRR_{fd} = FAR_{fd}) will be reported as a performance indicator. Note that FAR_{fd} and FRR_{fd} do not include verification/identification errors and must be combined with them to characterize the overall system errors.

In order to evaluate the proposed approach, a database of image sequences was collected. The database was acquired in the Biometric System Laboratory of the University of Bologna from 20 volunteers. Two fingers (thumb and forefinger of the right hand) were collected from each volunteer and two additional fingers (thumb and forefinger of the left hand) were collected from six of them; five image sequences were recorded for each finger. 12 fake fingers were manufacted (four made of RTV silicone, four of gelatine and four of latex) starting from fingers of three cooperating volunteers; five image sequences were recorded for each of them. The image sequences were acquired using the optical fingerprint scanner “TouchView II” by Identix, which produces 420×360 fingerprint images at 500 DPI. A Matrox Meteor frame grabber was used to acquire frames at 30 fps). The database was divided into two disjoint sets: a *validation set* (12 real fingers and 6 fake fingers) used for tuning the various parameters of the approach and a *test set* (40 real fingers and 6 fake fingers), used to measure the performance. The following transactions were performed on the test set:

- 400 genuine attempts (each sequence was matched against the remaining sequences of the same finger, excluding the symmetric matches to avoid correlation, thus performing 10 attempts for each of the 40 real fingers);
- 1200 impostor attempts (each of the 30 fake sequences was matched against the first sequence of each real finger). Note that, since only fake-detection performance was evaluated (not combined with identity verification) and considering that the proposed approach is based only on the elastic properties of real/fake fingers, it is

not necessary that a fake finger corresponding to the real finger is used in the impostor attempts: any fake finger can be matched against any real finger without adding any bias to the results.

The EER_{fd} of the proposed approach measured in the above described experimentation was 4.9%.

4 Conclusions and Future Work

We believe the results obtained are very promising: the method achieved a reasonable EER_{fd} (4.9%), proved to be very efficient (on a Pentium IV at 3.2Ghz, the average processing and matching time is less than eight ms) and not too annoying for the user (the whole fake-detection process, including the acquisition of the fingerprint sequence, takes about two seconds). The proposed approach has also the advantage of being software-based (i.e. no additional hardware is required to detect the fake fingers: the only requirement for the scanner is the capability of delivering frames at a proper rate). We are currently acquiring a larger database to perform additional experiments and investigating other alignment techniques for the DistortionCode sequences.

References

[1] N.K. Ratha, J.H. Connell, and R.M. Bolle, "An analysis of minutiae matching strength", Proc. AVBPA 2001, Third International Conference on Audio- and Video-Based Biometric Person Authentication, pp. 223-228, 2001.

[2] T Matsumoto, H. Matsumoto, K. Yamada, S. Hoshino, "Impact of Artificial 'Gummy' Fingers on Fingerprint Systems", Proceedings of SPIE, vol. 4677, January, 2002.

[3] T. Putte and J. Keuning, "Biometrical fingerprint recognition: don't get your fingers burned", Proc. IFIP TC8/WG8.8, pp. 289-303, 2000.

[4] Umut Uludag and Anil K. Jain, "Attacks on biometric systems: a case study in fingerprints", Proceedings of SPIE – v. 5306, Security, Steganography, and Watermarking of Multimedia Contents VI, June 2004, pp. 622-633.

[5] R. Derakhshani, S.A.C. Schuckers, L.A. Hornak, and L.O. Gorman, "Determination of vitality from a non-invasive biomedical measurement for use in fingerprint scanners", Pattern Recognition, vol. 36, pp. 383-396, 2003.

[6] PD Lapsley, JA Less, DF Pare, Jr., N Hoffman, "Anti-Fraud Biometric Sensor that Accurately Detects Blood Flow", SmartTouch, LLC, US Patent #5,737,439.

[7] D. Maltoni, D. Maio, A.K. Jain, and S. Prabhakar, Handbook of Fingerprint Recognition, Springer, 2003.

[8] R. Cappelli, D. Maio and D. Maltoni, "Modelling Plastic Distortion in Fingerprint Images", in proceedings 2nd International Conference on Advances in Pattern Recognition (ICAPR2001), Rio de Janeiro, March 2001, pp.369-376.

[9] A. K. Jain, S. Prabhakar and L. Hong, "A Multichannel Approach to Fingerprint Classification", in IEEE Transactions on Pattern Analysis and Machine Intelligence, vol. 21 no. 4, April 1999, pp. 348-359.

Model-Based Quality Estimation of Fingerprint Images

Sanghoon Lee, Chulhan Lee, and Jaihie Kim

Biometrics Engineering Research Center(BERC),
Department of Electrical and Electronic Engineering, Yonsei University, Seoul, Korea
{hoony, devices, jhkim}@yonsei.ac.kr

Abstract. Most automatic fingerprint identification systems identify a person using minutiae. However, minutiae depend almost entirely on the quality of the fingerprint images that are captured. Therefore, it is important that the matching step uses only reliable minutiae. The quality estimation algorithm deduces the availability of the extracted minutiae and allows for a matching step that will use only reliable minutiae. We propose a model-based quality estimation of fingerprint images. We assume that the ideal structure of a fingerprint image takes the shape of a sinusoidal wave consisting of ridges and valleys. To determine the quality of a fingerprint image, the similarity between the sinusoidal wave and the input fingerprint image is measured. The proposed method uses the 1-dimensional (1D) probability density function (PDF) obtained by projecting the 2-dimensional (2D) gradient vectors of the ridges and valleys in the orthogonal direction to the local ridge orientation. Quality measurement is then caculated as the similarity between the 1D probability density functions of the sinusoidal wave and the input fingerprint image. In our experiments, we compared the proposed method and other conventional methods using FVC-2002 DB I, III procedures. The performance of verification and the separability between good and bad regions were tested.

1 Introduction

The performance of any fingerprint recognition system is very sensitive to the quality of the acquired fingerprint images. There are three factors that lead to poor quality fingerprint images: 1) Physical skin injuries: scratches, broken ridges, and abrasions, 2) Circumstantial influences: wet or dry levels of humidity and dirty fingers, 3) Inconsistent contact: excessive or weak pressure. There are many previous works that deal with estimating the quality of fingerprint images. Hong et al. [1] modeled the ridge and valley pattern as a sinusoidal wave, and calculated amplitude, frequency and variance to determine the quality of fingerprint images. Michael [2] computed the mean and the variance of a sub-block of fingerprint images to measure the quality. Neither method was able to distinctly classify good regions and bad regions within the images. Bolle et al. [3] proposed a method that used the ratio of the directional region to the non-directional region. However, a limitation of this method is that the gray-level ridge and valley

D. Zhang and A.K. Jain (Eds.): ICB 2006, LNCS 3832, pp. 229–235, 2005.

structure of fingerprint images contains much more information. Shen, et al. [4] used the variance of the 8-directional Gabor filter response. The performance of this method depends on the number of Gabor filters, and the computational complexity is high. Ratha and Bolle [5] proposed a method for image quality estimation in the wavelet domain, which is suitable for WSQ-compressed fingerprint images. But it is unsuitable when dealing with uncompressed fingerprint images. Lim [6] observed both global uniformity and local texture patterns in fingerprint images. However, it is necessary to determine the weights for global and local quality measurements when using this method. In this paper, we propose model-based quality estimation of fingerprint images. The structure of an ideal fingerprint image takes the shape of a sinusoidal wave. To determine the quality of each sub-block image, we measure the similarity between the ideal fingerprint structure (sinusoidal wave) and the input fingerprint structure. In the following sections, we will explain model-based quality estimation of fingerprint images. Section 2 addresses the main steps of our algorithm and the method used to measure the similarity between the ideal fingerprint structure and the input fingerprint image. In section 3, the proposed method is compared to previous methods using the separability between good and bad regions and the performance of fingerprint verification. Section 4 shows the conclusions we arrived at in the course of our experiments.

2 Model-Based Quality Estimation

Fingerprint quality estimation divides a pixel (or a block) in an input fingerprint image into good regions and bad regions. Good regions are the regions where minutiae can be detected. Bad regions are the regions where minutiae cannot be detected or false minutiae are more prominent. The ideal fingerprint region can be shown by a mono-dimensional sinusoidal wave and the obscure region is represented by an arbitrary wave. The main idea of our proposed method is to measure the similarity of the structures between the sinusoidal wave and the input fingerprint image. This method is inspired by independent component analysis (ICA) that extracts a 1-dimensional independent signal from n-dimensional mixture signals [7]. Fig. 1 shows the overall procedure of our proposed method schematically.

2.1 Preprocessing

The preprocessing stage that is composed of normalization and Gaussian masking. We used normalization and Gaussian smoothing to remove the effects of sensor noise and finger pressure difference.

2.2 2D-Gradient Vectors

2D-gradient vectors of fingerprint images are obtained by gradient operators. Depending on computational requirements, either the Prewitt operator, the Sobel

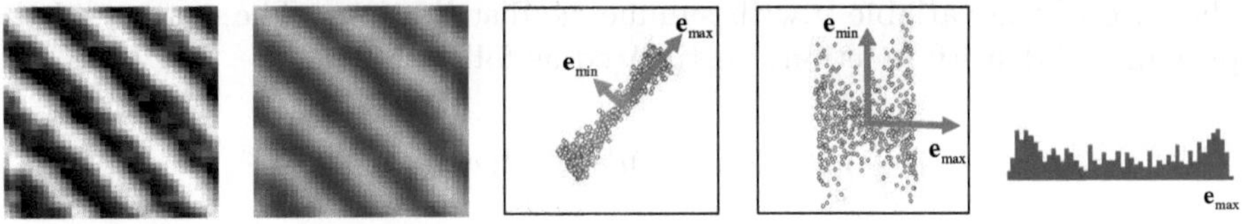

(a) (b) (c) (d) (e)

Fig. 1. Quality measurement block diagram: (a) Sub-block fingerprint image; (b) Pre-processing; (c) 2D-Gradient vectors; (d) Whitening; (e) 1D-Gradient PDF

operator, or the Marr-Hildreth operator [8] is chosen. In this paper, we used the Sobel operator. Fig. 1(c) shows the 2-channel gradient of a sub-block fingerprint image.

2.3 Whitening

Fig. 1(c) shows the 2D-gradient vectors of a sub-block fingerprint image. The 2D-gradient vector mixes up the orthogonal and parallel differential information to the ridge orientation. Because only the orthogonal differential information to the ridge is required to acquire the 1D-gradient PDF in order to estimate the quality of a sub-block of the fingerprint image, the mixed 2D-gradient vector must be separated. Fig. 1(d) indicates the whitened gradient vector that is rotated to align the horizontal axis ($\mathbf{e}_{\max}$) in the orthogonal direction of the ridge orientation. The whitening process separates the mixed 2D-gradient vector into two 1D-gradient vectors: the gradient vector G_v with only orthogonal differential information to the ridge orientation, and the gradient vector G_h with only parallel differential information to the ridge orientation. Since we have separated the mixed 2D-gradient vector, we can obtain the 1D-gradient PDF(Fig. 1(d)) by projecting the whitened gradient vector G_v to the $\mathbf{e}_{\max}$ axis.

2.4 Quality Measurement

In order to estimate the quality of the fingerprint image, we assume that the ideal structure of ridges and valleys shows a sinusoidal wave. At each sub-block of images, the 1D probability density function (PDF) is obtained by projecting the whitened 2D-gradient vectors in the orthogonal direction to the local ridge orientation. With finite samples, polynomial density expansion like Taylor expansion is used to estimate a PDF. However, two other expansions are usually used for PDF estimation: the Gram-Charlier expansion and the Edgeworth expansion. In this paper, we use the Gram-Charlier expansion with Chebyshev-Hermit polynomials to estimate the 1D-gradient PDF p_v as follows:

$$p_v(\xi) \approx \hat{p}_v(\xi) = \varphi(\xi)\{1 + \kappa_3(v)\frac{\mathrm{H}_3(\xi)}{3!} + \kappa_4(v)\frac{\mathrm{H}_4(\xi)}{4!}\}, \qquad (1)$$

where κ_3 and κ_4 are skewness and kurtosis, $\mathrm{H_i}$ represents the Chebyshev-Hermit polynomials of order i, and $\varphi(\xi)$ is the standardized Gaussian density. κ_3 is zero in the case of the variable v with symmetric distributions. The entropy of the approximated density function is estimated as follows:

$$H(v) \approx -\int \hat{p}_v(\xi) \log \hat{p}_v(\xi) d\xi = H(v_{gauss}) - \frac{\kappa_4^2(v)}{48}, \tag{2}$$

where v_{gauss} is the Gaussian variable of the zero mean and unit variance. The following equation is explicitly derived:

$$J(v) = H(v_{gauss}) - H(v) \propto \kappa_4^2(v), \tag{3}$$

where $J(v)$ is negentropy [7]. The 1D-gradient PDF of the ideal fingerprint region is sub-Gaussian and negentropy has a large value when the distribution of v is sub-Gaussian. Therefore we may define the quality measurement as follows:

$$Quality = \kappa_4^2(v) \approx J(v) \tag{4}$$

However, $J(v)$ also has a large value when the distribution of v is super-Gaussian. Because the 1D-gradient PDF of a dry or wet fingerprint region is super-Gaussian, the quality measurement must discriminate between images that are sub-Gaussian and super-Gaussian. Therefore, the quality measurement defined in equation (6) must be adjusted as follows:

$$Quality = sign(\kappa_4(v))\kappa_4^2(v) \tag{5}$$

Because expectations of polynomials like the fourth power ($\kappa_4(v) = E\{v^4\} - 3$) are much more strongly affected by data far from zero than by data close to zero, approximation kurtosis by a non-polynomial function G is used[7]:

$$\begin{aligned} &\kappa_4(v) = \mathrm{E}\{\mathrm{G}(v)\} - \mathrm{E}\{\mathrm{G}(v_{\mathrm{gauss}})\} \\ &\mathrm{G}(v) = \tfrac{1}{\mathrm{a}}\log(\cosh(\mathrm{a}v)),\ 1 \le \mathrm{a} \le 2 \end{aligned} \tag{6}$$

3 Experimental Results

The quality value procedure assigned validity to each 8x8 block and quantized 256 levels (with 255 the highest quality and 0 the lowest). Fig. 2(a) is a sample fingerprint image that includes a region of interest (ridges and valleys) and a background region. The block-wise quality value for the fingerprint image in Fig. 2(a) is shown in Fig. 2(b).

3.1 Separability of Quality Measurement: Separability Between High and Poor Quality Regions

We evaluated the proposed quality measurement using separability between values from good and bad regions. We first defined the quality of the sub-block by including minutiae as good and bad regions. The good regions are the sub-blocks around

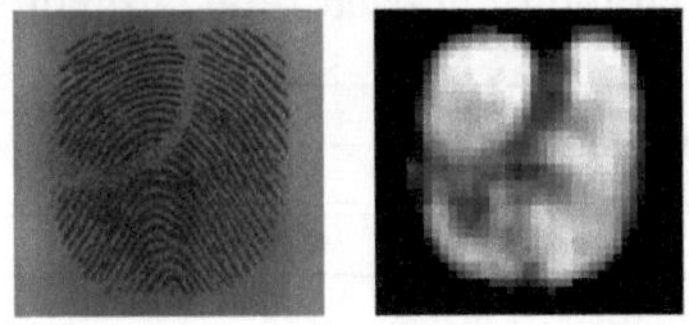

(a) (b)

Fig. 2. Quantized quality value: (a) Original image; (b) Block-wise quality value

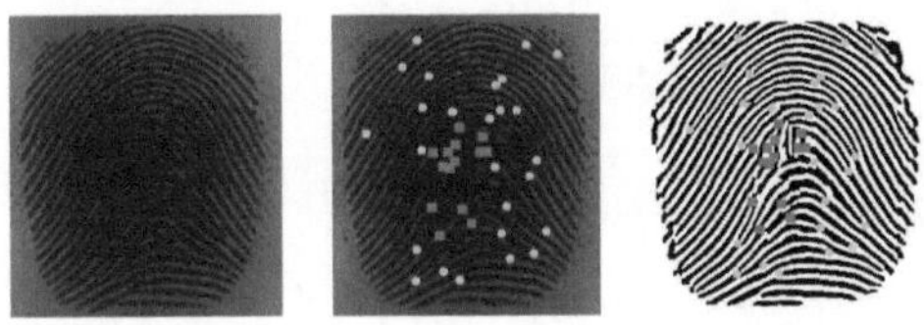

(a) (b) (c)

Fig. 3. Minutiae points of manually-defined quality (false minutiae: red rectangles, true minutiae: blue circles): (a) Original image; (b) Enhanced binary image; (c) Marked Region

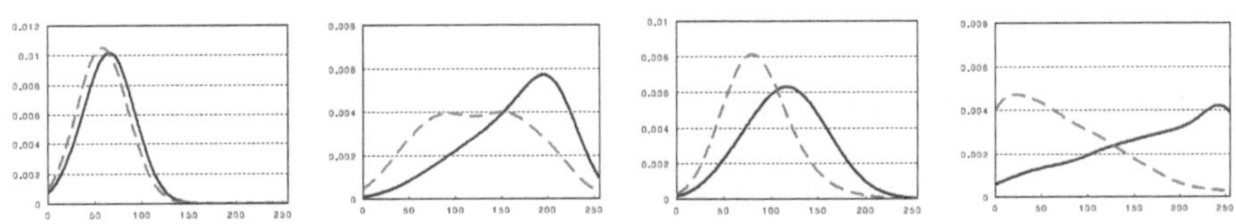

(a) (b) (c) (d)

Fig. 4. Probability density function of each type of quality measurement (good region: solid line, bad region: dotted line): (a)Standard deviation; (b)Coherence; (c)Gabor; (d)The proposed method

the true minutiae and the bad regions are the sub-regions around the false minutiae. True minutiae are determined if the minutiae extracted by the feature extraction algorithm are equal to the manually extracted minutiae, and if the minutiae are not equal, we determined the minutiae as false minutiae. The proposed quality definition method is more objective than the visual (subjective) assessments method. Fig. 3 shows the true and false minutiae. With 100 randomly selected fingerprint images that were separated into good and bad regions, we calculated the probability distribution of each corresponding quality measurement. Fig. 4 shows the distribution of four quality measurements and Table 1 shows the separability of each distribution using FVC2002 DB I, III. These clearly show that the distribution when using the proposed method is more separable than when using existing methods. The separability is calculated as follows:

$$\text{Separability} = |\mu_{\text{Good}} - \mu_{\text{Bad}}| \Big/ \sqrt{(\sigma^2_{\text{Good}} + \sigma^2_{\text{Bad}})} \quad (7)$$

Table 1. The separability of each type of quality measurement

Quality Measurement	Separability	
	DB I	DB III
Standard deviation	0.19	0.05
Coherence	0.64	0.88
Gabor filter	0.61	0.44
Proposed method	1.48	1.55

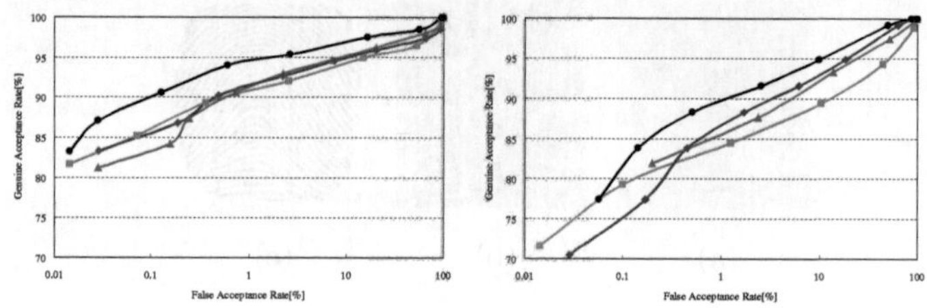

(a) (b)

Fig. 5. Receiver Operating Curves (s.d. : rectangle, coherence : diamond, gabor : triangle, proposed method : circle) : (a) FVC 2002 DB I; (b) FVC 2002 DB III

3.2 Verification Performance

We examined verification performance according to the quality methods. The verification system used the same algorithms (preprocessing, frequency estimation [10], enhancement [1] and matching [11]) with the exception of the quality estimation algorithm. The thresholds for each quality estimation algorithm were chosen at the point of minimum quality decision error using a Bayesian decision. In the experiment, we compared the proposed method and other conventional methods using FVC-2002 DB I, III. Fig. 5 shows the matching results with the ROC in order to compare the proposed algorithm with existing algorithms. From this experiment, we can observe that performance of the fingerprint verification system was significantly improved when our quality estimation algorithm was applied to the input fingerprint images.

4 Conclusions

In this paper, we proposed a method to determine the quality of a fingerprint image with similarity between the ideal fingerprint model and an estimated 1D-PDF. The ideal fingerprint image model has a monodimensional sinusoidal wave and uses a sub-Gaussian PDF when the project whitened 2D-gradient moves in the orthogonal direction of orientation of the sub-block. Quality estimation uses separability between high and poor quality regions and takes into account the performance of fingerprint verification. We compared the separability of each quality estimation method and the proposed method observed the highest separability using FVC-2002 DB I, III procedures. We also observed the lowest equal error rate (EER). The 1D-PDF is influenced not only by the quality of the fingerprint image but also by the projection axis. The projection axis corresponds to the orientation of the sub-block in the fingerprint image. In further research, we will continue to examine the robust orientation estimation method.

Acknowledgments

This work was supported by the Korea Science and Engineering Foundation (KOSEF) through the Biometrics Engineering Research Center at Yonsei University.

References

1. L. Hong, Y. Wan and A. K. Jain, "Fingerprint Image Enhancement: Algorithm and Performance Evaluation", IEEE Transactions on Pattern Analysis and Machine Intelligence, vol.20, pp. 777 -789, Aug.1998
2. Michael Yi-Sheng Yao, Sharath Pankanti, Norman Haas, Nalini Ratha, Ruud M. Bolle, "Quantifying Quality: A Case Study in Fingerprints", AutoID'02 Proceedings Workshop on Automatic Identification Advanced Technologies, pp.126-131, March 2002.
3. Bolle et al, "System and method for determining the quality of fingerprint images", United State Patent number, US596956, 1999.
4. L. L. Shen, A. Kot and W.M. Koo, "Quality Measures of Fingerprint Images", Third International Conference on AVBPA 2001, pp. 266-271, Jun. 2001.
5. N. K. Ratha, and M. Bolle, "Fingerprint Image Quality Estimation", IBM Computer Science Research Report RC 21622, 1999.
6. Lim. E.,Jiang XD, Yau WY, "Fingerprint Quality and Validity Analysis", IEEE 2002 International Conference on Image Processing.
7. Aapo Hyvärinen, Juha Karhunen, Erkki Oja, "Independent Component Analysis", John Wiley Sons. Inc, 2001.
8. D.Marr, Vision. San Francisco, Calif.:W.H. Freeman, 1982.
9. Richard O.Dula, Peter E.Hart, David G.Stork, "Pattern Classification", John Wiley Sons. Inc, 2001.
10. Maio D., Maltoni D.," Ridge-line Density Estimation in Digital Images", International Conference on Pattern Recognition, Australia, August 1998.
11. D. Lee, K. Choi and Jaihie Kim, "A Robust Fingerprint Matching Algorithm Using Local Alignment", International Conference on Pattern Recognition, Quebec, Canada, August 2002.

A Statistical Evaluation Model for Minutiae-Based Automatic Fingerprint Verification Systems

J.S. Chen and Y.S. Moon

Department of Computer Science and Engineering,
The Chinese University of Hong Kong, Shatin, N. T. Hong Kong
{jschen, ysmoon}@cse.cuhk.edu.hk

Abstract. Evaluation of the reliability of an Automatic Fingerprint Verification System (AFVS) is usually performed by applying it to a fingerprint database to get the verification accuracy. However, such an evaluation process might be quite time consuming especially for big fingerprint databases. This may prolong the developing cycles of AFVSs and thus increase the cost. Also, comparison of the reliability of different AFVSs may be unfair if different fingerprint databases are used. In this paper, we propose a solution to solve these problems by creating an AFVS evaluation model which can be used for verification accuracy prediction and fair reliability comparison. Experimental results show that our model can predict the performance of a real AFVS pretty satisfactorily.

1 Introduction

Minutia-based AFVS is widely used in numerous security applications. A common practice for evaluating the reliability of an AFVS is to apply it to a fingerprint database to get the FAR and FRR. Generally speaking, the experimental result can provide sufficient confidence only if the database is big enough. As one to one matching is usually adopted in such evaluations, experiment time required will grow very fast when the database becomes bigger. As AFVSs need to be repeatedly fine tuned during development, the rise in the evaluation time will prolong the developing cycles and thus increase the cost. Also, when comparing the reliability of two AFVSs, if different databases are used, the conclusion can be essentially unfair. To solve these problems, we propose an evaluation model for AFVSs. The model can be used to predict the reliability of AFVSs as well as compare different AFVSs on a fair basis.

Actually, the accuracy of an AFVS depends on the system properties as well as the inter-class variation of fingerprints, or fingerprint individuality. Fingerprint individuality study can be traced back to more than 100 years ago [2]. From then on, most related studies have focused on minutiae based representations [1, 3, 4], among which Pankanti's model [1] has been regarded as a very simple but effective one for solving fingerprint individuality problems. This model will serve as the basis for building our AFVS evaluation model. The objective of Pankanti's model is to quantify the amount of available minutiae information to establish a correspondence between TWO fingerprints.

D. Zhang and A.K. Jain (Eds.): ICB 2006, LNCS 3832, pp. 236 – 243, 2005.

The rest of this paper is organized as follows. Section 2 defines some necessary symbols and terminologies. Section 3 describes the idea of our fingerprint individuality model. Section 4 gives a formal presentation of our AFVS evaluation model. Experiments are reported in Section 5 in which a real AFVS system is used to test the validity of our model. The last section is a conclusion of our work.

2 Symbols and Terminologies

The following symbols and terminologies are adopted through the rest of this paper.

Genuine minutiae: The minutiae manually (carefully) extracted by a fingerprint expert from a fingerprint image of enough image quality; *False minutiae*: Any extracted minutiae which are not genuine minutiae; *Matching score*: Number of minutiae correspondences between a master template and a live template; *Genuine matching*: The matching templates are from the same finger tip; *Imposter matching*: The matching templates are from different finger tips; *Genuine matching score*: The score of a genuine matching; *Imposter matching score*: The score of an imposter matching; *Genuine minutiae correspondence*: A declared correspondence between a genuine minutia and its counterpart; *False minutiae correspondence*: A declared minutiae correspondence which is not a genuine minutiae correspondence.

t: Matching score; *FAR(t)*(*FRR(t)*): False acceptance(rejection) rate; *G(t)*: Discrete **P**robability **D**ensity **F**unction (PDF) of genuine matching score; *I(t)*: Discrete PDF of imposter matching score; *EER:* Equal error rate; *HG(x, M, K, N)*: PDF of hypergeometric distribution: $C_K^x C_{M-K}^{N-x} / C_M^N$; *b(x, n, p)*: Binomial distribution PDF: $C_n^x p^x (1-p)^{n-x}$; *chi2cdf(x, γ)*: **C**umulative **D**ensity **F**unction (CDF) of χ^2 distribution, where γ is the degrees of freedom; *poiss(x, λ)*: PDF of Poisson distribution: $\lambda^x e^{-\lambda}/x!$;. *round(x):* The integer closest to *x*; *erf(x)*: Error function for Gaussian integration: $2/\sqrt{\pi}\int_0^x e^{-t^2}dt$; *N(x, μ, σ)*: Normal distribution PDF: $\exp(-(x-\mu)^2/2\sigma^2)/\sigma\sqrt{2\pi}$.

3 Minutiae-Based Fingerprint Individuality Model

The following are the assumptions of our fingerprint individuality model.
A1) Only ridge terminations & bifurcations are considered; A2) Only locations & directions are considered for minutiae correspondence; A3) 2D fingerprint minutiae patterns follow the **C**omplete **S**patial **R**andomness (CSR) [5]; A4) Ridges have equal widths; A5) There is one and only one correct alignment between a master and a live template; A6) The minutiae correspondences are independent events and are equally important; A7) Only positive evidence from a minutiae correspondence is considered; A8) In an imposter matching, the minutiae direction difference between two minutiae matched in spatial position approximately follows the following distribution (PDF):

$$p_\theta(x) = 2/3\left(N(x,0,17\sqrt{2}) + N(180-x,0,17\sqrt{2})\right) + 1/(3\times180) \quad ,(0\le x\le180) \quad (1)$$

Our model differs from Pankanti's model in assumptions A3) and A8). Assumption A3) ensures that we can describe both the spatial minutiae distribution in one single fingerprint as well as the distribution of minutiae number among many different fingerprints. Assumption A8) is a strict mathematical description of the minutiae direction distribution. The fp383 database [6] which contains 1149 fingerprint images from 383 user finger tips was used to test the validity of these two assumptions.

For assumption A3), the hypothesis of CSR asserts: (i) the number of events (points) in any planar region A with area $|A|$ follows a Poisson distribution with mean $\lambda |A|$; (ii) given n events x_i in a region A, the x_i are independent random samples from the uniform distribution on A [5]. The test of hypothesis (i) is quite straightforward. The minutiae templates of fp383 were extracted using an AFVS which can achieve more than 95% verification accuracy on fp383 [6]. For each fingerprint, a rectangle R was placed randomly inside its effective region. The **E**mpirical **D**istribution **F**unction (EDF) of the minutiae number inside R was calculated. This EDF was then compared to a Poisson distribution with mean $\lambda |R|$, where λ was set to 54/65536pixel^2, the average minutiae density of fp383. $|R|$ varies from 2304 to 9216pixel^2. CSR hypothesis (i) is strongly supported by the test results. Fig. 1 shows one typical case.

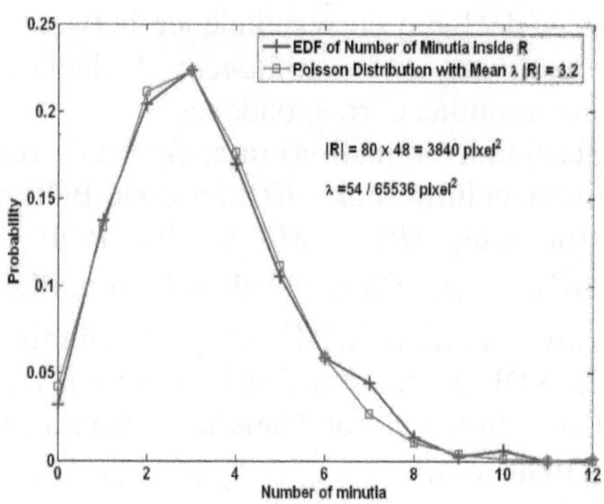

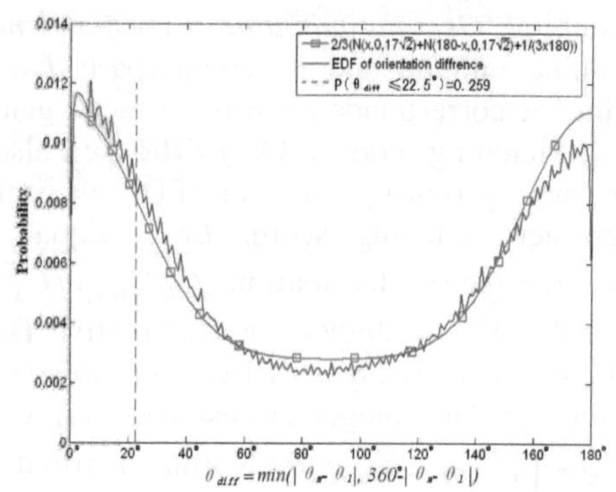

Fig. 1. Minutiae number distribution **Fig. 2.** Minutiae direction differences distribution

The "nearest neighbor distances" method [5] was used to test CSR hypothesis (ii). Minutiae of 40 fingerprints from fp383 were manually marked. Nearest neighbor distance test was then applied to them. Experimental results reveal that 39 fingerprints can pass the test. Boundary effect seems to be the main reason for the only fail case. In any event, for most of the test cases, uniform distribution is confirmed.

Assumption A8) is actually based on the empirical observation that minutiae directions are NOT uniformly distributed [1]. We further observe that in certain areas (~2/3) of the fingerprints, minutiae directions tend to cluster, while uniform distribution dominates in other areas (~1/3). Let θ_m denotes the direction of a master template minutia and θ_l denotes that of a live template minutia. The direction difference between these two minutiae is defined as $min(|\theta_m - \theta_l|, 360° - |\theta_m - \theta_l|)$ [1]. We calculated the EDF of the direction differences of minutiae pairs matched in position for imposter matching in fp383. Equation (1) is obtained by fitting the observation to the experimental result, as shown in Fig. 2. Although equation (1) is based on the experiment on fp383 only, it seems to have considerable generality. In [1], Pankanti et al claim that the possibility that the direction difference is ≤ 22.5° is 0.267 on their database, while equation (1) suggests 0.259 ($\int_0^{22.5} p_\theta(x)dx$).

4 A Minutiae-Based AFVS Evaluation Model

In this section we will apply our fingerprint individuality model to build an AFVS evaluation model with a capability of describing the characteristics of AFVSs as well as the intra-class variation of fingerprints. We will focus on modeling the three major components of a typical AFVS: fingerprint collector, minutia extractor and matcher. The following are the assumptions for our minutiae-based AFVS evaluation model.

E1) The minutia extractor can extract minutiae in a fingerprint image, which has "enough" image quality, with the following parameters (registration & verification):
a) Missing a genuine minutia is an independent event with probability p_{miss}
b) The extracted false minutiae form a CSR pattern with density λ_{false}
c) For a genuine minutia, the extracted position follows a bivariate normal distribution with equal standard deviation σ_{pos} in both dimensions; and the extracted direction follows a normal distribution with standard deviation σ_{ori}. This assumption actually tolerates the possible fingerprint intra-class variation caused by distortion.
E2) The master template covers all areas of the corresponding finger tip. In most AFVSs, a common mechanism for ensuring high reliability is to intentionally put more control on registration to make master templates' information more complete.
E3) The fingerprint collector can always capture fingerprint images with "enough" image quality; in the verification phase, the effective fingerprint area is $|S|$.
E4) The genuine minutia density of the fingerprint set to be verified is λ.
E5) The matcher declares a correspondence between a mater template minutia and a live template minutia if and only if the following three conditions are all fulfilled:
a) The Euclidean distance between these two minutiae is $\leq D$
b) The direction difference between these two minutiae is $\leq \theta_0$
c) No duplicated correspondence of one minutia is allowed.
E6) The matching score equals to the number of minutiae correspondences.

Combining the fingerprint individuality model defined in Section 3, we can formulate *G(t),I(t), FRR(t)* and *FAR(t)* of our AFVS evaluation model. *I(t)* is more related to the fingerprint individuality model. Considering assumptions E1a&b) and E4), we can see that the AFVS extracted minutiae patterns still comply with our fingerprint individuality model besides the overall minutiae density is equation (2).

$$\lambda_{ovr} = \lambda(1 - p_{miss}) + \lambda_{false}, \quad p_{match}(m,n,t) = \sum_{x=t}^{\min(m,n)} HG(x, |S|/2\omega D, m, n) \times b(t,x,l) \qquad (2, 3)$$

Consider an imposter matching situation X in which m minutiae exist in a master template and n minutiae in a live template. According to [1], the probability that there are exactly t minutiae correspondences between these two templates is equation (3), where ω is the ridge period and $l = \int_0^{\theta_0} p_\theta(x)dx$. According to assumption A3) and E3), the probability of the occurrence of situation X can be expressed by equation (4). Combining equations (3) and (4), we can have equation (5).

$$p_{mn} = poiss(m, \lambda_{ovr}|S|) \times poiss(n, \lambda_{ovr}|S|), \quad I(t) = \sum_{m=t}^{+\infty}\sum_{n=t}^{+\infty} p_{mn} p_{match}(m,n,t) \qquad (4, 5)$$

G(t) is relatively more difficult since genuine and false minutiae coexist in the templates. We simply assume that false minutiae correspondences are declared after all genuine minutiae correspondences have been declared (*). Let $\{x_m, y_m, \theta_m\}$, $\{x_l, y_l, \theta_l\}$ denote the occurrences of a genuine minutia existing in the master and live template respectively. According to assumptions E1c) and properties of normal distribution, independent random variables $X=(x_m- x_l)$ and $Y=(y_m- y_l)$ both follow $N(x, 0, \sqrt{2}\sigma_{pos})$; and $\Theta=(\theta_m-\theta_l)$ follows $N(x, 0, \sqrt{2}\sigma_{ori})$. Let $Z=(x_m- x_l)^2+(y_m- y_l)^2$. It can be shown that $Z/2\sigma^2_{pos}$ follows a χ^2 distribution with the degrees of freedom 2. Thus, $chi2cdf(D^2/\sigma^2_{pos},2)$ is the probability that the Euclidean distance between these two minutiae is $\leqslant D$. Also, by applying the property of normal distribution to Θ, we get $P(\Theta \leqslant \theta_0)=erf(\theta_0/2\sigma_{ori})$. Therefore the probability that these two minutiae match is

$$p_{ggm} = chi2cdf(D^2/2\sigma^2_{pos},2) \times erf(\theta_0/2\sigma_{ori})) \qquad (6)$$

Consider a genuine matching situation *X*, in which the number of genuine minutiae in the effective fingerprint area is α. Assume there are m_g genuine minutiae and m_f false minutiae in the master template and there are n_g genuine minutiae and n_f false minutiae in the live template. Equation (7) represents the probability that there are exactly t_g genuine minutiae correspondences and t_f false minutiae correspondences.

$$P_{pm}(\alpha, m_g, n_g, m_f, n_f, t_g, t_f) = \left(\sum_{\varphi=t_g}^{\min(m_g,n_g)} HG(\varphi,\alpha,m_g,n_g) \times b(t_g,\varphi,p_{ggm})\right) \times p_{match}(m_g+m_f-t_g, n_g+n_f-t_g, t_f) \qquad (7)$$

The probability of the occurrence of situation *X* can be expressed as:

$$p_{\alpha mn} = poiss(\alpha,\lambda|S|) \times b(m_g,\alpha,(1-p_{miss})) \times b(n_g,\alpha,(1-p_{miss})) \times poiss(m_f,\lambda_{false}|S|) \times poiss(n_f,\lambda_{false}|S|) \qquad (8)$$

Combining equations (7) and (8), we have

$$G(t) = \sum_{\alpha=0}^{+\infty}\sum_{m_g=0}^{\alpha}\sum_{n_g=0}^{\alpha}\sum_{m_f=t-m_g}^{+\infty}\sum_{n_f=t-n_g}^{+\infty}\sum_{t_g=0}^{t} p_{\alpha mn} \times p_{pm}(\alpha,m_g,n_g,m_f,n_f,t_g,t-t_g) \qquad (9)$$

Equation (9) is prohibitively complicated. Simplification can be achieved by replacing the summations with mean values for some variables. The expectation of false minutiae number is $f=round(\lambda_{false}|S|)$. The mean value of the number of genuine minutiae is $g=\alpha(1-p_{miss})^2$. By introducing these two mean values into (9), we have

$$\hat{G}(t) = \sum_{\alpha=\max(0,t-f)}^{+\infty}\sum_{\varphi=\max(0,t-f)}^{\alpha}\sum_{t_g=\max(0,t-f)}^{\min(t,k)} poiss(\alpha,\lambda|S|) \times b(\varphi,\alpha,(1-p_{miss})^2) \times b(t_g,\varphi,p_{ggm}) \times p_{match}(g+f-t_g, g+f-t_g, t-t_g) \qquad (10)$$

Three sets of numerical simulation were performed on equation (9) and (10) with different parameters. The biggest difference between the value of *G(t)* and $\hat{G}(t)$ is

0.004. Therefore, we can conclude that equation (10) is an accurate approximation of equation (9) in case that the error tolerance is higher than 0.01.

FAR(t) and *FRR(t)* can then be directly deduced as (11) and (12). According to our AFVS evaluation model, the matching scores *t* can only take discrete values, so *EER* is defined as equation (13).

$$FAR(t) = 1 - \sum_{i=0}^{t-1} I(i), \qquad FRR(t) = \sum_{i=0}^{t-1} G(i) \tag{11, 12}$$

$$EER = \{(FAR(t_0) + FRR(t_0))/2 \mid |FAR(t_0) - FRR(t_0)| = \min(|FAR(t) - FRR(t)|)\} \tag{13}$$

Equations (5), (9) ~ (13) depict the verification performance of an AFVS under our evaluation model. It is obvious that these equations are too complicated to be solved algebraically so that numerical simulations are used for all the experiments.

5 Experimental Results and Discussions

To test the validity of our model, the AFVS mentioned in Section 3 was used. The AFVS was first applied to fp383 to get the practical verification performances ($G_1(t)$, $I_1(t)$, $FAR_1(t)$,$FRR_1(t)$ and EER_1). Then, model parameters were evaluated for this AFVS and numerical simulation was performed to achieve its theoretical verification performance ($G'(t)$, $I'(t)$, $FAR'(t)$,$FRR'(t)$ and EER') under our evaluation model.

Kingston's estimation on the genuine minutiae density of 0.246minutiae/mm^2 [1] was adopted here so that $\lambda=51/65536pixel^2$. $\omega=8.2pixels/ridge$ for 450 dpi images [1]. D and θ_0 were set to 20pixels and 22.5° respectively. Core points were used as reference points. During the matching process, only the minutiae whose distances from the core point lie between 80pixels and 16pixels were considered. This leads to $|S|=19300pixels^2$. The automatic minutiae extraction results of 40 fingerprints were compared to their manually extracted templates which gives out $P_{miss}=0.3$, and $\lambda_{false}=18/65536pixel^2$. σ_{pos} and σ_{ori} were estimated by fitting $Z/2\sigma^2_{pos}$ to a χ^2 distribution and Θ to a normal distribution respectively which leads to $\sigma_{pos}=2.5$ and $\sigma_{ori}=5.0$.

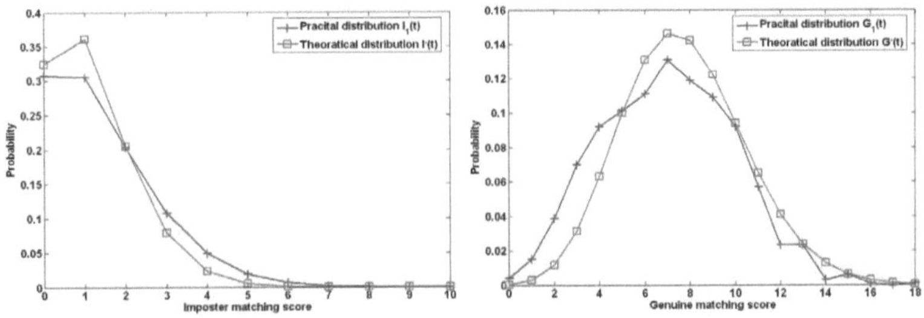

Fig. 3. Comparisons of theoretical and practical distributions of *G(t)* and *I(t)*

Fig. 3 compares the practical and theoretical distribution of *I(t)* and *G(t)*. There are mainly three reasons for the overestimation of *G(t)*: a) The core points of around 2.7% fingerprints in fp383 could not be consistently extracted [6]. Deviation in the reference point locations will surely degrade the genuine matching score. b) The

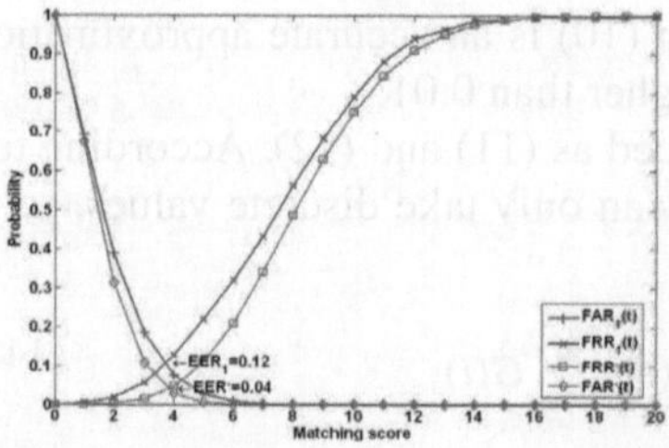

Fig. 4. Comparison of the ROC curves

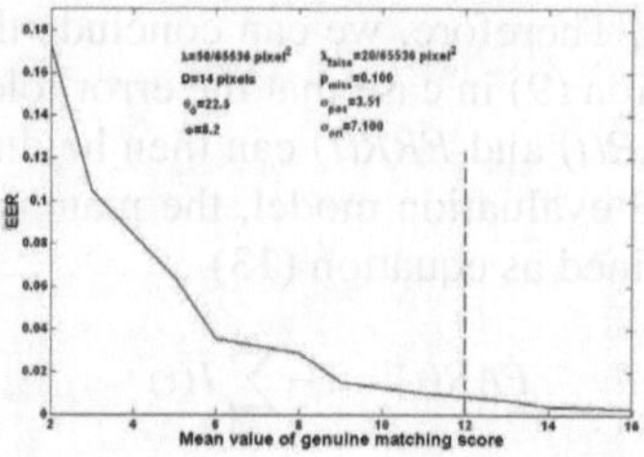

Fig. 5. *EER* values under different |*S*| values

overestimate of the effective fingerprint area as different fingerprints has different core point location. c) The assumption (*) made in section 4 is not always true.

The ROC curves are shown in Fig. 4. We can see that our model can predict the distribution of *I(t)* and *G(t)* satisfactorily. The overestimation of *G(t)* which directly leads to an obvious underestimate of *EER* is probably caused by inconsistency between the model assumptions and the experimental settings as discussed above. In addition, the quaternion $\{p_{miss}, \lambda_{false}, \sigma_{pos}, \sigma_{ori}\}$ actually decides the intrinsic reliability of a extraction process, making it possible to separate the extractor and the matcher when evaluating an AFVS. Clearly, our model can help AFVS developers to improve their systems by analyzing how different parameters can affect the system reliability. Fig. 5 and Fig. 6 show the relationship between *EER* and |*S*|, *D* and θ_0 respectively. The conclusion made in [6] that "when |*S*| is big enough, the increasing of |*S*| will not lead to an obvious improvement in *EER*" can be easily observed from Fig. 5. Fig. 6 shows that best system accuracy that can be achieve when $D \approx 3\sigma_{pos}$ and $\theta_0 \approx 3\sigma_{ori}$.

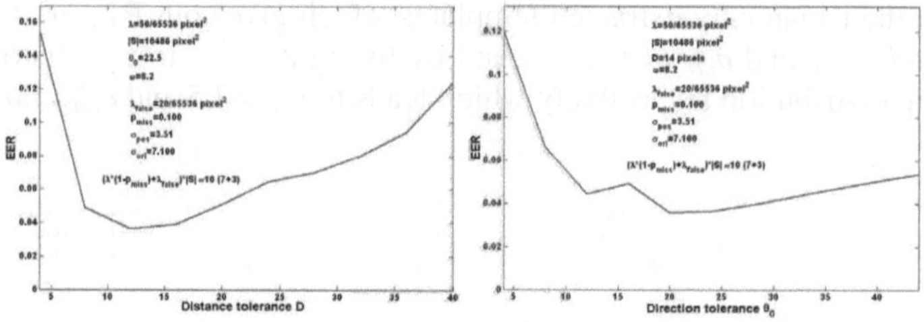

Fig. 6. The relationship between *EER* and distance/direction tolerance

6 Conclusion and Acknowledgement

We have proposed an evaluation model for minutiae-based AFVSs. We first adopt Pankanti's model with some strengthening assumptions to describe the fingerprint individuality. Then we parameterize the three major components of an AFVS. Equations are then derived to describe the verification performance under the model assumptions. Experimental results show that our model can predict the distribution of the *G(t)* and *I(t)* of an AFVS satisfactorily. Furthermore, our model can serve as an assistant for AFVS developers to improve their system reliability since (a) our model makes it possible to analyze different components in an AFVS separately; (b) how different model parameters will affect the system reliability can be used as a guidance for the developers to fine tune their systems.

This work was partially supported by the Hong Kong Research Grants Council Project 2300011, "Towards Multi-Modal Human-Computer Dialog Interactions with Minimally Intrusive Biometric Security Functions".

References

[1] S. Pankanti, S. Prabhakar, A. K. Jain, On the Individuality of Fingerprints, IEEE Trans. on Pattern Analysis and Machine Intelligence, pp. 1010-1025, vol. 24, no. 8, August 2002
[2] F. Galton, Finger Prints, London: McMillan, 1892
[3] M. Trauring, Automatic Comparison of Finger Ridge Patterns, Nature, pp. 938-940, 1963
[4] D. A. Stoney, J. I. Thornton, A Critical Analysis of Quantitative Fingerprint Individuality Models, J. Forensic Sciences, pp. 1187-1216, vol.31, no. 4, October 1986
[5] P. J. Diggle, Statistical Analysis of Spatial Point Patterns, Oxford University Press, 2003
[6] K. C. Chan, Y. S. Moon, P. S. Cheng, Fast Fingerprint Verification Using Sub-regions of Fingerprint Images, IEEE Trans. On Circuits and Systems for Video Technology, pp. 95-101, vol. 14, issue 1, January 2004

The Surround Imager™: A Multi-camera Touchless Device to Acquire 3D Rolled-Equivalent Fingerprints

Geppy Parziale, Eva Diaz-Santana, and Rudolf Hauke

TBS North America Inc. 12801,
Worldgate Drive, Herndon, VA 20170, USA
{geppy.parziale, eva.diaz-santana, rudolf.hauke}@tbsinc.com

Abstract. The Surround Imager™, an innovative multi-camera touchless device able to capture rolled-equivalent fingerprints, is here presented for the first time. Due to the lack of contact between the elastic skin of the finger and any rigid surface, the acquired images present no deformation. The multi-camera system acquires different finger views that are combined together to provide a 3D representation of the fingerprint. This new representation leads to a new definition of minutiae bringing new challenges in the field of fingerprint recognition.

1 Introduction

The current fingerprinting technologies rely upon either applying ink (or other substances) to the finger tip skin and then pressing or rolling the finger onto a paper surface or touching or rolling the finger onto a glass (silicon, polymer, proprietary) surface (platen) of a special device. In both cases, the finger is placed on a hard or semi-hard surface, introducing distortions and inconsistencies on the images [1, 2].

Touchless Biometric Systems[1], formally TBS, has developed the Surround Imager™, an innovative live-scan device able to capture a rolled-equivalent (nail-to-nail) fingerprint without the need of touching any surface. The intrinsic problems of the touch-based technology, also known as *inconsistent, non-uniform* and *irreproducible contacts* [2], are definitively overcome with this new device.

The paper describes this new acquisition technology that, besides the above mentioned advantages, introduces also a novel representation of fingerprints. In fact, the multi-camera system acquires different finger views that are combines to generate a 3D representation of the fingerprint. This implies the design and development of new algorithms that are able to manage the 3D information provided by the new device and bring new challenges in the field of fingerprtin recognition.

The paper is organized as follows. In the next Section 2, the main functionalities of the Surround Imager™ are reported. Section 3 provides an overview of the image processing algorithms involved with the 3D reconstruction. The new representation and a new definition of minutiae is provided in Section 4. In the same Section, the problem of matching the new fingerprint against *traditional* representation and a possible approach to match minutiae in 3D are discusses. Finally, concluding remarks and future activities are presented in Section 5.

[1] http://www.tbsinc.com

D. Zhang and A.K. Jain (Eds.): ICB 2006, LNCS 3832, pp. 244–250, 2005.

2 The Surround Imager™

The left-hand side of Fig. 1 highlights a schematic view of the Surround Imager™. The device is a cluster of 5 cameras[2] located on a semicircle and pointing to its center, where the finger has to be placed during the acquisition. The size of the acquired images is 640 × 480 pixels.

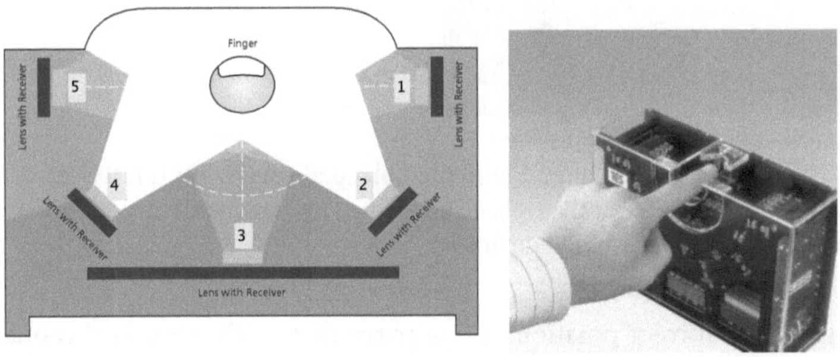

Fig. 1. The Surround Imager™ (on the right-hand side) and its schematic view (left-hand side)

The Surround Imager™ has currently the size of 15 cm × 24 cm × 10 cm. This size (large compared with other fingerprint devices) is mainly due to our choice of a reasonable quality-price ratio. Since the finger has to be far away from the 5 sensors with a distance depending on the sensor size and dot-pitch, the lens system and the required optical resolution, we chose the best solution in term of image quality, resolution and final costs of the device. The chosen distance has been fixed to 50 mm. Moreover, the device contains a set of 16 green LED arrays and the large size has also been chosen to dissipate the heat generated by the light system.

The LED intensities can be individually controlled during each acquisition. In previous experiments, we demonstrated that the green light produces a better contrast on the fingerprint structure than the red and the blue lights. The advantage of the use of a green light is illustrated in Fig. 2. The touchless approach combined with the green LEDs allows the acquisition of fingerprints with a very dry or a very wet skin. These kinds of fingers are very difficult to acquire by touch-based devices.

Due to the large distance between the camera and the object (with respect to their size), the image resolution is not constant within the image and decreases from the center to the image extremities. The optical system has been designed to ensure a resolution of 700 dpi in the center and a minimum of 500 dpi on the image borders.

During a capture, the finger is placed on a special support (right-hand side of Fig. 1) to avoid trembling that could create motion blur. The portion of the finger that has to be captured does not touch any surface. Moreover, the finger has to be placed in a correct position so that it is completely contained in the field-of-views of the 5 cameras at the same time. A realtime algorithm helps the user during the finger placement. Once

[2] Since the Surround Imager™ is a modular device, versions with 1 or more (up to five) cameras are also available on request.

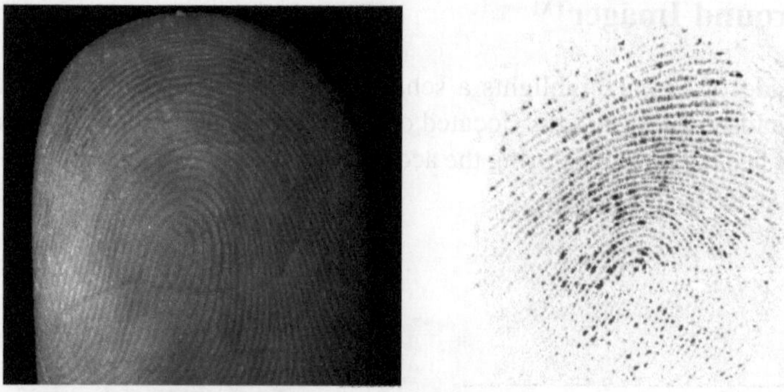

Fig. 2. The same fingerprint acquired with the Surround Imager ™ (on the left-hand side) and a touch-based optical device (on the right-hand side). The finger skin is very dry and thus, has a very low contrast on the touch-based device

the finger is in the correct position, the user receive a *'Don't move'* request from the device and the capture can start automatically. During an acquisition, each LED array is set to a specific light intensity and the 5 cameras capture synchronously a picture of the finger. This procedure is repeated 16 times in only 120 ms, ensuring that eventual finger movements are negligible for the following computation steps.

Each camera captures 16 times the same portion of the finger skin with different light conditions. Since the following 3D reconstruction steps are very complex and computationally expensive, the different illuminations are used to help these algorithms in extracting special image features.

In Fig. 3, a comparison of the same fingerprint acquired by the touchless device (on the left-hand side) and a touch-based optical sensor (on the right-hand side) is highlighted. Observing the two images, one can immediately notice that the Surround Imager™ provides a *negative polarity* representation of the fingerprint, i.e. the ridges appears to be brighter than the valleys. Besides, the image obtained by the TBS device contains also the structure of the valleys. This information is completely inexistent in other technologies where the valleys belong to the image background.

Fig. 3. The same portion of a fingerprint skin acquired with the Surround Imager ™ (on the left-hand side) and a touch-based optical device (on the right-hand side)

3 3D Reconstruction Algorithm

A detailed description of the used 3D reconstruction algorithms goes beyond the scope of this paper, but an overview of them is here reported for completeness.

The Surround Imager™ has been designed to provide a precise deformation-free representation of the fingerprint skin. The 3D reconstruction procedure is based on stereovision and photogrammetry algorithms. Thus, the exact position and orientation of each camera (*camera calibration*) with respect to a given reference system are needed for the following processing steps [5, 6]. The calibration is done off-line, using a 3D target on which points with known positions are marked.

The position of the middle camera (camera 3 in Fig. 1) has been chosen so that it could capture the central portion of the fingerprint, where the *core* and the *delta* are usually located. Then, the other cameras have been placed so that their field-of-views partially overlap. In this way, the images contain a common set of pixels (homologous pixels) representing the same portion of the skin. To compute the position of each pixel in the 3D space (3D reconstruction), the correspondences between two image pixels must be solved (*image matching*). This is done computing the cross-correlation between each adjacent image pair. Before that, the distortions generated by the mapping of a 3D object (the finger) onto the 2D image plane have to be minimized. This reduces errors and inconsistencies in finding the correspondences between the two neighbor image pair. Using shape-from-silhouette algorithms, it is possible to estimate the finger volume. Then, each image is unwrapped from the 3D model to a 2D plane obtaining the corresponding *ortho-images*.

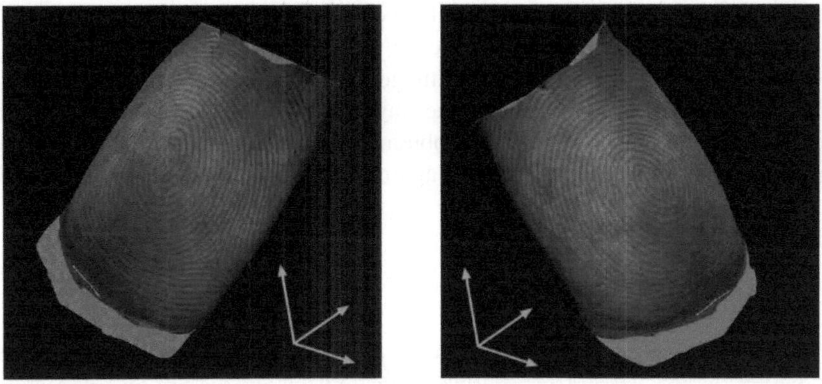

Fig. 4. Two views of a fingerprint reconstructed with the approach described in Section 3

The unwrapped images are used to search for homologous pixels in the image acquired by each adjacent camera pair. To improve the image matching, a multiresolution approach [4] has been chosen and an image pyramid is generated from each image [7]. Then, starting from the lower resolution level, a set of features is extracted for every pixel, obtaining a feature vector that is used to search the homologous pixel in the other image. When this is completed, the search is refined in the higher levels, until the original image resolution is reached. Once the pixel correspondences have been resolved, the third dimension of every image pixel is obtained using the camera geometry [6]. In Fig. 4, an example of the 3D reconstruction is highlighted.

4 A New Representation of Fingerprints

The image processing shortly described in Section 3 provides a new representation model for fingerprints. Since each image pixel can be described in a 3D space, a new representation of minutiae has to be adopted.

In the 2D image domain, a minutia may be described by a number of attributes, including its location in the fingerprint image, orientation, type (e.g. ridge termination or ridge bifurcation), a weight based on the quality of the fingerprint image in the minutia neighborhood, and so on [2, 3, 8]. The most used representation considers each minutia as a triplet $\{x, y, \theta\}$ that indicates the (x,y) minutia location coordinates and the minutia orientation θ. Considering this simple representation and adapting it to the 3D case (Fig. 5), a minutia point M_i may be represented by the t-upla $\{x, y, z, \theta, \phi\}$ that indicates the x, y and z coordinates and the two angles θ and ϕ representing the orientation of the ridge in 3D space.

Besides the coarse 3D representation of the fingerprint shape, the Surround Imager™ provides also a more fine 3D description of the ridge-valley structure. Since during the acquisition the finger does not touch any surface, the ridges are free of deformation. Besides, as shown in Section 2, this technology is also able to capture the information related to the fingerprint valleys. Thus, the entire 3D ridge-valley structure captured with a specific illumination can be well represented by the image gray-levels, mapping each image pixel into a 3D space $\{x, y, I(x, y)\}$, where $I(x, y)$ represents the value of the gray-level of the fingerprint image I at position (x, y). An example of this mapping is illustrated in Fig. 6, where the fingerprint portion of Fig. 3 is reported using a 3D representation.The fingerprint obtained by the Surround Imager™ would be useless if it was not possible to match it against fingerprints acquired with *traditional*

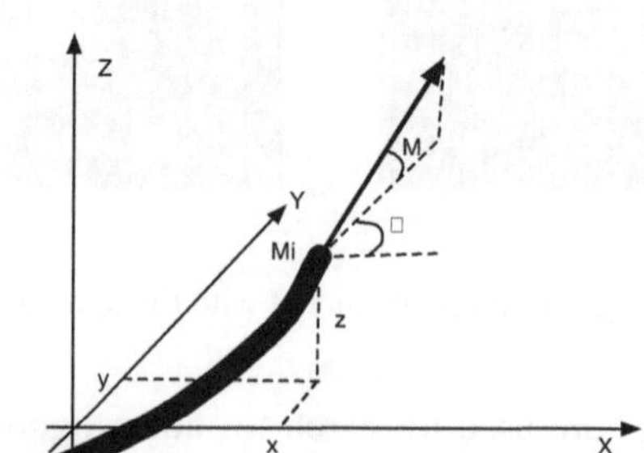

Fig. 5. 3D representation of a minutia M_i (ridge ending). The feature point is uniquely represented by the t-upla $\{x, y, z, \theta, \phi\}$.

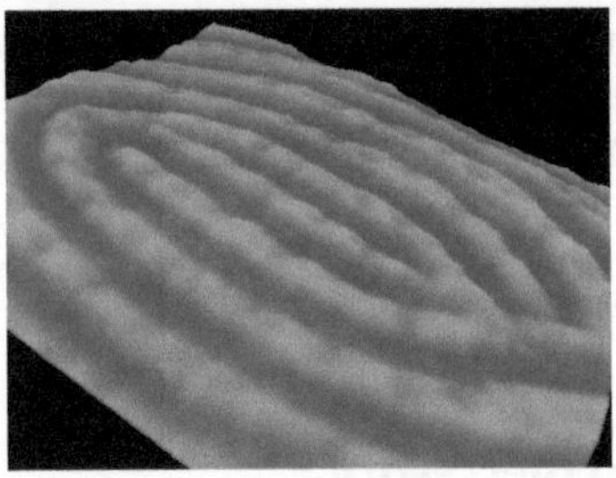

Fig. 6. A detail of the 3D ridge-valley structure

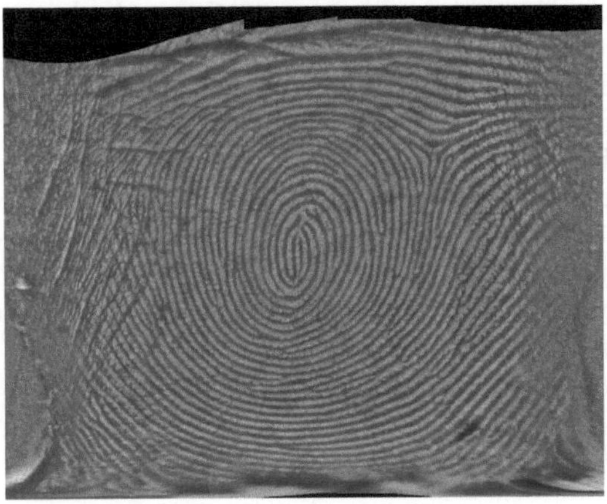

Fig. 7. A detail of the 3D ridge-valley structure

technologies. Besides, since large fingerprint databases are already available, it is inconvenient or/and impossible to build them up again using this new device. Thus, to facilitate the integration of the Surround Imager™ into existing systems, a 2D version of the reconstructed fingerprint is also provided after the reconstruction. The computed 3D finger geometry can be used to *virtually roll* the fingerprint onto a plane, obtaining a complete rolled-equivalent fingerprint of the the acquired finger (Fig. 7).

The presented 3D representation brings new challenges in field of fingerprint recognition and new algorithms to match fingerprints directly in the 3D space have been designed. This has many advantages with respect to the 2D matching. In fact, since fingerprints acquired by the Surround Imager™ do not present any skin deformation, the relative position of the minutia points is always maintained[3] during each acquisition. In this case, the minutiae matching problem can be considered as a rigid 3D point-matching problem [2].

[3] In reality, a small change in the water content of the skin can modify the relative distance among minutiae. These small variations can be corrected directly on the 3D reconstructed model.

The approach used to matching fingerprints in the 3D space is a generalization to the 3D case of the algorithm presented in [3]. Once the minutiae have been localized on the fingerprint skeleton, a 3D Delaunay triangulation is applied to the point clouds. From each triangle, many features are computed (length of the triangle sides, internal angles, angles between the minutia orientation and the triangle side, and so on) and then used to match the triangles in the other fingerprint.

5 Conclusion and Further Work

A novel device to acquire fingerprints has been here presented. The Surround Imager™ is a touchless device using 5 calibrated cameras that provide a 3D representation of the captured fingerprints. This novel representation leads also to a new definition of minutiae in 3D space, here given for the first time.

Because the different nature of the finger image with respect to the traditional approaches new methods for image quality check, analysis, enhancement and protection can be implemented to provide additional flexibility for specific applications. Besides, new forensic and pattern-based identification can also be developed and exploited to surpass the existing fingerprint methods. Also, due to this flexibility, the provided finger images are compatible with existing Automated Fingerprint Identification System (AFIS) and other fingerprint matching algorithms, including the ability to be matched against legacy fingerprint images.

References

1. D. R. Ashbaugh: Quantitative-Qualitative Friction Ridge Analysis. An Introduction to Basic and Advanced Ridgeology, CRC Press LLC, USA, 1999.
2. D. Maltoni, D. Maio, A. K. Jain, S. Prabhakar: Handbook of Fingerprint Recognition, Springer Verlag, June 2003.
3. G. Parziale, A. Niel: A Fingerprint Matching Using Minutiae Triangulation, on Proc. of International Conference on Biometric Authentication (ICBA), LNCS vol. 3072, pp. 241-248, Hong Kong, 15-17 July 2004.
4. M. del Pilar Caballo-Perucha, Development and analysis of algorithms for the optimisation of automatic image correlation, Master of Advanced Studies of the Post-graduate University Course Space Sciences, University of Graz, Austria, Dec. 2003.
5. M. Sonka, V. Hlavac, R. Boyle: Image Processing, Analysis, and Machine Vision, Second Edition, Brooks/Cole Publishing, USA, 1999.
6. R. Hartley, A. Zisserman: Multiple View Geometry in Computer Vision, Cambridge University Press, UK, 2003.
7. R. C. Gonzalez, R. E. Woods: Digital Image Processing, Prentice Hall, New Jersey, USA, 2002.
8. A. K. Jain, L. Hong, R. Bolle: On-Line Fingerprint Verification, PAMI, Vol. 19, No. 4, pp. 302-313, 1997.

Extraction of Stable Points from Fingerprint Images Using Zone Could-be-in Theorem

Xuchu Wang[1], Jianwei Li[1], Yanmin Niu[2], Weimin Chen[1], and Wei Wang[1]

[1] Key Lab on Opto-Electronic Technique of State Education Ministry, Chongqing University, 400044, Chongqing, P.R. China
Seadrift@sina.com, jwli@cqu.edu.cn
[2] College of Physics and Information Techniques, Chongqing Normal University, 400047, Chongqing, P.R. China

Abstract. This paper presents a novel zone Could-be-in theorem, and applies it to interpret and extract singular points (cores and deltas) and estimate directions of cores in a fingerprint image. Where singular points are regarded as stable points (attracting points and rejecting points just according to their clockwise or anticlockwise rotation), and pattern zones are stable zones. Experimental results validate the theorem. The corresponding algorithm is compared with popular Poincaré index algorithm under two new indices: reliability index (RI) and accuracy cost (AC) in FVC2004 datasets. The proposed algorithm are higher 36.49% in average RI, less 2.47 in average AC, and the advantage is more remarkable with the decrease of block size.

1 Introduction

Singular points (SPs) are global features in fingerprint images and play an important role in fingerprint identification/authentication [1]. Henry defined two early types of singular points, where a core is the topmost point of the innermost curving ridge and a delta is the center of triangular regions where three different direction flows meet [2]. Since the directional field around SPs is discontinuous, many approaches intended to solve the problem by orientation distribution [3][4][5][6][7][8]. Now the popular and elegant detecting method is Poincaré index based approach [9][10], and point orientation is often replaced by block orientation due to efficiency. Ref. [9] made some useful improvements to quicken detection. Nevertheless, little attention was focused on the definition of SPs and the direction estimation of core points in previous research on this topic.

SP is more regarded as a region than a point and it can be represented by a barycenter of the region. Different methods lead to different positions while they are situated in a similar region, so the reliability must be considered firstly and then be the accuracy. A limitation of Poincaré index method is the contradiction of reliability and accuracy. Another limitation is that when the noise is heavy, more pseudo SPs will be gotten or right points will be omitted due to increasing false orientations [1][4], so [9] proposed to refuse pseudo points by iterative smoothing method and which reduced accuracy. The third is the method can't estimate directions of core

D. Zhang and A.K. Jain (Eds.): ICB 2006, LNCS 3832, pp. 251 – 257, 2005.

points. Hence, it is an especially challengeable problem to improve the reliability of SPs without extra expenditure.

In this paper, we present new idea to interpret and detect SPs. The fingerprint orientations are interpreted by some original definitions in a dynamic viewpoint, where SPs are regarded as stable points surrounded by shortest stable boundary. When the stable points are rotating clockwise, they are assumed to get an ability of attracting ridges and other stable points, we call them as attracting points. Similarly, when they are rotating anticlockwise they are rejecting points with rejecting ability. The pattern zones around the stable points are regarded as stable zones. We propose a zone Could-be-in theorem to extract the stable points and estimate the directions of core points simultaneously by analyzing the property of the shortest stable boundary. (All of them are included in the fingerprint growth model proposed by the author.) We also define reliability index (RI) and accuracy cost (AC) to evaluate different performances of extraction algorithms. Experimental results show that our algorithm is higher 36.49% in AI and less 2.47 in AC than Poincaré index algorithm. When the block size is decreased, the advantage of our algorithm is more remarkable.

2 Zone Could-be-in Theorem

According some statistical analysis about orientations of ridges in fingerprint images, some results in nonlinear dynamic system, we present some definitions as follows:

Discrete orientation field *O*: A support set in 2-dimension plane composed by a serial of directional particles in square meshes. The term is written as $O=\{K_i \mid \theta_i \in [0,\pi), i \in Z\}$, we use orientation to describe directionality in images for distinction, so θ_i is the orientation of particle K_i ;.

Could-be-in: If the orientation of K_1, K_2, K_3 in ***O*** can be described as $|\theta_1 - \theta_2| > \theta_3$,we call K3 is Could-be-in to K1,K2. Suppose $\theta_1 \le \theta_2$, the term is written as $K_3 \mapsto \langle \hat{K}_1, \hat{K}_2 \rangle$.

Zone Could-be-in: There are K_p and a sequence of $\{K_i \mid i=0,1,...L-1\}$ in ***O***, $K_p \notin \{Ki\}$, if the term $K_p \mapsto \langle K_s, K_{(s+1) \bmod L} \rangle; s \in \{0,1,...L-1\}$ is true, then we regard that the loop $L\{K_i \mid i=0,1,...L-1\}$ composed by $\{K_i \mid i=0,1,...L-1\}$ is Zone Could-be-in to K_p. The term is written as $K_p \mapsto LOOP\langle K_1, K_2, ..., \hat{K}_s, \hat{K}_{(s+1) \bmod L} ... K_{L-1} \rangle$. The "^" symbolizes the entrance position of K_p. We can get Entrance Times N according to the number of "^". If N is equal to L, the zone is Could-be-in to K_p everywhere. If N is equal to 2, the zone is Could-be-1-in to K_p. If N is less than 2, the Zone is not Could-be-in to K_p. Apparently Could-be-in is a special case of Zone Could-be-in.

Monotone zone Could-be-in: There are K_p and $\{K_i \mid i=0,1,...L-1\}$ where $K_p \notin \{Ki\}$, $K_p \mapsto LOOP\langle K_1, K_2, ..., \hat{K}_s, \hat{K}_{(s+1) \bmod L} ... K_{L-1} \rangle$, if $\{K_i\}$ is monotone and $L\{K_i\}$ is

Monotone zone Could-be-in to K_p, the term is written as $K_p \mapsto \overrightarrow{LOOP}\langle K_0, K_1, ..., \hat{K}_s, \hat{K}_{(s+1) \bmod L} ... K_{L-1}\rangle$. If only one s can let the term true, we call $L\{K_i\}$ is **Monotone Zone Could-be-1-in** to K_p, so $K_p \mapsto \overrightarrow{LOOP}\langle \hat{K}_0, K_1, ..., \hat{K}_{L-1}\rangle$.

Stable Zone: There are K_p and $\{K_i \mid i = 0,1,...L-1\}$ where $K_p \notin \{Ki\}$, if K_p is in loop $L\{K_i \mid i = 0,1,...L-1\}$ and term $K_p \mapsto \overrightarrow{LOOP}\langle \hat{K}_0, K_1, ..., \hat{K}_{L-1}\rangle$ is true, then $L\{K_i \mid i = 0,1,...L-1\}$ can be regarded as a gradual stable zone to K_p. The shortest one of all $L\{K_i\}$ is regarded as the stable zone to K_p. It is reasonable to call $\{K_i\}$ as the boundary sequence of the zone. If the length of $\{K_i\}$ is in range [4, 8], we further call it the shortest stable boundary to K_p and K_p is a stable point.

Term zone Could-be-in describes relationship of an orientation and a sequence of orientation. It can be interpreted from two aspects, one is that the entrance orientation is accord with the orientation loop and it can be a part of the zone surrounded by orientation loop, the other is that there is a directional particle which can attract or reject the orientation loop. By the mutual function, both of them get a stable status. That's the reason we call them "stable point" and "stable boundary". Like some handedness phenomenon in particle physics field, we assume the attracting or rejecting ability is the property of a particle and determined just by rotating direction of the particle. As Fig.1 depicts, the shortest stable boundary is convex or concave in order to get a harmony It's apparent that the orientation loop around a stable point must satisfy some conditions to get a kind of harmony, which will be discussed by the following theorem.

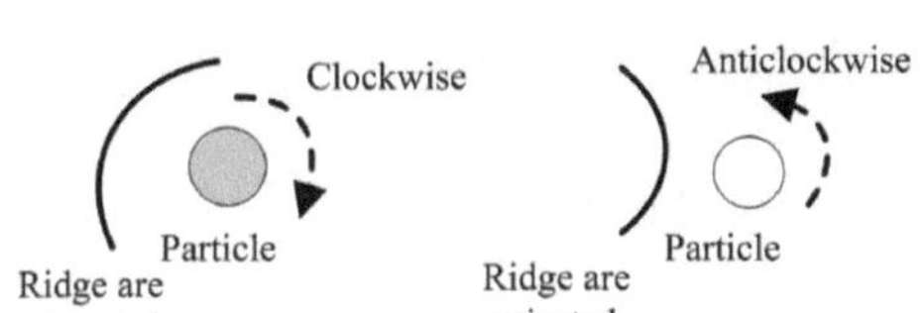

Fig. 1. Attracting or rejecting ability of a particle

Theorem. If a sequence is monotone zone Could-be-1-in to a directional particle, the entrance position must be between the extremums of the sequence.

Proof. Let K_p be the directional particle and $\{K_i \mid i = 0,1,...L-1\}$ be the sequence, since $\{K_i\}$ is monotone, it can be arranged as a loop called $L\{K_i\}$ in which the maximum and minimum is neighbor (max, min represent their positions in same period of the loop and $\theta_{\max}, \theta_{\min}$ are the corresponding orientations). Assume the entrance position of K_p is mid (mid is not equal to max or min), hence MIN{max, min}<mid<MAX{max, min}, MAX and MIN are getting extremums operation,

respectively. Since $\theta_{\min} < \theta_{mid} < \theta_{\max}$ and we can disconnect $L\{K_i\}$ at mid position and get sequence $\{K'_i\}$, which is descending from mid to min, and ascending from max to mid, so $\{K'_i\}$ is not a monotone sequence. This means mid position cannot be gotten, the one and only entrance position is between min and max unless the orientations of all directional particle are equal, which contradicts the monotony of the sequence. Q.E.D.

Further suppose the length of $L\{K_i\}$ is L, disconnect it to a monotone ascending sequence $\{K_i\}$, let $\Delta\theta_i = \theta_{(i+1) \bmod L} - \theta_i;\ i = 0,1...L-1$, then

$$\theta_p > \Delta\theta_i > 0; \quad i = 0,1...L-2. \tag{1}$$

$$\theta_p < \Delta\theta_{L-1} \tag{2}$$

The theorem, especially the two inequations above, qualifies the relationships of K_p, $\{K_i\}$, $L\{K_i\}$ and supplies a criterion of existence of stable boundary. Note that there are many methods to detect the existence of a stable boundary and we just provide one way here, for some case, Poincaré index is a detecting method, too. When L is 8, $\{K_i\}$ becomes an eight-direction stable boundary. It also provides a way to extract stable points through detecting the eight-direction stable boundary. By the way, the entrance position is a clue to estimate the direction of a core point in a fingerprint image and we will discuss it in the following section.

3 Extraction Methodology

As a description of orientation changing rule of discrete orientation, zone Could-be-in theorem provides some perspectives to detect some special zone and zone distribution. It can be used to extract fingerprint SPs and the algorithm procedure is as follows:

Step.1 —Segmenting background by variance threshold method.
Step.2 —Building discrete orientation field. Divide fingerprint image M into blocks of size $W_i \times W_i$ (16×16) and use the least square method [11] to estimate a directional image. The result is regarded as a discrete orientation field O_{all} of the whole image. If the input is a part of fingerprint image, the corresponding O is O_{part_i}.
Step.3—Detecting stable zones by zone Could-be-in theorem. Locate the 8-connected zones of the stable zones and divide them into O_{core} and O_{delta}.
Step.4—Overlapping and location. Map the regions of M by O_{core}, O_{delta} and decrease the block size $W_i \times W_i$ as 12×12, 8×8, return to step 2,3 to get $O_{part\_1}; O_{part\_2}; \cdots$
Step.5—Break out when W_i is less than a presetting threshold.

The direction of a core point can provide very useful information for fingerprint classification and fingerprint matching even though it is not very accurate. A little literature discussed this topic such as [6], and the method is computationally expensive. While in zone Could-be-in theorem, this problem can be easily solved. Firstly, we formulate the entrance position *i* and directional range *Li*:

$$L_i = [\lfloor (i+1)/2 \rfloor \times \pi/2 + (-1)^i \beta, (1 + \lfloor (i+1)/2 \rfloor) \times \pi/2 + (-1)^i \beta); i = 0,1,...7 \tag{3}$$

where $\beta = arctg(1/2)$ and the length of every range is $\pi/2$, $\lfloor \bullet \rfloor$ means floor integer operation. Note that when i is 6 or 7, the range is composed by two sub-ranges. Secondly, let θ_p be the orientation of a stable point and $\theta_i, \theta_{(i+1) \bmod 8}$ be the extremums of its eight-direction shortest stable boundary, considering they are orientations in range $[0, \pi)$, we map an orientation to the direction range *Li* by a function $f(\theta_i)$ .Lastly, we consider three elements dominate a core point direction β_{core} together:

$$f(\theta_i) = \begin{cases} \theta_i + \pi; & \theta_i + \pi \in L_i \\ \theta_i + \pi/2; & \theta_i + \pi/2 \in L_i \\ \theta_i - \pi/2; & \theta_i - \pi/2 \in L_i \\ \theta_i; & \theta_i \in L_i \end{cases} \tag{4}$$

$$\beta_{core} = \lambda_1 f(\theta_i) + \lambda_2 f(\theta_{(i+1) \bmod 8}) + \lambda_3 f(\theta_p) \tag{5}$$

where i is entrance position, $\lambda_1, \lambda_2, \lambda_3$ are weighted coefficients (0.3, 0.3, 0.4 empirically).

4 Experimental Results

Some detecting results and the comparision with popular Poincaré index algorithm under different block sizes are shown in Fig.2. In order to emphasis locations, some directions of the core points in our algorithm are omitted.

Apparently the locations of singular points in both methods are similar and overlapped in some portions. We define two indices: *reliability index (RI)* and *accuracy cost (AC)* to evaluate the performance of different algorithms:

$$RI = RZ/TZ \times 100\% \; ; AC = RN/TN \tag{6}$$

where *TZ* is the number of total zones detected according to 8-connectness, *RZ* is the number of right zones determined by human experts, *TN* is the total number of detected SPs, and *RN* is the number of SPs in right zones. The ideal performance of a singular point extraction algorithm is that *RI* is near to 100% and *AC* is near to 1. Table 1 reports the comparison matrix about average value of every image in FVC2004 and induces some conclusions: (i) average *RI* and *AC* of Alg.ZC are higher 36.49%, less 2.47 than those of Alg.P. (ii) advantage of Alg.ZC is more remarkable with decrease of block size.

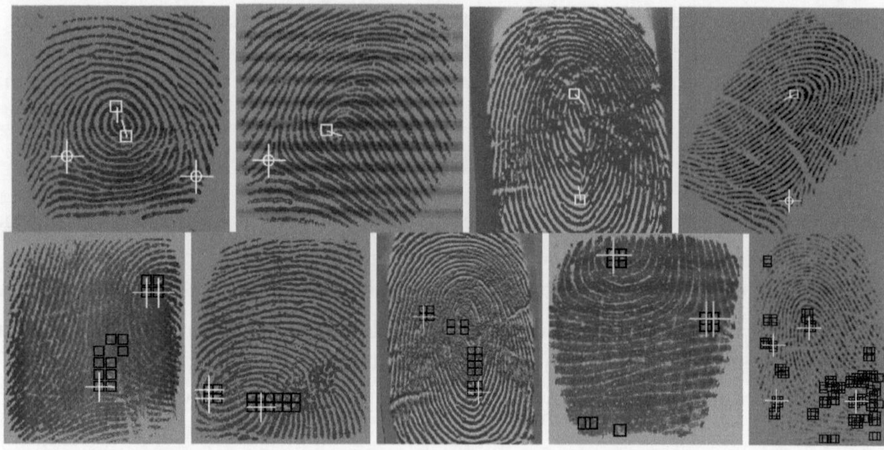

Fig. 2. Up row is some results of Alg.ZC, and low row is comparion by Alg.ZC ("+") and Alg.P ("□") under different block size

Table 1. Extraction matrix by zone Could-be-in algorithm (Alg.ZC) and by Poincaré index algorithm (Alg.P) with block sizes 16,12 and 8 in FVC2004 four datasets

DB	*Alg.*	*TZ*	*RZ*	*TN*	*RN*	*RI(%)*	*AC*
1	Alg.ZC	1.41,1.43,1.49	1.35	1.62,1.67, 1.72	1.46,1.52, 1.64	95.74, 94.40, 90.60	1.08, 1.13, 1.21
	Alg.P	2.29,2.46,3.02	1.35	7.87,8.03, 10.94	5.28,5.33 , 5.37	58.95, 54.88,44.70	3.91, 3.95, 3.98
2	Alg.ZC	1.56,1.61,1.73	1.25	1.87,1.95, 2.03	1.52,1.78 , 1.91	80.13,83.85, 72.25	1.22, 1.42, 1.53
	Alg.P	2.89,3.90,5.18	1.25	10.04,12.06, 8.11	4.93,4.95 , 4.98	43.25, 32.05, 24.13	3.94, 3.96, 3.98
3	Alg.ZC	2.37,2.38,2.41	2.24	4.17,4.21,4.40	4.06,4.13 , 4.35	94.51, 94.12,92.95	1.81, 1.84, 1.94
	Alg.P	2.55,2.76,3.47	2.24	9.88,10.10, 13.22	8.69,8.71 , 8.72	87.84,81.16,64.55	3.88, 3.89, 3.89
4	Alg.ZC	1.70,1.77,1.85	1.61	2.07,2.14, 2.34	1.92,2.03 , 2.28	94.70,90.96,87.03	1.19, 1.26,1.42
	Alg.P	2.98,3.36,4.04	1.61	8.23,8.92, 13.26	6.06,6.09 , 6.11	54.03,47.92,39.85	3.76, 3.78, 3.80
Avg	Alg.ZC	1.81	1.61	2.51	2.39	89.27	1.42
	Alg.P	3.24	1.61	10.89	6.27	52.78	3.89

5 Conclusions and Future Work

The contribution of this paper lies in three points:

(i) Define the singular points as stable points (attracting points and rejecting points just by their rotation) and pattern zones are stable zones from a new viewpoint.

(ii) Propose some innovative definitions and a theorem called zone Could-be-in theorem to extract the stable points and their directions.

(iii) Define two indices: reliability index (RI) and accuracy cost (AC) to evaluate the performance of different extraction algorithms. The average RI, AC of our proposed algorithm are higher 36.49%, less 2.47 than those of Poincaré index based algorithm in four FVC2004 datasets, the advantages are more remarkable when block size decreases.

In further research we will apply these ideas to enhance and classify fingerprints.

References

1. D. Maltoni, D. Maio, A. K. Jain, S. Prabhakar: Handbook of Fingerprint Recognition. Springer, New York, (2003)96-99
2. E. R. Herny: Classification and Uses of Finger Prints. George Routledge & Sons, London, 1900
3. V.S.Srinivasan, N.N.Murthy: Detection of singular points in fingerprint images. PR, 25(1992)139-153
4. Marius Ticu, Pauli Kuosmanen: A multiresolution method for singular points detection in fingerprint images, Proc. 1999 IEEE ISCS, 4(1999)183-186
5. X. Wang, J. Li, and Y. Niu: Fingerprint Classification Based on Curvature Sampling and RBF Neural Networks, Lecture Notes in Computer Science, 3497(2005)171-176
6. Asker M. Bazen and Sabih H. Gerez: Systematic methods for the computation of the directional fields and dingular points of fingerprints. IEEE Trans. PAMI, 24(2002)905-919, 2002
7. Maio D, Maltoni D: A structural approach to fingerprint classification. Proc. 13th ICPR, (1996)578-585
8. M. Kawagoe, A. Tojo: Fingerprint pattern classification. PR, 17(1984)295-303
9. K. Karu, A. K Jain: Fingerprint classification, PR, 29(1996)389-404
10. Nojun Kwak, Chong-Ho Choi: Input feature selection by mutual information based on Parzen window. IEEE Trans. PAMI,24(2002)1667-1771
11. L.Hong, Y.Wan, and A. Jain: Fingerprint image enhancement: algorithm and performance evaluation. IEEE Trans. PAMI, 20(1998)777-789

Fingerprint Image Enhancement Based on a Half Gabor Filter

Wonchurl Jang, Deoksoo Park, Dongjae Lee, and Sung-jae Kim

Samsung Electronics, SoC R&D Center, Korea
{wc7.jang, deoksoo.park, djae.lee, sungjae.kim}@samsung.com

Abstract. The performance of a fingerprint recognition system relies on the quality of the input fingerprint images. Several researches have been studied on the enhancement of fingerprint images for fingerprint recognition. The representative enhancement is the adaptive filtering method based on Gabor filter (GF). However, this method is computationally expensive due to the large mask size of GF. In this paper, we propose a half Gabor filter (HGF), which is suitable for fast implementation in spatial domain. The HGF is a modified filter which preserves the frequency property of a GF and reduces the mask size of the GF. Compared with the GF, the HGF not only reduces the processing time approximately by 41% but also enhances the fingerprint image which is as reliable as the GF.

Keywords: Gabor Filter, Gabor Enhancement, Fingerprint Image Enhancement, Adaptive Filter.

1 Introduction

Fingerprint patterns consist of ridges and valleys. These structures provide essential information for recognition. Conventionally, most fingerprint recognition systems use minutiae, a group of ridge end points and bifurcations, as the features of fingerprint patterns. The clearness of the extracted minutiae relies on the quality of the acquired fingerprint image. For this reason, fingerprint recognition systems heavily depend on the quality of the acquired fingerprint image. Hence, we need the image enhancing technique to improve the quality of the fingerprint image. Basically, the fingerprint image enhancement algorithm ought to satisfy two conditions. The first condition is to improve the clarity of ridge and valley structures of the fingerprint images. The second condition is to remove noise within ridge and valley pattern. The GF has the properties of spatial localization, orientation selectivity, and spatial-frequency selectivity [3]. With these properties, the GF satisfies the conditions of the fingerprint image enhancement algorithm [1]. Therefore the GF has been popularly used to enhance the fingerprint image. However this algorithm suffers from a major drawback which is a large computation cost. To solve this problem, we propose a HGF and a half Gabor stabilization filter (HGSF). The HGF is a modified filter which reduces the mask size of a GF and preserves the frequency property of a GF. The HGSF

D. Zhang and A.K. Jain (Eds.): ICB 2006, LNCS 3832, pp. 258–264, 2005.

is a low pass filter which equalizes the frequency domain property of the HGF and GF. The proposed algorithm is faster than the conventional enhancement algorithm based on the GF and saves the memory space for filters. In addition, this algorithm extracts the ridge patterns as reliably as the GF.

2 General Gabor Filter

The GF has been used as a very useful tool to enhance a fingerprint image [1,2,4]. The configurations of the parallel ridges and valleys with well defined frequency and orientation in a fingerprint image provide useful information which helps removing the undesired noise. The sinusoidal-shaped waves of ridges and valleys vary slowly in a local constant orientation. Gabor filters have both frequency-selective and orientation-selective properties in frequency domain [2]. Therefore, it is appropriate to use GF as a bandpass filter to remove the noise and preserve true ridge/valley structures.

The 2-Dimensional GF is a harmonic oscillator, composed of a sinusoidal plane wave of a particular frequency and orientation within a Gaussian envelope. In [1], the general even-symmetric 2D GF is defined as

$$h(x,y,\theta,f_0) = \exp\left\{-\frac{1}{2}\left[\frac{(x\cos\theta)^2}{\delta_x^2} + \frac{(y\sin\theta)^2}{\delta_y^2}\right]\right\}\cos(2\pi f_0 x_\theta) \tag{1}$$
$$x_\theta = x\cos\theta + y\sin\theta$$

where θ stands for an orientation of the GF and f_0 is the frequency of the sinusoidal plane wave (or the center frequency of the GF). Additionally, δ_x and δ_y represent the space constants of the Gaussian envelope along x and y axes. The frequency f_0 and the orientation θ are computed by a inter ridge distance and ridge orientation information [1].

3 Half Gabor Filter and Fingerprint Image Enhancement

In the previous section, we explained the general GF method based on the local ridge orientation and ridge frequency estimated from the input image. Although this algorithm can obtain a reliable ridge structures even for corrupted images, it is unsuitable for an embedded identification system because it spends a significant amount of efforts for GF computation. To improve the efficiency of GF, we propose the HGF and the HGSF algorithms. Figure 1 shows the block diagram (Fig. 1-b) of HGF and the image enhancement module based on HGF (Fig. 1-a). The frequency passband of HGF consists of the general GF term $G(u,v)$ and the phase shifted GF term $G(u-\pi, v-\pi)$. In order to reliably enhance the ridge patterns using the HGF, it is necessary to remove the noise passed by the phase shifted GF term and the filter mask normalization method to prevent the type changing of the enhanced ridge pattern. For this reason, we propose the HGSF, it is a low pass filter, which has a passband defined by the equation (13). Also,

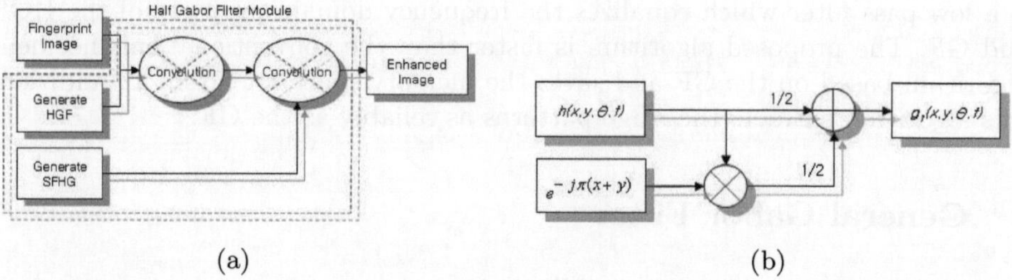

Fig. 1. Fingerprint image enhancement module based on the HGF: (a) Image enhancement module based on the HGF, (b) HGF generator

to prevent the type changing of the enhanced ridge pattern, we normalize the filter mask using the mask coefficient α (6).

The ridge pattern extraction steps are as follows :

Stage 1: Compute the mask coefficient α of HGF using the following equations.

$$p(x,y,\theta_i,f_0)=\begin{cases} h(x,y,\theta_i,f_0) & \text{if} \quad h(x,y,\theta_i,f_0)>0 \\ 0 & \text{otherwise.} \end{cases} \tag{2}$$

$$n(x,y,\theta_i,f_0)=\begin{cases} h(x,y,\theta_i,f_0) & \text{if} \quad h(x,y,\theta_i,f_0)<0 \\ 0 & \text{otherwise.} \end{cases} \tag{3}$$

$$pSum=\sum_{y=0}^{N-1}\sum_{x=0}^{N-1} p(x,y,\theta_i,f_0) \tag{4}$$

$$nSum=\sum_{y=0}^{N-1}\sum_{x=0}^{N-1} n(x,y,\theta_i,f_0) \tag{5}$$

$$\alpha=\frac{|nSum|}{pSum} \tag{6}$$

Here, θ_i is a quantified orientation ($\theta_i = 0, \frac{\pi}{16}, \ldots, \frac{15\pi}{16}$), and f_0 is a local ridge frequency ($f_0 = 0.12$).

Stage 2: Generate a half Gabor mask $g_h(x,y,\theta_i,f_0)$ of $N \times N$(N=15) sizes. However the effective mask size of HGF is $N \times \frac{N}{2}$ because only non-zero elements are used. Figure 2 shows the masks of GF and HGF.

$$m(x,y,\theta_i,f_0)=\begin{cases} \alpha \cdot p(x,y,\theta_i,f_0) & \text{if} \quad h(x,y,\theta_i,f_0)>0 \\ n(x,y,\theta_i,f_0) & \text{otherwise.} \end{cases} \tag{7}$$

$$g_h(x,y,\theta_i,f_0)=\frac{1}{2}\{m(x,y,\theta_i,f_0)+(e^{-j\pi})^{x+y}m(x,y,\theta_i,f_0)\} \tag{8}$$

Stage 3: Convolute a fingerprint image $t(x,y)$ with the HGF mask $g_h(x,y,\theta_i,f_0)$. We get the enhanced image $o(x,y)$. The discrete Fourier transform(DFT) of the image $o(x,y)$ is expressed by the $O(u,v)$.

1	2	3	4	5	6	7	7	7	6	5	4	3	2	1
0	0	-1	-1	-2	-3	-3	-3	-3	-3	-2	-1	-1	0	0
-3	-6	-10	-14	-19	-24	-27	-28	-27	-24	-19	-14	-10	-6	-3
-6	-11	-18	-26	-35	-43	-49	-51	-49	-43	-35	-26	-18	-11	-6
-5	-9	-15	-22	-30	-36	-41	-43	-41	-36	-30	-22	-15	-9	-5
1	1	2	3	4	5	5	5	5	5	4	3	2	1	1
9	16	25	36	48	59	67	70	67	59	48	36	25	16	9
14	23	36	52	69	85	96	100	96	85	69	52	36	23	14
9	16	25	36	48	59	67	70	67	59	48	36	25	16	9
1	1	2	3	4	5	5	5	5	5	4	3	2	1	1
-5	-9	-15	-22	-30	-36	-41	-43	-41	-36	-30	-22	-15	-9	-5
-6	-11	-18	-26	-35	-43	-49	-51	-49	-43	-35	-26	-18	-11	-6
-3	-6	-10	-14	-19	-24	-27	-28	-27	-24	-19	-14	-10	-6	-3
0	0	-1	-1	-2	-3	-3	-3	-3	-3	-2	-1	-1	0	0
1	2	3	4	5	6	7	7	7	6	5	4	3	2	1

(a)

1	0	2	0	5	0	7	0	7	0	5	0	2	0	1
0	0	0	-2	0	-3	0	-4	0	-3	0	-2	0	0	0
-3	0	-10	0	-20	0	-28	0	-28	0	-20	0	-10	0	-3
0	-12	0	-27	0	-44	0	-52	0	-44	0	-27	0	-12	0
-6	0	-16	0	-31	0	-43	0	-43	0	-31	0	-16	0	-6
0	1	0	2	0	4	0	5	0	4	0	2	0	1	0
9	0	25	0	48	0	67	0	67	0	48	0	25	0	9
0	23	0	52	0	84	0	100	0	84	0	52	0	23	0
9	0	25	0	48	0	67	0	67	0	48	0	25	0	9
0	1	0	2	0	4	0	5	0	4	0	2	0	1	0
-6	0	-16	0	-31	0	-43	0	-43	0	-31	0	-16	0	-6
0	-12	0	-27	0	-44	0	-52	0	-44	0	-27	0	-12	0
-3	0	-10	0	-20	0	-28	0	-28	0	-20	0	-10	0	-3
0	0	0	-2	0	-3	0	-4	0	-3	0	-2	0	0	0
1	0	2	0	5	0	7	0	7	0	5	0	2	0	1

(b)

Fig. 2. Examples of the 15x15 GF's mask and the HGF's mask (for $\theta = 0$ and $f = 0.12$): (a) GF mask, (b) HGF mask (A coefficient in colored element is not an effective value)

$$o(x,y) = \sum_{b=0}^{N-1} \sum_{a=0}^{N-1} g_h(a,b,\theta_i,f_0) \cdot t(a-x, b-y) \tag{9}$$

$$O(u,v) = T(u,v) \cdot \frac{1}{2}\{M(u,v,\theta_i,f_0) + M(u-\pi, v-\pi, \theta_i, f_0)\} \tag{10}$$

where $T(u,v)$ and $M(u,v,\theta_i,f_0)$ are the DFT of $t(x,y)$ and $m(x,y,\theta_i,f_0)$.
Stage 4: Apply the HGSF $l(x,y)$ to an enhanced image $o(x,y)$.

$$o_{lpf}(x,y) = \sum_{j=0}^{M-1} \sum_{i=0}^{M-1} l(i,j) \cdot o(i-x, j-y) \tag{11}$$

where l(x,y) is the $M \times M (M = 3)$ sized gaussian filter having the passband which is defined by the equation (13).
Stage 5: Binarize the filtered image $o_{lpf}(x,y)$.

$$b(x,y) = \begin{cases} 1 & \text{if } \; o_{lpf}(x,y) > T_b \\ 0 & \text{otherwise.} \end{cases} \tag{12}$$

From the stage 2, we generate a HGF mask which is half the size of GF mask. If we convolute a $S_x \times S_y$ sized fingerprint image with a $N \times N$ sized Gabor mask $h(x,y,\theta_i,f_0)$, then the computation power is $S_x \times S_y \times N \times N$. On the other hand, if we convolute a fingerprint image $S_x \times S_y$ with a $N \times N/2$ sized half Gabor mask $g_h(x,y,\theta_i,f_0)$, then the computation power is $S_x \times S_y \times N \times N/2$. The half Gabor filtered image $O(u,v)$ consists of the original GF passing image $I(u,v)H(u,v)$ and the phase shifted image $I(u,v)H(u-\pi, v-\pi)$, as shown in figure 3. To get an image such as the original Gabor filtered image, we have to remove the phase shifted image $I(u,v)H(u-\pi, v-\pi)$. For this reason, we apply the HGSF to the half Gabor filtered image. If the HGSF $l(x,y)$ satisfies the condition of the equation (13), then the $o_lpf(x,y)$ is expressed by the general Gabor filtered image as the equation (14).

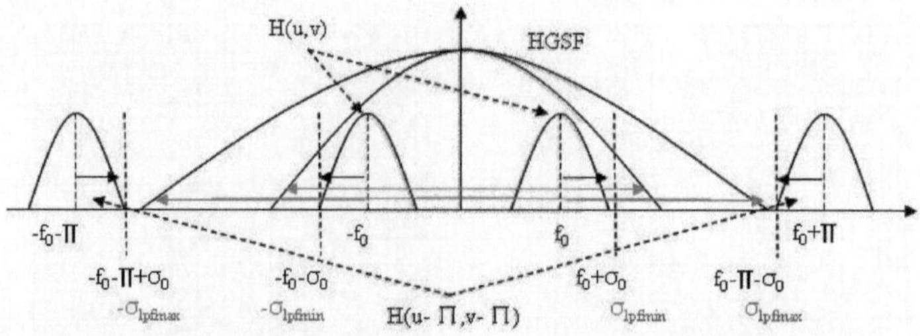

Fig. 3. The frequency property of HGF and the passband of HGSF

$$\sqrt{(f_0+\delta_0)^2} < \sqrt{(u^2+v^2)_{max}} < \sqrt{(f_0+\pi-\delta_0)^2} \tag{13}$$

$$o_{lpf}(x,y) = i(x,y) \otimes \frac{1}{2}h(x,y,\theta,f) \tag{14}$$

Where, $\delta_0 = \delta_x = \delta_y (\delta_0 = 4.0)$, f_0 is a ridge frequency($f_0 = 0.12$), $\sqrt{(u^2+v^2)_{max}}$ is bandwidth of H(u,v).

4 Experimental Results

We evaluated the efficiency and robustness of our algorithm using FVC2002 Database1(DB2) and our collected fingerprint images (DB1), which were captured

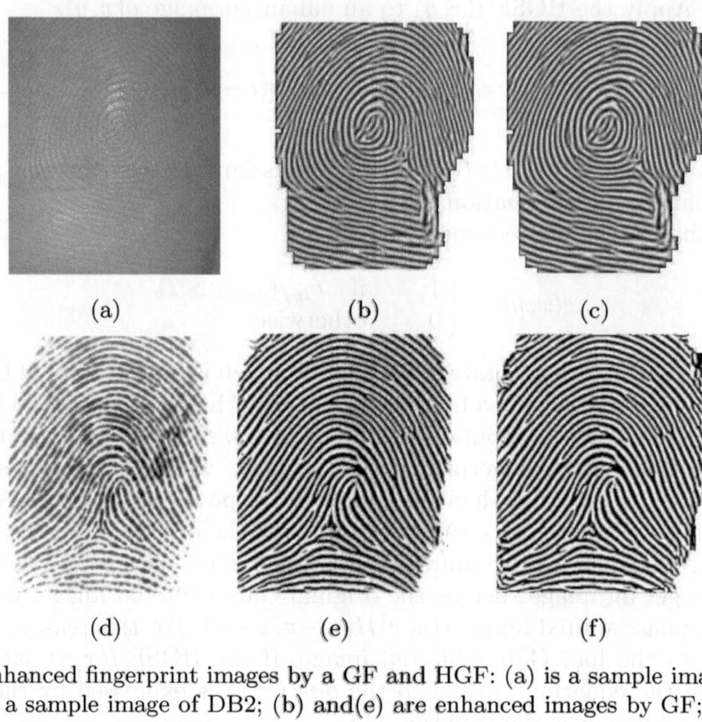

Fig. 4. Enhanced fingerprint images by a GF and HGF: (a) is a sample image of DB1 and (d) is a sample image of DB2; (b) and(e) are enhanced images by GF; (c) and(f) are enhanced images by HGF

Table 1. The performance of minutiae extraction : DMR(Dropped minutiae ratio), EMR(Exchanged minutiae ratio), TMR(True minutiae ratio), and FMR(False minutiae ratio)

Filter	DMR		EMR		FMR		TMR	
	DB1	DB2	DB1	DB2	DB1	DB2	DB1	DB2
GF	7%	3%	2%	3%	7%	4%	91%	94%
HGF	8%	3%	2%	5%	9%	5%	90%	92%

Table 2. The matching performance under the enhanced fingerprint images by HGF and GF

Filter \ DB Type		DB1			DB2		
		FAR		EER	FAR		EER
		0.1%	1.0%		0.1%	1.0%	
GF	FRR	5.24%	2.78%	2.32%	3.38%	1.53%	1.25%
HGF	FRR	5.41%	2.83%	2.41%	3.52%	1.59%	1.36%

Table 3. The time cost of image enhancement and the memory size for filter mask (Gabor orientations : 16 step, Gabor frequencies : 20 step, Gabor mask size : 15x15 pixels, Total number of Gabor masks : 320)

Filter	Time Cost (msec)	Memory Size (Kbyte)
GF	286	1033
HGF	170	557

by a 1.3 mega pixel digital camera. The DB2 consists of 840 fingerprint images (10 fingerprint-images are given by each 84 individuals) with various image qualities. Our experimental results show that our HGF is more efficient than the GF. Figure 4 shows the enhancement results obtained with the HGF and GF. In order to evaluate the performance, we examined the minutiae extracting rate, feature matching rate and time cost of the fingerprint image enhancement. In the examination of the minutiae extracting rate, we compared the minutiae manually taken by the experts with the minutiae automatically extracted using HGF and GF. Table 1 shows the minutiae extraction rate of HGF and GF. The difference between HGF and GF is less than 2% in TMR and FMR (Table 1). In the evaluation of matching performance, the difference between HGF and GF is less than about 0.1% in EER (Table 2). In the embedded system based on ARM-9, the GF takes 286 msec, but the HGF consumes 170 msec reducing 41% of time cost. Also, we can save the memory size for filter mask generation around 46% (Table 3).

5 Conclusions

Generally, the GF is used to enhance the fingerprint image. However the enhancement method based on the GF is computationally very expensive due to the large mask size. In this paper, we proposed an enhancement algorithm based on the HGF and HGSF which reliably improves the clarity of the ridge and valley patterns as well as permits a very efficient implementation in the spatial domain. We developed the HGF which reduces the mask size of the GF by using a frequency domain property of the GF in a fingerprint image. And we designed the HGSF which maintains a frequency domain property of the GF and HGF. The performance of our algorithm was evaluated using the minutiae extracting rate, feature matching rate, time cost and memory consumption. According to the experiment results, our algorithm is more suitable for an embedded system than the presented method based on the general GF.

References

1. L. Hong, Y. Wan, and A.K. Jain, *Fingerprint Image Enhancement: Algorithm and Performance Evaluation*, IEEE Trans. 1998, PAMI-20, (8), pp. 777-789 .
2. Chil-Jen Lee, Sheng-De Wang, and Kuo-Ping Wu, *Fingerprint Recognition Using Principal Gabor Basis Function*, Proceedings of 2001 International Symposium on Intelligent Multimedia, pp. 393-396.
3. J.G. Daugman, *Uncertainty relation for resolution in space, spatial frequency, and orientation optimized by tow-dimensional visual cortical filter*, J. Opt. Soc. Amer. A, vol. 2, no. 7, pp. 1160-1169, 1985.
4. Jianwei Yang, Lifeng Liu, Tianzi Jiang and Yong Fan, *A modified Gabor filter design method for fingerprint image enhancement*, Pattern Recognition Letters vol. 24, pp. 1805-1817, 2003.

Fake Fingerprint Detection by Odor Analysis*,**

Denis Baldisserra, Annalisa Franco, Dario Maio, and Davide Maltoni

DEIS, Università di Bologna,
Viale Risorgimento 2, 40136 Bologna, Italy
{baldisse, franco, maio, maltoni}@csr.unibo.it

Abstract. This work proposes a novel approach to secure fingerprint scanners against the presentation of fake fingerprints. An odor sensor (electronic nose) is used to sample the odor signal and an ad-hoc algorithm allows to discriminate the finger skin odor from that of other materials such as latex, silicone or gelatin, usually employed to forge fake fingerprints. The experimental results confirm the effectiveness of the proposed approach.

1 Introduction

Although the recognition performance of state-of-the-art biometric systems is nowadays quite satisfactory for most applications, much work is still necessary to allow convenient, secure and privacy-friendly systems to be designed. Fingerprints represent today one of the most used biometric characteristics in human recognition systems, due to its uniqueness and reliability. Some recent studies [6] [5] have shown that most of the fingerprint-based recognition systems available on the market can be fooled by presenting to the sensing device a three-dimensional mold (such as a rubber membrane, glue impression, or gelatin finger) that reproduces the ridge characteristics of the fingerprint. While manufacturing a fake finger with the cooperation of the finger owner is definitely quite easy, producing a sufficient quality clone from a latent fingerprint is significantly more difficult; in any case adequate protections have to be studied and implemented to secure the new generation of fingerprint sensing devices.

In the literature, some approaches have been recently presented to deal with the above problem which is often referred to as "fingerprint aliveness detection", i.e. the discrimination of a real and live fingerprint from a fake or deceased one. Some approaches use ad-hoc extra-hardware to acquire life signs such as the epidermis temperature [6], the pulse oximetry and the blood pressure [7], or other properties such as the electric resistance [6], optical characteristics (absorption, reflection, scattering and refraction) or dielectric permittivity [5]. Unfortunately, the performance achieved by most of these methods is not satisfactory, due to the inherent variability of such characteristics. Another aliveness detection method has been recently proposed in [1] where a sequence of fingerprint images is analyzed to detect the perspiration process that typically does not occur in cadaver or artificial

* This work was partially supported by European Commission (BioSec - FP6 IST-2002-001766).

** Patent Pending (IT #BO2005A000398).

D. Zhang and A.K. Jain (Eds.): ICB 2006, LNCS 3832, pp. 265 – 272, 2005.

fingerprints. It is worth noting that, since the only aim of the aliveness detection module is to verify if the fingerprint is real, and not to verify/identify the user, the module is usually integrated into a more complete verification/identification system where aliveness detection is often executed before user recognition.

In this work a new aliveness detection approach based on the odor analysis is presented. The paper is organized as follows: in section 2 a brief introduction to electronic odor analysis is given, in section 3 the hardware system designed for odor acquisition is presented; section 4 describes the odor recognition approach while section 5 reports the experimental results; finally, in section 6, some concluding remarks are given.

2 Electronic Odor Analysis

Everything that has an odor constantly evaporates tiny quantities of molecules, the so called odorants; a sensor able to detect these molecules is called *chemical sensor*. An *electronic nose* is an array of chemical sensors designed to detect and discriminate several complex odors. Odor stimulation to the sensing system produces the characteristic pattern of an odor. Since the strength of the signal in most sensors is proportional to the concentration of a compound, quantitative data can be produced for further processing. Electronic noses are equipped with hardware components to collect and transport the different odors to the sensor array, as well as electronic circuits to digitize and store the sensor response for subsequent signal processing.

Several *electronic noses* are nowadays available on the market [2]. The main applications where electronic noses are employed are [3]: medical diagnosis, environmental applications to identify toxic and dangerous escapes, systems aimed to assess quality in food production and pharmaceutical applications.

Although "odor recognition" is not a novel modality in the biometric system arena (see for example [4]), to the best of our knowledge this is the first approach where the finger odor is used to detect fake fingerprints.

3 The Odor Acquisition System

3.1 The Odor Sensors and the Acquisition Board

Different odor sensors, based on metal-oxide technology (MOS), have been tested in our experiments. Some of these sensors are available in the market (Figaro TGS 2600, Figaro TGS 822, FIS SB-31, FIS SB-AQ1A), other sensors are prototypes produced by an Italian company (SACMI) which is currently developing electronic noses for the food industry. Each of these sensors reacts to some odors while ignoring others: some of them are designed to detect gaseous air contaminants, other are designed to detect organic compounds, etc. All the sensors can be miniaturized enough (few mm^2) to be embedded into very small devices, and the sensor cost is quite small for volume productions (few €).

An electronic board has been developed[1] to drive the different odor sensors and to acquire the odor signals through a PC; the board allows to: 1) heat the sensors to make them working at the proper temperature (200 – 400 °C); 2) tune and modify the sensors operating point, the offset and to compensate for thermal deviation; 3) pre-amplify and pre-elaborate the signals provided by the MOS sensors; 4) convert (A/D) the pre-amplified analog signals into (10-bit resolution) digital signals; 5) sample the odor signal (of the pre-selected sensor) every few ms and send it to a PC via RS-232 interface. It is worth noting that embedding MOS odor sensors into a fingerprint scanner is not straightforward and special care must be taken to guarantee that the same part of skin which is sensed for identity verification is also sensed for odor analysis.

3.2 The Acquisition Process

The acquisition of an odor pattern consists of sampling the data coming from an odor sensor during a given time interval, usually few seconds. A typical acquisition session is composed of three different stages: calibration, recording and restoration.

When the system is idle (i.e., there are no fingers placed on the sensor surface), it periodically read data from the electronic board to establish (and update) a baseline response, denoted as "response in fresh air". This operation, called *calibration*, is continuously performed in background since the prototype version of the system works in an open environment and the sensors are thus exposed to environmental changes (e.g. breathing close to the odor sensors or accidental sprinkling of particular substances). The *recording* stage measures the sensor response when a finger is placed on the sensor surface. The user's finger has to be placed on the odor sensor surface for a few seconds and then lifted. Finally, the *restoration* stage starts when the finger is lifted from the sensor surface and is aimed at restoring the sensor to its initial conditions. The time necessary to restore the sensor response may vary depending on the sensor characteristic and environmental condition (a typical time interval for the sensors used is 10-15 seconds).

4 Odor Recognition

4.1 Data Processing

Let X be an acquisition sequence consisting of n sensor readings X={$\mathbf{x}_1$, $\mathbf{x}_2$,...,$\mathbf{x}_n$}; each reading is represented by a two dimensional vector $\mathbf{x}_i=[x_i^t, x_i^v]^T$ where x_i^t denotes the elapsed time since the beginning of the acquisition and x_i^v the recorded voltage ($x_i^v \in [0,V]$, where V=5 in our acquisition system). The first sample is acquired at the beginning of the acquisition stage; the acquisition covers all the recording stage (5 seconds) and the first 8 seconds of the restoration stage. The

[1] The electronic board has been developed by the Italian company Biometrika, which is one of the DEIS (University of Bologna) partners in the BioSec project (IST-2002-001766).

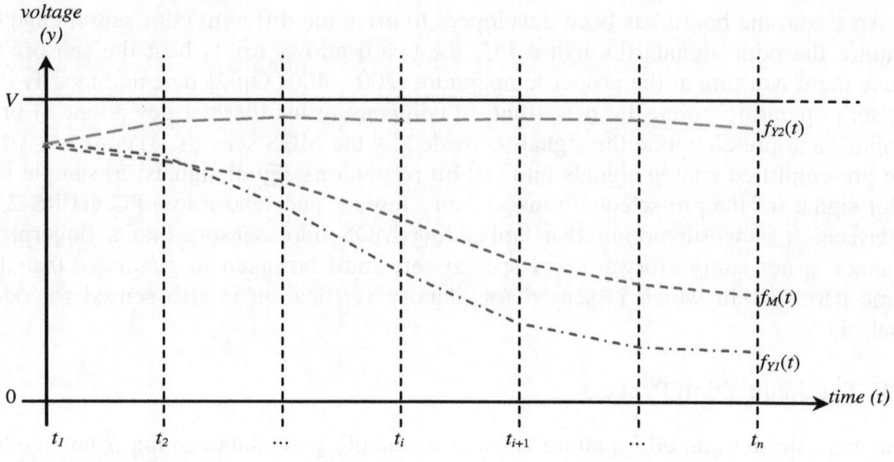

Fig. 1. Three piecewise linear functions $f_M(t)$, $f_{Y1}(t)$ and $f_{Y2}(t)$ representing the stored user's template M and the acquisition sequences of two artificial fingerprints (Y_1 and Y_2) forged using gelatine and silicone respectively

sampling frequency is about 100 Hz. The acquired sequence is then interpolated and downsampled in order to: 1) obtain the voltage values at predefined and regular intervals of width Δt (200 ms in our experiment); 2) partially smooth the data and reduce noise. The processed sequence $Y=\{y_1, y_2,\ldots,y_n\}$ has length n and each element y_i represents the voltage value at time $t_i = t_1 + i \cdot \Delta t$. We indicate with $f_Y(t)$ the piecewise linear function interpolating the sequence Y, obtained by connecting each couple of consecutive points (y_i, y_{i+1}) by a straight line (see **Fig. 1**).

A template, consisting of an acquisition sequence $M=\{m_1, m_2,\ldots, m_n\}$, represented by the piecewise linear function $f_M(t)$, is created for each new user enrolled into the system. The aliveness verification of a user fingerprint is carried out by comparing the function $f_Y(t)$ and $f_M(t)$ representing the newly acquired data Y and the user's stored template M respectively. The comparison between the two functions is based on the fusion of three different features extracted from the sequences: the function trend, the area between the two functions and the correlation between the two data sequences. The three similarity values are combined to produce the final decision.

4.1.1 Function Trend

Some preliminary experiments showed that, when the odor sensors are exposed to skin or gelatin, the acquired voltage gradually decreases, while when exposed to other substances such as silicone or latex the voltage increases (see Fig. 1); analyzing the trend of the curve, allows a first distinction between these two groups of compounds to be made. The trend is analyzed on the basis of the angle between the two functions and the horizontal axis. The angle α_i between $f_M(t)$ and the horizontal axis, in the interval $[t_i, t_{i+1}]$, is calculated as: $\alpha_i = \arctan\left(\frac{f_M(t_i) - f_M(t_{i+1})}{\Delta t}\right)$

The angle β_i of $f_Y(t)$ in the interval $[t_i, t_{i+1}]$ is computed analogously. Intuitively the similarity value should be higher if the two functions are concordant (both increasing of both decreasing in the considered interval), and lower otherwise. The similarity s_i^{trend} is thus calculated as follows:

$$s_i^{trend} = \begin{cases} 1-\left(|\alpha_i-\beta_i|+\pi\right)/2\pi & \text{if } ((\alpha_i>0)\text{ and }(\beta_i<0))\text{ or }((\alpha_i<0)\text{ and }(\beta_i>0)) \\ 1-\left(|\alpha_i-\beta_i|\right)/2\pi & \text{if } ((\alpha_i>0)\text{ and }(\beta_i>0))\text{ or }((\alpha_i<0)\text{ and }(\beta_i<0)) \end{cases}$$

The overall trend similarity is given by a simple average of the similarity values s_i^{trend} over all the intervals: $s^{trend} = \sum_{i=1}^{n} s_i^{trend} \Big/ n$. Please note that, since $s_i^{trend} \in [0,1]$, the overall similarity s^{trend} is a value in the interval [0,1] as well.

4.1.2 Area Between the Two Functions

For a single interval $[t_i, t_{i+1}]$ the area between $f_Y(t)$ and $f_M(t)$ is defined as:

$$d_i = \int_{t_i}^{t_{i+1}} |f_Y(t) - f_M(t)|\, dt$$

The piece-wise form of the two functions (see Fig. 1) allows a simple expression to be derived for d_i: $d_i = \left|\frac{\Delta t}{2}\cdot\left(f_Y(t_i)+f_Y(t_{i+1})\right)-\frac{\Delta t}{2}\cdot\left(f_M(t_i)+f_M(t_{i+1})\right)\right|$

Since the voltage values are constrained to the interval $[0,V]$, a local upper bound d_i^{UB} to the distance from the template function $f_M(t)$ in the interval $[t_i, t_{i+1}]$ can be estimated as the maximum area between $f_M(t)$ and the two horizontal axis of equation $f(t)=0$ and $f(t)=V$ (maximum voltage value) respectively:

$$d_i^{UB} = \int_{t_i}^{t_{i+1}} \max\left(f_M(t),\ V - f_M(t)\right)\, dt$$

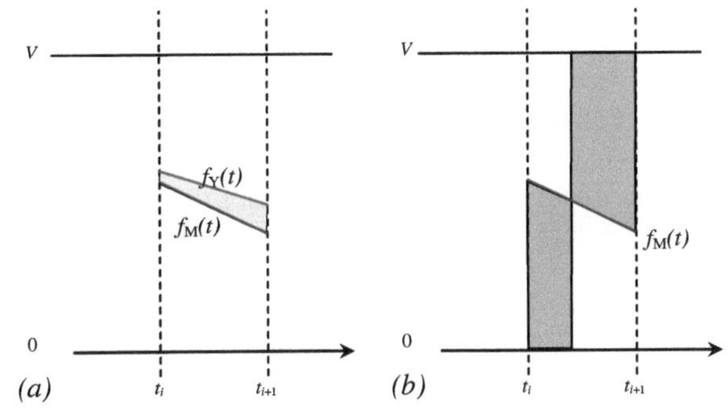

Fig. 2. *(a)* Distance in terms of area between the user's template M, approximated by the function $f_M(t)$, and the current input Y represented by $f_Y(t)$; *(b)* local upper bound d_i^{UB} (grey area) to the distance from the template function $f_M(t)$ in the interval $[t_i, t_{i+1}]$

In Fig. 2a an example of the distance between the user's template and the current input is given; in Fig. 2b the area representing the normalization factor is highlighted. The similarity in terms of area between the two functions in a generic interval $[t_i, t_{i+1}]$ is then simply defined as: $s_i^{area} = 1 - \frac{d_i}{d_i^{UB}}$. The overall similarity in the interval $[t_1, t_n]$ is calculated by averaging the similarity values s_i^{area} over all the intervals: $s^{area} = \sum_{i=1}^{n} s_i^{area} \Big/ n$.

4.1.3 Correlation

The correlation is a useful statistical indicator that measures the degree of relationship between two statistical variables represented in this case by the two data sequences Y and M. Let $\bar{y}$ ($\bar{m}$) and σ_Y (σ_M) be the mean value and the standard deviation of the data sequence Y (M) respectively. The correlation between the two data sequences, considering the whole interval $[t_1, t_n]$ is simply defined as:

$$\rho_{Y,M} = \frac{\frac{1}{n}\sum_{i=1}^{n}(y_i - \bar{y})(m_i - \bar{m})}{\sigma_Y \cdot \sigma_M}$$

Since the correlation value $\rho_{Y,M}$ lies in the interval [-1,1], a similarity value in the interval [0,1] is derived by the simple formula $s^{corr} = (\rho_{Y,M} + 1)/2$.

4.1.4 Final decision

Let w^{trend}, w^{area} and w^{corr} be the weights assigned to the trend, the area and the correlation similarities respectively. The final score is calculated as the weighted average of the three values: $s = w^{trend} \cdot s^{trend} + w^{area} \cdot s^{area} + w^{corr} \cdot s^{corr}$

The fingerprint is accepted as a real one if the final score s is higher than a predefined threshold *thr*.

5 Experimental Results

In this section the experiments carried out in order to evaluate the fake fingerprint detection approach are presented. Though several odor sensors have been considered in this work, for the sake of brevity only the results obtained by one of the most promising sensors (FIGARO TGS 2600) are here detailed. The database used for testing consists of 300 acquisitions of real fingerprints obtained by capturing 10 odor samples of 2 fingers for each of the 15 volunteers, and 90 acquisitions of artificial fingerprints obtained by capturing 10 odor samples of 12 fingerprints forged using different compounds (3 using the bi-component silicone Prochima RTV 530, 3 using natural latex and 3 using gelatine for alimentary use). An additional validation set, whose acquisitions have not been subsequently used for testing, has been acquired to tune the parameters of the algorithm. It consists of 50 acquisitions of real fingerprints, obtained by capturing 5 odor samples of 2 fingers for each of the 5 volunteers, and 30 acquisitions of artificial fingerprints obtained by capturing 10 odor samples of 3 artificial fingerprints forged each using one of the materials described above. The system was tested by performing the following comparisons:

- *genuine recognition attempts*: the template of each real fingerprint is compared to the remaining acquisitions of the same finger, but avoiding symmetric matches;
- *impostor recognition attempts*: the template of the first acquisition of each finger is compared to all the artificial fingerprints.

Then the total number of genuine and impostor comparison attempts is 1350 and 2700, respectively. The parameters of the method, tuned on the validation set, have been fixed as follows: w^{trend}=0.3, w^{area} = 0.5, w^{corr} = 0.2. The equal error rate (EER) measured during the experiments is 7.48%, corresponding to a threshold *thr*=0.9518. In **Fig. 3** the ROC curve, i.e. false rejection rate (FRR) as a function of false acceptance rate (FAR), is reported. An analysis of the results show that, while it's relatively easy to detect fake fingerprints forged using some materials such as silicone, some problems persist in presence of other compounds (e.g. gelatine) for which the sensor response is similar to that obtained in presence of human skin. Since different sensor present different responses to a particular material, a possible solution to this problem is the combination of data acquired by different odor sensors to obtain a more robust system.

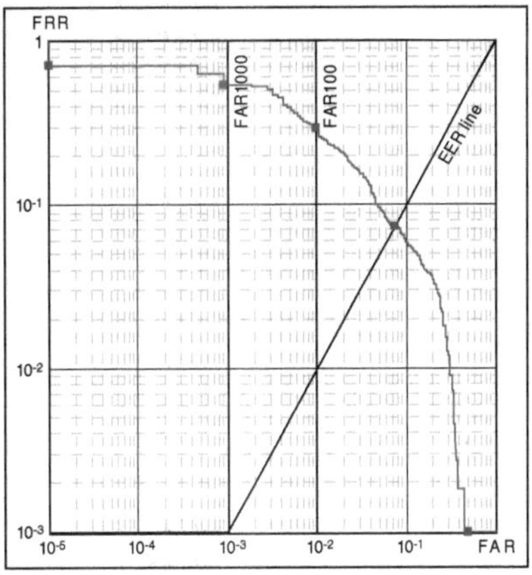

Fig. 3. ROC curve of the proposed approach

6 Conclusions

In this work a new approach to discriminate between real and fake fingerprints is proposed. The method is based on the acquisition of the odor by means of an electronic nose, whose answer in presence of human skin differs from that obtained in presence of other materials, usually employed to forge artificial fingerprints. The experimental results confirm that the method is able to effectively discriminate real fingerprints from artificial reproductions forged using a wide range of materials.

As to future research, we intend to investigate other similarity measures to compare the user's template with the current input. Moreover the creation a single model of human skin, instead of a template for each user, will be evaluated.

References

[1] Derakhshani R., Scuckers S., Hornak L., O'Gorman L., "Determination of Vitality From A Non-Invasive Biomedical Measurement for Use in Fingerprint Scanners", *Pattern Recognition*, vol. 17, no. 2, pp. 383-396, 2003.

[2] Harwood D., "Something in the air", *IEE Review*, vol. 47, pp. 10-14, 2001.

[3] Keller, P. E., "Electronic noses and their applications", *IEEE Technical Applications Conference and Workshops Northcon*, pp. 116- 120, 1995.

[4] Korotkaya Z., "Biometric Person Authentication: Odor", available at http://www.it.lut.fi/kurssit/03-04/010970000/seminars/Korotkaya.pdf

[5] Matsumoto T., Matsumoto H., Yamada K., Hoshino S., "Impact of Artificial "Gummy" Fingers on Fingerprint Systems", in *Proc. SPIE*, pp. 275-289, 2002.

[6] Putte T.v.D., Keuning J., "Biometrical Fingerprint Recognition: Don't Get Your Fingers Burned", in *Proc. Working Conference on Smart Card Research and Advanced Applications*, pp. 289-303, 2000.

[7] Schuckers S.A.C., "Spoofing and anti-spoofing measures", *Information Security Technical Report*, vol. 7, pp. 56-62, 2002.

Ridge-Based Fingerprint Recognition

Xiaohui Xie, Fei Su, and Anni Cai

Abstract. A new fingerprint matching method is proposed in this paper, with which two fingerprint skeleton images are matched directly. In this method, an associate table is introduced to describe the relation of a ridge with its neighbor ridges, so the whole ridge pattern can be easily handed. In addition, two unique similarity measures, one for ridge curves, another for ridge patterns, are defined with the elastic distortion taken into account. Experiment results on several databases demonstrate the effectiveness and robustness of the proposed method.

Keywords: fingerprint recognition, point-pattern matching, ridge sampling, ridge matching.

1 Introduction

Minutiae (fingerprint ridges' bifurcations and ends) are commonly employed as the basic features in most fingerprint recognition algorithms. In such circumstances, fingerprint recognition can be regarded as a point-set matching problem, where the best match with the maximal number of corresponding point pairs in the two point sets is searched under certain error restriction. Many solutions have been proposed to solve this problem [1][2][3][4][5]. Most of the proposed methods are based on a rigid-body model, and do not have a proper way to handle the elastic distortion problem in fingerprint matching. In addition, there always exist some quality problems on fingerprint images collected, and fake minutiae may be generated during feature extraction process because of noise on fingerprint images. Most of the current algorithms could not do well at these circumstances.

In order to solve the problems mentioned above, in addition to minutiae, more fingerprint features such as global features (center and triangle points) or ridge features (ridge flow and ridges count between two minutiae) are introduced by some researchers to decrease the possibility of error occurred during matching. However, the features newly introduced also have elastic distortion, and thus these methods could not solve the problems ultimately. Looking for more robust and more efficient fingerprint matching algorithms is still a challenge problem.

Usually we can obtain skeleton images through enhancement, segmentation, binarization, and thinning stages of common fingerprint image preprocessing, and ridges in the skeleton image are single-pixel-wide curves. The skeleton image contains not only all of the minutiae information but also the whole ridge pattern. There has been few work on ridge-pattern-based fingerprint matching published in the literature. In this paper, we propose a novel fingerprint matching method with which two fingerprint ridge images are directly matched. The main

D. Zhang and A.K. Jain (Eds.): ICB 2006, LNCS 3832, pp. 273–279, 2005.

contributions of this work are two folds: First, an associate table is introduced to describe the relation of a ridge with its neighbor ridges, and consequently the whole ridge pattern can be easily handled; secondly, by taking the elastic distortion into account, two unique similarity measures, one for ridge curves, another for ridge patterns, are defined. These make this algorithm effective and robust.

The rest of the paper is organized as follows: In section II, we introduce a way to obtain skeleton ridge images from the gray-scale fingerprint images; In section III, the proposed method is presented; Experiment results are given in section IV; Section V provides the conclusions and the future work.

2 Fingerprint Skeleton Image

Fingerprint skeleton image can be obtained through the common preprocess procedure which includes segmentation, filtering, binarization and thinning stages. However, this preprocess procedure exists some problems when used for ridge extraction since it was tuned to minutiae extraction. Also the filtering stage are often time consuming.

Maio and Maltoni [6] presented a novel approach to extract minutiae directly from gray-level fingerprint images. With their algorithm ridges can be extracted by following the ridges until they terminate or intersect with other ridges. As the fingerprint image need not be filtered at every pixel, the computational complexity of the algorithm is low. We modified Maio's method in the following way to obtain skeleton images. First, ridges are extracted in high-quality image areas with Maio's method, and then more paths are searched and a strict stop criterion is adopted during ridge following in blurred image areas. Finally we employ the method proposed by Chen [7] to connect the broken ridges caused by scars, dryness or other reasons. A sample skeleton ridge image is shown in Fig. 1.

(a) Origin fingerprint image (b) Skeleton ridge image

Fig. 1. A skeleton image

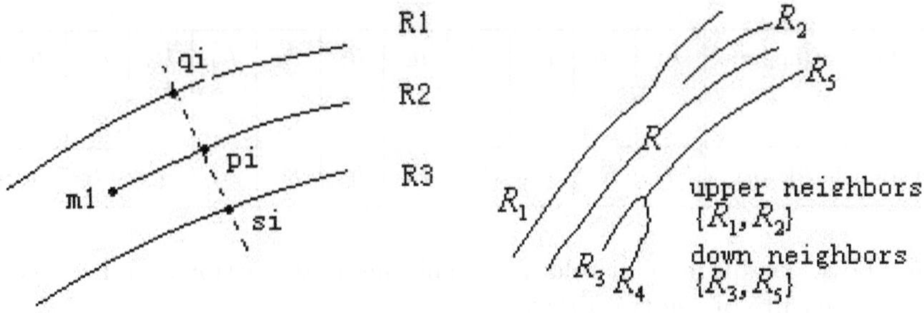

(a) Associate points (b) Ridge neighbors

Fig. 2. The neighborhood of ridges

3 Ridge Matching

As shown in Fig.2(a), ridges $R1$ and $R3$ are neighbor ridges of ridge $R2$. A ridge curve may have more than one neighbor on its each side in the skeleton image. The neighborhood relationships among ridges are invariant during one's life time and are robust to elastic distortions of fingerprint images. These steady relationships make the base of the ridge-based fingerprint matching method proposed by us.

Define a direction for a ridge along which the ridge following procedure is performed. Then the left-hand-side neighbors of the ridge are called its upper neighbors and the right-hand-side neighbors are called its down neighbors (see Fig.2(b)). Suppose to draw a line at point p_i normal to ridge $R2$, the line intersects $R1$ at q_i and $R3$ at s_i, and q_i and s_i are called p_i 's associate points.

3.1 Similarity Measure of Two Ridge Curves

Suppose P_m and P_n are respectively the starting point and the ending point of ridge f , and P_m and P_n could be ridge end, ridge bifurcation or ridge broken points. The curvature γ of curve f is defined as:

$$\gamma = \int_{P_m}^{P_n} |d^2 f| \tag{1}$$

γ describes a curve's winding degree, and it's an invariant to image rotation and translation.

Suppose the lengths of two ridges f_1 and f_2 are d_1 and d_2 respectively, and the starting and ending points of f_1 and f_2 are not ridge broken points, we say these two ridges pre-matched to each other if the following conditions are satisfied:

$$\begin{cases} |(d_2 - d_1)/d_2| \leq th_1 \\ \varsigma = 1 - |(1-\kappa)/(1+\kappa)| \cdot |(\gamma_{f_1} - \gamma_{f_2})/(\gamma_{f_1} + \gamma_{f_2})| \geq th_2 \end{cases} \tag{2}$$

Where κ is the stretch factor of ridge f_1 and f_2, and is defined as:

$$\kappa = d_1/d_2 \tag{3}$$

Table 1. Associate table

Associate point(upper)	R_1	R_1	R_1	R_1	R_2	R_2	R_2	R_3	R_3	R_3	R_3	...
Sampling point	p_0	p_1	p_2	p_3	p_4	p_5	p_6	p_7	p_8	p_9	p_{10}	...
Associate point(down)	R_4	R_4	R_4	R_4	R_4	R_4	R_4	R_4	R_5	R_5	R_5	...

The above conditions can tolerate small elastic distortions, and ς gives the similarity measure of the two ridges.

3.2 Associate Table

As shown in Fig.2(b), there may exist more than one upper neighbor and down neighbor for one ridge. We will describe the relationships of a ridge with its neighbors by using a table, which is called associate table. The associate table is constructed with the following way.

We sample ridge R with interval d from its starting point to end point, and obtain one sampling point-set Θ and its associate point-sets Ψ_{up} and Ψ_{down}. All the points in Ψ_{up} and Ψ_{down} are labelled by their corresponding ridges (NULL for empty). The labels and the sampling point set Θ make up of ridge R's associate table. A typical ridge associate table is shown in table I.

Assume that the length of the shortest ridge is not less than 7 pixels in our system, and ridges shorter than 7 pixels are always generated by noise. Thus we choose the sampling interval of 7 pixels, although using dynamic sampling interval according to the ridge stretch factor can depict the neighborhood relationships of the ridge more accurate. The associate tables of all ridges contain all information and features the image has.

3.3 Ridge Matching Procedure

Ridge matching is performed by using ridge associate tables and travelling all the ridges. Suppose $R_{I_1} = \{r_i \,|i \le M\}$ and $R_{I_2} = \{r_j \,|j \le N\}$ are the skeleton ridge sets of fingerprint images I_1 and I_2 respectively. The procedure of matching I_1 and I_2 can be described as below:

1. Calculate each ridge's curvature in R_{I_1} and R_{I_2}, and compare ridge pairs which have the same type of starting and ending points. If the pair of ridges satisfies the conditions stated in section III part A, the pair of ridges is pre-matched. Arrange the matched ridge pairs in descending order according to their similarity measures. These pairs of ridges will be used as the initial pairs for matching. Multiple initial pairs may be needed for proper alignment of the two images.
2. Choose the first ridge pair of the initial set and record their starting points into the task queue.
3. Get one task point pair from the task queue, and sample the corresponding ridges (R_a and R_b).

4. Construct the associate tables of R_a and R_b, and put the associate points of the starting points of R_a and R_b into the task queue.
5. Check the associate tables of the two ridges and find the maximal matched length m of R_a and R_b. This is done in the following way. First set $m = 0$, and then:
   (a) Check the ridge labels of the consecutive upper associate points starting from the mth sampling points of R_a and R_b. If the ridge labels of the upper associate points of the $(m+i)^{th}$ sampling point $(i \geq 3)$ in either of the two tables is changed, update $m = m + i$ and $i = 0$; Put the starting point pair of the new neighbor ridges into the task queue, and go to (b);
   (b) Check the ridge labels of the consecutive down associate points starting from the mth sampling point of R_a and R_b. If the ridge labels of the down associate points of the $(m+j)^{th}$ sampling point $(j \geq 3)$ in either of the two tables is changed, update $m = m + j$ and $j = 0$; Put the starting point pair of the new neighbor ridge into the task queue, and go to (a);
   The above loop stops if no further match can be found.
6. According to the result obtained at step 5), we obtain the newly matched relation of *Ra* and *Rb* from the starting point to the mth sampling point.
7. According to the result obtained at step 5), suppose ridge labels of the consecutive associate points do not have changes from i to j, $R^{'}$ and $R^{''}$ are ridges labels of the corresponding associate points respectively, we obtain the newly matched relation of $R^{'}$ and $R^{''}$ from sampling point i to j when $(j - i) \geq 3$ is satisfied.
8. If the newly matched ridges conflict with the previous matching results, i.e. if there already exists a ridge segment (longer than 3 times of sampling intervals) in R_{I_1} matched with the newly matched ridge segment in R_{I_2}, or vice versa, stop the matching procedure, and return to step 2) to restart the matching procedure by choosing a new initial ridge pair.
9. If there is no matching confliction, return to step 3). Matching goes on until the task queue is empty.
10. Calculate the matching score according to Eq.(4) presented in the next subsection. If the score is larger than a threshold, the whole matching procedure stops; if not, return to step 2) to restart the matching procedure by choosing a new initial ridge pair. The maximal matching score resulted from the different initial pairs gives the final result.

3.4 Similarity Measure of Two Ridge Patterns

The similarity measure of two fingerprints is defined as:

$$score = N/(C \times distortion) \tag{4}$$

Where N is the total length of all matched ridges, more ridges matched would achieve higher score; C is a scaling constant, and the distortion is defined as follows:

$$distortion = \sum_{i,j}^{|P|} |(|p_i p_j| - |q_i q_j|)|/(|P| \cdot (|P| - 1)) \tag{5}$$

Where $p_i, p_j \in P$, $q_i, q_j \in Q$, P and Q are two point sets containing the termination points of all the matched ridge pairs, $|P|$ denotes the number of elements in P. The distortion describes the distortion between the ridge structures formed by matched ridge pairs, wrong matched ridge pair always leads to higher distortion value and lower score.

4 Experiment Results and Performance Evaluation

We tested our algorithm on the database of FVC2002[8], which contains 4 testing data sets and every set has 800 gray-level fingerprint images. The images in one set came from 100 different fingers with 8 sample images each. We matched every two fingerprints for each data set, which means 2800 times true matching and 36000 times false matching. The average matching time is 0.025s to 0.33s per

Table 2. Results on database of FVC2002

EER	DB1	DB2	DB3	DB4
Matching Ridges	0.35	0.63	1.45	0.7
Matching minutiae[9]	0.78	0.95	3.1	1.15

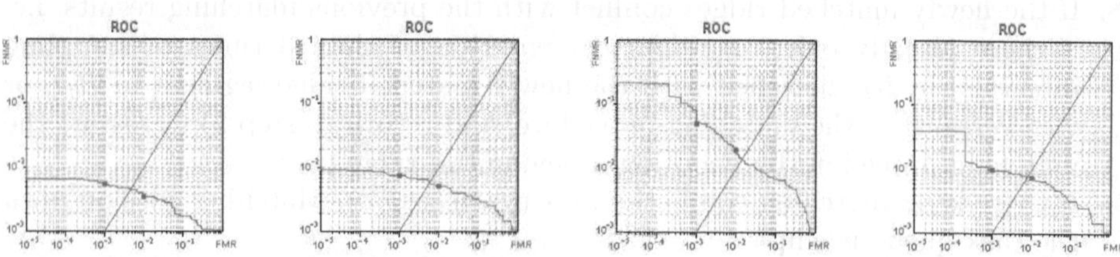

(a) ROC curve on DB1 (b) ROC curve on DB2 (c) ROC curve on DB3 (d) ROC curve on DB4

Fig. 3. ROC curves on FVC2002 databases

(a) Image I_1 of Sample A (b) Image I_2 of Sample A (c) Image I_1 of Sample B (d) Image I_2 of Sample B

Fig. 4. Ridge based fingerprint matching results

match by using a laptop with a PIII 866 CPU. Comparisons between this method and minutiae based method proposed in paper [9] on the four data sets are given in table II, The result shows that the algorithm has better performance than that in [9]. Fig.3 gives ROC curves on the four databases, and Fig.4 shows two examples of the matched images from the same finger. From figure 4, we can find that the method proposed in this paper not only handles the elastic distortion problem well but also helps to eliminate the matching uncertainty (such as caused by not having enough minutiae) since it fully utilizes the ridge information.

5 Summary and Future Work

In this paper, we have presented a novel fingerprint matching algorithm based on ridge structures. The method matches fingerprint skeleton images directly. Associate tables are introduced in this method to describe the neighborhood relations among ridge curves. Also two unique similarity measures, which properly handle the elastic distortions, are defined. Thus better performance is achieved by this method compared to minutiae-based matching method. However, future research is still needed on this method: match ridges more effectively, find fast ways to construct ridge associate tables, find more effective rules to follow matched or unmatched ridges. Blurred image area could generate fake ridges, and how to introduce fuzzy theory in ridge extraction stage is also important.

References

[1] A.K.Jain, L.Hong, and R.M.Bolle. *On-line Fingerprint Verification.* IEEE Trans. on Pattern Analysis and Machine Intelligence. 19 (4): 302-313, April 1997.

[2] N.K.Ratha, K.Karu, S.Chen, and A.K.Jain. *A Real-time Matching System for Large Fingerprint Database.* IEEE Trans. on Pattern Analysis and Machine Intelligence. 18 (8): 799-813. Aug 1996

[3] N.K.Ratha, R.M.Bolle, V.D.Pandit, V.Vaish. *Robust Fingerprint Authentication Using Local Structural Similarity.* Applications of Computer Vision, 2000, Fifth IEEE Workshop on., 4-6 Dec. 2000 Page(s) : 29-34

[4] Z.Chen, C.H.Kou, *A Toplogy-based Matching Algorithm for Fingerprint Authentication.* Security Technology. 1991. Proceedings. 25th Annual 1991 IEEE International Carnahan Conference on , 1-3 Oct. 1991 Page(s): 84-87

[5] D.K.Isenor and S.G.Zaky. *Fingerprint Identification Using Graph Matching. Pattern Recognition.* 19(2): 113-122, 1986

[6] D.Maio and D.Maltoni, Direct Gray-Scale Minutiae Detection in Fingerprints. IEEE Trans. PAMI 19(1):27-40, 1997

[7] Chen PH and Chen XG, A New Approach to Healing the Broken Lines in the Thinned Fingerprint Image. Journal of China Institute of Communications. 25(6):115-119, June 2004

[8] D.Maio, D.Maltoni, R.Cappelli, J.L.Wayman, A.K.Jain. *FVC2002: Second Fingerprint Verification Competition.* Pattern Recognition, 2002, Proceedings. 16th International Conference on., 11-15 Aug. 2002 Page(s): 811-814 vol.3

[9] Xiaohui Xie, Fei Su, Anni Cai and Jing'ao Sun, "A Robust Fingerprint Matching Algorithm Based on the Support Model". Proc. International Conference on Biometric Authentication (ICBA), Hong Kong, China, July 15-17, 2004

Fingerprint Authentication Based on Matching Scores with Other Data

Koji Sakata[1], Takuji Maeda[1], Masahito Matsushita[1], Koichi Sasakawa[1], and Hisashi Tamaki[2]

[1] Advanced Technology R&D Center, Mitsubishi Electric Corporation, 8-1-1, Tsukaguchi-Honmachi, Amagasaki, Hyogo, 881-8661, Japan
[2] Faculty of Engineering, Kobe University, 1-1, Rokkodai, Nada, Kobe, Hyogo, 657-8501, Japan

Abstract. A method of person authentication based on matching scores with the fingerprint data of others is proposed. Fingerprint data of others is prepared in advance as a set of representative data. Input fingerprint data is verified against the representative data, and the person belonging to the fingerprint is confirmed from the set of matching scores. The set of scores can be thought of as a feature vector, and is compared with the feature vector already enrolled. In this paper, the mechanism of the proposed method, the person authentication system using this method are described, and its advantage. Moreover, the simple criterion and selection method of the representative data are discussed. The basic performance when general techniques are used for the classifier is FNMR-3.6% at FMR-0.1%.

1 Introduction

Generally, biometric authentication systems either use the biometric data as is or use some processed version of the biometric as feature data. There is a real danger with this kind of authentication that if the enrolled data is leaked, the leaked data could be used to impersonate the legitimate user for illegitimate purposes. When a password is used for authentication, all you need do is to change the password in the event that the password is leaked, but biometric data cannot generally be changed. Then, the method of making former data not restorable is proposed as a protection method of the enrolled data. Biometric data is transformed by one way function or geometrical conversion[1]. Moreover, biometric data is protected by using the cryptology, and there is a method of correcting swinging of the input image by using helper data[2].

Now we use fingerprint authentication scheme using features extracted from fingerprint images[3]. We propose a method of fingerprint matching based on matching scores with other data[4]. A set of representative data is prepared in advance, and the set of scores obtained by verifying the input data against the set is regarded as a feature vector.

First we will provide an overview of conventional matching and proposal methods. Moreover, the person authentication system using this method is described,

D. Zhang and A.K. Jain (Eds.): ICB 2006, LNCS 3832, pp. 280–286, 2005.

and it explains the advantage. Next, we consider what feature data is suitable for the representative data, and a simple criterion is discussed. Finally, general techniques are applied to the classifier, and the basic performance of the correlation matching is clarified.

2 Conventional Matching and Correlation Matching

In this section we describe differences of conventional matching and correlation matching.

2.1 Conventional Fingerprint Matching

In conventional matching, features extracted from fingerprint image or the image are verified. Important to note is that, while there are some differences in the data being verified, there is no difference in the enrollment of his fingerprint data to the system (Fig. 1).

Since having the user's biometric data is required for conventional matching, this means that the data has to be stored somewhere in an authentication system. If the user's biometric data is retained, then there is an inherent risk that the data could be leaked. Various schemes have been proposed for encrypting or somehow transforming the enrollment data to reduce the risk, but that does not alter the fact that the individual's biometric data enrolled in the system. Since biometric data cannot readily be changed, a user whose data had been leaked might be compelled to use a different finger for authentication or some other equally inconvenient tactic.

2.2 Correlation Matching

Here we present an overview of correlation matching, a fingerprint matching technique that does not require enrollment of biometric data. Fig. 2 shows a schematic overview of correlation matching. Correlation matching requires that a number of fingerprint data used for matching are prepared in advance. This

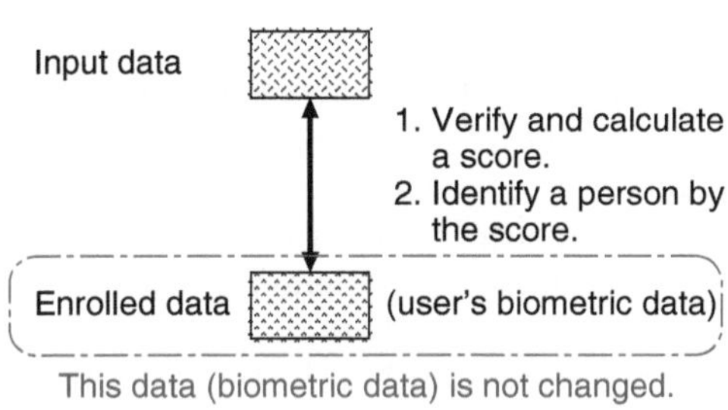

Fig. 1. Conventional matching method. The individual's own data is necessary for verification.

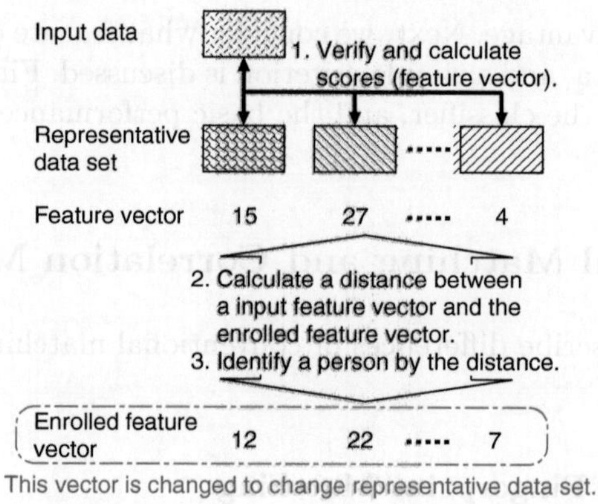

Fig. 2. Overview of correlation matching. Input data is verified with individual data items in the representative data set to derive a feature vector. The most simple matching method is to calculate the distance between the input feature vector and previous enrolled feature vector.

set of fingerprint data is called a set of representative data. Input data of a user is not verified with his enrolled biometric data, but rather is verified against his representative data items. The set of scores obtained by verifying the input data against his representative data items can be thought of as a feature vector. The distance is then calculated between this feature vector and other enrolled feature vectors derived previously by the same procedure, and the person is identified by the distance.

Here, it explains the calculation time. In this method, it is assumed to the verification of input data and representative data to use the conventional matching. If it takes n second for the verification in the conventional matching, it will take the $M \times n$ second in the correlation matching to calculate the feature vector. M is assumed to be a number of representative data. In addition, the time that hangs in the comparison between the input vector and the enrolled vector will add.

An advantage of correlation matching is that it does not require the enrollment of users' biometric data. Rather, the information that is enrolled in the system is feature vectors indicating the relationship with the representative data items. The risk of a leak thus comes to focus on the sets of representative data and feature vectors. Note however that the set of representative data readily changed by transposing the data items themselves or by changing the number of data items. The feature vectors are determined by the number and type of representative data. Though the searching method for the enrolled data by the steetest descent method in a face recognition system[5] is reported, it might be difficult to search for the number of elements and the element value at the same time.

Here, one example of the person authentication system that uses the correlation matching is shown in Fig.3. In this authentication system, user's fingerprint data is enrolled nowhere and doesn't flow in the network. This is an advantage of the correlation matching.

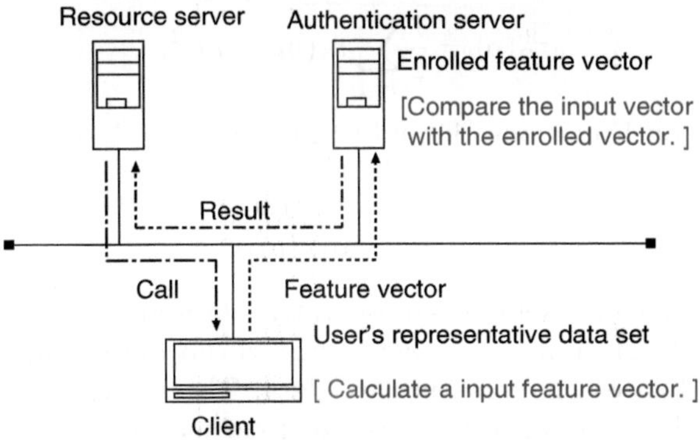

Fig. 3. Overview of the authentication system. A fingerprint data is taken on the client and a feature vector is calculated for the data. In the Authentication server, the input feature vector and the enrolled feature vector are compared.

3 Correlation Matching Scheme

Next let us consider the criterion for selecting the representative data items. Moreover, it thinks about the classifier when the fingerprint is matching.

3.1 Representative Data

We observed earlier that correlation matching requires that representative data be prepared in advance. Here we will consider the criterion by which this representative data should be prepared.

we assume here that representative data sets are set up for each enrollee. Consider that the set of representative data is selected for a fingerprint F_{i^*}. Here we assume that each representative data item incorporates fingerprint data enabling F_{i^*} to be distinguished from $\mathrm{F}_{j \neq i^*}$. Thus the group of scores x_{p,d_1} yielded by verifying fingerprint data $p \in \mathcal{D}$ with $d_1 \in \mathcal{D}_{i^*}$ is called class ω_1, and the group of scores x_{p,d_2} obtained by verifying with $d_2 \in \mathcal{D}_{j \neq i^*}$ is called class ω_2. Here $\mathcal{D}$ is a fingerprint data set, and $\mathcal{D}_i \subset \mathcal{D}$ is the fingerprint data set for fingerprint F_i.

The value of p is based on the within-class variance between-class variance ratio between these two classes ω_1 and ω_2. The within-class variance between-class variance ratio J_σ represents the extent or degree of separation between the classes. In other words, the bigger the J_σ score, the greater the distance between classes. Here x_{p,d_i} belonging to ω_i is $\mathcal{X}_i$, the $\mathcal{X}_i$ element number is n_i, and the average score is m_i. The total number of element is n and the total average score is m. Here the within-class variance is represented by σ_W^2, the between-class variance is represented by σ_B^2, and can be written

$$\sigma_W^2(p) = \frac{1}{2} \sum_{i=1}^{2} \sum_{x_{p,d_i} \in \mathcal{X}_i} (x_{p,d_i} - m_i)^2 \tag{1}$$

$$\sigma_B^2(p) = \frac{1}{2}\sum_{i=1}^{2} n_i(m_i - m)^2. \tag{2}$$

Therefore, based on Equations (1) and (2), the p score $J_\sigma(p)$ is given by

$$J_\sigma(p) = \frac{\sigma_B^2(p)}{\sigma_W^2(p)}. \tag{3}$$

Next, we consider how sets of representative data are constructed using representative data. First, a large number of fingerprint data samples are prepared to serves as candidates of representative data. The values of these candidates are derived based on the criterion described earlier. If a set of representative data consists of M number of representative data samples, then M number of samples are chosen from among these candidates, and arranged in the order to highest value first to make up the set of representative data.

3.2 Adopting Classifier

In this section we consider the procedure for identifying fingerprints. A set of representative data can be prepared by above method. The following problem is a classifier. In a word, the method of matching the fingerprint from the feature vector. The easiest method is to match the fingerprint from the distance of the input vector and the enrolled vector. Moreover, there is the one using the KL expansion and the linear discriminant method. The one using the neural net work is a superior method, too. In addition, there is a method of the combination of these classifier. For example, there is bagging[6] that studies the data set where distribution is different, and there is boosting[7] that increases the weight of the instance of the mis-classification and repeats study. Moreover, there are cascading[8] and stacking[9, 10] that controls the combination of the classification machine by study.

Here, the KL expansion and the linear discriminant method that is the standard way will be used. Moreover, the method of combining these two methods is applied. Therefore, the basic performance of the correlation matching is confirmed.

4 Computer Experiments

In this section, the basic performance of the correlation matching is confirmed.

4.1 Experimental Procedures

Four basic experiments are conducted in which the matching is done

(a) using feature space,
(b) using space whose dimensionality is reduced by KL expansion (KL),
(c) using discriminate space based on the linear discriminant method (LD), and
(d) using discriminate space based on a combination of KL and LD.

In the experiments we use a database of 30,000 fingerprints compiled by scanning 2,000 fingers 15 times each. Essentially, the 2,000 fingers are divided into three groups as follows: 500 fingers are used to calculate the performance (Group A), 500 different fingers are used to calculate the values of the candidates (Group B), and the remaining 1,000 fingers are used as candidates for the representative data (Group C).

The experiments are conducted in the following order:

(1) A set of representative data is defined for each finger in Group A. Using the first 10 data samples out of 15 and the data in Group B, values are derived for the candidate data in Group C. M number of candidates are selected forming his set of representative data in the order of highest values.
(2) Enrolled feature vectors are calculated for each finger in Group A. Ten feature vectors are derived from his set of representative data defined in (1) and from the first ten data samples. This average vector is regarded as his enrolled feature vector.
(3) When KL and LD are applied, conversion matrix and vector were calculated. These matrix and vector are derived using his feature vectors calculated in (2) and another-person feature vectors calculated from the Group B data and his set of representative data defined in (1).
(4) For each finger in Group A we obtain a genuine distribution calculated from the distance and frequency between the enrolled feature vectors calculated in (2) and the feature vectors derived from the remaining 5 data. Imposter distributions are then obtained calculating the distance and frequency of the feature vectors derived from the remaining 5 data of the other fingers. In other words, the distance calculations are performed between 2,500 pairs of the same finger, and between 1,247,000 pairs of different fingers, and the frequencies are derived from these calculations.

4.2 Experimental Results

In experiment (a), we change M from 100 to 1000. In experiment (b), the results are obtained when 1000-dimension feature space is converted to L reduced-dimension subspace by KL expansion. L is changed from 100 to 1000. In experiment (c), we show the results for discriminant space derived by the linear discriminant method for M-dimension feature space. The range of M is from 100 to 900. In the last experiment (d), we show the results when matching is done using discriminate space derived by applying the linear discriminant method to the L-dimension subspace. Here $L = 100$ to 900. The result of each experiment is

Table 1. The best FNMR when a threshold is set up in FMR = 1% and FMR = 0.1%

Experiment	(a)	(b)	(c)	(d)
FMR= 1%	12.2%	10.9%	3.2%	1.6%
FMR= 0.1%	33.7%	27.0%	7.7%	3.6%

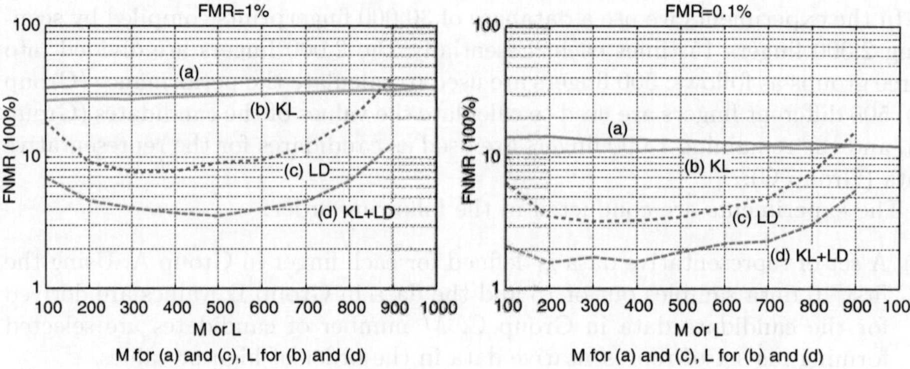

Fig. 4. FNMR in the each experiment is shown. The left figure is a result in FMR= 1%, and a right figure is a result in FMR= 0.1%.

shown in Fig. 4. And, the best result is shown in Table. 1. The best performance is FNMR= 3.6% at FMR= 0.1% when the combining classifier is applied.

5 Conclusions

In this paper, we showed the overview of the correlation matching and examined the basic performance. To realize better performance, we will improve the method to prepare representative data, construct them, and adopt more advanced classifiers in the future study.

References

1. Ratha N., Connell J., Bolle R., "Enhancing security and privacy in biometrics based authentication systems", IBM Systems Journal40, pp.61-634, 2001.
2. Soutar C., Roberge D., Stoianov A., Gilroy R., Kumar V.,"Biometric Encryption", http://www.bioscrypt.com/assets/Biometric_Encryption.pdf
3. K. Sasakawa, F. Isogai, S. Ikebata, "Personal Verification System with High Tolerance of Poor Quality Fingerprints", in Proc. SPIE, vol. 1386, pp. 265-272, 1990.
4. M. Matsushita, T. Maeda, K. Sasakawa, "Personal verification using correlation of score sets calculated by standard biometrics data", Technical Paper of the Inst. of Electronics and Communication Engineers of Japan, PRMU2000-78, pp. 21-26, 2000.
5. Adler A., "Sample images can be independently restored from face recognition template", Can. Conf. Electrical Computer Eng., pp.1163-1166, 2003.
6. Breiman, L "Bagging Predictors", Machine Learning, 24(2), pp. 123-140, 1996.
7. Freund, Y. Schapire, R. E. "Experiments with a new boosting algorithm", in Proc. of Thirteenth International Conference on Machine Learning, pp. 138-156, 1996.
8. Gama, J. and Brazdil, P. "Cascade Generalization", Machine Learning, 41(3), Kluwer Academic Publishers, Button, pp. 315-343, 2000.
9. Wolpert, D. "Stacked Generalization", Neural Network 5(2), pp.241-260, 1992.
10. Dzeroski S., and Zenko B., "Is combining classifiers better than selecting the best one?", Machine Learning, 54, pp.255-273, 2004.

Effective Fingerprint Classification by Localized Models of Support Vector Machines

Jun-Ki Min, Jin-Hyuk Hong, and Sung-Bae Cho

Department of Computer Science, Yonsei University,
Biometrics Engineering Research Center,
134 Shinchon-dong, Sudaemoon-ku, Seoul 120-749, Korea
{loomlike, hjinh}@sclab.yonsei.ac.kr, sbcho@cs.yonsei.ac.kr

Abstract. Fingerprint classification is useful as a preliminary step of the matching process and is performed in order to reduce searching time. Various classifiers like support vector machines (SVMs) have been used to fingerprint classification. Since the SVM which achieves high accuracy in pattern classification is a binary classifier, we propose a classifier-fusion method, multiple decision templates (MuDTs). The proposed method extracts several clusters of different characteristics from each class of fingerprints and constructs localized classification models in order to overcome restrictions to ambiguous fingerprints. Experimental results show the feasibility and validity of the proposed method.

1 Introduction

Fingerprint classification is a technique that classifies fingerprints into the predefined categories according to the characteristics of the image. It is useful for an automated fingerprint identification system (AFIS) as a preliminary step of the matching process and is performed in order to reduce searching time. Fig. 1 shows the examples of fingerprint classes.

Various classifiers, such as neural networks, k-nearest neighbors, and SVMs, have been widely used in fingerprint classification [1]. Since the SVM which shows good performance in pattern classification was originally designed for binary classification, it requires a combination method in order to classify multiclass fingerprints [2].

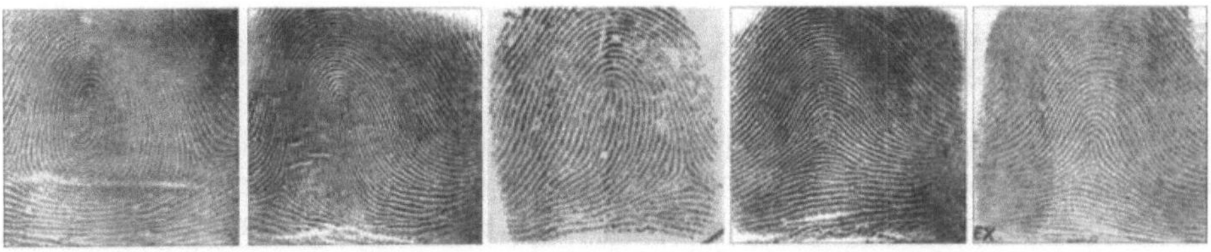

(a) (b) (c) (d) (e)

Fig. 1. Five fingerprint classes in the NIST database 4. (a) Whorl, (b) Right loop, (c) Left loop, (d) Arch, (e) Tented arch.

D. Zhang and A.K. Jain (Eds.): ICB 2006, LNCS 3832, pp. 287 – 293, 2005.

Many classifier-fusion methods have been investigated for the purpose of extending binary classification to multiclass classification or for improving classification accuracy [4]. Especially, the decision templates (single-DTs) have produced good performance in recent applications [5]. Since this method abstracts the outputs of the classifiers to a template, there is a limitation of applying it to complex problems with ambiguous samples such as fingerprints [6]. For the effective combination of SVMs in order to classify fingerprints, we propose multiple decision templates (MuDTs) that localized fusion models with clustering algorithm. The MuDTs decompose one class into several clusters to produce decision templates of each cluster. The proposed method is validated on the NIST database 4 using FingerCode features.

2 Related Works

2.1 The FingerCode

The FingerCode, as proposed by Jain in 1999, was extracted from NIST database 4 using a filter-based method. The algorithm set a registration point in a given fingerprint image and tessellated it into 48 sectors. Then, it transformed the image using the Gabor filter of four directions (0°, 45°, 90°, and 135°). Ridges parallel to each filter direction were accentuated, and ridges not parallel to the directions were blurred (Fig. 2). Standard deviations were computed on 48 sectors for each of the four transferred images in order to generate the 192-dimensional feature vector called FingerCode. Jain achieved 90% accuracy at a 1.8% rejection rate with two stage classification of K-NN/neural networks using these features [3].

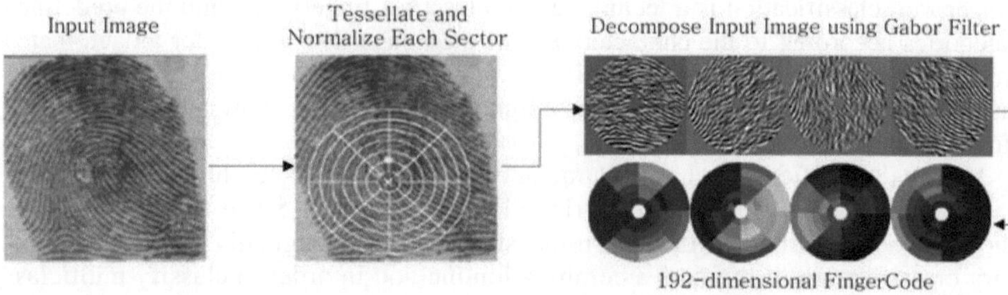

Fig. 2. Flow diagram of the FingerCode feature vector [3]

2.2 Support Vector Machines

The SVM is a technique for binary classification in the field of pattern recognition. This technique maps an input sample to a high-dimensional feature space and finds the optimal hyperplane that minimizes the recognition error for the training data using the non-linear transformation function.

Let n be the number of training samples. For the i th sample x_i with class label $c_i \in \{1, -1\}$, the SVM calculates

$$f(x) = \sum_{i=1}^{n} \alpha_i c_i K(x, x_i) + b, \quad K(x, x_i) = \Phi(x) \cdot \Phi(x_i). \tag{1}$$

Coefficient α_i in Eq. (1) is non-zero when x_i is a support vector that composes the hyperplane. Under all other conditions, it is zero. The kernel function $K(x, x_i)$ is easily computed by defining an inner product of the non-linear mapping function.

To classify fingerprints using SVMs, decomposition strategies such as one-vs-all, pairwise, and complete-code are needed [7].

2.3 The Decision Templates

The decision templates (single-DTs) generate templates of each class by averaging the decision profiles (DPs) for the training samples. For the M-class problem with L classifiers, $DP(x_i)$ of the i th sample is

$$DP(x_i) = \begin{bmatrix} d_{1,1}(x_i) & \cdots & d_{1,M}(x_i) \\ \vdots & d_{y,z}(x_i) & \vdots \\ d_{L,1}(x_i) & \cdots & d_{L,M}(x_i) \end{bmatrix}, \tag{2}$$

where $d_{y,z}(x_i)$ is the degree of support given by the y th classifier for the sample x_i of the class z. When DPs are generated from the training data, Eq. (3) estimates the decision template DT_c of the class c.

$$DT_c = \begin{bmatrix} dt_c(1,1) & \cdots & dt_c(1,M) \\ \vdots & dt_c(y,z) & \vdots \\ dt_c(L,1) & \cdots & dt_c(L,M) \end{bmatrix}, \quad dt_c(y,z) = \frac{\sum_{i=1}^{n} ind_c(x_i)\, d_{y,z}(x_i)}{\sum_{i=1}^{n} ind_c(x_i)} \tag{3}$$

$Ind_c(x_i)$ has a value of 1 if x_i's class is c, otherwise it has a value of zero.

In the test stage, it computes the distance between the DP of a new sample and the decision templates of each class. The class label is decided as the class of the most similar decision templates [5].

3 Multiple Decision Templates

In order to construct the MuDTs, we composed decision profiles with 5 one-vs-all SVMs (whorl, right loop, left loop, arch, and tented arch versus all). Decision profiles of each class $DP_{whorl}(x), \ldots, DP_{tented\ arch}(x)$ were clustered with a SOM algorithm (Eq. (4)). Each $DP(x)$ mapped a sample to the cluster (k, l) using Euclidean distance, with $w_{i,j}$ as the weight of the (i, j) th cluster [8].

$$\left\| DP(x) - w_{k,l} \right\| = \min_{i,j=1,\ldots,N} \{ \left\| DP(x) - w_{i,j} \right\| \} \tag{4}$$

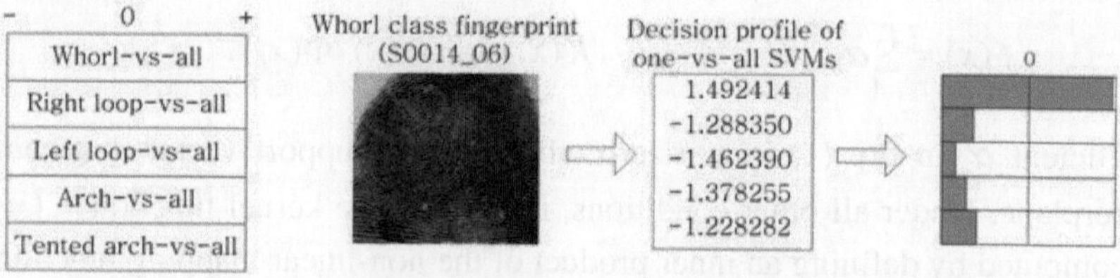

Fig. 3. A template of one-vs-all SVMs with its graphical representation

A decision template $DT_c^{k,l}$, which is the template of a cluster (k,l) of class c, was computed by Eq. (5). $Ind_c^{k,l}(x_i)$ refers to an indicator function of 1 if x_i belongs to the (k,l) th cluster of class c. If this is not the case, it refers to zero.

$$DT_c^{k,l} = \begin{bmatrix} dt_c^{k,l}(1,1) & \cdots & dt_c^{k,l}(1,M) \\ \vdots & dt_c^{k,l}(y,z) & \vdots \\ dt_c^{k,l}(L,1) & \cdots & dt_c^{k,l}(L,M) \end{bmatrix}, \quad dt_c^{k,l}(y,z) = \frac{\sum_{i=1}^{n} ind_c^{k,l}(x_i)\, d_{y,z}(x_i)}{\sum_{i=1}^{n} ind_c^{k,l}(x_i)} \tag{5}$$

Since the SVM is a binary classifier, we represented the output of a classifier to one column with positive and negative signs (Fig. 3). Sixteen decision templates of a class were estimated by clustering 4×4 SOM as shown in Fig. 4.

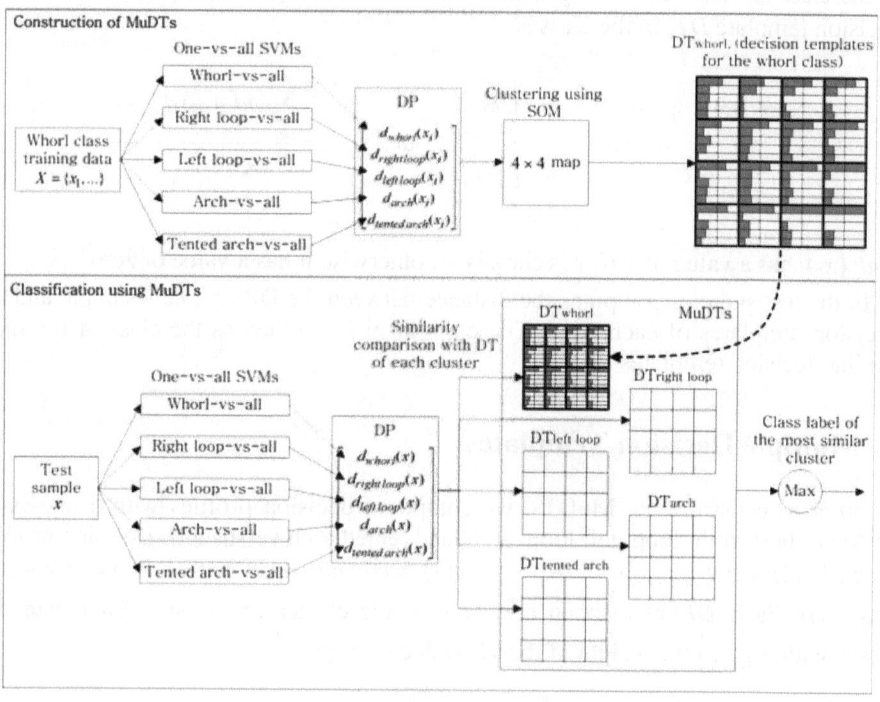

Fig. 4. Construction and classification of 4×4 MuDTs (case of whorl class)

The classification process of the MuDTs is similar to that used with single-DTs'. The distance between the decision profile of a new sample and each decision template of clusters is calculated (Fig. 4), and then the sample is classified into the class that contains the most similar clusters. In this paper, the Euclidean distance (Eq. (6)) is used to measure the similarity for its simplicity and good performance [5].

$$dst_c^{i,j}(x) = \sqrt{\sum_{y=1}^{L}\sum_{z=1}^{M}(dt_c^{i,j}(y,z) - d_{y,z}(x))^2}\ , \quad \min_{c=1,\dots,M}\left(\min_{i,j=1,\dots n} dst_c^{i,j}(x)\right) \tag{6}$$

4 Experimental Results

4.1 Experimental Environments

We have verified the proposed method on the NIST database 4. The first set of impressions of the fingerprints (F0001~F2000) were used as the training set while the second set of impressions of the fingerprints (S0001~S2000) were used as the test set. Jain's FingerCode features were used after normalization $(+1 \sim -1)$. The FingerCode rejected a few fingerprint images in both the training set (1.4%) and the test set (1.8%) [3]. The LIBSVM package (available at http://www.csie.ntu.edu.tw/~cjlin/libsvm) was used for the SVM classifiers. The Gaussian kernel with $\sigma^2 = 0.0625$ was selected based on the experiment.

4.2 MuDTs Versus DTs

The MuDTs of the one-vs-all (OVA) SVMs yielded an accuracy of 90.4% for the 5-class classification task. For the 4-class classification task, 94.9% was achieved. The confusion matrices of the one-vs-all SVMs combined with the single-DTs and MuDTs with the Euclidean distance are shown in Table 1 and Table 2. Because the MuDTs produce multiple classification models for one class, they classify ambiguous fingerprint images more accurately than single-DTs (Fig. 5).

Table 1. Confusion matrix for the single-DTs of OVA SVMs

	W	R	L	A	T
W	380	6	8	0	0
R	7	357	1	6	21
L	7	0	363	13	13
A	1	2	1	347	60
T	1	8	9	37	316

Table 2. Confusion matrix for the MuDTs of OVA SVMs

	W	R	L	A	T
W	380	6	7	0	1
R	9	369	1	5	17
L	8	0	366	14	10
A	1	4	1	356	50
T	1	10	6	38	304

4.3 Comparison with Other Methods

The winner-takes-all, ECCs, BKS, and single-DTs methods were compared with the MuDTs. The Euclidean distance was used for ECCs, single-DTs, and MuDTs. For the

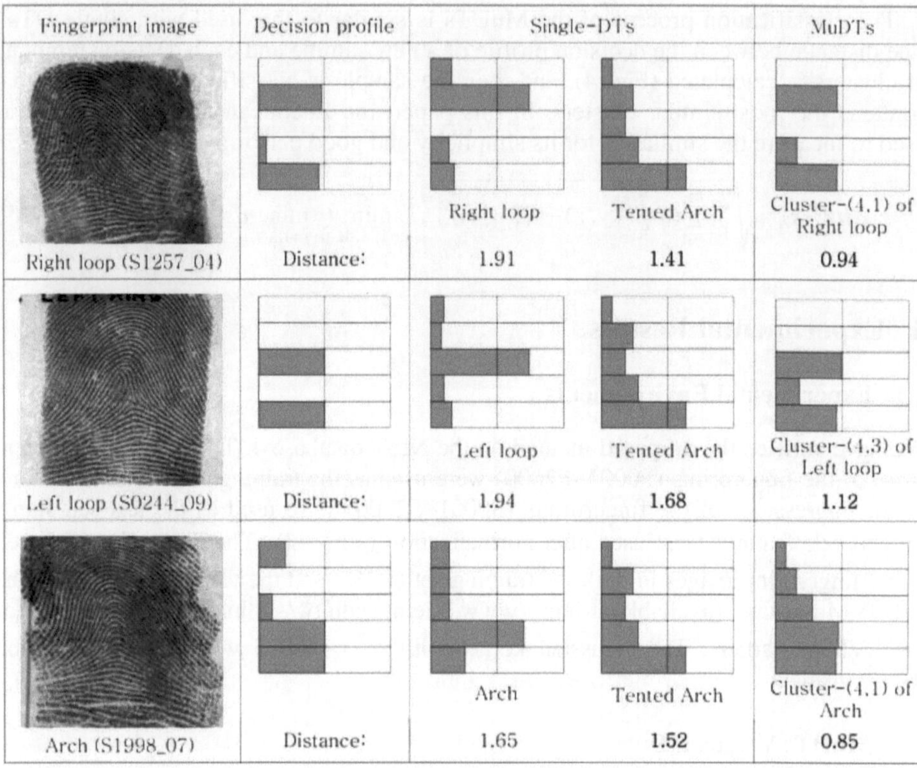

Fig. 5. Classification of ambiguous fingerprints

BKS method, when ties or new output patterns occurred, the winner-takes-all method was alternatively used. As shown in Table 3, the MuDTs achieved the highest accuracy of 89.5%~90.4%. Given the simplicity of the SOM algorithm with the low-dimension vector, despite the additional step for clustering at the training phase, there is nearly no difference between the classification times of the MuDTs and single-DTs. It took about 60ms on a Pentium 4 (2.4 GHz) machine to train the SOM with 2,000 fingerprints which can be ignored, compared to the training time of the SVMs.

Table 3. The accuracies of various classifier fusion schemes (%)

Fusion methods	One-vs-all	Pairwise	Complete-code
Winner-takes-all	90.1	87.7	90.0
ECCs	90.1	88.6	90.0
BKS	88.8	89.4	89.3
Single-DTs	89.8	88.3	89.5
MuDTs	90.4	89.5	90.3

5 Conclusion

This paper has proposed an effective classifier fusion method (MuDTs) to classify ambiguous fingerprint images which show more than one characteristic in terms of fingerprint class. The outputs of one-vs-all SVMs for the training data were clustered by the SOM to decompose the class into several clusters to separate and examine diverse characteristics. The localized decision templates were estimated for each cluster, and then the MuDTs were constructed. Experiments were performed on the NIST database 4 using FingerCodes. We achieved 90.4% for 5-class classification with 1.8% rejection, and 94.9% for 4-class classification. Experimental results show the effectiveness of the multiple templates method with higher accuracy than other methods. In future work, we will investigate effective classifier decomposition methods with appropriate cluster maps to maximize the effectiveness of the MuDTs.

Acknowledgements. This work was supported by the Korea Science and Engineering Foundation (KOSEF) through the Biometrics Engineering Research Center (BERC) at Yonsei University. We would like to thank Prof. Anil Jain and Dr. Salil Prabhakar for providing the FingerCode data.

References

1. A. Senior, "A combination fingerprint classifier," *IEEE Trans. Pattern Analysis and Machine Intelligence*, vol. 23, no. 10, pp. 1165-1174, 2001.
2. Y. Yao, et al., "Combining flat and structured representations for fingerprint classification with recursive neural networks and support vector machines," *Pattern Recognition*, vol. 36, no. 2, pp. 397-406, 2003.
3. A. K. Jain, et al., "A multichannel approach to fingerprint classification," *IEEE Trans. Pattern Analysis and Machine Intelligence*, vol. 21, no. 4, pp. 348-359, 1999.
4. L. I. Kuncheva, *Combining Pattern Classifiers*, Wiley-Interscience, 2004.
5. L. I. Kuncheva, et al., "Decision templates for multiple classifier fusion: An experimental comparison," *Pattern Recognition*, vol. 34, no. 2, pp. 299-314, 2001.
6. R. Cappelli, et al., "Fingerprint classification by directional image partitioning," *IEEE Trans. Pattern Analysis and Machine Intelligence*, vol. 21, no. 5, pp. 402-421, 1999.
7. R. M. Rifkin and A. Klautau, "In defence of one-vs-all classification," *Jnl. of Machine Learning Research*, vol. 5, pp. 101-141, 2004.
8. K. Obermayer and T. J. Sejnowski, *Self-Organizing Map Formation Foundations of Neural Computation*, The MIT Press, 2001.

Fingerprint Ridge Distance Estimation: Algorithms and the Performance*

Xiaosi Zhan[1], Zhaocai Sun[2], Yilong Yin[2], and Yayun Chu[1]

[1] Computer Department, Fuyan Normal College, 236032, Fuyang, China
xiaoszhan@263.net, chuyayun.fync@163.com
[2] School of Computer Science & Technology, Shandong University, 250100, Jinan, China
sunnykiller@126.com, ylyin@sdu.edu.cn

Abstract. Ridge distance is one important attribute of the fingerprint image and it also is one important parameter in the fingerprint enhancement. It is important for improving the AFIS's performance to estimate the ridge distance correctly. The paper discusses the representative fingerprint ridge distance estimation algorithms and the performance of these algorithms. The most common fingerprint ridge distance estimation algorithm is based on block-level and estimates the ridge distance by calculating the number of cycle pattern in the block fingerprint image. The traditional Fourier transform spectral analysis method has been also applied to estimate the fingerprint ridge distance. The next kind of method is based on the statistical window. Another novel fingerprint ridge distance estimation method is based on the region-level which regards the region with the consistent orientation as the statistical region. One new method obtains the fingerprint ridge distance from the continuous Fourier spectrum. After discussing the dominant algorithm thought, the paper analyzes the performance of each algorithm.

1 Introduction

The fingerprint images vary in quality. It is important to enhance effectively the fingerprint image with low quality for improving the performance of the automatic fingerprint identification system [1,2,3]. As one key attribute of the fingerprint image, most fingerprint enhancement algorithms regard the ridge distance as one essential parameter for enhancing the fingerprint image effectively. It is important to estimate the accurate ridge distance for improving the performance of the AFIS. In recent years, fingerprint ridge estimation method is the research focus and many methods for estimating the ridge distance have been brought forward in the correlative literatures.

D. C. Douglas Hung estimated the average distances of the all ridges on the whole fingerprint image [4]. Mario and Maltoni did mathematical characterization of the local frequency of sinusoidal signals and developed a 2-Dmodel of the ridge pattern in

* Supported by the National Natural Science Foundation of China under Grant No. 06403010, Shandong Province Science Foundation of China under Grant No.Z2004G05 and Anhui Province Education Department Science Foundation of China under Grant No.2005KJ089.

D. Zhang and A.K. Jain (Eds.): ICB 2006, LNCS 3832, pp. 294 – 301, 2005.

order to obtain the ridge density [5]. Lin and Dubes attempted to count ridge number in one fingerprint image automatically and assumed the ridge distance is a constant value on the whole fingerprint image [6]. L. Hong et al proposed the direction window method to estimate the ridge frequency [3]. O'Gorman and Nickerson acquired the ridge distance using the statistics mean value and used the ridge distance as a key parameter in the design of filters [7]. Z. M. Kovace-Vajna et al brought out two kinds of ridge distance estimation methods: the geometry approach and the spectral analysis approach, which are both based on the block fingerprint image to estimate the ridge distance [8]. Y. Chen et al proposed two kinds of methods to estimate the ridge distance: the spectral analysis approach and the statistical window approach [9]. In addition, Y.L. Yin et al proposed the ridge distance estimation method based on regional level. The method divides the fingerprint image into several regions according to the consistency of the orientation information on the whole fingerprint image and calculates the ridge distance to every region respectively [10].

The paper chooses four representative fingerprint ridge estimation methods to analyze. After introducing the dominant realization steps, the paper analyzes the performance of the four methods mainly.

2 The Primary Fingerprint Ridge Distance Algorithms

Up to now, we can sum up the fingerprint ridge distance estimation algorithms into the following four primary kinds: (1) the statistical window method; (2) the region-level method; (3) the discrete Fourier spectral method and (4) the continuous spectrum analysis method.

2.1 Method for Fingerprint Ridge Distance Estimation Based on Statistical Window

The method defined the statistical window and the base line firstly. After dividing the fingerprint image into block image with the size of 32×32, the method estimated the ridge distance of each block image by detecting the distributing of the gray histogram. The definitions of the statistical window and the base line are showed as the following Fig.1 and the key steps of the method can be described as:

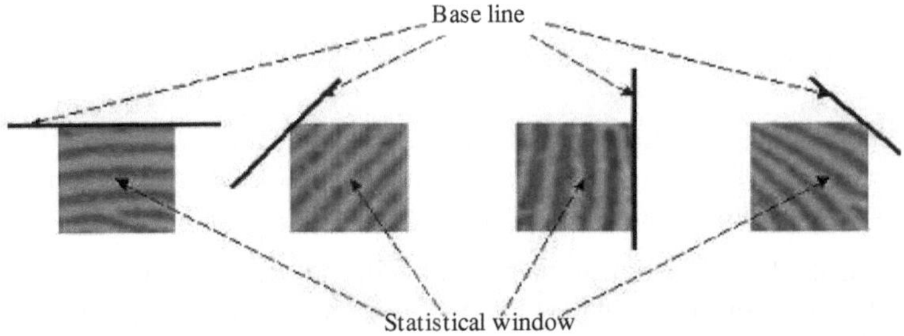

Fig. 1. Definitions of the statistical window and base line of different fingerprint image region

Step 1: Calculate the fingerprint image orientation field based on block-level. Here, we can adopt the ridge orientation estimation method put forward by L. Hong et al. or other method.

Step 2: Translate the gray fingerprint image into the binary fingerprint image adopting the local self-adapt segmentation method.

Step 3: Define the base line and the statistical window of each block image according to the Fig.1. And then obtain the ridge distributing histogram in each block image.

Step 4: Detect and memorize the locality of all local peak value in the ridge distributing histogram. Obviously, every local peak value is corresponding to one ridge and the distance between the two adjacent points with peak value is the ridge distance between the two adjacent fingerprint ridges in the block image.

Step 5: Calculate the dependability degree of the ridge distance values in all fingerprint image region and adjust ridge distance value with low dependability degree.

2.2 Method for Fingerprint Ridge Distance Estimation Based on Region Level

The selecting of the window size is the key issue in the method in the statistical window method. It is the precondition for selecting the right window size to confirm the ridge distance. In theory, it is impossible. Consequently, Y. Long et al. put forward the method for estimating the fingerprint ridge distance based on the region level.

Based on the orientation field of block image, the method clustered the region with the close ridge orientation as one region by the region increasing method. Obviously, one fingerprint image can be segmented into several regions with the close ridge orientation. The following Fig.2 showed the segmentation results:

Fig. 2. Segmentation results of the directional images about three typical fingerprints

After segmenting the directional image into several regions, the method regarded every region with the same ridge orientation as one unit to estimate the ridge distance. The method can be described as the following steps.

Step 1: Calculate the area of each region. In the method, the region area can be defined as the number of the image blocks of the corresponding region. Go to step 2 if the number is more than or equal with the threshold value R_{min} (R_{min}=8 in the paper).

Step 2: Define the statistical window and the base line of each region. Here, the statistical window and the base line is same with the definitions in 2.1.

Step 3: Translate the gray fingerprint image into the binary fingerprint image adopting the local self-adapt segmentation method.

Step 4: Calculate and memorize the distance between the ridge pixel in the region and the base line. Obtain the ridge distributing histogram on the reference frame by defining the distance to the base line and the number of the ridge pixel in the same distance as the x-axle and y-axle respectively.

Step 5: Estimate the ridge distance in one region by obtaining the peak values of the corresponding histogram.

Step 6: The ridge distance of one region with the area smaller than Rmin can be defined the average value of the ridge distances in the circumjacent regions.

2.3 Fingerprint Ridge Distance Estimation Based on the Discrete Fourier Spectrum

Spectral analysis method that transforms the representation of fingerprint images from spatial field to frequency field is a typical method of signal processing in frequency field. It is a traditional method for ridge distance estimation in fingerprint images. Generally, if $g(x, y)$ is the gray-scale value of the pixel (x,y) in an $N \times N$ image, the DFT of the $g(x, y)$ is defined as follows:

$$G(u,v) = \frac{1}{N^2}\sum_{i=1}^{N}\sum_{j=1}^{N} g(x,y)e^{2\pi j/N<(x,y)(u,v)>} \tag{1}$$

Where j is the imaginary unit, $u, v \in \{1, \cdots N\}$ and $<(x,y)(u,v)>=xu+yv$ is the vector pot product. In theory, the module $|G_{(u,v)}|$ of $G(u,v)$ describes the cycle character of signal. We can acquire the dominant cycle of the signal in one region by calculating the values of $|G_{(u,v)}|$ of each pixel in the region, which can be defined the ridge frequency in the fingerprint ridge distance estimation. For obtaining the right ridge distance value, Y. Chen et al. define the radial distribute function in [10] as follows:

$$Q(r) = \frac{1}{\# C_r} \sum_{(u,v) \in C_r} |G_{(u,v)}| \tag{2}$$

Where C_r is the set of the whole pixels that satisfy the function $\sqrt{u^2+v^2}=r$, $\#C_r$ is the number of the element of the set C_r. Then define the $Q(r)$ as the distribution intensity of the signal with the cycle N/r in the $N \times N$ image. The value of r corresponding to the peak of $Q(r)$ can be defined as the cycle number of the dominant signal in the $N \times N$ image. Search for the value of r_0 that enables the value of $Q(r_0)$ is the local maximum. At this moment, the ridge distance of the block image can be estimated with $d=N/r_0$. The dominant steps of the method can be described as the following:

Step 1: Divide the fingerprint image into non-overlap block image with the N×N size (N is 32 generally).

Step 2: Calculate $|G_{(u,v)}|$ of each pixel (x, y) ($x, y \in \{0, \cdots, 31\}$) in one block image by adopting the 2-D fast Fourier transform.

Step 3: Calculate the value of $Q(r)$ $(0 \le r \le N-1)$.

Step 4: Search the appropriate r which make each r ($0 \le r_{min} \le r \le r_{max} \le N-1$ and $r \ne r'$) having $Q(r') \ge Q(r)$.

Step 5: Cannot estimate the ridge distance of one block image if the following two conditions cannot be satisfied: $Q(r') > Q(r-1)$ and $Q(r') > Q(r+1)$. Here, we think that $Q(r')$ is not the local peak value. Otherwise, search the appropriate r'' that makes each r ($0 \le r_{min} \le r \le r_{max} \le N-1, r \ne r'$ and $r \ne r''$) having $Q(r'') \ge Q(r)$.

Step 6: Calculate the dependability degree according to the following formula:

$$\frac{\alpha}{Q(r')}\min\{Q(r') - Q(r''), Q(r') - Q(r'-1), Q(r') - Q(r'+1)\} \tag{3}$$

Estimate the ridge distance of one block image with the formula $\overline{d} = N/r'$ when the dependability degree is larger than 0.4. Otherwise, the ridge distance of one block image can' t be estimated.

2.4 Ridge Distance Estimation Method Based on the Continues Spectrum

The precision does not meet the requirement if we carry through the discrete Fourier transform. At the same time, the speed can't meet the requirement of the real time disposal if we make the continuous Fourier transform. So, the method adopts the 2-D sampling theorem to transform the 2-D discrete Fourier spectrum into 2-D continuous Fourier spectrum and estimates the ridge distance based on the continuous spectrum.

Suppose the Fourier transform function $F(s_1, s_2)$ about the function f (x1, x2) of L2 (R2) is tight-supported set (namely that the function F is equal to zero except the boundary region D and the boundary region D can be defined as the rectangle region $\{(s_1, s_2) \mid |s_1| \le \Omega \text{ and } |s_2| \le \Omega\}$ in the paper. Here, we firstly assume $\Omega = \pi$ in order to simplify the function. Then the Fourier transform function about the function $f(x_1, x_2)$ can be denoted as follows:

$$F(s_1, s_2) = \sum_{n_1}\sum_{n_2} C_{n_1,n_2} e^{-jn_1s_1 - jn_2s_2} \tag{4}$$

Here, the C_{n_1,n_2} is defined as follows:

$$C_{n_1,n_2} = \frac{1}{(2\pi)^2}\int_{-\infty}^{+\infty}\int_{-\infty}^{+\infty} ds_1 ds_2 e^{jn_1s_1 + jn_2s_2} F(s_1, s_2) = \frac{1}{2\pi} f(n_1, n_2) \tag{5}$$

Then, we can acquire the following function as:

$$f(x_1, x_2) = \sum_{n_1}\sum_{n_2} C_{n_1,n_2} \frac{\sin \pi(x_1 - n_1)}{\pi(x_1 - n_1)} \frac{\sin \pi(x_2 - n_2)}{\pi(x_2 - n_2)} \tag{6}$$

In this way, the discrete signal $C_{n1,n2}$ can be recovered for the continuous signal $f(x_1, x_2)$ through the sampling theorem. Then, the discrete frequency spectrum of each block fingerprint image can be recovered for the continuous frequency spectrum.

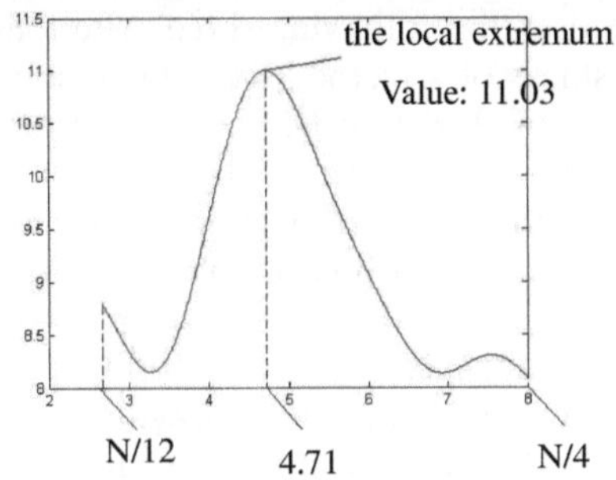

Fig. 3. The cutaway view of the continuous spectrum in the normal orientation

We can try to obtain accurately the local extreme value (that is the "light spot" position we care about) in random small step in the continuous frequency spectrum. Thus we can calculate the ridge distance accurately. But it is a long period course that we search the continuous spectrum image that is recovered from one $N \times N$ point matrix in a small step for the local peak value. Thus we need to search the continuous spectrum purposefully.

Suppose that the ridge orientation is θ and then the normal orientation of the ridge is $\theta+\pi/2$. We can obtain the position of the local extreme point in the continuous spectrum if we search the region, which is confirmed by the radius: N/12-N/4 and the direction: $\theta+\pi/2$, in the step length as 0.01. As Fig.3 shows, the local extreme points are 11.03, the corresponding radius is 4.71, and the ridge distance of the image is 32/4.71=6.79.

Step 1: Divide the fingerprint image into non-overlap block with the size $N \times N$, the N is equal to 32 generally.

Step 2: To each block fingerprint image $g(i,j)$, carry on two-dimension fast Fourier transform and get the corresponding discrete spectrum $G(u,v)$.

Step 3: To each discrete spectrum $G(u,v)$, apply the sampling theorem to get the continuous spectral function G(x,y).

Step 4: Adopt Rao method to obtain the ridge orientation θ.

Step 5: Search the region confirmed by the radius N/12-N/4 and the direction $\theta+\pi/2$ in a small step length L for finding the radius r corresponding the local extreme point. Generally, the value of L is 0.01.

Step 6: If don't find the local extreme point then think that the ridge distance of the fingerprint image region can't be obtained. Else estimate the fingerprint image ridge distance from d =N/r.

4 Performance Analysis and Conclusion

To evaluate the performance of the methods, we use 30 typical images (10 good quality, 10 fair quality, 10 poor quality) selected from NJU fingerprint database (1200 live-scan images; 10 per individual) to estimate ridge distance with the four ridge distance estimation methods respectively. In order to describe the performance in the same criterions, the paper chooses the following three criterions DER, EA and TC. Here, DER indicates the robustness of a method for ridge distance estimation in fingerprint images, EA is the degree of deviation between the estimation result and the actual value of the ridge distance and TC is the time needed for handing a fingerprint image. A high DER value means that the method is flexible and insensitive to a variety of image quality and ridge directions. A high EA value indicates that the estimation result is close to the actual value of the ridge distance. A lower TC value means that the method is faster. The following table 1 is the performance of the four methods.

From the Table 1, we can obtain the following conclusion. (1) The statistical method has the middle DER value, the EA value and the TC value. For the method, the sixty-four dollar question is that the method cannot estimate the ridge distance in a good deal of region. It doesn't perform well in these regions where there is acute variation of ridge directions. But, the obvious advantage of the statistical window method is that it is simple and can estimate the right ridge distance in good quality image region. (2) The region-level method has the highest DER value with the lowest EA and TC values. The method divides the fingerprint image into several regions and the ridge distance of each region can be estimated generally. But, the ridge distance is not accurate in most block images because there is only one ridge distance value in one big region and the ridge distance is diverse in the same region. (3) The discrete spectrum method has the lowest DER value with the middle TC value and EA value. For this method, the biggest problem is how to determine e r' accurately and reliably. If we can acquire the value of r' accurately and reliably, the performance will improve significantly. (4) The continuous spectrum has the highest EA value and TC value with the middle DER value. The method can obtain the ridge distance of most regions in a fingerprint image except the pattern region and the strong-disturbed region because the sub-peak is not obvious in these regions. In addition, the processing time of our method is more that the other two methods because our method is based on the two-dimension continuous spectrum. It shows that the method has higher performance except the processing time.

In order to illustrate the performance of the four methods farther, the paper chooses 10 representative fingerprint images (5 fingerprint images with good quality, 3 fingerprint images with fair quality and 2 fingerprint images with low quality) to test the effect on the minutiae exactness method. Firstly, we extract the right minutia artificially and consider the minutia set as the standard minutia set. Then, we extract the minutiae with the same processing methods except the ridge estimation method. Here, we define TMN, RMN, LMN, EMN and Rate as the total minutiae number, the lost minutiae number, the right minutiae number, the error minutiae number and the accurate rate respectively. Here, we define the accurate rate as the ratio between the RMN and the sum of the TMN and the LMN. The test results is showed as the following Table.2:

Table 1. The three performance indexes of the four methods

Method	DER (%)	EA (%)	TC (second)
Statistical window method	63.8	93	0.31
Region-level method	100	68	0.28
Discrete spectrum method	44.7	84	0.42
Continuous spectrum method	94.6	95	0.63

Table 2. The minutiae exactness results of the four methods

Method	TMN	LMN	RMN	EMN	Rate (%)
Statistical window method	484	32	448	36	86.8
Region-level method	512	26	440	72	81.8
Discrete spectrum method	501	24	445	56	84.8
Continuous spectrum method	487	15	459	28	91.4

From Table 2 we can find that the continuous spectrum method has the highest performance with the lowest LMN EMN values and the highest RMN and Rate values. For region-level method, the processing result is affected by the fingerprint image quality. The method can't process well for these fingerprint images with low quality. Generally, the statistical window method can process well except some strive noised fingerprint images.

For the fingerprint ridge estimation, we should combine the availability of the methods based on spatial field and frequency field. Continuous spectrum analysis method has its merits and has the highest performance. But, the key issues are that the time consuming is very high and we should search better method for transforming the spatial fingerprint image into two-dimension continuous frequency spectrum and making certain the more appropriate step length in order to find the two sub-peak points faster and accurately.

References

[1] L. Hong, A.K. Jain, R. Bolle et al. Identity authentication using fingerprints. Proceedings of FirstInternational Conference on Audio and Video-Based Biometric Person Authentication, Switzerland, 1997:103-110.

[2] L. Yin, X. Ning, X. Zhang. Development and application of automatic fingerprint identification technology.Journal of Nanjing University(Natural Science), 2002, 38(1):29-35.

[3] L. Hong, Y. Wan, A. K. Jain. Fingerprint image enhancement: algorithm and performance evaluation, IEEE Transactions on Pattern Analysis and Machine Intelligence, 1998, 20(8): 777-789.

[4] D. C. Douglas Hung, Enhancement feature purification of fingerprint images, Pattern Recognition, 1993, 26(11) : 1661-1671.

[5] D. Maio and D. Maltoni. Ridge-line density estimation in digital images. Proceedings of 14th International Conference on Pattern Recognition, Brisbane, Australia, 1998: 534-538.

[6] W. C. Lin and R. C. Dubes. A review of ridge counting in dermatoglyphics, Pattern Recognition, 1983, 16(2): 1-8.

[7] L. O'Gorman, J. V. Neckerson. An approach to fingerprint filter design, Pattern Recognition, 1989, 22(1): 29-38.

[8] Z. M. Kovacs-Vajna, R. Rovatti, and M. Frazzoni, Fingerprint ridge distance computation methodologies, Pattern Recognition, 33 (2000) 69-80.

[9] Y. Chen, Y. Yin, X. Zhang et al, A method based on statistics window for ridge distance estimation, Journal of image and graphics, China, 2003, 8(3): 266-270.

[10] Y. Yin, Y Wang, F Yu, A method based on region level for ridge distance estimation, Chinese computer science, 2003, 30(5): 201-208.

Enhancement of Low Quality Fingerprints Based on Anisotropic Filtering*

Xinjian Chen, Jie Tian**, Yangyang Zhang, and Xin Yang

Center for Biometrics and Security Research,
Key Laboratory of Complex Systems and Intelligence Science, Institute of Automation,
Chinese Academy of Sciences, Graduate School of the Chinese Academy of Science,
P.O. Box 2728, Beijing 100080, China
tian@doctor.com, jie.tian@mail.ia.ac.cn
http://www.fingerpass.net

Abstract. The enhancement of the low quality fingerprint is a difficult and challenge task. This paper proposes an efficient algorithm based on anisotropic filtering to enhance the low quality fingerprint. In our algorithm, an orientation filed estimation with feedback method was proposed to compute the accurate fingerprint orientation. The gradient-based approach was firstly used to compute the coarse orientation. Then the reliability of orientation was computed from the gradient image. If the reliability of the estimated orientation is less than pre-specified threshold, the orientation will be corrected by the mixed orientation model. And an anisotropic filtering was used to enhance the fingerprint, with the advantages of its efficient ridge enhancement and its robustness against noise in the fingerprint image. The proposed algorithm has been evaluated on the databases of Fingerprint verification competition (FVC2004). Experimental results confirm that the proposed algorithm is effective and robust for the enhancement of the low quality fingerprint.

1 Introduction

There are still many challenging tasks in fingerprint recognition. One of them is the enhancement of low quality fingerprints. The effect of enhancement of poor quality fingerprints is seriously affects the performance of the whole recognition system.

Many image enhancement techniques have been developed for poor quality images. Shi et al[1] proposed a new feature Eccentric Moment to locate the blurry boundary using the new block feature of clarified image for segmentation. Zhou et.al [2] proposed a model-based algorithm which is more accurate and robust to dispose the degraded fingerprints. Lin et al [3] made use of Gabor filter banks to enhance

* This paper is supported by the Project of National Science Fund for Distinguished Young Scholars of China under Grant No. 60225008, the Key Project of National Natural Science Foundation of China under Grant No. 60332010, the Project for Young Scientists' Fund of National Natural Science Foundation of China under Grant No.60303022, and the Project of Natural Science Foundation of Beijing under Grant No.4052026.

** Corresponding author. Tel: 8610-62532105; Fax: 8610-62527995, *Senior Member, IEEE.*

D. Zhang and A.K. Jain (Eds.): ICB 2006, LNCS 3832, pp. 302 – 308, 2005.

fingerprint images and reported to achieve good performance. Yang et al.[4] proposed a modified Gabor filter to enhance fingerprints, specified parameters deliberately through some principles instead of experience, preserved fingerprint image structure and achieved image enhancement consistency. Willis et al [5] proposed a Fourier domain based method that boosts up a low quality fingerprint image by multiplying the frequency spectrum by its magnitude.

This paper proposes an efficient algorithm based on anisotropic filtering to enhance the low quality fingerprint. The main steps of the algorithm include: normalization, orientation field estimation, reliability of orientation computing, orientation correction, region mask estimation and filtering. In our algorithm, an orientation filed estimation with feedback method was proposed to compute the accurate fingerprint orientation, and an anisotropic filtering was used to enhance the fingerprint

This paper is organized as follows. Section 2 indicates out the details of the enhancement of fingerprint images. Section 3 shows the performance of the proposed algorithm by experiments. Section 4 gives out our conclusion.

2 Fingerprint Enhancement Algorithm

The flowchart of the proposed fingerprint enhancement algorithm is shown in figure 1.

2.1 Normalization

Normalization is performed to decrease the dynamic range of the gray scale between ridges and valleys of the image, which facilitates the subsequent enhancement steps. In this paper, Lin et al [3]'s method has been used to process the normalization. The image intensity values were standardized by adjusting the range of grey-level values so that it lies within a desired range of values.

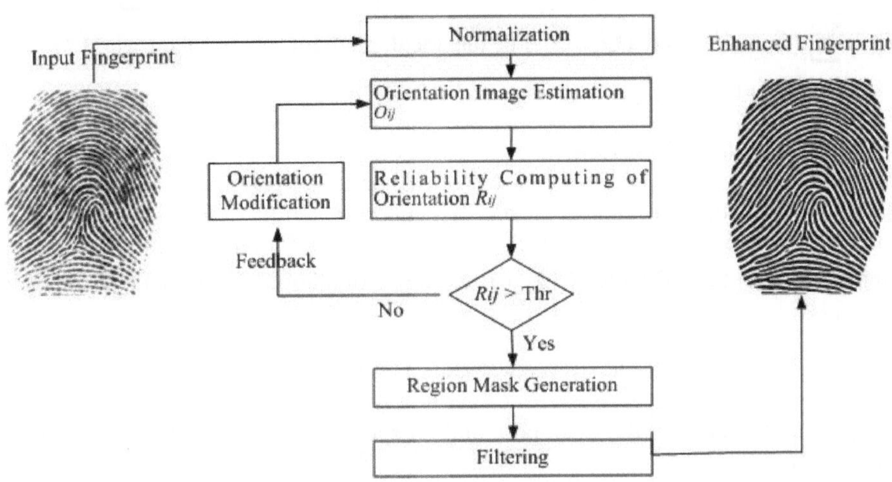

Fig. 1. The flowchart of the proposed enhancement algorithm

2.2 Orientation Field Estimation with Feedback

We proposed an orientation field estimation with feedback method to get the accurate fingerprint orientation. First, the gradient-based approach was used to compute the coarse orientation. Then we compute the reliability of orientation from the gradient image. If the reliability of the estimated orientation *rij* is less than threshold *thr*, the orientation will be corrected by the proposed mixed orientation model, otherwise the estimated orientation was taken as the true orientation.

2.2.1 The Gradient-Based Approach

In our algorithm, the gradient-based approach proposed by Lin et al [3] was used to compute the coarse orientation. But in our algorithm we divide the normalized image into an odd block of size (15*15) instead of (16*16).

2.2.2 Reliability of Orientation Computing

An additional value rij is associated with each orientation element Oij to denote the reliability of the orientation. The value rij is low for noise and seriously corrupted regions and high for good quality regions in the fingerprint image. The reliability rij is derived by the coherence of the gradient Gij within its neighborhood. It is defined as follows:

$$r_{ij} = \frac{\left|\sum_W (G_{i,x}, G_{j,y})\right|}{\sum_W \left|(G_{i,x}, G_{j,y})\right|} = \frac{\sqrt{(G_{xx} - G_{yy})^2 + 4G_{xy}{}^2}}{G_{xx} + G_{yy}} \tag{1}$$

Where $(G_{i,x}, G_{j,y})$ is the squared gradient, $G_{xx} = \sum_W G_x{}^2$, $G_{yy} = \sum_W G_y{}^2$, $G_{xy} = \sum_W G_x \cdot G_y$ and (G_x, G_y) is the local gradient. W is taken as 11*11 block around (i,j).

2.2.3 Orientation Correction

The mixed orientation model is consisted of two parts, polynomial model and singular model. Due to the smoothness of the original orientation field, we could choose proper polynomial curves to approach it. We map the orientation field to a continuous complex plane [2]. Denote $\theta(x, y)$ as the orientation field. The mapping is defined as:

$$\mathrm{U} = \mathrm{R} + \mathrm{iI} = \cos(2\theta) + \mathrm{i}\sin(2\theta) \tag{2}$$

where R and I denote the real part and image part of the unit-length complex respectively.

To globally approach the function R and I, a common bivariate polynomial model is chosen for them respectively, which can be formulated as:

$$(1\ x \cdots x^n) \cdot \begin{bmatrix} p_{00} & p_{01} & \cdots & p_{0n} \\ p_{10} & p_{11} & \cdots & p_{1n} \\ \vdots & \vdots & \ddots & \vdots \\ p_{n0} & p_{n1} & \cdots & p_{nn} \end{bmatrix} \cdot \begin{pmatrix} 1 \\ y \\ \vdots \\ y^n \end{pmatrix} \quad (3)$$

where the order n can be determined ahead.

It is difficult to be modeled with polynomial functions near the singular points. The orientation model proposed by Sherlock and Monro [6] is added at each singular point, and we name it as the singular model. The model allows a consistent directional map to be calculated from the position of the cores and deltas only. In this model the image is located in the complex plane and the orientation is the phase of the square root of a complex rational function with the fingerprint macro-singularities Let $c_i\ (i = 1..n_c)$ and $d_i\ (i = 1..n_d)$ be the coordinates of the cores and deltas respectively; the orientation O' at each point (x,y) is calculated as:

$$O'(z) = O_0 + \frac{1}{2}\left[\sum_{i=1}^{n_d} \arg(z - d_i) - \sum_{i=1}^{n_c} \arg(z - c_i)\right] \quad (4)$$

where O_0 is the background orientation (we set O_0 =0), and the function arg(z) returns the argument of the complex number z (x,y).

To combine the polynomial model with singular model smoothly, a weight function is defined for singular model, its weight at (x, y) is defined as:

$$w = \begin{cases} 0 & if\ (\sum_{i=1}^{k} w_i > 1) \\ 1 - \sum_{i=1}^{k} w_i & otherwise \end{cases} \quad (5)$$

$$w_i = \begin{cases} 0 & if\ (D_i(x, y) > r_i) \\ 1 - D_i(x, y) / r_i & otherwise \end{cases} \quad (6)$$

where k is the number of singular points, i is the ordinal number of singular points, $D_i(x, y)$ is the distance between point (x,y) and i-th singular point, r_i is i-th singular point's effective radius.

Finally, the mixed model for the whole fingerprint's orientation field can be formulated as:

$$O_m = (1 - w) \cdot \theta + w \cdot O' \quad (7)$$

In order to implement the orientation correction algorithm, the position and type of singular points are need to detected. In our algorithm, the Poincare index method is used to detect the singular points. On the other hand, many parameters need to be ascertained. Some of them are initiated and modified based on the experiments while others are computed by least square method.

2.3 Region Mask Generation

In this step, we will classify each pixel in an input fingerprint image into a recoverable region or an unrecoverable region. In our algorithm, an optimal linear classifier has been trained for the classification per block and the criteria of minimal number of misclassified samples are used. Morphology has been applied as post-processing to reduce the number of classification errors. The detailed algorithm can be seen from our previous work [7].

2.4 Fingerprint Filtering

In the proposed algorithm we replaced the Gabor filter[3] with an anisotropic one, which was proved to be robust and efficient for the filtering of the fingerprint ridges. The structure adaptive anisotropic filtering [8] was modified for fingerprint image filtering. We use both a local intensity orientation and an anisotropic measure to control the shape of the filter. The filter kernel applied to fingerprint image at each point (x,y) is defined as follows:

$$h(x, y, \psi) = c_1 + c_2 \cdot \exp(-\frac{x_\psi^2}{2\sigma_1^2} - \frac{y_\psi^2}{2\sigma_2^2}) \cdot \frac{\sin f \cdot x_\psi}{f \cdot x_\psi} \tag{8}$$

$$x_\psi = x\cos\psi + y\sin\psi \tag{9}$$

$$y_\psi = -x\sin\psi + y\cos\psi \tag{10}$$

c1, c2, σ_1^2, σ_2^2 are empirical parameters,c1=-1, c2=2, σ_1^2=4, σ_2^2=2 in our algorithm. f is a parameter related to the ridge frequency. Applying a 2D Fourier transform to Equation (8), we obtain the filter's frequency response:

$$H(u, v, \psi) = c_1 \cdot 4\pi^2 \delta(u, v) + 2\pi \cdot c_2 \sigma_1 \sigma_2 \cdot \exp(-\frac{u_\psi^2}{2\sigma_u^2} - \frac{v_\psi^2}{2\sigma_v^2}) * G(u_\psi) \tag{11}$$

$$G(u_\psi) = \begin{cases} \frac{1}{2f} & |u_\psi| < 2\pi f \\ 0 & otherwise \end{cases} \tag{12}$$

$$u_\psi = u\cos\psi + v\sin\psi \tag{13}$$

$$v_\psi = -u \sin\psi + v \cos\psi \tag{14}$$

Where * stands for convolution, $\sigma_u = 1/2\pi\sigma_1, \sigma_v = 1/2\pi\sigma_2$

Let G be the normalized fingerprint images, O be the orientation image and R be the recoverable mask, the enhanced image F(i,j) is obtained as follows:

$$F(i,j) = \begin{cases} 255 & \text{if } R(i,j) = 0 \\ \sum_{u=-w_f/2}^{w_f/2} \sum_{v=-w_f/2}^{w_f/2} h(u,v:O(i,j)) \cdot G(i-u, j-v) & \text{otherwise} \end{cases} \tag{15}$$

where $w_f = 13$ specifies the size of the filters.

3 Experimental Results

The proposed algorithm has been evaluated on the databases of FVC2004 [9]. As the limits of pages, only the results on FVC2004 DB2 were listed in this paper.

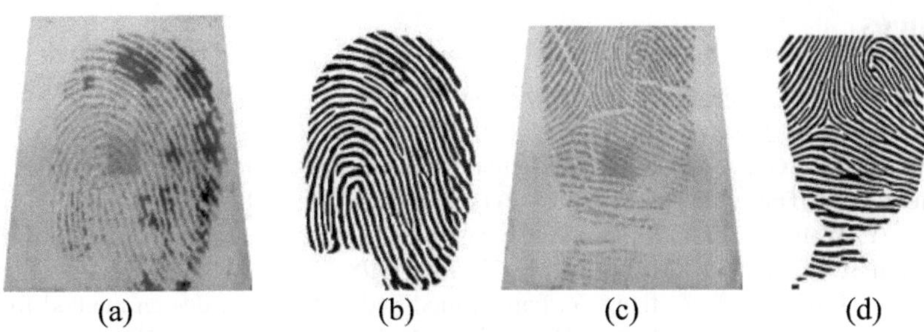

Fig. 2. Some examples of low quality fingerprints and their enhanced results in FVC2004 DB2. (a) Original image, very dry, (b) Enhanced image of (a), (c) Original image, with scars, (d) Enhanced image of (c).

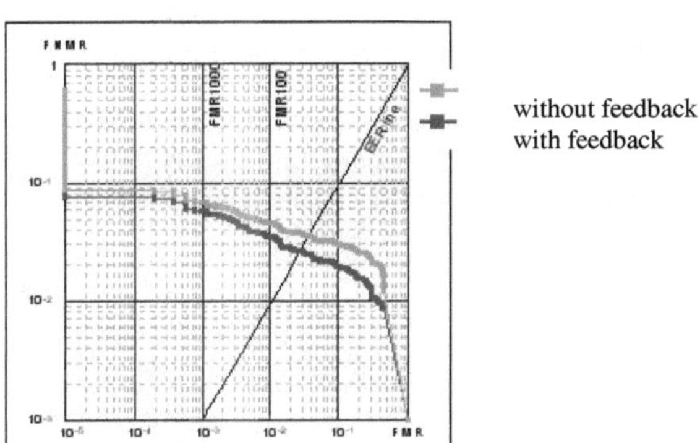

Fig. 3. The comparison of the algorithm with and without feedback method on FVC2004 DB2

Figure 2 show some examples of low quality fingerprints and their enhanced results in FVC2004 DB2. It can be seen form figure that these poor fingerprints (very dry, with many scars) are enhanced well. The average time for enhancing a fingerprint is about 0.32 second on PC AMD Athlon 1600+ (1.41 GHz).

Experiments were also done to compare the orientation estimation algorithm with and without feedback method. The comparison results on FVC2004 DB2 are shown in figure 3. The EER was 2.59 for the algorithm with feedback method, while 3.49 for the algorithm without feedback method. It is clear that the performance the recognition algorithm was improved by the feedback method.

4 Conclusion

In this paper, an orientation filed estimation with feedback method was proposed to compute the accurate fingerprint orientation. And an anisotropic filtering was applied to enhance the fingerprint, with the advantages of the efficient ridge enhancement and robustness against noise in the fingerprint image. Experimental results confirm that our algorithm is effective and robust for the enhancement of the low quality fingerprint.

References

1. C. Shi, Y.C. Wang, J. Qi, K. Xu, A New Segmentation Algorithm for Low Quality Fingerprint Image, ICIG 2004, pp.314-317.
2. J. Zhou and J. W. Gu, A Model-based Method for the computation of Fingerprints' Orientation Field, IEEE Trans. On Image Processing , Vol. 13, No. 6, pp.821-835, 2004
3. L. Hong, Y. Wan, A. K Jain, Fingerprint Image Enhancement: Algorithm and Performance Evaluation. IEEE Trans. PAMI, 20(8), pp.777–789, 1998.
4. J. W. Yang, L. F. Liu, T. Z. Jiang, Y. Fan, A modified Gabor filter design method for fingerprint image enhancement, Pattern Recognition, Vol.24, pp.1805-1817, 2003.
5. A.J. Willis, L. Myers, A Cost-effective Fingerprint Recognition System for Use with Low-quality Prints and Damaged Fingertips. Pattern Recognition, 34(2), pp.255–270, 2001
6. B. Sherlock and D. Monro, A Model for Interpreting Fingerprint Topology, Pattern Recognition, v. 26, no. 7, pp. 1047-1095, 1993.
7. X. J. Chen, J. Tian, J. G. Cheng, X. Yang, Segmentation of Fingerprint Images Using Linear Classifier. EURASIP Journal on Applied Signal Processing, Vol. 2004, No. 4, pp.480–494, Apr.2004
8. G.Z. Yang, P. Burger, D.N. Firmin and S.R. Underwood, Structure Adaptive Anisotropic Filtering. Image and Vision Computing 14: 135–145, 1996.
9. Biometric Systems Lab, Pattern Recognition and Image Processing Laboratory, Biometric Test Center, http://bias.csr.unibo.it/fvc2004/.

K-plet and Coupled BFS: A Graph Based Fingerprint Representation and Matching Algorithm

Sharat Chikkerur, Alexander N. Cartwright, and Venu Govindaraju

Center for Unified Biometrics and Sensors, University at Buffalo, NY, USA
{ssc5, anc, govind}@buffalo.edu

Abstract. In this paper, we present a new fingerprint matching algorithm based on graph matching principles. We define a new representation called K-plet to encode the local neighborhood of each minutiae. We also present CBFS (Coupled BFS), a new dual graph traversal algorithm for consolidating all the local neighborhood matches and analyze its computational complexity. The proposed algorithm is robust to non-linear distortion. Ambiguities in minutiae pairings are solved by employing a dynamic programming based optimization approach. We present an experimental evaluation of the proposed approach and showed that it exceeds the performance of the NIST BOZORTH3 [3] matching algorithm.

1 Introduction

Clearly the most important stage of a fingerprint verification system is the matching process. The purpose of the matching algorithm is to compare two fingerprint images or templates and return a similarity score that corresponds to the probability of match between the two prints. Minutiae features are the most popular of all the existing representation for matching and also form the basis of the process used by human experts [7]. Each minutiae may be described by a number of attributes such as its position (x,y) its orientation θ, its quality etc. However, most algorithms consider only its position and orientation. Given a pair of fingerprints, their minutiae features may be represented as an unordered set given by

$$I_1 = \{m_1, m_2....m_M\} \text{ where } m_i = (x_i, y_i, \theta_i) \quad (1)$$

$$I_2 = \{m'_1, m'_2....m'_N\} \text{ where } m'_i = (x'_i, y'_i, \theta'_i) \quad (2)$$

Usually points in I_2 is related to points in I_1 through a geometric transformation $T()$. Therefore, the technique used by most minutiae matching algorithms is to recover the transformation function T() that maps the two point sets . While there are several well known techniques for doing this, several challenges are faced when matching the minutiae point sets. The fingerprint image is obtained by capturing the three dimensional ridge pattern on the finger on to a two-dimensional surface. Therefore apart from skew and rotation assumed under most distortion models, there is also considerable stretching. Most matching algorithms assumed the prints to be rigidly transformed(strictly rotation and displacement) between different instances and therefore perform poorly under such situations. (See Figure 1).

D. Zhang and A.K. Jain (Eds.): ICB 2006, LNCS 3832, pp. 309–315, 2005.

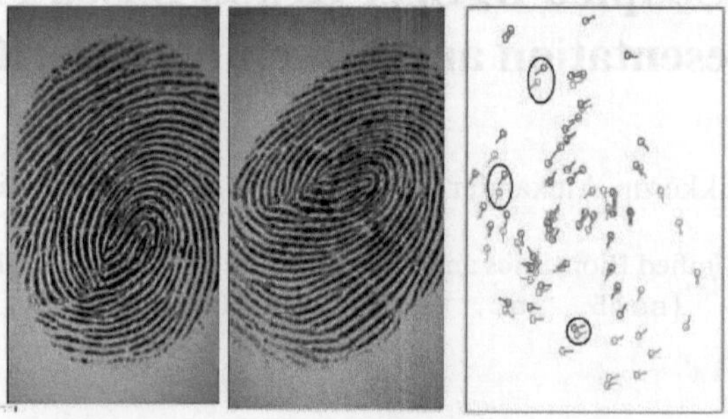

Fig. 1. An illustration of the non-linear distortion

1.1 Prior Related Work

A large number of recognition algorithms have been proposed in literature to date. The problem of matching minutiae can be treated as an instance of generalized *point pattern matching* problem. It is assumed that the two points sets are related by some geometrical relationship and the problem reduces to finding the most optimal geometrical transformation that relates these two sets. Most existing algorithms can be broadly classified as follows

1. **Global Matching:** In this approach, the matching process tries to simultaneously align all points at once. The global matching approach can be further categorized into
   (a)*Implicit Alignment:* Here the process of finding the point correspondences and finding the optimal alignment are performed simultaneously. This includes the iterative approach proposed by Ranade and Rosenfield [8] and the generalized Hough Transform based approach of Ratha et al. [9]
   (b)Explicit Alignment In this approach, the optimal transformation is obtained after explicitly aligning one of more corresponding points. The alignment may be absolute (based on singular point such as core and delta) or relative(based on a minutiae pair). Absolute alignment approaches are not very accurate since singular point location in poor quality prints is unreliable. Jain et al [4] proposed a relative alignment approach based on alignment of ridges.
2. **Local Matching:** In local matching approaches, the fingerprint is matched by accumulating evidence from matching local neighborhood structures. Each local neighborhood is associated with structural properties that are invariant under translation and rotation. Therefore, local matching algorithms are more robust to non-linear distortion and partial overlaps when compared to global approaches. However, local neighborhood do not sufficiently capture the global structural relationships making false accepts very common. Thefore in practice, matching algorithms that rely on local neighborhood information are implemented in two stages (a) *Local structure matching*: In this step, local structures are compared to derive candidate matches for each structure in the reference print. (b) *Consoldiation*: In this step, the candidate matches are validated based on how it agrees to the global match and a score is

generated by consolidating all the valid matches. Examples of matching algorithm based on local properties can be found in Jian and Yau [6],Jea and Govindaraju [5] and Ratha et al. [10].

2 Proposed Approach: Graph Based Matching

We propose a novel graph based algorithm for robust fingerprint recognition. We define a new representation called *K-plet* to represent local neighorhood of a minutiae that is invariant under translation and rotation. The local neighborhoods are matched using a dynamic programming based algorithm. The consolidation of the local matches is done by a novel *Coupled Breadth First Search* algorithm that propagates the local matches simultaneously in both the fingerprints. In the following section, we describe our approach using the following three aspects (i)Representation, (ii)Local Matching and (iii)Consolidation.

Table 1. Left: An illustration of K-plets defined in a fingerprint, Right:Local co-ordinate system of the K-plet

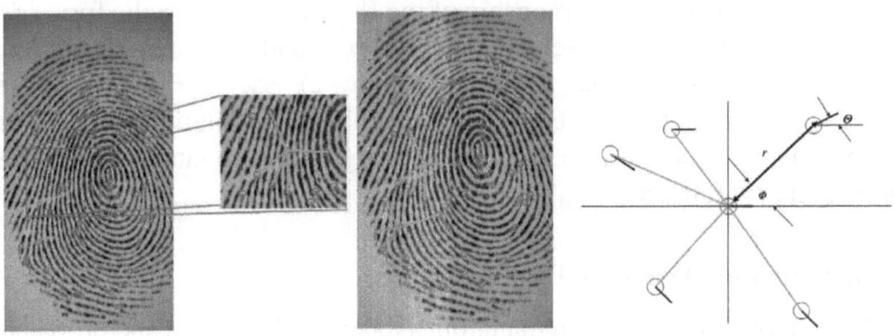

2.1 Representation: K-plet

The *K-plet* consists of a central minutiae m_i and K other minutiae $\{m_1, m_2...m_K\}$ chosen from its local neighborhood. Each neigbhorhood minutiae is defined in terms of its local radial co-ordinates $(\phi_{ij}, \theta_{ij}, r_{ij})$ (See Table 1) where r_{ab} represents the Euclidean distance between minutiae m_a and m_b. θ_{ij} is the relative orientation of minutia m_j w.r.t the central minutiae m_i. ϕ_{ij} represents the direction of the edge connecting the two minutia. The angle measurement is made w.r.t the X-axis which is now aligned with the minutia direction of m_i. Unlike the *star* representation, the K-plet does not specify how the K neighbors are chosen. We outline two different approaches of doing this althought this is not meant to be an exhaustive enumeration of ways to construct the K-plet. (i)In the first approach we construct the K-plet by considering the K-nearest neighbors of each minutia. This is not very effective if the minutia are clustered since it cannot propagate matches globally. (ii) In the second approach, in order to maintain high connectivity between different parts of the fingerprint, we chose K neighboring minutia such that a nearest neighbor is chosen in each of the four quadrant sequentially. Our results are reported based on this construction.

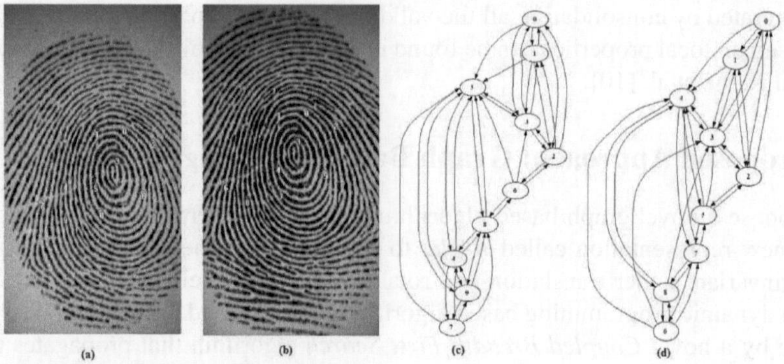

Fig. 2. Illustration of two fingerprints of the same user with marked minutiae and the corresponding adjacency graph based on the K-plet representaion. It is to be noted that the topologies of the graphs are different due to an extra unmatched minutiae in the left print.

2.2 Graphical View

We encode the local structural relationship of the K-plet formally in the form of a graph $G(V, E)$. Each minutiae is represented by a vertex v and each neighboring minutiae is represented by a directed edge (u, v) (See Figure 2). Each vertex u is colored with attributes $(x_u, y_u, \theta_u, t_u)$ that represents the co-ordinate, orientation and type of minutiae(ridge ending or bifurcation). Each directed edge (u, v) is labelled with the corresponding K-plet co-ordindates $(r_{uv}, \phi_{uv}, \theta_{uv})$

2.3 Local Matching: Dynamic Programming

Our matching algorithm is based on matching a local neighborhood and propagating the match to the K-plet of all the minutiae in this neighborhood successively. The accuracy of the algorithm therefore depends critically on how this local matching is performed. We convert the unordered neighbors of each K-plet into an ordered sequence by arranging them in the increasing order of the radial distance r_{ij}. The problem now reduces to matching two ordered sequences $S\{s_1, s_2...s_M\}$ $T\{t_1, t_2...t_N\}$. We utilize a dynamic programming approach based on string alignment algorithm [2].

Formally, the problem of string alignment can be stated as follows: Given two strings or sequences S and T, the problem is two determine two auxiliary strings S' and T' such that

1. S' is derived by inserting spaces (_) in S
2. T' is derived by inserting spaces in T
3. $length(S') = length(T')$
4. The cost $\sum_{i=1}^{|S'|} \sigma(s'_i, t'_i)$ is maximized.

For instance, the result of aligning the sequences $S = \{acbcdb\}$ and $T = \{cadbd\}$ is given by

$$S' = ac\_\_bcdb \quad (3)$$

$$T' = \_cadb\_d\_ \quad (4)$$

A trivial solution would be to list all possible sequences S' and T' and select the pair with the least/most alignment cost. However, this would require exponential time. Instead we can solve this using dynamic programming in O(MN) time as follows. We define D[i,j]($i \in \{0,1...M\}, j \in \{0,1...N\}$) as the cost of aligning substrings S(1..i) and T(1..j). The cost of aligning S and T is therefore given by D[M,N]. Dynamic programming uses a recurrence relation between D[i,j] and already computed values to reduce the run-time substantially. It is assumed ofcourse that D[k,l] is optimal $\forall k < i, l < j$. Given that the previous sub-problems have been optimally defined, we can match s_i and t_j in three ways

1. the elements s[i] and t[j] match with cost $\sigma(s[i], t[j])$,
2. a gap is introduced in t (s[i] is matched with a gap) with cost $\sigma(s[i], \_)$
3. a gap is introduced in s (t[j] is matched with a gap) with cost $\sigma(\_, t[j])$

Therefore, the recurrence relation to compute D[i,j] is given by

$$D[i,j] = max \left\{ \begin{array}{c} D[i-1,j-1] + \sigma(s[j], t[i]) \\ D[i-1,j] \quad + \quad \sigma(s[i], \_) \\ D[i,j-1] \quad + \quad \sigma(\_, t[j]) \end{array} \right\} \tag{5}$$

2.4 Consolidation: Coupled Breadth First Search

The most important aspect of the new matching algorithm is a formal approach for consolidating all the local matches between the two fingerprints without requiring explicit

```
Algorithm:  CBFS
Inputs   :  Graphs G(V,E) and H(V,E) corresponding to reference and
            test fingerprints
            i: source node in graph G
            j: source node in graph H
Outputs  :  No. of vertices that can be matched from the given sources

1. Let G(V,E) and H(V,E) represent the graphs corresponding to the two
   prints
2. Let GQ and HQ represent a FIFO queue
3. Let M represent a set of matched vertex pairs <g,h>
4. Initialize
      a. For each vertex g in G(V,E) and h in H(V,E)
           i. color[g] = WHITE //unvisited node
          ii. color[h] = WHITE
      b. color[i] = GRAY
      c. color[j] = GRAY
      d. M = M + <g[i],h[j]>
      e. ENQUEUE(GQ,g[i])
      f. ENQUEUE(HQ,h[j])
5. While (GQ is not empty and HQ is not empty)
      a. gu = DEQUEUE(GQ)
      b. hu = DEQUEUE(HQ)
      c. Find matching neighbors of gu,hu using dynamic programming
      d. For each matching neighbor gv (of gu) and hv (of gv)
      e. If (color[gv] == WHITE and color[hv] == WHITE)
           i. M = M + <gv,hv>
          ii. ENQUEUE(GQ,gv)
         iii. ENQUEUE(HQ,hv)
6. Return M (the size of M gives the match count)
```

Fig. 3. An overview of the CBFS algorithm

alignment. We propose a new algorithm called *Coupled BFS algorithm*(CBFS) for this purpose. CBFS is a modification of the regular breadth first algorithm [2] except for two special modifications. (i) The graph traversal occurs in two directed graphs G and H corresponding to reference and test fingerprints simultaneously. (The graphs are constructured as mentioned in Section 2.2) (ii) While the regular BFS algorithm visits each vertex v in the adjacency list of , CBFS visits only the the vertices $v_G \in V$ and $v_H \in H$ such that v_G and v_H are locally matched vertices. The overview of the CBFS algorithm is given in Figure 3

2.5 Matching Algorithm

It is to be noted that the CBFS algorithm requires us to specify two vertices as the source nodes from which to begin the traversal. Since the point correspondences are not known apriori, we execute the CBFS algorithm for all possible correspondence pairs $g[i], h[j]$). We finally consider the maximum number of matches return to compute the matching score. The score is generated by using [1] $s = \frac{m^2}{M_R M_T}$. Here m represents the number of matched minutiae and M_R and M_T represent the number of minutiae in the reference and template prints respectively.

3 Experimental Evaluation

In order to measure the objective performance, we run the matching algorithm on images from FVC2002 DB1 database. The database consists of 800 images (100 distinct fingers, 8 instances each). In order to obtain the performance characterists such as EER (Equal Error Rate) we perform a total of 2800 genuine comparision and 4950 impostor comparisons .We present the comparative results in Table 2. The improvement in the ROC characteristic can be seen from Figure 4.

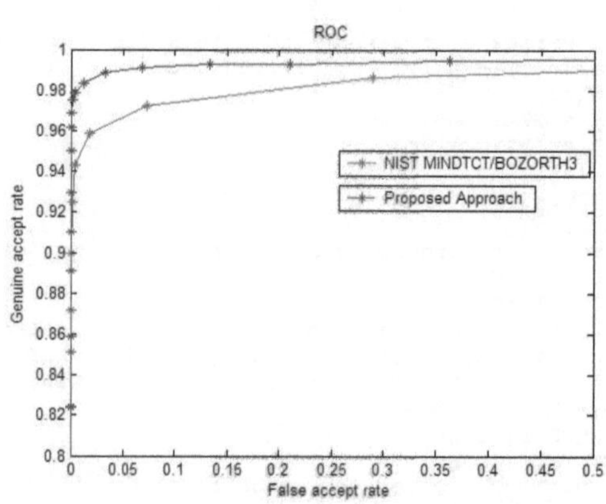

Fig. 4. A comparision of ROC curves for FVC2002 DB1 database

Table 2. A summary of the comparative results

Database	NIST MINDTCT/BOZORTH3		Proposed Approach	
	EER	FMR100	EER	FMR100
FVC2002 DB1	3.6%	5.0%	1.5%	1.65%

4 Summary

We presented a novel minutia based fingerprint recognition algorithm that incorporates three new ideas. Firstly, we defined a new representation called K-plet to encode the local neighborhood of each minutia. Secondly, we also presented a dynamic programming approach for matching each local neighborhood in an optimal fashion. Lastly, we proposed CBFS (Coupled Breadth First Search), a new dual graph traversal algorithm for consolidating all the local neighborhood matches and analyze its computational complexity. We presented an experimental evaluation of the proposed approach and showed that it exceeds the performance of the popular NIST BOZORTH3 matching algorithm.

References

1. Asker M. Bazen and Sabih H. Gerez. Fingerprint matching by thin-plate spline modeling of elastic deformations. *Pattern Recognition*, 36:1859–1867, 2003.
2. Thomas H. Cormen, Charles E. Leiserson, and Ronald L. Rivest. *Introduction to algorithms*. McGraw-Hill Book Company, 1998.
3. M. D. Garris, C. I. Watson, R. M. McCabe, and C. L. Wilson. Users guide to nist fingerprint image software (nfis). Technical Report NISTIR 6813, National Institute of Standards and Technology, 2002.
4. A. Jain, L. Hong, and R. Bolle. On-line fingerprint verification. In *Pattern Analysis and Machine Intelligence*, volume 19, pages 302–313, 1997.
5. Tsai-Yang Jea and Venu Govindaraju. A minutia-based partial fingerprint recognition system. *Submitted to Pattern Recognition*, 2004.
6. Xudong Jiang and Wei-Yun Yau. Fingerprint minutiae matching based on the local and global structures. In *International Conference on Pattern Recognition*, pages 1038–1041, 2000.
7. D. Maio, D. Maltoni, A. K. Jain, and S. Prabhakar. *Handbook of Fingerprint Recognition*. Springer Verlag, 2003.
8. A. Ranade and A. Rosenfeld. Point pattern matching by relaxation. *Pattern Recognition*, 12(2):269–275, 1993.
9. N. K. Ratha, K. Karu, S. Chen, and A. K. Jain. A real-time matching system for large fingerprint databases. *Transactions on Pattern Analysis and Machine Intelligence*, 18(8):799–813, 1996.
10. N. K. Ratha, V. D. Pandit, R. M. Bolle, and V. Vaish. Robust fingerprint authentication using local structure similarity. In *Workshop on applications of Computer Vision*, pages 29–34, 2000.

A Fingerprint Recognition Algorithm Combining Phase-Based Image Matching and Feature-Based Matching

Koichi Ito[1], Ayumi Morita[1], Takafumi Aoki[1], Hiroshi Nakajima[2], Koji Kobayashi[2], and Tatsuo Higuchi[3]

[1] Graduate School of Information Sciences, Tohoku University, Sendai 980–8579 Japan
ito@aoki.ecei.tohoku.ac.jp
[2] Yamatake Corporation, Isehara 259–1195, Japan
[3] Faculty of Engineering, Tohoku Institute of Technology, Sendai 982–8577, Japan

Abstract. This paper proposes an efficient fingerprint recognition algorithm combining phase-based image matching and feature-based matching. The use of Fourier phase information of fingerprint images makes possible to achieve robust recognition for weakly impressed, low-quality fingerprint images. Experimental evaluations using two different types of fingerprint image databases demonstrate efficient recognition performance of the proposed algorithm compared with a typical minutiae-based algorithm and the conventional phase-based algorithm.

1 Introduction

Biometric authentication has been receiving extensive attention over the past decade with increasing demands in automated personal identification. Biometrics is to identify individuals using physiological or behavioral characteristics, such as fingerprint, face, iris, retina, palm-print, etc. Among all the biometric techniques, fingerprint recognition [1, 2] is the most popular method and is successfully used in many applications.

Major approaches for fingerprint recognition today can be broadly classified into feature-based approach and correlation-based approach. Typical fingerprint recognition methods employ feature-based matching, where minutiae (i.e., ridge ending and ridge bifurcation) are extracted from the registered fingerprint image and the input fingerprint image, and the number of corresponding minutiae pairs between the two images is used to recognize a valid fingerprint image [1]. Feature-based matching is highly robust against nonlinear fingerprint distortion, but shows only limited capability for recognizing poor-quality fingerprint images with low S/N ratio due to unexpected fingertip conditions (e.g., dry fingertips, rough fingertips, allergic-skin fingertips) as well as weak impression of fingerprints.

On the other hand, as one of the efficient correlation-based approaches[3], we have proposed a fingerprint recognition algorithm using phase-based image matching [4] — an image matching technique using the phase components

D. Zhang and A.K. Jain (Eds.): ICB 2006, LNCS 3832, pp. 316–325, 2005.

in 2D Discrete Fourier Transforms (2D DFTs) of given images —, and developed commercial fingerprint verification units for access control applications [5]. Historically, the phase-based image matching has been successfully applied to high-accuracy image registration tasks for computer vision applications [6, 7, 8]. The use of Fourier phase information of fingerprint images makes possible highly reliable fingerprint matching for low-quality fingerprints whose minutiae are difficult to be extracted as mentioned above. However, the performance of the phase-based fingerprint matching is degraded by nonlinear distortions in fingerprint images.

In order to improve matching performance for both fingerprint images with poor image quality and with nonlinear shape distortions, we propose a novel fingerprint recognition algorithm combining phase-based image matching and feature-based matching. In this algorithm, two approaches are expected to play a complementary role and may result in significant improvements of recognition performance. Experimental evaluations using two different types of fingerprint image databases demonstrate efficient recognition performance of the proposed algorithm compared with a typical minutiae-based algorithm and the conventional phase-based algorithm.

2 Phase-Based Fingerprint Matching

In this section, we introduce the principle of phase-based image matching using the Phase-Only Correlation (POC) function (which is sometimes called the "phase-correlation function") [6, 7, 8]. We also describe the POC-based fingerprint matching algorithm.

2.1 Fundamentals of Phase-Based Image Matching

Consider two $N_1 \times N_2$ images, $f(n_1, n_2)$ and $g(n_1, n_2)$, where we assume that the index ranges are $n_1 = -M_1 \cdots M_1$ $(M_1 > 0)$ and $n_2 = -M_2 \cdots M_2$ $(M_2 > 0)$ for mathematical simplicity, and $N_1 = 2M_1 + 1$ and $N_2 = 2M_2 + 1$. Let $F(k_1, k_2)$ and $G(k_1, k_2)$ denote the 2D DFTs of the two images. $F(k_1, k_2)$ is given by

$$F(k_1, k_2) = \sum_{n_1, n_2} f(n_1, n_2) W_{N_1}^{k_1 n_1} W_{N_2}^{k_2 n_2} = A_F(k_1, k_2) e^{j\theta_F(k_1, k_2)}, \tag{1}$$

where $k_1 = -M_1 \cdots M_1$, $k_2 = -M_2 \cdots M_2$, $W_{N_1} = e^{-j\frac{2\pi}{N_1}}$, $W_{N_2} = e^{-j\frac{2\pi}{N_2}}$, and $\sum_{n_1, n_2}$ denotes $\sum_{n_1=-M_1}^{M_1} \sum_{n_2=-M_2}^{M_2}$. $A_F(k_1, k_2)$ is amplitude and $\theta_F(k_1, k_2)$ is phase. $G(k_1, k_2)$ is defined in the same way. The cross-phase spectrum $R_{FG}(k_1, k_2)$ is given by

$$R_{FG}(k_1, k_2) = \frac{F(k_1, k_2)\overline{G(k_1, k_2)}}{|F(k_1, k_2)\overline{G(k_1, k_2)}|} = e^{j\theta(k_1, k_2)}, \tag{2}$$

where $\overline{G(k_1, k_2)}$ is the complex conjugate of $G(k_1, k_2)$ and $\theta(k_1, k_2)$ denotes the phase difference $\theta_F(k_1, k_2) - \theta_G(k_1, k_2)$. The POC function $r_{fg}(n_1, n_2)$ is the 2D Inverse DFT (2D IDFT) of $R_{FG}(k_1, k_2)$ and is given by

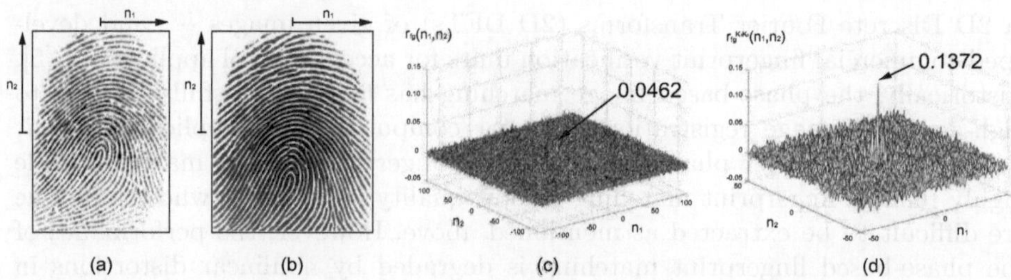

Fig. 1. Example of genuine matching using the original POC function and the BLPOC function: (a) registered fingerprint image $f(n_1, n_2)$, (b) input fingerprint image $g(n_1, n_2)$, (c) POC function and (d) BLPOC function with $K_1/M_1 = K_2/M_2 = 0.48$

$$r_{fg}(n_1, n_2) = \frac{1}{N_1 N_2} \sum_{k_1,k_2} R_{FG}(k_1, k_2) W_{N_1}^{-k_1 n_1} W_{N_2}^{-k_2 n_2}, \tag{3}$$

where $\sum_{k_1,k_2}$ denotes $\sum_{k_1=-M_1}^{M_1} \sum_{k_2=-M_2}^{M_2}$. When two images are similar, their POC function gives a distinct sharp peak. When two images are not similar, the peak drops significantly. The height of the peak gives a good similarity measure for image matching, and the location of the peak shows the translational displacement between the images.

We modify the definition of POC function to have a BLPOC (Band-Limited Phase-Only Correlation) function dedicated to fingerprint matching tasks. The idea to improve the matching performance is to eliminate meaningless high frequency components in the calculation of cross-phase spectrum $R_{FG}(k_1, k_2)$ depending on the inherent frequency components of fingerprint images [4]. Assume that the ranges of the inherent frequency band are given by $k_1 = -K_1 \cdots K_1$ and $k_2 = -K_2 \cdots K_2$, where $0 \le K_1 \le M_1$ and $0 \le K_2 \le M_2$. Thus, the effective size of frequency spectrum is given by $L_1 = 2K_1 + 1$ and $L_2 = 2K_2 + 1$. The BLPOC function is given by

$$r_{fg}^{K_1 K_2}(n_1, n_2) = \frac{1}{L_1 L_2} {\sum_{k_1,k_2}}' R_{FG}(k_1, k_2) W_{L_1}^{-k_1 n_1} W_{L_2}^{-k_2 n_2}, \tag{4}$$

where $n_1 = -K_1 \cdots K_1$, $n_2 = -K_2 \cdots K_2$, and $\sum'_{k_1,k_2}$ denotes $\sum_{k_1=-K_1}^{K_1} \sum_{k_2=-K_2}^{K_2}$. Note that the maximum value of the correlation peak of the BLPOC function is always normalized to 1 and does not depend on L_1 and L_2. Figure 1 shows an example of genuine matching using the original POC function r_{fg} and the BLPOC function $r_{fg}^{K_1 K_2}$. The BLPOC function provides the higher correlation peak and better discrimination capability than that of the original POC function.

2.2 Fingerprint Matching Algorithm Using BLPOC Function

This section describes a fingerprint matching algorithm using BLPOC function. The algorithm consists of the three steps: (i) rotation and displacement alignment, (ii) common region extraction and (iii) matching score calculation with precise rotation.

(i) Rotation and displacement alignment
We need to normalize the rotation and the displacement between the registered fingerprint image $f(n_1, n_2)$ and the input fingerprint image $g(n_1, n_2)$ in order to perform the high-accuracy fingerprint matching. We first normalize the rotation by using a straightforward approach as follows. We first generate a set of rotated images $f_\theta(n_1, n_2)$ of the registered fingerprint $f(n_1, n_2)$ over the angular range $-50° \le \theta \le 50°$ with an angle spacing 1°. The rotation angle Θ of the input image relative to the registered image can be determined by evaluating the similarity between the rotated replicas of the registered image $f_\theta(n_1, n_2)$ $(-50° \le \theta \le 50°)$ and the input image $g(n_1, n_2)$ using the BLPOC function. Next, we align the translational displacement between the rotation-normalized image $f_\Theta(n_1, n_2)$ and the input image $g(n_1, n_2)$. The displacement can be obtained from the peak location of the BLPOC function between $f_\Theta(n_1, n_2)$ and $g(n_1, n_2)$. Thus, we have normalized versions of the registered image and the input image, which are denoted by $f'(n_1, n_2)$ and $g'(n_1, n_2)$. In practical situation, we store in advance a set of rotated versions of the registered image into a memory in order to reduce the processing time.

(ii) Common region extraction
Next step is to extract the overlapped region (intersection) of the two images $f'(n_1, n_2)$ and $g'(n_1, n_2)$. This process improves the accuracy of fingerprint matching, since the non-overlapped areas of the two images become uncorrelated noise components in the BLPOC function. In order to detect the effective fingerprint areas in the registered image $f'(n_1, n_2)$ and the input image $g'(n_1, n_2)$, we examine the n_1-axis projection and the n_2-axis projection of pixel values. Only the common effective image areas, $f''(n_1, n_2)$ and $g''(n_1, n_2)$, with the same size are extracted for the use in succeeding image matching step.

(iii) Matching score calculation with precise rotation
The phase-based image matching is highly sensitive to image rotation. Hence, we calculate the matching score with precise correction of image rotation. We generate a set of rotated replicas $f''_\theta(n_1, n_2)$ of $f''(n_1, n_2)$ over the angular range $-2° \le \theta \le 2°$ with an angle spacing 0.5°, and calculate BLPOC function $r^{K_1K_2}_{f''_\theta g''}(n_1, n_2)$. If the rotation and displacement between two fingerprint images are normalized, the correlation peak can be observed at the center of the BLPOC function. The BLPOC function may give multiple correlation peaks due to elastic fingerprint deformation. Thus, we define the matching score between the two images as the sum of the highest P peaks of the BLPOC function $r^{K_1K_2}_{f''_\theta g''}(n_1, n_2)$, where search area is $B \times B$-pixel block centered at $(0, 0)$. In this paper, we employ the parameters $B = 11$ and $P = 2$. The final score $S_P (0 \le S_P \le 1)$ of phase-based matching is defined as the maximum value of the scores computed from BLPOC function $r^{K_1K_2}_{f''_\theta g''}(n_1, n_2)$ over the angular range $-2° \le \theta \le 2°$.

3 Feature-Based Fingerprint Matching

The proposed feature-based fingerprint matching algorithm extracts the corresponding minutiae pairs between the registered image $f(n_1, n_2)$ and the input image $g(n_1, n_2)$, and calculates the matching score by block matching using BLPOC. This algorithm consists of four steps: (i) minutiae extraction, (ii) minutiae pair correspondence, (iii) local block matching using BLPOC function, and (iv) matching score calculation.

(i) Minutiae extraction
We employ the typical minutiae extraction technique [1], which consists of the following four steps: (a) ridge orientation/frequency estimation, (b) fingerprint enhancement and binarization, (c) ridge thinning, and (d) minutiae extraction with spurious minutiae removal. Each extracted minutia is characterized by a feature vector $\boldsymbol{m}_i$, whose elements are its (n_1, n_2) coordinates, the orientation of the ridge on which it is detected, and its type (i.e., ridge ending or ridge bifurcation). Let $\boldsymbol{M}^f$ and $\boldsymbol{M}^g$ be sets of minutiae feature vectors extracted from $f(n_1, n_2)$ and $g(n_1, n_2)$, respectively.

(ii) Minutiae pair correspondence
A minutia matching technique based on both the local and global structures of minutiae is employed to find corresponding minutiae pairs between $f(n_1, n_2)$ and $g(n_1, n_2)$ [9]. For every minutia $\boldsymbol{m}_i$, we calculate a local structure feature vector $\boldsymbol{l}_i$, which described by the distances, ridge-counts, directions and radial angles of the minutia relative to each of two nearest-neighbor minutiae and the types of these minutiae. Let $\boldsymbol{L}^f$ and $\boldsymbol{L}^g$ be sets of local structure feature vectors calculated from $\boldsymbol{M}^f$ and $\boldsymbol{M}^g$, respectively. We perform minutiae matching between $\boldsymbol{M}^f$ and $\boldsymbol{M}^g$ by using their local structure information $\boldsymbol{L}^f$ and $\boldsymbol{L}^g$, and find the best matching minutiae pair $(\boldsymbol{m}^f_{i_0}, \boldsymbol{m}^g_{j_0})$, which is called *reference minutiae pair*. All other minutiae are aligned based on this reference minutiae pair by converting their coordinates to the polar coordinate system with respect to the reference minutia. Thus, we have the aligned minutiae information $\boldsymbol{M'}^f$ and $\boldsymbol{M'}^g$. For every aligned minutia $\boldsymbol{m'}^f_i \in \boldsymbol{M'}^f$ (or $\boldsymbol{m'}^g_j \in \boldsymbol{M'}^g$), we calculate a global feature vector $\boldsymbol{g}^f_i$ (or $\boldsymbol{g}^g_j$), which is described by the distance, direction and radial angle of the minutia relative to the reference minutia $\boldsymbol{m}^f_{i_0}$ (or $\boldsymbol{m}^g_{j_0}$). Based on the distance $|\boldsymbol{g}^f_i - \boldsymbol{g}^g_j|$, we can now determine the correspondence between the minutiae pair $\boldsymbol{m'}^f_i$ and $\boldsymbol{m'}^g_j$. As a result, we obtain a set of the corresponding minutiae pairs between $\boldsymbol{M'}^f$ and $\boldsymbol{M'}^g$ as well as the matching score S_{minutiae} $(0 \leq S_{\text{minutiae}} \leq 1)$ defined as

$$S_{\text{minutiae}} = \frac{(\#\text{ of corresponding minutiae pairs})^2}{|\boldsymbol{M'}^f| \times |\boldsymbol{M'}^g|}. \tag{5}$$

(iii) Local block matching using BLPOC function
When the number of corresponding minutiae pairs is greater than 2, we extract local binary images, from $f(n_1, n_2)$ and $g(n_1, n_2)$, centered at the corresponding

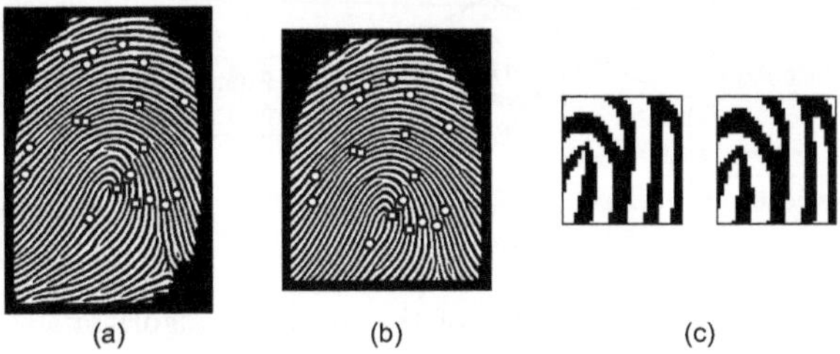

Fig. 2. Example of local block matching using BLPOC function for a genuine pair ($S_{\mathrm{minutiae}} = 0.41$ and $S_{\mathrm{block}} = 0.57$): (a) binarized registered image, (b) binarized input image, (c) a pair of block around corresponding minutiae (the score of local block matching is 0.59). The symbols $\circ$ and $\square$ denote the corresponding minutiae.

minutiae. The size of local binary image is $l \times l$ pixels, where we use $l = 31$ in our experiments. For every pair of local binary images, we align image rotation using the information of minutiae orientation, and calculate the BLPOC function between the local image blocks to evaluate the local matching score as its correlation peak value. The score of block matching S_{block} ($0 \leq S_{\mathrm{block}} \leq 1$) is calculated by taking an average of the highest three local matching scores. On the other hand, when the number of corresponding minutiae pairs is less than 3, we set $S_{\mathrm{block}} = 0$. Figure 2 shows an example of local block matching using BLPOC function for a genuine pair.

(iv) Matching score calculation
The combined score S_F ($0 \leq S_F \leq 1$) of feature-based matching is calculated from S_{minutiae} and S_{block} as follows:

$$S_F = \begin{cases} 1 & \text{if } S_{\mathrm{minutiae}} \times S_{\mathrm{block}} > T_F \\ S_{\mathrm{minutiae}} \times S_{\mathrm{block}} & \text{otherwise,} \end{cases} \tag{6}$$

where T_F is a threshold.

4 Overall Recognition Algorithm

In this section, we describe a fingerprint recognition algorithm combining phase-based image matching and feature-based matching. Figure 3 shows the flow diagram of the proposed fingerprint recognition algorithm.

(I) Classification
In order to reduce the computation time and to improve the recognition performance, we introduce the rule-based fingerprint classification method [1] before matching operation. In our algorithm, we classify the fingerprints into 7 categories: "Arch", "Left Loop", "Right Loop", "Left Loop or Right Loop", "Arch or Left Loop", "Arch or Right Loop", and "Others". If the two fingerprints to be verified fall into different categories, we give the overall score $S = 0$, otherwise matching operation is performed to evaluate the overall score.

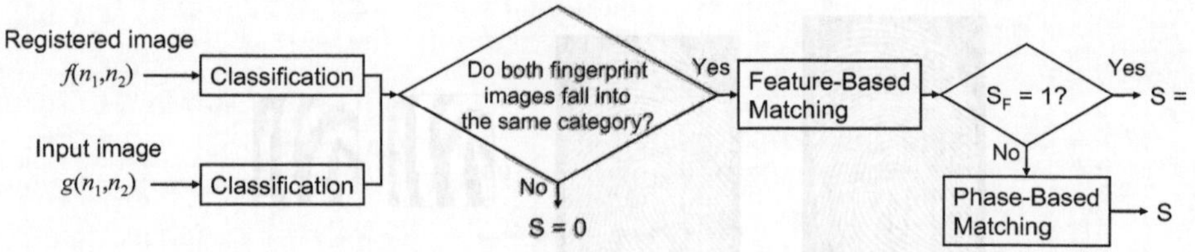

Fig. 3. Flow diagram of the proposed algorithm

(II) Feature-based matching
This stage evaluates the matching score S_F of feature-based matching as described in section 3. If $S_F = 1$, then we set the overall score as $S = 1$ and terminate matching operation, otherwise we proceed to the stage (III).

(III) Phase-based matching
This stage evaluates the matching score S_P of phase-based fingerprint matching as described in section 2. Then, the overall matching score S is computed as a linear combination of S_F and S_P, given by

$$S = \alpha \times S_F + (1 - \alpha) \times S_P, \tag{7}$$

where $0 \leq \alpha \leq 1$. In our experiments, we employ $\alpha = 0.5$.

5 Experimental Results

This section describes a set of experiments, using our original database (DB_A) collecting low-quality fingerprint images and the FVC 2002 DB1 set A [10] (DB_B), for evaluating fingerprint matching performance of the proposed algorithm. The following experiments are carried out for the two databases.

A set of fingerprint images in DB_A is captured with a pressure sensitive sensor (BLP-100, BMF Corporation) of size 384 × 256 pixels, which contains 330 fingerprint images from 30 different subjects with 11 impressions for each finger. In the captured images, 20 of subjects have good-quality fingerprints and the remaining 10 subjects have low-quality fingerprints due to dry fingertips (6 subjects), rough fingertips (2 subjects) and allergic-skin fingertips (2 subjects). Thus, the test set considered here is specially designed to evaluate the performance of fingerprint matching under difficult condition. We first evaluate genuine matching scores for all possible combinations of genuine attempts; the number of attempts is ${}_{11}C_2 \times 30 = 1650$. Next, we evaluate impostor matching scores for impostor attempts: the number of attempts is ${}_{30}C_2 = 435$, where we select a single image (the first image) for each fingerprint and make all the possible combinations of impostor attempts.

A set of fingerprint images in DB_B is captured with an optical sensor (Touch View II, Identx Incorporated) of size 388 × 374 pixels, which contains 800 fingerprint images from 100 different subjects with 8 impressions for each finger. We first evaluate genuine matching scores for all possible combinations of genuine attempts; the number of attempts is ${}_{8}C_2 \times 100 = 2800$. Next, we evaluate impostor

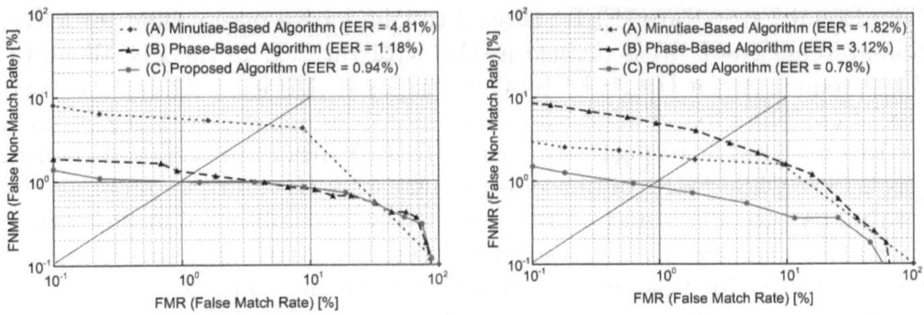

(a) (b)

Fig. 4. ROC curves and EERs: (a) DB_A and (b) DB_B

matching scores for impostor attempts: the number of attempts is ${}_{100}C_2 = 4950$, where we select a single image (the first image) for each fingerprint and make all the possible combinations of impostor attempts.

We compare three different matching algorithms: (A) a typical minutiae-based algorithm (which is commercially available), (B) a phase-based algorithm described in section 2, and (C) the proposed algorithm. In our experiments, the parameters of BLPOC function are $K_1/M_1 = K_2/M_2 = 0.40$ for DB_A and $K_1/M_1 = K_2/M_2 = 0.48$ for DB_B. The threshold value for feature-based matching is $T_F = 0.046$ for DB_A and $T_F = 0.068$ for DB_B.

The performance of the biometrics-based identification system is evaluated by the Receiver Operating Characteristic (ROC) curve, which illustrates the False Match Rate (FMR) against the False Non-Match Rate (FNMR) at different thresholds on the matching score. Figures 4 (a) and (b) show the ROC curve for the three algorithms (A)–(C) for DB_A and DB_B, respectively. In both cases, the proposed algorithm (C) exhibits significantly higher performance, since its ROC curve is located at lower FNMR/FMR region than those of the minutiae-based algorithm (A) and the phase-based algorithm (B).

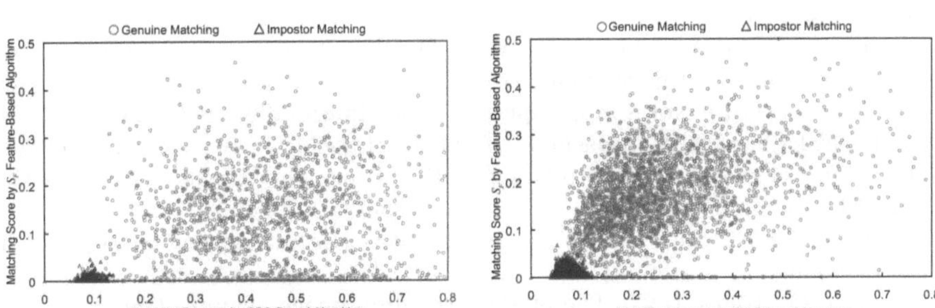

(a) (b)

Fig. 5. Overall joint distribution of matching scores for phase-based matching S_P and feature-based matching S_F: (a) DB_A and (b) DB_B

The Equal Error Rate (EER) is used to summarize performance of a verification system. The EER is defined as the error rate where the FNMR and the FMR are equal. As for DB_A, EER of the proposed algorithm (C) is 0.94%, while EERs of the phase-based algorithm (B) and the minutiae-based algorithm (A) are 1.18% and 4.81%, respectively. As for DB_B, EER of the proposed algorithm (C) is 0.78%, while EERs of phase-based algorithm (B) and the minutiae-based algorithm (A) are 3.12% and 1.82%, respectively. As is observed in the above experiments, the combination of phase-based matching and feature-based matching is highly effective for verifying low-quality difficult fingerprints.

Figure 5 (a) and (b) show the joint distribution of matching scores for phase-based matching S_P and feature-based matching S_F. Although we can observe weak correlation between S_P and S_F, both figures (a) and (b) show wide distributions of matching scores. This implies independent matching criteria used in phase-based and feature-based approaches can play a complementary role for improving overall recognition performance.

6 Conclusion

This paper has proposed a novel fingerprint recognition algorithm, which is based on the combination of two different matching criteria: (i) phase-based matching and (ii) feature-based matching. Experimental results clearly show good recognition performance compared with a typical minutiae-based fingerprint matching algorithm. In our previous work, we have already developed commercial fingerprint verification units for access control applications [5], which employs specially designed ASIC [11] for real-time phase-based image matching. The algorithm in this paper could be easily mapped onto our prototype hardware, since the computational complexity of feature-based matching algorithm is not significant.

References

1. Maltoni, D., Maio, D., Jain, A.K., Prabhakar, S.: Handbook of Fingerprint Recognition. Springer (2003)
2. Wayman, J., Jain, A., Maltoni, D., Maio, D.: Biometric Systems. Springer (2005)
3. Venkataramani, K., Vijayakumar, B.V.K.: Fingerprint verification using correlation filters. Lecture Notes in Computer Science **2688** (2003) 886–894
4. Ito, K., Nakajima, H., Kobayashi, K., Aoki, T., Higuchi, T.: A fingerprint matching algorithm using phase-only correlation. IEICE Trans. Fundamentals **E87-A** (2004) 682–691
5. `http://www.aoki.ecei.tohoku.ac.jp/poc.html`. Products using phase-based image matching
6. Kuglin, C.D., Hines, D.C.: The phase correlation image alignment method. Proc. Int. Conf. on Cybernetics and Society (1975) 163–165
7. Takita, K., Aoki, T., Sasaki, Y., Higuchi, T., Kobayashi, K.: High-accuracy subpixel image registration based on phase-only correlation. IEICE Trans. Fundamentals **E86-A** (2003) 1925–1934

8. Takita, K., Muquit, M.A., Aoki, T., Higuchi, T.: A sub-pixel correspondence search technique for computer vision applications. IEICE Trans. Fundamentals **E87-A** (2004) 1913–1923
9. Jiang, X., Yau, W.Y.: Fingerprint minutiae matching based on the local and global structures. International Conference on Pattern Recognition. **2** (2000) 1038–1041
10. `http://bias.csr.unibo.it/fvc2002`. Fingerprint verification competition 2002
11. Morikawa, M., Katsumata, A., Kobayashi, K.: An image processor implementing algorithms using characteristics of phase spectrum of two-dimensional Fourier transformation. Proc. IEEE Int. Symp. Industrial Electronics **3** (1999) 1208–1213

Fast and Robust Fingerprint Identification Algorithm and Its Application to Residential Access Controller

Hiroshi Nakajima[1], Koji Kobayashi[2], Makoto Morikawa[3], Atsushi Katsumata[3], Koichi Ito[4], Takafumi Aoki[4], and Tatsuo Higuchi[5]

[1] Building Systems Company, Yamatake Corporation, 54 Suzukawa, Isehara, Kanagawa 259-1195 Japan
[2] Building Systems Company, Yamatake Corporation, 2-15-1 Kounan, Minato, Tokyo 108-6030 Japan
[3] Research and Development Center, Yamatake Corporation, 1-12-2 Kawana, Fujisawa, Kanagawa 251-8522, Japan
[4] Graduate School of Information Science, Tohoku University, 6-6 Aoba, Aramaki, Aoba, Sendai, Miyagi 980-8579, Japan
[5] Faculty of Engineering, Tohoku Institute of Technology, 35-1 Kasumi, Yagiyama, Taihaku, Sendai, Miyagi 982-8577, Japan

Abstract. A novel fingerprint recognition algorithm suitable for poor quality fingerprint is proposed, and implementation considerations to realize fingerprint recognition access controllers for residential applications are discussed. It is shown that optimizing spatial sampling interval of fingerprint image has equivalent effect of optimizing high limit frequency of low-pass filter in the process of phase based correlation. The processing time is 83% shorter for the former than the latter. An ASIC has been designed, and it is shown that fingerprint matching based access controller for residential applications can be successfully realized.

1 Introduction

Biometrics has been recognized as indispensable means to attain security in various areas of social life. Fingerprint is the most frequently used, because it exhibits higher performance by smaller size at lower cost than other biometrics [1,2,3].

It is widely recognized that there are some percentage of people whose fingerprint is difficult for automatic recognition. Typical cases include senior citizens whose finger skin tend to be flat, house wives who uses fingertip hard, or those who suffer skin diseases such as atopic dermatitis. In general, pressure sensitive fingerprint sensor [4] produces better images than optical sensors or various types of semiconductor fingerprint sensors in cases when fingertip is dry or wet. However, when the problem stems from structure of finger surface itself, some other approaches have to be taken.

The authors have been studying a pattern-matching algorithm named Phase-Only Correlation [5]. POC is not only good for biometrics such as fingerprint, but also for sub-pixel precision translation measurements for industrial applications [6]. Band-Limited POC (BLPOC) is modified POC in that high frequency components are eliminated in the process of POC calculations [7].

Typical fingerprint recognition algorithm extracts lineal structure from the image. Such kinds of methods are referenced as minutiae algorithms in this paper. The struc-

D. Zhang and A.K. Jain (Eds.): ICB 2006, LNCS 3832, pp. 326 – 333, 2005.

tural reproducibility is especially important for minutiae algorithms in order to reduce false rejections for genuine attempts. It has been shown that BLPOC improves fingerprint recognition performance especially when numbers of images from those who have poor quality fingerprints are included.

On the other hand, POC based algorithms require more computational resources than minutiae algorithms in general, because the algorithms are based on two-dimensional discrete Fourier transformation (DFT). It is too much burden for typical microprocessors to process a fingerprint image in a moment. However, the algorithm is suited for hardware implementation such as ASIC, because DFT is calculated by repetitive executions of sum-of-products arithmetic.

In this paper, a novel fingerprint recognition algorithm that has as good recognition performance as BLPOC is described. The effect of eliminating high frequency components of BLPOC is now realized by optimizing spatial sampling interval of fingerprint image. The computational time for the proposed algorithm is 83% shorter than that for BLPOC. The recognition performance is evaluated using fingerprint database in comparison with BLPOC and a typical minutiae algorithm.

The CPU burden for the algorithm is still high, and therefore an ASIC has been implemented. The architecture of the ASIC is based on pipelining. Required functions such as re-sampling and scaling are executed in pipeline fashion with DFT calculation, therefore, the time for those functions can be eventually neglected. The processing time is 110 times faster for the ASIC than a typical personal computer. As a result, a prototype of compact access controller for residential applications that uses the algorithm, the ASIC, and a pressure sensitive fingerprint sensor can be realized.

2 Phase-Based Fingerprint Recognition Algorithm

2.1 Proposed Fingerprint Recognition Algorithm

The fingerprint recognition algorithm using BLPOC is described as following steps of processes. Refer [7] for more details of definitions of POC and BLPOC.

(a) Rotation Alignment

Let f be an input fingerprint image and g be a registered image. For each image f_θ rotated by θ in 1 degree step, $-20^\circ \le \theta \le 20^\circ$, of f, compute POC function $\hat{r}_{fg}$ with g. Θ is the angle of f_θ that produces the highest peak value of the POC function. f_Θ is defined as the rotationally aligned image of f.

(b) Translation Alignment

$\hat{r}_{fg}$ also gives the amount of two-dimensional translation displacement δ as the location of the peak. Align f_Θ and g by using δ. Let f' and g' be the resultant translation aligned images.

(c) Conjuncture Area Extraction

Let f'' and g'' be the part of f' and g' where the fingerprint image is common.

(d) Upper Limit Frequency Calculation

Calculate upper limit frequencies of (K_1, K_2) as inherent frequency band by using two-dimensional DFT.

(e) BLPOC Calculation
Calculate BLPOC function $\hat{r}_{f''g''}^{K_1K_2}$ from f'' and g'' using (K_1, K_2).
(f) Score Calculation
The BLPOC score is defined as sum of two largest peak values of BLPOC function.

The essential part of BLPOC is in step (e) above, where K_1 and K_2 are adaptively determined per individual fingerprint image pair. Hardware implementation of BLPOC may not be straightforward because the size of the images varies.

In our experiments using pressure sensitive fingerprint sensor [4], BLP-100 384×256 pixels, and 0.058×0.058 mm pixel pitch, the optimum values of K_1 and K_2 ranges roughly 0.4 to 0.6. It is expected that selecting value of 0.5 may not produce significant performance differences.

Widening spatial re-sampling interval of an original image has similar effects as of lowering cutoff frequency of low-pass filter. It is assumed that the effect of aliasing stemmed from re-sampling can be neglected. Setting high limit cut-off frequencies of BLPOC is replaced by wider spatial re-sampling interval. Indices for DFT and inverse DFT are selected to be constants. Conjuncture area extraction and improving score calculation function are simplified as well.

The processes of the proposed algorithm significantly simplify aforementioned BLPOC processes as follows.

(a) Re-sampling
Images f and g are re-sampled by scaling factor of S. The resultant image is defined as constant size of 128×128 pixels, because DFT calculation is faster for 2's power indices than arbitrary indices. The center of re-sampled image is moved to the gravity center of the original image instead of adjusting translation deviation. This is considered to be simplified version of BLPOC steps (b) and (c).
(b) Rotational Alignment
For each images f_θ rotated by θ in one degree step, $-20^\circ \le \theta \le 20^\circ$, of f, compute POC function $\hat{r}_{f_\theta g}$ with g. This process corresponds to step (a) of BLPOC.
(c) Score Calculation
The three largest peaks within 5×5 pixels from the maximum peak are evaluated. The evaluation function to get score value is either the value of the maximum peak, or the sum of peak values weighed by the inverse of the distance from the maximum peak. The distance has offset value of 1, therefore weight is 1 for the maximum peak. The reason of the weight function is that POC function of imposter calculations tends to produce large peaks in far location from the maximum peak.

2.2 Performance Evaluations

The ratio of those who have difficult fingerprint pattern is intentionally increased to create fingerprint database for performance evaluation. Total of 12 subjects, 8 males and 4 females, are participated. Seven of them have fine fingerprint condition, three dry finger, one rough finger skin, and one atopic dermatitis skin lesion. The typical ratio of difficult fingerprint person of some percent is intentionally higher here, 41.6%, for this database.

Ten fingerprint images are taken from each subject. The genuine match combinations are ${}_{10}C_2 \times 12 = 540$, and imposter combinations ${}_{120}C_2 - 540 = 6600$.

The first experiment is to test POC recognition performance by varying spatial sampling interval in order to verify that widening of spatial sampling interval has equivalent effect of lowering cutoff frequency of low-pass filter by BLPOC. The results are shown in Figure 1. The original image from BT-100 is re-sampled by the factor of 100% to 30% in 5% steps. Note that the sampling interval is converted to bits per inch (BPI) by using the sensor's 0.058 micrometer dot pitch. EER and zero FMR values are plotted per sampling interval by two evaluation functions. Zero FMR values may be less significant, because the size of database is small for this evaluation. The first evaluation function simply uses the value of the largest peak. The second one uses aforementioned weighed and averaged peak values. The EER and zero FMR of BLPOC is also shown in the figure as references. 200 DPI sampling produces the best performance, and it is equivalent to that of BLPOC as shown in the figure. The result also implies that the cost of fingerprint sensor can be further reduced by realizing, possibly low-cost, low-resolution sensor.

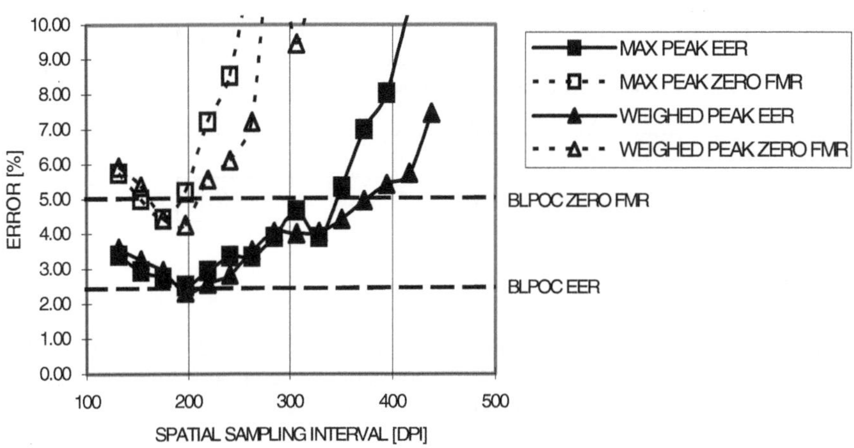

Fig. 1. Characteristics of Spatial Sampling Interval

The second experiment is to compare the performance of the algorithm with that of a minutiae algorithm and BLPOC. The EER and zero FMR values are summarized in Table 1, and ROC characteristics are shown in Figure 2. Again, zero FMR values may be less significant for this small database. The proposed algorithm shows as good performance as that of BLPOC, and both are superior to the minutiae algorithm.

The proposed algorithm can be processed considerably faster than BLPOC. The CPU time to calculate the proposed algorithm using a personal computer of Pentium 4, 3.06GHz, using MATLAB 7.01 is 19.07s and 2.45s for BLPOC and the proposed algorithm, respectively.

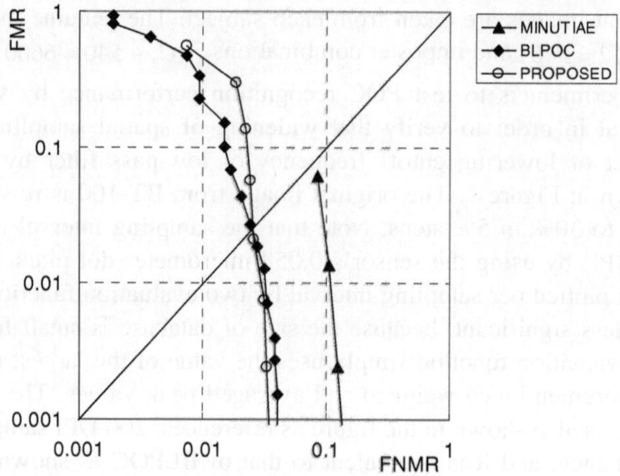

Fig. 2. ROC Comparison

Table 1. Summery of Performance Comparison

	EER [%]	ZeroFMR [%]
MINUTIAE	7.34	17.41
BLPOC	2.46	5.00
PROPOSED	2.34	4.26

3 LSI Implementation

There have been POC dedicated LSI implementations reported [8, 9, 10]. ASIC approach is very important for residential applications, because it reduces number of components while processing POC algorithm in a moment.

An ASIC has been developed. The picture is shown in Figure 3, and the block diagram in Figure 4. The pipeline architecture is fully adopted. The fingerprint image signal is re-sampled, and the output image is 128×128 pixels. The image goes through the internal memory bus, and fed into the local memory through the post-processing controller. The controller calculates the image parameters such as average brightness, and maximum brightness. The image interface, resizing, and image parameter measurements are processed in pipeline fashion with data transfer, and therefore the processing time for those functions can be neglected effectively. Image in the local memory are next read to internal memory through the pre-processing controller, and it eliminates offset and converts real data to complex data for succeeding DFT calculation again in pipeline fashion. The internal memory is divided into four blocks, each of which is for two pairs of horizontal lines. One pair is for input image, and the other pair for registered image. As soon as a line of data transfer is completed and DFT conversion has started, transfer of the next line to the other buffer is started. Therefore, the data transfer time can be neglected. The output data of the DFT unit goes to the local memory through post-processing controller, and the data can be scaled by the multiplexer, or converted to phase in order to minimize registration data size for storage. In this way, the ASIC removes most of heavy POC related burdens from CPU.

Fig. 3. Picture of the ASIC

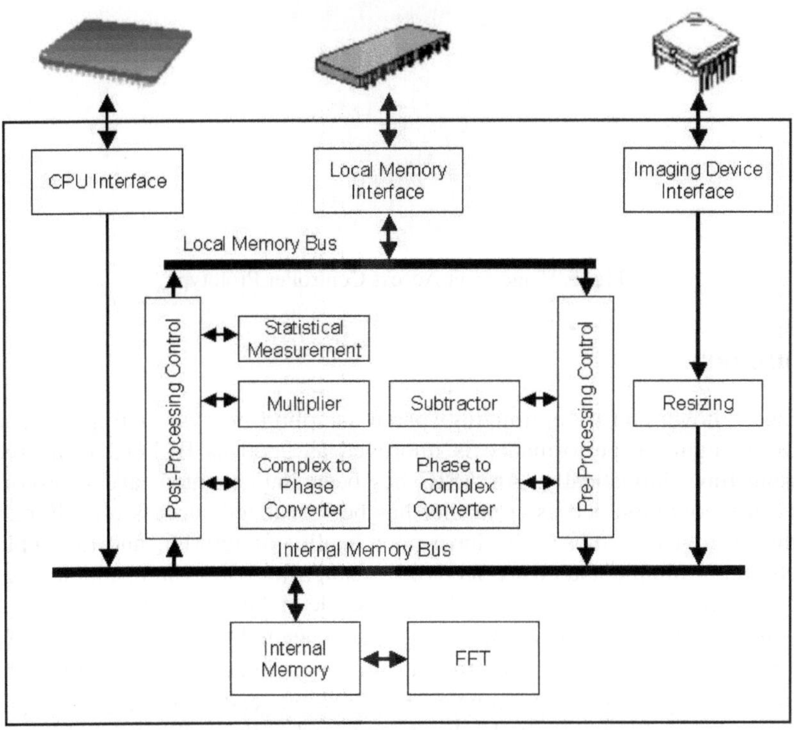

Fig. 4. ASIC Block Diagram

The throughput of the ASIC is compared with that of a typical personal computer. The time for fundamental 128×128 POC calculation is 8.8ms at 57MHz clock, whereas the same calculation takes average of 28ms by aforementioned PC. The performance of the LSI is $\frac{28}{8.8} \times \frac{3060}{57} \cong 171$ times higher than PC, if the performance is compared in a normalized clock frequency.

4 Fingerprint Access Controller for Residential Applications

The most important feature of the fingerprint recognition access controller for residential applications is to realize a good product for ordinary people, especially for senior citizens or housewives who tends to have poor quality fingerprints and frequently at rough conditions. Pressure sensitive fingerprint sensor is applied, because it is insensitive to wet or dry fingers. The ASIC processes a verification calculation at 0.3 second.

The prototype has a graphical LCD display unit, and it provides various user-friendly interface capabilities. Fingerprint image is displayed in case when fingertip is mistakenly placed and the sensor cannot take adequate image. Figure 5 shows the picture of the prototype.

Fig. 5. Fingerprint Access Controller Prototype

5 Summary

It has been shown that by optimizing spatial sampling interval of fingerprint image, the POC recognition performance is improved as good as BLPOC while reducing processing time dramatically. An ASIC has been implemented, and a prototype of fingerprint recognition access controller has been realized successfully. Because the algorithm is robust to those who have poor quality fingerprint, and the application products can be simple and cost-effective by using the ASIC, the resultant fingerprint recognition access controller can be ideal for residential applications.

It is noticeable that the POC based algorithms, including the one described in this paper, are less dependent on the structure of target images, and therefore they are good for other biometrics. For examples, POC exhibits excellent recognition performance for iris recognition [11]. It also has been tested as three-dimensional human face measurements [12, 13, 14]. POC also calculates parallax in one-hundredth resolution using a pair of images taken by cameras set in parallel for the case.

References

1. Wayman, J., Jain, A., Maltoni, D., Maio, D.: Biometric Systems. Springer (2005)
2. Maltoni, D., Maio, D., Jain, A. K., Prabhakar, S.: Handbook of Fingerprint Recognition. Springer (2003)

3. Jain, A. K., Hong, L., Pankanti, S., Bolle, R.: An Identity Authentication System Uusing Fingerprints. Proc. IEEE, Vol.85, No.9 (1997) 1365-1388
4. http://www.bm-f.com/
5. Nakajima, H., Kobayashi, K., Kawamata, M., Aoki, T., Higuchi, T.: Pattern Collation Apparatus based on Spatial Frequency Characteristics. US Patent 5,915,034 (1995)
6. Takita, K., Aoki, T., Sasaki, Y., Higuchi, T., Kobayashi, K.: High-accuracy Subpixel Image Registration Based on Phase-only Correlation," IEICE Trans. Fundamentals, Vol.E86-A, No.8 (2003) 1925-1934
7. Ito, K., Nakajima, H., Kobayashi, K., Aoki, T., Higuchi, T.: A Fingerprint Matching Algorithm Using Phase-only Correlation. IEICE Trans. Fundamentals, Vol.E87-A, No.3 (2004) 682-691
8. Morikawa, M., Katsumata, A., Kobayashi, K.: Pixel-and-Column Pipeline Architecture for FFT-based Image Processor. Proc. IEEE Int. Symp. Circuit and Systems, Vol.3 (2003) 687-690
9. Morikawa, M., Katsumata, A., Kobayashi, K.: An Image Processor Implementing Algorithms using Characteristics of Phase Spectrum of Two-dimensional Fourier Transformation. Proc. IEEE Int. Symp. Industrial Electronics, Vol.3. (1999) 1208-1213
10. Miyamoto, N., Kotani, K., Maruo, K., Ohmi, T.: An Image Recognition Processor using Dynamically Reconfigurable ALU. Technical Report of IEICE, ICD2004-123. (2004) 13-18 (in Japanese).
11. Miyazawa, K., Ito, K., Aoki, T., Kobayashi, K.: A Design of an Iris Matching Algorithm based on Phase-only Correlation. Int. Conf. Image Processing (2005) (in press)
12. Takita, K, Muquit, M. A., Aoki, T, Higuchi, T.: A Sub-Pixel Correspondence Search Technique for Computer Vision Applications. IEICE Trans. Fundamentals, Vol.E87-A, No.8. (2004) 1913-1923
13. http:/www.aoki.ecei.tohoku.ac.jp/poc/
14. Uchida, N., Shibahara, T., Aoki, T., Nakajima, H, Kobayashi, K.: 3D Face Recognition using Passive Stereo Vision. Int. Conf. Image Processing (2005) (in press)

Design of Algorithm Development Interface for Fingerprint Verification Algorithms

Choonwoo Ryu, Jihyun Moon, Bongku Lee, and Hakil Kim

Biometrics Engineering Research Center (BERC),
School of Information and Communication Engineering, INHA Unversity, Incheon, Korea
{cwryu, jhmoon, bklee, hikim}@vision.inha.ac.kr

Abstract. This paper proposes a programming interface in order to standardize low-level functional modules that are commonly employed in minutiae-based fingerprint verification algorithms. The interface, called FpADI, defines the protocols, data structures and operational mechanism of the functions. The purpose of designing FpADI is to develop a minutiae-based fingerprint verification algorithm cooperatively and to evaluate the algorithm efficiently. In a preliminary experiment, fingerprint feature extraction algorithms are implemented using FpADI and an application program, called FpAnalyzer, is developed in order to evaluate the performance of the implemented algorithms by visualizing the information in the FpADI data structures.

1 Introduction

Biometrics of different modality requires different techniques of data processing, and a certain biometric technique can be implemented by various approaches. Therefore, standardization of biometric techniques is not a simple task. If biometrics modality and its technical approach are fixed, then the design of the standards is much easier. However, there are still many problems to be solved. For example, a certain fingerprint verification algorithm has a unique logical sequence of functional modules, some of which are not necessary in other verification algorithms.

The purpose of this study is to design a programming interface, so called *Fingerprint Verification Algorithm Development Interface* (FpADI) in order to standardize low-level functional modules that are commonly employed in minutia-based fingerprint verification algorithms [1]. FpADI focuses on function protocols, data structures and operational mechanism of the functional modules. In particular, FpADI must be differentiated from BioAPI [2] in the sense that it deals with low-level functions and data structures as listed in Table 1 and 2. BioAPI focuses on the interfaces between a biometric sensing device and an application program leaving the detailed algorithm for processing biometric data to algorithm developers. Meanwhile, FpADI defines the specification of the detailed algorithm for fingerprint verification in terms of the function protocols and data structures. In particular, the data structures are designed by referring to ISO standard committee's literatures [3-5].

D. Zhang and A.K. Jain (Eds.): ICB 2006, LNCS 3832, pp. 334 – 340, 2005.

Conventional methods of performance evaluation in biometrics are only able to compare the recognition results of overall algorithms consisting of numerous low-level functions such as segmentation, binarization, and thinning. They cannot compare the performance of different low-level functions for a specific data processing inside a recognition algorithm. They even fail to identify which function mainly deteriorates the performance of the recognition algorithm. The proposed standardization, however, facilitates both the comparison of the performance of different schemes for a specific low-level function and the improvement of the performance by easy modification of the algorithm. Furthermore, this standard specification will encourage several developers to invent interoperable algorithms or even a single algorithm.

2 Definition of Function Protocols and Data Structures

There are three types of data structures for FpADI as listed in Table 1. *Image* is either gray or binary while *Map* is a block-wise structure where the size of block is arbitrary. They contain the typical information produced as intermediate results by most of minutiae-based fingerprint recognition algorithms. Moreover, *Feature* contains a list of minutiae and singular points as the final result of a minutiae-based fingerprint recognition algorithm. It also has user-defined areas for algorithms generating extended features for fingerprint matching so that FpADI can cope with proprietary fingerprint verification algorithms. Table 1 describes various data for each data structure of FpADI in minutiae extraction.

Table 1. Data structure for feature extraction FpADI

Data Structure		Comments
Image	Input Image	Captured fingerprint image by a fingerprint sensor. It is the only image data provided by the FpADI calling function.
	Gray Image	Intermediate gray image output by FpADI functions.
	Binary Image	Intermediate binary image output by FpADI functions.
	Thinned Image	Binary image containing curves of one pixel width which represents fingerprint ridge or valley.
Map	Orientation	Map containing local orientation information which represents the direction of ridge flow in each block.
	Segmentation	Map containing local information of fingerprint foreground or background region.
	Frequency	Map containing local ridge frequency information representing the ridge distance between neighboring ridges in each block.
	Quality	Map containing global fingerprint image quality as well as local image quality.
Feature	Singular Points	User defined features as well as core/delta information.
	Minutiae	User defined features as well as ridge ending and bifurcation information.

Table 2 describes the functionality and typical output data type of the low-level functions employed by most of minutiae-based fingerprint recognition algorithms. Except the opening and the closing functions (*FPADI_SetInputImage* and *FPADI_FeatureFinalization*), the FpADI functions can be called at any order inside the feature extraction algorithm, which makes it possible for FpADI to develop feature extraction algorithms in different logical sequences.

Table 2. FpADI functions for feature extraction

Function	Comments
FPADI_SetInputImage	Input a fingerprint image to the feature extraction algorithm. This function is the first function to be called in the extraction algorithm.
FPADI_Preprocessing	Pre-process an *Input Image*. Typical output data: *Gray Image* in *Image*
FPADI_LocalOrientation	Compute local orientation. Typical output data: *Orientation* in *Map*
FPADI_QualityEvaluation	Compute global and local fingerprint quality. Typical output data: *Quality* in *Map*
FPADI_Segmentation	Segment an image into foreground and background regions. Typical output data: *Segmentation* in *Map*
FPADI_RidgeFrequency	Compute local ridge frequency. Typical output data: *Frequency* in *Map*
FPADI_Enhancement	Enhance a gray or binary image by noise removal. Typical output data: *Gray image* or *Binary image* in *Image*
FPADI_Binarization	Produce a binary image from a gray image. Typical output data: *Binary image* in *Image*
FPADI_Skeletonization	Generate a thinned image Typical output data: *Thinned image* in *Image*
FPADI_MinutiaeDetection	Generate minutiae and their extended features Typical output data: *Minutiae* in *Feature*
FPADI_MinutiaeFiltering	Post-process to eliminate noise in minutiae information Typical output data: *Minutiae* in *Feature*
FPADI_SingularityDetection	Generate singular points and their extended features Typical output data: *Singular Points* in *Feature*
FPADI_FeatureFinalization	Release all internal memory blocks in the feature extraction. This is the last function to be called by the request of either user or the algorithm itself.

As shown in Fig. 1, the FpADI manipulation module in an application calls all FpADI functions. FpADI functions are not allowed to call any other FpADI functions. However, the FpADI compliant algorithm, called FpADI SDK, specifies the order of FpADI function calls. In detail, the FpADI manipulation module calls the opening function (*FPADI_SetInputImage*) by providing a fingerprint image as the *Input image*. *FPADI_SetInputImage* mainly performs initializations of the feature extraction algorithm, and its return value indicates the next function to be called by the FpADI manipulation module. In the same fashion, the FpADI manipulation module calls all the FpADI functions in the SDK until the closing function (*FPADI_FeatureFinalization*) is called.

FPADI_FeatureFinalization resets the internal memory blocks and prepares for the next feature extraction. Normally, *FPADI_FeatureFinalization* is called by the FpADI manipulation module according to the request from a certain FpADI function in the SDK. However, it also can be called directly from the application-specific module in the middle of the feature extraction process. In this case, it has to clean up all unnecessary memory blocks and prepares for the next feature extraction.

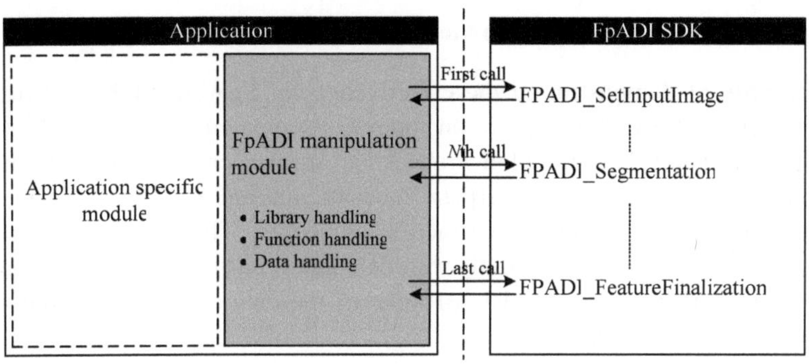

Fig. 1. Mechanism of FpADI function call

Except *FPADI_FeatureFinalization*, each FpADI function has four input parameters which correspond to *Image*, *Map*, *Feature* and *Calling order*, respectively. The data corresponding to the first three parameters are generated and referred to by FpADI functions themselves, while *Calling order* is a number starting from one and increases by one as a next function is called. Therefore, *Calling order* is a unique number associated with each FpADI function called. It distinguishes the functions especially when a certain function is called multiple times and each time performs different tasks. Fig. 2 shows an example of the FpADI function protocol.

The return value of all FpADI functions contains three types of information, function status, data-updating indicator, and the next calling function. The function status indicates the function's completion status, *success*, *failure*, or *bad parameter*. The data-updating indicator informs which input data have been updated by the function itself. And, the next calling function contains the name of the FpADI function which must be called in the next step.

```
UINT32 FPADI_QualityEvaluation(LPIMAGE Image, LPMAP Map, LPFEATURE Feature, UINT32 CallingOrder);
```

Fig. 2. Example of the FpADI function protocol

In summary, FpADI has following characteristics to encompass minutiae-based fingerprint verification algorithms of various logical sequences and data:

- Data structure for both pre-defined and algorithm-defined (extendable) fingerprint features
- Algorithm-defined sequence of calling functions
- Omission or multiple calls of a function

3 Implementations

3.1 Common Visual Analyzer: FpAnalyzer

For the purpose of demonstrating the effectiveness of FpADI, SDKs for fingerprint feature extraction, FpADI manipulation module (implemented in C++ class), and a visual algorithm analysis tool called *FpAnalyzer* are implemented. Firstly, the SDKs implemented in this study are fingerprint local orientation estimation, image quality estimation and fingerprint feature extraction algorithm. They observe the proposed FpADI specification. The first two algorithms contain partial functionality compared to the third algorithm which consists of most of the data and the functions listed in Table 1 and 2, respectively.

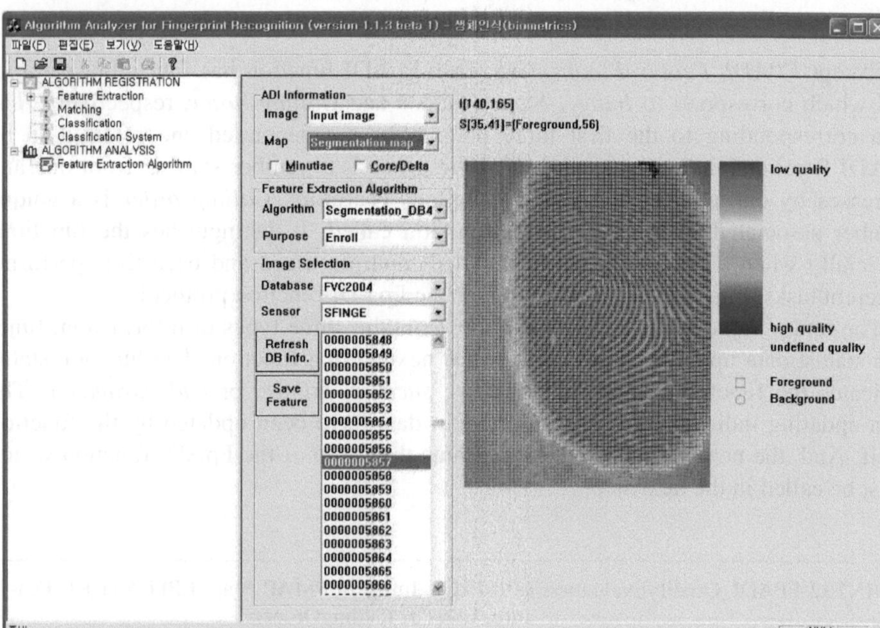

Fig. 3. *FpAnalyzer* - Visual algorithm analysis tool for fingerprint minutiae extraction

Secondly, the FpADI manipulation class, called *CFeatureADI*, can load and execute any FpADI compliant algorithms. It calls FpADI functions in the FpADI compliant SDK and performs data management such as memory allocation according to the requests of the called FpADI functions.

Finally, *FpAnalyzer* is an application tool for analyzing the algorithms under MS-Windows as shown in Fig. 3. It utilizes the *CFeatureADI* class for handling any FpADI compliant algorithms and displays all the data in the FpADI data structures listed in Table 1. It also provides a linkage between FpADI compliant algorithms and fingerprint databases.

3.2 FpADI Compliant Fingerprint Feature Extraction Algorithms

As mentioned in the previous section, three FpADI compliant algorithms have been implemented, fingerprint local orientation estimation, image quality estimation and fingerprint feature extraction algorithm, in order to show FpADI's characteristics under various programming requirements such as various block sizes and different sequences of FpADI function calls. Technical analysis of these algorithms is out of the scope of this study. Therefore, this paper will describe only the structural features of the algorithms.

The fingerprint local orientation estimation produces an orientation map in pixel, *i.e.*, 1×1 pixel block, where the orientation angle is in degree from 0 to 179. As shown in Table 3, this algorithm is the simplest one consisting of only three FpADI functions: *FPADI_SetInputImage*, *FPADI_LocalOrientation* and *FPADI_FeatureFinalization*. The second algorithm, image quality estimation, has six FpADI functions. Unlike in the first algorithm, *FPADI_LocalOrientation* is called at the fourth and *FPADI_QualityEvaluation* produces a map of 32×32 pixel blocks.

The third algorithm is a typical fingerprint feature extraction algorithm. Therefore, it generates minutiae information from the input image. Further, the block size of the orientation map in this algorithm is 8×8 pixels and its angle is represented in 8-directions. As listed in Table 3, this algorithm has 11 out of 13 FpADI functions. The function *FPADI_RidgeFrequency* and *FPADI_SingularityDetection* are not implemented because the algorithm does not utilize the information of local ridge frequency and singular points. Figure 4 shows an experimental example of local orientation of the first and third algorithm for the same input image.

Table 3. Calling functions of the implemented algorithms

Calling order	Local orientation estimation	Image quality estimation	Feature extraction
1	FPADI_SetInputImage	FPADI_SetInputImage	FPADI_SetInputImage
2	FPADI_LocalOrientation	FPADI_Segmentation	FPADI_Preprocessing
3	FPADI_FeatureFinalization	FPADI_Preprocessing	FPADI_LocalOrientation
4		FPADI_LocalOrientation	FPADI_Segmentation
5		FPADI_QualityEvaluation	FPADI_QualityEvaluation
6		FPADI_FeatureFinalization	FPADI_Enhancement
7			FPADI_Binarization
8			FPADI_Skeletonization
9			FPADI_MinutiaeDetection
10			FPADI_MinutiaeFiltering
11			FPADI_FeatureFinalization

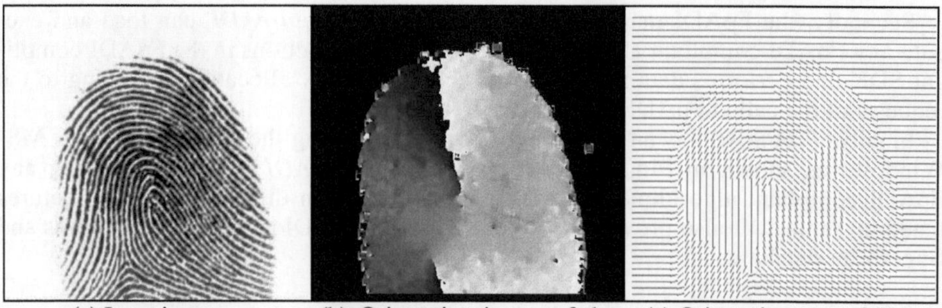

(a) Input image (b) Orientation image of the first algorithm (c) Orientation map of the third algorithm

Fig. 4. Input and output data of the implemented algorithms

4 Conclusions and Future Works

We designed and implemented FpADI, a programming interface for development of minutiae-based fingerprint feature extraction algorithms. The function protocols and the data structures are defined in order to be able to cope with flexibility in various minutiae-based feature extraction algorithms. FpADI can provide technical benefits, for example, easy co-working with several algorithm developers and easy modification of an algorithm. In the near future, the implemented products including the sample SDK, *CFeatureADI* and *FpAnalyzer* will be available to public with the FpADI specification. One of our future works includes the design of FpADI specification for fingerprint matching algorithms.

Acknowledgement

This work was supported by the Korea Science and Engineering Foundation (KOSEF) through the Biometrics Engineering Research Center (BERC) at Yonsei University.

References

1. D. Maltoni, D. Maio, A. K. Jain and S. Prabhakar, *Handbook of Fingerprint Recognition*, Springer, 2003.
2. Biometric Consortium, *BioAPI Specificaion Version 1.1*, March 2001.
3. ISO/IEC FDIS 19794-2:2004, *Information Technology - Biometric data interchange Formats-Part 2: Finger minutiae data*, ISO/IEC JTC 1/SC 37 N954, January 2005.
4. ISO/IEC FDIS 19794-4:2004, *Information Technology - Biometric data interchange Formats-Part 4: Finger image data*, ISO/IEC JTC 1/SC 37 N927, November 2004.
5. ISO/IEC FCD 19785-1:2004, *Information Technology - Common Biometric Exchange Formats Framework - Part 1: Data Element Specification*, ISO/IEC JTC 1/SC 37 N628, October 2004.
6. B.M. Mehtre and B.Chatterjee, "Segmentation of fingerprint images-a composite method," *Pattern Recognition*, vol.22, no.4, pp.381-385, 1989.
7. A.M. Bazen and S.H. Gerez, "Segmentation of Fingerprint Images," in Proc. Workshop on Circuits Systems and Signal Processing(ProRISC2001), pp.276-280, 2001.

The Use of Fingerprint Contact Area for Biometric Identification

M.B. Edwards, G.E. Torrens, and T.A. Bhamra

Extremities Performance Research Group, Department of Design and Technology, Loughborough University, Loughborough, LE11 3TU, UK
m.b.edwards@lboro.ac.uk

Abstract. This paper details the potential use of finger contact area measurement in combination with existing fingerprint comparison technology for the verification of user identity. Research highlighted includes relationships between finger contact area, pressure applied and other physical characteristics. With the development of small scale fingerprint readers it is starting to be possible to incorporate these into a wide range of technologies. Analysis of finger pressure and contact area can enhance fingerprint based biometric security systems. The fingertip comprises a range of biological materials which give it complex mechanical properties. These properties govern the way in which a fingertip deforms under load. Anthropometric measurements were taken from 11 males and 5 females along with fingerprint area measurements. Strong correlations were found between fingerprint area and many other measurements, including hand length. Notably there were more strong correlations for the female group than for the male. This pilot study indicates the feasibility of fingerprint area analysis for biometric identification. This work is part of a long term program of human physical characterization.

1 Introduction

This paper details the potential use of finger contact area measurement in combination with existing fingerprint comparison technology for the verification of an individual's identity. Details of current knowledge in the field provide an indication of the feasibility of using enhanced fingerprint technology in this way. The information highlighted includes relationships between finger contact area, pressure applied and other physical characteristics.

2 Fingerprinting Technology

Fingerprinting is a known technology with well established protocols for fingerprint comparison. With the development of small scale fingerprint readers it is starting to be possible to incorporate these into security systems and products. The small silicone-based sensors are now compact enough to fit into hand-held devices.

However the use of fingerprint matching technology opens up the possibility of abuse of the system. A number of techniques have been developed that can be used in

D. Zhang and A.K. Jain (Eds.): ICB 2006, LNCS 3832, pp. 341 – 347, 2005.

conjunction with fingerprinting to improve its accuracy. These techniques use metrics such as temperature, conductivity and pulse measurement to check the finger placed upon a sensor is from a living person [1]. While these do improve the fallibility of fingerprinting, all of these methods can be circumvented. For example, checks upon the temperature of the finger can be falsified using a thin silicone rubber cast of the desired fingerprint placed upon a finger. This will be kept at the correct temperature by the underlying finger and have the correct pattern of ridges to deposit the required fingerprint.

3 Fingertip Deformation Prediction

Consideration of the tissues of the fingertip shows that the analysis of finger pressure and contact area can prevent the use of fake fingerprints for the accessing of protected system. The different tissues within the fingertip give it complex mechanical characteristics which are dependent upon a number of different factors, including size, rate and direction of force application [2]. This allows the fingertip to attenuate small forces applied and transmit larger forces to the underlying bones so making it an effective tool for both exploratory and manipulative tasks. Deformation of these tissues occurs when the fingertip is pressed against a surface and the amount of deformation dictates the size of the fingerprint deposited.

Non-linear viscoelastic theory has been used by a number of researchers to model the deformation of the finger. These models do not accurately predict the changes to the separate materials within the fingertip, instead considering it as one homogeneous material. These models have been found to be accurate in predicting a variety of factors such as plastic distortion of the skin [3], force displacement during tapping [4] and the mechanical responses of the fingertip to force application during dynamic loading [5,6].

All of these fingertip models use information about the physical properties of the finger including its size, elasticity and viscosity to predict the manner in which the fingertip deforms. These physical properties are treated as constants while size and force applied are variables in the models. As such, knowledge of finger size and applied force should allow for the prediction of fingerprint area. The force applied can be measured using transducers placed within a fingerprint scanner. This leaves the deposited fingerprint size as a variable through which one person can be distinguished from another.

4 Fingertip Size

Anthropometric surveys conducted in the UK have shown fingertip dimensions vary across the population. Index finger depth at the distal joint has been found to vary between 12.5mm and 15.1mm while its breadth varies between 16.5mm to 17.1mm [7]. No link has been found between the pattern of ridges on a fingertip and body size. The range of sizes across the population makes finger size a useful measurement for validating a deposited fingerprint. As fingertip size influences the contact area between the fingertip and a object, it will be a component factor in a model predicting fingerprint area. If finger size is measured when a fingerprint is first entered into a database, the deposited fingerprint area can be calculated using a suitable model each time the fingerprint is read and used for validation of the entered print.

For the validation of a model for fingertip deformation, Serina *et al* [8] performs some preliminary tests of finger contact areas for a range of finger forces. In this testing, all forces were subject generated at specified forces between 0.25N and 5N. The forces were held for 2 seconds and the contact area measured by inking the finger before the test and measuring the resultant fingerprint. The author then nondimensionalised the data by dividing the contact areas by the square of the finger width. The non-dimensionalised data shows a rapid increase in contact area below 1N, after which the area steadily in relation to force. This shows that the contact area of the finger for a set force is repeatable and should be modelable. While the authors own predictive model appears to be a poor fit for the data, it purpose is mainly to model finger displacement, with contact are as a extra output. By focusing purely upon contact area, it should be possible to produce a better model. Dividing the data by the square of the finger width removes the main effects of finger size within this figure. This illustrates the basic relationship between force and contact area and gives a indication of the importance of finger size.

5 Body Size Proportionality

Another possible further application of this idea is the use of the proportionality of the human body to attempt to predict an approximate body size or weight from a fingerprint. This can then be compared to other measurement of the individual whose fingerprint is being taken, such as height or weight.

Attempts to define the proportion of the human body have been made for centuries, many by artists in order to produce realistic figure drawings. These art-based methods often define the proportion of the body using a limb length as a unit of distance through which the rest of the body can be measured, essentially defining the body as being proportional in size. For example, stature is often defined as being eight times the distance from the chin to the top of the head. More recent anthropometric studies have shown that many individual anatomical measures of the body are correlated and that the human body does indeed have a degree of proportionality. Roebuck [9] gives the correlation values for a range of anthropometric measurements of both U.S. civilian males and females. For both of these groups there are a number of strong correlation coefficients, which indicate proportionality within the human body. For example many bone lengths are strongly correlated as are many limb girths with weight.

6 Fingerprint Area Investigation

In order to investigate the relationship between fingerprint area and other body characteristics a survey of 17 (n=11 male, n=5 female) students at Loughborough University, UK, was conducted. This study measured both male and female students although analysis of these was conducted separately as size and geometry differences have been found between male and female hands [10].

Fingerprint area was measured by applying a 10N load to the back of an inked finger, which pressed the finger against a sheet of photocopy paper. The load was applied by a moving platen held within a guiding frame which ensured the force was applied perpendicularly to the back of the finger. The area of the resultant fingerprint was then measured using a planimeter. For comparison with the fingerprint area, each of the participants had nine anthropometric measurements taken. All length measurements were taken with either digital calipers or a Holtain anthropometer depending on the size of measurement to be taken. Height was taken using a portable stadiometer to the nearest millimeter and weight using digital weighing scales, accurate to the nearest half kilogram.

6.1 Results

Correlations were produced for all measurements against fingerprint area (see Table 1) and those with a high correlation (Pearsons $r > 0.65$) were noted. These illustrated a correlation of fingerprint area with a number of measurements, including fingertip length,

Table 1. Correlation coefficients between various anthropometric measurements and fingerprint area

	Fingerprint areas	
	Male	Female
Stature	0.70	-0.22
Weight	0.64	0.85
Arm length	0.68	0.64
Hand length	0.83	0.81
Hand width	0.76	0.90
Finger tip length	0.76	0.79
Finger tip width	0.52	0.95
Finger tip Depth	0.28	0.88
Finger tip diameter	0.26	0.81

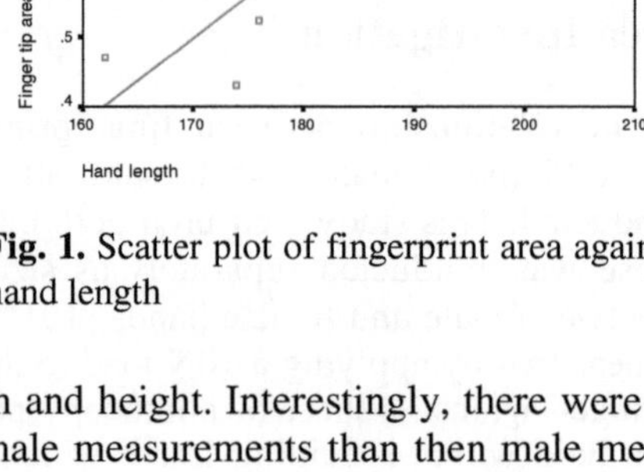

Fig. 1. Scatter plot of fingerprint area against hand length

hand length, arm length and height. Interestingly, there were a larger number of high correlations for the female measurements than then male measurements with the notable exception of height, which is the only negative coefficient. It is thought that this is due to an erroneous height measurement which will have a large effect on the small sample group. Scatter plots of the high correlations were created to confirm that these correlations were not erroneous, and example of which can be seen in Figure 1.

7 Discussion

Fingerprinting is the most commonly used biometric security method; however it is not without its problems. The consideration of fingertip structure shows there is a relationship between finger contact area, pressure applied and finger size. This knowledge can be used to enhance current fingerprint security by incorporating it into existing fingerprinting technology. In addition to this, possible links between fingerprint area and body size may allow for a larger increase in the security of fingerprint protected devices.

In order for fingerprint area measurement to become a successful security system, it is important to have an accurate method of measuring the contact area of a finger placed upon a sensor.

A number of laboratory based area measurement techniques have been evaluated by the authors. These all measured the area of an inked fingerprint and included manual techniques involving graph paper, different types of planimeters and a computer program written specifically for the task which was used a scanner to digitise the fingerprint. All of these methods were found to be reliable and repeatable apart from the fully automatic program. This was due to the variability of the amount of ink deposited by the finger. An excess of ink upon the finger makes a much darker fingerprint and this influenced the measurement made by this computer system. The other techniques were not affected by this as they all involved human judgment being used to define the edges of the fingerprint.

The influence of the amount of ink on the finger upon the automatic measurement illustrates some of the problems which may be encountered with a system that is to be used outside of a controlled laboratory. Environmental factors, such as dirt, oil and moisture may have a similar influence to ink for an automatic system, making the fingerprint appear bigger. These are examples of a few environmental factors that require consideration.

The physiological condition of the finger is also a matter that requires consideration. A number of factors can change the mechanical properties of finger tissues and this will affect its deformation. Temperature affects the rigidity of many of the tissues in the body, sweat will make the skin more flexible and stress will affect the levels of sweat produced upon the palm.

From existing literature and the development of the procedure for the tests described in the previous section, a number of different issues were found to be important. These are shown in Figure 2. Many of the issues identified were kept constant, however preliminary testing was done to acertain the effects of variations in the angle of the finger how far along the length of the finger was considered a print of the fingertip. These both were found to have a large effect upon the results. To remove these effects, they were controlled by keeping hand posture the same for each measurement and ensuring only the fingertip above the distal interphalangeal joint was in contact with the paper. These factors all require further investigation before fingerprint area measurement can be used to.

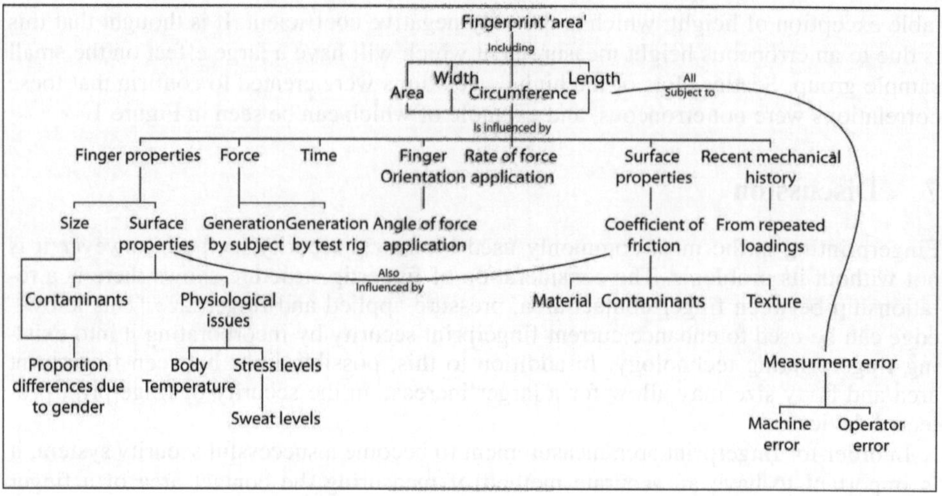

Fig. 2. Issues found to be relevant for fingerprint area deposition

As these factors are addressed, it should be possible to begin to use fingerprint area measurement to enhance biometric security systems through the development of an accurate model predicting fingertip deformation. In order to take this idea from being a concept, to a proven method, for fingerprint-based security augmentation a number of stages of research are planned.

8 Conclusions

The use of fingerprint area measurement provides a new method for augmenting fingerprint recognition. This can be potentially applied within numerous security systems due to the size of the sensors required. Before it can be applied, there are a number of issues that need to be addressed, including the effects of a number of factors upon fingerprint area, the production of a model predicting fingerprint deformation and the accuracy of the method used for fingerprint area measurement. Work is currently being performed to address these issues and bring this concept closer to being a usable technique for augmenting fingerprint based security.

A more in-depth investigation into the relationship between fingertip size and deposited fingerprint is currently planned. This will involve the use of a range of sizes of fingertips, a range of force applications and rates of force applications. With these relationships known, a pragmatic model of the fingertip and its deposition area is to be developed. This model will attempt to allow for the determination of fingertip size from a deposited fingerprint at a known load and so not model the deformation of the fingertip. Once this is completed other factors shown in Figure 2 will be investigated to broaden the model.

References

1. Biometric Technology Today (2001). Forging Ahead. **9**, 9-11
2. Serina, E. R., Mote Jr, C. D. and Rempel, D. (1997). Force response of the fingertip pulp to repeated compression - effects of loading rate, loading angle and anthropometry. Journal of biomechanics, **30,** 1035-1040.
3. Cappelli, R., Maio, D. and Maltoni, D. (2001). Modelling Plastic Distortion in Fingerprint Images. In Second International Conference on Advances in Pattern Recognition (ICAPR2001)Rio de Janeiro, pp. 369-376.
4. Jindrich, D., Zhou, Y., Becker, T. and Dennerlein, J. (2003). Non-linear viscoelastic models predict fingertip pulp force-displacement characteristics during voluntary tapping. Journal of Biomechanics, **36,** 497-503.
5. Wu, J. Z., Dong, R. G., Smutz, W. P. and Rakheja, S. (2003a). Dynamic interaction between a fingerpad and a flat surface: experiments and analysis. Medical Engineering & Physics, **25,** 397-406.
6. Wu, J. Z., Dong, R. G., Smutz, W. P. and Schopper, A. W. (2003b). Modelling of time-dependant force response of fingertip to dynamic loading. Journal of Biomechanics, **36,** 383-392.
7. Department of Trade and Industry (1998) Adultdata: The handbook of Adult Anthropometric and strength measurements - Data for Design Safety. Institute for Occupational Ergonomics, Nottingham.
8. Serina, E. R., Mockenstrum, E., Mote Jr, C. D. and Rempel, D. (1998). A structural model of the forced compression of the finger pulp. Journal of Biomechanics, **31,** 639-646.
9. Roebuck, J. A. (1995) Anthropometric Methods: Designing to Fit the Human Body. Human Factors and Ergonomics Society, Santa Monica.
10. Rahman, Q. and Wilson, G. D. (2003). Sexual orientation and the 2nd to 4th finger length ratio: evidence for organising effects of sex hormones or developmental instability? Psychoneuroendocrinology, **28,** 288-303.

Preprocessing of a Fingerprint Image Captured with a Mobile Camera

Chulhan Lee[1], Sanghoon Lee[1], Jaihie Kim[1], and Sung-Jae Kim[2]

[1] Biometrics Engineering Research Center,
Department of Electrical and Electronic Engineering,
Yonsei University, Seoul, Korea
devices@yonsei.ac.kr
[2] Multimedia Lab., SOC R&D center,
Samsung Electronics Co., Ltd, Gyeonggi-Do, Korea

Abstract. A preprocessing algorithm of a fingerprint image captured with a mobile camera is proposed. Fingerprint images from a mobile camera are different from images from conventional or touch-based sensors such as optical, capacitive, and thermal sensors. For example, images from a mobile camera are colored and the backgrounds or non-finger regions can be very erratic depending on how the image captures time and place. Also, the contrast between the ridges and valleys of images from a mobile camera is lower than that of images from touch-based sensors. Because of these differences between the input images, a new and modified fingerprint preprocessing algorithm is required for fingerprint recognition when using images captured with a mobile camera.

1 Introduction

Mobile products are used in various applications such as communication devices, digital cameras, schedule management devices, and mobile banking. Due to the proliferation of these products, privacy protection is becoming more important. Fingerprint recognition has been the most widely exploited because of stability, usability, and low cost. There are already a few commercial mobile products equipped with fingerprint recognition systems. However, these products require additional fingerprint sensors. This leads to weakening durability and increasing price. Fortunately, almost all modern mobile products have high computational power and are already equipped with color cameras. These cameras are comparable in quality to commercial digital cameras, with features such as zooming, auto-focusing, and high resolution. Because of hardware contributions(high computational power and camera) and privacy protection problems in mobile environments, a new fingerprint recognition system which uses these kinds of mobile camera devices is realizable in near future. There are many challenges when developing fingerprint recognition systems which use a mobile camera. First, the contrast between the ridges and the valleys in the images is lower than that in images obtained with touch-based sensors. Second, because the depth of field of

D. Zhang and A.K. Jain (Eds.): ICB 2006, LNCS 3832, pp. 348–355, 2005.

the camera is small, some parts of the fingerprint regions are in focus but some parts are out of focus. Third, the backgrounds, or non-finger regions, in mobile camera images are very erratic depending on how the image captures place and time. For these reasons, a new and modified preprocessing algorithm is required. In Section 2, we explain how we obtained the fingerprint image for our work and explain the segmentation algorithm. Section 3 presents the orientation estimation. Experimental results are shown in Section 4, followed by a conclusion and future work in Section 5.

2 Fingerprint Segmentation

Firstly, we explain how we obtained the fingerprint images for our work. We used an acquisition device composed of a 1.3M pixel CMOS camera used on a mobile phone and a LED (Light Emitting Diode). The working distance was set as 5cm in front of the camera, and a finger was positioned here to get fingerprint images with a additional holder. Because of the LED, we were able to obtain fingerprint images which are less affected by outside light conditions. After acquiring a fingerprint image with a mobile camera, the first step is fingerprint segmentation. This process divides the input image into a foreground (fingerprint) region and a background region. When a fingerprint image is obtained from a touch-based sensor such as a capacity, optical, or thermal sensor, the background or non-finger region is easy to segment from the fingerprint region because the region has similar patterns depending on sensor types. However, when a fingerprint is captured by a mobile camera, the background regions are very erratic depending on how the image captures place and time.

2.1 Fingerprint Segmentation Using Color Information

In order to segment fingerprint regions using color information, we compare each pixel in the input image with the distribution of the fingerprint color model in the normalized color space. [1] shows that even though the skin color of humans is different from that of the entire human species according to each melanin, the skin color of the palm (including the fingers) is mainly influenced by an absorption spectrum of oxygenated hemoglobin because of the absence of melanin at the palm. Therefore, the fingers of all humans show similar reflection rates according to visual wavelengths. With this characteristic, the normalized color distribution which we determined with our sample images can be applied to all humans. In this paper, we model fingerprint color distribution with a non-parametric modeling method using a lookup table (LUT) [2]. We produced 100 training images and 400 test images by manually segmentation. One of the training images is shown in Fig. 1(a),(b). With the training images, the foreground regions are transferred to normalized rgb color space and then accumulate the normalized r, b information in order to make the distribution of fingerprint color(Fig 1(c)). To create the LUT, the color space (rb space) is quantized into a number of cells according to a predefined resolution and the value of each

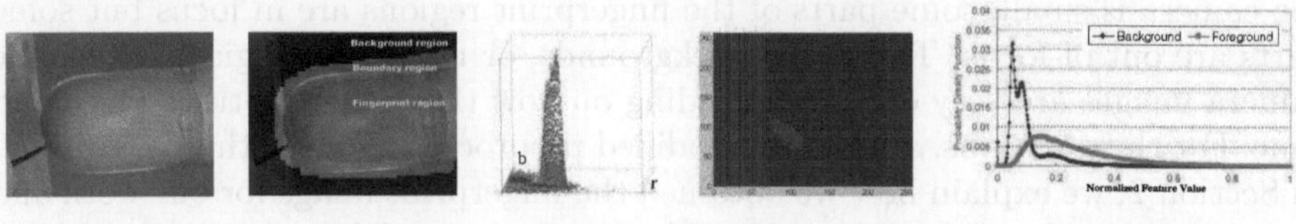

(a) (b) (c) (d) (e)

Fig. 1. (a) Original Image, (b) Manually Segmented Image (c)The distribution of the fingerprint color model, (d) The LUT with 256×256 Resolution (e) The distribution of Tenengrad-based measurement

cell is divided by the largest value. We categorize the cells as fingerprint region cells if the divided value is larger than T_{LUT}. If not, the cells represent background region cells. We experimentally define T_{LUT} as 0.1. Fig. 1(d) shows the LUT with 256×256 resolution. With the LUT, each pixel $x(i,j)$ is segmented as follows:

$$x(i,j) = \begin{cases} \text{fingerprint\_region\_if\_LUT}[r(i,j)][b(i,j)] = \text{fingerprint\_cell} \\ \text{background\_region\_if\_LUT}[r(i,j)][b(i,j)] = \text{background\_cell} \end{cases}, \quad (1)$$

where r(i,j) and b(i,j) are the normalized r and b values of a pixel $x(i,j)$. To reduce noise, we apply this process to each block. Each block is represented by the average r and b values within the blocks of predefined size (8×8).

2.2 Fingerprint Segmentation Using Frequency Information

In order to capture a fingerprint image with a camera, a close-up shot is required. This makes the depth of field (Dof) small. This means that the fingerprint region is in focus and the background region is out of focus, which produces clear ridge-pattern images in the fingerprint region and blurred-pattern images in the background region. Our method is based on this difference between the two image regions. We consider the Tenegrad based method that has been exploited in the auto-focusing technique [3]. In the Tenengrad based method, using the Sobel operator, we calculate the horizontal(G_H) and vertical(G_V) gradients of the images. Our Tenengrad based measurements are determined as follows:

$$\text{Tenengrad}(i,j) = \frac{1}{(2n+1)^2} \sum_{k=i-n}^{i+n} \sum_{l=i-n}^{i+n} G_V^2(k,l) + G_H^2(k,l) \quad (2)$$

Fig. 1(e) show the distributions of the measurement of the fingerprint region and the background region with manually segmented images (the training images in Section 3.1). The distribution shows that the measured values of the background region are concentrated on low values and the values of the fingerprint region are spread out wildly. Taking advantage of these characteristics, segmentation is achieved through the simple threshold method. The threshold is determined by Bayesian theory with two distributions, the background distribution and the foreground distribution. In Bayesian theory, we assume that a priori probabilities are the same.

2.3 Fingerprint Segmentation Using the Region Growing

The final fingerprint segmentation algorithm is conducted with the region growing method. In the region growing algorithm [4], the seed region and the similarity measurement (which merges neighboring pixels) must be determined. To determine the seed region, we combine the results of color (Section 2.1) and frequency (Section 2.2) with the AND operator. This is because the fingerprint region should be well focused, and also because it should show the finger color. From the determined seed region, we estimate the color distribution of each input finger as the color distribution of the seed region. With the color distribution, the similarity measurement is defined as follows:

$$\begin{aligned} &\mathbf{D}(i,j) = (\mathbf{x}(i,j) - \mathbf{m})^{\mathbf{T}} \Sigma^{-1} (\mathbf{x}(i,j) - \mathbf{m}) \\ &(i,j)\text{\_is\_fingerprint\_region} : \text{if\_}\mathbf{D}(i,j) < \mathrm{T_S} \\ &(i,j)\text{\_is\_background\_region} : \text{otherwise} \end{aligned} \quad (3)$$

,where $\mathbf{x(i,j)}$ is the normalized r and b value of a neighbor pixel that will be merged. $\mathbf{m}$ and Σ are the means of the normalized r and b values and the covariance matrix calculated within the seed region. Fig. 2 shows the resulting images of color, frequency, combining color and frequency, and final segmentation ($T_s = 4$). In Section 5, the proposed segmentation algorithm is evaluated by manually segmented images.

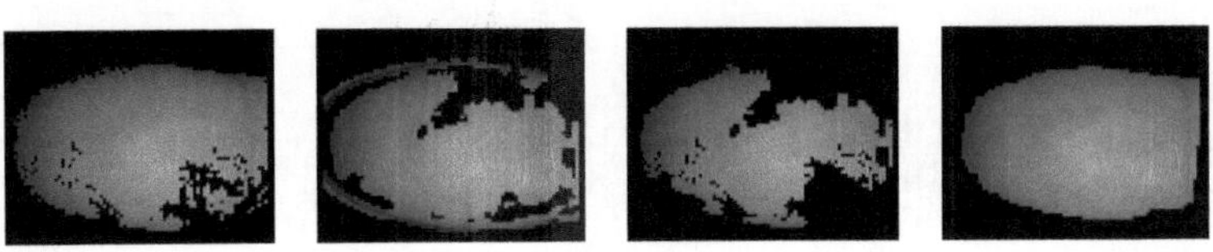

(a) (b) (c) (d)

Fig. 2. The resulting images: (a) Color (b) frequency (c) Combining (d) Final segmentation ($T_s = 4$)

3 Fingerprint Orientation Estimation

Many algorithms have been proposed for orientation estimation. Among these, gradient-based approaches[5][6] are the most popular because of low computational complexity. However, gradient-based approaches are very sensitive to noise, especially non-white Gaussian noise in the gradient field because it is based on least square method. In this section, we propose the robust orientation estimation method based on iterative regression method.

3.1 Orientation Estimation Based on the Iterative Robust Regression Method

In fingerprint images captured with a mobile camera, since the contrast between ridges and valleys is low, outliers are caused by not only scars on specific fingerprints but also by camera noise. To overcome the problem of outliers, we apply the robust regression method. This method tends to ignore the residuals associated with the outliers and produce essentially the same results as the conventional gradient based method when the underlying distribution is normal and there are no outliers. The main steps of the algorithm include:

i) 2-Dimensional gradients ($\mathbf{x}_i = [G_x, G_y]$): An input image is divided into sub-blocks, and the 2-Dimensional gradients are calculated using the Sobel operator.

ii) Orientation estimation: Using the calculated 2-Dimensional gradients, the orientation of the sub-block is estimated by the conventional gradient method.

iii) Whitening: The gradients ($\mathbf{x}_i$) are whitened to measure a norm in the Euclidean space.

iv) Removing outliers: In the whitened 2-Dimensional gradients field, a gradient $\mathbf{x}_i$ is removed if the Euclidean norm of the whitened gradient ($\|\mathbf{x}_i\|_w$) is larger than 2σ, where σ is 1 because of whitening.

v) Orientation re-estimation: Using the 2-Dimensional gradients from step 4, the orientation ($\theta(n+1)$) of the sub-block is re-estimated by the conventional gradient method.

vi) Iterative procedure: If $|\theta(n+1) - \theta(n)|$ is less than T_θ, the procedure is stopped. If not, we revert to step 3. The T_θ is defined according to quantized Gabor filter orientation that is used on ridge enhancement.

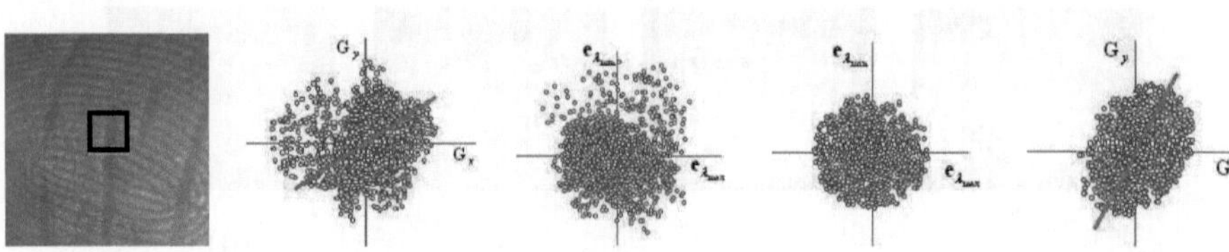

(a) (b) (c) (d) (e)

Fig. 3. (a) A sub-block image (b) A 2D gradient field with outliers (c) A whitened 2D gradient field (d) whitened 2D gradient field without outliers (e) A 2D gradient field without outliers

Since the gradient elements corresponding to the outliers have an influence on the orientation estimation, they have relatively larger Euclidean norm values than those corresponding to the ridges in the whitened gradient field. So, the gradient elements corresponding to the outliers are removed by comparing the norms of the gradient elements on step iv). Fig. 3 shows the result of our proposed algorithm schematically. The ridge orientation in the sub-block is represented by the orthogonal direction to the line shown in (b) and (e). The line in (b) is pulled by the outliers caused by the scar. After removing the outliers in the gradient field, the line in (e) represents the reliable direction.

4 Experimental Result

4.1 Segmentation

400 test images from 150 different fingers were evaluated in terms of segmentation. Each test image was manually separated into fingerprint regions and background regions. To evaluate the segmentation algorithm, we compared the output of the proposed segmentation method with the manually labeled results. We created 4 different resolution LUTs (256×256, 128×128, 64×64, 32×32) and calculated the error according to merging-threshold T_s. There are two types of error: a type I error which misjudges the fingerprint region as the background region, or a type II error which misjudges the background region as the fingerprint region. Fig. 4(a) shows the total error (type I + type II) curve. Here, the horizontal axis represents the value of merging-threshold T_s, and the vertical axis is the error rate. Fig. 4(a) indicates that we get the best segmentation performance when T_s is between 4 and 5, and better segmentation performance when larger resolution LUTs are used. When T_s is less than 4, the type I error increases and the type II error decreases. When T_s is greater than 5, the type I error decreases and the type II error increases.

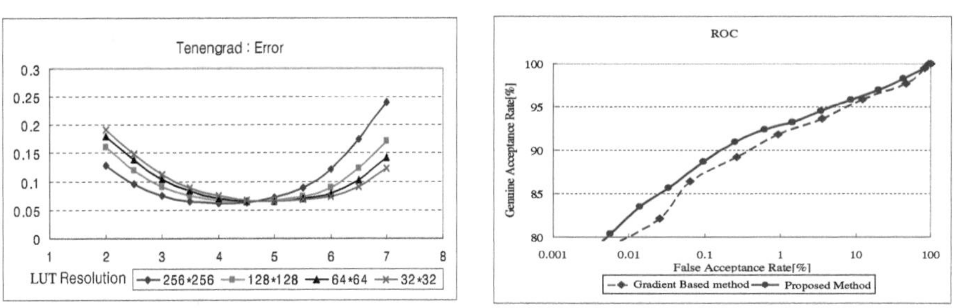

(a) (b)

Fig. 4. (a)Fingerprint segmentation total error curve (b)The ROC curve of the gradient-based method and the proposed method

4.2 Orientation Estimation

We compared the orientation estimation methods with verification performance. To evaluate verification performance, we applied the proposed segmentation algorithm and implemented a minutia extraction [7] and a matching algorithm [8]. In this experiment, we used a fingerprint database of 840 fingerprint images from 168 different fingers with 5 fingerprint images for each finger. We compared the verification performance after applying a conventional gradient-based method and the proposed method for orientation estimation. Fig. 4(b) shows the matching results with the ROC curve. We can observe that the performance of the fingerprint verification system is improved when the proposed orientation method is applied.

5 Conclusion and Future Work

In this paper, we propose a fingerprint preprocessing algorithm using a fingerprint image captured with a mobile camera. Since the characteristics of fingerprint images acquired with mobile cameras are quite different from those obtained by conventional touch-based sensors, it is necessary to develop new and modified fingerprint preprocessing algorithms. The main contributions of this paper are the method of fingerprint segmentation and the robust orientation estimation algorithm when using a mobile camera. In future work, we will develop the matching algorithm that is invariant to 3D camera viewpoint change in mobile camera images and compare fingerprint recognition system in images captured with mobile cameras and touch-based sensors. In this comparison, we will compare not only verification performance but also image quality, the convenience of usage and the number of true minutiae.

Acknowledgements

This work was supported by Samsung Electronics Co. Ltd. and Korea Science and Engineering Foundation (KOSEF) through the Biometrics Engineering Research Center at Yonsei University.

References

1. Angelopoulou Elli, “Understanding the Color of Human Skin. ”Proceedings of the 2001 SPIE conference on Human Vision and Electronic Imaging VI, SPIE Vol. 4299, pp. 243-251, May 2001.
2. Zarit, B. D., Super, B. J., and Quek, F.K.H. “Comparison of Five Color Models in Skin Pixel Classification” International Workshop on Recognition, Analysis, and Tracking of Faces and Gestures in Real-Time Systems, pp. 58-63, 1999.
3. NK Chern, PA Neow, MH Ang Jr. “Practical issues in pixel-based autofocusing for machine vision”, Int. Conf. On Robotics and Automation, pp. 2791- 2796, 2001.
4. R. C. Gonzalez, R. E. Woods, “Digital Image Processing”, Addison-Wesley, Second Edition, pp. 613, 2002.
5. Nalini K., Ratha, Chen Shaoyun, Anil K. Jain, “Adaptive flow orientation-based feature extraction in fingerprint images”, Pattern Recognition, Vol. 28, Issue 11, pp. 1657-1672, November 1995.

6. A.M. Bazen and S.H. Gerez, "Directional field computation for fingerprints based on the principal component analysis of local gradients", in Proceedings of ProRISC2000, 11th Annual Workshop on Circuits, Systems and Signal Processing, Veldhoven, The Netherlands, November 2000.
7. L. Hong, Y. Wan and A.K. Jain, "Fingerprint Image Enhancement: Algorithms and Performance Evaluation", IEEE Transactions on PAMI, Vol. 20, No. 8, pp.777-789, August 1998.
8. D. Lee, K. Choi and Jaihie Kim, "A Robust Fingerprint Matching Algorithm Using Local Alignment", International Conference on Pattern Recognition, Quebec, Canada, August 2002.

A Phase-Based Iris Recognition Algorithm

Kazuyuki Miyazawa[1], Koichi Ito[1], Takafumi Aoki[1], Koji Kobayashi[2], and Hiroshi Nakajima[2]

[1] Graduate School of Information Sciences, Tohoku University, Sendai 980–8579, Japan
miyazawa@aoki.ecei.tohoku.ac.jp
[2] Yamatake Corporation, Isehara 259–1195, Japan

Abstract. This paper presents an efficient algorithm for iris recognition using phase-based image matching. The use of phase components in two-dimensional discrete Fourier transforms of iris images makes possible to achieve highly robust iris recognition with a simple matching algorithm. Experimental evaluation using the CASIA iris image database (ver. 1.0 and ver. 2.0) clearly demonstrates an efficient performance of the proposed algorithm.

1 Introduction

Biometric authentication has been receiving extensive attention over the past decade with increasing demands in automated personal identification. Among many biometrics techniques, iris recognition is one of the most promising approaches due to its high reliability for personal identification [1–8].

A major approach for iris recognition today is to generate feature vectors corresponding to individual iris images and to perform iris matching based on some distance metrics [3–6]. Most of the commercial iris recognition systems implement a famous algorithm using iriscodes proposed by Daugman [3]. One of the difficult problems in feature-based iris recognition is that the matching performance is significantly influenced by many parameters in feature extraction process (eg., spatial position, orientation, center frequencies and size parameters for 2D Gabor filter kernel), which may vary depending on environmental factors of iris image acquisition. Given a set of test iris images, extensive parameter optimization is required to achieve higher recognition rate.

Addressing the above problem, as one of the algorithms which compares iris images directly without encoding [7, 8], this paper presents an efficient algorithm using phase-based image matching – an image matching technique using only the phase components in 2D DFTs (Two-Dimensional Discrete Fourier Transforms) of given images. The technique has been successfully applied to high-accuracy image registration tasks for computer vision applications [9–11], where estimation of sub-pixel image translation is a major concern. In our previous work [12], on the other hand, we have proposed an efficient fingerprint recognition algorithm using phase-based image matching, and have developed commercial fingerprint verification units [13]. In this paper, we demonstrate that the

D. Zhang and A.K. Jain (Eds.): ICB 2006, LNCS 3832, pp. 356–365, 2005.

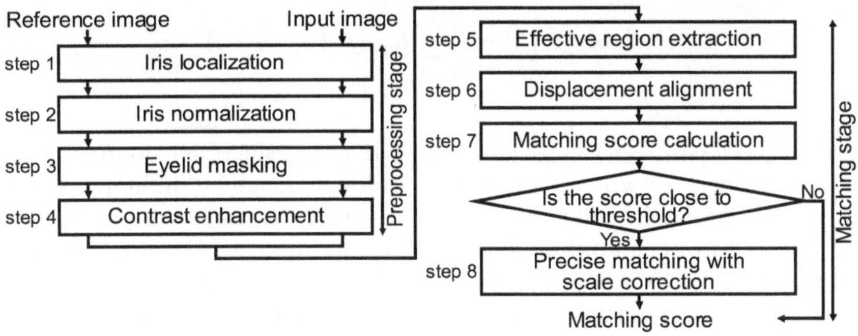

Fig. 1. Flow diagram of the proposed algorithm

same technique is also highly effective for iris recognition. The use of Fourier phase information of iris images makes possible to achieve highly robust iris recognition in a unified fashion with a simple matching algorithm. Experimental performance evaluation using the CASIA iris image database ver. 1.0 and ver. 2.0 [14] clearly demonstrates an efficient matching performance of the proposed algorithm.

Figure 1 shows the overview of the proposed algorithm. The algorithm consists of two stages: (i) preprocessing stage (step 1 – step 4) and (ii) matching stage (step 5 – step 8). Section 2 describes the image preprocessing algorithm (stage (i)). Section 3 presents the iris matching algorithm (stage (ii)). Section 4 discusses experimental evaluation.

2 Preprocessing

An iris image contains some irrelevant parts (eg., eyelid, sclera, pupil, etc.). Also, even for the iris of the same eye, its size may vary depending on camera-to-eye distance as well as light brightness. Therefore, before matching, the original image needs to be preprocessed to localize and normalize the iris.

2.1 Iris Localization

This step is to detect the inner (iris/pupil) boundary and the outer (iris/sclera) boundary in the original image $f_{org}(m_1, m_2)$ shown in Figure 2(a). Through a set of experiments, we decided to use an ellipse as a model of the inner boundary. Let (l_1, l_2) be the lengths of the two principal axes of the ellipse, (c_1, c_2) be its center, and θ be the rotation angle. We can find the optimal estimate $(l_1, l_2, c_1, c_2, \theta)$ for the inner boundary by maximizing the following absolute difference:

$$|S(l_1 + \Delta l_1, l_2 + \Delta l_2, c_1, c_2, \theta) - S(l_1, l_2, c_1, c_2, \theta)| . \tag{1}$$

Here, Δl_1 and Δl_2 are small constants, and S denotes the N-point contour summation of pixel values along the ellipse and is defined as

$$S(l_1, l_2, c_1, c_2, \theta) = \sum_{n=0}^{N-1} f_{org}(p_1(n), p_2(n)), \quad (2)$$

where $p_1(n) = l_1\cos\theta \cdot \cos(\frac{2\pi}{N}n) - l_2\sin\theta \cdot \sin(\frac{2\pi}{N}n) + c_1$ and $p_2(n) = l_1\sin\theta \cdot \cos(\frac{2\pi}{N}n) + l_2\cos\theta \cdot \sin(\frac{2\pi}{N}n) + c_2$. Thus, we will detect the inner boundary as the ellipse on the image for which there will be sudden change in luminance summed around its perimeter. In order to reduce computation time, the parameter set $(l_1, l_2, c_1, c_2, \theta)$ can be simplified depending on iris images. For example, in our experiments using the CASIA iris image database ver. 1.0 and ver. 2.0, assuming $\theta = 0$ causes no degradation on its performance. The outer boundary, on the other hand, is detected in a similar manner, with the path of contour summation changed from ellipse to circle (i.e., $l_1 = l_2$).

2.2 Iris Normalization and Eyelid Masking

Next step is to normalize iris to compensate for the deformations in iris texture. We unwrap the iris region to a normalized (scale corrected) rectangular block with a fixed size (256×128 pixels). In order to remove the iris region occluded by the upper eyelid and eyelashes, we use only the lower half (Figure 2(a)) and apply a polar coordinate transformation (with its origin at the center of pupil) to obtain the normalized image shown in Figure 2(b), where n_1 axis corresponds to the angle of polar coordinate system and n_2 axis corresponds to the radius.

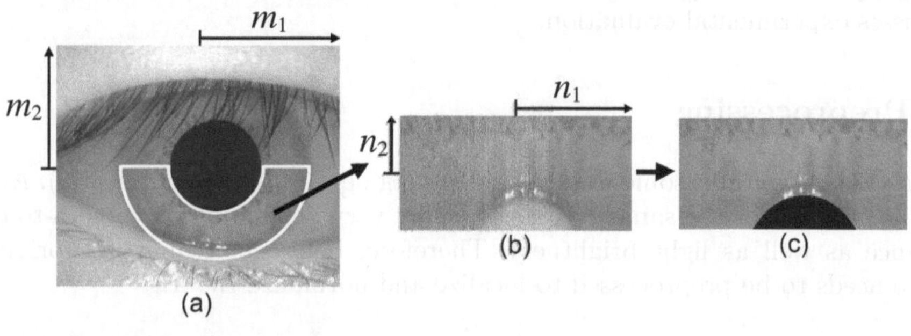

Fig. 2. Iris image: (a) original image $f_{org}(m_1, m_2)$, (b) normalized image, and (c) normalized image with eyelid masking $\tilde{f}(n_1, n_2)$

In general, the eyelid boundary can be modeled as an elliptical contour. Hence the same method for detecting the inner boundary can be applied to eyelid detection. The detected eyelid region is masked as shown in Figure 2(c).

2.3 Contrast Enhancement

In some situation, the normalized iris image has low contrast. Typical examples of such iris images are found in the CASIA iris image database ver. 2.0. In such a case, we improve the contrast by using local histogram equalization technique [4]. Figure 3 shows an example of contrast enhancement.

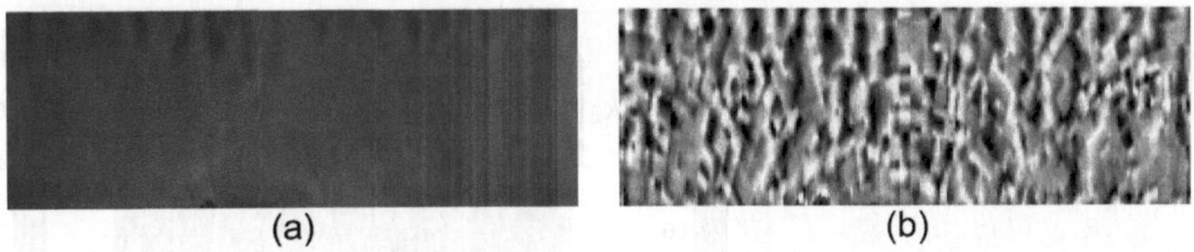

Fig. 3. Contrast enhancement: (a) normalized iris image, and (b) enhanced image

3 Matching

In this section, we describe the detailed process of effective region extraction (section 3.2), image alignment (section 3.3) and matching score calculation (section 3.4 and section 3.5). The key idea in this paper is to use phase-based image matching for image alignment and matching score calculation. Before discussing the algorithm, section 3.1 introduces the principle of phase-based image matching using the Phase-Only Correlation (POC) function [10–12].

3.1 Fundamentals of Phase-Based Image Matching

Consider two $N_1 \times N_2$ images, $f(n_1, n_2)$ and $g(n_1, n_2)$, where we assume that the index ranges are $n_1 = -M_1 \cdots M_1$ $(M_1 > 0)$ and $n_2 = -M_2 \cdots M_2$ $(M_2 > 0)$ for mathematical simplicity, and hence $N_1 = 2M_1 + 1$ and $N_2 = 2M_2 + 1$. Let $F(k_1, k_2)$ and $G(k_1, k_2)$ denote the 2D DFTs of the two images. $F(k_1, k_2)$ is given by

$$F(k_1,k_2) = \sum_{n_1=-M_1}^{M_1} \sum_{n_2=-M_2}^{M_2} f(n_1,n_2) W_{N_1}^{k_1 n_1} W_{N_2}^{k_2 n_2} = A_F(k_1,k_2) e^{j\theta_F(k_1,k_2)}, \quad (3)$$

where $k_1 = -M_1 \cdots M_1$, $k_2 = -M_2 \cdots M_2$, $W_{N_1} = e^{-j\frac{2\pi}{N_1}}$, and $W_{N_2} = e^{-j\frac{2\pi}{N_2}}$. $A_F(k_1, k_2)$ is amplitude and $\theta_F(k_1, k_2)$ is phase. $G(k_1, k_2)$ is defined in the same way. The cross-phase spectrum $R_{FG}(k_1, k_2)$ between $F(k_1, k_2)$ and $G(k_1, k_2)$ is given by

$$R_{FG}(k_1,k_2) = \frac{F(k_1,k_2)\overline{G(k_1,k_2)}}{|F(k_1,k_2)\overline{G(k_1,k_2)}|} = e^{j\theta(k_1,k_2)}, \quad (4)$$

where $\overline{G(k_1, k_2)}$ is the complex conjugate of $G(k_1, k_2)$ and $\theta(k_1, k_2)$ denotes the phase difference $\theta_F(k_1, k_2) - \theta_G(k_1, k_2)$. The POC function $r_{fg}(n_1, n_2)$ is the 2D inverse DFT of $R_{FG}(k_1, k_2)$ and is given by

$$r_{fg}(n_1,n_2) = \frac{1}{N_1 N_2} \sum_{k_1=-M_1}^{M_1} \sum_{k_2=-M_2}^{M_2} R_{FG}(k_1,k_2) W_{N_1}^{-k_1 n_1} W_{N_2}^{-k_2 n_2}. \quad (5)$$

When two images are similar, their POC function gives a distinct sharp peak. When two images are not similar, the peak value drops significantly. The height

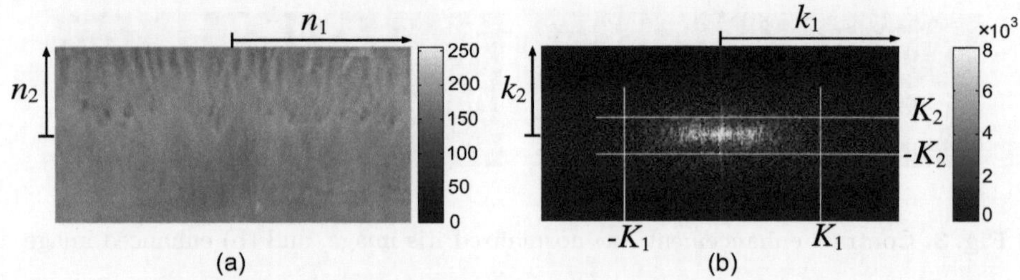

Fig. 4. Normalized iris image in (a) spatial domain, and in (b) frequency domain (amplitude spectrum)

of the peak can be used as a similarity measure for image matching, and the location of the peak shows the translational displacement between the two images.

In our previous work on fingerprint recognition [12], we have proposed the idea of BLPOC (Band-Limited Phase-Only Correlation) function for efficient matching of fingerprints considering the inherent frequency components of fingerprint images. Through a set of experiments, we have found that the same idea is also very effective for iris recognition. Our observation shows that (i) the 2D DFT of a normalized iris image sometimes includes meaningless phase components in high frequency domain, and that (ii) the effective frequency band of the normalized iris image is wider in k_1 direction than in k_2 direction as illustrated in Figure 4. The original POC function $r_{fg}(n_1, n_2)$ emphasizes the high frequency components, which may have less reliability. We observe that this reduces the height of the correlation peak significantly even if the given two iris images are captured from the same eye. On the other hand, BLPOC function allows us to evaluate the similarity using the inherent frequency band within iris textures.

Assume that the ranges of the inherent frequency band are given by $k_1 = -K_1 \cdots K_1$ and $k_2 = -K_2 \cdots K_2$, where $0 \leq K_1 \leq M_1$ and $0 \leq K_2 \leq M_2$. Thus, the effective size of frequency spectrum is given by $L_1 = 2K_1 + 1$ and $L_2 = 2K_2 + 1$. The BLPOC function is given by

$$r_{fg}^{K_1 K_2}(n_1, n_2) = \frac{1}{L_1 L_2} \sum_{k_1=-K_1}^{K_1} \sum_{k_2=-K_2}^{K_2} R_{FG}(k_1, k_2) W_{L_1}^{-k_1 n_1} W_{L_2}^{-k_2 n_2}, \qquad (6)$$

where $n_1 = -K_1 \cdots K_1$ and $n_2 = -K_2 \cdots K_2$. Note that the maximum value of the correlation peak of the BLPOC function is always normalized to 1 and does not depend on L_1 and L_2. Also, the translational displacement between the two images can be estimated by the correlation peak position.

In our algorithm, K_1/M_1 and K_2/M_2 are major control parameters, since these parameters reflect the quality of iris images. In our experiments, $K_1/M_1 = 0.6$ and $K_2/M_2 = 0.2$ are used for the CASIA iris image database ver. 1.0, and $K_1/M_1 = 0.55$ and $K_2/M_2 = 0.2$ are used for the CASIA iris image database ver. 2.0. It is interesting to note that iris images in both databases have effective frequency band of only 20% in k_2 direction (radius direction of iris).

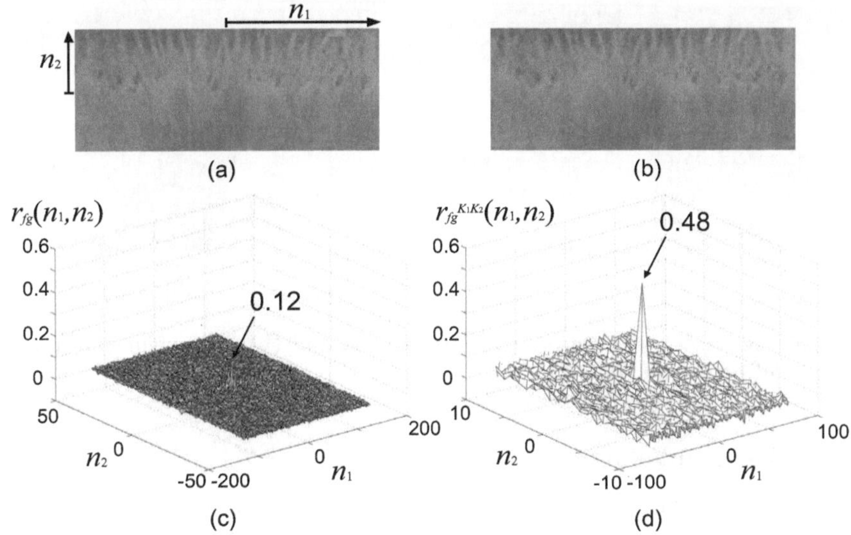

Fig. 5. Example of genuine matching using the original POC function and the BLPOC function: (a) iris image $f(n_1, n_2)$, (b) iris image $g(n_1, n_2)$, (c) original POC function $r_{fg}(n_1, n_2)$, and (d) BLPOC function $r_{fg}^{K_1K_2}(n_1, n_2)$ ($K_1/M_1 = 0.6$, $K_2/M_2 = 0.2$).

Figure 5 shows an example of genuine matching, where the figure compares the original POC function r_{fg} and the BLPOC function $r_{fg}^{K_1K_2}$ ($K_1/M_1 = 0.6$ and $K_2/M_2 = 0.2$). The BLPOC function provides a higher correlation peak than that of the original POC function. Thus, the BLPOC function exhibits a much higher discrimination capability than the original POC function.

In the following, we explain the step 5 – step 8 in Figure 1. The above mentioned BLPOC function is used in step 6 (displacement alignment), step 7 (matching score calculation) and step 8 (precise matching with scale correction).

3.2 Effective Region Extraction

Given a pair of normalized iris images $\tilde{f}(n_1, n_2)$ and $\tilde{g}(n_1, n_2)$ to be compared, the purpose of this process is to extract effective regions of the same size from the two images, as illustrated in Figure 6(a). Let the size of two images $\tilde{f}(n_1, n_2)$ and $\tilde{g}(n_1, n_2)$ be $\tilde{N}_1 \times \tilde{N}_2$, and let the widths of irrelevant regions in $\tilde{f}(n_1, n_2)$ and $\tilde{g}(n_1, n_2)$ be $w_{\tilde{f}}$ and $w_{\tilde{g}}$, respectively. We obtain $f(n_1, n_2)$ and $g(n_1, n_2)$ by extracting effective regions of size $\tilde{N}_1 \times \{\tilde{N}_2 - \max(w_{\tilde{f}}, w_{\tilde{g}})\}$ through eliminating irrelevant regions such as masked eyelid and specular reflections.

On the other hand, a problem occurs when the extracted effective region becomes too small to perform image matching. In this case, by changing the parameter w, we extract multiple effective sub-regions from each iris image as illustrated in Figure 6(b). In our experiments, we extract at most 6 sub-regions from a single iris image by changing the parameter w as 55, 75 and 95 pixels.

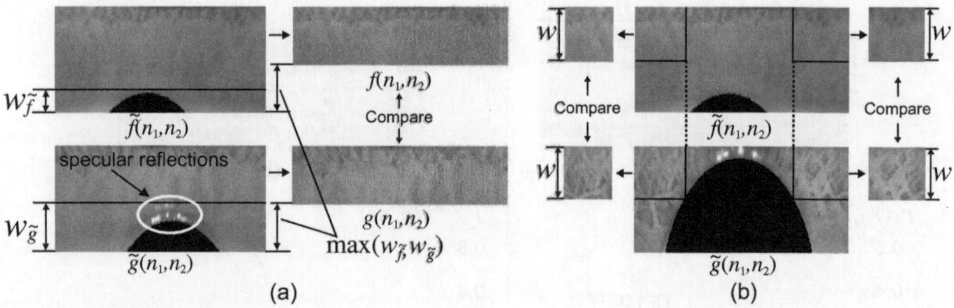

Fig. 6. Effective region extraction: (a) normal case, and (b) case when multiple sub-regions should be extracted

3.3 Displacement Alignment

This step is to align the translational displacement τ_1 and τ_2 between the extracted images $f(n_1, n_2)$ and $g(n_1, n_2)$. Rotation of the camera, head tilt and rotation of the eye within the eye socket may cause the displacements in normalized images (due to the polar coordinate transformation). The displacement parameters (τ_1, τ_2) can be estimated from the peak location of the BLPOC function $r_{fg}^{K_1K_2}(n_1, n_2)$. The obtained parameters are used to align the images.

3.4 Matching Score Calculation

In this step, we calculate the BLPOC function $r_{fg}^{K_1K_2}(n_1, n_2)$ between the aligned images $f(n_1, n_2)$ and $g(n_1, n_2)$, and evaluate the matching score. In the case of genuine matching, if the displacement between the two images is aligned, the correlation peak of the BLPOC function should appear at the origin $(n_1, n_2) = (0, 0)$. So, we calculate the matching score between the two images as the maximum peak value of the BLPOC function within the $r \times r$ window centered at the origin, where we choose $r = 11$ in our experiments. When multiple sub-regions are extracted at the "effective region extraction" process, the matching score is calculated by taking an average of matching scores for the sub-regions.

3.5 Precise Matching with Scale Correction

For some iris images, errors take place in estimating the center coordinates of the iris and the pupil in the preprocessing. In such a case, slight scaling of the normalized images may occur. And the matching score drops to a lower value even if the given two iris images are captured from the same eye. Then, if the matching score is close to threshold value to separate genuine and impostor, we generate a set of slightly scaled images (scaled in the n_1 direction), and calculate matching scores for the generated images. We select their maximum value as the final matching score.

4 Experiments and Discussions

This section describes a set of experiments using the CASIA iris image database ver. 1.0 and ver. 2.0 [14] for evaluating matching performance.

- **CASIA iris image database ver. 1.0.** This database contains 756 eye images with 108 unique eyes and 7 different images of each unique eye. We first evaluate the genuine matching scores for all the possible combinations of genuine attempts; the number of attempts is ${}_{7}C_{2}\times 108 = 2268$. Next, we evaluate the impostor matching scores for all the possible combinations of impostor attempts; the number of attempts is ${}_{108}C_{2}\times 7^2 = 283122$.
- **CASIA iris image database ver. 2.0.** This database contains 1200 eye images with 60 unique eyes and 20 different images of each unique eye. We first evaluate the genuine matching scores for all the possible combinations of genuine attempts; the number of attempts is ${}_{20}C_{2}\times 60 = 11400$. Next, we evaluate the impostor matching scores for ${}_{60}C_{2}\times 4^2 = 28320$ impostor attempts, where we take 4 images for each eye and make all the possible combinations of impostor attempts.

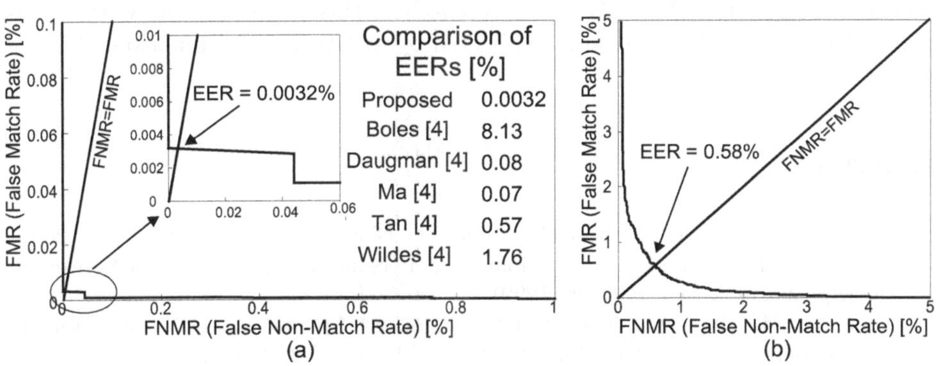

Fig. 7. ROC curve and EER: (a) CASIA iris image database ver. 1.0, and (b) ver. 2.0

Figure 7(a) shows the ROC (Receiver Operating Characteristic) curve of the proposed algorithm for the database ver. 1.0. The ROC curve illustrates FNMR (False Non-Match Rate) against FMR (False Match Rate) at different thresholds on the matching score. EER (Equal Error Rate) shown in the figure indicates the error rate where FNMR and FMR are equal. As is observed in the figure, the proposed algorithm exhibits very low EER (0.0032%). Some reported values of EER from [4] using the CASIA iris image database ver. 1.0 are shown in the same figure for reference. Note that the experimental condition in [4] is not the same as our case, because the complete database used in [4] is not available at CASIA [14] due to the limitations on usage rights of the iris images.

Figure 7(b) shows the ROC curve for the database ver. 2.0. The quality of the iris images in this database are poor, and it seems that the recognition task is difficult for most of the reported algorithms. Although we cannot find any reliable official report on recognition test for this database, we believe that our result (EER=0.58%) may be one of the best performance records that can be achieved at present for this kind of low-quality iris images. All in all, the above mentioned two experimental trials clearly demonstrate a potential possibility of phase-based image matching for creating an efficient iris recognition system.

5 Conclusion

The authors have already developed commercial fingerprint verification units [13] using phase-based image matching. In this paper, we have demonstrated that the same approach is also highly effective for iris recognition task. It can also be suggested that the proposed approach will be highly useful for multimodal biometric system having iris and fingerprint recognition capabilities.

Acknowledgment. Portions of the research in this paper use the CASIA iris image database ver 1.0 and ver 2.0 collected by Institute of Automation, Chinese Academy of Sciences.

References

1. Wayman, J., Jain, A., Maltoni, D., Maio, D.: Biometric Systems. Springer (2005)
2. Jain, A., Bolle, R., Pankanti, S.: Biometrics: Personal Identification in a Networked Society. Norwell, MA: Kluwer (1999)
3. Daugman, J.: High confidence visual recognition of persons by a test of statistical independence. IEEE Trans. Pattern Analy. Machine Intell. **15** (1993) 1148–1161
4. Ma, L., Tan, T., Wang, Y., Zhang, D.: Efficient iris recognition by characterizing key local variations. IEEE Trans. Image Processing **13** (2004) 739–750
5. Boles, W., Boashash, B.: A human identification technique using images of the iris and wavelet transform. IEEE Trans. Signal Processing **46** (1998) 1185–1188
6. Tisse, C., Martin, L., Torres, L., Robert, M.: Person identification technique using human iris recognition. Proc. Vision Interface (2002) 294–299
7. Wildes, R.: Iris recognition: An emerging biometric technology. Proc. IEEE **85** (1997) 1348–1363
8. Kumar, B., Xie, C., Thornton, J.: Iris verification using correlation filters. Proc. 4th Int. Conf. Audio- and Video-based Biometric Person Authentication (2003) 697–705
9. Kuglin, C.D., Hines, D.C.: The phase correlation image alignment method. Proc. Int. Conf. on Cybernetics and Society (1975) 163–165
10. Takita, K., Aoki, T., Sasaki, Y., Higuchi, T., Kobayashi, K.: High-accuracy subpixel image registration based on phase-only correlation. IEICE Trans. Fundamentals **E86-A** (2003) 1925–1934

11. Takita, K., Muquit, M.A., Aoki, T., Higuchi, T.: A sub-pixel correspondence search technique for computer vision applications. IEICE Trans. Fundamentals **E87-A** (2004) 1913–1923
12. Ito, K., Nakajima, H., Kobayashi, K., Aoki, T., Higuchi, T.: A fingerprint matching algorithm using phase-only correlation. IEICE Trans. Fundamentals **E87-A** (2004) 682–691
13. http://www.aoki.ecei.tohoku.ac.jp/poc/
14. http://www.sinobiometris.com

Graph Matching Iris Image Blocks with Local Binary Pattern

Zhenan Sun, Tieniu Tan, and Xianchao Qiu

Center for Biometrics and Security Research,
National Laboratory of Pattern Recognition, Institute of Automation,
Chinese Academy of Sciences, P.O. Box 2728, Beijing, 100080, P.R. China
{znsun, tnt, xcqiu}@nlpr.ia.ac.cn

Abstract. Iris-based personal identification has attracted much attention in recent years. Almost all the state-of-the-art iris recognition algorithms are based on statistical classifier and local image features, which are noise sensitive and hardly to deliver perfect recognition performance. In this paper, we propose a novel iris recognition method, using the histogram of local binary pattern for global iris texture representation and graph matching for structural classification. The objective of our idea is to complement the state-of-the-art methods with orthogonal features and classifier. In the texture-rich iris image database UPOL, our method achieves higher discriminability than state-of-the-art approaches. But our algorithm does not perform well in the CASIA database whose images are less textured. Then the value of our work is demonstrated by providing complementary information to the state-of-the-art iris recognition systems. After simple fusion with our method, the equal error rate of Daugman's algorithm could be halved.

1 Introduction

Iris-based identity authentication has many important applications in our networked society. Since the last decade, much research effort has been directed towards automatic iris recognition. Because the distinctive information of iris pattern is preserved in the randomly distributed micro-textures, constituted by freckles, coronas, stripes, furrows, etc., most of the state-of-the-art iris recognition algorithms are based on the local features of iris image data. Typical iris recognition methods are Gabor-based phase demodulation [1], local intensity variations [2] and wavelet zero-crossing features [3], etc. However, the minutiae-based iris representation is sensitive to noise, such as the occlusions of eyelids and eyelashes, non-linear deformations, imperfect localization or alignment, etc. So it is a straightforward idea to complement local features based methods with global structural features.

In our early attempt [4], blobs of interest are segmented from the iris images for spatial correspondence. Experimental results demonstrated the effectiveness of combining local statistical features and global structural features. But the segmentation of foreground regions in some poor quality images, e.g. defocused iris images, is a difficult problem. In addition, both the feature extraction and matching of blob patterns [4] were not very efficient.

D. Zhang and A.K. Jain (Eds.): ICB 2006, LNCS 3832, pp. 366 – 372, 2005.

We think the distinctiveness of an iris pattern relies on the statistical features of local image regions and the spatial relationship between these regions. Motivated by the fact that the literature has ignored the global topological information in iris data, the iris features are represented from both local and global aspects in this paper: local binary pattern (LBP operator) is adopted to characterize the iris texture in each image block, and all localized image blocks are used to construct a global graph map. Then the similarity between two iris images is measured by a simple graph matching scheme.

The novelty of this paper is that both LBP and image blocks based graph matching are introduced for the first time to iris recognition and in a fusion manner. Another contribution is that our method is good complement of the state-of-the-art iris recognition systems with orthogonal features and classifiers.

The remainder of this paper is organized as follows. Section 2 introduces the LBP-based attribute graph representation scheme. The graph matching method, aiming to find the correspondence between two iris images, is provided in Section 3. Experimental results on two publicly available iris databases are reported in Section 4. Section 5 concludes this paper.

2 LBP-Based Iris Feature Representation

LBP describes the qualitative intensity relationship between a pixel and its neighborhoods, which is robust, discriminant, and computationally efficient so it is well suited to texture analysis [5]. We choose LBP to represent iris image blocks'

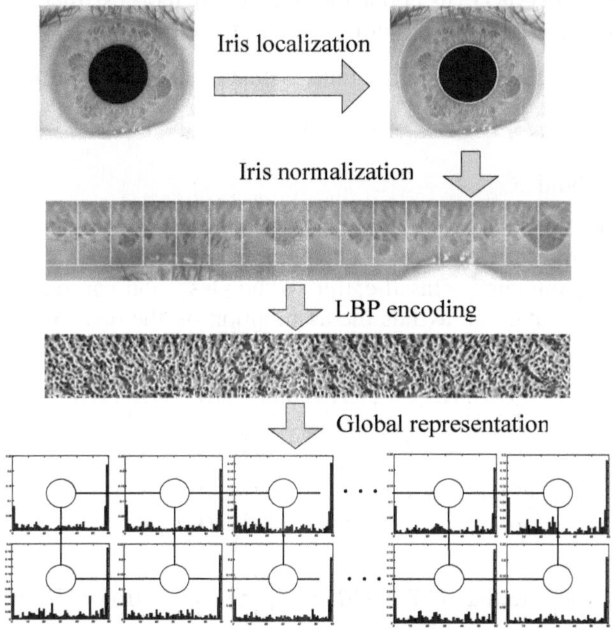

Fig. 1. The flowchart of the LBP-based iris graph representation

distinctive information because iris pattern could be seem as texture constituted by many minute image structures. This is the first attempt in literature to use LBP for iris recognition.

The whole procedure of iris feature extraction is illustrated in Figure 1. Firstly, the input iris image should be preprocessed and normalized to correct the position and scale variations before iris feature extraction and matching. In our paper, the resolution of the normalized iris image is 80 by 512. To exclude the possible occlusions of eyelids and eyelashes, we divide the upper region of the normalized iris image into 2*16=32 blocks, and each block has the size 32 by 32. For each block in the normalized iris image, an eight-neighborhood uniform LBP histogram with radius 2 (59 bins) [5] may be obtained. In our labeled graph representation of iris pattern, each manually divided image block is regarded as a graph node, associated with the attributes of the local region's LBP histogram. And the spatial layout of these image blocks is used to model the structural relations among the nodes. Finally, a graph with 32 nodes is constructed as the template of each iris image (Figure 1).

3 Graph Matching Iris Features

Because an iris pattern has randomly distributed minute features, varying from region to region, the basic idea underlying our graph matching scheme is qualitative corresponding theory. For each block of an iris image, it should be the most similar to the corresponding block in another image if these two iris images (*A* and *B*) are from the same eye. So we only need to count the number of the best matching block pairs, which are required to satisfy two conditions:

1) The matching blocks have the minimal distance based on a similarity metric, i.e. $\min_j Distance(A^i, B^j)\ \forall i, j = 1,2,\cdots,32$. In addition, their distance should be lower than a given threshold C_{Th}.

2) The matching blocks have the same topological layout, i.e. the corresponding blocks have the same spatial position in the graph representation.

Compared with parametric classification principles, non-parametric classification strategy is more flexible and avoids the assumption on the distribution of input data. In this paper, the Chi-square statistic is used to evaluate the dissimilarity between two LBP histograms $HA^i\{HA^i_1, HA^i_2, \cdots, HA^i_{59}\}$ and $HB^j\{HB^j_1, HB^j_2, \cdots, HB^j_{59}\}$:

$$\chi^2(HA^i, HB^j) = \sum_{k=1}^{59} \frac{(HA^i_k - HB^j_k)^2}{HA^i_k + HB^j_k} \tag{1}$$

Because it is possible that $HA^i_k + HB^j_k = 0$, the summation only includes the nonzero bins. Suppose the LBP features of the two iris images are $HA\{HA^1, HA^2, \cdots, HA^{32}\}$ and $HB\{HB^1, HB^2, \cdots, HB^{32}\}$ respectively, so their matching score *S* is computed as follows:

```
Set S=0
for i=1 : 32
    compute χ²(HA^i, HB^i)
    if χ²(HA^i, HB^i) < C_Th
        set Flag=0
        for j=1 : 32
            if (j ≠ i)
                compute χ²(HA^i, HB^j)
                if χ²(HA^i, HB^j) < χ²(HA^i, HB^i)
                    Flag=1;
                    break;
                end
            end
        end
        if Flag=0
            S= S+1
        end
    end
end
```

Fig. 2. The pseudo code of the graph matching of LBP features

C_{Th} is a constant value learned from the training set. For genuine corresponding block pairs, the probability of their Chi Square lower than the C_{Th} should be more than 0.8. The matching score *S* has the range from 0 to 32, and could be normalized as *S*/32 to obtain a uniform output for fusion. The higher the matching score, the higher the probability of the two images being from the same eye.

4 Experiments

To evaluate the effectiveness of our method for iris recognition, two publicly available iris databases, UPOL [6] and CASIA [7] are used as the test datasets. The first one is constituted by European volunteers, captured under visible lighting. And the second one mainly comes from Chinese volunteers, captured under infrared illumination.

The UPOL iris database [6] includes 384 iris images from 64 persons. All possible intra-class and inter-class comparisons are made to estimate the genuine distribution and imposter distribution respectively, i.e. totally 384 genuine samples and 73,152 imposter samples. The distribution of these matching results is shown in Figure 3. For the purpose of comparison, two state-of-the-art iris recognition algorithms, Daugman's [1] and Tan's [2], are also implemented on the same dataset. Although these three methods all achieve perfect results, i.e. without false accept and false reject, our method obtains higher discriminating index ($DI = \frac{|m_1 - m_2|}{\sqrt{(\delta_1^2 + \delta_2^2)/2}}$, where m_1 and δ_1^2 denote the mean and variance of intra-class Hamming distances, and m_2 and δ_2^2 denote the mean and variance of inter-class Hamming distances.) [1] (See Fig. 3).

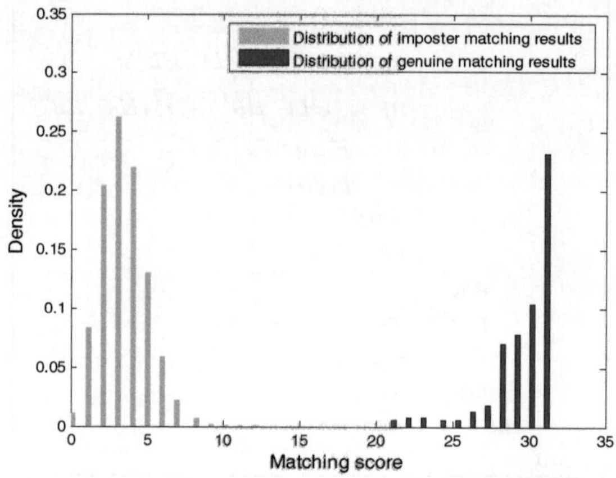

Fig. 3. The distribution of matching results of our method on the UPOL database. The DI is 15.2. In contrast, the DI of Daugman's method [1] is 7.9 and that of Tan's [2] is 8.6.

The CASIA database is the largest open iris database [7] and we only use the subset described in [2] for performance evaluation. There are totally 3,711 intra-class comparisons and 1,131,855 inter-class comparisons. The distribution of the matching results of our method is shown in Fig. 4. The maximal inter-class matching score is 12. We can see that the comparison results of genuine and imposter are well separated by our method although they overlap each other in a minor part. The ROCs (receiver operating curve) of the three methods are shown in Fig. 5. It is clear that our method does not perform as well as the state-of-the-art methods on this dataset. We think the main reason is that the texture information of Asian subjects is much less than that of the Europeans, especially on the regions far from the pupil, but the effectiveness of LBP histogram heavily depends on the abundant micro-textures. The main purpose of this paper is to develop the complementary global features, along with the commonlyused local features, to improve the accuracy and robustness of an iris recognition system. The score-level fusion results based on Sum rule are shown in Fig. 5 and Table 1. After introducing the matching results of LBP features and structural classifier, the equal error rate (EER) of Daugman's method [1] is halved. Similarly, about 30% EER is reduced from Tan's method [2] (Table 1). Comparatively, combining two local features based methods does not show significant improvement (Table 1). The disadvantage of our method is that the graph matching diagram is time consuming because of many iterations, but it still could be implemented in real time. In addition, if we adopt a cascading scheme like that described in [4], the computational complexity could be considerably reduced.

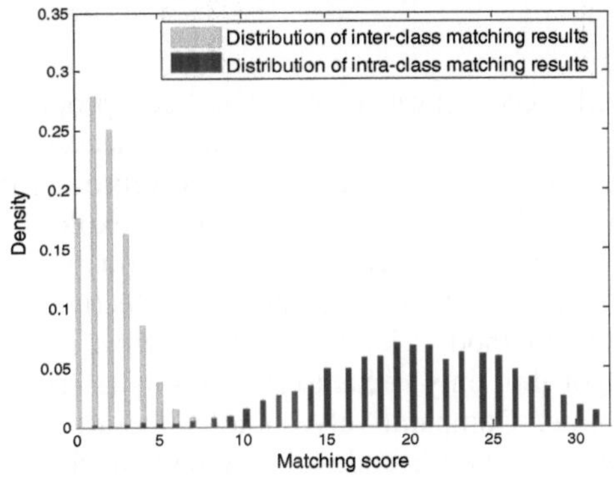

Fig. 4. The distribution of matching results of our method on CASIA database

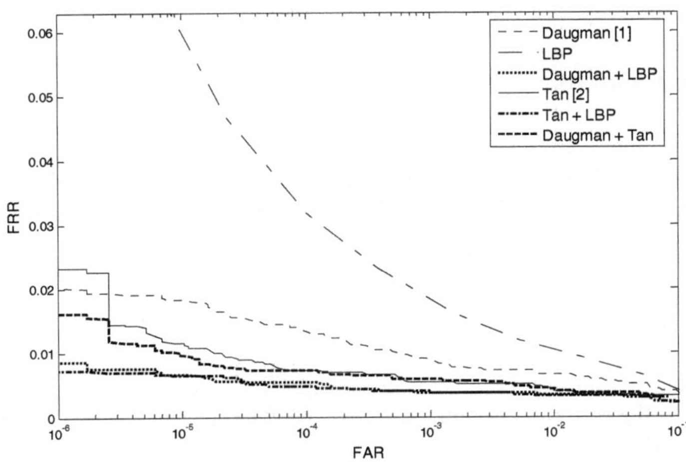

Fig. 5. Comparison of ROC curves of different iris recognition methods on CASIA database

Table 1. Comparison of recognition accuracy of various recognition schemes

Recognition scheme	DI	EER
Daugman [1]	4.74	0.70%
Tan [2]	5.36	0.51%
LBP	4.46	0.86%
Daugman + LBP	5.31	0.37%
Tan + LBP	5.51	0.32%
Daugman + Tan	5.23	0.49%

5 Conclusions

In this paper, a new iris recognition method has been proposed to complement the state-of-the-art approaches. LBP operator, which is successfully applied to texture analysis and face recognition, is firstly employed to represent the robust texture features of iris images. A novel graph matching scheme is exploited to measure the similarity between two iris images. Experimental results on two publicly available iris image databases, UPOL and CASIA, illustrated the effectiveness of our method. The largest advantage of our method is its robustness against noise or occlusions in iris images because our algorithm only needs to match only a fraction of all image blocks to authenticate a genuine. Comparatively, state-of-the-art iris recognition methods [1][2][3] require that most of the iris codes should be matched.

How to define suitable global features to strengthen the robustness of local features based methods is not well addressed before, and it should be an important issue in future works. In addition, we think the global features should play a defining role in indexing of large scale iris databases.

Acknowledgement

This work is funded by research grants from the National Basic Research Program (Grant No. 2004CB318110), Natural Science Foundation of China (Grant No. 60335010, 60121302, 60275003, 60332010, 69825105) and the Chinese Academy of Sciences.

References

1. J. Daugman, “High Confidence Visual Recognition of Persons by a Test of Statistical Independence”, IEEE Trans. Pattern Analysis and Machine Intelligence, Vol.15, No.11, pp.1148-1161, 1993.
2. L. Ma, T. Tan, Y. Wang, and D. Zhang, “Efficient Iris Recognition by Characterizing Key Local Variations”, IEEE Trans. Image Processing, Vol. 13, No. 6, pp.739–750, 2004.
3. C. Sanchez-Avila, R. Sanchez-Reillo, "Two different approaches for iris recognition using Gabor filters and multiscale zero-crossing representation", Pattern Recognition, Vol. 38, No. 2, pp. 231-240, 2005.
4. Zhenan Sun, Yunhong Wang, Tieniu Tan, Jiali Cui, “Improving Iris Recognition Accuracy via Cascaded Classifiers”, IEEE Transactions on Systems, Man, and Cybernetics—Part C: Applications and Reviews, Vol. 35, No. 3, pp.435-441, August 2005.
5. Topi Mäenpää, Matti Pietikäinen, “Texture analysis with local binary patterns”, Chapter 1, in C. Chen and P. Wang (eds) Handbook of Pattern Recognition and Computer Vision, 3rd ed, World Scientific, pp.197-216, 2005.
6. Michal Dobeš and Libor Machala, UPOL Iris Database, http://www.inf.upol.cz/iris/.
7. CASIA Iris Image Database, http://www.sinobiometrics.com..

Localized Iris Image Quality Using 2-D Wavelets

Yi Chen, Sarat C. Dass, and Anil K. Jain

Michigan State University, East Lansing, MI, 48823
{chenyi1, jain}@cse.msu.edu,
{sdass}@stt.msu.edu

Abstract. The performance of an iris recognition system can be undermined by poor quality images and result in high false reject rates (FRR) and failure to enroll (FTE) rates. In this paper, a wavelet-based quality measure for iris images is proposed. The merit of the this approach lies in its ability to deliver good spatial adaptivity and determine local quality measures for different regions of an iris image. Our experiments demonstrate that the proposed quality index can reliably predict the matching performance of an iris recognition system. By incorporating local quality measures in the matching algorithm, we also observe a relative matching performance improvement of about 20% and 10% at the equal error rate (EER), respectively, on the CASIA and WVU iris databases.

1 Introduction

Iris recognition is considered the most reliable form of biometric technology with impressively low false accept rates (FARs), compared to other biometric modalities (e.g., fingerprint, face, hand geometry, etc) [1]. However, recent studies on iris recognition systems have reported surprisingly high false reject rates (FRRs) (e.g., 11.6% [3], 7% [4] and 6% [5]), due to poor quality images. Causes of such poor quality include occlusion, motion, poor focus, non-uniform illumination, etc. (see Figure 1(a)) [2].

There have been several efforts in iris image quality analysis in the past. Daugman [7] measured the energy of high frequency components in Fourier spectrum to determine the focus. Zhang and Salganicoff [8] analyzed the sharpness of the pupil/iris boundary for the same purpose. Ma et al. [9] proposed a quality classification scheme to categorize iris images into four classes, namely clear, defocused, blurred and occluded.

We propose a novel iris quality measure based on local regions of the iris texture. Our argument is that the iris texture is so localized that the quality varies from region to region. For example, the upper iris regions are more often occluded than lower regions, and the inner regions often provide finer texture compared to the outer regions (see Figure 1(b)). Sung et al. have shown that by simply weighting the inner (respectively, outer) iris regions with the weight 1(0), the matching performance can be improved [12]. To estimate the local quality, we employ 2D wavelets on concentric bands of a segmented iris texture. By weighting the matching distance using the local quality, we observe a relative improvement of about 20% and 10% at the equal error rate (EER) in the matching performance, respectively, on CASIA1.0 [16] and WVU databases. Further, we combine the local quality into a single image quality index, Q, and demonstrate its capability of predicting the matching performance.

D. Zhang and A.K. Jain (Eds.): ICB 2006, LNCS 3832, pp. 373–381, 2005.

The rest of the paper is organized as follows: Section 2 describes the iris segmentation algorithms. In Section 3, localized quality are derived using 2D wavelets. In Section 4, an overall quality index Q is computed. Two experiments are conducted in Section 5 to predict and improve the matching performance using the quality derived. Summary and conclusions are provided in Section 6.

2 Image Preprocessing

The iris region, consisting of the annulus band between the pupil and sclera (see Figure 1(b)), is the essential feature used in iris biometric systems. The segmentation of iris region involves two steps, (i) iris boundary detection, and (ii) eyelid detection.

The iris/sclera boundary and the pupil/iris boundary (see Figure 1(b)) can be approximated by two circles using the following method.

1. The grayscale morphological opening is conducted on a given image to remove noise (e.g., eyelashes). Intensity thresholding is used to locate the pupil area and approximate the pupil center (c) and radius (r).
2. To approximate the pupil/iris boundary, Canny edge detection is performed onto a circular neighborhood centered at c and with radius ($r + 20$). Noise-like edges are removed and the edge map is down-sampled before circular Hough transform is applied to detect the pupil/iris boundary.
3. To detect the iris/sclera boundary, Step 2 is repeated with the neighborhood region replaced by an annulus band (of width R, say) outside the pupil/iris boundary. The edge detector is tuned to the vertical direction to minimize the influence of eyelids.

The upper and lower eyelids are oval-shaped and can be approximated by second-order parabolic arcs, as shown below:

1. The original image is decomposed into four sub bands (HH, HL, LH, LL) using Daubechies wavelets [15]. The LH image, which contains details in the vertical direction is processed through Canny edge detection. Here, the Canny edge detector is tuned to the horizontal direction to minimize the influence of eyelashes.
2. To detect the upper eyelid, edges outside the upper iris/sclera boundary neighborhood are removed. The remaining edge components that are located close to each other within a certain distance are connected.
3. The longest connected edge is selected and fit with a second-order parabolic curve
$$f(x) = ax^2 + bx + c, \tag{1}$$
where a, b, c are the parameters to be estimated. The estimation is carried out by minimizing the sum of squared error $\frac{1}{N}\sum_{i=1}^{N}(f(x_i) - y_i)^2$, where $(x_i, y_i)_{i=1,2,\ldots,N}$ represent N points on the selected edge.
4. To detect the lower eyelid, Steps 2 and 3 are repeated with the rectangular neighborhood in Step 2 taken around the lower iris/sclera boundary.

A simple intensity thresholding operation is implemented to remove eyelashes in the CASIA1.0 database, but not in the WVU database (Note that the two databases used different iris image capture devices). Figure 2(I) illustrates the segmentation results using the algorithms discussed above on several iris images from the CASIA1.0 database.

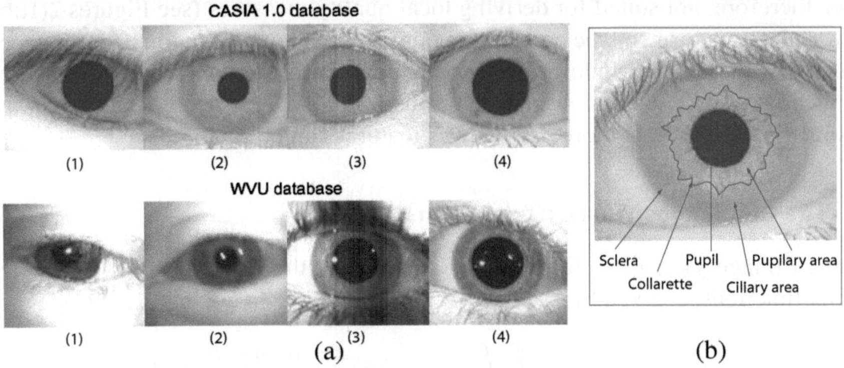

Fig. 1. (a) Poor quality of iris images caused by (1) occlusion, (2) poor focus and eye motion, (3) non-uniform illumination, and (4) large pupil area. The top (respectively, bottom) panels are images from the CASIA1.0 (WVU) databases. (b) Components of the eye and iris pattern. The inner iris (pupillary) area and the outer iris (ciliary) area are separated by the collarette boundary.

3 Localized Quality Assessment

Ma et al. [9] used the energy of low, moderate and high frequency components in 2D Fourier power spectrum to evaluate iris image quality. However, it is well known that Fourier transform (or Short Time Fourier Transform (STFT)) does not localize in space,

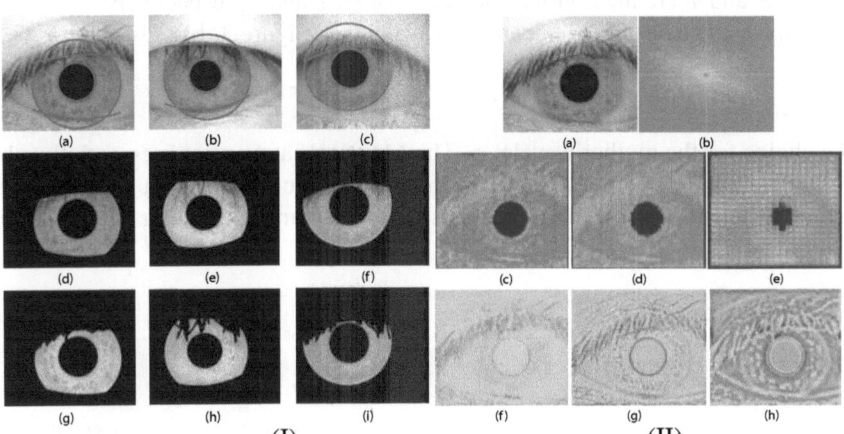

Fig. 2. (I) Three iris images from CASIA1.0 database with (a-c) iris boundaries and eyelids detected; (d-f) The extracted iris pattern; (g-i) The extracted iris pattern after eyelash removal. (II) Demonstrating the effectiveness of the wavelet transform in achieving better space-frequency localization compared to Fourier transform and STFT: (a) Original eye image; (b) Fourier transform of the image; (c-e) STFT using rectangular windows with sizes of $2 \times 4, 4 \times 6$, and 14×16, respectively; (f-h) Wavelet transform using Mexican hat with scales of 0.5, 1.0, 2.0, respectively.

and is, therefore, not suited for deriving local quality measures (see Figures 2(II:b-e)). The wavelet transform, on the contrary, obtains smooth representation in both space and frequency with flexible window sizes varying up to a scale factor (see Figures 2(II:f-h)). Specifically, we use continuous wavelet transform (CWT) instead of discrete wavelet transform (DWT) so that more detailed iris features can be captured.

3.1 The Continuous Wavelet Transform (CWT)

Given an image $f(x, y) \in R^2$, its CWT, defined as the convolution with a series of wavelet functions, is given by

$$w(s,a,b) = \frac{1}{\sqrt{s}} \int\int_{R^2} f(x,y)\phi(\frac{x-a}{s}, \frac{y-b}{s})dxdy, \quad (2)$$

where s is the dilation (scale) factor and (a, b) denotes the translation (or, shift) factor. To simplify computations, the convolution in equation (2) can be converted into multiplication in the Fourier frequency domain. For a function g, we denote by G the corresponding 2D Fourier transform of g, given by

$$G(\omega_1, \omega_2) = \int\int_{R^2} g(x,y)e^{-i2\pi(\omega_1 x + \omega_2 y)}dxdy. \quad (3)$$

Then, equation (2) can be re-written in the frequency domain as

$$W(s, \omega_1, \omega_2) = \sqrt{s}F(\omega_1, \omega_2)Phi(s\omega_1, s\omega_2), \quad (4)$$

where W, F and Φ are the Fourier transforms of w, f and ϕ, respectively.

We employ the isotropic Mexican hat wavelet (see Figure 3 (b)), given by:

$$\Phi(s\omega_1, s\omega_2) = -2\pi((s\omega_1)^2 + (s\omega_2)^2)e^{-\frac{1}{2}((s\omega_1)^2+(s\omega_2)^2)} \quad (5)$$

as the choice for the mother wavelet ϕ. The Mexican hat wavelet is essentially a band pass filter for edge detection at scales s. In addition, the Mexican hat wavelet has two vanishing moments and is, therefore, sensitive to features exhibiting sharp variations

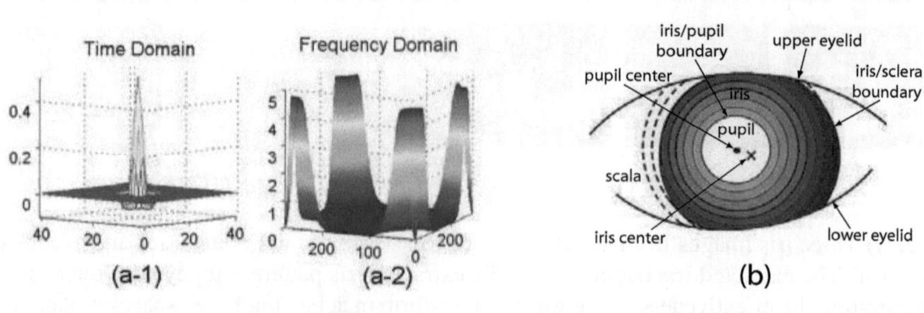

Fig. 3. (a) A Mexican hat wavelet illustrated (a-1) in the space domain, and (a-2) in the frequency domain. (b) Partitioning the iris texture into local regions. Multiple concentric annulus bands with fixed width are constructed and local quality is measured based on the energy in each band.

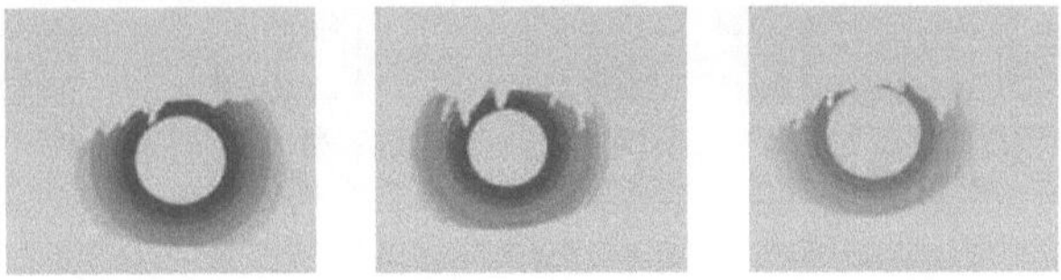

(a) (b) (c)

Fig. 4. The local quality measures based on the energy concentration in the individual bands. The estimated quality indices Q for these three images are 10, 8.6, 6.7, respectively.

(e.g., pits and freckles) and non-linearity (e.g., zigzag collarette, furrows). In order to capture various features at multiple scales, we obtain the product responses given by

$$w^{mul}(s_1, s_2, s_3) = w(s_1) \times w(s_2) \times w(s_3), \quad (6)$$

where s_1, s_2, s_3 are the three scales introduced in Figures 2(II:f-h), namely 0.5, 1.0, 2.0.

To obtain the local quality measure of an iris texture, we partition the region into multiple concentric (at the pupil center) bands with a fixed width until the iris/sclera boundary is reached (see Figure 3(b)). Let T be the total number of bands. The energy E_t of the t-th $(t = 1, 2, ...T)$ band is defined as

$$E_t = \frac{1}{N_t} \sum_{i=1}^{i=N_t} |w_{t,i}^{mul}|^2, \quad (7)$$

where $w_{t,i}^{mul}$ represents the i-th product-based wavelet coefficient in the t-th band, and N_t is the total number of wavelet coefficients in the t-th band. The energy, E_t, is a good indicator of the distinctiveness of the iris features, and hence, a reliable measure of local quality; high values of E_t indicate good quality and vice versa (see Figure 4).

The quality index Q is defined as a weighted average of the band-wise local quality

$$Q = \frac{1}{T} \sum_{t=1}^{T} (m_t \times \log E_t), \quad (8)$$

where T is the total number of bands and m_t is the weight [17]

$$m_t = \exp\{-\|l_t - l_c\|^2/(2q)\}, \quad (9)$$

with l_c denoting the center of the pupil, and l_t denoting the mean radius of the t-th band to l_c. The justification for using weights m_t is that inner iris regions provide more texture [12] and is less occluded by eyelashes compared to outer iris regions.

4 Iris Matching

Before incorporating local quality measures, there are several difficulties in matching two iris images: (i) the iris region may vary due to dilations of the pupil caused by changes in lighting conditions; (ii) the iris size may vary since the capturing distance

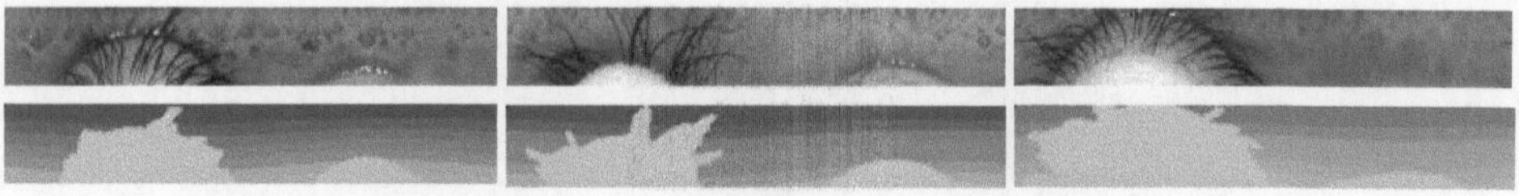

Fig. 5. The normalized iris patterns (top row) associated with Figures 2(I:a-c) and their corresponding normalized quality map (bottom row). The normalization introduces nonlinear distortion when the iris and pupil centers do not coincide.

from the camera is not strictly controlled; and (iii) genuine iris images may have slight rotation due to variability in the acquisition process.

To account for these variations, the Daugman's rubber sheet model [7] is applied to normalize both the iris texture and the local quality measures. Although this nonlinear mapping introduces distortion (Figure 5), it is essential for compensating for pupil dilation and size variability of the iris. Then, Daugman's matching algorithm based on Gabor wavelets is applied to generate the IrisCode for any iris patterns [6]. To measure the similarity of two IrisCodes, X and Y, we compute the Hamming distance, given by

$$HD = \frac{1}{B}\sum_{i=1}^{B} X_i \bigotimes Y_i \ , \tag{10}$$

where X_i and Y_i represent the i-th bit in the sequence X and Y, respectively, and N is the total number of bits in each sequence. The symbol $\otimes$ is the "XOR" operator. To account for rotational variability, we shift the template left and right bit-wise (up to 8 bits) to obtain multiple Hamming distances, and then choose the lowest distance.

To incorporate local quality measures into the matching stage, we modify Daugman's matching algorithm by deriving a weighted Hamming distance, given by

$$HD_w = \frac{1}{B}\frac{\sum_{i=1}^{B}\sqrt{E^X_{g(i)} \times E^Y_{g(i)}} \times (X_i \bigotimes Y_i)}{\sum_{i=1}^{B}\sqrt{(E^X_{g(i)} \times E^Y_{g(i)})}} \ , \tag{11}$$

where $g(i)$ is the index of the band that contains the i-th bit of the IrisCode. The symbols $E^X_{g(i)}$ and $E^Y_{g(i)}$ are the associated local quality measures of the $g(i)$-th band in X and Y, respectively. The weighting scheme is such that regions with high quality in both X and Y contribute more to the matching distance compared to regions with poor quality.

5 Experimental Results

Our proposed local quality and the overall quality index Q are derived for two iris databases. The CASIA1.0 database [16] contains 756 greyscale images from 108 different eyes. The West Virginia University (WVU) Iris Database has a total of 1852 images from 380 different eyes. The number of acquisitions for each eye ranges from 3-6 in this database. The images were captured using an OKI IrisPass-H hand-held device.

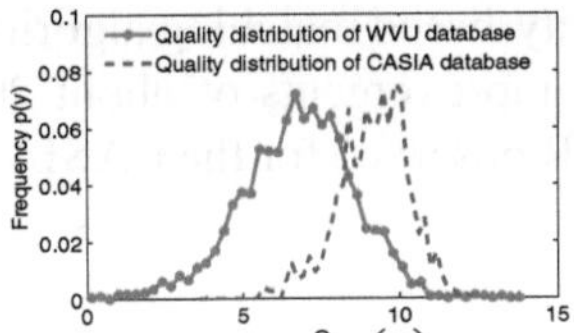

(a)

Accuracy	Iris boundary	Upper eyelid	Lower eyelid
Daugman [6]	98.6%	N/A	N/A
Wildes [11]	99.9%	N/A	N/A
Cui et al. [12]	99.34%	97.35%	93.39%
Proposed	99.47%	98.41%	99.07%

(b)

Fig. 6. (a) Image quality distribution of CASIA1.0 (dotted line) and WVU (solid line) databases. (b) Performance comparison of different segmentation algorithms on CASIA1.0 database.

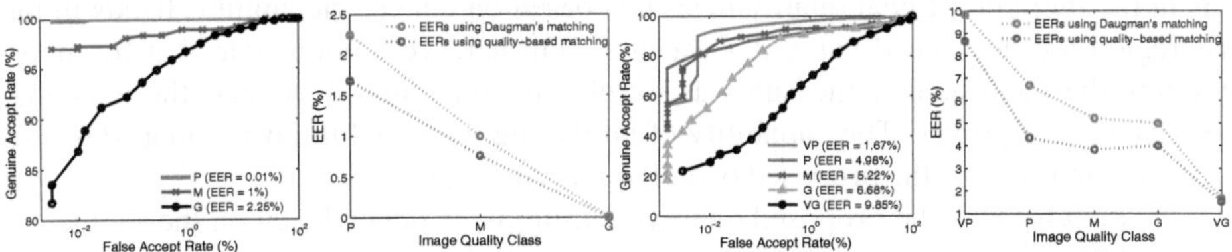

Fig. 7. Demonstrating the improvement in matching performance using the proposed quality measures on the CASIA1.0 database: (a) ROC curves of the P, M, and G image quality classes. (b) Improvement in the matching performance (in terms of EER) using the proposed quality-based matching algorithm. Similar results on the WVU database: (c) ROC curves of the VP, P, M, G, VG quality classes. (d) Improvement in the matching performance (in terms EER).

Figure 6(a) shows distribution of the overall quality index Q for the two databases. Note the longer left tail of the WVU database, indicating lower quality compared to CASIA1.0. In fact, images in the WVU database were captured without any quality control and were heavily affected by lighting conditions. Further, size of the iris exhibits high variability due to inconsistencies in capture distance during image acquisition.

Since segmentation results on CASIA1.0 are available in the literature [11], we compare them with the performance of our proposed method in Figure 6(b). We can see the proposed method is highly comparable with the others, particularly for lower eyelid detection. Results of Daugman's and Wildes's algorithms were also reported in [11].

Two experiments are conducted to evaluate the proposed quality measures. In the first experiment, we classify images in CASIA1.0 into three quality classes based on Q, namely, Poor (P), Moderate (M), and Good (G). The matching performance for each class is obtained using Daugman's matching algorithm and the corresponding ROC curves are shown in Figure 7 (a). Note that the proposed quality index Q are effective in predicting the matching performance. Higher values of Q indicate better matching performance. In the second experiment, Daugman's matching algorithm was modified by equation (11) and the corresponding ROC curves are obtained. We compare the ERRs of the modified algorithm with those of the Daugman's algorithm. As shown in Figure 7 (b), quality-based matching reduces EERs for all three classes with the greatest improvement on the poor class. Similar experiments were conducted on WVU database (see Figure 7(c-d)). Due to the large size, we classify images in WVU into five classes, namely, Very Poor (VP), Poor (P), Moderate (M), Good (G), and Very Good (VG).

The improvement of matching performance using quality-based matching algorithm is also studied across the entire database, with relative improvements of about 20% (from 1.00% to 0.79%) and 10% (7.28% to 6.55%) in EER observed for the CASIA1.0 and WVU databases, respectively.

6 Summary and Conclusions

In this paper, we study the effects of iris image quality on the matching performance of iris recognition. Two segmentation algorithms are proposed and compared with methods in the literature. Local quality measures based on concentric annulus bands in the iris region are developed using 2D wavelets. Further, we demonstrate that by incorporating the local quality measures as weights for matching distances, the matching performance improves. The capability of predicting the matching performance is also evaluated in terms of the proposed overall quality index Q.

One drawback of the proposed quality measure is its dependency on the segmentation performance, since segmentation itself is affected by poor image quality. In future work, we want to solve this by conducting the two modules in parallel to optimize both.

Acknowledgements

This work is supported by a contract from the Lockheed-Martin corporation. Thanks to Dr. Arun Ross at West Virginia University and Dr. Yunhong Wang at Chinese Academy of Science for providing the iris databases. Thanks are also due to Mr. Libor Masek for sharing MATLAB code of Daugman's matching algorithm as public resource [18].

References

1. T. Mansfield, G. Kelly, D. Chandler, and J. Kane, "Biometric Product Testing Report," CESG/BWG Biometric Test Programme, National Physical Laboratory, UK, 2001
2. Committee Draft, "Biometric Data Interchange Formats - Part 6: Iris Image Data," *International Organization for Standarization (ISO)*, 2003
3. H. Wang, D. Melick, R. Vollkommer and B. Willins, "Lessons Learned From Iris Trial," *Biometric Consortium Conference*, 2002
4. D. Thomas, "Technical Glitches Do Not Bode Well For ID Cards, Experts Warn," *Computer Weekly*, May, 2004
5. S. King, H. Harrelson and G. Tran, "Testing Iris and Face Recognition in a Personal Identification Application," *Biometric Consortium Conference*, 2002
6. J. Daugman, "Recognizing Persons By Their Iris Patterns", in *Biometric Systems: Technology, Design and Performance Evaluation,* J. Wayman, A.K. Jain, etc. (Eds.), Springer, 2004
7. J. Daugman, "Statistical Richness of Visual Phase Information: Update on Recognizing Persons by Iris Patterns", *Int'l Journal on Computer Vision*, Vol. 45, no. 1, pp. 25-38, 2001
8. G. Zhang and M. Salganicoff, "Method of Measuring the Focus of Close-Up Image of Eyes," *United States Patent, no. 5953440*, 1999
9. L. Ma, T. Tan, Y. Wang and D. Zhang, "Personal Identification Based on Iris Texture Analysis," *IEEE Trans. on Pattern Analysis and Machine Intelligence*, Vol. 25, no. 12, 2003

10. R. Wildes, "Automated Iris Recognition: An Emerging Biometric Technology," *Proc. of the IEEE*, Vol. 85 no. 9, pp. 1348-1363, 1997
11. J. Cui, Y. Wang, etc., "A Fast and Robust Iris Localization Method Based on Texture Segmentation," *SPIE Defense and Security Symposium*, Vol. 5404, pp. 401-408, 2004
12. H. Sung, J. Lim, J. Park and Y. Lee, "Iris Recognition Using Collarette Boundary Localization," *Proc. of the 17th Int'l Conf. on Pattern Recognition*, Vol. 4, pp. 857-860, 2004
13. N. Graham, "Breaking the Visual Stimulus Into Parts," Current Directions in Psychologial Science, Vol. 1, no. 2, pp. 55-61, 1992
14. J. Antoine, L. Demanet, etc., "Application of the 2-D Wavelet Transform to Astrophysical Images," *Physicalia magazine*, Vol. 24, pp. 93-116, 2002
15. C. Burrus, R. Gopinath, and H. Guo, "Introduction to Wavelets and Wavelet Transforms," *Prentice Hall*, New Jersy, 1998
16. Chinese Academy of Sciences - Institute of Automation Iris Database 1.0, available online at: *http://www.sinobiometrics.com*, 2003
17. N. Ratha, R. Bolle, "Fingerprint Image Quality Estimation," IBM RC21622, 1999
18. L. Masek, *http://www.csse.uwa.edu.au/ pk/studentprojects/libor/* , 2003

Iris Authentication Using Privatized Advanced Correlation Filter

Siew Chin Chong, Andrew Beng Jin Teoh, and David Chek Ling Ngo

Faculty of Information Science and Technology (FIST), Multimedia University,
Jalan Ayer Keroh Lama, Bukit Beruang, Melaka 75450, Malaysia
{chong.siew.chin, bjteoh, david.ngo}@mmu.edu.my

Abstract. This paper proposes a private biometrics formulation which is based on the concealment of random kernel and the iris images to synthesize a minimum average correlation energy (MACE) filter for iris authentication. Specifically, we multiply training images with the user-specific random kernel in frequency domain before biometric filter is created. The objective of the proposed method is to provide private biometrics realization in iris authentication in which biometric template can be reissued once it was compromised. Meanwhile, the proposed method is able to decrease the computational load, due to the filter size reduction. It also improves the authentication rate significantly compare to the advance correlation based approach [5][6] and comparable to the Daugmant's Iris Code [1].

1 Introduction

Nowadays, security is in critical demand of finding reliable and cost-effective alternatives to passwords, ID cards or PIN due to the increasing of financial losses from computer-based fraud such as computer hacking and identity theft. Biometric solutions address these fundamental problems due to the fact that the biometric data is unique and cannot be transferred. However, the traditional biometrics system does not completely solve the security concerns. One critical issue is the cancelability or replaceability of the biometric template once it is compromised by an attacker.

Some authors like Bolle et. al. [2] and Davida et al. [3] have introduced the terms cancelable biometrics and private biometrics to rectify this issue. These terms are used to denote biometrics data that can be cancelled and replaced, as well as is unique to every application. The cancelability issue of biometrics was also addressed by Andrew et al. [4]. They introduced the freshness into the authenticator via a randomized token. The revocation process is essentially the inner-product of a tokenized pseudo-random pattern and the biometrics information iteratively. Most recently, Savvides et al. [5] proposed a cancelable biometrics scheme which encrypted the training images used to synthesize the correlation filter for biometrics authentication. They demonstrated that convolving the training images with any random convolution kernel prior to building the biometric filter does not change the resulting correlation output peak-to-sidelobe ratios, thus preserving the authentication performance. In other word, their work does not show any improvement in terms of performance.

D. Zhang and A.K. Jain (Eds.): ICB 2006, LNCS 3832, pp. 382–388, 2005.

In this paper we propose a private or cancelable biometric formulation method based on Savvides et al. advance correlation filter formulation. We multiply training images with the user-specific random kernel in frequency domain instead of convolving the training images with random kernel in spatial domain that done by Savvides et al. The objectives of the proposed method are three fold: to provide private biometrics realization in iris authentication in which biometric template can be reissued by replacing the random kernel if it was compromised. Secondly, it helps to decrease the computational load during the enrollment as filter size is greatly reduced. In terms of authentication rate, the proposed method shows better performance than the advance correlation based approach.

The outline of the paper is organized as follow: Section 2 briefly explains MACE filter. Section 3 introduces the proposed method. Experiments and results are reported in Section 4. Conclusion is presented in Section 5.

2 Overview of Minimum Average Correlation Energy (MACE) Filter

Kumar et al [6] [7] has proposed many types of advanced correlation filters for biometrics authentication purpose. Minimum average correlation energy (MACE) filter is one of the advanced correlation filters. MACE is designed such as correlation function levels at all points can be reduced except at the origin of the correlation plane and thereby obtained a very sharp correlation peak [8].

During the enrollment stage, multiple training images are being used to form a MACE filter. Let $\mathbf{D}_i$ be a d x d diagonal matrix containing the power spectrum of training image i along its diagonal, and let diagonal matrix $\mathbf{D}$ be the average of all $\mathbf{D}_i$. Also, $\mathbf{X} = [\mathbf{x}_1, \mathbf{x}_2, \ldots, \mathbf{x}_N]$ is a d x N matrix with N training image vectors, $\mathbf{x}$ as its columns. MACE filter is given as follows:

$$\mathbf{h} = \mathbf{D}^{-1}\mathbf{X}(\mathbf{X}^{+}\mathbf{D}^{-1}\mathbf{X})^{-1}\mathbf{u} \tag{1}$$

In general, $\mathbf{u} = [\mathbf{u}_1, \mathbf{u}_2, \ldots, \mathbf{u}_N]^T$ and $\mathbf{u}_i$ is user defined. All $\mathbf{u}_i$ belonging to an authentic class are set to 1; otherwise they are set to 0. The superscript + denotes the complex conjugate transpose. On the other hand, the test image will be cross-correlated with the MACE filter to produce the correlation output in the authentication stage.

3 The Proposed Method

During the enrollment phase, we multiply normalized iris training images, $\mathbf{x}$ with the user-specific random kernel, $\mathbf{R}$ in the frequency domain before biometric filter is created:

$$\mathbf{e}(\mathbf{x}, \mathbf{R}) = \mathbf{R}_{dm}^{T}\mathbf{x}_d \quad \text{where } m < d \tag{2}$$

where d is the original template size and m is the size after the concealment. The concealed patterns are used to synthesize a minimum average correlation energy (MACE) filter. Meanwhile, for the authentication stage, a testing iris image with its associated random kernel will be also gone through the concealment operation to generate the concealed iris pattern and will then convolute with the trained MACE filter to produce a correlation output. Fig.1 shows the idea of the proposed method.

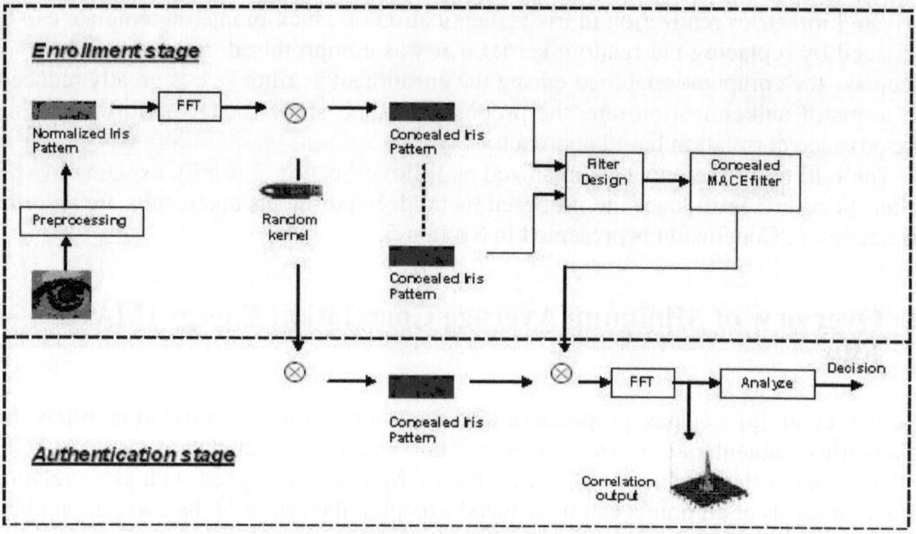

Fig. 1. Block diagram of the proposed method

In practice, random kernel can be generated from a physical device, for example smartcard or USB token. There is a seed which stores in USB token or smartcard microprocessor to generate **R** using a random number generator. Different user will have different seeds for different applications and these seeds are recorded during the enrollment process. A lot of pseudo-random bit/number algorithms are publicly available, such as ANSI X9.17 generator or Micali-Schnorr pseudo-random bit generator [9].

The process flow of the enrollment phase is as follow:

1) Perform Fast Fourier transform (FFT) to each normalized iris patterns, $I \in \Re^{d_1 \times d_2}$.
2) Convert each of the FFTed iris patterns into the column vector, **x** with dimension d (d_1 x d_2) through column-stacking.
3) Then, multiply **x** with random kernel, **R**, thus $\mathbf{e}(\mathbf{x},\mathbf{R}) = \mathbf{R}^T_{dm}\mathbf{x}_d$, where $m \leq d$.
4) Then $\mathbf{E} = [\mathbf{e}_1, \mathbf{e}_2, ..., \mathbf{e}_N]$ will be used to synthesize the MACE filter as follow:

$$\mathbf{h} = \mathbf{D}^{-1}\mathbf{E}(\mathbf{E}^{+}\mathbf{D}^{-1}\mathbf{E})^{-1}\mathbf{u} \tag{3}$$

where **D** is a m_1 x m_2 diagonal matrix containing the average power spectrum of all the training images along its diagonal. Also, $\mathbf{u} = [\mathbf{u}_1, \mathbf{u}_2, ..., \mathbf{u}_N]^T$ is a N x 1 column vector containing the desired peak values for N training images. The resulting **h** is a column vector with m entries that need to be re-ordered into matrix to form the MACE filter.

From the above description, a concealed iris size, **e** is either equal or less than the **x** original iris template, $m \leq d$; hence the MACE filter size can be greatly trimmed down if m is small. This helps increase the computation speed, especially the calculation of inversion matrix **D** in eq(3).

In order to ascertain how similar a test image to a MACE filter, a corresponding metric is needed. Kumar [6] suggested the Peak-to-Sidelobe Ratio (PSR) as a "summary" of the information in each correlation plane. Thus, the PSR is used to evaluate the degree of similarity of correlation planes. The PSR is defined as follows:

$$PSR = \frac{mean(mask) - mean(sidelobe)}{\sigma(sidelobe)} \tag{4}$$

First, the correlation peak is located and the mean value of the central mask (e.g., of size 3 x 3) centered at the peak is determined. The sidelobe region is the annular region between the central mask and a larger square (e.g., of size 10 x 10), also centered at the peak. The mean and standard deviation of the sidelobe are calculated.

4 Experimental Results

The experiments were conducted by using Chinese Academy of Sciences-Institute of Automation (CASIA) Iris image database [10], which consists of 756 grey scale eye images with i=108 individuals and 7 images each. In the experiment, 3 images of each person are randomly selected as training images while other j=4 images are used as testing images. For the False Accept Rate (FAR) test and imposter population distribution, the specific MACE filter of each iris is cross-correlated against all other testing iris images, leading to 46224 imposter attempts $(((i-1)\times j)\times i)$. For the False Reject Rate (FRR) test and genuine population distribution, the specific MACE filter of each iris is cross-correlated against all images of the same iris, leading to 432 genuine attempts $(i \times j)$.

In the experiment, the performance of MACE, the proposed method (RMACE) and Daugmant's Iris Code. (For a detailed study of Daugman's Iris Code see [1]) are examined. During the authentication phase, the filter is cross-correlated with the testing images to generate correlation outputs which will be used for calculating the PSR. Fig. 2 shows the correlation plane of RMACE-20x50, from a person during the authentication phase. As demonstrated by the figure, the correlation output will exhibit a sharp peak for authentics but no such peak for imposters.

As illustrated in Fig. 3 and Table 1, the performance of the original and the proposed method are tested. The proposed method, RMACE is tested with different size of m. For the original MACE filter, its original size is 20x240 and the EER achieved is 14.78%. If compare to RMACE, the authentication of RMACE-m where m = 20x20, 20x40 and 20x50 are far better than MACE. The best authentication rate can attained from RMACE-20x50 in which the EER is 0.0726%. For Daugman's Iris Code, we can see that the EER achieved is 0.43% which is better than MACE but poorer than RMACE-20x50.

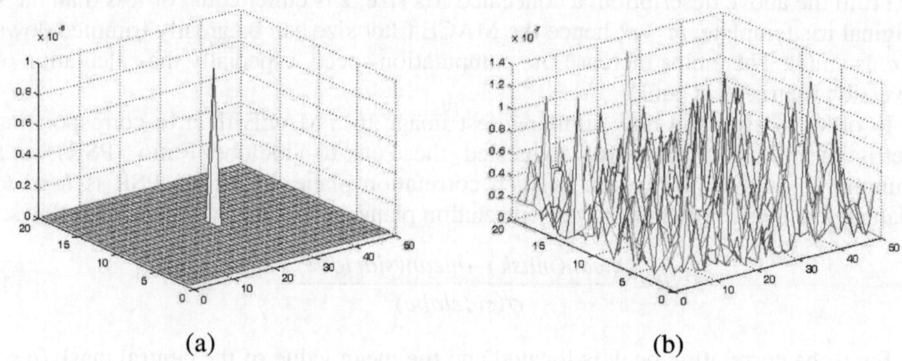

Fig. 2. Correlation plane of RMACE-1000 of a person: (a) Genuine class (b) Imposter class

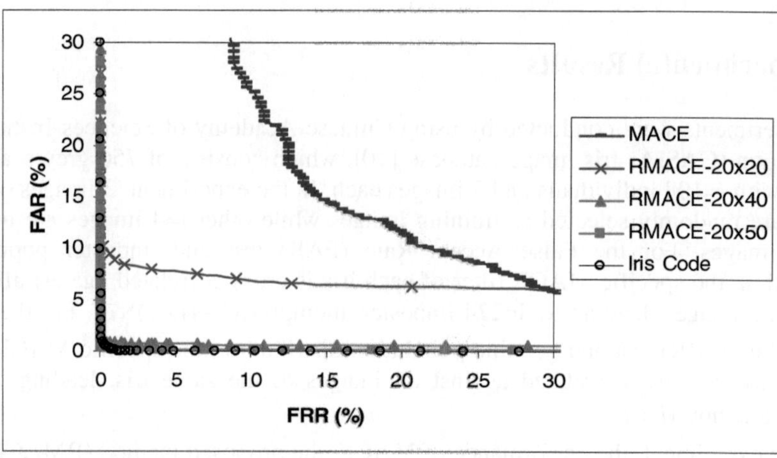

Fig. 3. Receiver operating curve for MACE, RMACE and Iris Code

Table 1. Performance evaluation of genuine class and imposter class of CASIA Iris Image Database using MACE and RMACE tested on different size of concealed template

Method	Concealed template size, *m*	FAR (%)	FAR (%)	EER (%)
MACE	20x240 (=4800)	14.7456	14.8148	14.7802
RMACE	20x20 (=400)	7.4831	6.4815	6.9823
	20x40 (=800)	0.8589	0.9259	0.8924
	20x50 (=1000)	0.0715	0.0729	0.0722
Iris Code	2048 bit binary code	0.4253	0.4409	0.4331

In addition, from the result obtained, it is obviously that the size of the iris templates is greatly reduced if compared to the original MACE methodology and Daugman's Iris Code. MACE's template has 20x240 and Iris Code's template has 2048 bit binary code whereas RMACE can provide the best EER with size 20x50. Among these three methods, our proposed method is able to generate the best EER with smaller template size. Intuitively, smaller size is less accurate in performing authentication task. However, our proposed method shows that the size reduction does not weaken the accuracy in authentication task but somehow improve the authentication rate. Meanwhile, the size reduction also helps to reduce the computational load.

Fig. 4 shows the PSRs of RMACE-20x50 for the first 400 comparisons of genuine and imposter class. A clear separation is found between the genuine and the imposter plots. This implies that RMACE can recognize the genuine and imposter perfectly.

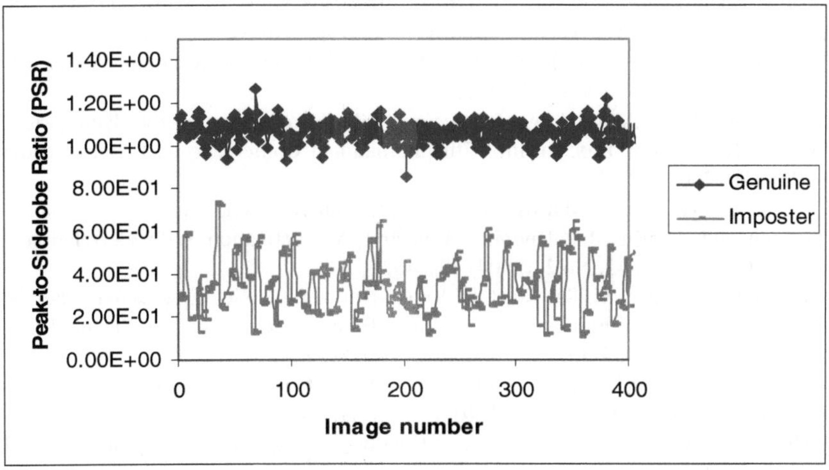

Fig. 4. PSR plots using RMACE-1000 for the first 400 comparisons of Genuine and Imposter class

5 Conclusion and Future Works

In this paper, a promising method for private iris authentication is presented. The privatization of biometrics is done based on the concealment of random kernel and the iris images to synthesize a minimum average correlation energy (MACE) filter for the iris authentication. Specifically, we multiply training images with the user-specific random kernel in frequency domain before biometric filter is created. Therefore, new private biometrics filter can be easily reissued if his/her possession has been lost/stolen.

In terms of authentication rate, it improves the performance significantly compare to the advance correlation based approach and comparable to the

Daugmant's Iris Code. Besides that, the filter synthesizing speed during the enrollment is notably increased due to the size reduction of the concealed iris template.

The research presented here will be further investigated by considering more challenging conditions such as noise contaminated, rotated and random occlusion iris images. Besides, it is interesting to look at the theoretical aspect on the proposed method.

References

1. J.G Daugman,: Recognizing Persons by their Iris Patterns In Biometrics: Personal Identification in Networked Society. Kluwer, (1998) 103-121.
2. R. M. Bolle, J. H. Connel and N. K. Ratha.: Biometric Perils and Patches. Pattern Recognition, Vol. 35, (2002) 2727 2738.
3. Davida, G., Frankel, Y., & Matt, B. J.: On enabling secure applications through off-line biometric identification. Proceeding Symposium on Privacy and Security, (1998) 148-157
4. Andrew Teoh Beng Jin, David Ngo Chek Ling and Alwyn Goh.: An Integrated Dual Factor Verification Based On The Face Data And Tokenised Random Number. LNCS, Springer-Verlag, 3072, (2004)117-123.
5. Marios Savvides, B.V.K. Vijaya Kumar and P.K. Khosla.: Cancelable Biometric Filters For Face Recogntion. Proc. of the 17th International Conference on Pattern Recognition (ICPR'04), (2004).
6. B. V. K. Vijaya Kumar, Marios Savvides, Chunyan Xie, Krithika Venkataramani, Jason Thornton and Abhijit Mahalanobis.: Biometric Authentication With Correlation filters. Applied Optics, Vol. 43, No.2, (2004) 391-402.
7. B.V.K Vijaya Kumar, M. Savvides, K. Venkataramani, C. Xie.: Spatial frequency domain image processing for biometric recognition. Proc. of Int. Conf. On Image Processing (ICIP), Vol.1, (2002) 55-56.
8. A.Mahalanobis, B.V.K Vijaya Kumar, and D.Casasent.: Minimum average correlation filters. Appl, Opt 26, (1987) 3633-3640.
9. A.Menezes, P.V. Oorschot, S. Vanstone.: Handbook of Applied Cryptography. CRC Press, Boca Raton, (1996).
10. CASIA Iris Image Database, Version 1.0. From: http://www.sinobiometrics.com

Extracting and Combining Multimodal Directional Iris Features

Chul-Hyun Park[1] and Joon-Jae Lee[2]

[1] School of Electrical and Computer Engineering, Purdue University,
West Lafayette, Indiana 47907-2035, USA
park95@purdue.edu
[2] Dept. of Computer and Information Engineering,
Dongseo University, Busan, Korea
jjlee@dongseo.ac.kr

Abstract. In this paper, we deal with extracting and combining multimodal iris features for person verification. In multibiometric approaches, finding reasonably disjoint features and effective combining methods are crucial. The proposed method considers the directional characteristics of iris patterns as critical features, and first decomposes an iris image into several directional subbands using a directional filter bank (DFB), then generates two kinds of feature vectors from the directional subbands. One is the binarized output features of the directional subbands on multiple scales and the other is the blockwise directional energy features. The former is relatively robust to changes in illumination or image contrast because it uses the directional zero crossing information of the directional subbands, whereas the latter provides another form of rich directional information though it is a bit sensitive to contrast change. Matching is performed separately between the same kind of feature vectors and the final decision is made by combining the matching scores based on the accuracy of each method. Experimental results show that the two kinds of feature vectors used in this paper are reasonably complementary and the combining method is effective.

1 Introduction

Though human irises have been successfully used in some applications as a means for human identification [1], finding a method robust to various environmental situations such as changes in illumination or image contrast is still a challenging issue. Actually, the local and global brightness values of an iris image change according to the positions of various light sources and the image contrast also varies due to different focusing of the camera. To accomplish robustness to such changes, most conventional approaches use the quantized values of the transformed data or multi-resolution features [2-4]. However, the approaches do not utilize significant components of rich discriminatory information available in iris patterns.

Therefore, in order to extract rich distinctive iris features robust to contrast and brightness differences in an image or between images, the proposed method attempts to combine the two separate approaches, in which one is robust to changes in

D. Zhang and A.K. Jain (Eds.): ICB 2006, LNCS 3832, pp. 389 – 396, 2005.

illumination and contrast; and the other represents rich information of iris patterns in another form. Since combing two matchers increases the complexity of the system, it is important to design an efficient way of sharing the common information from the two feature extractors as much as possible and to find a combining method that maximizes the advantage of each method.

The two methods used in this paper consider the directionality of iris patterns as a key feature and both methods decompose an iris image into 8 directional subband images using a directional filter bank (DFB) [5]. Thereafter, one of them generates a feature vector consisting of the sampled and binarized subband outputs [6], and the other takes the normalized energy values of the tessellated directional subband blocks as a feature vector [7].

Matching is performed separately between the input and template feature vectors extracted from the same feature extractor and the final decision is made by combining the two matching scores based on the accuracy of each method. Since both the two matchers extract iris features from the subband outputs decomposed by the same DFB, the complexity of the entire system does not increase so much though two matchers are combined, whereas the accuracy (or reliability) of the system increases reasonably.

2 Iris Region Detection

An iris is a ring shaped area surrounding the pupil of the eye as shown in Fig. 1(a). Since the pupil area has little discriminatory information, only the iris region is used for verification. Fortunately, the iris region is darker than the (white) sclera and brighter than the pupil except for eyes with cataract, thus the iris region can be easily detected by the circular edge detector [1]. Among the detected region, only the inner half regions of the left and right 90 degree cone-shaped areas are used for feature extraction in order to simply exclude the region commonly occluded by the eyelids (refer to Fig. 1(b)). The detected ROI (region of interest) is converted again into polar coordinates to facilitate the following feature extraction as illustrated in Fig. 1(c).

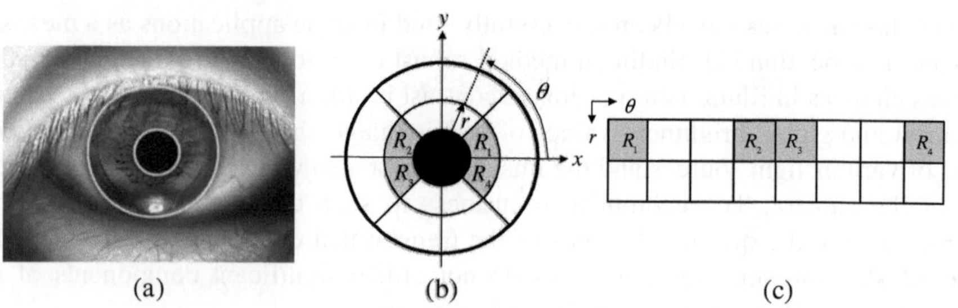

Fig. 1. Iris region detection and ROI extraction. (a) Detected inner and outer boundaries of an iris, (b) ROI in Cartesian coordinate system, and (c) ROI (R1, R2, R3, R4) in polar coordinate system.

3 Multimodal Directional Iris Feature Extraction

Irises include various (directional) patterns such as arching ligament, crypts, ridges, and a zigzag collarette, thus the information on how much components of a certain direction exist according to the image location can be exploited as a good feature. For this reason, the DFB that effectively and accurately decomposes an image into several subband images is suitable for extracting directional features of iris images. The proposed method attempts to accomplish the higher accuracy by extracting and combining the two different forms of directional features (complementary features) from the directional subband outputs decomposed by the DFB.

3.1 Directional Decomposition

In the proposed method, the ROI images R_1, R_2, R_3, and R_4 (See Fig. 1) are decomposed into 8 directional subband outputs separately using the 8-band DFB. Since the DFB partitions the two-dimensional spectrum of an image into wedge-shaped directional passband regions accurately and efficiently as shown in Fig. 2(a), each directional component or feature can be captured effectively in its subband image. The decomposed subband images have a downscaled rectangular shape whose width and height are different and this is due to the post sampling matrices used to remove frequency scrambling [5]. Fig. 2 shows an example of the ROI images and the directional subband images decomposed by the 8-band DFB.

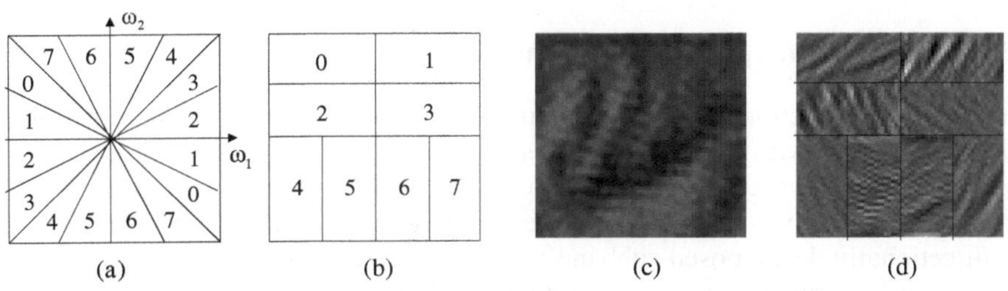

Fig. 2. Directional decomposition by the DFB. (a) Frequency partition map of the 8-band DFB, (b) positions of 8 subband outputs, (c) sample ROI image, and (d) decomposed outputs of (c).

3.2 Binary Directional Feature Extraction

Since the iris images are acquired by a digital camera under various internal and external illumination conditions, they have contrast and brightness differences in an image or between images. Therefore, robust features to such differences need to be extracted for reliable verification or identification. To extract the iris features that represent well the directional diversity of an iris pattern and have robustness to various brightness or contrast changes at the same time, the proposed method binarizes the directional subband outputs by making all the outputs with the positive value a binary 1, all other outputs a binary 0 [6]. Since each decomposed subband output value has an average value of almost 0, those values thresholded by 0 preserve the directional linear features and are robust to changes in illumination or brightness.

The method uses an additional low-pass filter to extract the iris features on multiple scales [8]. The extracted ROI is low-pass filtered and decomposed by an 8-band DFB. The resultant subband outputs are then thresholded at either 1 or 0 according to their signs and sampled at regular intervals. For the subband outputs of an image filtered by a low-pass filter with a cut-off frequency of π/n, sampling is performed every n pixels. The used method extracts the features from the two different scales, and the procedures for the feature extraction are illustrated in Fig. 3. The feature values are graphically displayed and enlarged to the original image scale to make the feature extraction procedure understandable.

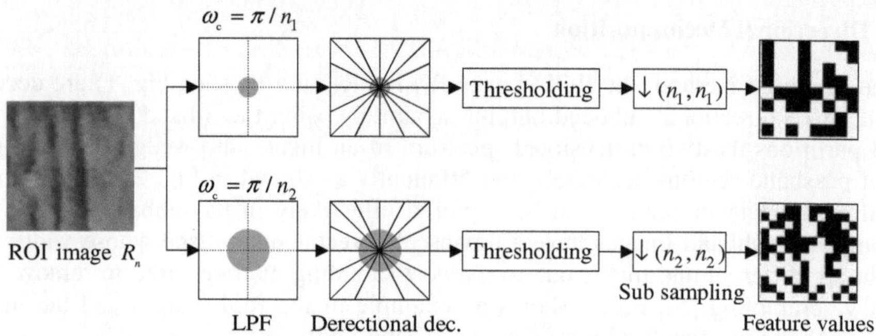

Fig. 3. Procedure for extracting the thresholded directional subband output feature

3.3 Directional Energy Feature Extraction

The binarized directional subband output features are robust to contrast or illumination change, however it dose not represent enough the rich information of iris patterns. Accordingly, the second method extracts another complementary feature from the directional subband outputs [7]. The intuitive feature that can be extracted from the directionally decomposed subband images is a directional energy. This directional energy can be a good feature in case that the illumination or contrast conditions are similar, but it changes severely according to illumination or contrast. Therefore, image normalization is necessary in order to use a directional energy as the iris feature, yet this is not easy in the iris image in which the brightness or contrast differences in an image or between images exist. To solve this problem, the proposed method first enhances the iris image using the method in [9] and employs the ratio of the directional energy in each block instead of the directional energy itself.

Let $e_{k\theta}^{(n)}$ denote the energy value of subband θ (which we call $S_{k\theta}^{(n)}$). More specifically, $S_{k\theta}^{(n)}$ corresponds to kth block $B_k^{(n)}$ of the nth ROI image R_n; $\hat{e}_{k\theta}^{(n)}$ is the normalized energy value of $e_{k\theta}^{(n)}$; and $c_{k\theta}^{(n)}(x, y)$ is the coefficient value at pixel (x, y) in subband $S_{k\theta}^{(n)}$. Now, $\forall\ n \in \{0, 1, 2, 3\}$, $k \in \{0, 1, 2, \ldots, 35\}$, and $\theta \in \{0, 1, 2, \ldots, 7\}$, the feature value, $v_{k\theta}^{(n)}$, can be given as

$$v_{k\theta}^{(n)} = [v_{\max} \times \hat{e}_{k\theta}^{(n)}] \tag{1}$$

where

$$\hat{e}_{k\theta}^{(n)} = \frac{e_{k\theta}^{(n)}}{\sum_{\theta=0}^{7} e_{k\theta}^{(n)}} \tag{2}$$

$$e_{k\theta}^{(n)} = \sum_{x,y \in S_{k\theta}} \left| c_{k\theta}^{(n)}(x,y) - \bar{c}_{k\theta}^{(n)} \right| \tag{3}$$

$[x]$ is the function that returns the nearest integer to x, $\bar{c}_{k\theta}^{(n)}$ is the mean of pixel values of $c_{k\theta}^{(n)}(x, y)$ in the subband $S_{k\theta}^{(n)}$, and v_{max} is a positive integer normalization constant. In this method, the high frequency components are removed by the low pass filter to reduce the effect of noise, and then the normalized directional energy features are extracted from the low pass filtered image (see Fig. 4).

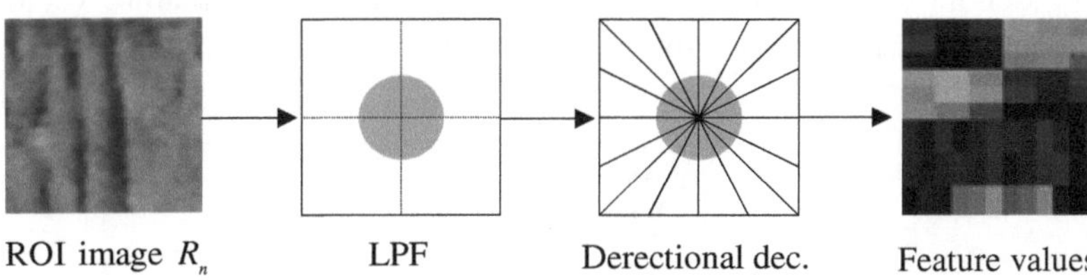

Fig. 4. Procedure for extracting the normalized directional energy feature

4 Matching

The two kinds of feature vectors are obtained for the single input image. One is the feature vector that consists of the binarized and sampled directional subband outputs on multiple scales, and the other is the feature vector of which elements are the blockwise normalized directional energy values. We call the former the binary feature vector, and the latter the energy feature vector for convenience' sake in this paper. In multibiometric approaches, the information presented by multiple traits can be fused at various levels such as feature extraction, matching score, and decision [10], but since the binary and energy feature vectors have different size and characteristics, combing the two feature vectors at the matching score level is one of the most effective and simplest ways. In the database, two kinds of feature vectors are also enrolled. Matching is performed between the input and template feature vectors extracted from the same feature extractor, and the final decision is made based on combining the matching scores from the two matchers.

To achieve the rotational alignment between the input and template feature vectors, the proposed method generates the additional feature vectors, in which various rotations are considered, by shifting the directional subband images and recalculating the feature values. Thereafter the method finds the minimum distances between the corresponding feature vectors for the rotational alignment [6, 7].

The matching between the binary feature vectors of the input and template iris images is based on finding the Hamming distance. Let ${V_{Bj}}^R$ denote the jth feature value of the input binary feature vector considering $R \times 45 \times (4/N)$ degree rotation and let T_{Bj} denote the jth feature value of the template binary feature vector, then the Hamming distance between input and template binary feature vectors, D_B, is given by

$$D_B = \min_R \frac{1}{N_B} \sum_{j=1}^{N_B} V_{Bj}^R \oplus T_{Bj} \tag{4}$$

where $R \in \{-10, -9, \ldots, -2, -1, 0, 1, 2, \ldots, 9, 10\}$, N_B is the size of the binary feature vector, and $\oplus$ is an exclusive-OR operator that yields one if ${V_{Bj}}^R$ is not equal to T_{Bj}, and zero otherwise.

The matching between the energy feature vectors of the input and template iris images is based on finding the Euclidean distance. Let ${V_{Ej}}^R$ denote the jth feature value of the input energy feature vector considering $R \times 45 \times (4/N)$ degree rotation and let T_{Ej} denote the jth feature value of the template energy feature vector, then the Euclidean distance between input and template energy feature vectors, D_E, is given by

$$D_E = \min_R \sqrt{\sum_{j=1}^{N_E} (V_{Ej}^R - T_{Ej})^2} \tag{5}$$

where N_E is the size of the energy feature vector. Once the two matching distances (D_B, D_E) are obtained, the final distance D_T is calculated using the following equation:

$$D_T = \alpha \cdot D_B + \beta \cdot D_E \tag{6}$$

where α and β are weighting factors and their sum is 1. These weighting parameters were determined considering the EER (equal error rate), a compact measure of accuracy for biometric systems, of each method. If the final distance is below a certain threshold the input iris is accepted, otherwise it is rejected.

5 Experimental Results

For the experiments, we acquired a total of 434 iris images from 10 persons using a digital movie camera and 50W halogen lamp. The iris images were captured from a distance about 15-20cm and the light was located below the camera so that the glint only appeared in the lower 90° cone of the iris. The acquired iris images were 256 grayscale images with the size of 640×480.

In order to estimate the performance as a personal verification, an EER, which is the error rate at which a FAR (false accept rate) is equal to a FRR (false reject rate), is calculated and the result was compared with that of the Gabor filter bank-based method [1]. Table 1 shows the EER for each method.

The performance of a verification system can also be evaluated using a receiver operator characteristic (ROC) curve, which graphically demonstrates how the genuine acceptance rate (GAR) changes with a variation in FAR. The ROC curve for the proposed method is shown in Fig. 5. We can see that the verification performance can be effectively improved by the combining the multiple matchers.

Table 1. Decidability index and equal error rate for each method

Features	Gabor	Binary	Energy	Binary & Energy
EER	4.25%	5.45%	3.80%	2.60%

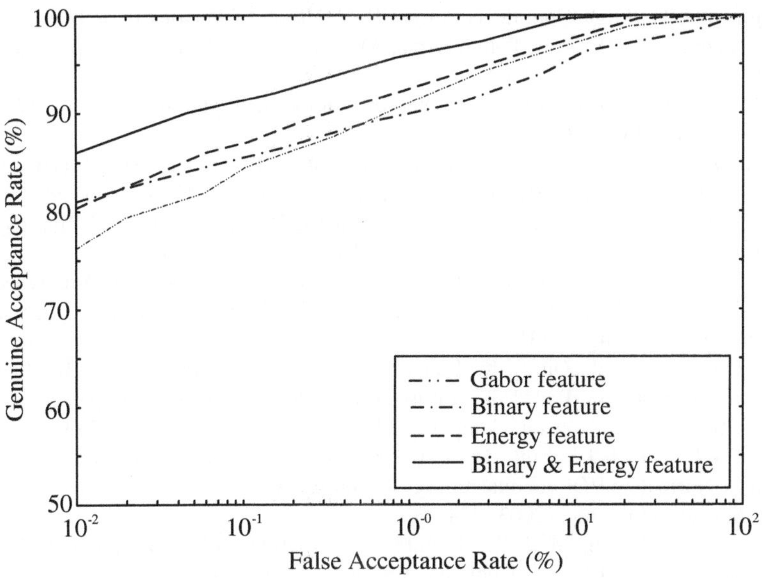

Fig. 5. ROC curve for the proposed method

6 Conclusion

We have presented an iris-based personal authentication method based on combining the multiple matchers. The proposed method represents the diverse directionality of the iris pattern into two forms using the same DFB: One is the binarized directional subband outputs at multiple scales, and the other is the blockwise normalized directional energy values. The former captures the multiscale and directional features that are robust to contrast or brightness differences between images, and the latter extracts another form of discriminatory iris features. Those two feature vectors are generated from the input iris image, and these feature vectors are compared with the enrolled template feature vectors, which consist of two sorts of feature vectors as in the input feature vectors. The final distance is obtained combing the matching distances from the two matchers. The experimental results show that the proposed multimodal approach based on combing the multiple matchers is effective in extracting robust and discriminatory iris features.

Acknowledgements

This work was supported by the IT postdoctoral fellowship program of the Ministry of Information and Communication (MIC), Republic of Korea.

References

1. Daugman, J. G.: High Confidence Visual Recognition of Persons by a Test of Statistical Independence. IEEE Trans. Pattern Anal. Machine Intell., Vol. 15, No. 11 (1993) 1148-1161
2. Wildes, R. P.: Iris Recognition: An Emerging Biometric Technology. Proc. IEEE, Vol. 85, No. 9 (1997) 1348-1363
3. Boles, W. W., Boashash, B.: A Human Identification Technique Using Images of the Iris and Wavelet Transform. IEEE Trans. Signal Processing, Vol. 46, No. 4 (1998) 1185-1998
4. Lim, S., Lee, K., Byeon, O., Kim, T.: Efficient Iris Recognition through Improvement of Feature Vector and Classifier. ETRI Journal, Vol. 23, No. 2 (2001) 61-70
5. Park, S., Smith, M. J. T., Mersereau, R. M.: Improved Structures of Maximally Decimated Directional Filter Banks for Spatial Image Analysis. IEEE Trans. Image Processing, Vol. 13, No. 11 (2004) 1424-1431
6. Park, C.-H., Lee, J.-J., Oh, S.-K., Song, Y.-C., Choi, D.-H., Park, K.-H.: Iris Feature Extraction and Matching Based on Multiscale and Directional Image Representation. Scale Space 2003, Lecture Notes in Computer Science, Vol. 2695 (2003) 576-583
7. Park, C.-H., Lee, J.-J., Smith, M. J. T., Park, K.-H.: Iris-Based Personal Authentication Using a Normalized Directional Energy Feature. AVBPA 2003, Lecture Notes in Computer Science, Vol. 2688 (2003) 224-232
8. Rosiles, J. G., Smith, M. J. T.: Texture Classification with a Biorthogonal Directional FilteBank. Proc. IEEE Intl. Conf. on Acoustics, Speech, and Signal Processing, Vol. 3 (2001) 1549-1552
9. Ma, L., Tan, T., Wang, Y., Zhang, D.: Personal Identification Based on Iris Texture Analysis. IEEE Trans. Pattern Anal. Machine Intell., Vol. 25, No. 12 (2003) 1519-1533
10. Jain, A. K., Ross, A.: Multibiometric Systems. Communications of the ACM, Special Issue on Multimodal Interfaces, Vol. 47, No. 1 (2004) 34-40

Fake Iris Detection by Using Purkinje Image

Eui Chul Lee[1], Kang Ryoung Park[2], and Jaihie Kim[3]

[1] Dept. of Computer Science, Sangmyung University,
7 Hongji-dong, Jongro-Ku, Seoul, Republic of Korea,
Biometrics Engineering Research Center (BERC)
oryong@smu.ac.kr
[2] Division of Media Technology, Sangmyung University,
7 Hongji-dong, Jongro-Ku, Seoul, Republic of Korea,
Biometrics Engineering Research Center (BERC)
parkgr@smu.ac.kr
[3] Department of Electrical and Electronic Engineering, Yonsei University,
Biometrics Engineering Research Center (BERC), Seoul, Republic of Korea,
jhkim@yonsei.ac.kr

Abstract. Fake iris detection is to detect and defeat a fake (forgery) iris image input. To solve the problems of previous researches on fake iris detection, we propose the new method of detecting fake iris attack based on the Purkinje image. Especially, we calculated the theoretical positions and distances between the Purkinje images based on the human eye model and the performance of fake detection algorithm could be much enhanced by such information. Experimental results showed that the FAR (False Acceptance Rate for accepting fake iris as live one) was 0.33% and FRR(False Rejection Rate of rejecting live iris as fake one) was 0.33%.

1 Introduction

Counterfeit iris detection is to detect and defeat a fake (forgery) iris image. In previous research, Daugman proposed the method of using FFT (Fast Fourier Transform) in order to check the printed iris pattern [1][3][7]. That is, the method checks the high frequency spectral magnitude in the frequency domain, which can be shown distinctly and periodically from the print iris pattern because of the characteristics of the periodic dot printing. However, the high frequency magnitude cannot be detected in case that input counterfeit iris image is defocused and blurred purposely and the counterfeit iris may be accepted as live one in such case. The advanced method of counterfeit iris detection was introduced by iris camera manufacturer. They use the method of turning on/off illuminator and checking the specular reflection on a cornea. However, such method can be easily spoofed by using the printed iris image with cutting off the printed pupil region and seeing through by attacker's eye, which can make corneal specular reflection [6].

To overcome such problems, we propose the new method of detecting fake iris attack based on the Purkinje image by using collimated IR-LED (Infra-Red Light Emitting Diode). Especially, we calculated the theoretical positions and distances between the Purkinje images based on the human eye model and the performance of fake detection algorithm could be much enhanced by such information.

D. Zhang and A.K. Jain (Eds.): ICB 2006, LNCS 3832, pp. 397–403, 2005.

2 Proposed Method

2.1 The Overview of the Proposed Method

The overview of the proposed method is as following. At first, we capture an iris image and calculate the focus value of input image by the Daugman's method [7]. If the calculated focus value is bigger than the predefined threshold (as 50), we regard the input image as focused one and perform the iris recognition. However, the focus value is smaller than threshold, our system capture an iris image again until the well focused image is acquired enough to recognition. Then, if the user's identification is completed, our system turns on two 'collimated IR-LEDs', alternatively. The collimated IR-LED has the property of smaller illumination angle than the conventional IR-LED for iris recognition. One of the collimated IR-LEDs is used for measuring Z-distance between a camera and an eye, another is used for making the Purkinje image. Two 'collimated IR-LEDs' turns on, alternatively, synchronized with image frame and we can obtain two images. Then we capture a bright iris image with the 760nm + 880nm IR-LED and detect the regions of pupil and iris in image. At the next, we measure Z-distance between a camera lens and an eye. The measured Z-distance is used for calculating a theoretical distance between the Purkinje images. In detail, we define three 'the Purkinje image searching boxes' based on the measured Z-distance and the Purkinje image model (Fig. 1). The Purkinje image model is obtained by using the Gullstrand eye scheme [2]. Due to detecting Purkinje images in the searching boxes, we can reduce the processing time. Then we detect the 1st, 2nd and 4th Purkinje images in the searching boxes. From that, we check whether the 1st and 2nd Purkinje images exist in the searching box of the iris area (because of our system configuration of collimated IR-LED, the 1st and 2nd Purkinje images exist in iris region). If so, we also check whether the 4th Purkinje image exists in the searching box of the pupil area (because of our system configuration of collimated IR-LED, the 4th Purkinje image exists in pupil region). If so, we determine the input image as the live iris and accept the user. If not, we reject the input image as the fake iris.

2.2 The Proposed Iris Camera Structure

Our iris recognition camera (made by our Lab.) uses dual IR-LEDs for iris recognition and two collimated IR-LEDs. For camera, we use a conventional USB camera (Quickcam-Pro 4000 [9]) with CCD sensor (in which IR-cut filter is removed).

The wavelength of dual IR-LEDs for recognition is 760nm and 880nm. In this camera, we also use two collimated IR-LEDs. The illumination (divergence) angle of collimated IR-LED is about 2.9 degrees.

2.3 Detecting Purkinje Images

Conventional human eye has four optical surfaces, each of which reflects bright lights: the front and back surface of the cornea, and the front and back surface of the lens. In this case, the 4 reflected images of incident light on each optical surface are mentioned as Purkinje images. The positions of these four Purkinje reactions depend on the geometry of the light sources [5]. Fig.1 is the Purkinje image shaping model which is designed based on the Gullstrand eye model [2].

To overcome the vulnerable problems of the Daugman method using the Purkinje images [4], we consider the shaping model of the Purkinje image. Since this model is designed with the Gullstrand eye model, the theoretical distances between the Purkinje images can be obtained. Because such distances are determined by human eye model (such as refraction rate, diameter of cornea and lens, etc), the distances from live iris are different from those from fake one. So, it is difficult to make fake iris showing the Purkinje images having same distances to those of live eye, because the material characteristics (such as refraction rate, diameter of cornea and lens, etc) of fake iris is different to that of live iris. In Fig.1, we show the method of calculating the theoretical distances between the Purkinje images.

In Fig.1, the radius and the focal length of an each optical surface (anterior cornea, posterior cornea, anterior lens, posterior lens) are shown. C_{ac} is the center of an anterior cornea's curvature and the radius of anterior cornea is 7.7 mm. F_{ac} (= 3.85 mm) is a focal point (a half of radius). Similarly, C_{pc} is the center of a posterior cornea's curvature and the radius of anterior cornea is 6.8 mm. F_{pc} (= 3.4) is the focal point of posterior cornea's curvature. C_{pl} is the center of a posterior lens's curvature and the radius of posterior lens is -6.0 mm. F_{pl} (= -3.0 mm) is the focal point of posterior lens's curvature [2].

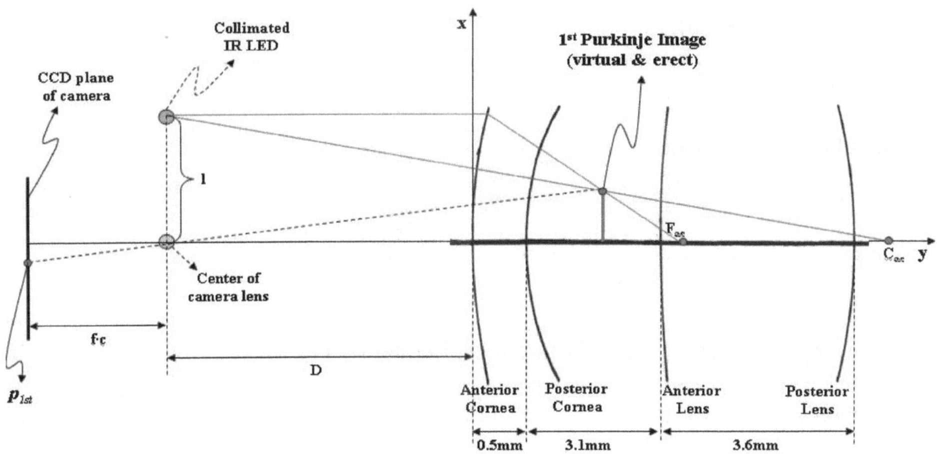

Fig. 1. The Purkinje image shaping model

Since the 1st, 2nd and 3rd Purkinje image are shaped by reflecting from a convex mirror, images are virtual and erect. But the 4th Purkinje image is real and inverted since it is shaped by reflecting from a concave mirror. Due to these facts, we can know the 1st and 2nd Purkinje images exist in symmetric position to 4th Purkinje image about the center of iris. Actually, there can be 3rd Purkinje image made from anterior lens. But, the 3rd Purkinje image is not seen in image. That is because the 3rd Purkinje image happens on the behind position of an iris from the camera. Generally, a diameter of pupil is reported to be 2mm~8mm [3] and its size is changed according to environmental light. The stronger the light is, the smaller the pupil size becomes. In our case, since we use collimated IR-LED and its light is entered into the pupil area, the pupil size becomes the smallest (2mm). So, the iris area is enlarged consequently and the 3rd Purkinje image is hidden by an iris area in captured eye image (cannot be seen).

Now, we introduce the method of calculating the distances between the 1st, 2nd and 4th Purkinje images, theoretically. As seen in Fig. 1, we can suppose the surfaces of anterior and posterior corneas as convex mirror models. In addition, we can do the surfaces of posterior lens as concave mirror model. So, we can use the camera lens model [8].

The 1st Purkinje image :

$$y_{1st} = \frac{D \cdot F_{ac}}{D - F_{ac}} \quad x_{1st} = \frac{l \cdot (7.7 - y_{1st})}{D + 7.7} \tag{1}$$

(because the radius of anterior cornea is 7.7, D is the distance between camera lens and anterior cornea surface. l is that between camera lens and collimated IR-LED as shown in Fig. 1)

The 2nd Purkinje image :

$$y_{2nd} = \frac{F_{pc} \cdot (D + 0.5)}{(D + 0.5) - F_{pc}} + 0.5 \qquad x_{2nd} = \frac{l \cdot (7.3 - y_{2nd})}{D + 7.3} \tag{2}$$

(because the depth of cornea is 0.5 and the radius of posterior cornea is 6.8. 7.3 = 6.8+ 0.5)

The 4th Purkinje image :

$$y_{4th} = 7.2 + \frac{F_{pl} \cdot (D + 7.2)}{(D + 7.2) - F_{pl}} \qquad x_{4th} = \frac{l \cdot (y_{4th} - 7.2)}{D + 7.2} \tag{3}$$

(because the distance between the anterior cornea and the posterior lens is 7.2)
(In all cases, l is 50mm as shown in Fig. 1)

According to the similarity of triangle and equation (1), an each Purkinje image's axis values on coordinate x, y is as follows. By using x_{1st}, x_{2nd}, x_{4th} and the perspective transformation, we can obtain the corresponding position of the 1st, 2nd and 4th Purkinje images in input image as shown in Eq. (4)(5)(6). Experimental results (from 100 test images) show that the x image positions of the 1st, 2nd and 4th Purkinje images are 42.9, 38.1 and -33.7 pixels, respectively. And the measured x range of iris region is -37.7 ~ +37.7 pixels in image. From that, we can know that the 1st and 2nd Purkinje image exist in the iris area, but the 4th Purkinje image does in the pupil area in a captured image. And we can obtain the distance between the 1st and 4th Purkinje image in Eq.(7). Also, we can measure the distance between the 1st and 2nd Purkinje image in Eq.(8).

1st Purkinje image in CCD plane : $$p_{1st} = \frac{fc \cdot X_{1st}}{D + Y_{1st}} \tag{4}$$

2nd Purkinje image in CCD plane : $$p_{2nd} = \frac{fc \cdot X_{2nd}}{D + Y_{2nd}} \tag{5}$$

4th Purkinje image in CCD plane : $$p_{4th} = \frac{fc \cdot X_{4th}}{D + Y_{4th}} \quad (7)$$

$$d_1 = p_{1st} - p_{4th}$$
$$d_2 = p_{1st} - p_{2nd} \quad (8)$$

2.4 Finding the Purkinje Image in the Searching Box

Based on Eq.(7) and (8), we can know the theoretical distance between the 1st and 4th Purkinje image, and that between the 1st and 2nd Purkinje image. So, we first detect the 1st Purkinje image in input image by the information of p_{1st} in Eq. (4). Then, we define the 2nd and 4th Purkinje image searching boxes (for the 2nd Purkinje image, the size of searching box is 20*20 pixels. For the 4th Purkinje image, the size of searching box is 37*37 pixels.) by the information of d_1 and d_2 in Eq. (7)(8) and detect the 2nd and 4th Purkinje images in the searching boxes. To detect the Purkinje images in the searching box, we perform the binarization (threshold of 190), component labeling and size filtering [8]. The sizes of Purkinje images are largest in each searching box. From that, we can detect the exact positions of the Purkinje images excluding the noise by eyebrows, etc.

3 Experimental Result

For experiments, the live iris samples were acquired from 30 persons (10 persons without glasses (no contact lens), 10 persons without glasses (contact lens) and 10 persons with glasses (no contact lens)). Each person tried to recognize 10 times and total 300 eye images were acquired to test our algorithm. In addition, we acquired total 15 counterfeit samples for testing. They were composed of 10 samples for 2D printed iris image on planar or on / with convex surface. Also, 2 samples were acquired for 3D artificial eye. And 3 samples were for 3D patterned contact lens. With each sample, we tried to 20 times to spoof our counterfeit iris detecting algorithm.

At first test, we measure the accuracy of our fake detection algorithm with FAR and FRR. Here, the FAR means the error rate of accepting the counterfeit iris as the live one. And the FRR does the error rate of rejecting the live iris as the counterfeit one. Experimental result shows that the FAR is about 0.33% (1/300) and FRR is 0.33% (1/300), but the FRR becomes 0% allowing for the second trial of fake checking. In this case, FRR does not happen in case of live iris with normal contact lens. In detail, about 2D printed iris image on planar or on / with convex surface, the FAR is 0% (0/200). About 3D artificial eye, the FAR is also 0% (0/40). However, the FAR is increased to 1.67% (1/60) about 3D patterned contact lens. In case of fake contact lens, the attacker uses his live pupil and 1 cases of FAR (that the 1st, 2nd and 4th Purkinje images happen like live iris) happen.

At second test, we measure the error rate according to Z distance between eye and camera. As shown in Table 1, the FAR and FRR are almost same irrespective of Z distance.

Table 1. Z Distance vs. the FAR and the FRR

Distance(mm)	80	100	120	140	160	180	200	220	240	250
FAR(%)	0.11	0.10	0.09	0.11	0.11	0.09	0.12	0.11	0.09	0.11
FRR(%)	0.34	0.31	0.33	0.30	0.32	0.33	0.33	0.34	0.30	0.34

At third test, we measure the accuracy according to the size of searching boxes for the 2nd and 4th Purkinje images. Experimental results show that when the size of searching box is increased, the FAR is increased and FRR is decreased, vice versa. From that, we can know that if we use the size of searching box for the 2nd and 4th Purkinje images as 20*20 and 37*37 pixels respectively, we can obtain the performance of minimum EER (FAR = FRR = 0.33%)).

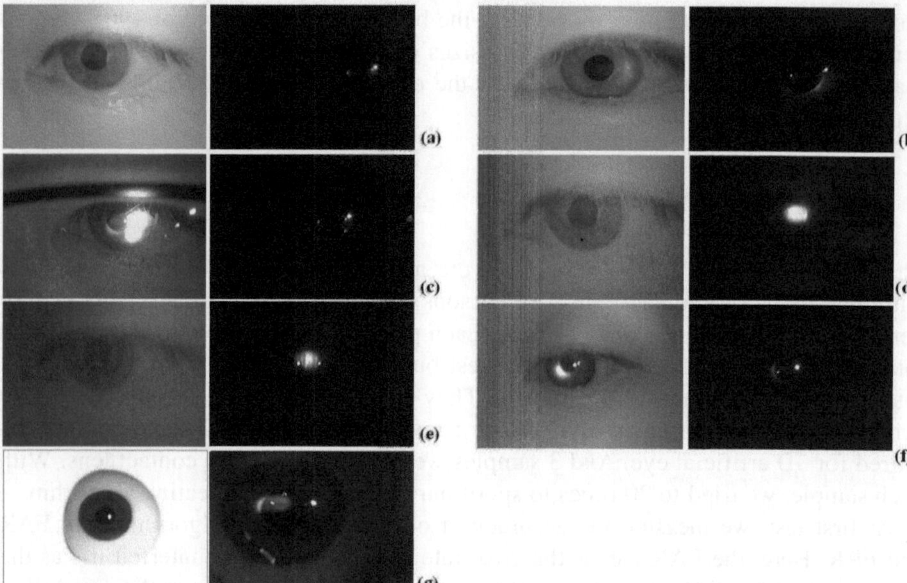

Fig. 2. The test examples of live and fake iris.(a) Live eye. (b) Live eye with a normal contact lens. (c) Live eye with glasses. (d) 2D printed eye. (e) 3D print eye with a contact lens. (f) Eye with 3D fake patterned lens. (g) 3D artificial eye. (The left of each part image is normal image and the right of each part is the Purkinje image).

The processing time of detecting Purkinje images is so small as 11ms in PC of Pentium-4 2GHz CPU. Fig. 2 is the test examples of live and fake eyes. As shown in Fig.2 (a), (b), (c), we can know that the 1st and 2nd Purkinje images exist in an iris area and 4th Purkinje image does in pupil area. In case of (c), though the specular reflection on glasses surface happen, such 3 Purkinje images still happen. As shown in Fig.2 (d), (e), (f), we can know that the fake eye shows the different characteristics from the live eye about Purkinje image. Especially, in case of (d) and (e), we can find that a big bright spot happens in the pupil region different from live iris. That is because the pupil area of such fake iris is not a hole and a big bright spot reflected on surface happens. In case of (f), we can't find the 2nd Purkinje image because the refraction factor of patterned lens is different from that of live iris. In case of (g), the 3D artificial eye shows also big bright spot. And though it shows the 1st, 2nd and 3rd Purkinje images, the distances between them are different from those of live iris.

4 Conclusions

For higher security level of iris recognition, the importance for detecting iris is much highlighted recently. In this paper, we propose the new method of detecting fake iris attack based on the Purkinje image. Experimental results show that the FRR and FAR are 0.33%, respectively. To enhance the performance of our algorithm, we should have more field tests and consider more countermeasures against various situations and counterfeit samples in future.

Acknowledgements

This work was supported by the Korea Science and Engineering Foundation (KOSEF) through the Biometrics Engineering Research Center (BERC) at Yonsei University.

References

[1] John G. Daugman, "High confidence visual recognition of personals by a test of statistical independence". IEEE Trans. PAMI, vol.15, no.11, pp.1148-1160, 1993
[2] Gullstrand A, "Helmholz's physiological optics", Optical Society of America, App.pp 350–358, 1924
[3] http://www.iris-recognition.org, accessed on 2005.6.1
[4] John Daugman, "Recognizing persons by their iris patterns", (http:// www.cse. msu.edu/~cse891/)
[5] Konrad P. Ko¨rding *, Christoph Kayser, Belinda Y. Betsch, Peter Ko¨nig, "Non-contact eye-tracking on cats" Journal of Neuroscience Methods, June 2001
[6] http://www.heise.de/ct/english/02/11/114/, accessed on 2005.6.1
[7] Daugman J, "How Iris Recognotion Works" IEEE Transactions on Circuit and Systems For Video Technology, Vol. 14, No. 1, January 2004
[8] Rafael C. Gonzalez, et al., "Digital Image Processing" Second Edition, Prentice Hall
[9] http://www.logitech.com, accessed on 2005.8.18

A Novel Method for Coarse Iris Classification

Li Yu[1], Kuanquan Wang[1], and David Zhang[2]

[1] Department of Computer Science and Technology,
Harbin Institute of Technology, Harbin, 150001, China
{lyu, wangkq}@hit.edu.cn
[2] Department of computing, The Hong Kong Polytechnic University,
Hung Hom, Kowloon, Hong Kong
csdzhang@comp.polyu.edu.hk

Abstract. This paper proposes a novel method for the automatic coarse classification of iris images using a box-counting method to estimate the fractal dimensions of the iris. First, the iris image is segmented into sixteen blocks, eight belonging to an upper group and eight to a lower group. We then calculate the fractal dimension value of these image blocks and take the mean value of the fractal dimension as the upper and the lower group fractal dimensions. Finally all the iris images are classified into four categories in accordance with the upper and the lower group fractal dimensions. This classification method has been tested and evaluated on 872 iris cases and the accuracy is 94.61%. When we allow for the border effect, the double threshold algorithm is 98.28% accurate.

1 Introduction

Biometrics is one of the most important and reliable methods for computer aided personal identification. The fingerprint is the most widely used biometric feature, but the most reliable feature is the iris and it is this that accounts for its use in identity management in government departments requiring high security.

The iris contains abundant textural information which is often extracted in current recognition methods. Daugman's method, based on phase analysis, encodes the iris texture pattern into a 256-byte iris code by using some 2-dimensional Gabor filters, and taking the Hamming distance [1] to match the iris code. Wildes [2], matches images using Laplacian pyramid multi-resolution algorithms and a Fisher classifier. Boles et al, extract iris features using a one-dimensional wavelet transform [3], but this method has been tested only on a small database. Ma et al. construct a bank of spatial filters whose kernels are suitable for use in iris recognition [4]. They have also developed a preliminary Gaussian-Hermite moments-based method which uses local intensity variations of the iris [5]. They recently proposed an improved method based on characterizing key local variations [6].

Although these methods all obtain good recognition results, all iris authentication methods require the input iris image to be matched against a large number of iris images in a database. This is very time consuming, especially as the iris databases being used in identity recognition growing ever larger. To reduce both the search time and computational complexity, it would be desirable to be able to classify an iris

D. Zhang and A.K. Jain (Eds.): ICB 2006, LNCS 3832, pp. 404 – 410, 2005.

image before matching, so that the input iris is matched only with the irises in its corresponding category, but as yet the subject of iris classification has received little attention in the literature.

This paper is intended to contribute to the establishment of meaningful quantitative indexes. One such index can be established by using box-counting analysis to estimate the fractal dimensions of iris images with or without self-similarity. This allows us to classify the iris image into four categories according to their texture and structure.

2 Counting Boxes to Estimate the Fractal Dimension of the Iris

The concept of the fractal was first introduced by Mandelbrot [7], who used it as an indicator of surface roughness. The fractal dimension has been used in image classification to measure surface roughness where different natural scenes such as mountains, clouds, trees, and deserts generate different fractal dimensions. Of the wide variety of methods for estimating the fractal dimension that have so far been proposed, the box-counting method is one of the more used widely [8], as it can be computed automatically and can be applied to patterns with or without self-similarity.

In the box-counting method, an image measuring size $R \times R$ pixels is scaled down to $s \times s$, where $1 < s \leq R/2$, and s is an integer. Then, $r = s / R$. The image is treated as a 3D space, where two dimensions define the coordinates (x, y) of the pixels and the third coordinate (z) defines their grayscale values. The (x, y) is partitioned into grids measuring $s \times s$. On each grid there is a column of boxes measuring $s \times s \times s$. If the minimum and the maximum grayscale levels in the $(i, j)^{th}$ grid fall into, respectively, the kth and l^{th} boxes, the contribution of n_r in the $(i, j)^{th}$ grid is defined as:

$$n_r(i, j) = l - k + 1 \tag{1}$$

In this method N_r is defined as the summation of the contributions from all the grids that are located in a window of the image:

$$N_r = \sum_{i,j} n_r(i, j) \tag{2}$$

If N_r is computed for different values of r, then the fractal dimension can be estimated as the slope of the line that best fits the points $(\log(1/r), \log N_r)$.

The complete series of steps for calculating the fractal dimension are follows. First, the image is divided into regular meshes with a mesh size of r. We then count the number of square boxes that intersect with the image N_r. The number N_r is dependent on the choice of r. We next select several size values and count the corresponding number N_r. Following this, we plot the slope D formed by plotting $\log(N_r)$ against $\log(1/r)$. This indicates the degree of complexity, or the dimensions of the fractal. Finally, a straight line is fitted to the plotted points in the diagram using the least square method. In accordance with Mandelbrot's view, the linear regression equation used to estimate the fractal dimension is

$$\log(N_r) = \log(K) + D\log(1/r) \tag{3}$$

where K is a constant and D denotes the dimensions of the fractal set.

3 Iris Classification

3.1 The Calculation of the Fractal Dimension

The calculation of the fractal dimension begins with preprocessing the original image to localize and normalize the iris. In our experiments, the preprocessed images were transformed into images measuring 256×64.

Because all iris images have a similar texture near the pupil, we do not use the upper part of the iris image when classifying an iris. Rather we make use only of the middle and lower part of the iris image. Preliminarily, we use the box-counting method to calculate the fractal dimension. To do this, we first divide a preprocessed iris image into sixteen regions. Eight regions are then drawn from the middle part of the iris image, as shown in Fig. 1. We call these the upper group. The remaining eight regions are drawn from the bottom part of iris image. These are referred to as the lower group. From these sixteen regions we obtain sixteen 32×32 image blocks. We then use the box-counting method to calculate the fractal dimensions of these image blocks. This produces sixteen fractal dimensions, FD_i (i=1,2...16). The mean values of the fractal dimensions of the two groups are taken as the upper and lower group fractal dimensions, respectively.

$$FD_{upper} = \frac{\sum_{i=1}^{8} FD_i}{8}, \quad FD_{lower} = \frac{\sum_{i=9}^{16} FD_i}{8} \tag{4}$$

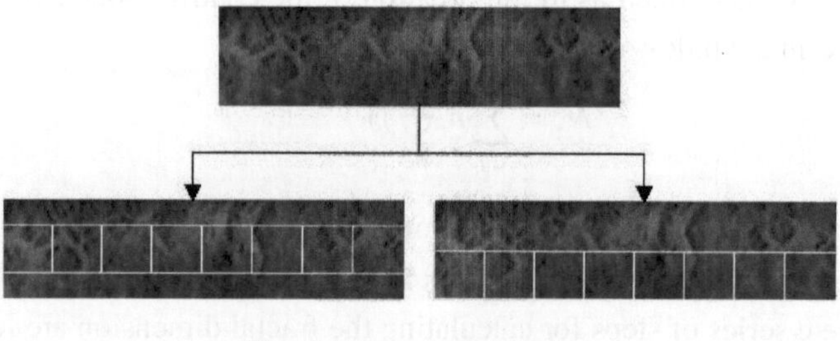

Fig. 1. Image segmentation

3.2 Classifying an Iris Using the Double Threshold Algorithm

The double threshold algorithm uses two thresholds to classify the iris into the following four categories, according to the values of the upper and lower group fractal dimensions.

Category 1 (net structure): The iris image appears loose and fibrous. The fibers are open and coarse, and there are large gaps in the tissue. The values of both the upper and lower group fractal dimensions are less than the first threshold E_I .

$$\{(FD_{upper}, FD_{lower}) \mid FD_{upper} < E_I \textbf{ AND } FD_{lower} < E_I\} \tag{5}$$

Category 2 (silky structure): The iris image appears silky. It displays few fibers and little surface topography. The Autonomic Nerve Wreath (also known as the Ruff and Collarette) is usually located less than one-third the distance from the pupil to the iris border. The values of the upper and lower group fractal dimensions are more than the second threshold E_{II} .

$$\{(FD_{upper}, FD_{lower}) \mid FD_{upper} < E_I \textbf{ AND } FD_{lower} < E_I\} \tag{6}$$

Category 3 (linen structure): The iris image appears to have a texture between those of Category 1 and Category 2. The Autonomic Nerve Wreath usually appears one-third to halfway between the pupil and the iris border, and the surface of ciliary zone is flat. (The Autonomic Nerve Wreath divides the iris into two zones, an inner pupillary zone, and an outer ciliary zone.) The value of lower group fractal dimension is more than the second threshold E_{II} and the value of upper group fractal dimension is less than the second threshold E_{II} .

$$\{(FD_{upper}, FD_{lower}) \mid FD_{upper} < E_I \textbf{ AND } FD_{lower} < E_I\} \tag{7}$$

Category 4 (hessian structure): The iris image appears to have a similar texture to Category 3 but with a few gaps (Lacunae) in the ciliary zone. When the upper and lower group fractal dimension values of an iris fail to satisfy the rules of Categories 1, 2, or 3, they are classified into Category 4.

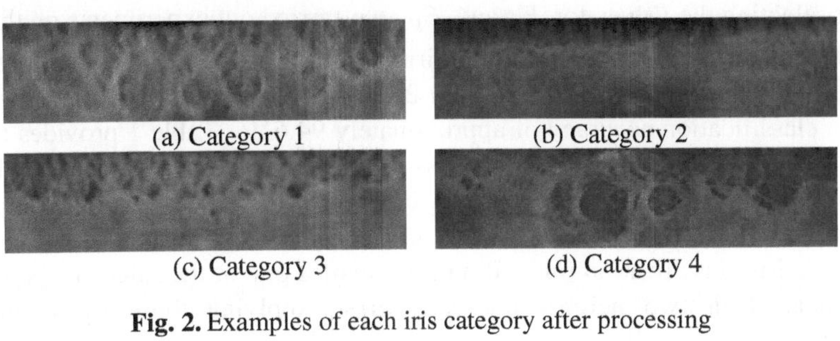

(a) Category 1 (b) Category 2

(c) Category 3 (d) Category 4

Fig. 2. Examples of each iris category after processing

Fig. 2 shows the range of possible textures. Categories 3 and 4 are both in a range between Categories 1 and 2. Category 3 is more like Category 2 and Category 4 is more like Category 1.Because the value of a fractal dimension is continuous, when classifying we must take into account the border effect. For the value near the threshold, we cannot simply classify the iris image into one category. Therefore, the nearby categories should be considered at the same time. The complementary rules for classifying the image are as follows:

Rule 1. If $\{(FD_{upper}, FD_{lower}) \mid FD_{upper} \le E_I$ AND $FD_{lower} \le E_I + \Delta E)\}$ or $\{(FD_{upper}, FD_{lower}) \mid (E_I - \Delta E \le FD_{upper} \le E_I + \Delta E)$ AND $FD_{lower} \le E_I\}$, the image belongs to Category 1 or Category 4, so Category 1 and Category 4 should be matched. Here ΔE is a small value.

Rule 2: If $\{(FD_{upper}, FD_{lower}) \mid (E_{II} - \Delta E \le FD_{upper} \le E_{II} + \Delta E)$ AND $E_{II} \le FD_{lower}\}$ or $\{(FD_{upper}, FD_{lower}) \mid E_{II} \le FD_{upper}$ AND $(E_{II} - \Delta E \le FD_{lower} \le E_{II} + \Delta E)\}$, the image belongs to Category 2 or Category 3, so Category 2 and Category 3 should be matched.

Rule 3: If $\{(FD_{upper}, FD_{lower}) \mid FD_{upper} < E_{II} - \Delta E$ AND $(E_{II} - \Delta E < FD_{lower} < E_{II} + \Delta E)\}$ the image belongs to Category 3 or Category 4, so Category 3 and Category 4 should be matched.

4 Experimental Results

Extensive experiments on a large image database were carried out to evaluate the effectiveness and accuracy of the proposed methods.

An iris image is correctly classified when the label of its category is the same as that of the iris. When there is no such match, the iris has been misclassified. The following subsections detail the experiments and their results.

Our iris classification algorithm was tested on a database containing 872 iris images captured from 436 different eyes. There are two images of each eye. The images measure 758×568 with eight bits per pixel and the irises have been labeled manually. Of the 872 irises in the database, 48 samples belong in Category 1, 336 belong in Category 2, 190 belong in Category 3 and 298 belong in Category 4.

After selecting the values for E_I and E_{II}, we carried out experiments on these two thresholds to classify the iris. Of the 872 irises in the database, 47 samples were misclassified: 6 in Category 1, 5 in Category 2, 20 in Category 3 and 16 in Category 4. This is a classification accuracy of approximately 94.61%. Table 1 provides the confusion matrix. It shows that many misclassified irises are to be found in neighboring categories.

To reduce the influence of the border effect on classification accuracy, we have added three iris classification rules. If an iris satisfies one of the rules, it is simultaneously matched in two neighboring categories. Applying these rules, and with $\Delta E = 0.0050$, the classification was 98.28% accurate. Clearly, this is a great improvement on the method which did not take into account the border effect.

Using coarse iris classification can reduce the time in searching. Table 2 shows the search time with and without coarse iris classification. As shown in Table 2, the search time of our iris recognition system can be reduced almost 70% of the original search time by using coarse iris classification, if taking into account of the border effect, the search time is less than half of the original search time.

Table 1. Classification results of the double threshold algorithm

Assigned Category No	True Category No. 1	2	3	4
1	**48**	0	0	6
2	0	**321**	5	0
3	0	9	**175**	11
4	0	6	10	**281**

Table 2. The search time of the system with and without coarse classification

without coarse classification(ms)	Using coarse classification	
	Without border effect (ms)	Consider border effect (ms)
81	25	**32**

According to this rule, we can evaluate the size of the database to decide when to use the coarse classification method. Suppose N is the database size, T is the original search time without coarse classification and T_c is the search time with coarse classification. The time used for iris coarse classification is T_f. If the computational cost for coarse classification is less than the reduced matching time, then this identification system can use coarse iris classification. That is:

$$T_f < T - T_c \tag{8}$$

As presented before, the reduced search time is about half of the original search time, So $T - T_c = T/2$, and $T = N*t$. Here t is the match time (from a pair of feature vectors to the match result). Therefore, we obtained:

$$T_f < N*t/2 \tag{9}$$

In our method, T_f and t are about 98 ms and 0.2 ms, so $N > 980$. It shows that when the database size N becomes bigger than 980, the coarse classification can reduce the computational time of the identification system.

5 Conclusion

Among the biometrics approaches, iris recognition is known for its high reliability, but as databases grow ever larger, an approach needed that can reduce matching times. Iris classification can contribute to that. As the first attempt to classify iris images, this paper presents a novel iris classification algorithm based on the box-counting method of fractal dimension. The approach uses the fractal dimension of the iris image to classify the iris image into four categories according to texture. The classification method has been tested and evaluated on 872 iris cases. After taking the border effect into account, the best result was obtained using the double threshold algorithm, which was 98.28% accurate.

In the future, we will modify the image preprocessing method to reduce the influence of light and eyelids. There is also much work to be done on the selection of classification methods. We will also try other approaches to the improvement of classification accuracy.

Acknowledgment

This work is partially supported by PhD program foundation of the Ministry of Education of China, (20040213017), the central fund from The Foundation of the H.L.J Province for Scholars Return from Abroad (LC04C17) and the NSFC fund (90209020).

References

1. J.G. Daugman.: High Confidential Visual Recognition by Test of Statistical Independence. In: IEEE Trans. PAMI, Nov. vol.15, No.11, (1993) 1148-1161.
2. R. P. Wildes.: Iris Recognition: an Emerging Biometric Technology. In: Proc. IEEE, Sep. vol.85, (1997) 1348-1363.
3. W. W. Boles and B. Boashash.: A Human Identification Technique Using Images of the Iris and Wavelet Transform. In: IEEE Trans. Signal Processing, Apr, vol.46, No.4, (1998) 1185-1188.
4. L. Ma, T. Tan, Y. Wang and D. Zhang.: Personal Identification Based on Iris Texture Analysis. In: IEEE Trans. PAMI, Dec, vol.25, No.12, (2003) 1519-1533.
5. L. Ma, T. Tan, Y. Wang and D. Zhang.: Local Intensity Variation Analysis for Iris Recognition. In: Pattern Recognition, vol.37, (2004) 1287-1298.
6. L. Ma, T. Tan, Y. Wang and D. Zhang.: Efficient Iris Recognition by Characterizing Key local Variations. In: IEEE Trans. Image Processing, Jun, vol.13, No.6, (2004).739-749
7. B. B. Mandelbrot and J. W. Van Ness.: Fractional Brownian motions, fractional noises and applications. In: SIAM Rev., vol.10, no.4, (1968) 422–437.
8. H.O. Peitgen, H. Jurgens and D. Saupe.: Chaos and Fractals New Frontiers of Science. Berlin, Germany: Springer-Verlag, (1992) 202–213.

Global Texture Analysis of Iris Images for Ethnic Classification

Xianchao Qiu, Zhenan Sun, and Tieniu Tan

Center for Biometrics and Security Research,
National Laboratory of Pattern Recognition, Institute of Automation,
Chinese Academy of Sciences, P.O. Box 2728, Beijing, P.R. China, 100080
{xcqiu, znsun, tnt}@nlpr.ia.ac.cn

Abstract. Iris pattern is commonly regarded as a kind of phenotypic feature without relation to the genes. In this paper, we propose a novel ethnic classification method based on the global texture information of iris images. So we would argue that iris texture is race related, and its genetic information is illustrated in coarse scale texture features, rather than preserved in the minute local features of state-of-the-art iris recognition algorithms. In our scheme, a bank of multichannel 2D Gabor filters is used to capture the global texture information and AdaBoost is used to learn a discriminant classification principle from the pool of the candidate feature set. Finally iris images are grouped into two race categories, Asian and non-Asian. Based on the proposed method, we get an encouraging correct classification rate (CCR) of 85.95% on a mixed database containing 3982 iris samples in our experiments.

1 Introduction

Iris texture is a distinct and stable biometric trait for personal identification. Some examples are shown in Fig. 1, which are from three different iris databases: CASIA[1] version 2, UPOL[2], and UBIRIS[3]. The iris of human eye is the annular part between the black pupil and the white sclera, in which texture is extremely rich.

Since Daugman's[4] iris recognition algorithm, many studies have been conducted on the randomness and uniqueness of human iris texture. Many people regard iris texture as phenotypic feature[4, 5, 6]. That is to say, the iris texture is the result of the developmental process and is not dictated by genetics. Even the genetically identical irises, the right and left pair from any given person have different textural appearance.

However, through investigating a large number of iris images of different races, Asian and non-Asian, we found that these iris patterns have different characteristics on the overall statistical measurement of the iris texture. At small scale, the details of iris texture are not dictated by genetics, but at large scale, the overall statistical measurement of iris texture is correlated with genetics. Motivated by this assumption, we try to do ethnic classification based on iris texture.

D. Zhang and A.K. Jain (Eds.): ICB 2006, LNCS 3832, pp. 411–418, 2005.

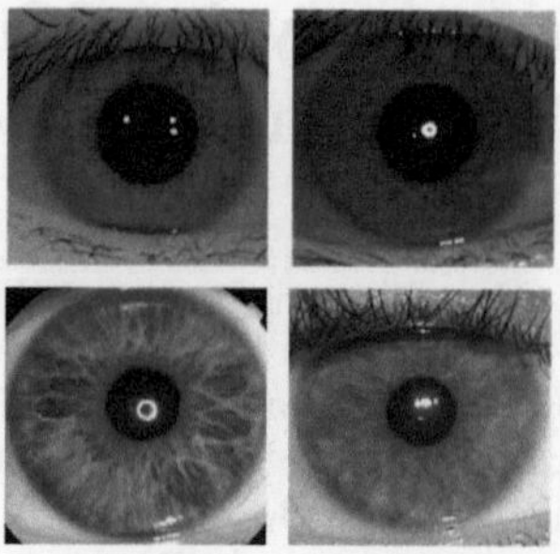

Fig. 1. Iris examples from different databases

So far, no work on ethnic classification with iris texture has been introduced in the public literature. In this paper, we propose a novel method for ethnic classification based on global texture analysis of iris images. Because the main purpose of this paper is to find the relationship between iris texture and race, only gray iris images are adopted in our experiments.

The remainder of this paper is organized as follows. Related work is presented in Section 2. The proposed method is discussed in Section 3. Experimental results are presented and discussed in Section 4 prior to conclusions in Section 5.

2 Related Work

Ethnic classification is an old topic in social science. It is often assumed to be a fixed trait based on ancestry. But in natural science, few attempts have been made to perform automatic ethnic classification based on images of human. One example is Gutta et al.[7] with hybrid RBF/decision-trees. Using a similar architecture with Quinlan's C4.5 algorithm, they were able to achieve an average accuracy rate of 92% for ethnic classification based on face images. Recently, Shakhnarovich, Viola and Moghaddam[8] used a variant of AdaBoost to classify face images as Asian and non-Asian. Their approach yields a classifier which attains accuracy rate of 78%. Lu and Jain[9] presented a Linear Discriminant Analysis (LDA) based scheme for two-class (Asian vs. non-Asian) ethnic classification from face images. Their reported accuracy is about 96%.

3 Global Texture Analysis

In this paper, an ethnic classification algorithm includes three basic modules: image preprocessing, global feature extraction, and training. Fig. 2 shows how the proposed algorithm works. Detailed descriptions of these steps are as follows.

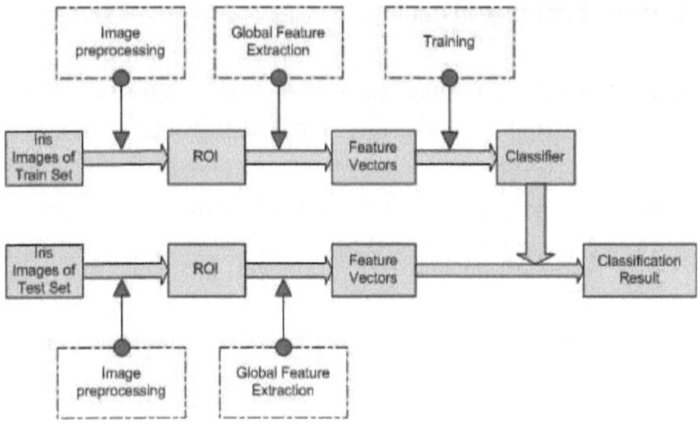

Fig. 2. The flowchart of our approach

3.1 Image Preprocessing

A typical iris recognition system must include image preprocessing. Fig. 3 illustrates the preprocessing step involving localization, normalization and enhancement. More details can be found in our previous work[6]. To exclude the eyelids and eyelashes, only the inner 3/4 of the lower half of an iris region is used as the region of interest (ROI) for feature extraction, as shown in Fig. 3 (c). In our experiment, the size of ROI is 60×256 and it is divided into two equal regions, region A and region B, as shown in Fig. 3 (d).

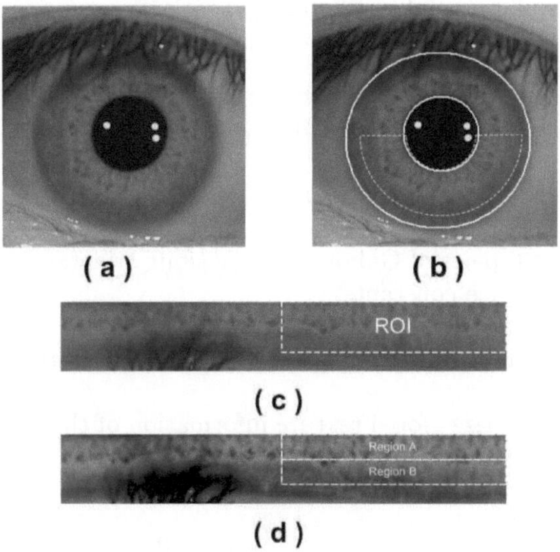

Fig. 3. Image preprocessing. (a) Original image. (b) Iris Localization. (c) Normalized image. (d) Normalized image after enhancement.

3.2 Global Feature Extraction

Once ROI has been created, we can proceed with feature extraction based on multichannel Gabor filtering[10, 11]. Gabor Energy[12] of each image point is used to represent texture features.

An input image (ROI) $I(x, y)$, $(x, y) \in \Omega$ (Ω denotes the set of image points), is convolved with a 2D Gabor filter to obtain a Gabor filtered image $r(x, y)$.

$$r(x, y) = \iint I(x_1, y_1) h_i(x - x_1, y - y_1) dx_1 dy_1 \; ; \; i = e, o. \tag{1}$$

where h_e and h_0 denote the even- and odd- symmetric Gabor filter.

The outputs of the even- and odd- symmetric Gabor filter in each image point can be combined into a single quantity called the Gabor energy[12]. This feature is defined as follows:

$$e_{f,\theta,\sigma}(x, y) = \sqrt{r^2_{even_{f,\theta,\sigma}}(x, y) + r^2_{odd_{f,\theta,\sigma}}(x, y)} \tag{2}$$

where $r_{even_{f,\theta,\sigma}}(x, y)$ and $r_{odd_{f,\theta,\sigma}}(x, y)$ are the responses of even- and odd- symmetric Gabor filter respectively.

For Asians, region A has rich texture, but region B often has less texture. However for non-Asians, region A and region B nearly have the same rich texture. Thus, high-pass filtering could extract the discrimination between different races.

We design a bank of Gabor filters to extract Gabor energy features. Since the Gabor filters we use are of central symmetry in the frequency domain, only half of the frequency plan is need. Four values of orientation θ are used: 0, $\frac{\pi}{4}$, $\frac{\pi}{2}$, and $\frac{3\pi}{4}$. Because we are interested in the higher spatial frequencies in the frequency domain, for each orientation, we choose six spatial frequencies and ten space constants as follows:

$$f = 0.25 + 2^{(i-0.5)}/256, i = 1, 2, \ldots, 6. \tag{3}$$

$$\sigma = 3 + i * 0.25, i = 0, 1, \ldots, 9. \tag{4}$$

It gives a total of 240 pairs of Gabor channels (four orientations, six frequencies combined with ten space constants).

For each pair of Gabor filters, we can get the Gabor energy image by Eqn. 2. Then the average Gabor energy values of region A and region B, m_A and m_B, are calculated.

In order to characterize global texture information of the ROI, two statistical features of the Gabor energy image, Gabor Energy (GE) and Gabor Energy Ratio (GER), are extracted.

$$GE = m_B, \quad GER = \frac{m_A}{m_B}. \tag{5}$$

These features are combined to form the pool of candidate classifiers.

3.3 Training

Many features have been extracted for each iris image, but our final application requires a very aggressive process which would discard the vast majority of features. For the sake of automatic feature selection, the AdaBoost algorithm[8] is used in our experiment to train the classifier.

4 Experimental Results

4.1 Image Database

Three iris databases are used in our experiments to evaluate the performance of the proposed method. They are CASIA[1], UPOL[2] and UBIRIS[3] iris image databases.

Because the iris images of the CASIA database are all from Asians in this version and the images of the UPOL and UBIRIS databases are mainly from Europeans, we divide all images into two categories, the Asian and the non-Asian. The Asian set includes 2400 images (all images from CASIA database), and the non-Asian set includes 1582 images (384 images from UPOL database and 1198 images from session-1 of UBIRIS database except 16 images without iris). All images from UPOL database and UBIRIS database are converted into 8 bit depth gray images as those in CASIA database. Then the images are separated into two sets: a training set of 1200 images (600 images randomly selected from the Asian and 600 images randomly selected from the non-Asian) and a testing set of 2782 images (the remaining images).

4.2 Performance Evaluation of the Proposed Algorithm

Statistical test is carried out to measure the accuracy of the algorithm. Correct Classification Rate (CCR) of the algorithm is examined.

Fig. 4 shows the distribution of Gabor Energy on the training set. The parameters of the Gabor filters used in this test were carefully selected to get the best peformance. When $f = 0.338$, $\theta = \frac{\pi}{4}$, $\sigma = 4$ and the threshold is set to 600, the value of CCR is 77.92%.

Fig. 5 shows the distribution of Gabor Energy Ratio on the training set. The parameters of the Gabor filters used in this test were $f = 0.427$, $\theta = \frac{\pi}{4}$, $\sigma = 4.5$ and the threshold is set to 0.93. The value of CCR is 83.75%.

For the sake of automatic feature selection, the AdaBoost algorithm was used in our experiment to learn a classification function. Given different feature sets, we get different results of classification, as shown in Table 1.

From Table 1, we can draw a conclusion that Gabor Energy Ratio is better than Gabor Energy in representing texture features. But the highest Correct Classification Rate (CCR) is achieved when both Gabor Energy and Gabor Energy Ratio are used.

As mentioned before,Shakhnarovich et al. get a Correct Classification Rate (CCR) of 79.2% with 3500 images of human faces collected from the World Wide

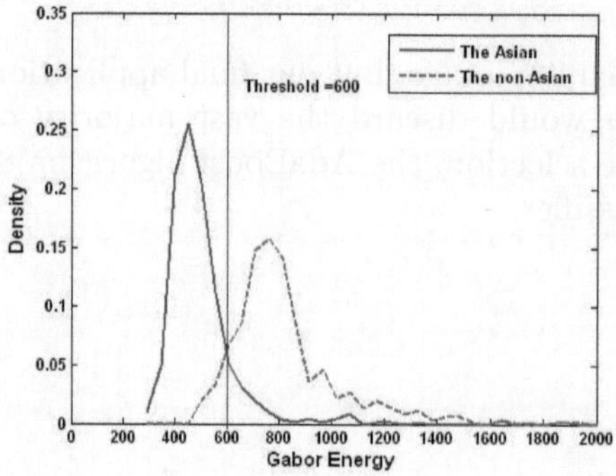

Fig. 4. Distribution of GE ($f = 0.338$, $\theta = \frac{\pi}{4}$, $\sigma = 4$)

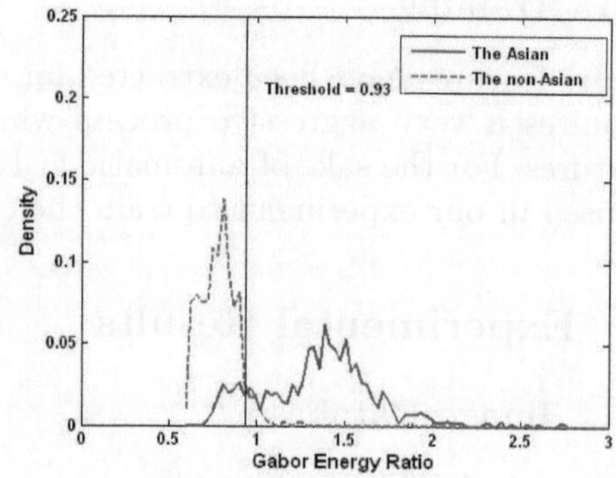

Fig. 5. Distribution of GER ($f = 0.427$, $\theta = \frac{\pi}{4}$, $\sigma = 4.5$)

Web. From the ethnic classification point of view, our method gets higher CCR of 85.95% than theirs.

Most of classification errors in our experiments are caused by three factors. Firstly, UBIRIS is a noisy iris image database and it includes many defocused images, which lacked of higher spatial frequencies. Secondly, the occlusions of eyelids and eyelashes in ROI may affect the classification result. Thirdly, there are some outliers in both classes. For example, an iris image from the Asian (CASIA database) may have very high Gabor energy in region B, while an iris image from the non-Asian (UPOL and UBIRIS database) may have very low Gabor energy in region B.

Images used in our experiments are acquired in different illumination. The UPOL database and UBIRIS database were acquired using visible light(VL) illumination, and CASIA database acquired in near infrared (NIR) illumination. In order to measure the influence of illumination conditions on the classification result, we conduct another experiment on a relatively small database. This database contains 480 iris images, 240 images are randomly selected from the CASIA database, and the other 240 images of 12 subjects were acquired using the same cameras but in visible light(VL) illumination, and all 480 images were from the Asian. All images are taken as the illumination testing set, it was divided into two classes, the VL and the NIR. Then three classifiers we had trained before were used for classification.

Table 1. Correct Classification Rate(CCR) resulted from the proposed method

Feature Type	Number of features	Number of selected features	Correct Classification Rate(%)		
			Training Set	Testing Set	Overall
GE	240	4	80.36	78.52	79.44
GER	240	6	84.17	85.73	84.95
GE&GER	480	6	85.42	86.48	85.95

Table 2. Correct Classification Rate on the illumination testing set

Feature Type	Number of features	Number of selected features	CCR on Illumination Testing Set (%)
GE	240	4	57.50
GER	240	6	53.62
GE&GER	480	6	49.17

As Table 2 shows, the classification result is only a little better than random guess as there are only two classes. The result demonstrates that the classifiers we had trained before were not tuned to classify the iris images in different illumination. The difference between iris images from different races is due to the inherent characteristics of iris texture.

5 Conclusion

In this paper, we have presented a novel method for automatic ethnic classification based on global texture analysis of iris images. A bank of multichannel 2D Gabor filters is used to capture the global texture information in some iris regions. An AdaBoost learning algorithm is used to select the features and train the classifier. Using the proposed method, we get an encouraging correct classification rate (CCR) of 85.95% in our experiments. Based on the analytical and experimental investigations presented in this paper, the following conclusion may be drawn: 1)At a small scale, the local features of the iris are unique to each subject, whereas at a large scale, the global features of the iris are similar for a specific race, and they seem to be dependent on the genes; 2)and the global texture features of iris are efficient for ethnic classification.

Acknowledgement

This work is funded by research grants from the National Basic Research Program (Grant No. 2004CB318110), the Natural Science Foundation of China (Grant No. 60335010, 60121302, 60275003, 60332010, 69825105) and the Chinese Academy of Sciences.

References

1. Chinese Academy of Sciences Institute of Automation. *CASIA iris image database, http://www.sinobiometrics.com.* 2003.
2. Michal Dobes and Libor Machala. *Upol iris image database, http://phoenix.inf.upol.cz/iris/.* 2004.
3. Hugo Proenca and Luis A. Alexandre. *Ubiris iris image database, http://iris.di.ubi.pt.* 2004.
4. John Daugman. High confidence visual recognition of persons by a test of statistical independence. *IEEE TRANS. PAMI*, 15(11):1148–1161, 1993.

5. R.P. Wildes. Iris recognition: An emerging biometric technology. *Proceedings of the IEEE*, 85(9):1348–1363, 1997.
6. Li Ma, Tieniu Tan, Yunhong Wang, and Dexin Zhang. Personal identification based on iris texture analysis. *IEEE TRANS. PAMI*, 25(12):1519–1533, 2003.
7. S. Gutta, H. Wechsler, and P. J. Phillips. Gender and ethnic classification. In *International Conference on Automatic Face and Gesture Reconition*, pages 194–199, 1998.
8. Gregory Shakhnarovich, Paul A. Viola, and Baback Moghaddam. A unified learning framework for real time face detection and classification. In *International Conference on Automatic Face and Gesture Reconition*, 2002.
9. Xiaoguang Lu and Anil K. Jain. Ethnicity identification from face images. In *Proc. SPIE Defense and Security Symposium*, April 2004.
10. Yong Zhu, Tieniu Tan, and Yunhong Wang. Font recognition based on global texture analysis. *IEEE TRANS. PAMI*, 23(10):1192–1200, 2001.
11. Tieniu Tan. Rotation invariant texture features and their use in automatic script indentification. *IEEE TRANS. PAMI*, 20(7):751–756, 1998.
12. Simona E. Grigorescu, Nicolai Petkov, and Peter Kruizinga. Comparison of texture features based on gabor filters. *IEEE Transactions on Image Processing*, 11(10):1160–1167, 2002.

Modeling Intra-class Variation for Nonideal Iris Recognition*

Xin Li

Lane Dept. of Computer Science and Electrical Engineering,
West Virginia University, Morgantown WV 26506-6109

Abstract. Intra-class variation is fundamental to the FNMR performance of iris recognition systems. In this paper, we perform a systematic study of modeling intra-class variation for nonideal iris images captured under less-controlled environments. We present global geometric calibration techniques for compensating distortion associated with off-angle acquisition and local geometric calibration techniques for compensating distortion due to inaccurate segmentation or pupil dilation. Geometric calibration facilitates both the localization and recognition of iris and more importantly, it offers a new approach of *trading* FNMR with FMR. We use experimental results to demonstrate the effectiveness of the proposed calibration techniques on both ideal and non-ideal iris databases.

1 Introduction

Inter-class and intra-class variations are at the heart of any pattern recognition problem. They jointly determine the receiver operating characteristics (ROC) performance measured by false matching rate (FMR) and false non-matching rate (FNMR). Inter-class variation is largely determined by the "randomness" of a pattern itself - for example, since the iris pattern appears to be more random than the fingerprint pattern, iris recognition can easily achieve an extremely low FMR [2], [6], [7], [8]. However, the con side of randomness is large intra-class variation and accordingly high FNMR.

For iris images, intra-class variation is caused by various uncertainty factors (e.g., eyelid/eyelash occlusion, pupil dilation/constriction, reflection of lights). Although it is possible to use quality control at the system level to alleviate the problem to some extent (e.g., in [6] an iris image is suggested to be rejected if the eye is overly blurred or occluded), such strategy is often bad for the ergonomics of biometric systems. Moreover, there is increasing evidence that less-controlled iris acquisition might be inevitable in practice. For instance, it is not always feasible to capture the iris images at the front angle and the level position due to varying height, head tilting and gaze direction. Such class of "nonideal iris images" raise new challenges to the existing iris recognition systems since none of them can handle geometric distortion caused by off-angle acquisition (refer to Fig. 1).

* This work was partially supported by NSF Center for Identification Technology Research.

D. Zhang and A.K. Jain (Eds.): ICB 2006, LNCS 3832, pp. 419–427, 2005.

In this paper, we present geometric calibration techniques for reducing intra-class variation. Given a pair of nonideal images, we first globally calibrate them by geometric transformations (rotation and scaling) to recover the circular shape of pupil. To the best of our knowledge, this is the first study on compensating geometric distortion of off-angle images in the open literature. After standard iris localization, unwrapping into polar coordinate and enhancement, we propose to locally calibrate enhanced images by constrained-form deformation techniques before matching. Local calibration is shown to dramatically reduce intra-class variation at the cost of slightly increased inter-class variation. Due to global and local calibration, we can even directly match two enhanced images without any spatial or frequency filtering (for feature extraction) and still obtain good recognition performance.

2 Nonideal Iris Acquisition

Due to the small physical size of human iris, its acquisition is not as easy as other biometrics such as face and fingerprint. Even under a controlled environment, the acquired images are seldom perfect - various uncertainty factors could give rise to severe intra-class variation, which makes the matching difficult. We structure those factors into two categories: sensor-related and subject-related.

A. Sensor-related
The first assumption we make is that the camera is sufficiently close to the subject such that iris region with enough spatial resolution is acquired. Empirical studies have shown that it is desirable to have the resolution of above 100dpi for iris recognition. In addition to camera distance, the angle of camera is the other dominating factor in the acquisition. When the camera is located at an off-angle position, nearly-circular structure of human pupil would become elliptic (refer to Fig. 1). Most existing iris recognition algorithms can not handle such nonideal (off-angle) images.

There are two different off-angle scenarios under our investigation. In the first case, the camera and the eyes are at the same height and the following scaling transformation relates the front-angle image to its off-angle counterpart:

$$\begin{bmatrix} x' \\ y' \end{bmatrix} = \begin{bmatrix} cos\theta & 0 \\ 0 & 1 \end{bmatrix} \begin{bmatrix} x \\ y \end{bmatrix}. \tag{1}$$

It simply compresses the horizontal direction - for instance, a circle in $f(x, y)$ becomes an ellipse in $f(x', y')$ whose long and short axes are parallel to vertical and horizontal directions. In the second case, the camera and the eyes are not in the same horizontal plane and the projection of iris onto imaging plane becomes slightly complicated. Instead of an ellipse at the straight position, we observe a rotated ellipse with the angle being determined by the tilting of the camera.

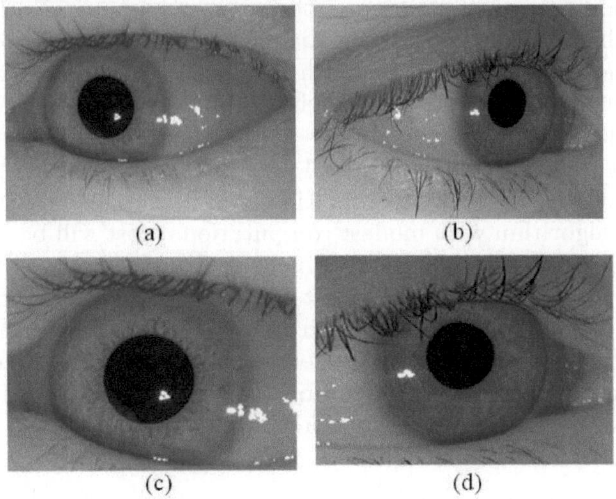

Fig. 1. Examples of nonideal iris images: a) off-angle but the same level; b) off-angle and different level; c) calibrated image of a); d) calibrated image of b).

In addition to geometric distortions, sensor also introduces photometric distortions such as out-of-focus, reflection and shading. We usually assume that iris images are acquired with good focus; but in practice manual adjustment of the focus is only possible when images are captured by well-trained personnel. Reflection of light source often gives rise to bright spots in iris images, which need to be treated as occlusions. Another potential reflection source is the contact lens, though such issue has been largely ignored in the literature of iris recognition so far. Shading could also affect the intensity values of iris images especially during off-angle acquisition, which often makes robust detection of limbus boundary more difficult.

B. Subject-related

The fundamental cause of subject-related uncertainty factors is *motion*. For iris recognition, three levels of motion could interfere with the acquisition: head movement, eye movement and pupil motion. Head movement can often be avoided by verbal commands; but even when the head remains still, its varying height and tilting position could give rise to different projections. Eye movement consists of eye open/close and saccadic eyeball movement. Both eyelid and eyelashes could render occlusions; gaze direction interacts with camera angle, which makes captured iris images seldom ideal except when the camera is extremely close to eye (e.g., CASIA database).

There are two kinds of pupillary motion: hippus and light reflex. Hippus refers to spasmodic, rhythmical dilation and constriction of the pupil that are independent of illumination, convergence, or psychic stimuli. The oscillation frequency of hippus is around 0.5Hz and its origin remains elusive. Light reflex refers to pupillary dilation and constriction in response to the change in the amount of light entering the eye. It is known that the diameter of human pupil can change as much as nine times ($1 - 9mm$). Such dramatic variation leads to complex elastic deformation of iridal tissues, which can only be partially handled by the existing normalization technique.

One might argue that quality control at the system level can solve all the problems caused by uncertainty factors. However, it is our opinion that a robust iris recognition algorithm with modest computational cost will be more effective than redoing the acquisition. Note that in the real world, it is nontrivial to take all those uncertainty factors into account and even more frustrating for human operators to figure out what is wrong with an innocent-looking image. Therefore, the main objective of this paper is to present geometric calibration techniques for improving the robustness of iris recognition algorithms (the issue of photometric distortion is outside the scope of this work).

3 Geometric Calibration

A. Global Calibration

Global calibration of nonideal iris images refers to the compensation of geometric distortion caused by off-angle cameras. The key motivation behind global calibration is to make the shape of pupil in an iris image as *circular* as possible. Although slightly non-circular pupils exist, they won't cause us any problem as long as we perform the calibration to both the enrolled and inquiry iris images. Therefore, we suggest that the pursuit of circular shape is an effective strategy for globally calibrating iris images even if both the enrolled and inquiry image are off-angle.

Detecting the pupil boundary in an off-angle image can use standard Least-Square (LS) based ellipse fitting techniques such as [3]. However, the accuracy of ellipse fitting degrades in the presence of outliers. Though it is often suggested that RANSAC can lead to improved robustness, we argue that it is more efficient to exploit our a priori knowledge about the outlier than the power of randomness. For example, outliers to ellipse detection in iris images are mainly attributed to light reflection and eyelashes. Light reflection often shows up as small round balls with high-intensity values, which can be masked during ellipse detection. Eyelashes have similar intensity values to pupil but highly different morphological attributes. Therefore, morphological filtering operations can effectively suppress the interference from eyelashes.

Ellipse fitting returns five parameters: the horizontal and vertical coordinates of pupil center (c_x, c_y), the length of long and short axes (r_x, r_y), and the orientation of the ellipse ϕ. Our global calibration consists of two steps: 1) rotate the image around (c_x, c_y) by $-\phi$ to restore the straight position of ellipse; 2) apply the inverse of scaling transformation defined by Eq. (1) to restore the circular shape of pupil. The parameter in scaling transformation is given by $cos\theta = \frac{r_x}{r_y}$ (assume r_x, r_y correspond to the short and long axes respectively). One tricky issue in the practical implementation is the periodicity of orientation parameter

ϕ. Since [3] does not put any constraint on the range of ϕ (e.g., ϕ and $\phi + \pi$ generates exactly the same ellipse), we need to further resolve the ambiguity among the set $\{\phi + \frac{k\pi}{2}, k \in Z\}$.

B. Local Calibration

After global calibration, we assume the compensated images are first unwrapped into polar coordinate based on the estimated parameters of inner(pupil) and outer(limbus) boundaries. Iris localization problem has been well studied in the literature (e.g., the coarse-to-fine integro-differential operator suggested by Daugman in [2]). The detection of non-iris structures (eyelid, eyelashes and reflections) has also been studied in [1] and [5]. However, two major challenges remain.

First, it has been experimentally found in [7] that excessive pupil dilation often gives rise to large intra-class variation. Unwrapping into polar coordinate partially alleviates the problem due to normalization along the radial axis; but it is cannot completely account for nonlinear elastic deformation of iridal tissues when dilation ratio is large. Second and more importantly, pupil dilation often interacts with erroneous estimate of inner and outer boundaries (due to poor contrast or eyelash occlusion), which gives rise to inaccurate alignment along the radial axis. We propose to compensate the remaining geometric distortions by local calibration techniques.

Our local calibration is decomposed of two steps. In the first step, enhanced image is structured into eight nonoverlapping blocks along the angular coordinate and block matching is applied to linearly compensate translational displacement (e.g., due to head tilting). In the second step, nonlinear elastic deformation is approximated by Horn's optical flow field (v_1, v_2) [4]. Specifically, Horn's method targets at the minimization of

$$E = E_{of}^2 + \alpha^2 E_s^2. \tag{2}$$

where E_{of} is the error of optical flow equation, $E_s^2 = ||\nabla v_1||^2 + ||\nabla v_2||^2$ measures the smoothness of optical flow field. By selecting a fairly large regularization parameter α (suggested value is 200), we enforce the optical flow model to only accommodate small and localized deformation. Fig. 2 shows an example of deformed sampling lattice after local calibration.

Although local calibration effectively reduces intra-class variation, its impact on inter-class variation can not be ignored. If iris patterns were truly random, our calibration should have no effect because of the constraints enforced above. Neither linear shifting nor regularized optical flow can deform a random pattern into another. However, in practice iris patterns are still characterized by notable structures such as flower, jewelry, shake and stream. Therefore, the impact of local calibration on inter-class variation is structure-dependent. For structures with less discriminating capability (e.g., stream), its optimal recognition performance is fundamentally worse than other's (e.g., flower). As we will see next, the proposed local calibration technique is also often more effective on high-texture iris images than low-texture ones.

4 Experimental Results

We have incorporated the proposed calibration techniques into the well-known Daugman's algorithm as shown in Fig. 3. In our current implementation, we have search for the largest bounding boxes for upper and lower eyelid respectively based on an approximate estimate of locations. Fig. 2b) shows several examples of different occlusion scenarios. In this section, we report our experimental results with both ideal (front-angle) and non-ideal (off-angle) iris databases.

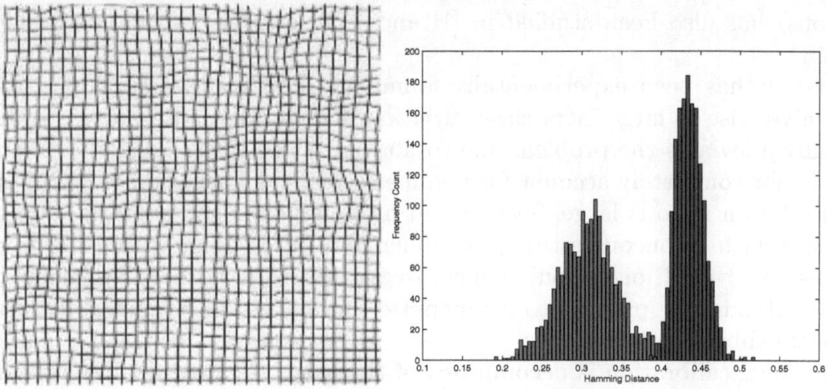

Fig. 2. An example of deformed mesh obtained by local calibration (left) and HD distributions of simply thresholding enhanced images (right)

A. Ideal Iris Database

For ideal database such as CASIA, no global calibration is needed. Therefore, we first demonstrate how local calibration facilitates iris recognition - an iris code can be obtained by simply thresholding the enhanced image. Fig. 2b) shows the distribution of Hamming distance (HD) for the whole 108 images (1620 intra-class and 1600 inter-class comparisons). It can be observed that without any sophisticated feature extraction technique, our plain iriscode already achieves reasonably good separation of intra-class and inter-class distributions. Empirical studies show that among the 2% intra-class comparisons whose HD is above 0.4, about 80% occur with two difficult subjects (No. 41 and 101, one example is shown as the bottom image in Fig. 3b) whose iris contain little texture and is severely occluded.

To further illustrate the impact of iris type on recognition performance, we manually pick out 30 subjects with high-texture (e.g., the middle image in in Fig. 3b) and low-texture (e.g., the top image in in Fig. 3b) iris respectively. The HD distributions for these two classes are shown in Fig. 4. For high-texture iris images, the separation of intra-class and inter-class distributions is nearly optimal regardless of the occlusion (on the average, 20% pixels are occluded in CASIA database). Low-texture iris is more challenging especially when occlusion also occurs. How to improve the performance for low-texture iris is left for our future study.

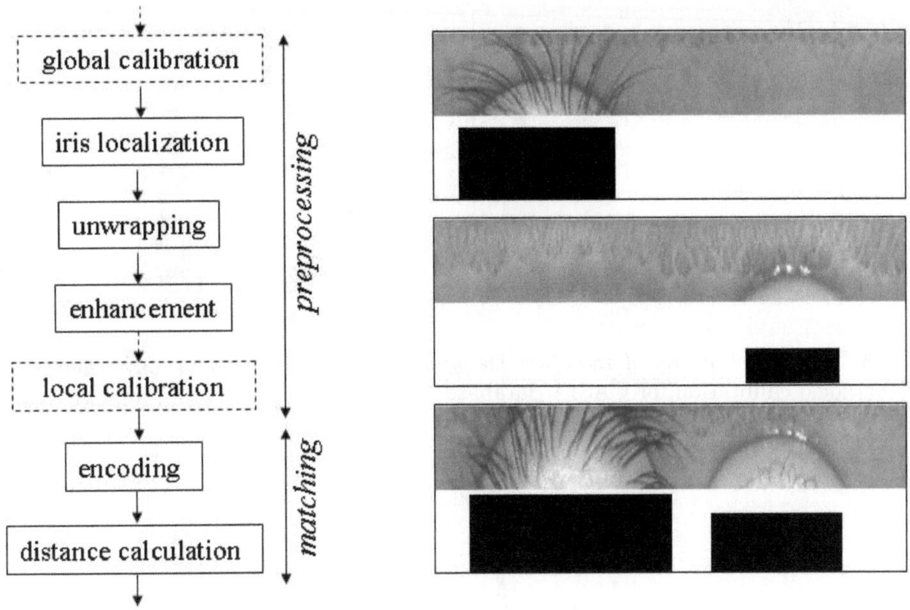

(a) (b)

Fig. 3. a) The diagram of the proposed iris recognition system; b) examples of ROIs

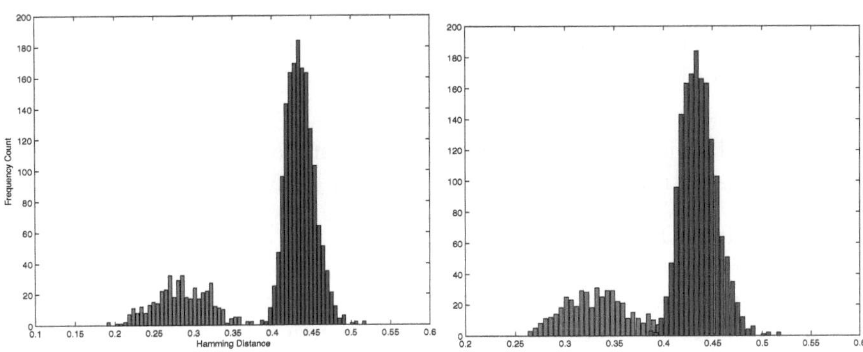

Fig. 4. HD distributions for high-texture iris (left) and low-texture iris (right)

We have also tested the proposed local calibration technique with our own implementation of Daugman's algorithm. The distributions of HD before and after calibration are shown in Fig. 5. It can be observed that local calibration effectively reduces intra-class variation at the price of slightly increased inter-class variation. Though more extensive experiments are required to evaluate the impact on ROC performance, it seems that local calibration at least suggests a new way of trading FNMR with FMR - i.e., in order to satisfy the accuracy require-

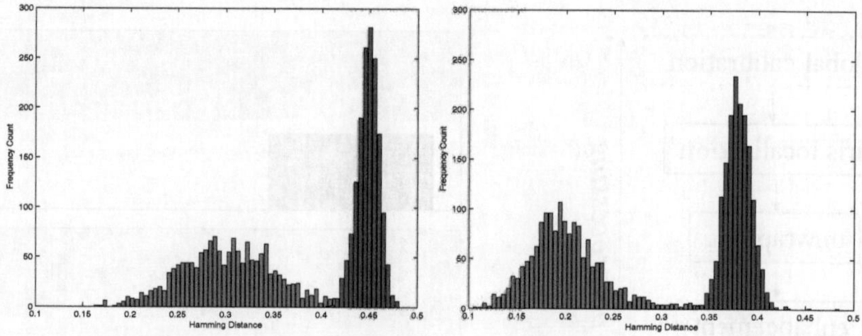

Fig. 5. HD distributions of modified Daugman's algorithm without (left) and with (right) local calibration for CASIA database

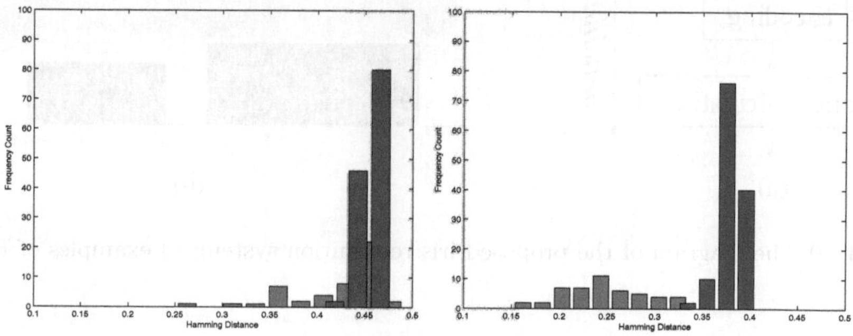

Fig. 6. HD distributions of modified Daugman's algorithm without (left) and with (right) local calibration for EI database

ments imposed by biometric applications, we might want to slightly sacrifice the FMR (since it is extremely low) in order to lower FNMR.

B. Nonideal Iris Database

We have also collected a database of nonideal images for about 100 people in collaboration with the Eye Institute (EI) of West Virginia University in the past year. For each eye of a person, two images are acquired at the front and off-angle respectively; the total number of images in EI database is around 800. Although the off-angles are preset to be 15° and 30°, we have found that those parameters cannot be directly used for global calibration due to varying gaze and head positions. We have also found that acquiring well-focused iris images is not easy for people without sufficient experience on operating cameras (e.g., auto-focus does not work properly for iris acquisition). Out-of-focus images can still be used for testing global calibration and iris localization techniques; but not for iris matching. Therefore, we can only perform our experiments with nonideal iris recognition on a small set of images (8 subjects) that are reasonably focused.

Experimental results have shown that ellipse-fitting based calibration works very well. By manually inspecting 80 calibrated images randomly selected from the database, we do not observe any error - pupils all appear circular after the calibration, which implies that nonideal iris recognition is transformed back to the ideal case. For the small set of focused iris images after global calibration, we have compared the results of modified Daugman's algorithm with and without local calibration. Fig. 6 shows the distributions of HD for 48 intra-class and 128 inter-class comparisons, from which we can again see the effectiveness of local calibration. .

References

[1] J. Cui, Y. Wang, T. Tan, L. Ma, and Z. Sun. A fast and robust iris localization method based on texture segmentation. In Proc. SPIE on Biometric Technology for Human Identification, 2004.

[2] J. Daugman. How iris recognition works? IEEE Transactions on Circuits Syst. Video Tech., 14:21–30, 2004.

[3] A. W. Fitzgibbon, M. Pilu, and R. B. Fisher. Direct least-squares fitting of ellipses. IEEE Trans. on Pattern Anal. Mach. Intell., 21:476–480, 1999.

[4] B. Horn and B. Schunck. Determining optical flow. Artif. Intell., 17:185–203, 1981.

[5] W. Kong and D. Zhang. Accurate iris segmentation based on novel re.ection and eyelash detection model. In Int. Sym. on Intell. Multimedia, Video and Speech Proc., 2001.

[6] L. Ma, T. Tan, and Y. W. D. Zhang. Personal identi.cation based on iris texture analysis. IEEE Trans. on Pattern Analysis and Machine Intelligence, 25(12):1519 – 1533, 2003.

[7] L. Ma, T. Tan, and Y. W. D. Zhang. E.cient iris recognition by characterizing key local variations. IEEE Trans. on Image Processing, 13(6):739 – 750, 2004.

[8] R. Wildes. Iris recognition: an emerging technology. Proc. of IEEE, 85:1348– 1363, 1997.

A Model Based, Anatomy Based Method for Synthesizing Iris Images

Jinyu Zuo and Natalia A. Schmid

Lane Department of Computer Science and Electrical Engineering,
West Virginia University, Morgantown, WV 26506, USA
{jinyuz, natalias}@csee.wvu.edu

Abstract. Popularity of iris biometric grew considerably over the past 2-3 years. It resulted in development of a large number of new iris encoding and processing algorithms. Since there are no publicly available large scale and even medium size databases, neither of the algorithms has undergone extensive testing. With the lack of data, two major solutions to the problem of algorithm testing are possible: (i) physically collecting a large number of iris images or (ii) synthetically generating a large scale database of iris images. In this work, we describe a model based/anatomy based method to synthesize iris images and evaluate the performance of synthetic irises by using a traditional Gabor filter based system and by comparing local independent components extracted from synthetic iris images with those from real iris images. The issue of security and privacy is another argument in favor of generation of synthetic data.

1 Introduction

Popularity of iris biometric grew considerably over the past 2-3 years. It resulted in development of a large number of new iris encoding and processing algorithms. Most of developed systems and algorithms are claimed to have exclusively high performance. However, since there are no publicly available large scale and even medium size datasets, neither of the algorithms has undergone extensive testing. The largest dataset of frontal view infrared iris images presently available for public use is CASIA-I dataset [1]. It consists of 108 classes, 7 images per class. With the lack of data, two major solutions to the problem of algorithm testing are possible: (i) physically collecting a large number of iris images or (ii) synthetically generating a large scale dataset of iris images. In this work, we describe a model based, anatomy based method to synthesize iris images and evaluate the performance of synthetic irises by using a traditional Gabor filter based system. The issue of security and privacy is another argument in favor of generation of synthetic data.

The first methodology for generating synthetic irises has been proposed by Cui et al. [2], where a sequence of small patches from a set of iris images was collected and encoded by applying Principle Component Analysis (PCA) method. Principle components were further used to generate a number of low resolution iris images from the same iris class. The low resolution images were combined in a single high resolution iris image using a superresolution method. A small set of random parameters was

D. Zhang and A.K. Jain (Eds.): ICB 2006, LNCS 3832, pp. 428 – 435, 2005.

used for generation of images belonging to different iris classes. Another method for generation of synthetic irises based on application of Markov Random Field has been recently developed at WVU [3] and offered as an alternative to the model based, anatomy based method described in this paper.

When generating synthetic iris images, the problem that one faces is to define a measure of "realism." What is the set of requirements that synthetic iris has to satisfy to be recognized and treated as a physically collected iris image? The conclusion could be: (i) it should look like a real iris; (ii) it should have the statistical characteristics of a real iris.

We have conducted extensive anatomical studies of the iris including study of ultra-structure images and high-resolution images [4, 5], structure and classification of irises due to iridology [6], and models available for the iris. As a result, a few observations on common visual characteristics of irises have been made: (i) most iris images used in biometrics research are infrared images; (ii) the information about iris texture is mainly contained in the structure, not in the color; (iii) radial fibers constitute the basis for the iris tissue; (iv) a large part of iris is covered by a semitransparent layer with a bumpy look and a few furrows; (v) the collaret part is raised; (vi) the top layer edge contributes to the iris pattern.

Thus, the main frame of the iris pattern is formed by radial fibers, raised collaret, and partially covered semitransparent layer with irregular edge. The difference of pixel values in an infrared iris image is not only the result of the iris structure information. It is related to the material that the iris is composed of, surface color, and lighting conditions.

2 Methodology

In this work, the generation of iris image can be subdivided into five major steps:

1. Generate continuous fibers in cylindrical coordinates (Z, R, and Θ), where the axis Z is the depth of the iris, R is the radial distance, and Θ is the rotational angle measured in degrees with a 0 value corresponding to the 3 o'clock position and values increasing in the counter-clockwise direction. Each fiber is a continuous 3D curve in this cylindrical coordinates. Currently 13 random parameters are used for generation of each continuous fiber. The curve is further sampled in R direction to obtain matrices of Z and Θ coordinates.
2. Project 3D fibers into a 2D flat image space. Then shape the pupil and iris. Generated 3D fibers are projected into a 2D polar space to form a 2D frontal view fiber image. Only the top layer of fibers can be seen. The gray value of each pixel in 2D space is determined by the Z value of the top layer at that point in the 3D cylindrical space. A set of basic B-spline functions in the polar coordinate system (R, Θ) is used to model shapes of the pupil and iris, that is, their deviation from a circular shape.
3. Transform the basis image to include the effect of collaret. Add a semitransparent top layer with an irregular edge. The edge of the top layer is modeled

using cosine functions. The top layer is then blurred to make it semitransparent. The area of collaret is brightened to create the effect of a lifted portion of the iris.

4. Blur the iris root and add a random bumpy pattern to the top layer. Blur the root of the iris to make the area look continuous. Then add a smoothed Gaussian noise layer.
5. Add the eyelids at a certain degree of opening and randomly generated eyelashes. Based on a required degree of eyelid opening, draw two low frequency cosine curves for eyelids. Then randomly generate eyelashes.

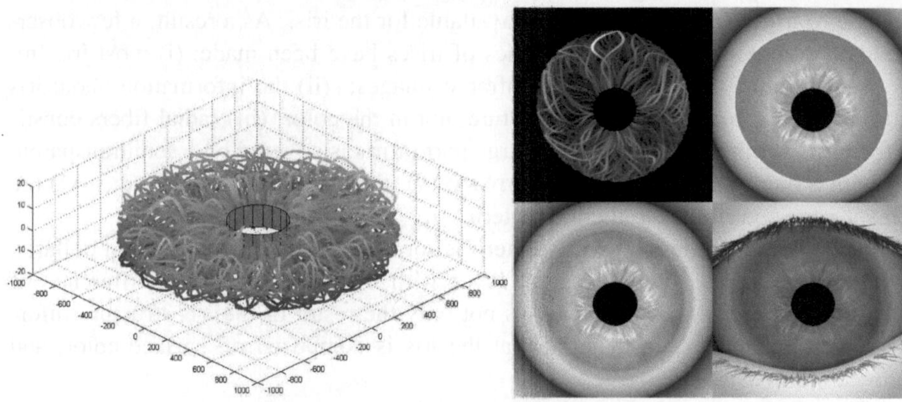

Fig. 1. Shown are the steps of iris image synthesis

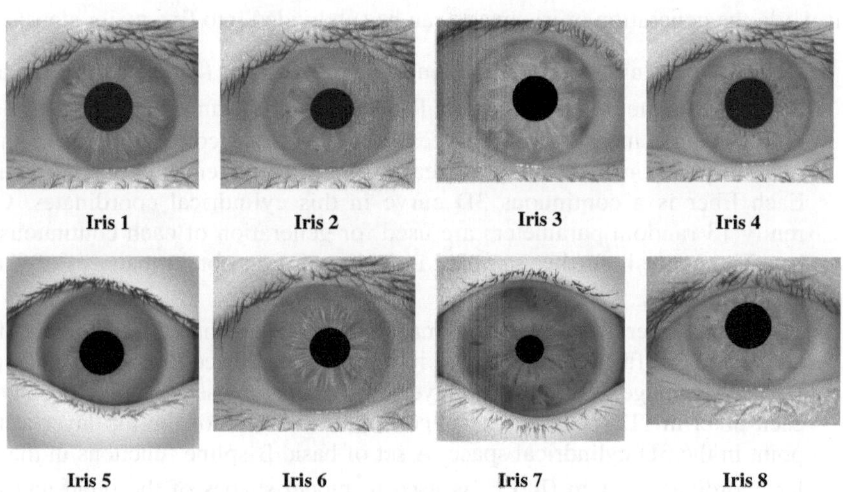

Fig. 2. A gallery of synthetic iris images generated using model based, anatomy based approach. Iris 4 is a real iris image borrowed from CASIA dataset

The generation of iris images is based on other 40 controllable random parameters including fiber size, pupil size, iris thickness, top layer thickness, fiber cluster degree, iris root blur range, the location of the collaret, the amplitude of the collaret, the range of the collaret, top layer transparency parameter, net structure parameter, eye angle, eye size, eye horizontal location, number of crypts, number of eyelashes. If we also account for the random variables used in the calculation of the fiber shape, the resulting number of random parameters is of the order of several thousands. Most of the parameters are uniformly distributed on a prescribed interval. The range of intervals is selected to ensure the appearance close to the appearance of real irises. Fig. 1 demonstrates our generation procedure. Other effects influencing the quality of iris image including noise, off-angle, blur, specula reflections, etc. can be easily incorporated.

3 Real and Synthetic Iris Images: Similarity Measures

We identified three levels at which similarity of synthetic and real iris images can be quantified. They are as follows: (i) global layout, (ii) features of fine iris texture, and (iii) recognition performance.

3.1 Visual Evaluation

A gallery of synthetic iris images generated using our model based approach is shown in Fig. 2. To ensure that generated irises look like real irises, we borrowed a few eyelids from CASIA dataset. Note that only one image in Fig. 2 is a real iris image, a sample from CASIA dataset. It is placed among synthetic irises for the purpose of comparison. To further demonstrate that our synthetic iris images look similar to real iris images, we displayed three normalized enhanced iris images in Fig. 3. The samples on the upper and middle panels are unwrapped images from CASIA and WVU non-ideal iris datasets. The sample on the lower panel is an unwrapped image from our dataset of synthetic irises. Although it looks slightly oversmoothed on the bottom portion of the image, the unwrapped synthetic iris image has all major features of real iris images.

(a) (b) (c)

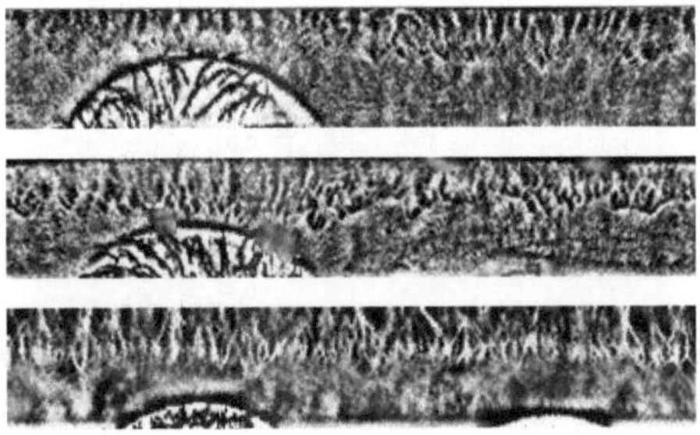

Fig. 3. Shown are three segmented unwrapped and enhanced iris images. The images are samples from (a) CASIA dataset, (b) WVU non-ideal iris dataset, and (c) dataset of synthetic irises generated using our model based approach.

3.2 Comparison of Local ICA Functions

To evaluate similarity of iris images at a fine feature level, we encode iris images using local Independent Component Analysis (ICA) [7, 8, 9] and compare local ICA functions extracted from synthetic iris images with the ICA functions extracted from real iris images. We find the best matching pairs of local ICA functions using normalized Euclidean distance. ICA functions are obtained using FastICA MATLAB package [10].

To extract ICA basis functions for each of three datasets, within each dataset we randomly selected 50,000 patches from 100 iris classes, with 3 segmented unwrapped and enhanced iris images per class in CASIA dataset, with one segmented unwrapped and enhanced iris image per class in synthetic dataset and with 2 segmented unwrapped and enhanced iris images per class in WVU non-ideal iris dataset. We ensured that patches contain no occlusions (eyelids and eyelashes). Each segmented unwrapped image has the size 64×360 pixels. The selected patch size is 5×5. We repeated this procedure 20 times, which resulted in the total 480 local ICA functions. We found the best matching pairs of local ICA basis functions, based on the minimum Euclidean distance between two local ICA functions, for the following pairs of datasets: CASIA-synthetic, WVU-synthetic, and CASIA-WVU. To summarize the results of comparison, Fig. 4 and Fig. 5 show distributions of the minimum Euclidean distance for best matching pairs of ICA functions. The left panel in Fig. 4 is the distribution of the minimum Euclidean distance when local ICA functions extracted from CASIA and synthetic datasets are compared. The right panel in Fig. 4 is the distribution of the minimum Euclidean distance when local ICA functions extracted from WVU and synthetic datasets are compared. The left panel in Fig. 5 shows the results when local ICA functions extracted from CASIA and WVU datasets are compared. To provide a baseline, we also plot the distribution of the minimum Euclidean distance for best matching pairs of ICA functions extracted for two non-overlapping sets of iris images from CASIA dataset . This distribution is shown

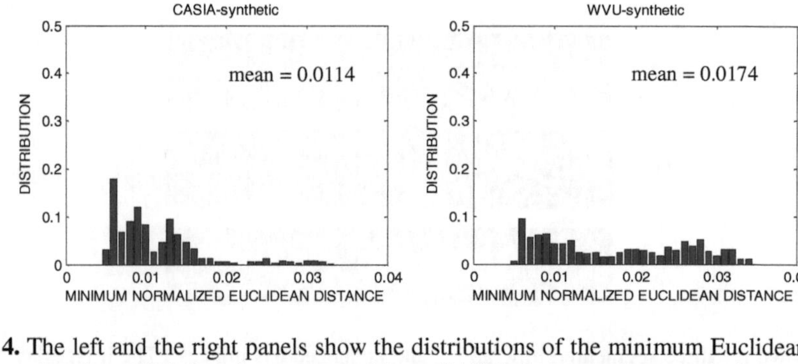

Fig. 4. The left and the right panels show the distributions of the minimum Euclidean distance scores when local ICA functions extracted from CASIA dataset and synthetic dataset are compared and when ICA functions are extracted from WVU and synthetic datasets, respectively

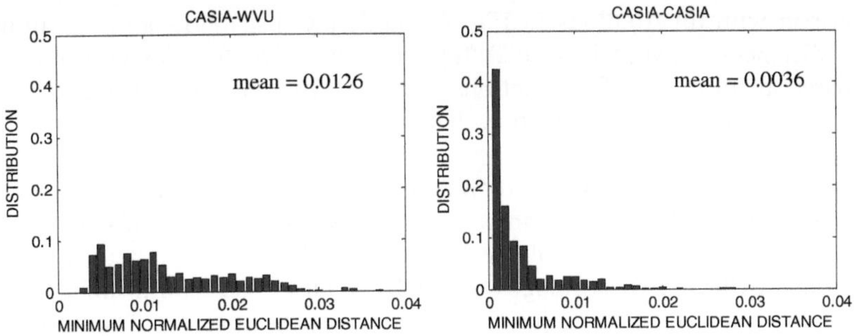

Fig. 5. The left and the right panels show the distributions of the minimum Euclidean distance scores when local ICA functions extracted from CASIA dataset and WVU datasets are compared and when ICA functions are extracted from two different subsets of CASIA dataset

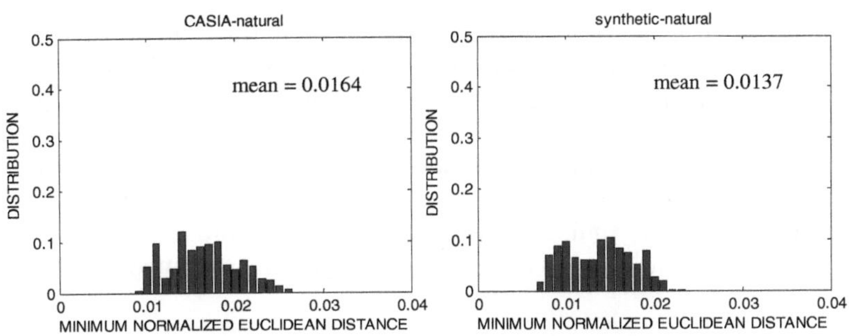

Fig. 6. The left and the right panels show the distributions of the minimum Euclidean distance scores when local ICA functions extracted from CASIA dataset and natural images are compared and when ICA functions are extracted from synthetic dataset and natural images, respectively.

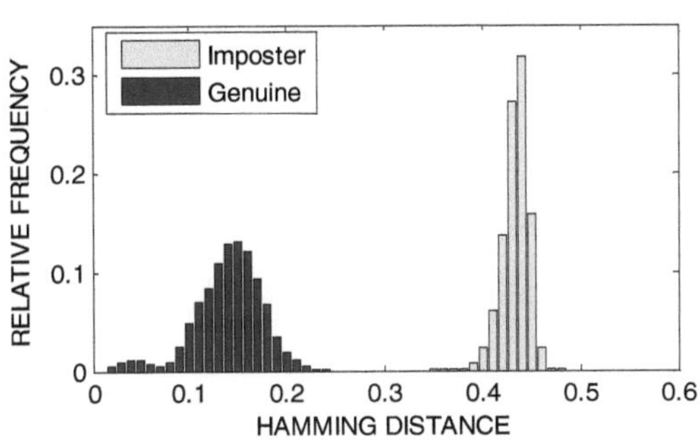

Fig. 7. Verification performance

on the right panel in Fig. 5. Note that the score distributions in Fig. 4 (CASIA – synthetic) and (WVU – synthetic) and Fig. 5 (CASIA – WVU) look and perform similar.

In comparison with distributions in Fig. 4 and 5, the distributions of the minimum Euclidean distances between local ICA functions extracted from natural images [11] and compared against local ICA functions extracted from synthetic or real iris images have a compact support and do not achieve 0.005 of the minimum distance (see Fig. 6).

When the patch size is increased (for instance, to the size 12-by-12 pixels) the similarity between the ICA basis functions extracted from images in CASIA dataset and from synthetic iris images will decrease while the similarity between the ICA basis functions extracted from images in CASIA dataset and natural images will increase. We conjecture that the major reason for this is the absence of multi-level texture (results from tissues having fibers of different size and thickness) in synthetic irises. We are currently enhancing our generator to incorporate this feature into synthetic iris images.

3.3 Verification Performance

To evaluate the performance of synthetic iris images from recognition perspective, we used a Gabor filter based encoding technique (our interpretation of Daugman's algorithm [12]). We generated iris images that could belong to 204 individuals, 2 eyes per individual, 6 iris images per iris class including one frontal view, two rotated, and three blurred and rotated. No False Acceptance and False Rejection are reported, that is, the genuine score and imposter score histograms do not overlap. D-prime, a measure of separation between genuine and imposter matching score distributions, is equal to 11.11. Fig. 7 shows the plot of two distributions, genuine and imposter.

4 Summary

We proposed a model based, anatomy based method for synthesizing iris images with the major purpose to provide the academia and industry with a large database of generated iris images to test newly designed iris recognition algorithms. Since synthetic data are known to introduce a bias that is impossible to predict [13, 14], the data have to be used with caution. We believe, however, that the generated data provide an option to compare efficiency, limitations, and capabilities of newly designed iris recognition algorithms through their testing on a large scale dataset of generated irises. We anticipate that synthetic data because of their excessive randomness and limited number of degrees of freedom compared to real iris images will provide overoptimistic bound on recognition performance.

References

1. CASIA Iris Image Dataset (ver. 1.0), http://www.sinobiometrics.com/casiairis.htm
2. Cui, J., Wang, Y., Huang, J., Tan, T., Sun, Zh.: An Iris Image Synthesis Method Based on PCA and Super-resolution. In Proc. of the 17th Intern. Conf. on Pattern Recognition (2004) 471-474.
3. Makthal, S., Ross, A.: Synthesis of Iris Images using Markov Random Fields. Proc. of 13th European Signal Processing Conference (EUSIPCO), (Antalya, Turkey), September 2005. To appear.

4. Miles Research: Iris Pigmentation Research Info. http://www.milesresearch.com/iris/
5. Miles Research: Iris Images from Film Camera.
6. http://www.milesresearch.com/download/exampleirisimages.ppt
7. Sharan, F.: Iridology - a complete guide to diagnosing through the iris and to related forms of treatment. HarperCollins, Hammersmith, London (1992).
8. Hyvärinen, A., Karhunen, J., Oja, E.: Independent Component Analysis, John Wiley and Sons (2001).
9. Noh, S., Pae, K., Lee, C., Kim, J.: Multiresolution Independent Component Analysis for Iris Identification. In Proc. of the Intern. Technical Conf. on Circuits / Systems, Comp. and Commun., Puket, Thailand (2002) 1674-1677.
10. Bae, K., Noh, S., Kim, J.: Iris Feature Extraction Using Independent Component Analysis. In Proc. of the 4th Intern. Conf. on Audio-and Video-Based Biometric Person Authentication, Guildford, UK, June (2003) 838-844.
11. FastICA MATLAB Package. Available online at http://www.cis.hut.fi/projects/ica/fastica
12. Natural images. Available online at http://www.cis.hut.fi/projects/ica/imageica/
13. Daugman, J.: High Confidence visual Recognition of Persons by a test of Statistical Independence. In IEEE Trans. on Pattern Analysis and Machine Intelligence, vol. 15, no. 11 (1993) 1148-1161.
14. Mansfield, A. J., Wayman, J. L.: Best Practices in Testing and Reporting Performance of Biometric Devices (2002). Available online at:
15. http://www.cesg.gov.uk/site/ast/biometrics/media/BestPractice.pdf
16. Wayman, J., Jain, A., Maltoni, D., Maio, D. (Eds): Biometric Systems: Technology, Design, and Performance Evaluation, Springer, New York (2005).

Study and Improvement of Iris Location Algorithm

Caitang Sun, Chunguang Zhou*, Yanchun Liang, and Xiangdong Liu

College of Computer Science and Technology,
Jilin University, Changchun, 130012, China
cgzhou@jlu.edu.cn

Abstract. Iris location is a crucial step in iris recognition. Taking into consideration the fact that interior of the pupil, there would have some lighter spots because of reflection, this paper improves the commonly used coarse location method. It utilizes the gray scale histogram of the iris graphics, first computes the binary threshold, averaging the center of chords to coarsely estimate the center and radius of the pupil, and then finely locates it using the algorithm of circle detection in the binary graphic. This method could reduce the error of locating within the pupil. After that, this paper combines Canny edge detector and Hough voting mechanism to locate the outer boundary. Finally, a statistical method is exploited to exclude eyelash and eyelid areas. Experiments have shown the applicability and efficiency of this algorithm.

Keywords: Iris Location, Circle Detection, Canny Edge Detection, Hough Voting Mechanism.

1 Introduction

Iris recognition has become an important solution for individual identification. As an emerging biometric recognition technology, it has some advantages compared to others: (1) It is impossible that any two individual's texture of iris is complete the same, and even the left and the right one of the same individual are also different from each other; (2) The features of the iris are changeless during one's lifetime without any accident; (3) Unlike other information such as face and password, it is difficult to change or simulate. All these advantages make it a hot topic.

Iris location aims at locating the inner boundary (pupil) and outer one (sclera) of the iris, providing valid areas for iris feature extraction, which could directly influence the effect of iris recognition. There are two most commonly used iris location algorithms. One is the circle-detection algorithm proposed by J.Daugman[1-4], which uses circular edge detecting operator to detect the inner and outer boundary of the iris, exploiting the geometrical characteristic that the iris is approximately a circle. And the other one is the two-step method proposed by P.Wildes[5]. Cui Jiali etc[6] combine SVM and LDA for iris location, but it may be influenced if the eyelashes are heavy; Yuan Weiqi etc[7] present an active contour method-Snake-Daugman, and it could also be influenced by eyelashes. Most of the iris location algorithms coarsely locate pupil by finding the minimum of the sum of gray value

* Corresponding author.

D. Zhang and A.K. Jain (Eds.): ICB 2006, LNCS 3832, pp. 436–442, 2005.

before fine location, because the gray level of pupil is lower than that of all the other areas in the iris image. But the disadvantage is also obvious, i.e., if the gray level of some pixels in the horizontal or vertical direction of the true pupil center is made higher or others made lower because of some factors such as lightness, the result will be far away from the actual position. Xue Bai etc [8] use histogram to compute the threshold for binarization, and it improves the effect of iris location in certain extent, but in some conditions, the gray level of eyelash or eyebrow could be lower than that of the pupil, so the threshold would be so low that the result is not ideal. This paper improves the method in coarse location of the pupil, uses binary image circle detector, and combines edge detection and Hough voting mechanism in outer boundary detection [5,9,10]. Experiments show that the effect is satisfactory.

The remainder of this paper is organized as follows. Section 2.1 mainly introduces the coarse location method. Section 2.2 depicts the fine location method of the inner boundary. Section 2.3 describes the fine location method for the outer boundary. Section 3 introduces how to exclude eyelid and eyelash area from the result above. Section 4 presents experimental results and concludes with some remarks.

2 Iris Location

2.1 Inner Boundary (Pupil) Coarse Location

The objective of pupil coarse location is to approximately estimate the center and radius of the pupil, that is, to determine the pseudo center and pseudo radius of it. In general, in an iris image, the gray values inside the pupil are the lowest in the image. Whereas, the gray values of eyelashes and eyebrows are often near to those of the pupil, or even lower than them in some conditions.

In this paper, the image is binarized first. Selection of the threshold is crucial, which influences the following steps. If the threshold is too low, the area of pupil would be reduced, vice versa. Based on this analysis, this paper proposed the following method.

① Make the gray scale histogram of the image, then filter it, and compute the valley between the first two wave peaks, whose gray value is marked as T_0 . In some instances, it could obtain good result if T_0 is directly used as the threshold. But, in many images, the gray level of eyelash or eyebrow is lower than that of the pupil, in these cases, the calculated threshold will be lower than the needed one, and pupil would be mistaken for background, so, further judgement is necessary;

② Calculate the difference between T_0 and the first wave peak, if it is larger than 6, T_0 could be taken as T_1, the ultimate threshold; otherwise, continue to search for the valley after the third wave peak, and select the correspond gray value as T_1 (Fig.1(b));

③ Binarize the image, set the values of the pixels whose gray level below T_1 to 0, others to 255;

In some cases, there will still have some noises in the binary image because of the existence of eyelash or eyebrow (Fig.1(c)), but most of them could be removed by the Open and Close operation of morphology.

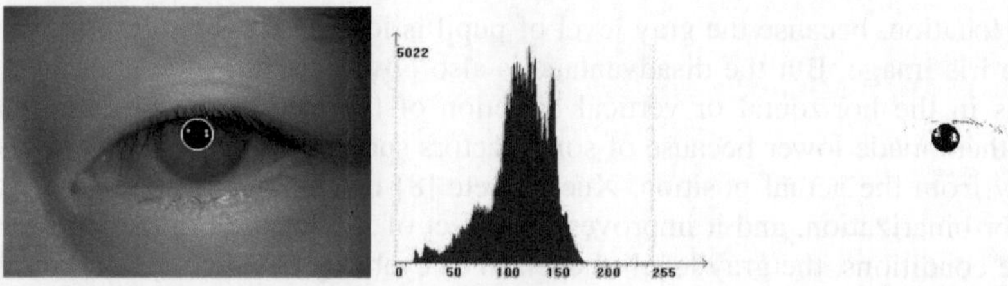

Fig. 1. (a) Location result of the inner boundary (b) histogram of the image with T_1=25 (c) binary image

Then summarize the gray value in the direction of x and y respectively, and find the point correspond to the minimum of them. This point may be near to the center of pupil, but it is also possible to be far away from it, so, it is necessary for further determination (In fact, because the most time consuming in iris location is fine location of the inner and outer boundary, while the result of coarse location will determine the time and effect of fine location, it is worthy of a little more time on coarse location). The algorithm is described as follows:

① Search for the x-coordinate of the pseudo center: take each point of (x_0, $y_0\pm10$) as a temporary center and search for the first white pixel on its left and right side, recording the x-coordinate of the midpoint as the new value x, then take the average of all the new values of x as the possible x-coordinate of the pseudo center of pupil(x_1).
② Search for the y-coordinate of the pseudo center: take each point of ($x_1\pm10$, y_0) as a temporary center and do as step ① does, then the coordinate of the pseudo center(x_1,y_1) could be attained.
③ Estimate the length of radius of pupil: Take (x_1,y_1) as the center to calculate the length of chords in some arbitrary directions, and the longest one is thought as the pseudo radius r_1.

The above method could efficiently reduce the searching range in the following inner boundary fine location, so as to speed it up.

2.2 Fine Location of the Inner Boundary

Based on the estimated result, the pupil could be finely located, and the most commonly used formula is (1)

$$\max_{r,x_0,y_0}\left| G_\sigma(r) \otimes \frac{\partial}{\partial r} \oint_{r,x_0,y_0} \frac{I(x,y)}{2\pi r} ds \right| \tag{1}$$

Formula(1) is a detector of circular edge with σ as the scalar, which searches for the optimal solution by iteration of the three-parameter space (r,x_0,y_0) to locate the pupil. In this formula, (x_0,y_0) is the center of the circle; r is the length of radius of it which ranges from (r_1-10) to (r_1+10); $G\sigma(r)$ is a filter usually in the form of Gaussian; and ⊗ is a convolution operation. The essence of the formula is to calculate the average of gray value of every pixel on the circumference of the circles with all the possible radii, then to filter the difference between two adjacent circles. Finally, the parameters correspond to the maximum difference is taken as the center and radius of the pupil. The discrete form of the formula is

$$\max_{n\Delta r, x_0, y_0} \left| \frac{1}{\Delta r} \sum_k [(G((n-k)\Delta r) - G((n-k-1)\Delta r)) \sum_m I(x_{m,k}, y_{m,k})] \right| \quad (2)$$

In real images, even the gray values of the pixels within the pupil may not be the same, especially in cases that some lighter areas are made because of reflection inside of it under the source of light. The gray values of these areas may be remarkably larger than those of others (Fig. 1(a)). In these conditions, if formula (2) is used, the gray value difference made by these pixels could be more than that of up to 10 regular ones, and so it may lead to the error of locating inside the pupil. This paper detects the circle in binary image, so all the points contribute equally, and this could effectively avoid the problem. The result could be seen in Fig.1(a).

2.3 Fine Location of the Outer Boundary

In this paper, the fine location of the outer boundary is based on the inner one. Most of the algorithms utilize circle detectors like formula (2), but in fact, the contrast between the gray values of the outer boundary of iris and sclera (the near to white area outside the iris) are not so remarkable, and the iris has resourceful texture, so it is difficult to locate the outer boundary accurately by those detectors. This paper first uses Canny for edge detection [11,12], and then imposes Hough Voting algorithm on the result to determine its radius and center. Canny algorithm is widely accepted as the best edge detector, which could eliminate the influence of noise more effectively without much loss of true edge information.

In practice, the inner and outer boundary of an iris are not homocentric, and experiments on the CASIA iris database, show that, the vertical difference is within 3 pixels, while the horizontal may be up to 6 pixels. In this paper, all the pixels in the range of [x±6 y±3] are taken as candidate centers of the outer boundary in Hough voting for circle detection. The detail of Hough voting algorithm is as follows:

(1) Set up an array A, with the dimension of (maximal length of radius of the outer boundary-radius of the inner boundary)*(number of candidates of center of circle, 91 in this paper), and initialize all the elements to 0;
(2) Scan the result of Canny edge detection, if a pixel is on the edge, calculate the distance r between it and the candidates of center of circle, all the values of elements correspond to r-1, r, r+1 in array A plus 1;
(3) Scan array A, and the subscripts correspond to the element with the maximal value is taken as the center and radius of the outer boundary respectively, see Fig. 2.

3 Exclude Non-iris Areas

In most cases, the result area after fine location would contain areas of eyelashes or (and) eyelids, if these areas are not removed from the actual ones, the accuracy of iris recognition would be reduced greatly. Many researchers do this by Hough transform, modeling eyelids as two parabolic arcs, but this method is very time-consuming and sometimes it is difficult to find the arcs. Basing on the observation that the gray values of the pixels near the outer boundary distribute uniformly, this paper presents a gray value statistical approach on the circumferences of a series of consecutive homocentric circles to obtain two thresholds: $T_1<T_2$, then scans all over the circumference of the outer boundary for further decision. The details of the algorithm are as follows.

① Build gray scale histogram of the outer boundary H: From the outer boundary to inside about 4 pixels, select an angle in the left and the right part respectively as the object for statistics, and build its histogram;
② Search for the maximum value (N_0) in the histogram, and record the correspond gray value as G_0, then calculate the total number $N=\sum H(i)$;
③ Search for the upper and lower threshold: Initialize α to a positive real number less than 1.0, then search from G_0 to left and right respectively in the histogram, if there are successively 4 gray values correspond to which the number of pixels is less than N0*α, stop and record the gray values $T_1<T_2$, calculate the number $N_1=\sum_{i=T_1}^{T_2} H(i)$, if $N_1/N \geq 85\%$ or $N0*\alpha<1$, go to ④; Else, decrease α, repeat ③;

T1 and T2 obtained by ③ are taken as the gray level distribution range of the outer boundary of the iris.

④ Exclude eyelash and eyelid areas: Scan the bottom right circumference from the point of 0°, if there are two consecutive points whose gray values are not in $[T_1, T_2]$, stop and record the angle as $-90° \leq \theta_1 \leq 0°$; So do the top right, the top left and the bottom left parts, suppose that the angles are θ_2, θ_3 and θ_4 respectively;

⑤ The circular arcs between $[\theta_1, \theta_2]$ and $[\theta_3, \theta_4]$ are selected as iris areas for following iris recognition, and the remainders are considered to be non-iris areas, the result could be seen in Fig.2(c).

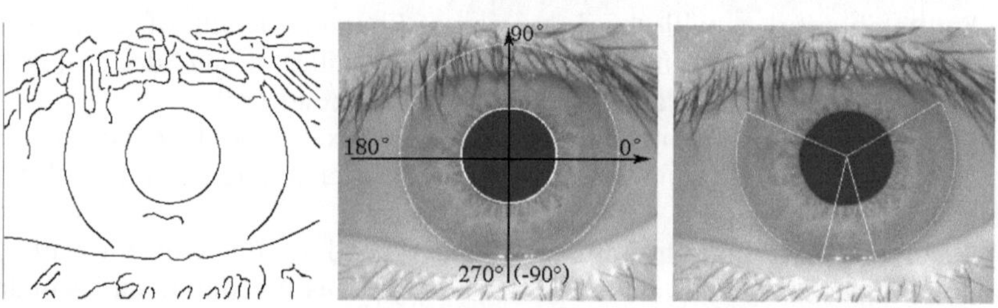

Fig. 2. (a) Result of Canny edge detection (b) fine location result (c)excluding eyelash and eyelid areas

This method could remove certain false areas using the outer boundary information, but how to remove other false inner areas still need careful study, motivated by Canny edge detection, this paper experimented of selecting two thresholds and it works well in some cases but is not so perfect.

4 Analysis of Experimental Result

The iris location algorithm is implemented in VC++6.0. Experiments were conducted using the images in CASIA Iris Image Database 1.0, and they show that the result is satisfactory. In the database, the images are all 320*280, and all the pseudo centers of circle acquired by coarse location are all in the searching range of the next step. Table 1 shows the comparison of the proposed and two commonly used algorithms practiced under the condition of 1.7GHz CPU 256M RAM. Figure 3 shows some results in experiments.

Table 1. Comparison of accuracy and time of location with other algorithms

Algorithms	accuracy	Average time
Daugman[1]	98.4%	6.76s
Wildes[10]	99.8%	8.94s
proposed	99.2%	0.23s

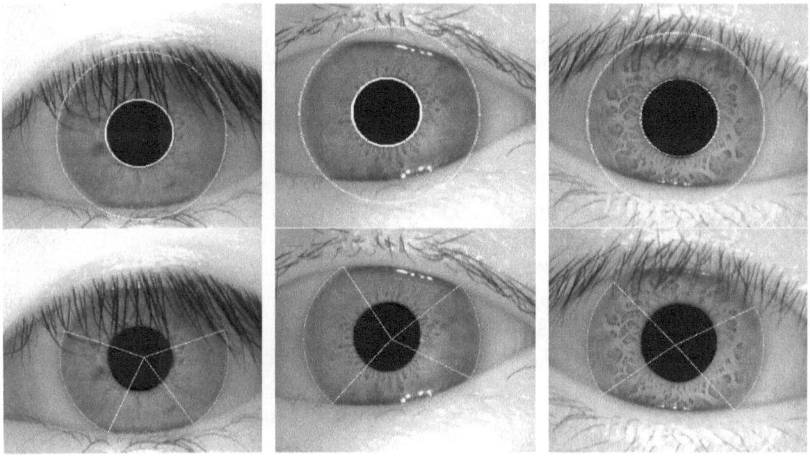

Fig. 3. Some samples of iris location. The above are the results of fine location, and the below ones are correspond results that excluding the eyelash and eyelid areas.

The reduction of time can be contributed to the following factors:

(1) The improved performance of coarse location, so that the searching range could be reduced in the following steps;
(2) Set up an array in advance, storing the offset of x and y correspond to certain length of radii and angles used in fine location, and it could reduce time to 1/7—1/5 of previous methods.

In the iris image database (CASIA 1.0), there are about 5 percent of images in which the gray levels of eyelash are lower than those of iris, using the method presented in [8] for coarse location would lead to such result that the located center of pupil could be far away from the real one, while uses the method proposed by this paper, the problem could be resolved, and the fine location time could also be decreased. Compared to the commonly used J.Daugman's iris location algorithm, it could effectively avoid the error location problem caused by reflection inside the pupil. And combines Canny edge detection and Hough voting mechanism for location of the outer boundary could ensure satisfactory performance under the condition of very clear texture.

Acknowledgements

This work was supported by the Natural Science Foundation of China (Grant No. (60433020)) and the Key Laboratory for Symbol Computation and Knowledge Engineering of the National Education Ministry of China. (93K-17).

All the images appeared in this paper were supplied by the Institute of Automation, Chinese Academy of Sciences, the authors would like to thank them.

References

[1] J.Daugman: High confidence visual recognition of persons by a test of statistical independence. IEEE Transactions On Pattern Analysis and Machine Intelligence, 15 (1993) 1148-1161

[2] Wang Chengru, Hu Zhengping: Iris Location Algorithm Based on Geometric Features. Journal of Image and Graphics A, 8 (2003) 683-685

[3] Fan Kefeng, Zeng Qingning: A Research on Iris Location Algorithm, Computer Engineering and Applications, 40 (2004) 60-61

[4] Yang Wen, Yu Li etc: A Fast Iris Location Algorithm. Computer Engineering and Applications, 40 (2004) 82-84

[5] Richard P. Wildes: Iris Recognition: An Emerging Biometric Technology. Proceedings of the IEEE, 85 (1997) 1348-1363

[6] Jiali Cui, Li Ma, etc: An Appearance-based Method for Iris Detection, ACCV2004, 2: 1091-1096

[7] Yuan Weiqi, Ma Junfang etc: A New Method of Iris location based on Active Contour. Computer Engineering and Applications, 39 (2003) 104-107

[8] Xue Bai, Liu Wenyao etc: Research on Iris Image Preprocessing Algorithm. Journal of Optoelectronics • Laser, 14 (2003) 741-744

[9] Xiaoyan Yuan And Pengfei Shi: An Iris Segmentation Procedure for Iris Recognition. Advances in Biometric Person Authentication, 5th Chinese Conference on Biometric Recognition, SINOBIOMETRICS 2004.12:546-553

[10] Chen Gong, Zhou Youling: Iris Location Based on Hough Transform. Journal of East China University of Science and Technology, 30 (2004) 230-233

[11] Wan Li, Yi Ang, Fu Ming: An improved edge-detection method based on Canny algorithm. Computing Technology and Automation, 22 (2003) 24-26

[12] Zhang Xiaohong, Yang Dan, Liu Yawei: Improved edge detection algorithm based on Canny operator. Computer Engineering and Applications, 39 (2003) 113-115

[13] CASIA Iris Image Database, http://www.sinobiometrics.com

Applications of Wavelet Packets Decomposition in Iris Recognition

Junying Gan and Yu Liang

School of information, Wuyi University,
Jiangmen, Guangdong, P.R.C. 529020
jygan@wyu.cn

Abstract. The method of Wavelet Packets Decomposition (WPD) originating from wavelet transform is more accurate in signal analysis, with the predominance of analyzing high-frequency information. Combined with the trait of WPD, an algorithm for iris recognition is presented in this paper. Firstly, iris image is divided into several windows, and WPD is done to them. At the same time, some of the subband images from each window are selected, which contain most information of iris image. Secondly, the farther feature extraction and compression are applied to these subband images by way of Singular Value Decomposition (SVD), and iris recognition features are obtained. Finally, Weighted Euclidean Distance (WED) classifier is utilized in recognition. Experimental results on CASIA (Chinese Academy of Sciences, Institute of Automation) iris image database show the method is valid in iris recognition.

1 Introduction

Iris recognition has the best potential of development and the bright future due to its advantages of invariability, stability, acquirability, un-intrusion and so on, though its history is about 20 years. Statistical data reveals that compared with face, voice and other untouched method of identity authentication, iris has higher veracity and becomes a hot study field these years [1]. Daugman is the one who first studied iris recognition, and proposed the method of iris texture phase encoding based on Gabor wavelet [2]. Wildes presented the approach for decomposing iris image using Gaussian filter at different resolution level [3]. Boles introduced a zero-crossing detection way based on wavelet transform [4]. All these methods are based on wavelet transform. Development of iris recognition technology was accompanied by the application of wavelet transform altogether. In recent years, the method of Wavelet Packets Decomposition (WPD) gradually draws attention. Chen ji et al investigated a quality evaluation method of iris image based on WPD [5]. Emine Krichen et al contrasted the recognition results of iris image acquired under the conditions of visible light by a standard camera with ones under the conditions of near infrared illumination by a monochrome CCD camera. He verified that under the conditions of visible light illumination, the use of classical wavelet transform is not so satisfactory compared with WPD [6].

D. Zhang and A.K. Jain (Eds.): ICB 2006, LNCS 3832, pp. 443 – 449, 2005.

Iris possesses abundant detailed information in texture, and one of the key problems in iris recognition is how to extract iris features effectively. It is well known high-frequency information can be analyzed by WPD. Integrated with the trait of WPD, an approach for iris recognition based on WPD is presented in this paper. Firstly, division is done to each iris image, and several windows with the same size are acquired. Secondly, WPD is applied to these windows, and then a set of subband images from each window are gained. In order to extract iris features adequately, some of subband images from each window need be selected. For making farther feature extraction and compression, Singular Value Decomposition (SVD) is applied in the selected subband images with the help of the properties of Singular Value (SV). Finally, Weighted Euclidean Distance (WED) classifier is used in iris recognition. Experimental results on CAISA (Chinese Academy of Sciences, Institute of Automation) iris image database show that, correct recognition rate in this paper reaches 100% and a new approach in iris recognition is supplied.

2 Wavelet Packets Decomposition

WPD originating from wavelet transform is more accurate in signal analysis. Wavelet transform is one kind of mathematical tool, which is used for function decomposition in multilevel. After wavelet transform, a signal can be described by its wavelet coefficients. Because of this, wavelet transform becomes more and more attractive in the fields of texture recognition, signal processing, image processing, pattern recognition and so on.

In the process of wavelet transform, wavelet function $\psi(x)$ called wavelet base is introduced, which is made b translation and then convolved with $f(x)$ analyzed in different scale a. That is

$$\mathrm{WT}_f(a,b) = \frac{1}{\sqrt{a}} \int_{-\infty}^{\infty} f(x) \overline{\psi\left(\frac{x-b}{a}\right)} dx, \qquad a > 0 \tag{1}$$

Two-dimensional wavelet transform can be obtained by the generalization of one-dimensional wavelet transform. $\mathrm{WT}_f(a,b)$ is regarded as wavelet transform only in direction x. On the basis of it, wavelet transform in direction y is introduced, in which b and c denote translation parameters in direction x and direction y respectively. Here, $\psi(\frac{x-b}{a})$ is defined as wavelet function in direction x; whereas $\psi(\frac{y-c}{a})$ is wavelet function in direction y. Suppose $f(x,y)$ express two-dimensional signal analyzed, two-dimensional wavelet transform can be deduced as

$$\mathrm{WT}_f(a,b,c) = \frac{1}{\sqrt{a}} \int_R \int_R f(x,y) \overline{\psi_x\left(\frac{x-b}{a}\right) \psi_y\left(\frac{y-c}{a}\right)} dx dy \tag{2}$$

In image processing, binary wavelet is usually used. Suppose scales are discretized as the power of 2, which is expressed as $a = 2^0,\ 2^1,\ 2^2,\ \cdots,\ 2^j,\quad j = 1,\ 2,\ \cdots,\ N$. Tiny change of j can result in a great change in scale, and it shows that binary wavelet is an effective strategy from coarse to fine signal analysis. If $f(x, y)$ represents iris image, two-dimensional wavelet transform can be done respectively by different one-dimensional filters in direction x and direction y. Then a set of subband images in low and high frequency can be gained, as shown in Fig.1(a), where *LL* represents horizontal and vertical low-frequency information, *HL* horizontal high-frequency and vertical low-frequency information, *LH* horizontal low-frequency and vertical high-frequency information, *HH* horizontal high-frequency and vertical high-frequency information which is named as diagonal detail part. Through wavelet transform, the more evident image information features in one frequency or direction are, the more powerful the energy of the corresponding subband image will be. That is to say, features of iris image concentrates mainly on a few wavelet coefficients.

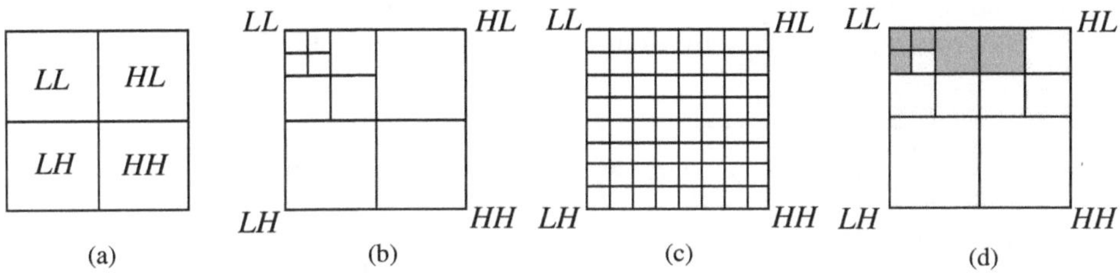

Fig. 1. Wavelet transform and WPD of iris image

In practical application, a signal can be divided into low-frequency and high-frequency information by wavelet analysis. Low-frequency information describes approximation part; whereas high-frequency information describes detail part. Then, decomposition is done to the approximation part. Thus its approximation and detail parts can be gained again. This operation can happen to the approximation part frequently if necessary, but never to the detail part, as shown in Fig.1(b). While each subband is divided into multilevel by WPD, it's more powerful in signal description and becomes compensation for wavelet analysis insufficiency in high-frequency information. In a word, decomposition is not only done to the approximation part, but also to the detail part, as shown in Fig.1(c). When WPD is used in texture image analysis, besides the analysis of approximation part, orthogonal decomposition is done to the selected detail part, and most information of iris image can be gained.

3 Wavelet Packets Decomposition in Iris Recognition

The flow char on WPD in iris recognition is shown in Fig.2. A series of operations to the acquired iris image are iris image processing, image division, WPD of windows, subband images selection, SVD and SV compression (SVC), and WED classifier respectively. The procedures of iris image processing and WPD in iris feature extraction will be discussed detailed.

3.1 Iris Image Preprocessing

Iris image preprocessing includes localization, normalization and image enhancement. Iris shape is approximately annular, in which inner round refers to the boundary between pupil and iris, and outer round refers to the boundary between iris and sclera. Therefore, localization of inner and outer round of iris image need be done respectively with two rounds un-homocentric. Firstly, threshold is utilized to pupil segmentation, then erosion and dilation are done to eliminate the useless spots, and projection is taken to carry out inner round localization, as shown in Fig.3(a). Because the gray between iris and sclera is not so evident, threshold method cannot be used to fix the two boundaries again. In this paper, Hough transform is utilized to iris outer round localization, in which Gaussian filter is used to iris image for smoothness firstly, then Canny operator is successively processed to the image for edge detection, and the outline of iris image is attained. According to the localization results of iris inner round, parameter ranges in Hough transform are limited. The center and radius of iris outer round can be fixed by Hough transform, as shown in Fig.3(a).

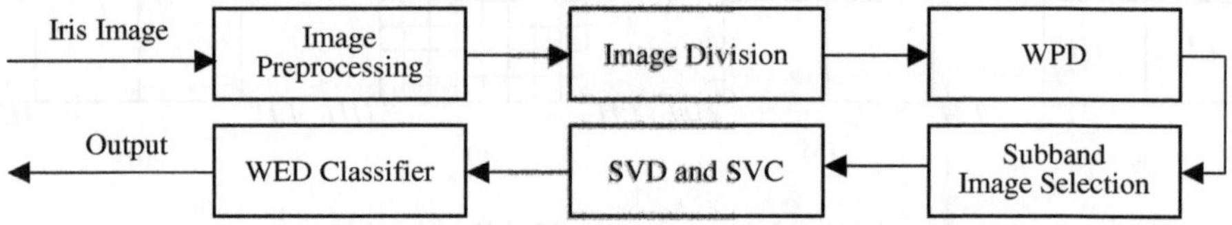

Fig. 2. Flow chart on iris recognition

To eliminate the influence of translation, proportion and rotation, it is necessary to adjust iris image to the same size after localization. Normalization is to map each pixel of iris image from orthogonal coordinates to polar coordinates, denoted by $I(x(r,\theta), y(r,\theta)) \to I(r,\theta)$, where $I(x,y)$ refers to the iris image after localization, and $I(r,\theta)$ the representation of iris image in polar coordinates. After normalization, iris image possesses translation invariance and changing invariance of iris inner and outer circle size. And iris image becomes a 64×512 rectangle in polar coordinates. For increasing recognition rate and lessening the influence of illumination, image enhancement is done to the iris rectangle by histogram equalization, as shown in Fig.3(b). Fig.3(b) demonstrates eyelid and eyelashes envelop iris partly, and can decrease recognition rate if feature extraction is done to the iris rectangle directly. In this paper, part of the enveloped iris is discarded, and a 64×256 iris rectangle is acquired as the object of feature extraction, as shown in Fig.3(c).

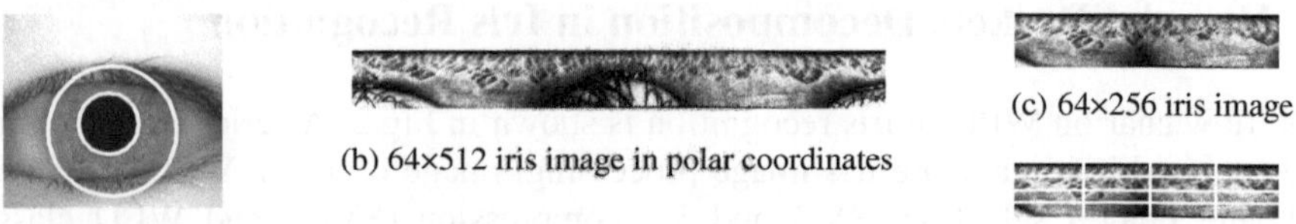

(a) Iris image localization

(b) 64×512 iris image in polar coordinates

(c) 64×256 iris image

(d) Iris image division

Fig. 3. Iris image preprocessing

3.2 Iris Feature Extraction by WPD

Iris image contains abundant information of texture details. If WPD is done to the whole iris image directly, a series of wavelet coefficients can be obtained. If the energy and variance of wavelet coefficients are regarded as iris image features, a great deal of image information can be abandoned and noise in high-frequency detail part can also be introduced. All these aspects can affect iris recognition rate. In order to extract local detail part of iris texture adequately, iris image is first divided into several windows with the same size, as shown in Fig.3(d). In Fig.3(a), the distribution of iris texture is radial. When iris image is mapped to polar coordinates, the distribution of iris texture is vertical. After done by wavelet transform, high-frequency information of iris image concentrates on horizontal detail part HL. While HH representing diagonal high-frequency information of iris image contains much noise, it is not suitable for feature extraction. In the process of wavelet transform, if the resolution levels are not enough, it can result in deficiency in classified information; whereas too many resolution levels can lead to huge calculations and the boosting up of boundary impact owing to small subbands. These aspects can affect iris recognition rate. Therefore, in this paper, Daubechies-4 wavelet decomposition is adopted, and each window of iris image is divided into 3 levels. After that, HL_1 containing most high-frequency information, whose subscript means sequence number of decomposition levels, is further decomposed according to the theory of WPD, and the corresponding low-frequency and high-frequency information are gained. In order to maintain effective information, get rid of the redundancy and diminish the effect of noise, LL_3, HL_3, LH_3, HL_2 and $HL_{1\_}LL$ altogether 5 subband images are held as the feature extraction objects, as shown in the shadows of Fig.1(d).

After WPD, SVD is applied to these 5 subband images respectively, and 5 SV vectors are acquired. Then, compression of SV vectors is necessary on account of redundant information and image noise. Reference [7] shows that the larger elements of SV play the key role in recognition. Therefore, the larger ones are maintained and the smaller ones omitted, which can help not only hold most information of iris image, but also weaken the influence of noise. In addition, the number of recognition features can be lessened greatly. Afterwards, the final recognition features are obtained, and WED classifier is utilized in iris recognition.

4 Experimental Results and Analysis

In this paper, experiments have been done on CASIA iris image database offered by the National Laboratory of Pattern Recognition (NLRP). This database (version 1.0) includes 756 iris images from 108 eyes of 80 subjects. They are gray images with resolution 320×280. For each eye, 7 images are captured in two sessions, where three samples are collected in the first session and the other four in the second session. In this paper, 280 images with 40 classes in CASIA iris image database are randomly selected in experiments.

In the process of image division, horizontal and vertical directions are divided into 4 and 4 respectively. So 16 windows are obtained, as shown in Fig.3(d). WPD is based on 3-level wavelet transform. First, All windows are processed respectively by

Daubechies-4 wavelet transform in 3 levels. After that, Daubechies-4 wavelet decomposition is utilized again to HL_1 where iris texture high-frequency detail part concentrates, and HL_1_LL retaining rich information of HL_1 is kept. Then LL_3, HL_3, LH_3, HL_2 and HL_1_LL 5 subband images of each window are chosen as the objects of further feature extraction, as shown in the shadows of Fig.1(d).

LL_3, HL_3, and LH_3 are three matrixes consisting of wavelet coefficients with the same size 8×14; whereas HL_2 and HL_1_LL are 9×21. Next, $[LL_3\ HL_3\ LH_3\ HL_2\ HL_1\_LL]$ 5 subband images are processed apart by SVD to obtain 5 SV vectors, the elements number of each SV vector are respectively [8 8 8 9 9]. With view of redundant information and noise contained in the SV vectors and the un-practicality of too many features, compression is done to the SV vectors. That is to preserve the larger elements of each SV vector, and then reshape the 5 SV vectors to a one-dimensional recognition feature vector. Finally, WED classifier is used in recognition.

Table 1 shows iris recognition rate based on WPD. In the experiments, the elements of 5 SV vectors are selected respectively in different ways. The number of training and testing samples is also changed to gain the results of iris recognition. In the meanwhile, several front images of each iris class are chosen as training samples, and the remainder ones of each iris class as testing samples. From Table 1, it can be known that, when the larger elements number of SV vectors corresponding to the $[LL_3\ HL_3\ LH_3\ HL_2\ HL_1\_LL]$ 5 subband images is [2 2 1 1 1], there are 7×16 feature elements with 112 in total, and the number selection of training and testing samples is 6_1, recognition rate reaches 100%. At the same time, increasing or decreasing the elements of SV vectors both results in the descending of recognition rate. But decreasing training samples and increasing testing samples results in the descending of recognition rate. When the number selection of training and testing samples is 3_4, recognition rate is 88.13%. On the whole, the selection of SV vectors in this case has better performance than the ones in other case. At the same time, increasing or decreasing the elements of SV vectors both results in the descending of recognition rate.

Table 1. Correct recognition rate of iris recognition on WPD

$[LL_3\ HL_3\ LH_3\ HL_2\ HL_1\_LL]\times 16$	6_1	5_2	4_3	3_4
$[3\ 3\ 2\ 1\ 1]\times 16 = 160$	95%	95%	93.33%	90%
$[3\ 3\ 1\ 2\ 1]\times 16 = 160$	97.5%	92.5%	91.67%	86.88%
$[3\ 3\ 1\ 1\ 1]\times 16 = 144$	97.5%	95%	92.5%	88.75%
$[3\ 2\ 1\ 1\ 1]\times 16 = 128$	100%	95%	92.5%	88.13%
$[2\ 2\ 1\ 1\ 1]\times 16 = 112$	100%	95%	92.5%	88.13%
$[2\ 1\ 1\ 1\ 1]\times 16 = 96$	100%	95%	90%	85.63%
$[1\ 1\ 1\ 1\ 1]\times 16 = 80$	95%	93.75%	87.5%	81.25%

*: The number '6' shown in '6_1'denotes the number of training samples in iris recognition, and '1' denotes the number of testing samples, the same to the others.

In addition, image division can also influence recognition rate. In this paper, the division of 4×4 is better than the other way of divisions. It is due to the proper consideration of local iris detail part in the vertical direction. If image is divided into more windows, the whole distribution of iris texture can be out of consideration.

5 Conclusions and Outlook

With the predominance of analyzing high-frequency information, WPD is utilized in iris image in this paper. Being a good algebraic description tool, SVD is applied to subband image, which contains most of iris texture detail part. Then SV compression is done to acquire final recognition features. Experimental results demonstrate that, the approach of iris recognition based on WPD is valid. But recognition rate of the approach is influenced by SV compression. Larger or smaller compression range can also result in the descending of iris recognition. Therefore, it's necessary to make further study on iris feature extraction and recognition.

Acknowledgement

This work is supported by NSF of Guangdong Province, P.R.C. (No.032356).

References

1. Anil K Jain, et al: Biometrics Personal Identification in Networked Society[M]. 101 Philip D river, A ssinippi Park, Norwell, Massachusetts 02061 USA, Kluwer Academic Publishers (1999) 103-121
2. Daugman J: High confidence visual recognition of persons by a test of statistical independence[J]. IEEE Trans. Pattern Analysis and Machine Intelligence, Vol. 15, No. 11, (1993)1148-1161
3. Woggen U: Optical properties of semiconductor quantum dots[M]. Germany: Springer, (1996)179-185
4. Bobles W W: A human identification technique using image of the iris and wavelet transform[J], Vol. 46. IEEE Transaction on Signal Processing (1998)1185 -1188
5. Chen Ji, Hu Guangshu, XU Jin: Iris image quality evaluation method based on wavelet packets decomposition[J], Vol.43, No.3. J Tsinghua Univ (Sci Tech)(2003)377-380
6. Emine Krichen, M. Anouar Mellakh, Sonia Garcia-Salicetti, Bernadette Dorizzi: Iris Identification Using Wavelet Packets[C], Vol. 4. Pattern Recognition, 17th International Conference, Proceeding of ICPR'04(2004)335-338
7. Gan Junying, Zhang Youwei: A New Approach for Face Recognition Based on Singular Value Features and Neural Networks[J], Vol. 32, No. 1. Acta Electronica Sinica(2004) 170-173
8. Yunhong Wang, Yong Zhu, Tieniu Tan: Biometrics Personal Identification Based on Iris Pattern[J], Vol. 28, No, 1. Acta Automatica Sinics(2002)1-10
9. Qing Qianqing, Yang Zongkai: Wavelet Analysis in Practical Applications[M]. Xi Dian Univerity Press (1998)

Iris Image Real-Time Pre-estimation Using Compound BP Neural Network

Xueyi Ye, Peng Yao, Fei Long, and Zhenquan Zhuang

Department of Electronic Science and Technology,
University of Science and Technology of China, HeFei 230026, P.R. China
xueyi_ye@ustc.edu, xueyiye@gmail.com

Abstract. A practical iris identification application system faces different types of bad iris images resulted from many reasons. Because previous image quality evaluation methods estimate an iris image whether bad or else by the resolution and the definition of the iris part, they just can deal with few types among them. For saving the time occupied by the localization in images real-time estimation, improving friendly interaction of an iris identification system, decreasing the localization failure on account of importing the bad-image, this paper proposes a method of real-time pre-estimation using the compound BP neural network. Multiple independent BP neural networks are used to extract both the overall contour feature and the local of an iris image and to calculate the pre-estimation output by different training weights. The experimental result is shown that the method can detects most types of the bad-image with comparatively low error rate and the pre-estimation network has fairly large throughput. It should satisfy the pre-estimation requirement of a real-time iris identification system.

1 Introduction

Biometrics technology using intrinsic physiological or behavioral characteristics of human beings to authenticate identities, so far, is the most prospective candidate displacing the traditional identification method (e.g., token-based or knowledge-based methods). Many different biometric methods, including fingerprint, iris, hand geometry, face, signature, and so on, have successfully deployed to provide useful value in practical applications. For distinctive abundant texture and good stability, under the precondition of having captured normal iris images, the iris identifier shows excellent performance (i.e., fairly low false acceptance rate, FAR, false reject rate, FRR, and friendly interaction). At one time, it also has relatively high failure to enrollment because it requires rigid sampling condition [1]. Users often feel uncomfortable at the time of failure to enrollment again and again specially in a real-time application system, which hampers the research community to broaden the iris identification to a large scale real application [2].

As shown in figure 1, there are many types of the bad-image (e.g., no iris part in an image, half-baked iris, excess distortion of iris, low definition for incorrect focus, losing much texture information for few pixels of iris part, eyes moving at the instant of capturing image.) because of different reasons, and therefore several factors

D. Zhang and A.K. Jain (Eds.): ICB 2006, LNCS 3832, pp. 450 – 456, 2005.

possibly result in which an iris sampler can not capture images according to the demand of identification process (i.e., appearing bad images).

Generally, the iris identification technology disposes the challenge of the bad-image with a strategy of feedback which can divide into three categories: 1) It evaluates each image captured by a camera, and then decides whether an image is bad or not by the output of the evaluation [3]. 2) From the hardware perspective, for instance, it use a peculiar mirror, so called the mono-direction-mirror, by which an user can directly see the approximate size and location of his iris in the mirror [4]. 3) A person controls the iris identifier and decides whether an input image satisfies the requirement of identification [4]. Apparently, the third category will be discarded gradually, because it is opposite to the trend of the automatic identification which is expected as little human supervision as possible.

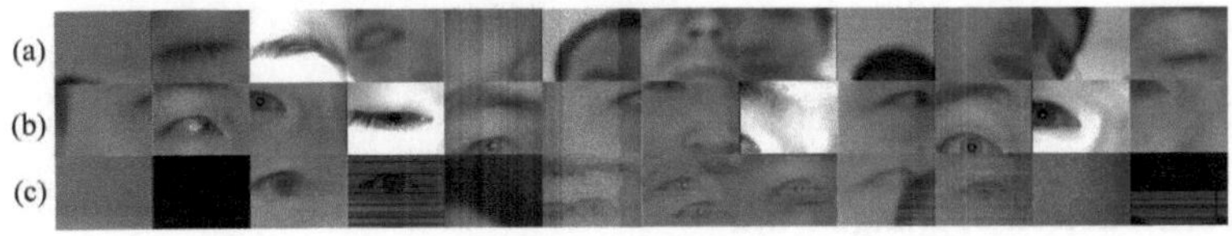

Fig. 1. Bad images possibly captured by iris identification system, (a): Iris absence in an image, (b): Incorrect view angle awing to head motion, (c): Errors of sampler-self

As shown in figure 1, these images are difficulties towards no matter the first category or the second. If iris image samples are pre-estimated automatically by a virtual watcher, and its output will decide whether an identifier continues the next step or captures images again, most types of the bad-image will be eliminated before the process of localization. This paper proposes a new idea of the pre-estimation accomplished by a compound BP neural network. That is, input samples are estimated in real time by the pre-estimation network before the localization step.

The remainder of this paper is organized as follows. Section 2 provides a detailed description of the pre-estimation method using the compound BP neural network including three subsections. Section 3 illuminates experimental items and their results. Analysis and discussion are reported in section 4.

2 Compound BP Neural Network

The pre-estimation of image samples is belonged to the forepart of the iris identification preprocess. In a real-time iris identification application system, it disposes an image or sequence images kept in the memory and exports an outcome parameter, by which the system decides to go to the next processing or back to the previous step again. This paper proposes a new compound BP neural network to pre-estimate iris image samples, in which it uses the back propagation algorithm with moment factors and expects to depress the adverse effect resulting from the BP neural network inextricable frauds recurring to its novel network topology and training modes [5]. The whole network comprises three parts (as shown in figure 2).

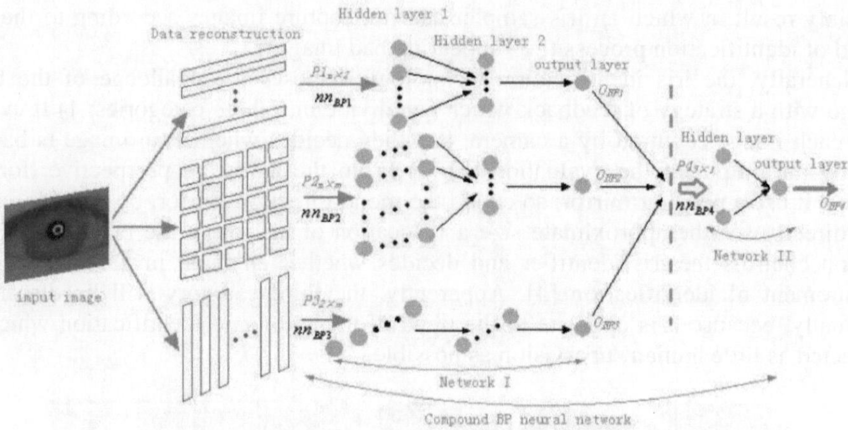

Fig. 2. Compound BP neural network for the pre-estimation of real-time images

2.1 Reconstruction of Image Data

Firstly, an iris image $\boldsymbol{I}$ is resized to be the image $I_{n\times m}$ and keep the same proportion of n to m of $I_{n\times m}$ as $\boldsymbol{I}$ (n and m is the number of rows and columns of the image $I_{n\times m}$, respectively) to ensure $I_{n\times m}$ not to be distorted. After converting the image $I_{n\times m}$ to the binary image by a threshold, the elements of this binary image is assembled to the input matrix $P2_{n\times m}$. Finally, $P1_{n\times 2}$ and $P3_{2\times m}$ are built by $P2_{n\times m}$ as shown in formulas (1) ~ (6). The matrix T is the training target of these BP neural networks and its size is equal to the number of input training samples. If an input training sample is a bad image, the corresponding element of T is -1; and if a normal image, it is 1.

$$I_{n\times m} = resize(I), \quad (n,m \in N), \quad P2_{n\times m} = binary(I_{n\times m}). \tag{1}$$

$$P1_{n\times 2}(i,1) = \mu_{i1} = \frac{1}{m}\sum_{j=1}^{m} P2_{n\times m}(i,j), \quad i \in (1,\cdots,n), j \in (1,\cdots,m). \tag{2}$$

$$P1_{n\times 2}(i,2) = \sigma_{i2} = \frac{1}{m}\sum_{j=1}^{m}[P2_{n\times m}(i,j) - \mu_{i1}]^2. \tag{3}$$

$$P3_{2\times m}(1,j) = \mu_{1j} = \frac{1}{n}\sum_{i=1}^{n} P2_{n\times m}(i,j). \tag{4}$$

$$P3_{2\times m}(2,j) = \sigma_{2j} = \frac{1}{n}\sum_{i=1}^{n}[P2_{n\times m}(i,j) - \mu_{1j}]^2. \tag{5}$$

$$P1_{n\times 2} = \begin{pmatrix} \mu_{11} & \sigma_{12} \\ \vdots & \vdots \\ \mu_{n1} & \sigma_{n2} \end{pmatrix}, \quad P2_{n\times m} = \begin{pmatrix} a_{11} & \cdots & a_{1m} \\ \vdots & \ddots & \vdots \\ a_{n1} & \cdots & a_{nm} \end{pmatrix}, \quad P3_{2\times m} = \begin{pmatrix} \mu_{11} & \cdots & \mu_{1m} \\ \sigma_{21} & \cdots & \sigma_{2m} \end{pmatrix}. \tag{6}$$

2.2 Part Ⅰ of the Network

The network part Ⅰ consists of three independent BP neural sub-networks nn_{BP1}, nn_{BP2} and nn_{BP3}. As shown in figure 2, they all include one input layer, two hidden layers and one output layer, but the corresponding layer node number of various networks are different except their output layers that are composed of single node and severally confirmed by the experience and the experimental effect [6][7][8]. Weight matrices of nn_{BP1}, nn_{BP2} and nn_{BP3} are respectively defined as $(W^1_{BP1}, W^2_{BP1}, W^3_{BP1})$, $(W^1_{BP2}, W^2_{BP2}, W^3_{BP2})$ and $(W^1_{BP3}, W^2_{BP3}, W^3_{BP3})$, and their bias matrices are $(b^1_{BP1}, b^2_{BP1}, b^3_{BP1})$, $(b^1_{BP2}, b^2_{BP2}, b^3_{BP2})$ and $(b^1_{BP3}, b^2_{BP3}, b^3_{BP3})$ as well. Their active functions select the consecutive derivative *Sigmoid* function. $P1_{n\times 2}$, $P2_{n\times m}$ and $P3_{2\times m}$ shown in the formula (6) form input vectors of those sub-networks, and then their outputs o_{BP1}, o_{BP2} and o_{BP3} respectively according to nn_{BP1}, nn_{BP2} and nn_{BP3} construct the matrix $P4_{3\times 1}$, as shown in formulas (7) ~ (10), regarded as the input vector of the network II (nn_{BP4}). The calculation ' ' in formulas means that $\sum_{i=1,j=1}^{k} X_{(1,i)} \times Y_{(j,1)}$ (X as a weight matrix and Y as an input matrix), and k is the dimension number of an input vector.

$$O_{BP1} = f^3_{Sigmoid}(W^3_{BP1}\ f^2_{Sigmoid}(W^2_{BP1}\ f^1_{Sigmoid}(W^1_{BP1}\ P1_{n\times 2} + b^1_{BP1}) + b^2_{BP1}) + b^3_{BP1}) . \tag{7}$$

$$O_{BP2} = f^3_{Sigmoid}(W^3_{BP2}\ f^2_{Sigmoid}(W^2_{BP2}\ f^1_{Sigmoid}(W^1_{BP2}\ P2_{n\times m} + b^1_{BP2}) + b^2_{BP2}) + b^3_{BP2}) . \tag{8}$$

$$O_{BP3} = f^3_{Sigmoid}(W^3_{BP3}\ f^2_{Sigmoid}(W^2_{BP3}\ f^1_{Sigmoid}(W^1_{BP3}\ P3_{2\times m} + b^1_{BP3}) + b^2_{BP3}) + b^3_{BP3}) . \tag{9}$$

$$P4_{3\times 1} = \begin{pmatrix} o_{BP1} \\ o_{BP2} \\ o_{BP3} \end{pmatrix}, \quad (O_{BP1}, O_{BP2}, O_{BP3} \in [-1,1]) \tag{10}$$

$$O_{BP4} = f^4_{Sigmoid}(W^2_{BP4}\ f^1_{Sigmoid}(W^1_{BP4}\ P4_{3\times 1} + b^1_{BP4}) + b^2_{BP4}), \quad (O_{BP4} \in [-1,1]) . \tag{11}$$

2.3 Part Ⅱ of the Network

The network Ⅱ also is a independent BP neural network nn_{BP4}. The connection between the network Ⅱ and the network Ⅰ is switched off at the time of training network Ⅰ. After having accomplished the training of the network Ⅰ, the output $P4_{3\times 1}$ of the network Ⅰ is regarded as the training input data of the network Ⅱ. According to the same training target matrix T, the network nn_{BP4} is trained. Finally, three parts of the compound BP neural network are connected to pre-estimate input samples and the pre-estimation final result is the network output (i.e. o_{BP4} shown in the formula 11). The sub-network nn_{BP4} consists of one hidden layer (including two nodes), one output layer (including one node) and one input layer $p4_{3\times1}$. In the training process of nn_{BP4}, the error back-propagation just take place in the bound of the network II, refreshing its weight matrices (W^1_{BP4}, W^2_{BP4}) and bias matrices (b^1_{BP4}, b^2_{BP4}) .

3 Experimental Results

Because the pre-estimation of image samples is in front of the previous processing of image quality evaluation, fast detecting whether an image sample includes the ingrate iris part is the unique task of the pre-estimation network. The whole experimental image samples are from the local iris database (Self-Databases), the bad-image database, CASIA-Open iris database [9] (CASIA-open) and the training iris database provide by the first contest of biometrics of China [10] (CASIA-contest-iris-DB1&DB2). Examples shown in figure 1 and figure 3 are selected from these databases.

There are 64 images demonstrated in figure 3 which include some typical image samples captured under different conditions, such as different sampling instruments, different illuminations, different imaging distances, different poses and different people.

There are sizeable discriminations between images captured by different instruments, as shown in figure 3. And, therefore, above four databases are not mixed at most of experiments. Finally, for proofing the robust of the pre-estimation network, mixed databases are done. As shown in Table 1, these classified experiments appear hortative result: Their false rates are lower than 1% and the throughput of the pre-estimation network mostly is 18000 samples per second. It should satisfy the requirement of a real-time iris identification application system.

Fig. 4a, 4b, 4c and 4d severally describe the output distribution of the pre-estimation network to four databases, corresponding the table 1, and fig. 4e and 4f demonstrate two types of mixing databases. In these figures, the horizon coordinate denotes the number of the testing sample and the vertical coordinate denotes the network output value to the input sample.

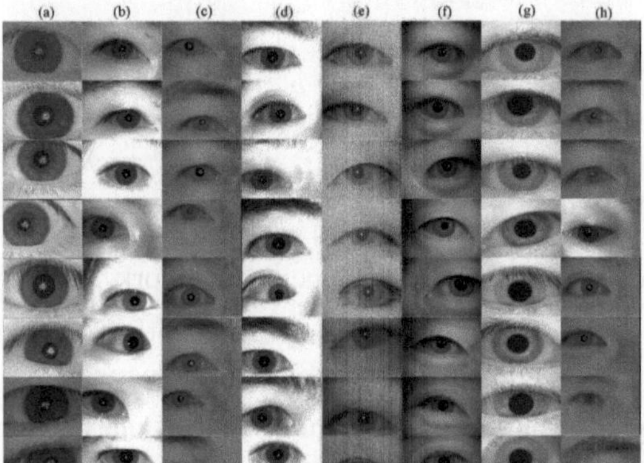

Fig. 3. Normal images accepted by pre-estimation network

Table 1. The experimental result

Databases	Training Samples Normal samples	Bad-image samples	Testing Samples Normal samples	Badimage samples	Gross testing time (s)	False Rate (%)
Self-Databases	540	244	1310	1104	0.125	0. 042
CASIA-open	256	244	500	1104	0.094	0.37
Contest-DB1	400	244	800	1104	0.109	0.58
Contest-DB2	400	244	800	1104	0.110	0.58
DB1&DB2	800	244	1600	1104	0.156	0.41
DB1,DB2& open	1056	244	2100	1104	0.172	0.41

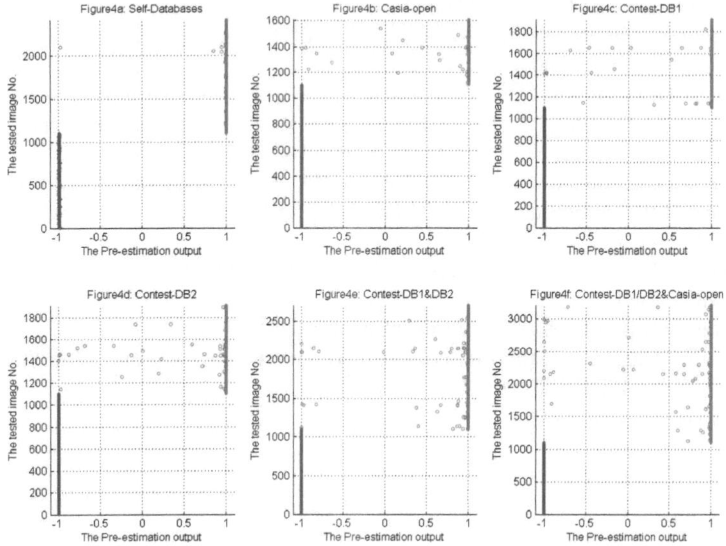

Fig. 4. The output distribution of the pre-estimation network to different databases

4 Analysis and Discussion

Addressing the purpose of the pre-estimation of iris images, It is easy to note on the experimental result that the pre-estimation method based on the compound BP neural network represents approving performances both the throughput and the false rate of the pre-estimation and can preferably satisfy the real-time estimation requirement of an iris identification system. By the novel network topology, the data-reconstruction, the independent training and associated testing tricks, sub-BP-neural-networks have been trained with lesser epochs. That is, depressing the possibility of the over-training, and meantime, leading the whole pre-estimation network to possess better generalization than an ordinary BP neural network. And that, three independent data channels of the network I can extract both overall and local features of the iris image geometric contour. Moreover, according to the figure 4, though the false rate of estimation is non-zero toward every database, the output of the pre-estimation to the bad-image belongs to the diminutive domain of -1 (i.e. the false acceptance rate of the pre-estimation network is zero).

References

1. Anil K.Jain, Sharath Pankanti, Salil Prabhakar, Lin Hong, Arun Ross, James L.Wayman , "Biometrics: A Grand Challenge", Proceedings of International Conference on Pattern Recognition, Cambridge, UK, Aug, 2004
2. NIST report to the United States Congress, "Summary of NIST Standards for Biometric Accuracy, Tamper Resistance, and Interoperability", Available at ftp://sequoyah.nist.gov/pub/nist_internal_reports/NISTAPP_Nov02.pdf, November 2002
3. HE Jia-feng, YE Hu-nian, YE Miao-yuan, "A Study on Iris Image Quality Evaluation", Journal of Image and Graphics, China,Vol.8(A), No.4, Apr.2003
4. J Daugman, "Iris Recognition: Current state of the art", The ASI'04 in HongKong, P.R.China, December 2004
5. Rumelhart D.E., Hinton G.E., Williams R.J., "Learning internal representations by error propagation in Parallel Distributed Processing", Vol.1, Rumelhart D.E. and McClelland J.L., EDS，Cambridge, MA: MIT Press, 1986, 318～362
6. Eberhart R.C., Dbbins R. W., "Neural Network PC tools", Academic Press, 1990
7. Lippmann R.P., "A Introduction to Computing with Neural Nets", IEEE ASSP Magazine, Nov. 1987, 4-22
8. Lippmann R.P., "Pattern Classification using Neutral Networks", IEEE Comm., Magazine, Nov. 1989, 47-64
9. Iris Database 1.0 of CASIA, www.sinobiometircs.com
10. Iris training Database of the first contest of Biometrics recognition and testing of China, 2004, www.sinobiometrics.com

Iris Recognition in Mobile Phone Based on Adaptive Gabor Filter

Dae Sik Jeong[1], Hyun-Ae Park[1], Kang Ryoung Park[2], and Jaihie Kim[3]

[1] Department of Computer Science, Sangmyung University,
7 Hongji-Dong, Jongro-ku, Seoul, Republic of Korea
Biometrics Engineering Research Center (BERC)
{jungsoft97, whitebbb}@smu.ac.kr
[2] Division of Media Technology, Sangmyung University,
7 Hongji-Dong, Jongro-ku, Seoul, Republic of Korea
Biometrics Engineering Research Center (BERC)
parkgr@smu.ac.kr
[3] Biometrics Engineering Research Center (BERC),
Department of Electrical and Electronic Engineering, Yonsei University,
Seoul, Republic of Korea
jhkim@yonsei.ac.kr

Abstract. As the security of personal information is becoming more important in mobile phones, we apply iris recognition technology to mobile device. Different from conventional iris recognition system used for access control, user puts the mobile phone by hands in this case. So, optical and motion blurring happens, frequently. In addition, most users have tendencies to use the mobile phone in outdoor and sunlight (which includes much amount of IR(Infra-Red) light) may have much effect on the input iris image in spite of the visible light cut filter attached in front of iris camera lens. To overcome such problems, we propose a new method of extracting the accurate iris code based on AGF (Adaptive Gabor Filter). The kernel size, frequency and amplitude of Gabor filter are determined by the amount of blurring and sunlight in input image, adaptively. Experimental results show that the EER by our propose method is 0.14 %.

1 Introduction

Compared to other biometric systems, the iris recognition system is reputed to the most reliable among all biometric methods [3][4][5][6][7]. Iris recognition is to recognize a person by using unique iris patterns, which exist in iris region between white sclera and black pupil [1][2].

Recently, with the increasing need of guaranteeing the security in case of using bank transaction service by using mobile phone, it is required to apply biometrics for the security of mobile phone. For example, mobile phone with fingerprint recognition (such as LG-KP3800 made by LG Electronic) has been already produced [9]. However, it requires additional fingerprint image acquisition sensor and DSP chip for fingerprint recognition. This cause the cost and size of mobile phone to be increased. So, we try to apply the iris recognition to mobile phone.

D. Zhang and A.K. Jain (Eds.): ICB 2006, LNCS 3832, pp. 457 – 463, 2005.

In general, for iris recognition, the magnified iris image (more than 200 pixels of diameter) is required for accurate recognition [13]. So we need a zoom and focus lenses for iris recognition camera, which cause the size and cost to be increased. However, in case of using mega pixel camera built in mobile phone, though face image is captured with the far distance between eye and camera (more than 30 to 40 cm), the iris region can include sufficient pixel information for recognition. So, we aim at developing the iris recognition system in mobile phone only by using a built-in mega-pixel camera and software without additional hardware component.

For iris recognition, iris feature extraction is critical step. Previous methods for iris feature extraction [3][4][7][8][14] require too much processing power to be used in ARM CPU in mobile phone (for example, ARM 9 has the processing power of about 200MHz [16] [17] and ARM CPU does not have an internal floating point processing). Otherwise, Noh's method [15] uses ICA based 1D Filter. However, it uses the determined ICA basis from training and cannot deal with the iris image affected by image blurring and sunlight in mobile phone. To overcome such problems, we proposed a new 1D Adaptive Gabor Filter apt for iris recognition in mobile phone.

2 Iris and Pupil Localization

To extract iris code from input iris image, we first detect iris and pupil region. Because the pupil has the hollow structure inside of cornea [18], the gray level of pupil in input image is very low compared to other region such as iris, sclera and skin. So, by simple binarization, we can discriminate the pupil and the other region such as iris, sclera and facial skin. However, some dark region such as shaded facial skin, iris and eyelash shows same gray level to pupil. So, we change decoder value of brightness and contrast, if we make the brightness and contrast value of A/D converter lower and higher, we can easily select the binarization threshold to detect the pupil region and bright corneal specular reflection [19]. From that information, we detect the pupil region and locate iris region by modified circular edge detection [19]. We then choose a region free from occlusions caused by eyelids or eyelashes. To detect eyelashes, we use the eyelash mask [25]. We also use the elliptical eyelid deformable template to detect the eyelid regions [26]. The iris codes generated in eyelash and eyelid are set to invalid codes and are not used for enrollment and recognition.

3 Adaptive Gabor Filter Considering Blurring and Sunlight

3.1 Overview of the Proposed Adaptive Gabor Filter and the Illuminator-On/Off Scheme

After detecting iris region in input image, we extract iris feature code by Adaptive Gabor Filter. The overview of the proposed method is shown in Fig. 1. As mentioned before, most users of mobile phone have tendencies to use the mobile phone in outdoor and sunlight (which includes much amount of IR(Infra-Red) light) may have much effect on the input iris image in spite of the visible light cut filter attached in front of iris camera lens. In addition, different from conventional iris recognition system used for access control, user puts mobile phone by hand and optical & motion blurring happen frequently. So, we measure the amount of blurring and sunlight from input image and change the kernel size, frequency and amplitude of Gabor filter according to the measure amount.

As shown in Fig. 1, user presses "the button for iris recognition" in the mobile phone and the camera micro-controller turns the IR (Infra-Red) illuminator on and off, successively, and six successive iris images are captured as shown in Fig. 1(1) and 2.

Then, we measure the amount of sunlight in input image ((2) in Fig. 1). For that, we calculate a average gray value of image frame #2 , #4 and #6 (of Fig.2) as Eq.(1). As shown Fig. 2, because image frame #2 , #4 and #6 are obtained with IR illuminator off and we use visible light cut filter in front of camera lens, the image is bright in case that environmental sunlight exist.

$$\boldsymbol{Br} = (\mathbf{A2} + \mathbf{B2} + \mathbf{C2})/3 \qquad (0 \le \boldsymbol{Br} \le 255), \tag{1}$$

where A2, B2 and C2 are a gray value in image frame #2 , #4 and #6, respectively. In case that ***Br*** exceeds in 50, we determine the outdoor environment in which sunlight exists. Otherwise, we do the indoor environment in which sunlight does not exist.

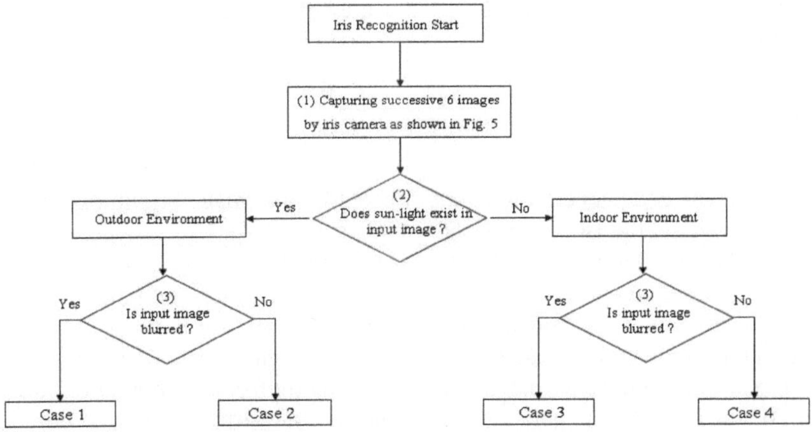

Fig. 1. Overview of proposed method

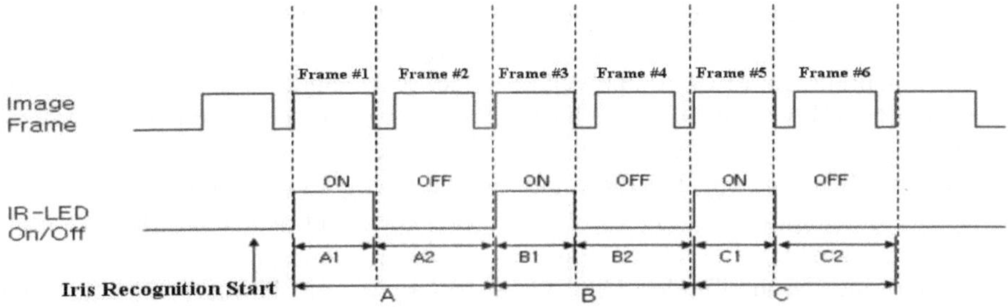

Fig. 2. Capturing successive 6 images by iris camera in mobile phone

Then we check whether the input image is blurred or not ((3) in Fig. 1). For that, we use the iris focus checking mask proposed by Kang [25], which checks high frequency component in input image. With that, we can obtain continuous focus value from 0 to 100 (100 means the most focused and 0 does the least focused). We apply the focus checking mask to the image frame #1 and #3 (of Fig.2), and obtain the focus value of **FV1** and **FV2** in Eq. (2). And final focus value in input image is calculated by ***Ft*** and in case that ***Ft*** exceeds in 50, we determine there exist optical & motion blurring in input image.

$$\boldsymbol{Ft} = (\mathbf{FV1} + \mathbf{FV2})/2 \qquad (0 \le \boldsymbol{Ft} \le 100), \tag{2}$$

where **FV1** and **FV2** are the measured focus value of A1 and B1. As shown in Fig. 1, we can classify 4 cases as such (Case 1 : Blurred and Sunlight, Case 2 : Focused and Sunlight, Case 3 : Blurred and No Sunlight, Case 4 : Focused and No Sunlight)

3.2 Adaptive Gabor Filter

Then, we use the measured ***Br*** and ***Ft*** for selecting the kernel size, frequency and amplitude of Gabor filter. In our case, we use adaptive Gabor Filter that can be defined as follows:

$$G(x) = A \cdot e^{-\pi[\frac{(x-x_0)^2}{\sigma^2}]}(\cos(2\pi[u_0(x-x_0)]), \quad \sum_{x_0=-N}^{N} G(x) = 0\,(\because DC = 0), \tag{3}$$

(To make iris code generation is not affected by the image brightness of iris texture, we set the DC component of Gabor filter to be 0)

where $\boldsymbol{A}$ is Gabor Filter's amplitude, σ and $\boldsymbol{u_0}$ are the kernel size and the frequency of Gabor Filter, respectively. $2\boldsymbol{N}$ is the number of Gabor Filter coefficient. Due to the limitation of processing power in mobile phone, we only use the real term of Gabor Filter. In addition, we do not use the scheme of iris region stretching to make rectangle iris region. Instead, we extract the iris code directly from iris region of polar coordinate. To consist with conventional Daugman iris code structure, we use 8 tracks and 256 sectors to extract iris code. After extracting iris code, we use the HD (Hamming Distance) to measure the dissimilarity between two iris codes [3].

The gray level change (contrast) in iris texture is diminished in case of sunlight. That is because sunlight increases the overall image brightness of iris texture. In addition, when the longer wavelength (more than 900nm) of IR illuminator is used by sunlight (the wavelength of our iris camera illuminator is 760 and 850nm, but sunlight includes much amount of IR component of longer wavelength), iris texture shows the characteristics of low contrast. In such case, the accurate iris code cannot be extracted by static Gabor filter. Especially, in case that user enrolls his iris in indoor and tries to identify himself in outdoor, the FRR may be increased. To overcome such problem, we increase the amplitude of Gabor filter in case of sunlight, which has the effect that increases the signal level of low contrast iris texture.

To determine the amplitude (***A***) of Gabor filter, we use the measured the image brightness by sunlight (***Br***) as shown in Eq. (4)

$$\boldsymbol{A} = (1/11)\boldsymbol{Br} + 180 \tag{4}$$

Eq.(4) was obtained by experiment. About each image brightness (***Br***), we measured the EER with Gabor filter (Eq.(3)) of various range of amplitude and selected the optimal amplitude(***A***) in each image brightness (***Br***). In this case, we used a static kernel size of 47 and frequency of $\pi/8$, which were obtained by our experiment with CASIA DB [23] of showing the minimum EER. Of course, the phase component of Gabor filter is not basically affected by the iris image contrast, but we use the offset margin for quantizing of iris code (0 or 1) instead of using threshold 0. So, in case that image contrast is low due to sunlight, the calculated phase component is very small and cannot be credible with offset margin. Due to such reason, Eq.(4) can be useful in case of iris texture of low image contrast. In similar method, we obtain the kernel size(σ) and frequency of Gabor filter (u_0) based on the focus value ***Ft*** of Eq. (2) as shown in Eq.(5). In general, the more the input image is blurred, the longer the kernel size and lower frequency of Gabor filter is required to deal with low frequency iris texture.

$$\sigma = (-1/16)\boldsymbol{Ft} + 54$$
$$u_0 = (-\pi/800)\boldsymbol{Ft} + \pi/4 \qquad (5)$$

From that, we can determine the amplitude and the kernel size & frequency of Adaptive Gabor Filter based on the measured image brightness and focus value and extract the accurate iris code for recognition.

4 Experimental Results

For experiment, we use the mobile phone (the model name is SPH-S2300 (by Samsung Electronics) with a 3.2 mega-pixel (2048*1536) CCD sensor). To capture detailed iris patterns, it is necessary to use an IR-illuminator and an IR pass filter [16][17]. We attached the IR pass filter (which only passes over the wavelength of 750 nm) in front of the camera lens and used the built-in Xenon flash lamp (which had the characteristics of including both visible and IR light wavelengths) for iris recognition. Tests were performed on 200 iris images captured from 20 persons. Test images were also obtained indoors (100 images) and outdoors (100 images). The average intensity of illumination in the indoor was 223 Lux. (measured by optical power meter) and that in the outdoor was 1,394 Lux. (including sunlight). Experimental results showed that in case of using Daugman's Gabor filter (with the kernel size of 47, the frequency of $\pi/8$ and amplitude of -120 ~ +170), the EER was 0.09% (***d'*** is 8.4 [8]) (all the enrollment and recognition were performed with the indoor images) and 0.11% (***d'*** is 8.1) (all the enrollment and recognition were performed with the outdoor images). However, the EER was much increased to 1.2 % (***d'*** is 1.5) (the enrollment was performed with indoor images, but the recognition was tried with outdoor images of sunlight). In this case, optimal kernel size, frequency and amplitude of Gabor filter were obtained using CASIA DB [23]. The CASIA DB consists of 8-bit gray-level images of 756 (108 eyes for 80 subjects) with 320x280 pixels. The EER was 0.088% (***d'*** is 8.5 [8]) (all the enrollment and recognition were performed with the CASIA DB) and we can know that iris image quality of our mobile camera is not degraded compared to CASIA DB excluding sunlight and severe blurring. In case of using Adaptive Gabor Filter based on the measured image brightness and focus value, the EER was 0.09% (*d'* is 8.4) (all the enrollment and recognition were performed with the indoor images) and 0.10% (*d'* is 8.2) (all the enrollment and recognition were performed with the outdoor images). However, the EER was somewhat increased to 0.14 % (*d'* is 7.7) (the enrollment was performed with indoor images, but the recognition was tried with outdoor images of sunlight). However, it is much smaller to that with static Gabor filter. In this case, we set the threshold for authentic and imposter distribution, with which the FAR is equal to the FRR, by experiment. Processing time is about 1,800ms in Mobile Phone (SPH-S2300) of ARM926EJ-STM (150 MHz).

5 Conclusions

In this paper, we propose a new method of extracting the accurate iris code based on AGF (Adaptive Gabor Filter). The kernel size, frequency and amplitude of Gabor filter are determined by the amount of blurring and sunlight in input image, adaptively. Experimental results show that the EER by propose method is 0.14 %. In future works, more field tests are required to enhance the performance of AGF. In addition, the eyelash shade region made by sunlight should be detected and excluded to extract iris code for better performance.

Acknowledgements

This work was supported by the Korea Science and Engineering Foundation (KOSEF) through the Biometrics Engineering Research Center (BERC) at Yonsei University.

References

[1] A. K. Jain, et al., "BIOMETRICS: Personal Identification in Networked Society," Kluwer Academics Publishers, Norwel, MA, 1999

[2] A. Pankanti, R. M. Bolle and A. K. Jain, "Biometrics: the future of identification," IEEE Computer, Vol. 33, No. 2, pp. 46-49, 2000.

[3] J. G. Daugman, "High Confidence Visual Recognition of Personals by a Test of Statistical Independence ", IEEE Trans. PAMI, Vol.15, No.11, pp.1148 - 1160, Nov., 1993.

[4] Richard P. Wildes, "Iris Recognition:An Emerging Biometrics Technology", Proceedings of the IEEE, Vol.85, No.9, pp.1348 - 1363, Sep., 1997.

[5] W.W. Boles and B. Boashash, "A Human Identification Technique Using Image of the Iris and Wavelet Transform", IEEE Trans. on SP, Vol.56, No.4, pp.1185-1188, 1998.

[6] Christel-loic, et al., "Person identification technique using human iris recognition", The 15th International Conference on Vision Interface, May, pp.294-299, 2002.

[7] Yong Zhu, et al., "Biometric personal identification based on iris patterns", Proceedings of 15th International Conference on Pattern Recognition, Sep., 2000, Vol. 2, pp. 805-808

[8] John G. Daugman, "The importance of being random: statistical principle of iris recognition", Pattren Recognition, vol. 36, no. 2, pp279-291

[9] http://www.lge.com (accessed on June 8, 2005)

[10] http://www.lgiris.com (accessed on June 8, 2005)
[11] http://www.iridiantech.com (accessed on June 8, 2005)
[12] http://www.panasonic.com/cctv/products/biometrics.asp (accessed on June 8, 2005)
[13] Kang Ryoung Park, Jaihie Kim, "A Real-time Focusing Algorithm for Iris Recognition Camera", Lecture Notes in Computer Science (ICBA 2004), Vol. 3072, pp.410~417, July 2004
[14] Li Ma, et al., "Personal Identification Based on Iris Texture Analysis", IEEE Trans. PAMI, Vol. 25, No. 12, Dec. 2003
[15] S.I.Noh, et al.,"A New Iris Recognition Method Using Independent Component Analysis", IEICE Transactions on Information and Systems, accepted for publication
[16] http;//www.arm.com (accessed on June 8, 2005)
[17] http://www.qualcom.com (accessed on June 8, 2005)
[18] Gregory A. Baxes, Digital Image Processing – Principles and Applications, Wiley
[19] Dal Ho Cho, et al., "Real-time Iris Localization for Iris Recognition in Cellular Phone", SNPD2005, Towson University, Maryland, USA, May 23 - 25, 2005
[20] http://www.biometrics.org (accessed on June 8, 2005)
[21] http://ww.iris-recognition.org (accessed on June 8, 2005)
[22] John. Dagman, "How Iris Recognition Work", IEEE Trans. on Circuits and Systems Video Technology, Vol. 14, No. 1, Jan. 2004
[23] CASIA iris DB of Chinese Academy of Science, http://www.sinobiometrics.com/resources.htm (accessed on June 8, 20)
[24] R. Jain, Machine Vision, McGraw-Hill
[25] Byung Joon Kang, Kang Ryoung Park, "A Study on Iris Image Restoration", Lecture Notes in Computer Science (AVBPA 2005), July 2005, accepted for publication
[26] Kang Ryoung Park, "Practical Gaze Point Computing Method by 3D Position Estimation of Facial and Eye Features", LNCS, Vol. 3339, pp.237~247, Dec., 2004

Robust and Fast Assessment of Iris Image Quality

Zhuoshi Wei, Tieniu Tan, Zhenan Sun, and Jiali Cui

National Laboratory of Pattern Recognition, Institute of Automation,
Chinese Academy of Sciences, P.O. Box 2728, Beijing, P.R. China, 100080
{zswei, tnt, znsun, jlcui}@nlpr.ia.ac.cn

Abstract. Iris recognition is one of the most reliable methods for personal identification. However, not all the iris images obtained from the device are of high quality and suitable for recognition. In this paper, a novel approach for iris image quality assessment is proposed to select clear images in the image sequence. The proposed algorithm uses three distinctive features to distinguish three kinds of poor quality images, i.e. defocus, motion blur and occlusion. Experimental results demonstrate the effectiveness of the algorithm. Clear iris images selected by our method are essential to subsequent iris recognition.

1 Introduction

Biometrics personal identification has been drawing extensive attention in recent years [1]. Among all the biometric traits, iris pattern has its distinctive advantages for its large inter-class and low intra-class variability. Therefore, iris recognition is no doubt a promising biometrics, which has great value in commercial and information security.

Iris image quality assessment is an important step in any iris recognition system. The existing iris recognition algorithms with good performance [2, 3, 4] are all based on certain quality images. Poor quality images will enlarge the intra-class variability and reduce the inter-class variability, and consequently increase FRR and FAR. So it is necessary to prevent poor quality images from entering subsequent processing (Fig.1).

Much attention has been paid on iris image quality assessment. Daugman [2] used a (8×8) convolution kernel to extract the high frequency of the image. Zhang et al. [5] used $\frac{1}{w} = \frac{gradient}{M_i - M_p}$ to measure how sharp the pupil/iris boundary is to distinguish the defocused images. Our earlier work [3] defined a descriptor $D = [(F_1 + F_2 + F_3), \frac{F_2}{F_1 + F_3}]$, $F_i = \int\int_{\Omega=(u,v|f_1^i<\sqrt{u^2+v^2}\leq f_2^i} \left|F(u,v)\right| dudv$ $(i = 1, 2, 3)$ to describe the iris image quality. However, some other differences between clear image and poor quality image are still expecting to be discovered, to make the iris recognition system faster and more robust.

To further explore this problem, an efficient iris image quality assessment algorithm is proposed in this paper. We attempt to distinguish three main cases of

D. Zhang and A.K. Jain (Eds.): ICB 2006, LNCS 3832, pp. 464–471, 2005.

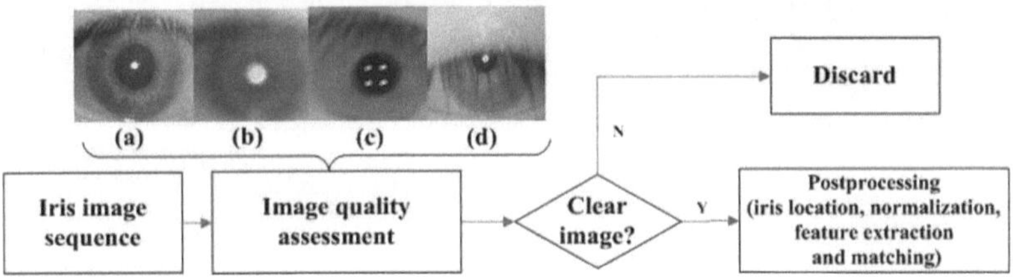

Fig. 1. Iris image quality assessment in a recognition system. (a) A clear image. (b) A defocused image. (c) A motion blurred image. (d) An occluded image.

poor quality iris images, i.e. defocus (Fig.1b), motion blur (Fig.1c) and occlusion (Fig.1d). The algorithm we proposed is conceptually simple but effective. Pupil location, which could be time-consuming, is avoided in our approach.

The remainder of this paper is organized as follows. Section 2 presents the main problems concerned. Section 3 goes into the details of our proposed method. Experimental results are provided in Section 4, and Section 5 gives the conclusion.

2 Problem Statement

There are three main problems concerned, i.e. defocus, motion blur and occlusion. A system that employs fixed-focus optical lens easily causes defocused iris images. Motion blurred images in our experiment are captured by a CCD sensor in interlaced scan mode, and a frame is combined by two fields with an interval of 20ms or less, the resulting image involves obvious interlacing lines in the horizontal direction [3]. Occluded image is the case that most area of the iris is covered by eyelid and eyelashes. It often happens if the client blinks while the images are being taken.

3 Feature Extraction

Our iris recognition system is supposed to assess image quality immediately after capturing the images (Fig.1). Based on such original data, no single solution can be made to separate the three kinds of images from clear images. As each of the three main problems has its own peculiarity, three features can be selected to classify them one by one. The quality of the image can be defined as $Q(p_1, p_2, p_3)$:

$$p_1 = \frac{1}{M \times N} \sum_x \sum_y HFP \tag{1}$$

$$p_2 = \frac{1}{M \times N} \sum_x \sum_y VHFP \tag{2}$$

$$p_3 = \frac{1}{M \times N} \sum_x \sum_y ROI \tag{3}$$

where M and N are the width and height of the image respectively; x, y are variables denote the pixel location; HFP is the high frequency power of the image; $VHFP$ is the vertical high frequency power; ROI is the Region of Interest.

-1	-1	-1	-1	-1	-1	-1	-1
-1	-1	-1	-1	-1	-1	-1	-1
-1	-1	+3	+3	+3	+3	-1	-1
-1	-1	+3	+3	+3	+3	-1	-1
-1	-1	+3	+3	+3	+3	-1	-1
-1	-1	+3	+3	+3	+3	-1	-1
-1	-1	-1	-1	-1	-1	-1	-1
-1	-1	-1	-1	-1	-1	-1	-1

(a)

-1	-1	-1	-1	-1
-1	+2	+2	+2	-1
-1	+2	0	+2	-1
-1	+2	+2	+2	-1
-1	-1	-1	-1	-1

(b)

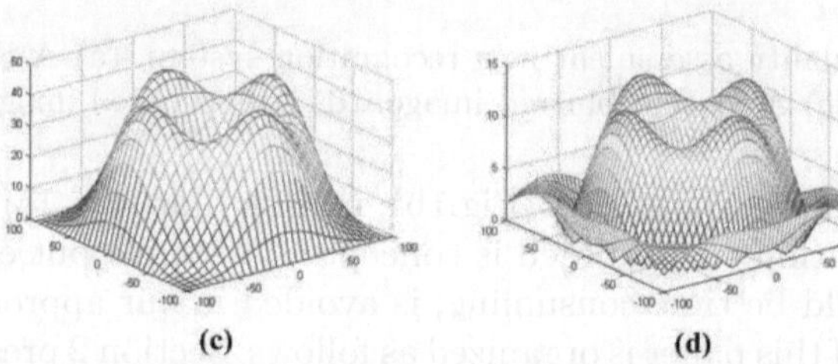

Fig. 2. (a) The (8×8) convolution kernel proposed by Daugman. (b) The proposed (5×5) convolution kernel $H_{(5\times5)}$. (c) The Fourier spectrum of (a). (d) The Fourier spectrum of (b).

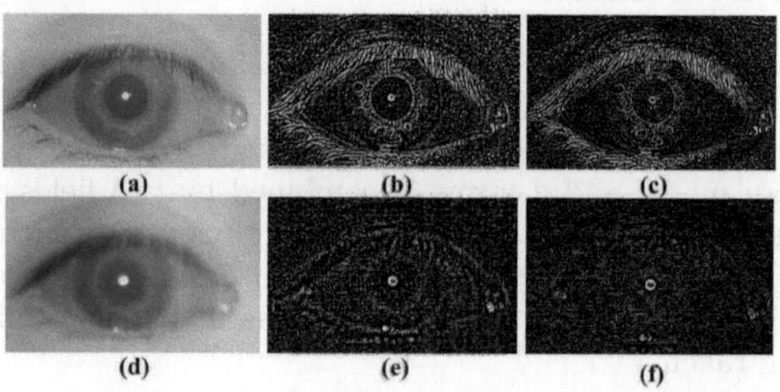

Fig. 3. (a) A clear image. (b) The result of image $a * (8 \times 8)$ kernel. (c) The result of image $a * H_{(5\times5)}$. (d) A defocused image. (e) The result of image $d * (8 \times 8)$ kernel. (f) The result of image $d * H_{(5\times5)}$.

3.1 For Defocused Images

A clear image has relatively uniform frequency distribution in the 2D Fourier spectrum. On the contrary, the energy of a defocused image concentrates on the lower frequency part. Therefore, using the high frequency power of the image to evaluate the degree of focus is a common method in previous research on image focus assessment.[2,6-9].

In order to obtain the high frequency power of the image, a proper high-pass convolution kernel is really important. Daugman [2] used a (8×8) convolution kernel (Fig.2a) to extract the high frequency power of the image. In this paper we propose a (5×5) high-pass filter $H_{(5\times5)}$, as shown in Fig.2b.

Here we give a short analysis of this operator. It is formed by three box functions, one of size (5×5) with amplitude -1, one of size (3×3) with amplitude

+3, and the last one of size (1×1) with amplitude -2. The overlap of these three boxes function constitutes the operator we use. We will describe $H_{(5\times5)}$ in 2 aspects. One important measurement is its 2D Fourier spectrum characteristic. The proposed kernel's 2D Fourier spectrum is shown in Fig.2d. It is a band-pass filter and its central frequency is around 0.4375, with a bandwidth (BW) of 0.3125 in which the attenuation is less than 3db with respect to the central frequency. Although both the proposed kernel $H_{(5\times5)}$ and the (8×8) kernel proposed by Daugman have a similar shape in 2D Fourier spectrum (Fig.2 c and d) and share the same principle of filtering image, compared with (8×8) convolution kernel, whose passband central frequency is about 0.28125 and BW is about 0.1875, $H_{(5\times5)}$ has a higher central frequency and larger BW. That is to say, it can select more high frequencies. Another measurement of an operator is its efficiency. In Daugman's method, the convolution results are squared by exploiting Parseval's theorem: $\int\int |I(x,y)|^2 dxdy = \int\int |F(u,v)|^2 dudv$, and accumulated by selecting every fourth row and fourth column [2]. Doing the same work, the proposed $H_{(5\times5)}$ operator would take less execution time because the total multiplication is less than the (8×8) one. Thus we can say that $H_{(5\times5)}$ is computationally more efficient.

Fig.3 shows two sample images, a clear image and a defocused image. Fig.3b and Fig.3e are the images after convolution with the (8×8) kernel, and Fig.3c and Fig.3f are the images after the convolution with $H_{(5\times5)}$. In order to get the same visual effect as the (8×8) one, we enlarge the image gray scale to 2.56 $((8\times8)/(5\times5))$ times. Since the $H_{(5\times5)}$ has a better performance in attenuating low frequencies, we can see that the clear images and defocused images become more discriminant under the proposed operator. The feature of defocus images is shown in (1).

3.2 For Motion Blurred Images

Motion blurred images are also undesired in iris recognition, since motion blur can severely degrade image quality. The two-field CCD sensor [12], one frame of whose image is formed by two fields, the odd field and the even field, is widely used for image capturing. The two fields are presented as adjacent rows in the resulting image. If the client moves when capturing, the two fields would present a very different scene.

As the adjacent rows are quite different in motion blurred image, the difference between every two rows of pixel is adopted as a measurement. Jarvis described SMD (Sum Modulus Difference) in his work [6] for obtaining the focus score. We find that SMD_x which denotes the row difference is suitable for distinguishing motion blurred images, but SMD_x is only the sum of single pixel difference. To achieve better performance, we build up an improved $(2 \times n)$ operator, the first row with amplitude -1 and the second row 1, which is a vertical highpass filter.

In order to select a proper n, we use 200 clear and 200 motion blurred images for training. We get the threshold by using Minimal Error Criterion and finally

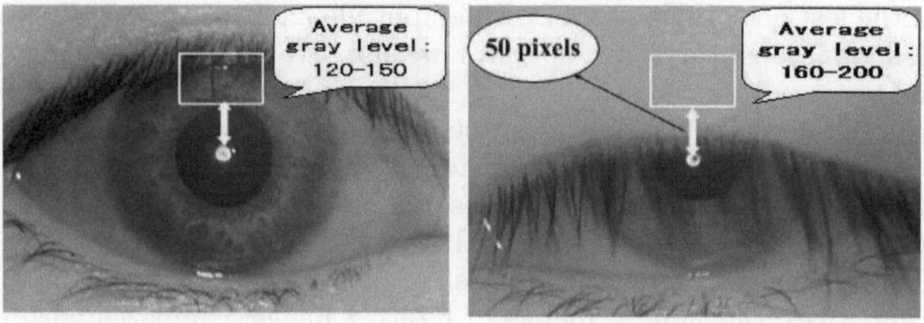

Fig. 4. Illustration of discriminating clear and occluded image

selected n=8 for the best tradeoff between time and performance. And the quality feature of motion blurred image is described in (2).

3.3 For Occluded Images

Occluded images are another big problem in iris image quality assessment. In an occluded image, at least 1/3 of the iris area is covered by eyelid, and the other part is often covered by eyelashes. So, usually we can not extract accurate iris information from occluded images. That is why we consider them poor quality images and have to discard them.

The most distinct difference between clear images and occluded images is the size of iris due to occlusion. Since the radius of iris is between 100 and 120 pixels and that of pupil is between 30 and 50 pixels, given the approximate location of the pupil, then at a certain distance along the vertical direction of the pupil, we can find iris or eyelashes in a clear image, but can only find eyelid in an occluded image. Considering that the gray level of the iris and eyelid is different, we can separate the occluded images and clear images in this way. The feature for occluded image is described in (3).

4 Experimental Results

4.1 Training and Testing Results

For the purpose of validating the effectiveness of the proposed algorithm, we constructed a training dataset which contains 152 clear images, 128 defocused images, 190 motion blurred images and 105 occluded images. The testing dataset contains 300 clear images, 271 defocused images, 287 motion blurred images and 335 occluded images. The positive samples (the clear images) come from the CASIA database Version 2.0. The negative samples come from those images that can not be recognized by our recognition algorithm presented in [4].

Simply taking clear/defocus images as samples, the proposed algorithm and the algorithm proposed by Daugman, Zhang et.al. are compared in Fig.5 and Table 1. The experiment is performed using Matlab 6.0 on a Pentium IV 1.3GHz processor with 256MB RAM.

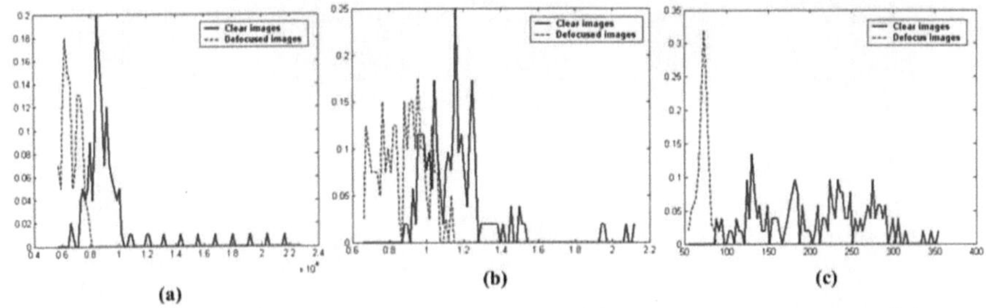

Fig. 5. Distribution of clear vs. defocused images (a)Daugman's algorithm. (b)Zhang's algorithm. (c) Proposed algorithm.

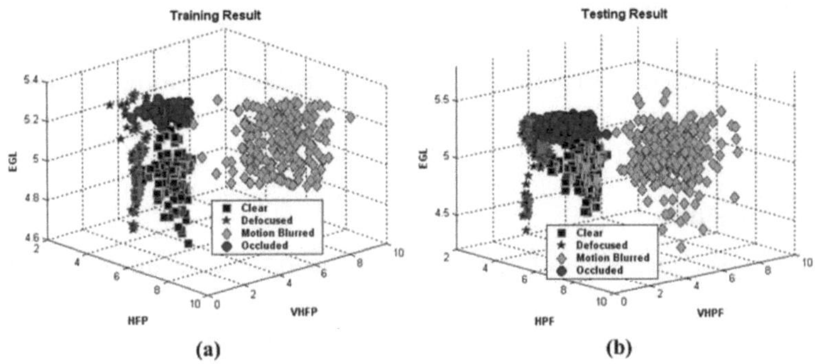

Fig. 6. (a)Training results. (b)Testing results.

Table 1. Comparison of three algorithms for detecting defocus image.

152 Clear/ 128 Defocus	Daugman's	Zhang's	Proposed
time(s)	0.1181	0.3431	0.0822
CCR	92.12%	85.62%	98.63%

Table 2. Performance of previous and proposed algorithm

	Previous	Proposed
Training	92.41%	98.80%
Testing	89.29%	97.68%

Fig.6 and Table 2 illustrates the training and testing results, showing that the proposed algorithm works well. The three axes in Fig.6 respectively denote three feature components. We choose SVM to characterize the distribution boundary, for it has good classification performance in high dimensional space. Table 2 also gives the results of our previous algorithm [3].

4.2 UBIRIS Database Assessment Results

To further evaluate the performance of the proposed algorithm, we did another experiment on the UBIRIS database [11]. UBIRIS section1 contains 1214 images which come from 241 eyes. It is a noisy database. We take 20 clear images and

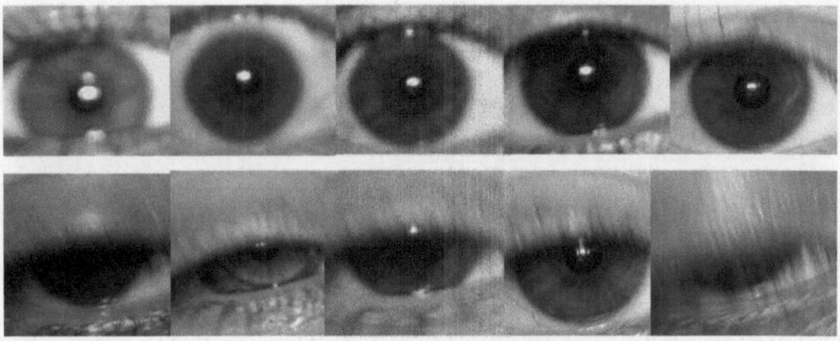

Fig. 7. Examples of poor quality images in the UBIRIS database identified by our algorithm

20 poor quality images as training samples, and the remaining images as testing samples. The results (shown in Fig.7) imply that the proposed algorithm can correctly identify the poor quality images.

4.3 Discussions

From the above analysis and results, a number of conclusions can be drawn as follows:

1. In detecting defocused image, Table 1 shows that the proposed algorithm is superior to the other two algorithms in terms of speed and accuracy.
2. The proposed algorithm has some distinct advantages compared with our previous algorithm. Experimental results demonstrated this conclusion. Pupil location is avoided in the proposed algorithm, which fits for a real-time recognition system.

5 Conclusion

Iris image quality assessment is an important step that can never be neglected in iris recognition systems. In this paper, a novel algorithm for iris image quality assessment has been proposed, which used three features to discriminate defocused, motion blurred and occluded images. It is robust and fast because we selected simple and stable features. By this method, clear images can be well selected for subsequent recognition process, and thus the intra-class distance can be reduced and the inter-class distance can be increased. It can help to further improve the performance of subsequent iris recognition and reduce the FRR.

Acknowledgement

This work is funded by research grants from the National Basic Research Program (Grant No. 2004CB318110), the Natural Science Foundation of China (Grant No. 60335010, 60121302, 60275003, 60332010, 69825105) and the Chinese Academy of Sciences.

References

[1] Anil K. Jain, R.M. Bolle, and S. Pankanti.*Biometrics: Personal Identification in Networked Society,* Norwell, MA: Kluwer, (1999)

[2] J. Daugman. *How Iris Recognition Works,* IEEE Trans. on Circuits and Systems for Video Technology, vol. 14, no.1 pp. 21-30, (2004)

[3] L. Ma, T. Tan, Y. Wang, D. Zhang. *Personal Identification Based on Iris Texture Analysis,* IEEE Trans. on Pattern Analysis Machine Intelligence, vol. 25, no.12 pp. 1519-1533, (2003)

[4] Zhenan Sun, T. Tan, Y Wang, *Robust Encoding of Local Ordinal Measures: A General Framework of Iris Recognition*, ECCV Workshop on Biometric Authentication, (2004)

[5] Zhang et al. *Method of Measuring the Focus of Close-up Image of Eyes,* United States Patent, No.5953440, (1999)

[6] R. A. Jarvis, *Focus Optimization Criteria for Computer Image Processing,* Microsope, vol.24(2), pp.163-180, (1976)

[7] S.K. Nayar and Y. Nakagawa,*Shape from focus,* IEEE Trans. on Pattern Analysis Machine Intelligence, vol. 15, no.8 pp.824-831, (1994)

[8] E. Krotkov, *Focusing,* International Journal of Computer Vision, Vol. 1, No. 3, October, pp. 223-237, (1987)

[9] Byung Jun Kang, Kang Ryoung Park, *A Study on Iris Image Restoration*, in Proc. of International Conference on Audio- and Video-based Biometric Person Authentication 2005

[10] *CASIA database*, http://www.sinobiometrics.com

[11] H.Proenca and L.A. Alexandre, *UBIRIS Iris Image Database, http://iris.di.ubi.pt.*

[12] Yuqing He, Yangsheng Wang, T. Tan, *Iris Image Capture System Design for Personal Identification*, Advances in Biometric Personal Authentication, Springer-Verlag, (2004)

Efficient Iris Recognition Using Adaptive Quotient Thresholding

Peeranat Thoonsaengngam, Kittipol Horapong,
Somying Thainimit, and Vutipong Areekul

Kasetsart Signal and Image Processing Laboratory (KSIP lab),
Department of Electrical Engineering, Faculty of Engineering,
Kasetsart University, Bangkok, 10900, Thailand
{g4765155, g4565244, fengsyt, fengvpa}@ku.ac.th

Abstract. This paper presents an intensity-based iris recognition system. The system exploits local intensity changes of the visible iris textures such as crypts and naevi. The textures are extracted using local histogram equalization and the proposed 'quotient thresholding' technique. The quotient thresholding partitions iris images in a database such that a ratio between foreground and background of each image is retained. By fixing this ratio, variations of illumination across iris images are compensated, resulting in informative and distinctive blob-like iris textures. An agreement of the two extracted textures is measured by finding spatial correspondences between the textures. The proposed system yields the 0.22 %EER and 100%CRR. The experimental results indicate encouraging and effective iris recognition system, especially when it is used in identification mode. The system is very robust to changes in decision ratio.

1 Introduction

The term 'biometrics' refers to the science and technology of authentication using physiological or behavioral characteristics of human such as fingerprints, signatures, faces, and irises. Among the biometrics, an iris pattern is a highly accurate and reliable characteristic. Flom and Safir [1] reported that the iris patterns are unique to individual and stable over time and across environments (e.g. occupations). The uniqueness of iris patterns is the product of dense collections of iris structures and textures such as pigment frills, furrows, freckles, and crypts. In addition, an iris is a protected organ located behind the cornea, but in front of the lens. This makes personal authentication possibilities life long.

Existing iris recognition systems were developed using several approaches. Major differences among the systems are methods used in analyzing and extracting iris features. Examples of these systems are briefly described as follows: Daugman [2] extracts iris textures using 2-D Gabor filter. Wides [3] analyses iris textures using 4-level Laplacian Pyramid. After these works, several iris recognition systems [4-5] have been proposed. However, few of them associate physical iris structures into their systems. One of the few is a system developed by Sun [6]. Sun applied zero-crossing wavelet transform to segment what he called blocks of interest-BOIs. From his experimental

D. Zhang and A.K. Jain (Eds.): ICB 2006, LNCS 3832, pp. 472 – 478, 2005.

results, only partial physical iris structures are segmented. He, then, cascades the approach with a local feature based classifier to obtain higher recognition accuracy.

Our iris recognition system also attempts to segment the blob-like structures of an iris. However, the proposed system is a non-cascading system. Our approach exploits local intensity variations of the iris textures. The blob-like structures are extracted using local histogram equalization and the proposed technique, quotient thresholding. Our system consists of three main steps: preprocessing, feature extraction and feature matching. The following sections describe our system in details.

2 Preprocessing

The first step of our approach is to locate boundaries of an iris in an input eye image: the pupil's boundary and the sclera's boundary. These boundaries are approximated using concentric circles, in our approach.

The pupil boundary is detected by initially segmenting the pupil using thresholding technique. The pupil's boundary is located by applying edge detection, followed by edge thinning and circular fitting. The fitting yields the pupil's radius and its center. This center serves as the reference point for the rest of the process.

The sclera boundary is detected by exploiting intensity difference between iris and sclera. The contrast enhancement algorithm in [7] is firstly applied to improve contrast of the image. Then, average intensity of pixels along virtual arcs covered $\pm 45°$ and moving outward from the center is calculated. The sclera boundary is located at the first abrupt change of the obtained average intensity.

Due to the iris size is varied depending on an amount of incoming light, the obtained iris images are normalized into the same fixed size.

3 Our Proposed Feature Extraction

Our feature extraction aims to extract visible and distinctive physical iris textures such as crypts, freckles, and moles. These textures occur randomly over the iris and have darker intensity level. They belong to regions of minimum of the image. Typically, the regions of minimum are segmented using conventional thresholding technique. One drawback of this technique is that it is sensitive to image illumination. Non-uniform illumination over and across images can significantly degrade its performance. Unfortunately, these non-uniformities can not be avoided during the iris acquisition process. In order to extract informative and discriminating iris textures, compensations for the non-uniform illumination must be accounted for. In this paper, we propose using local histogram equalization and quotient thresholding technique to compensate illumination changes over and across iris images.

3.1 Local Histogram Equalization

Local histogram equalization is used in our feature extraction in order to enhance iris textures and to compensate non-uniform illumination over the image. The local operation is chosen for two reasons. Firstly, local enhancement can compensate effects of non-uniform illumination of an image. Secondly, it is generally agreed that texture analysis should concern spatial distribution of pixels within a given neighborhood. This implies local operation is preferred over global operation for texture analysis.

Figure 1 illustrates effects of local versus global operation on segmenting the iris textures. The enhanced images of the same iris using global- and local- histogram equalization are shown in figure 1(a) and 1(c), respectively. Figure 1(b) and 1(d) show its corresponding segmented iris textures obtained by thresholding the enhanced images using threshold value of 150, respectively. The obtained results reveal the global operation is unable to compensate the non-uniform illumination over the image. The global operation provides indistinguishable iris textures. Contrarily, the local histogram equalization provides informative and distinguishable iris textures.

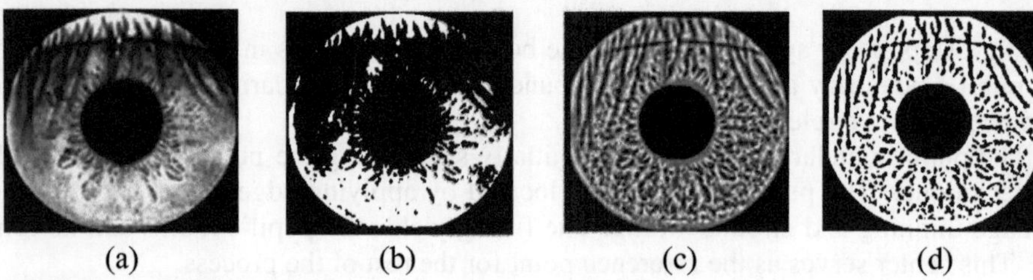

Fig. 1. (a) The enhanced iris image using global histogram equalization. (b) The segmented iris textures obtained by thresholding figure (a) using a threshold value of 150. (c) The enhanced image of the same iris using local histogram equalization. (d) The segmented iris textures obtained by thresholding figure (c) using a threshold value of 150.

From figure 1(d), does not only iris textures are enhanced, but also some irrelevant and unwanted textures. Since our approach utilizes local intensity changes of iris textures, noises caused from occlusions such as eyelashes can degrade our system performance dramatically. To reduce effects of these noises, only half radius of the iris ring covered 10° above and 50° below the iris's diameter is of our interest.

3.2 Adaptive Quotient Thresholding

In this paper, quotient thresholding is proposed to handle uneven illumination across iris images in a database. The motivation behinds the proposed thresholding is a human eye's adaptability characteristic. With our eye, a black-white pattern such as a checker lit differently is perceived similarly. If stationary scenes are assumed, this implies that a proportion between perceived black pixels and white pixels is the same. In term of image histogram, similar patterns can be achieved if a ratio between foreground and background is maintained. Maintaining this ratio is a principle of the proposed quotient thresholding technique.

The quotient thresholding partitions images in a database such that the ratio between foreground and background of each image, called decision ratio, is maintained. Figure 2 demonstrates our feature extraction scheme and its results. Two input images captured from the same iris are shown in the first row. The corresponding enhanced iris wedges using local histogram equalization overlaid over the iris ring are shown in the second row. The third row shows its corresponding histogram and the

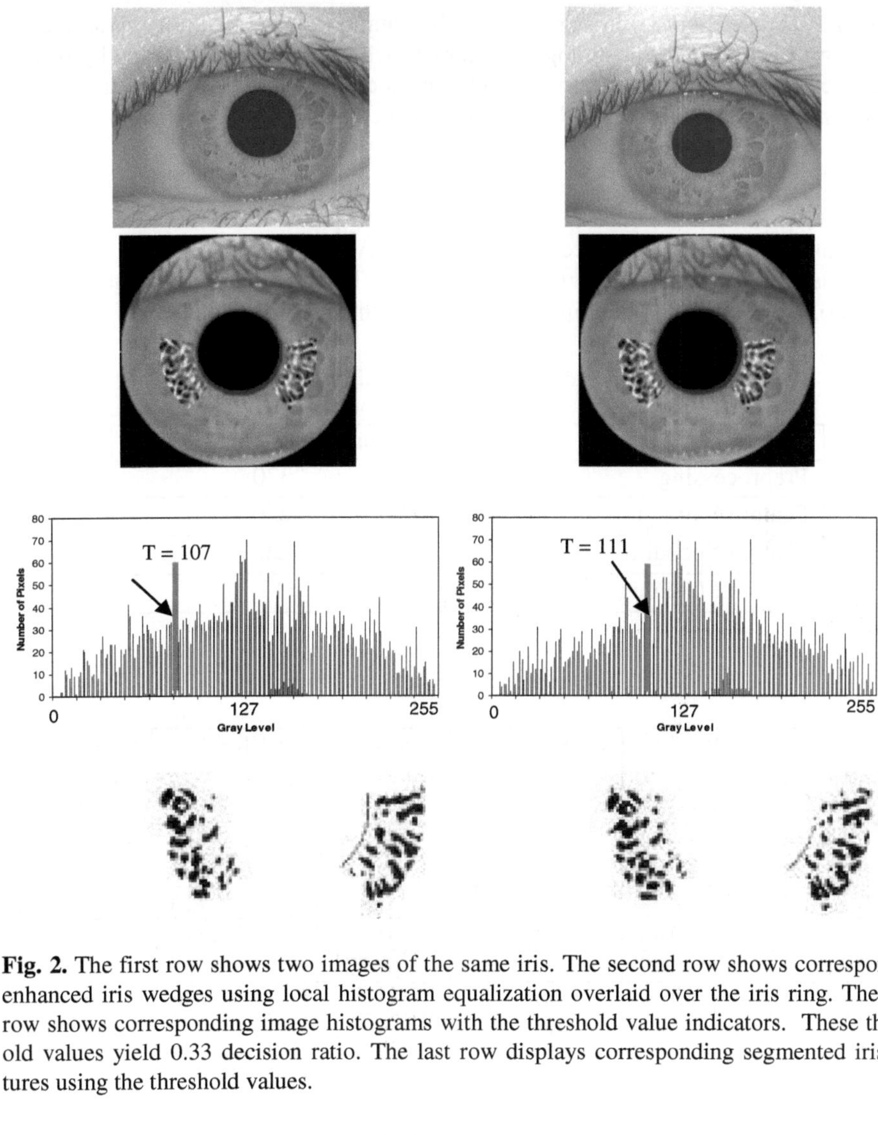

Fig. 2. The first row shows two images of the same iris. The second row shows corresponding enhanced iris wedges using local histogram equalization overlaid over the iris ring. The third row shows corresponding image histograms with the threshold value indicators. These threshold values yield 0.33 decision ratio. The last row displays corresponding segmented iris textures using the threshold values.

threshold value used to obtain a 0.33 decision ratio. The thresholded textures are displayed in the last row. It is clearly seen that the obtained textures well reflect the physical iris structures. Though the proposed feature extraction is very simple, it is very effective.

4 Iris Feature Matching

An agreement of two extracted iris textures is measured by finding spatial correspondences between the textures. The number of aligned pixels is counted and reflected in term of a matching score. In our system, the rotation and translation invariance is achieved by rotating and translating the template within a range of $\pm 10^{o}$ and ± 10 pixels, respectively. A matching score of each pair is calculated. The maximum matching score among them represents the matching score of the template.

5 Experimental Results

We validate our proposed iris recognition using CASIA iris database [8]. The database consists of 108 irises. Each iris has 7 images. The recognition system is implemented using VC++ on PC Pentium IV 2.4 GHz with 512 MB memory.

Table 1. Computational time of the proposed iris recognition system

Process	Computational Time (msec.)
Preprocessing	370
Feature Extraction	250
Feature Encoding	20
Feature Matching	17

Table 2. Our proposed system performance

Decision Ratio	EER (%)	CCR (%)
0.11	0.416487	100
0.13	0.344492	100
0.15	0.290305	100
0.17	0.293837	100
0.19	0.291188	100
0.21	0.288009	100
0.23	0.306199	100
0.25	0.297722	100

Decision Ratio	EER (%)	CCR (%)
0.27	0.260135	100
0.29	0.244447	100
0.31	0.259605	100
0.33	0.219723	100
0.35	0.264374	100
0.37	0.251335	100
0.39	0.342196	99.8677
0.40	0.348024	99.8677

Our first experiment is finding the decision ratio offered the best system performance. The system performance is measured in terms of Equal-Error-Rate (EER) and Correct-Recognition-Rate (CRR) using the leave-two-out method. Two images out of 756 images are randomly selected as test images. Table 1 indicates computational time of our system. Table 2 indicates the obtained EER and CRR for each decision ratio. From the experiment, decision ratio of 0.33 yields the best system performance with 0.22% EER and 100% CRR. From the results, our system is very well performed in an identification mode. Almost all decision ratios yield the 100% CRR. In addition, the results indicate that our system is quite insensitive to values of decision ratio.

Next, we examine the distribution of intra- and inter- class verification. The distribution of the matching distance is shown in figure 3. Comparing to existing iris

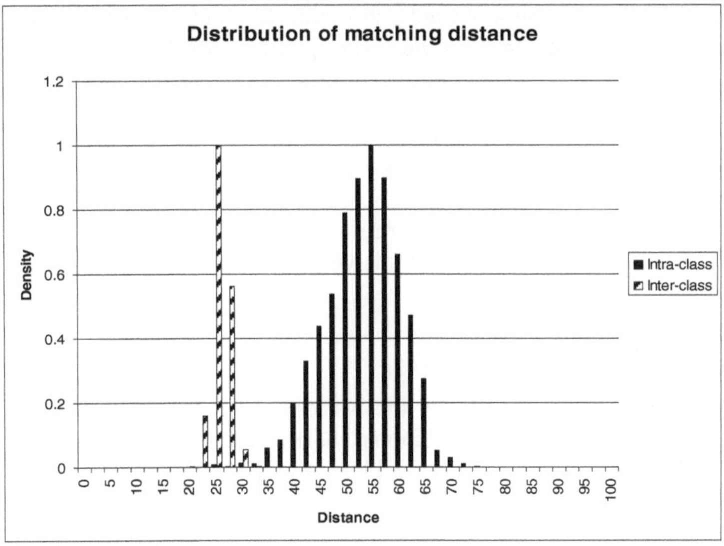

Fig. 3. The distribution of inter- and intra- class verification using 0.33 decision ratio

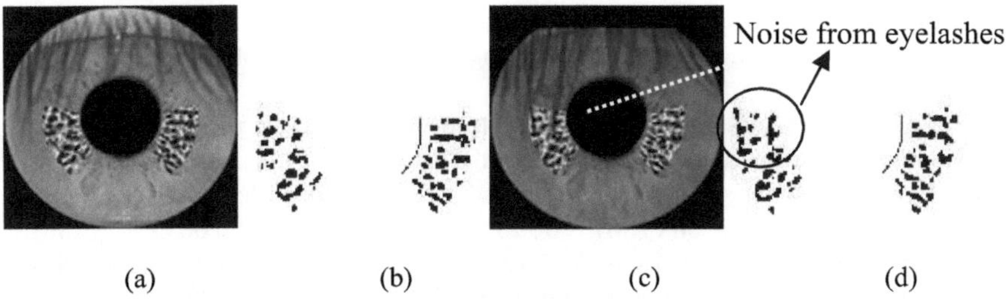

Fig. 4. (a) An iris image overlaid by its enhanced iris textures. (b) Its corresponding extracted textures. (c) Another image of the same iris captured during eyelashes are moving downward. (d) Its corresponding extracted textures. The circle in the figure indicates areas where the occlusions occur.

recognition systems [2-6], the proposed iris system performance in verification mode is lower. From our observations, main degradations of our system are caused from occlusion noises such as eyelashes. Since eyelashes have very low intensity, they are always included in the segmented textures. As a result, the segmented textures appear differently where the occlusions occur as shown in figure 4. Further system improvement can be achieved by integrating eyelids and eyelashes removal scheme into our system.

6 Conclusions and Future Works

An efficient intensity-based iris recognition algorithm is proposed in this paper. The local intensity variation of physical iris textures is exploited in the proposed method.

The physical iris textures are analyzed and extracted using local histogram equalization and adaptive quotient thresholding. The quotient thresholding partitions an image in a database into foreground and background such that a ratio between the foreground and background of all images in the database are maintained. The proposed algorithm yields informative and discriminating blob-like iris features. The matching between the input feature and the template is measured by finding spatial correspondences between the features. Our experimental results show encouraging iris recognition performance. With 0.33 decision ratio, our system yields the 0.22% EER and 100% CRR. Hence, the proposed system is very efficient but very simple.

Acknowledgement

This work was partially supported by the National Electronics and Computer Technology Center (NECTEC) under National Science and Technology Development Agency (NSTDA) under Grant NT-B-22-I3-12-47-07.

References

1. L. Flom, A. Safir,: Iris Recognition System. U.S. Patent No. 4 641 349, 1987.
2. J.Daugman: Biometric Personal Identification System based on Iris Analysis. U.S. Patent No. 5 291 560, 1994.
3. R. P. Wildes: Iris Recognition: An Emerging Biometric Technology. Proceeding of the IEEE, Vol. 85, No.9, pp.1348-1362, 1997.
4. W.W. Boles and B.Boashah: A Human Identification Technique using Images of the Iris and Wavelet Transform. IEEE Trans. on Signal Processing, Vol.46, pp1185-1188, 1998.
5. Lima, Tieniu Tan, Yunhong Wang, Dexie Zhang: Personal Identification Based on Iris Texture Analysis. IEEE Trans. on PAMI, Vol. 25, No.12, pp.1519-1533, 2003.
6. Zhenan Sun, Yunhong Wand, Tieniu Tan, and Jiali Cui: Cascading Statistical and Structural Classifiers for Iris Recognition. Proceedings of ICIP, pp.1261-1264, 2004.
7. L.Hong, Y.Wan, and A.K.Jain: Fingerprint Image Enhancement: Algorithm and Performance Evaluation. IEEE Trans. on PAMI, Vol. 20, No.8, pp.777-789, 1998.
8. http://www.sinobiometrics.com.

A Novel Iris Segmentation Method for Hand-Held Capture Device

XiaoFu He and PengFei Shi

Institute of Image Processing and Pattern Recognition,
Shanghai Jiao Tong University, Shanghai 200030, China
{xfhe, pfshi}@sjtu.edu.cn

Abstract. In this paper, a new iris segmentation method for Hand-held capture device is proposed. First, the pupil is binarized using the intensity threshold, then use morphologic method to denoise the eyelashes and eyelids noise. The geometrical method is used to calculate the coordinates of the pupil. Second, the outer (or limbus) boundary is localized using the shrunk image with the Hough transform and modified Canny edge detector in order to reduce computational cost. Third, the eyelids which are constrained to be within the outer boundary are estimated using the polynomial fitting method. The segmentation method was implemented and tested on iris database set which is captured by hand-held optical sensor device. Experimental results show that the proposed algorithm can separate the iris from the surrounding noises with good speed and accuracy.

1 Introduction

In recent years, iris recognition becomes one of the most reliable biometric technologies. The most important part for an iris recognition system is the iris segmentation. The goal of iris segmentation is to localize the iris from the surrounding noises, such noises include the pupil, the sclera, the eyelids, the eyelashes, the eyebrows and the reflections etc.

Whereas, it is difficult to localize the iris from the surrounding noises, especially for the hand-held capture device. The main reason is that the eyelid or eyelashes usually occlude the iris and the iris capture devices in application are mostly exposed to the natural scene, so the natural illumination or other variant conditions sometimes can greatly influence the iris images and further impact the segmentation result. All these factors will influence the subsequent processing since iris pattern represented improperly will inevitably result in poor recognition performance. To solve this problem, a robust segmentation method should be proposed to remove the influence of all these noises as much as possible.

It is widely accepted that an eye can be modeled by two circles, pupil and limbus, and two parabolas, upper and lower eyelids. In previous segmentation methods, most of which are based on integrodifferential operator or Hough transform. John G. Daugman [1][2][3] proposed an integrodifferential operator for localizing iris regions along with removing the possible eyelid noises. The Daugman system fits the circular contours via gradient ascent together with radial Gaussian. Wildes [4] processed iris segmentation through filtering and voting procedure, which is realized via Hough transform on

D. Zhang and A.K. Jain (Eds.): ICB 2006, LNCS 3832, pp. 479–485, 2005.

parametric definitions of the iris boundary contours, include pupil, limbus and the eyelids boundaries. W. K. Kong and D. Zhang [5] proposed an accurate iris segmentation method for reflection and eyelash detection. An edge detector is applied to detect separable eyelashes, and intensity variances are used to recognize multiple eyelashes. A threshold and statistical model is proposed to recognize the strong and weak reflection, respectively. Ma et al. [6][7] processed iris segmentation by edge detection and Hough Transform. Eyelids and eyelashes noises were not considered in his method. Huang et al. [8] proposes a new noise-removing approach based on the fusion of edge and region information. Edge information extraction was based on phase congruency. The iris is segmented using edge detection and Hough transform. Using the Hough transform segmentation technique, the circle parameters can be found out in the 3-dimensional parameter spaces. Direct application of the Hough transform for detecting circles and eyelid in images is not practical due to the expensive requirements of 3-dimensional parameter spaces. Space and time complexities are the main concerns in the application of Hough transform for detecting circles. Certainly the smaller of the searching region, the faster of the calculation speed. Moreover, if boundaries of iris pupil and sclera were clearly distinguished, most automatic segmentation methods will be proved to be successful. However, in practical application, it is very difficult to locate the boundary between the iris and the sclera. Difficulties arise from the fact that edge detection will fail to find the edges of the iris border.

So, in this paper, we develop a novel iris segmentation method for hand-held capture device. Since the pupil typically is darker than the iris, the pupil (or inner) boundary is detected using geometrical method together with the intensity threshold, which can reduce computational cost and still gets good results. We use the shrunk image to detect outer circle based on modified Canny edge detector [9][10] together with Hough transform, which has fewer noises, e.g spurious boundaries. The fewer noises are, the smaller of the searching region is. The upper and lower eyelids detection relies on the polynomial fitting method.

The remainder of this paper is organized as follows: Section 2 provides a description of the proposed method for pupil localization. Section 3 introduces the outer boundary detection. Eyelids, eyelashes and reflection detection are also given in this section. Section 4 reports experiments and results. Section 5 concludes this paper.

2 Pupil Localization

The iris is an annular part between the pupil (or inner) and the outer (or limbus) boundary. Both of them can approximately be taken as circles. Using the iris prior-knowledge, we first roughly determine the iris region in the original image and use intensity threshold to binarize the iris together with morphological operations. Then use the geometrical method to exactly calculate the parameters of the inner circle in the determined region after edge detection.

2.1 Rough Localization

To capture the rich details of iris patterns, an imaging system should resolve a minimum of 70 pixels in iris radius [3]. In most deployments of these algorithms to date, the resolved iris radius has typically been 80 to 130 pixels, though some companies have minute differences. So we can use the prior-knowledge to roughly locate the iris region. This will reduce the region for subsequent processing, which results in lower computational cost. Then we use intensity threshold to binarize the iris.

2.2 Edge Detection

From the above processing, we usually get a noised binary image, especially the eyelashes and top eyelid, since the intensity of the eyelashes and top eyelid are similar to the pupil in hand-held captured images. All these may affect the subsequent processing, so we use morphological operations to exclude unnecessary regions especially the eyelashes and eyelid in order to get a connected adjacent region. Then we use Sobel operator to extract the edges.

2.3 Pupil Localization

The geometrical method is used to locate the circle. We proposed that two perpendicular lines L1 and L2 intersect at P which is the inner of the circle. From the bottom point A, we draw several horizontal lines with interval $\triangle$. We then get two point sets A0 {A10, A20, A30…} and A1 {A11, A21, A31…} from these horizontal lines which intersect the circle, as shown in the diagram in Fig.1. Then use three points a0, a1 and A to calculate the radius and the center coordinates of the circle, since three points which are not on the same line can determine a circumcircle, where a0 and a1 comes from point set A0 and A1 respectively.

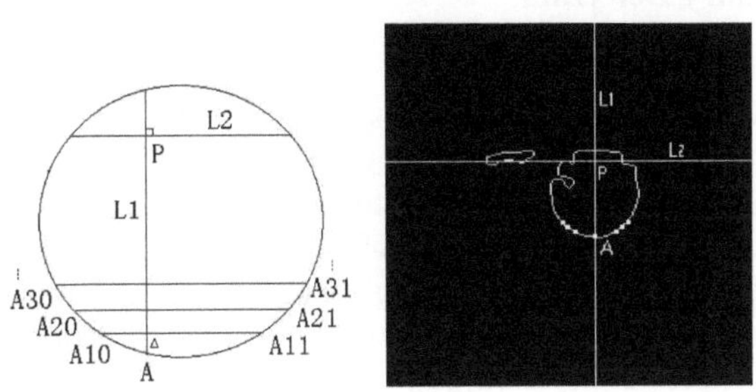

(a) (b)

Fig. 1. Circle localization. (a) Diagram. (b) Point sets overlaid on iris edge image.

We project the binary image in the vertical and horizontal direction to approximately estimate the centroid $P(x_p, y_p)$, which is usually not the center of the pupil. But at least we can see that $P(x_p, y_p)$ is the inner of the pupil since the pupil is usually darker than other areas. We draw two perpendicular lines L1 and L2 through the inner point $P(x_p, y_p)$, shown in Fig.1 (b). Then we get the lowest point A. From the point A, we draw several horizontal lines with interval $\triangle$ which result in two point sets. We use these two point sets and the bottom point A to calculate the radius and the center of the circle and get the average value of the radius and the center of the circle. The result of the pupil localization can be seen in Fig.2.

It is difficult to separate the pupil from the surrounding noises especially captured by Hand-held capture device. Since the pupil is partly occluded by the top eyelid and eyelashes. We use the bottom point A of the pupil as the reference point because the bottom point is insensitive to the pupil dilation and not affected by the top eyelid or eyelashes. We have found empirically that the bottom 3 or 4 lines with interval 3 pixels are enough to locate the pupil exactly.

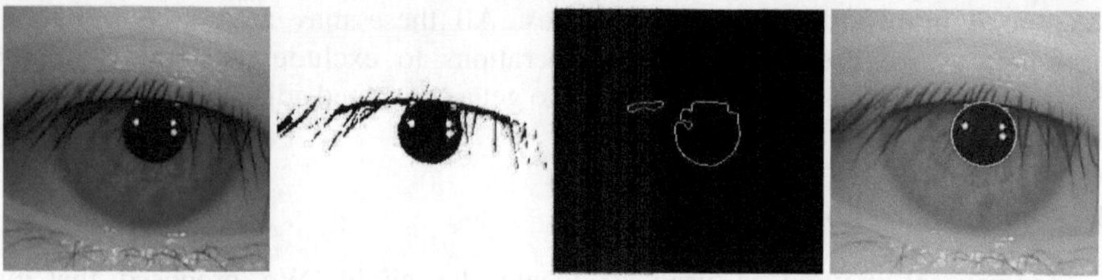

(a) (b) (c) (d)

Fig. 2. Pupil localization. (a) Source image. (b) Binary image. (c) Edge image after morphological operation and deleting small region. (d) Localized image.

3 Outer Boundary and Non-iris Areas Detection

3.1 Outer Boundary Detection

It is difficult to locate the outer boundary from the surrounding noises when there is little contrast between iris and sclera regions, especially the eyelids or eyelashes usually occlude the iris. So, Hough transform is used to detect outer boundary which is a standard machine vision technique for fitting simple contour models to images. Since space and time complexities are the main concerns in the application of Hough transform. In order to reduce the searching region we use the shrunk image to detect outer boundary together with modified Canny edge operator [9][10]. The output edge map is similar but has fewer noises, e.g spurious boundaries, using the improved Canny edge detector.

First, the image is shrunk with the rate of 30%. Then the shrunk image is filtered with modified Canny edge detector that is tuned in near vertical orientation since even in the face of occluding eyelids or eyelashes, the left and the right portions of the limbus should be clearly visible and oriented near the vertical. Whereas the detected edge usually is a noised image, especially affected by the eyelashes, shown in Fig.3(b). So we cut the top eyelid region and exclude pupil area, where top eyelid region is estimated through the pupil position, shown in Fig.3(c). Then we use 8-directional connection method to further exclude unnecessary edges using the certain threshold in order to save the connected adjacent edges, shown in Fig.3 (d). Second, the radius and the center coordinates of the outer circle are calculated using Hough transform on the denoised edge map, shown in Fig.3 (e).

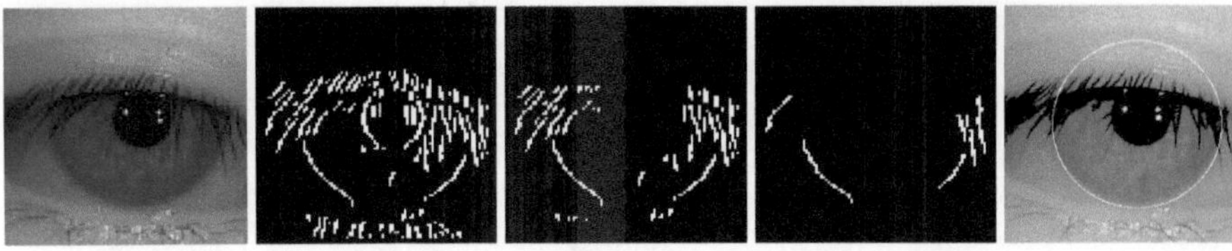

(a) (b) (c) (d) (e)

Fig. 3. Outer boundary detection. (a) Shrunk image. (b) Edge image. (c) Exclude top eyelid and pupil area. (d) Denoised edge image. (e) Outer circle localization result.

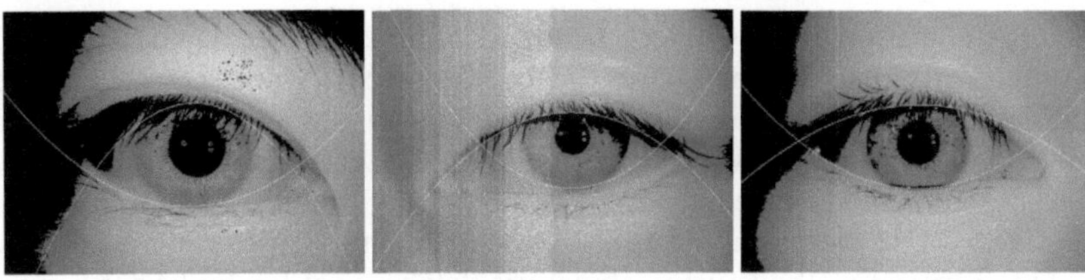

(a) (b) (c)

Fig. 4. Non-iris areas detection samples

3.2 Eyelids, Eyelashes and Reflection Detection

If the eyelids occlude part of the iris, then only that portion of the image below the upper eyelid and above the lower eyelid should be included. Here, we use least-squares sense to fit the parabola of eyelid together with modified Canny edge detector, i.e finds the coefficients of a polynomial P(X) of degree N that fits the data, P(X(I))~=Y(I) in a least-squares sense. Moreover, we additionally constrained the detected boundaries to be within the outer circle and above the pupil for upper eyelid and below the pupil for lower eyelid, respectively. Since the detected upper eyelid edge map is usually affected by the eyelashes, we use 8-directional connection method to further exclude unnecessary edges using the certain threshold in order to save the connected adjacent edges and exclude pupil areas.

Since the iris especially captured by hand-held capture device has been corrupted by the occlusion of the eyelashes, it is necessary to detect them as much as possible. We can see from Figure 3 (b) that the eyelashes area can be got from the binary image of the eye, so we detect the eyelashes areas using intensity threshold with a proper value. Since the reflection spots in most cases are much lighter than other part of the iris, it is also an effective way to detect them just by a certain threshold. The eyelids, eyelashes and reflection detection result can be seen from Fig.4.

4 Experimental Results

In this work, we use Hand-held captured iris database set, which comes from the National Laboratory of Pattern Recognition (NLPR) in China [11] collected by the

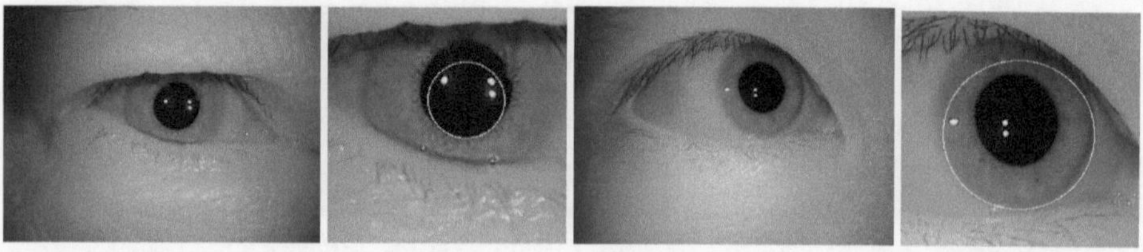

(a) (b) (c) (d)

Fig. 5. Failure localization samples. (a) Source image. (b) Pupil localization failure. (c) Source image. (d) Outer boundary detection failure.

Institute of Automation of the Chinese Academy of Science. The database set includes 1200 grayscale iris images from 60 eyes (hence 60 classes). For each eye, 20 images are captured in one session using the hand-held optical sensor. Image resolution is 640×480 and the distance between the device and the user is about 4-5cm.

The proposed methods have been implemented using Matlab 6.5 on a PC with Intel Pentium III 864MHz processor and 128MB system memory. We evaluated the success rate for the proposed methods using above iris database set for finding the inner and outer boundary. The correct segmentation rate can reach approximately 97%, only a few cases cannot be segmented correctly mainly due to the capture problem and the eye shape. The failed samples are shown in Fig.5. Moreover, we compared the time consuming of pupil localization with traditional Hough transform. The average time is about 0.7 seconds, whereas is about 76 seconds with traditional Hough method.

In general, under the same conditions and using the same iris database, the pupil localization of the proposed method speeds up with good localization accuracy. Whereas, in the case of that position of the iris is not correct or the pupil is distorted, our method will fail to localize the iris.

5 Conclusion

In this paper, a new iris segmentation method for hand-held capture device is presented. Pupil (or inner) boundary is detected using the geometrical method. In order to reduce computational cost, the shrunk image is used to detect outer boundary using Hough transform together with modified Canny edge detector. The polynomial fitting method is used to fit the parabola of eyelid together with modified Canny edge detector. Experimental results have illustrated the encouraging performance of the current method in accuracy.

In the future work, we will conduct experiments on a large number of iris databases in various environments for the proposed method to be more stable and reliable.

Acknowledgements

This work is funded by NNSF (No.60427002). Portions of the research in this paper use the CASIA iris image database collected by Institute of Automation, Chinese Academy of Sciences.

References

1. J. G. Daugman. High Confidence Visual Recognition of persons by a Test of Statistical Independence, IEEE Transaction on Pattern Analysis and Machine Intelligence, 15(11),(1993) 1148-1160
2. J.G. Daugman. The importance of being random: Statistical principles of iris recognition. Pattern Recognition,36(2), (2003) 279–291
3. J. G. Daugman. How iris recognition works. IEEE Transactions on Circuits and Systems for Video Technology,14(1),(2004) 21-30
4. R.Wildes. Iris recognition: An emerging biometric technology. Proceedings of the IEEE, 85(9), (1997) 1348–1363
5. W. K. Kong and D. Zhang. Detecting eyelash and reflection for accurate iris segmentation. International Journal of Pattern Recognition and Artificial Intelligence, 17(6),(2003)1025-1034
6. L. Ma, T. Tan, Y. Wang, D. Zhang. Efficient iris recognition by characterizing key local variations. IEEE Transactions on Image Processing,13(6),(2004) 739-750
7. L. Ma, T. Tan, Y. Wang, D. Zhang. Personal Recognition Based on Iris Texture Analysis. IEEE Transaction on Pattern Analysis and Machine Intelligence,25(12),(2003) 1519–1533
8. J.Z. Huang, Y.H. Wang, T. Tan, et al. A new iris segmentation method for recognition. Proceedings of the 17th International Conference on Pattern Recognition.Vol.3,(2004) 554-557
9. Fleck, M.M. Some defects in finite-difference edge finders. IEEE Transactions on Pattern Analysis and Machine Intelligence, 14(3),(1992)337-345
10. Canny, J.,A Computational Approach to Edge Detection, IEEE Transaction on Pattern Analysis and Machine Intelligence, 8,(1986)679-714
11. CASIA Iris Image Database, http://www.sinobiometrics.com

Iris Recognition with Support Vector Machines

Kaushik Roy and Prabir Bhattacharya

Concordia Institute for Information System Engineering,
Concordia University, Montreal, Quebec, Canada H3G 1M8
{kaush_ro, prabir}@ciise.concordia.ca

Abstract. We propose an iris recognition system for the identification of persons using support vector machines. Canny's edge detection and the Hough transform are used to find the iris/pupil boundary and a simple thresholding method is employed for eyelash detection. The Gabor wavelet technique is deployed in order to extract the deterministic features in the transformed iris of a person in the form of template. The extracted iris features are fed into a support vector machine (SVM) for classification. Our results indicate that the performance of SVM as a classifier is far better than the performance of a classifier based on the artificial neural network.

1 Introduction

To meet the increasing security requirements of the present-day society, personal identification is becoming more and more important. There has been a lot of interest recently for using the iris recognition technology as a tool in the fight against terrorism and other crimes [1][2][3]. The main concept behind this technique is to use the iris (the colored portion of the eye, around the pupil) as a means of uniquely identifying each person. Each iris is unique and does not change over the person's lifetime, therefore making their usage to identify people works even better than fingerprinting and human eye retina. Iris is a physiological biometric feature and it contains a unique texture which is complex enough to be used as a biometric signature. In this work, an iris recognition system is developed using supper vector machine (SVM) as a pattern classifier. A novel feature of this paper is the application of SVM for iris classification. Although SVM is widely used in pattern recognition applications [10], this paper is one of the few works on the application of SVM to iris classification except for [8][9].

2 Pre-processing

2.1 Iris Image Segmentation

The iris region isolation from the digital image of the eye is the first step of iris recognition. Lower and upper parts of the iris region are occluded by the eyelids and eyelashes. An automatic segmentation algorithm based on the circular Hough

D. Zhang and A.K. Jain (Eds.): ICB 2006, LNCS 3832, pp. 486 – 492, 2005.

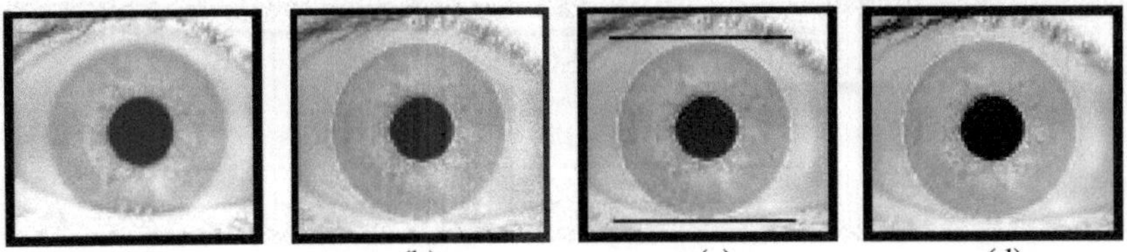

(a) (b) (c) (d)

Fig. 1. (a) Sample iris image (b) Iris and pupil boundary are detected from the sample iris image (c) Top and bottom eyelid detection (d) Eyelash has been detected using a thresholding method and denoted as black

transform is employed here. Firstly, an edge map is generated by calculating the first derivatives of intensity values in an image using canny edge detection and then thresholding the result. From the edge map, votes are cast in the Hough space for the parameters of circles passing through each edge point. A maximum point in the Hough space corresponds to the radius and centre coordinates of the circle best defined by the edge points. To perform the edge detection, the derivatives are biased in horizontal direction to detect the eyelids and outer circular boundary of iris in vertical direction since the eyelids are usually horizontally aligned, and also the eyelid edge map corrupts the circular iris boundary edge map, if using all gradient data. The range of radius values is set manually and the values of the iris radius range from 85 to 140 pixels, while the pupil radius ranges from 25 to 74 pixels. In order to make the circle detection process more efficient and accurate, the Hough transform for the iris/sclera boundary is performed first, then the Hough transform for the iris/pupil boundary is performed within the iris region, instead of the whole eye region, since the pupil is always within the iris region. Fig.1 (a) shows a sample iris image and iris/pupil regions are isolated with circular boundary from this image as shown in Fig.1 (b). Here eyelids are isolated by first fitting a line to the upper and lower eyelid using the linear Hough transform. A second horizontal line is then drawn, which intersects with the first line at the iris edge that is closest to the pupil. This process is illustrated in Fig. 1(c) and is done for both the top and bottom eyelids. The Canny edge detection is used to create an edge map, and only horizontal gradient information is taken. If the maximum in Hough space is lower than a threshold value then no line is fitted since this belongs to non-occluding eyelids. According to Kong and Zhang [6][7], separable eyelashes are detected using 1D Gabor filters, since a low output value is produced by convolution of a separable eyelash with the Gaussian smoothing function. Thus, if a resultant point is smaller than a threshold, it is noted that this point belongs to an eyelash. Multiple eyelashes are detected using the variance of intensity and if the values in a small window are lower than a threshold, the centre of the window is considered as a point in an eyelash as shown in Fig. 1(d).

2.2 Image Transformation

Once the iris region is successfully segmented, the next step is to transform the extracted iris region into a fixed dimension for further comparison. The transformation process will produce iris regions, which have the same constant

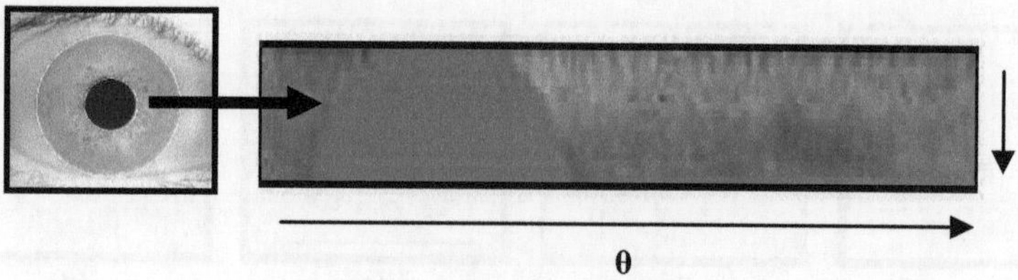

Fig. 2. Transformed Iris Image

dimensions, so that two images of the same iris under different conditions will have characteristic features at the same spatial location. The homogenous rubber sheet model devised by Daugman [2] is used here which remaps each point within the iris region to a pair of polar coordinates. The rubber sheet model takes into account pupil dilation and size inconsistencies to produce a normalized representation with constant dimension. In this way the iris region is modeled as a flexible rubber sheet anchored at the iris boundary with the pupil centre as the reference point. A constant number of points are chosen along each radial line, so that a constant number of radial data points are taken, irrespective of how narrow or wide the radius is at a particular angle. The transformed pattern is created by backtracking to find the Cartesian coordinates of data points from the radial and angular position in the normalized pattern. From the 'doughnut' iris region, it produces a 2D array with horizontal dimensions of angular resolution and vertical dimensions of radial resolution. Another 2D array is created for marking reflections, eyelashes, and eyelids detected in the segmentation stage. In order to prevent non-iris region data from corrupting the normalized representation, data points which occur along the pupil border or the iris border are discarded. Fig. 2 shows the transformed iris image.

2.3 Feature Extraction and Encoding

Feature encoding is implemented by convolving the normalized iris pattern with 1D Log-Gabor wavelets [4]. The 2D normalized pattern is broken up into a number of 1D signals, and then these 1D signals are convolved with 1D Gabor wavelets. The rows of the 2D normalized pattern are taken as the 1D signal, each row corresponds to a circular ring on the iris region. The angular direction is taken rather than the radial one, which corresponds to columns of the normalized pattern, since maximum independence occurs in the angular direction. The intensity values at known noise areas in the normalized pattern are set to the average intensity of surrounding pixels to prevent influence of noise in the output of the filtering. The output of filtering is then phase-quantized to four levels using the Daugman method, with each filter producing two bits of data for each phasor. The output of phase quantization is chosen to be a grey code, so that when going from one quadrant to another, only 1 bit changes. The encoding process produces a bitwise template containing a number of bits of information, and a corresponding noise mask that corresponds to corrupt areas within the iris pattern, and marks bits in the template as corrupt. Since the phase information will be meaningless at regions where the amplitude is zero, these regions are also marked in the noise mask. The total number of bits in the template will be twice of the angular resolution times the radial resolution multiplied by the number of filters used. Here a pattern of 300-bit sequence is selected for each iris image.

3 Support Vector Machine as a Classifier

Support Vector Machine (SVM) is a new technique for data classification. Given a training set of instance-label pairs (x_i, y_i), $i = 1, 2 \ldots, l$ where $x_i \in R^n$ and $y \in \{1,-1\}^l$ the support vector machine (SVM) requires the solution of the following optimization problem [10]:

$$\begin{aligned} \min_{w,b,\xi} \quad & \frac{1}{2} w^T w + C \sum_{i=1}^{l} \xi_i \\ \text{subject to} \quad & y_i (w^T \phi(x_i) + b) \geq 1 - \xi_i, \\ & \xi_i \geq 0. \end{aligned}$$

Here, the training vectors x_i are mapped into a higher (maybe infinite) dimensional space by the function ϕ. Then SVM finds a linear separating hyperplane with the maximal margin in this higher dimensional space. $C > 0$ is the penalty parameter of the error term. Furthermore, $K(x_i, xj) = \phi(x_i)^T \phi(x_j)$ is called the *kernel function*. The four basic kernel functions are given below [10]:

- Linear: $K(x_i, x_j) = x^T{}_i x_j$.
- Polynomial: $K(x_i, x_j) = (\gamma x_i^T x_j + r)^d, \lambda > 0.$
- Radial Basis Function (RBF): $K(x_i, x_j) = \exp(-\gamma \|x_i - x_j\|^2), \gamma > 0.$
- Sigmoid: $K(x_i, x_j) = \tanh(\gamma x_i^T x_j + r).$

In the present work, six iris image samples of each person from CASIA database are used to train the support vector machine. Here four kernel functions are used for experimentation and the most favorable one is selected for prediction. The remaining iris samples are used for recognition purpose.

4 Experimental Results and Discussion

We have used the CASIA (Chinese Academy of Sciences-Institute of Automation) iris image database and each iris class is composed of 7 samples taken in two sessions, three in the first session and four in the second. Sessions were taken with an interval of one month. Images are 320x280 pixels gray scale taken by a digital optical sensor designed by NLPR (National Laboratory of Pattern Recognition – Chinese Academy of Sciences). There are 108 classes of irises in a total of 756 iris images. Here the automatic iris and pupil detection procedure is proved as successful. For CASIA iris Database, iris region of 733 images out of 756 are detected properly

Table 1. Accuracy of the different parts of the iris recognition system

Number of sample Iris Images	Segmentation (Iris/Pupil Detection)	Transformation	Feature Extraction & Encoding
756	733 Accuracy = 96.95%	745 Accuracy =98.54%	741 Accuracy =98.01%

Table 2. Efficiency of various kernel functions

Kernel Type	No. of Support Vectors	Classification Accuracy (%)
Linear	678	93.67
Polynomial	643	92.56
Radial Basis Function (RBF)	728	97.34
Sigmoid	347	91.01

Table 3. Comparison of FAR and FRR for ANN and SVM

Type of errors	FAR (%)	FRR (%)
ANN	5.8	1.9
SVM	2.06	.60

Table 4. Average time consuming of different parts of iris recognition system

	Time Consuming (in ms)	% of Total Time
Iris Segmentation	433	95.16
Iris Image Transformation	17	3.73
Feature Extraction	4	0.87
Iris Classification	< 1	.21
Total	≈ 455	100

which corresponds to a success rate of 96.95%. The transformation procedure is also proved as successful with an accuracy rate of 98.54%. However, the same pattern from images with varying amounts of pupil dilation is not perfectly reconstructed by the transformation procedure, since deformation of iris results in small changes of its surface pattern. Feature encoding is performed by 1D Gabor wavelets. Table-1 shows the accuracy of the different parts of the system. In the current research work six iris image samples of each person from CASIA are used for classification by support vector machines and the remaining iris image samples are used for recognition purpose. Four kernel functions: Linear, polynomial, RBF and sigmoid are used here and classification efficiency of each kernel is measured which is shown in Table-2 with number of support vectors found during experimentation for RBF kernel. Since the highest classification accuracy is obtained by RBF kernel, this kernel is used in our system for classification and recognition purpose. To show the effectiveness of SVM as a classifier, extracted features are also given as input to Artificial Neural

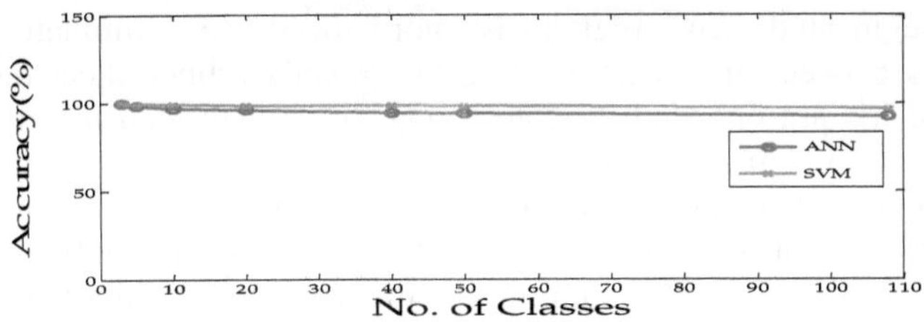

Fig. 3. Comparison of classification accuracy between ANN and SVM

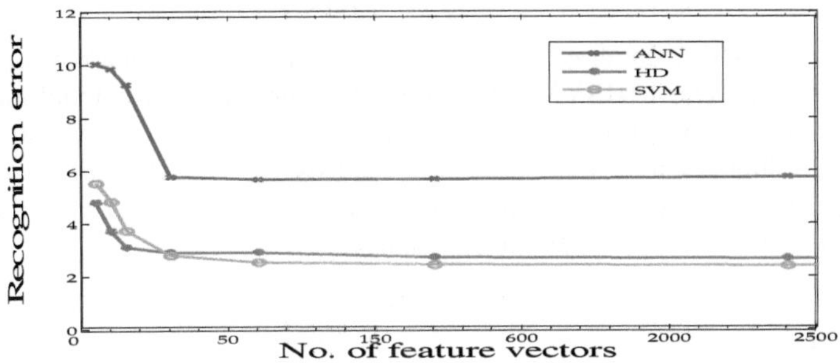

Fig. 4. Comparison between no. of feature vectors and recognition error

Network (ANN) for classification and accuracy of classification for various numbers of classes between ANN and SVM are shown in Fig.3. From this figure it is found that the performance of SVM as a classifier is far better than ANN though the classification accuracy is decreased as number of classes increased. Table-3 shows the False Accept rate (FAR) and False Reject Rate (FRR) for SVM and ANN. Finally, Table-4 indicates the time consumed by the different parts of the iris recognition system that is implemented on Pentium-4-3.00 GHz workstation. Since the total average execution time of verification process is not exceeded 455 ms, the system is suitable for a non-intrusive authentication process. Fig. 4 shows the comparison of the feature vectors vs. recognition error among Hamming Distance (HD), ANN and SVM. In this case only RBF kernel is considered due its reasonable classification accuracy. Here SVM provides satisfactory recognition rate in comparison with HD when number of feature vectors are increased. A pattern of 300 feature vectors for each iris image is chosen for its tolerable recognition accuracy. Fig.4 also shows that SVM outperforms ANN while the classification as well as recognition accuracy is concerned.

5 Conclusions

In this paper an iris recognition system is implemented using support vector machine. Firstly, an automatic segmentation algorithm is applied to localize the iris region from an eye image and isolate lower, upper eyelids, and eyelashes using Hough transform.

Next, the segmented iris region is normalized to eliminate dimensional inconsistencies between iris regions using Daugman's rubber sheet model. Finally, features of the iris are encoded by convolving the transformed iris region with 1D Log-Gabor filters and phase quantizing the output in order to produce a bit-wise biometric template. These extracted feature templates are then given to support vector machines for classification. In future, the speed of the system may be improved using the bootstrapping strategy. Also, non-symmetrical support vector machines may be used where the false accept/reject rate is treated differently.

Acknowledgements

The shared iris CASIA database is available on the web http://www.sinobiomatics.com/resources.htm. The LIBSVM (A Library for support vector machines) tool has been used in this paper for classification available at http://www.csie.ntu.edu.tw/~cjlin/libsvm/. This research has been supported by the NSERC, a Canada Research Chair grant and also by a grant from the Faculty of Engineering and Computer Science, Concordia University, Montreal, Canada.

References

1. J. Daugman, "How iris recognition works", Proc. of the IEEE Internat. Conf. on Image Processing, Vol. 1, pp. 33-36, 2002.
2. J. Daugman, "High Confidence Visual Recognition of Persons by a Test of Statistical Independence", IEEE Trans, on Pattern Analysis and Machine Intelligence, Vol. 15, No.11, pp.1148-1161, 1993.
3. J. Daugman, "Complete Discrete 2-D Gabor Transforms by Neural Network for Image Analysis and Compression", IEEE Trans. Acoust., Speech, Signal Processing, Vol. 36, No. 7, pp 1169-1179, 1988.
4. R. Wildes, "Iris Recognition: An Emerging Biometric Technology", Proc. of the IEEE, Vol. 85, No. 9, pp. 1348-1363, 1997.
5. R. Bolle, J. Connell, S. Pankanti, N. Ratha, and A. Senior, "Guide to Biometrics", Springer, New York, 2003.
6. W. Kong, and D. Zhang, "Accurate Iris Segmentation Based on Novel Reflection and Eyelash Detection Model". Proc. of Internat. Sympos. On Intelligent Multimedia, Video and Speech Processing, Hong Kong, 2001.
7. W. Kong, and D. Zhang, "Detecting Eyelash and Reflections for Accurate Iris Segmentation", Internat. Journal of Pattern Recognition and Artificial Intelligence, Vol. 17, No. 6 pp 1025-1034, 2003.
8. S. Yoon, S. Choi, S. Cha, Y. Lee, and C. Tappert, "On the Individuality of Iris Biometric", ICGST Internat. Journal. On Graphics, Vision and Image Processing, Vol. 5, 2005.
9. Y. Wang, and J. Han, "Iris Recognition Using Support Vector Machines", IEEE Internet. Sympos. on Neural Network, Dalian, China, 2004.
10. N. Cristianini and J. Shawe Taylor, "An Introduction to Support Vector Machines", Cambridge University press, New York, 2000.

Multi-level Fusion of Audio and Visual Features for Speaker Identification

Zhiyong Wu[1,2], Lianhong Cai[1], and Helen Meng[2]

[1] Department of Computer Science and Technology,
Tsinghua University, Beijing, China, 100084
john.zy.wu@gmail.com, clh-dcs@tsinghua.edu.cn
[2] Department of Systems Engineering and Engineering Management,
The Chinese University of Hong Kong, Shatin, N.T., Hong Kong SAR, China
hmmeng@se.cuhk.edu.hk

Abstract. This paper explores the fusion of audio and visual evidences through a multi-level hybrid fusion architecture based on dynamic Bayesian network (DBN), which combines model level and decision level fusion to achieve higher performance. In model level fusion, a new audio-visual correlative model (AVCM) based on DBN is proposed, which describes both the inter-correlations and loose timing synchronicity between the audio and video streams. The experiments on the CMU database and our own homegrown database both demonstrate that the methods can improve the accuracies of audio-visual bimodal speaker identification at all levels of acoustic signal-to-noise-ratios (SNR) from 0dB to 30dB with varying acoustic conditions.

1 Introduction

Human speech is produced by the movement of the articulatory organs. Since some of these articulators are visible, there are inherent correlations between audio and visual speech. There is also loose timing synchronicity between them, for instance, the mouth is opened before producing speech and closed after speech is produced.

While the audio is a major source of speech information, the visual component is considered to be a valuable supplementary information source in noisy environments because it remains unaffected by acoustic noise. Many studies have shown that the integration of audio and visual features leads to more accurate speaker identification even in noisy environments [1-3].

Audio-visual integration can be divided into three categories: feature fusion, decision fusion and model fusion [3-5]. In feature fusion, multiple features are concatenated into a large feature vector and a single model is trained [4]. However this type of fusion cannot easily represent the loose timing synchronicity between audio visual features. In decision fusion, audio and visual features are processed separately to build two independent models [5], which completely ignore the audio visual correlations. In model fusion, several models have been proposed, such as multi-stream hidden Markov model (HMM) [6], factorial HMM [6], coupled HMM [2], mixed DBN [7], etc. Multi-stream HMM and factorial HMM assume independence between audio

D. Zhang and A.K. Jain (Eds.): ICB 2006, LNCS 3832, pp. 493–499, 2005.

and visual features. Coupled HMM and mixed DBN force audio visual streams to be in strict synchrony at model boundaries by introducing "anchor-points".

This work attempts to capture the inter-correlations between audio and visual cues as well as the loose synchronicity between them for speaker identification. We propose a new audio-visual correlative model (AVCM) to describe the above relations, which is realized using the DBN. We also explore the fusion of audio and visual evidences through a multi-level hybrid fusion architecture based on DBN, which combines model level and decision level fusion to achieve higher performance.

The outline of this paper is as follows: Section 2 gives the details of the proposed audio-visual correlative model (AVCM). Then the multi-level audio visual fusion architecture is described in section 3. Section 4 presents the experimental results and analysis showing how the proposed approaches improve the speaker identification performance. Finally, section 5 concludes the paper.

2 DBN Based Audio-Visual Correlative Model (AVCM)

Dynamic Bayesian networks are a class of Bayesian networks designed to model temporal processes as stochastic evolution of a set of random variables over time [8]. A DBN is a directed acyclic graph whose topology structure can be easily configured to describe various relations among variables. DBN offers a flexible and extensible means of modeling the feature-based and temporal correlations between audio and visual cues for speaker identification.

We propose the AVCM model as depicted in figure 1. It illustrates a whole sentence model that consists of several words. The square nodes represent discrete variables. The round nodes represent continuous variables. The hollow nodes represent hidden variables and the shaded nodes are observed. The upper part of the model describes the audio stream (audio sub-model) and the lower part describes the video stream (video sub-model). The labeled nodes include:

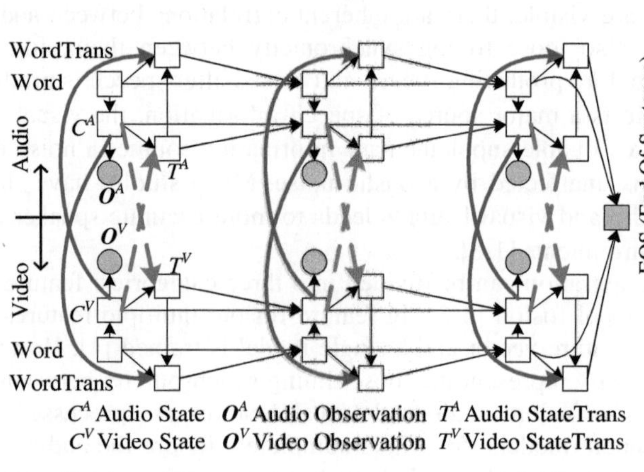

Fig. 1. Audio-visual correlative model (AVCM) based on DBN

- the "Word" node stands for the current word which is determined by the sentence;
- the "State" node (C^A, C^V) indicates the current state and is determined by "Word";
- the "Observation" node (O^A, O^V) represents the audio or visual observations;
- the "State Trans" node (i.e. state transition, T^A, T^V) indicates when the current state ends and switches to the next state;
- the "Word Trans" node may take the values *true* or *false* to respectively denote whether there is a word transition and is dependent on the "State Trans" node;
- the "EOS" node (End Of Sentence) represents the end of the whole sentence.

Inter-node dependencies modeled by the proposed AVCM include:

- the "State Trans" nodes of the audio and video streams are dependent on their "State" nodes from both two models, which describe the inter-correlations between the two streams. This is shown by the thick dashed arrows in figure 1;
- the "Word Trans" nodes are also dependent on the "State" nodes from both audio and video streams, which capture the loose timing synchronicity in between the two streams. This is shown by the thick solid arrows in figure 1.

The proposed AVCM model differs from previous approaches such as [2] and [7], where the loose timing synchronicity between audio and video streams is restricted by "anchor-points" at word boundaries. In AVCM, audio and video streams have their own independent "Word" and "Word Trans" nodes. Furthermore, the "Word Trans" node of the audio stream is dependent on the "State" node of the video stream, and vice versa. This models the loose synchronicity in between two streams and brings about performance improvements, as will be discussed later.

3 Multi-level Fusion of Audio and Visual Evidences

Incorporation of bimodal, correlated audio-visual features should achieve speaker identification performance that is superior to mono-modal systems. This is because the two modalities, if modeled properly, can complement and reinforce each other. However, under some conditions, the performance of AVCM is not as good as that of decision fusion (see point 4 in section 4.2). Similar observations are reported in [2].

In view of the advantages of model fusion and decision fusion, we proposed a multi-level fusion strategy via DBN, as illustrated in figure 2. There are three models altogether: the audio-only model, the video-only model and the AVCM model that performs model-based audio-video fusion. These three models are further combined by means of decision-level fusion to deliver the final speaker identification result.

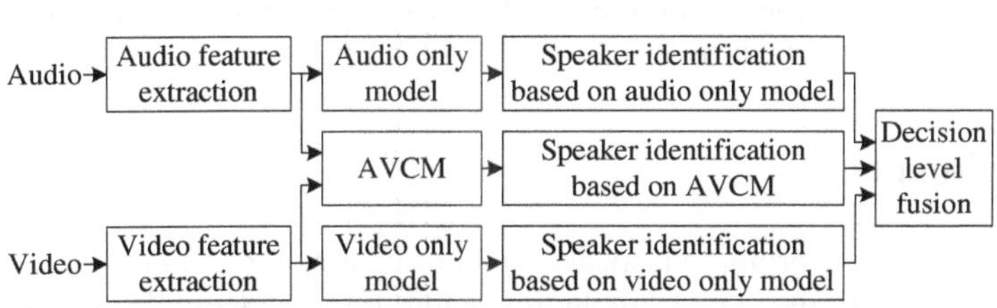

Fig. 2. Strategy for audio-visual multi-level fusion

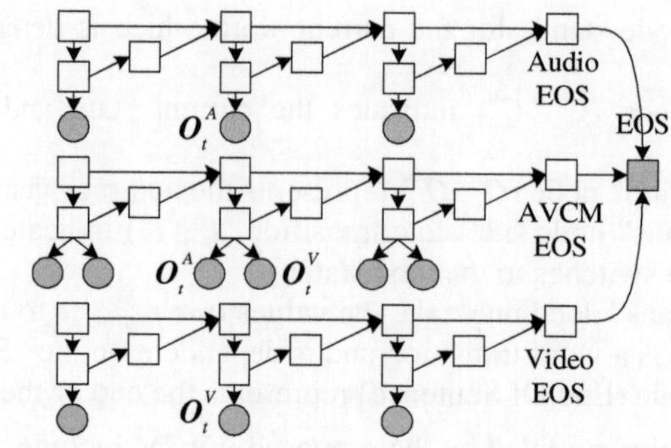

Fig. 3. DBN for audio-visual multi-level fusion

Decision-level fusion for the three models is also achieved through the use of DBN, by virtue of its extensibility. This is shown in figure 3, in which the EOS node of AVCM model (AVCM EOS), audio-only model (Audio EOS) and video-only model (Video EOS) are connected to the global "EOS" node. The EOS nodes of the three models are hidden and the global "EOS" node is observable.

Equation 1 shows the mathematical formula we use for multi-level fusion,

$$P(\boldsymbol{O}^A, \boldsymbol{O}^V \mid M^A, M^V, M^{AV}) = [P(\boldsymbol{O}^A \mid M^A)]^{\lambda_A} [P(\boldsymbol{O}^V \mid M^V)]^{\lambda_V} [P(\boldsymbol{O}^A, \boldsymbol{O}^A \mid M^{AV})]^{\lambda_{AV}}. \quad (1)$$

where $P(\boldsymbol{O}^A \mid M^A)$ is the recognition formula for audio-only model M^A of audio observation $\boldsymbol{O}^A$, and $P(\boldsymbol{O}^V \mid M^V)$ is the formula for video-only model M^V of video observation $\boldsymbol{O}^V$, and $P(\boldsymbol{O}^A, \boldsymbol{O}^V \mid M^{AV})$ is the formula for AVCM model M^{AV}. λ_A, λ_V and λ_{AV} are stream exponents (fusion weights) for audio-only, video-only and AVCM model, which encode the relative reliability of the models and can be varied according to ambient noise conditions (i.e. SNR). When the SNR is high, AVCM should be more reliable than mono-modal models and should carry higher weights. When SNR is low, the reliability of the audio-only and AVCM models degrade, hence the video-only model should carry the highest weight. The proposed multi-level fusion strategy combines the advantages of model fusion and decision fusion and has the potential of achieving performance improvement.

The estimation of the fusion weights is a key issue. We enforce the parameter constraints of $\lambda_A+\lambda_V+\lambda_{AV}=1$, $\lambda_A=\lambda_{AV}$ and $\lambda_A,\lambda_V,\lambda_{AV}\geq 0$. In addition we impose $\lambda_A=\lambda_{AV}$ by assuming that the performance of both audio-only and AVCM models are equally dependent on the quality of the acoustic speech. We then use a novel methodology known as support vector regression (SVR) to estimate the fusion weights directly from the original audio features [9].

4 Experiments

We perform the text-prompted speaker identification experiments to evaluate the performance of various models including audio-only, video-only, decision fusion, feature fusion, coupled HMM (CHMM), AVCM and multi-level fusion.

The experiments are conducted on the audio-visual bimodal database from Carnegie Mellon University (CMU database) [10] as well as our own homegrown database. The CMU database includes 10 subjects (7 males and 3 females) speaking 78 isolated words repeated 10 times. These words include numbers, weekdays, months, and others that are commonly used for scheduling applications. Our homegrown database includes 60 subjects (38 males and 22 females, aged from 20 to 65) with each subject speaks 30 connect-digit words (the digit length differs from 2 to 6), and each utterance is repeated three times at intervals of 1 month.

The acoustic front-end includes 13 Mel frequency cepstral coefficients (MFCCs) and 1 energy parameter (with frame window size of 25ms and frame shift of 11ms) together with their delta parameters. Hence the audio feature vector has 28 dimensions. The visual front-end includes mouth width, upper lip height, lower lip height [10] and their delta parameters. Thus the visual feature vector has 6 dimensions. The video frame rate is 30 frames per second (fps), which is up-sampled to 90fps (11ms) by copying and inserting two frames between each two original video frames.

Artificial white Gaussian noise was added to the original audio data (SNR=30dB) to simulate various SNR levels. The models were trained at 30dB SNR and tested under SNR levels ranging from 0dB to 30dB at 10dB intervals. We applied cross-validation for every subject's data, i.e. 90% of all the data are used as training set, and the remaining 10% as testing set. This partitioning was repeated until all the data had been covered in the testing set.

Table 1. Accuracies (%) of speaker identification under different SNR on CMU database

audio signal-to-noise-ratio (SNR)	30dB	20dB	10dB	0dB
video-only	77	77	77	77
audio-only	100	64	22	17
feature fusion	99	85	30	20
decision fusion	100	86	78	78
CHMM	100	88	79	60
AVCM	100	92	79	65
multi-level fusion	100	93	81	79
fusion weight for multi-level fusion (λ_A=λ_{AV})	0.4	0.32	0.1	0.01

Table 2. Accuracies (%) of speaker identification under different SNR on our own database

audio signal-to-noise-ratio (SNR)	30dB	20dB	10dB	0dB
video-only	74	74	74	74
audio-only	99	59	20	15
feature fusion	99	81	26	18
decision fusion	100	83	76	75
CHMM	100	85	77	57
AVCM	100	89	78	61
multi-level fusion	100	91	80	77

All the tested models are implemented as DBNs. A DBN is developed for each word, with a left-to-right no skipping topological structure. The number of the transited state is always equal to or 1 greater than the original state. The audio sub-model has 5 states, and video sub-model has 3 states, each state is modeled using the Gaussian mixture model (GMM) with 3 mixtures. During speaker identification, the words' DBNs are connected to form a whole sentence model, which is then used to identify the speakers. The DBNs are implemented using the GMTK toolkit [11].

The identification accuracies from all the testing data are averaged and reported as the final result. The results on CMU database are summarized in table 1. The experiments are also conducted on our own homegrown database with a larger number of speakers and results are summarized in table 2. Main observations include:

(1) Feature fusion performs worse than other fusion methods, even achieving accuracies lower than video-only model when SNR≤10dB. The main reason is due to the misalignment between audio and video streams;
(2) Decision fusion achieves lower accuracy than CHMM and AVCM when SNR≥10dB. However, when SNR<10dB, the accuracy is higher than CHMM and AVCM, as shown by the shaded table cells;
(3) The AVCM model proposed in this paper has higher accuracy than CHMM, because it describes both inter-correlations and loose timing synchronicity between audio and video features;
(4) Because of the misalignment during model training, when SNR<10dB, the performance of the AVCM and CHMM degrades and the accuracy is even lower than the video-only model;
(5) The audio-visual multi-level fusion strategy has solved the problem in (4) well. Best identification performance is obtained even when SNR=0dB. It can also be seen from both tables that better results are also obtained when SNR is 10dB and 20dB than the AVCM model. This is because that the multi-level fusion strategy combines the results from audio-only, video-only and AVCM model, and the results of audio and video model are supplementary to that of AVCM;
(6) From the results of table 1 and table 2, we can see that the performance of the speaker identification degrades a little with larger speaker numbers, but the conclusions from (1) to (5) can still be drawn. This indicates that the proposed model based on DBN has good extensibility for different databases.

5 Conclusions

This paper investigates the correlations between audio and visual features. A new audio-visual correlative model (AVCM) based on dynamic Bayesian network (DBN) is proposed, which describes both the inter-correlations and the loose synchronicity between audio and visual streams. The experiments on the audio-visual bimodal speaker identification demonstrate that the AVCM model improves the identification accuracies compared to the previous methods.

We also propose a DBN based audio-visual multi-level fusion strategy, which combines the results of audio-only, video-only and AVCM models through decision-level fusion. Experiments on both CMU database and our own homegrown database show that the proposed strategy integrates the advantages of both model level and decision level fusion and achieves the best accuracies of speaker identification at all levels of acoustic signal-to-noise ratio (SNR), ranging from 0dB to 30dB.

Acknowledgments

This work is supported by the research fund from the National Science Foundation of China (NSFC) under grant No. 60433030 and the joint fund of NSFC-RGC (Research Grant Council of Hong Kong) under grant No. 60418012 and N-CUHK417 /04.

References

1. Senior, A., Neti, C., Maison, B.: On the use of visual information for improving audio-based speaker recognition. In: Proc. Audio-visual Speech Processing Conf. (1999) 108–111
2. Nefian, A.V., Liang, L.H., Fu, T.Y., Liu, X.X.: A Bayesian approach to audio-visual speaker identification. In: Proc. 4th International Conf. Audio- and Video-based Biometric Person Authentication, Vol. 2688 (2003) 761–769
3. Chibelushi, C.C., Deravi, F., Mason, J.S.D.: A review of speech-based bimodal recognition. IEEE Trans. Multimedia 4 (2002) 23–37
4. Chibelushi, C.C., Mason, J.S.D., Deravi, F.: Feature-level data fusion for bimodal person recognition. In: Proc. 6th IEEE International Conf. Image Processing and its Applications. IEE, Stevenage (1997) 399–403
5. Chatzis, V., Bors, A.G., Pitas, I.: Multimodal decision-level fusion for person authentication. IEEE Trans. Syst. Man Cybern. A 29 (1999) 674–680
6. Dupont, S., Luettin, J.: Audio-visual speech modeling for continuous speech recognition. IEEE Trans. Multimedia 2 (2000) 141–151
7. Gowdy, J.N., Subramanya, A., Bartels, C., Bilmes, J.: DBN based multi-stream models for audio-visual speech recognition. In: Billene, M. (ed.): Proc. IEEE International Conf. Acoustics, Speech, and Signal Processing, Vol. 1. IEEE, Canada (2004) 993–996
8. Dean, T., Kanazawa, J.: Probabilistic temporal reasoning. In: Proc. 7th National Conf. Artificial Intelligence (1988) 524–528
9. WU, Z.Y.: Audio-visual bimodal modeling for speaker identification and visual-speech synthesis. Ph.D. Dissertation. Department of Computer Science and Technology, Tsinghua University, Beijing, China (2005)
10. Chen, T.: Audiovisual speech processing. IEEE Trans. Signal Processing 18 (2001) 9–21
11. Bilmes, J., Zweig, G.: The graphical models toolkit: An open source software system for speech and time-series processing. In: Proc. IEEE International Conf. Acoustics, Speech and Signal Processing, Vol. 4. IEEE, Florida (2002) 3916–3919

Online Signature Verification with New Time Series Kernels for Support Vector Machines

Christian Gruber, Thiemo Gruber, and Bernhard Sick

University of Passau, Institute of Computer Architectures
{gruberc, grubert, sick}@fmi.uni-passau.de

Abstract. In this paper, two new methods for online signature verification are proposed. The methods adopt the idea of the longest common subsequences (LCSS) algorithm to a kernel function for Support Vector Machines (SVM). The two kernels *LCSS-global* and *LCSS-local* offer the possibility to classify time series of different lengths with SVM. The similarity of two time series is determined very accurately since outliers are ignored. Consequently, LCSS-global and LCSS-local are more robust than algorithms based on dynamic time alignment such as Dynamic Time Warping (DTW). The new methods are compared to other kernel-based methods (DTW-kernel, Fisher-kernel, Gauss-kernel). Our experiments show that SVM with LCSS-local and LCSS-global authenticate persons very reliably.

1 Introduction

Authentication of a person's identity is an everydays issue. Often, signatures are used for verification. However, in most cases a signature is compared to a single reference signature with the naked eye only. Electronic, typeface-based techniques (so-called *offline signature verification*) can easily be outsmarted. Since authentication by signature is more widely accepted than any other technique (e.g., fingerprint or iris scan), a signature verification system that ensures a high level of security must be developed. Biometric signature verification systems that are based on the dynamics of a person's signature and not on its image are substantially more suitable for a reliable authentication (so-called *online signature verification*).

Support Vector Machines (SVM) are very popular since a few years. As they provide very good results for various pattern recognition problems, they also seem to be a good choice for online signature verification. Compared to most methods used for signature verification such as Hidden Markov Models (HMM) or Dynamic Time Warping (DTW), SVM, which are based on the principle of structural risk minimization, have various advantages such as a convex objective function with very fast training algorithms. On the other hand, SVM typically are applied to data sets containing feature vectors of fixed length and not to problems dealing with time series of variable length such as in online signature verification. In the following, the terms time series and sequence will be used equivalently.

D. Zhang and A.K. Jain (Eds.): ICB 2006, LNCS 3832, pp. 500–508, 2005.

2 Related Work

Only a few researchers in the area of online signature verification apply SVM as their usage poses a major problem: "Classic" kernel functions such as linear, polynomial, or Gauss-kernels require input vectors of fixed length. Since almost all signatures have different lengths, a way has to be found to deal with input vectors of different lengths.

One approach is to extract a fixed set of features, such as average velocity, pen-up time, etc., from each signature and to present these input vectors to an SVM utilizing kernels as mentioned above (so-called *static kernels*). This approach is described in [1]. SVM with classic kernels are also used in [2], but just for the fusion of two preceding classifiers' decisions (ensemble technique).

The other approach is to use kernels for sequences (so-called *dynamic kernels*). As far as we know, SVM with dynamic kernels have not been applied to online signature verification yet. Several publications such as [3, 4] provide a survey and a comparison of various kernel functions that are suitable for time series classification. Two categories of kernels can be distinguished: Within the first category are kernels based on distance measures. [5] introduced the *Time Alignment Kernel* for handwritten digit recognition. [6, 7, 8] developed kernels based on DTW for speech recognition and handwritten character recognition. The second category comprises kernel functions based on probabilistic models such as HMM or Gaussian Mixture Models (GMM). [9] introduced the *Fisher-kernel* which maps an input sequence onto a score vector with fixed length that is obtained from the parameters of the underlying probabilistic model. Fisher-kernels are used for classification of DNA-fragments, speaker-independent classification of spoken letters, or speaker identification. Another probabilistic kernel based on the Kullback-Leibler divergence of two GMM was described in [10]. This kernel was applied to speaker authentication as well as image classification.

3 LCSS-Based Kernels for SVM

3.1 Kernel Functions for Online Signature Verification

Given a binary classification problem $\mathcal{L} = \{(\mathbf{x}_i, y_i)\}$ with class labels $y_i \in \{-1, 1\}$, $i = 1 \ldots l$, an SVM (see, e.g., [14]) classifies a test sample $\mathbf{t}$ by $y_t = sign\left(\sum_{i=1}^{l} \alpha_i y_i K(\mathbf{t}, \mathbf{x}_i) + b\right)$with parameters $\alpha_i, b \in \mathbb{R}$. $K(\cdot, \cdot)$ is a kernel function which is the inner product of the samples transformed into a higher-dimensional space. In this feature space, an SVM tries to separate the two classes linearly. If a kernel function satisfies Mercer's conditions, the resulting kernel matrix is positive semidefinite, and the objective function that has to be optimized to determine the α_i and b is convex.

HMM are *probabilistic models* often used for online signature verification. Kernel functions based on HMM can either be applied to raw data or to features extracted from those data. In the former case a large number of hidden states is needed to get reasonable results. The consequence is a slow training phase. In the latter case complex pre-processing measures must be applied (e.g., segmentation of sequences and/or approximation with local models).

Approaches based on *distance measures* such as DTW can also be used as kernel functions. However, most of these methods are very sensitive to varying signal offsets or different ranges. Also, most of these methods do not deal appropriately with the type of outliers that appears within signatures (e.g., additional loops). We need kernel functions that do not require time-consuming pre-processing and meet our requirements concerning outliers.

3.2 LCSS-Global

In this section, a method that determines the similarity of two sequences with different lengths is introduced. In contrast to other methods such as DTW, non-matching subsequences (*gaps*) of these two sequences are ignored.

Subsequence Similarity: We are given two sequences $X = (x_1, \ldots, x_n)$, $Y = (y_1, \ldots, y_m)$ with $x_i, y_i \in \mathbb{R}$, $n, m \in \mathbb{N}$, $n \leq m$ and $\gamma, \epsilon \in \mathbb{R}^+$ with $\gamma \leq 1$. The sequences X and Y are called (γ, ϵ)-*similar*, if subsequences $X' = (x_{i_1}, \ldots, x_{i_{\lfloor \gamma \cdot n \rfloor}})$ and $Y' = (y_{j_1}, \ldots, y_{j_{\lfloor \gamma \cdot n \rfloor}})$ exist for all $k = 1, \ldots, \lfloor \gamma \cdot n \rfloor - 1$ with $i_k \leq i_{k+1}$ and $j_k \leq j_{k+1}$, such that $y_{j_k} - \epsilon \leq x_{i_k} \leq y_{j_k} + \epsilon$ for all $k = 1, \ldots, \lfloor \gamma \cdot n \rfloor$. The parameter γ determines the length of the subsequence of corresponding data points, ϵ controls how close (regarding the y-axis) matching points have to be. An example of two time series matched with LCSS-global is shown in Fig. 1. Similar points are matched, dissimilar subsequences (gaps) are ignored. The elements of X' and Y' are temporally ordered just as in X and Y, but they may consist of discontiguous fragments of the original sequences. Sketches for appropriate algorithms that determine such subsequences based on dynamic programming are set out in [12].

LCSS-Global Similarity: The overall similarity of two sequences X, Y based on LCSS-global with a user-defined $\epsilon \in \mathbb{R}^+$ is given by

$$Sim_\epsilon(X, Y) = \{max\ \gamma \mid X, Y \text{ are } (\gamma, \epsilon)\text{-similar}\}. \quad (1)$$

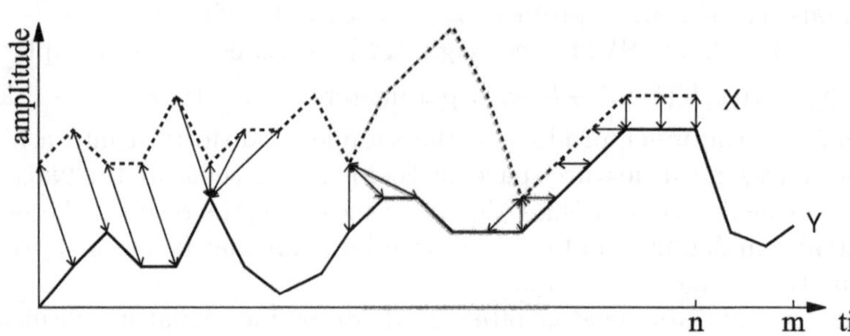

Fig. 1. Matching of two time series with LCSS-global

3.3 LCSS-Local

In LCSS-global, sequences must be rescaled to a common range in order to get suitable matchings. But, global rescaling may not reflect the similarity of two given sequences perfectly. LCSS-local allows the application of different local scaling functions to different subsequences of the time series. Additionally, very short discorresponding gaps are not ignored if they do not exceed a predefined length. Longer, discorresponding gaps are regarded as outliers and not matched. The algorithm is based on the method described in [13] which detects correlations of two sequences in three steps:

1. Computation of all ϵ-similar subsequences of a fixed length,
2. Iterative fusion of two subsequences to one longer subsequence, and
3. Determination of the longest common subsequence length.

Atomic Matchings: We are given two sequences $S = (s_1, \ldots, s_w)$ and $T = (t_1, \ldots, t_w)$ with equal length $w \in \mathbb{N}$ and $\epsilon \in \mathbb{R}^+$. S and T are ϵ-similar if $t_i - \epsilon \leq s_i \leq t_i + \epsilon$ for all $i = 1, \ldots, w$. Such a correlation is called an *atomic matching* of S and T.

In order to find all atomic matchings of two time series $X = (x_1, ..., x_n)$, $Y = (y_1, ..., y_m)$ with different lengths, they are split into all possible contiguous subsequences $\widetilde{S}_i = (x_i, \ldots, x_{i+w-1})$ and $\widetilde{T}_j = (y_j, \ldots, y_{j+w-1})$ with user-defined length w with $i = 1, \ldots, n - w + 1$ and $j = 1, \ldots, m - w + 1$. Subsequently, every subsequence $\widetilde{S}_i$ and $\widetilde{T}_j$ is rescaled to a specified interval by linear mapping functions f_i and g_j (with $f_i(x) = a_i x + b_i$ and $g_j(y) = c_j y + d_j$) in order to align the different ranges (e.g., rescaling to $[0, 1]$). With these local transformations, corresponding subsequences can now be found.

The rescaled subsequences S_i and T_j are compared pairwise and checked for ϵ-similarity. If two subsequences S_u and T_v are ϵ-similar, this is called an *atomic matching* (S_u, T_v) *of length* w of the two time series X and Y at points $u, \ldots, u + w - 1$ on X and $v, \ldots, v + w - 1$ on Y.

Longer Subsequences: Having computed all atomic matchings of a certain length w, the matching subsequences of X and Y must be fused to longer subsequences. Consider two atomic matchings (S_{i_1}, T_{j_1}) and (S_{i_2}, T_{j_2}) with $i_1 < i_2$ and $j_1 < j_2$. With *length*() denoting the length of a (sub-)sequence, these atomic matchings can be fused to longer subsequences if either the following conditions 1 and 3 or 2 and 3 hold (cf. Fig. 2):

1. The subsequences S_{i_1} and S_{i_2} are nonintersecting on X, that is, $i_1 + length(S_{i_1}) \leq i_2$. Moreover, the distance of two atomic matchings must be smaller than a given parameter ϱ: $i_2 - i_1 + length(S_{i_1}) \leq \varrho$. These conditions must also hold for T_{i_1} and T_{i_2}. It should be noticed that the gap between the two subsequences is included in the longer subsequence.
2. The two atomic matchings (S_{i_1}, T_{j_1}) and (S_{i_2}, T_{j_2}) intersect on both time series with the same length: $d = i_1 + length(S_{i_1}) - i_2 = j_1 + length(T_{j_1}) - j_2$.

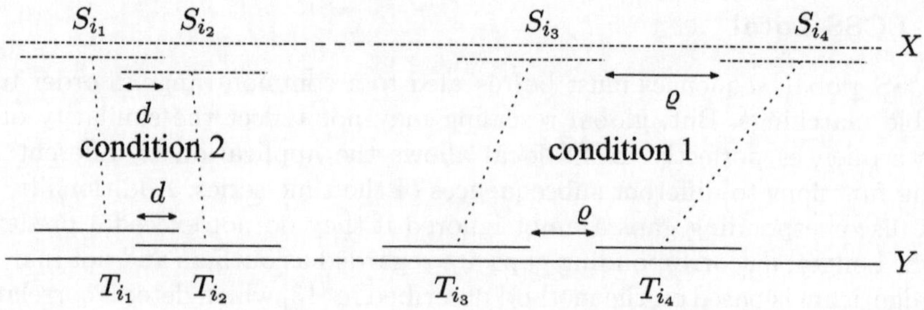

Fig. 2. Construction of longer subsequences

3. The parameters a_i, b_i, c_j and d_j used for scaling and translation are approximately equal: $|a_i - c_j| < \kappa_1$ and $|b_i - d_j| < \kappa_2$, with user-defined κ_1, κ_2.

Subsequently, longer subsequences can again be fused with other subsequences already fused as long as conditions 1 and 3 or 2 and 3 hold.

Longest Common Subsequences: Given a set with k pairs of matchings $\mathcal{S} = \{(S_1, T_1), \ldots, (S_k, T_k)\}$ that has been determined as described above. Now that subset $\mathcal{S}' = \{(S_{l_1}, T_{l_1}), \ldots, (S_{l_h}, T_{l_h})\}$ of $\mathcal{S}$ must be found for which

- The end points S_{l_i} and T_{l_i} precede the start points S_{l_j} and T_{l_j} on X and Y, respectively ($1 \leq i < j \leq h$). That is, the corresponding subsequences do not overlap.
- The total length of all subsequences in this subset $\sum_{i=1}^{h} length(S_{l_i}) + \sum_{i=1}^{h} length(T_{l_i})$ is maximal.

With $\circ$ denoting the sequential composition of two subsequences, the subsequences $X' = S_{l_1} \circ \ldots \circ S_{l_h}$ and $Y' = T_{l_1} \circ \ldots \circ T_{l_h}$ are called the *longest common subsequences* of X and Y. X' and Y' may have different lengths.

LCSS-Local Similarity: Let X' and Y' be the longest common subsequences of X and Y, then the similarity $Sim_{w,\epsilon,\varrho}(X, Y)$ of X and Y computed with LCSS-local is given by

$$Sim_{w,\epsilon,\varrho}(X, Y) = \frac{length(X') + length(Y')}{length(X) + length(Y)}. \tag{2}$$

3.4 Extension to Multivariate Time Series

Multivariate time series $X = (\mathbf{x}_1, \ldots, \mathbf{x}_n)$ and $Y = (\mathbf{y}_1, \ldots, \mathbf{y}_m)$ with $\mathbf{x}_i, \mathbf{y}_i \in \mathbb{R}^D$ with $D \in \mathbb{N}$ ($D \geq 2$) can be processed with LCSS-global and LCSS-local as well. Therefor, $Sim_\epsilon(X_l, Y_l)$ or $Sim_{w,\epsilon,\varrho}(X_l, Y_l)$ ($l = 1, \ldots, D$) are computed separately for each dimension l. Then, the similarity of the multivariate time series X and Y is computed by the average of the corresponding similarity measures.

3.5 LCSS-Global and LCSS-Local as Kernel Functions

As a kernel function $K(\cdot,\cdot)$ can be seen as a similarity measure of two samples, $Sim_\epsilon(X,Y)$ and $Sim_{w,\epsilon,\varrho}(X,Y)$ can be used as kernel functions:

$$K_{global}(X,Y) = \frac{2 \cdot Sim_\epsilon(X,Y) \cdot \min\{length(X), length(Y)\}}{length(X) + length(Y)}, \qquad (3)$$

$$K_{local}(X,Y) = Sim_{w,\epsilon,\varrho}(X,Y). \qquad (4)$$

Equations (3) and (4) do not define kernel functions that satisfy Mercer's conditions. I.e., $K(\cdot,\cdot)$ is not guaranteed to be positive semidefinite.

4 Experiments

The Biometric Smart Pen BiSP (see [15] for details) is a novel ballpoint pen for the acquisition of biometrical features based on handwriting movements which does not need a specific writing pad. For the verification of individuals by means of handwritten signatures the pen is equipped with sensors which measure the dynamics of pressure on the refill in three dimensions and the finger kinematics by means of tilt angels of the pen.

4.1 Database and Experimental Setup

For the following experiments, signatures of 71 persons have been recorded with the BiSP. The number of signatures available for a specific person varies from 6 to 40. In order to provide significant results, reference models are created only for those persons who provided at least 10 signatures (i.e., 63 persons).

The methods LCSS-global and LCSS-local presented in Section 3 are compared to other kernels, namely:

1. Fisher-kernel based on HMM [9] (Pre-processing: rescaling, 7 hidden states),
2. DTW-kernel [8] (Pre-processing: amplitude normalization and resampling to person-specific fixed length, standard deviation σ of the Gaussian function: person-specific), and
3. Gauss-kernel [14] (Pre-processing: amplitude normalization and resampling to person-specific fixed length, standard deviation σ of the Gaussian function: person-specific).

For LCSS-global all signatures are rescaled to $[0,1]$ and ϵ is set to 0.1. For LCSS-local no pre-processing is applied. The parameters are set to $w = 20$, $\epsilon = 0.2$, $\varrho = 15$ and local rescaling to $[0,1]$ is performed. These settings have been found empirically.

4.2 Results

The primary objectives of our experiments are the evaluation of SVM with LCSS-global and LCSS-local applied to online signature verification on the basis of two

Table 1. Experiment 1: Comparison with 3 originals for testing

Kernel	Error in %		
	FRR	FAR	TER
LCSS-global	3.81	0.24	0.57
LCSS-local	3.92	0.01	0.37
DTW-kernel	4.66	0.66	1.02
Fisher-kernel	14.18	3.79	4.73
Gauss-kernel	35.34	0.10	3.30

Table 2. Experiment 2: Comparison with 13 originals for testing

Kernel	Error in %		
	FRR	FAR	TER
LCSS-global	3.72	0.06	1.16
LCSS-local	3.08	0.00	0.93
DTW-kernel	5.38	0.00	1.63
Fisher-kernel	19.23	4.00	8.60
Gauss-kernel	37.44	0.00	11.32

signature sets with two different numbers of originals used for testing (experiment 1 and 2) and the performance of the proposed methods with varying numbers of original signatures used for the training (experiment 3).

In experiment 1, a training set containing 7 original signatures and 35 random forgeries selected randomly from the set of the 70 remaining persons is created for every person. Additionally, 3 originals and 30 random forgeries are selected for testing purposes and not used for training. Each of the five kernels is applied to each of the 63 data sets. Due to the fact that signatures used for training and testing are selected randomly, this experiment as well as the two following experiments are repeated 5 times in order to get a statistically more reliable result. Table 1 shows the average false rejection rates (FRR), the average false acceptance rates (FAR), and the average total error rates (TER). LCSS-local, LCSS-global, and the DTW-kernel yield the best results, with LCSS-local providing the lowest total error rate (TER = 0.37%). Fisher-kernel and Gauss-kernel seem not to be appropriate.

After this first experiment, the methods are evaluated with a second set of signatures (experiment 2): Therefor, 12 persons are selected who provided at least 20 signatures. 7 originals and 35 random forgeries are chosen for training as in experiment 1, but here, 13 originals and 30 random forgeries are taken for testing (see Table 2 for results). Due to the larger number of originals used for testing, the results are worse than in the first experiment. Again, LCSS-local provides the best results (TER = 0.93%) followed by LCSS-global and DTW-kernel.

In experiment 3, the number of originals used for training is varied between 3 and 17 and 35 random forgeries are used. For testing, 3 originals and 30 random forgeries are selected. Since LCSS-based methods provided the best results in experiments 1 and 2, only LCSS-global and LCSS-local are considered (see Fig. 3 for results). As the number of originals increases with LCSS-global and LCSS-local, FRR is decreasing rapidly. But in turn, as the variance of a person's signatures used for training increases due to the larger number of originals, FAR increases for both methods, but even faster for LCSS-global. It can be concluded that a number of seven originals effects a suitable compromise between FRR and FAR.

In a nutshell: LCSS-global, LCSS-local, and DTW-kernel provide the best results, with LCSS-local ranking first in each experiment. Fisher-kernels yield very high error rates just as Gauss-kernels. For a larger set of persons and only

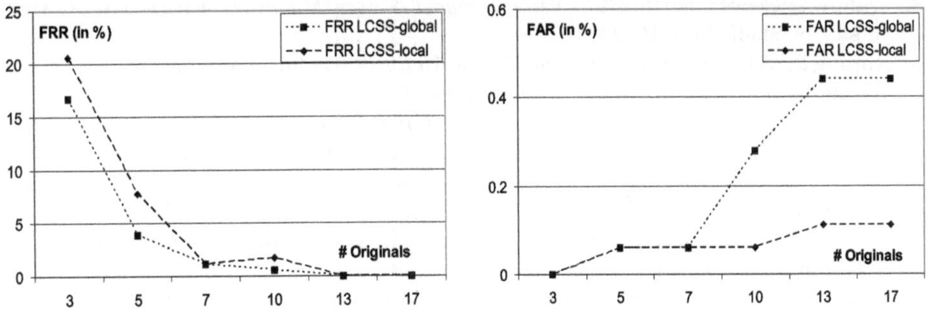

Fig. 3. Experiment 3: LCSS-global and LCSS-local with increasing number of originals used for training purposes

a few originals taken for the training of a reference model, LCSS-based methods are superior, whereas with an increasing amount of originals available for training purposes, LCSS-global and LCSS-local perform almost equally. But as the number of originals needed for very good authentication results is lowest with LCSS-local, it should be preferred.

5 Conclusion and Outlook

In this article, two new kernel functions (LCSS-global and LCSS-local) for SVM were proposed and applied to online signature verification. The experiments showed that these methods are capable to authenticate persons very reliable with LCSS-local providing the best results with only seven originals used for training purposes. It was also shown that in this particular application LCSS-local, LCSS-global, and DTW-kernel are superior to Gauss-kernel and Fisher-kernel.

Currently, we develop new time series segmentation and classification methods for online signature verification and intend to apply the new kernel functions to other time series classification tasks. To obtain a less biased evaluation of our methods, we also intend to use skilled forgeries. We also intend to improve our algorithms to reflect possible correlations between the individual dimensions of a multivariate time series and to compare the proposed kernels to a personalized version of the Fisher-kernel or related kernels (i.e., TOP-kernel [16]).

References

1. Thelusma F. and Muherjee S.: *Signature Verification*, Internal Report, Artificial Intelligence Laboratory, Massachusetts Institute of Technology, Cambridge, Massachusetts, URL: citeseer.ist.psu.edu/366946.html (06/03/2005)
2. Fuentes M., Garcia-Salicetti S., and Dorizzi B. *On-Line Signature Verification: Fusion of a Hidden Markov Model and a Neural Network via a Support Vector Machine* Proceedings of the Eighth International Workshop on Frontiers in Handwriting Recognition (IWFHR'02), Ontario, Canada, 2002, pp. 253–258

3. Wan V.: *Speaker Verification using Support Vector Machines*, Ph.D. Thesis, University of Sheffield, GB, 2003
4. Rüping S.: *SVM Kernels for Time Series Analysis*, Tagungsband der GI-Workshop-Woche LLWA 01, Dortmund, 2001, pp. 43–50
5. Chakrabartty S. and Deng Y.: *Dynamic Time Alignment in Support Vector Machines for Recognition Systems*, Internal Report, The John Hopkins University, Baltimore, URL: bach.ece.jhu.edu/gert/courses/774/2001/dtw.pdf (06/03/2005)
6. Shimodaira H., Noma K., Nakai M., and Sagayama S.: *Dynamic Time-Alignment Kernel in Support Vector Machine*, Advances in Neural Information Processing 14 (NIPS 2001), 2001, pp. 921–928
7. Shimodaira H., Noma K., Nakai M., and Sagayama S.: *Support Vector Machine with Dynamic Time-Alignment Kernel for Speech Recognition*, Proceedings of the European Conference on Speech Communication and Technology (Eurospeech), Aalborg, 2001, vol. 3, pp. 1841–1844
8. Bahlmann C., Haasdonk B., and Burkhardt H.: *On-line Handwriting Recognition with Support Vector Machines – A Kernel Approach*, 8th International Workshop on Frontiers in Handwriting Recognition (IWFHR '02), Ontario, 2002, pp. 49–54
9. Jaakkola T., Diekhans M., and Haussler D.: *Using the Fisher Kernel Method to Detect Remote Protein Homologies*, 7th International Conference on Intelligent Systems for Molecular Biology, Menlo Park, CA, 1999, pp. 149–158
10. Moreno P.J., Ho P.P., and Vasconcelos N.: *A Kullback-Leibler Divergence Based Kernel for SVM Classification in Multimedia Applications*, HP Laboratories Cambridge, Tech. Report HPL-2004-4, 2004
11. Das G., Gunopulos D. and Mannila H.: *Finding Similar Time Series*, 1st European Symposium on Principles of Data Mining and Knowledge Discovery, Trondheim, 1997, pp. 88–100
12. Bollobás B., Das G., Gunopulos D., and Mannila H.: *Time-Series Similarity Problems and Well-Separated Geometric Sets*, Proceedings of the 13th Annual ACM Symposium on Computational Geometry, Nice, 1997, pp. 454–456
13. Agrawal R., Lin K.-I., Sawhney H. S., and Shim K.: *Fast Similarity Search in the Presence of Noise, Scaling, and Translation in Time-Series Databases*, 21st International Conference on Very Large Databases, Zürich, 1995, pp. 490–501
14. Burges C. J. C.: *A Tutorial on Support Vector Machines for Pattern Recognition*, Data Mining and Knowledge Discovery, vol. 2 (2), 1998, pp. 121–167
15. Hook C., Kempf J., and Scharfenberg, G.: *A Novel Digitizing Pen for the Analysis of Pen Pressure and Inclination in Handwriting Biometrics*, LNCS, Springer, Berlin, Heidelberg, New York, vol. 3087, 2004, pp. 283–294
16. Tsuda K., Kawanabe M., Rötsch G., Sonnenburg S. and Müller K.-R.: *A new discriminative kernel from probabilistic models*, Neural Computation, vol. 14 (10), MIT Press, Cambridge, MA, 2002, pp. 2397–2414

Generation of Replaceable Cryptographic Keys from Dynamic Handwritten Signatures

W.K. Yip[1,2], A. Goh[2], David Chek Ling Ngo[1,2], and Andrew Beng Jin Teoh[1,2]

[1] Faculty of Information Science and Technology (FIST), Multimedia University, Jalan Ayer Keroh Lama, Bukit Beruang 75450, Melaka, Malaysia
{yip.wai.kuan04, david.ngo, bjteoh}@mmu.edu.my
[2] Corentix Technologies Sdn Bhd, B-S-06, Kelana Jaya, Petaling Jaya, 47301 Selangor, Malaysia
alwyn@corentix.com

Abstract. In this paper, we present a method for generating cryptographic keys that can be replaced if the keys are compromised and without requiring a template signature to be stored. The replaceability of keys is accomplished using iterative inner product of Goh-Ngo [1] Biohash method, which has the effect of re-projecting the biometric into another subspace defined by user token. We also utilized a modified Chang et al [2] Multi-state Discretization (MSD) method to translate the inner products into binary bit-strings. Our experiments indicate encouraging result especially for skilled and random forgery whereby the equal error rates are <6.7% and ~0% respectively, indicating that the keys generated are sufficiently distinguishable from impostor keys.

1 Introduction

In authentication systems, it is well known that password and public-key systems do not physically associate the user hence, identity frauds can be easily carried out. Therefore, there is a need to incorporate biometric factor (what you are) for authentication to provide better security. In this paper, we are interested in using dynamic hand-signatures as the biometric features because they are socially and generally well-accepted and are more cost effective in terms of capturing equipment (eg. PDAs, smartphones and mouse-pen). In particular, we are interested in deriving bit-strings from dynamic hand-signature data to be used as cryptographic keys in authentication protocols. The following issues are addressed in this paper: (1) biometrics is not exactly reproducible, (2) non-revocability of biometrics in that they are permanently associated with the users, and (3) non-secrecy nature of the biometric. Our solution to (1) is to use a modified MSD with Gray encoding to allow keys to be encoded as closely as possible within a permissible threshold bounded by the statistical deviation. Issue (2) is resolved using iterative inner product that causes the biometric feature to be projected into another random subspace dictated by the stored user random token which is an independent factor from the biometric. Lastly, the fact that our key statistics are linked to the mixed biometric with token randomness, and the inherent one-way transformation of the iterative inner product, guarantee the non-revelation the actual biometric even if the final keys are stolen.

D. Zhang and A.K. Jain (Eds.): ICB 2006, LNCS 3832, pp. 509 – 515, 2005.

2 Previous Works

The first biometrics hash on dynamic hand-signature was proposed by Vielhauer et al [3] which used a 24-feature-parameter set from dynamic hand-signature and an interval matrix, which stores the upper and lower threshold permissible for correct identification. Although the authors reported that the system has FAR 0% and FRR of 7.05% achieved for only 11 test subjects, it is not clear if the performance will be the same for a larger sample set. Similarly, Feng-Chan [4] also uses 43 features (but not all are published) and reported 8% EER but the uniqueness of the output vector is only 1 in 2^{40}, which is insufficiently long for cryptographic usages. Another scheme for face data, Chang et al [2] also uses user-specific boundaries information. The keys are generated from the biometric (permanent association) and hence, if compromised, the user needs to create a new biometric which is not feasible. This shortcoming is also observed in Davida et al [5] method of using error-correction codes directly on iris features.

Cancelable keys can be achieved by incorporating random tokens as in Soutar et al [6], Monrose et al [7-8], Juel-Wattenberg [9], Juels-Sudan [10] and Clancy et al [11]. Schemes [6-8] which utilized lookup tables, and [10-11] which require storing quite substantial number of additional chaff points, are not storage-efficient while Juel-Wattenberg is subjected to multiple key attack [12]. On the other hand, Goh-Ngo [1] is storage efficient, as only a randomized token is required, and is a secure one-way transform as the inner product cannot be reversed to recover the actual biometric.

For feature extraction, although most hand-signature verification methods [13-17] have reported the successful use of dynamic time warping (DTW) whereby the test signal is non-linearly aligned to a template signal, it is not suitable for our application due to the open storage of template. Another approach more suitable for our application is to process the signal using Fourier transformation as in Martinez et al [18] and Chan-Kamins [19]. We choose this approach as feature extraction as no template is needed and there exist a Fast Fourier Transform (FFT) that executes in n.log(n) time compared to common approaches DTW and linear discrimination methods that require at least n^2 computation.

3 Proposed Method

We adopt similar strategy as with Chan-Kamins [19] for feature extraction using FFT but using different combination of the dynamic signals derived from the input positional signals from the user devices. Our method then combines the Goh-Ngo iterative inner product step of mixing random token with the biometric data, with the discretizing scheme of Chang et al [2] to product multiple bits for each feature element, as outlined in Fig 1.

We assume a stylus-enabled PDA for capturing the signature in (x,y) coordinates and a timestamp (t) for each point. The signals are then pre-processed by cubic spline interpolation to derive the velocity (x1,y1) and acceleration (x2,y2), and re-sampled to obtain uniform signals length of 512 which is required for the optimum computation efficient for FFT. Finally, the signal is aligned to the origin by subtraction from the centroid.

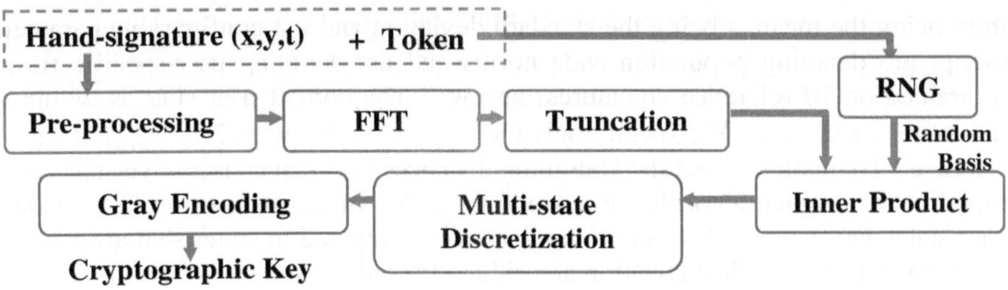

Fig. 1. Outline of the proposed method

Each signal x can be represented by a Fourier integral of form (implemented using the FFT algorithm) as $x(n) = \frac{1}{2\pi}\int_{-\pi}^{\pi} X(e^{j\omega})e^{j\omega n}d\omega$ with $j = \sqrt{-1}$. In general, the Fourier transformation is a complex-valued function of ω and can be expressed in rectangular form as $X(e^{j\omega}) = X_R(e^{j\omega}) + jX_I(e^{j\omega})$ and in the polar form as $X(e^{j\omega}) = |X(e^{j\omega})| e^{\measuredangle X(e^{j\omega})}$ where $|X(e^{j\omega})| = \sqrt{X_R^{\ 2} + X_I^{\ 2}}$ is the **mag**nitude and $\measuredangle X(e^{j\omega}) = \tan^{-1}\frac{X_I}{X_R}$ is the **angle** of the Fourier transform. Individual truncation by extracting the 20 most significant amplitudes of the Fourier transforms, followed by concatenation of the various truncated transforms (discussed in Section 5) is computed to obtain a compact biometric vector $\mathrm{B} \in \mathbb{R}^n$.

The biometric-token mixing stage (Table1) involves iterative inner products of biometric feature B and random basis vectors defined by user token (T). The random basis are orthonormalized vectors (using Gram-Schmidt algorithm) generated from a random number generator (RNG) eg. X917, that follows the Gaussian distribution of zero mean and unit standard deviation based on T as the seed.

At enrollment, the boundaries are specified for each feature element D_i:

- left boundary (LB_i) = $\min(m_{glo,i} - k_{glo} \cdot s_{glo,i}, m_{usr,i} - k_{usr} \cdot s_{usr,i})$ and
- right boundary (RB_i) = $\max(m_{glo,i} + k_{glo} \cdot s_{glo,i}, m_{usr,i} + k_{usr} \cdot s_{usr,i})$

Table 1. Iterative inner product and discretization steps

P := **inner product**(T, B):	D := **discretize** (P):
1. for i = 1..n-1	1. for i = 1..n-1
2. $r_i = \{RNG(T)_j\}_{j=1+(i-1).n}^{i.n}$	2. $D_i = \mathrm{gray}\left\lfloor \frac{P_i - LB_i}{w_i} \right\rfloor$
3. end for	3. end for
4. orthonormalize $\forall r_i$	
5. for i = 1..n-1	
6. $P_i = \langle r_i, B\rangle$	
7. end for	

with m being the mean, s being the standard deviation and k a configurable parameter, subscript glo denoting population-wide norms and usr denoting user-specific (in our case, trained on 10 reference signatures) norms. The correct user state is within the region of $m_{usr} \pm k_{usr}.s_{usr}$. We specify only the LB_i and width, w_i ($=2.k_{usr}.s_{usr,i}$) of each D_i for storage. To guarantee that the Hamming distances between consecutive states are 1, we modify the original algorithm to encode the index of the state using Gray code ie. distant states will have higher Hamming distances compared to states that are nearer to the authentic state. The discretization algorithm proceeds as shown in Table1.

4 Experiments and Discussion

The proposed method was tested on the Signature Verification Competition 2004 (SVC2004)[20] Task 1 database consists of 40 users with 20 genuine and 20 skilled forgery samples. The two error rates that we are interested to measure in our experiments are for: (1) skilled forgery where a non-genuine user replicates the genuine signature by imitation and (2) random forgery where a non-genuine user uses his signature. The different dynamic features from the positional, velocity and acceleration are then combined to extract longer bit-strings from (1) 80-feature vector V1= [mag(x),mag(y),mag(x1),mag(y1)], (2) 120-feature V2= [mag(x),mag(y),mag(x1), mag(y1),mag(x2),mag(y2)], and (3) 160-feature V3=[mag(x),mag(y),mag(x1), mag(y1),mag(x2),mag(y2),ang(x),ang(y)]. Fig 2-4 show the effect of varying k_{glo} and k_{usr} on the different combination of dynamic features used between the genuine and skilled forgery without any mixing of tokens. The optimal configuration is observed when k_{usr}=1.5 and k_{glo}=20. Using V1 provides the best result but shortest key length.

Using the optimal configuration found earlier, we obtain the results in Table 3 for the cases of using **no token**, forged signature with **stolen** token and with adversary own (**substitution**) token. The consistency of random forgery case EER2~0% confirms the clear separation between the genuine and random forgery distribution while

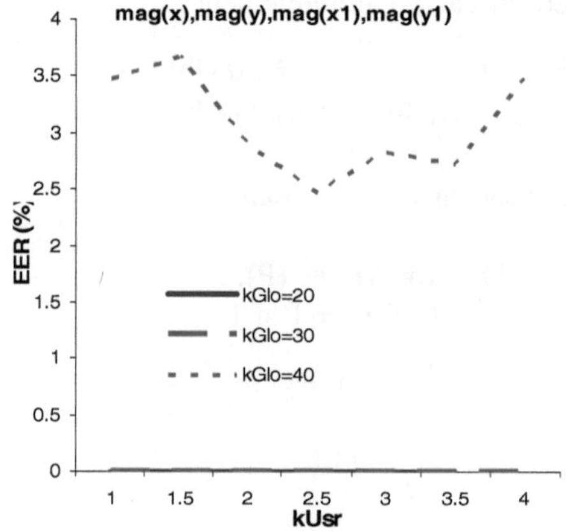

Fig. 2. Various k_{Glo} and k_{Usr} settings on V1

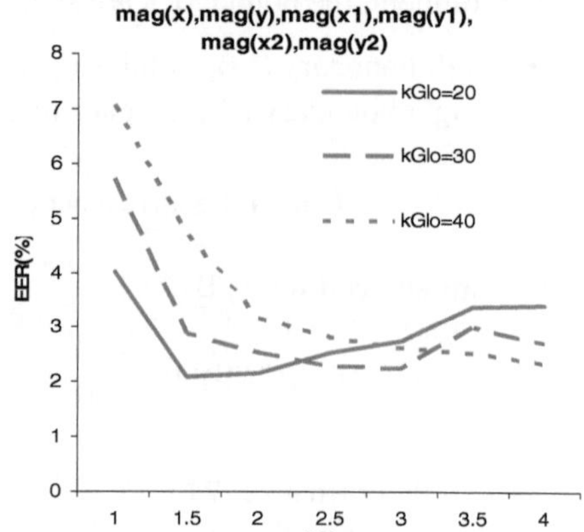

Fig. 3. Various k_{Glo} and k_{Usr} settings on V2

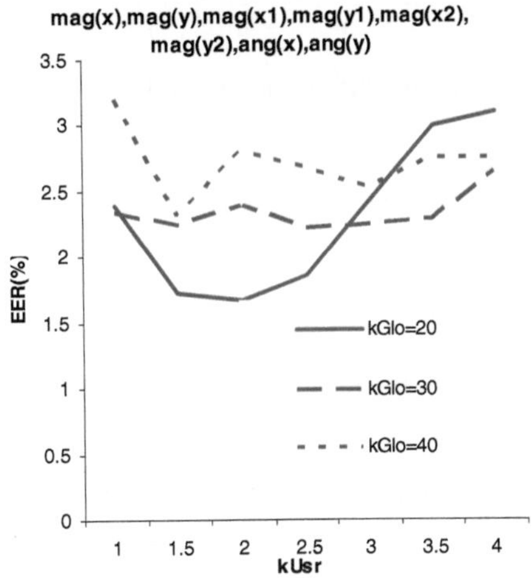

Fig. 4. Various k_{Glo} and k_{Usr} settings on

Table 3. EERs (in %) for no token, stolen token and substitution token case

Feature Vector		V1	V2	V3
Bits, N		320	480	640
No Token	EER1	0.00	2.44	1.61
	EER2	0.00	0.00	0.00
Stolen Token	EER1	4.25	6.64	4.66
	EER2	0.00	0.00	0.00
Substi-tution-Token	EER1	0.02	0.05	0.06
	EER2	0.00	0.00	0.00

results of skilled forgery EER1 <6.7% are also encouraging. The good separation for substitution token forgery is because each (including skilled forgery) user performs his own token-governed transformation, and hence the keys are all projected to different random subspaces, resulting in different keys. For stolen token scenario, the skilled forgery cases have higher EERs because similar signature vectors are projected onto the same subspace using the stolen token. We will discuss the security implication of these results in the Section 6.

5 Security Analysis and Discussion

5.1 Exhaustive Search

In this case, we assume that the attacker has no knowledge of the random token, key statistics or signature. Let N be the effective size of the bit-strings generated in our experiment. Using brute force attack, 2^N number of attempts is required.

5.2 Known Key Statistics Attack

In this scenario, the attacker has access to the statistical parameters used in MSD. The number of guesses he can make for each feature element is given by $\sum_{i=1}^{n-1} \frac{RB_i - LB_i}{w_i}$. For our experiment, each element can be represented by an average of 4 bits, hence the number of attempts made will be at least $2^{4(n-1)}$. It must be noted though the key statistics are not reflective of the actual biometric features but on the inner product values, hence are replaceable and will not pose a permanent security risk.

5.3 Stolen Token Attack

This is the case of an adversary using **stolen genuine token** and **forged signature** of the genuine user. The EER1 of <6.7% shows that it is of comparable performance to existing protocols. The best result is achieved using V1 signature features which provided EER1=4.25. In fact, our proposed scheme has a longer key length of effective bits as compared Feng-Chan (43 elements) and Vielheur et al (24 elements).

5.4 Substitution Token Attack

This is the case of an adversary using his **own token** and **forged signature** of the authentic user. EER1=0% in Table 2 confirmed that the scheme is extremely unyielding against such attacks.

In summary, the key security advantages our solution provides are (1) longer key space, (2) good separation between the genuine and skilled forgery curve and (3) perfect separation for the random forgery case. Another important improvement of our scheme as compared to DTW-based approaches is the non-requirement of template storage which could deter an adversary to reproduce the signature without even the actual signing action.

6 Concluding Remarks

Our experimental results have established that the proposed method of combining random token and biometric data is able to generate sufficiently long and distinguishing bit-strings. In particular, we have found that the method is comparable with existing schemes even for the more difficult case of a skilled forger using an authentic token. The use of MSD using user key statistics provides the error tolerance to accommodate intra-signal differences. By incorporating randomness via the iterative inner product, the keys generated are replaceable thus providing better management. The one-wayness of the inner product mixing, and the key statistics which are based on the token-based projected biometrics (instead on the plain biometrics) ensure that the biometric features are not compromised even if multiple keys stolen.

References

[1] Goh, A. & Ngo, D.C.L.: Computation of Cryptographic Keys from Face Biometrics, Seventh IFIP TC-6 TC-11 Conference on Communications and Multimedia Security, Springer-Verlag LNCS 2828 (2003)

[2] Chang, Y.C., Zhang, W. & Chen, T.: Biometric-based Cryptographic Key Generation, IEEE Conference on Multimedia and Expo, Taiwan (2004)

[3] Vielhauer, C., Steinmetz, R. & Mayerhorf, A.: Biometric Hash based on Statistical Features of Online Signatures, Proc. of the 16th Intl. Conference on Pattern Recognition (2002)

[4] Feng, H. & Chan, C.W.: Private Key Generation from On-line Handwritten Signatures, Information Management and Computer Security, MCB UP Limited (2000) 159-164.

[5] Davida, G., Frankel, Y., Matt, B.J. & Peralta, R.: On the Relation of Error Correction and Cryptography to an Off Line Biometric Based on Identification Scheme, WCC99, Workshop on Coding and Cryptography (1999)
[6] Soutar, C., Roberge, D., Stoianov, A., Gilroy, R. & Kumar, B.V.K.V.: Biometric Encryption Using Image Processing. SPIE 3314 (1998) 178-188
[7] Monrose, F., Reiter, M.K., Li, Q. & Wetzel, S.: Cryptographic Key Generation from Voice, Proc. of the 2001 IEEE Symp. on Security and Privacy (2001)
[8] Monrose, F., Reiter, M.K., Li, Q., Lopresti, D.P. & Shih, C.: Toward Speech-Generated Cryptographic Keys on Resource Constrained Devices, Proc. of the 11th USENIX Security Symposium (2002)
[9] Juels, A. & Wattenberg, M.: A Fuzzy Commitment Scheme, in Proc. 6th ACM Conf. Computer and Communications Security, G. Tsudik, Ed. (1999) 28-36
[10] Juels, A. & Sudan, M.: A Fuzzy Vault Scheme, in Proc. IEEE Int. Symp. Information Theory, A. Lapidoth & E. Teletar, Eds. (2002) 408
[11] Clancy, T.C., Kiyavash, N. & Lin, D.J.: Secure Smartcard-based Fingerprint Authentication, in Proc. ACM SIGMM 2993 Multimedia, Biometrics Methods and Applications Workshop (2003) 45-52
[12] Boyen, X.: Reusable Cryptographic Fuzzy Extractors, 11th ACM Conference on Computer and Communications Security (CCS 2004), ACM Press (2004) pp. 82-91
[13] Sakoe, H. & Chiba, S.: Dynamic Programming Algorithm Optimization for Spoken Word Recognition, IEEE Trans. on Acoustics, Speech & Signal Processing, Vol.ASSP-26, No.1 (1978)
[14] Hastie, T. & Kishon, E.: A model for Signature Verification, AT&T Bell Laboratories Technical Report (1992)
[15] Kholmatov, A.A.: Biometry Identity Verification Using On-line and Off-line Signature Verification, Master of Science Thesis, Sabanci University (2003)
[16] Plamondon, R. & Srihari, S.: On-line and Off-line Handwriting Recognition: A Comprehensive Survey, IEEE Trans. on Pattern Analysis and Machine Intelligence, Vol. 22, No. 1 (2000)
[17] Feng, H. & Chan, C.W.: Online Signature Verification using a New Extreme Points Warping Technique, Pattern Recognition Letters 24 (2003) 2943-2951
[18] Martinez, J.C.R., Lopez, J.J.V. & Rosas, F.J.L.: A Low-cost System for Signature Recognition, Int. Congress on Research in Electrical and Electronics Engineering (2002)
[19] Chan, F.L. & Kamins D.: Signature Recognition through Spectral Analysis, Pattern Recognition, Vol. 22, Issue 1 (1989) 39-44
[20] SVC 2004: First International Signature Verification Competition, http://www.cs.ust.hk/svc2004/

Online Signature Verification Based on Global Feature of Writing Forces*

ZhongCheng Wu[1], Ping Fang [1,2], and Fei Shen[1,2]

[1] Institute of Intelligent Machine, Chinese Academy of Science,
Hefei, Anhui Province, China 230031
[2] Department of Automation, University of Science & Technology of China
Hefei, Anhui Province, China 230026
{zcwu, shenfei}@iim.ac.cn, pingfang@ustc.edu

Abstract. Writing forces are important dynamics of online signatures and it is harder to be imitated by forgers than signature shapes. An improved DTW (Dynamic Time Warping) algorithm is put forward to verify online signatures based on writing forces. Compared to the general DTW algorithm, this one deals with the varying consistency of signature point, signing duration and the different weights of writing forces in different direction. The iterative experiment is introduced to decide weights for writing forces in different direction and the classification threshold. A signature database is constructed with F_Tablet and the experiments results are present in the end.

1 Introduction

Classical user authentication systems have been based in something that you have (like a key, an identification card, etc.) and/or something that you know (like a password, or a PIN). With biometrics, a new user authentication paradigm is added: something that you are (e.g., fingerprints or face) or something that you do or produce (e.g., handwritten signature or voice). The convenience for paper and pen in the electronic era is the reason why people still use handwriting as a mean to convey, retain, and facilitate communication. Together with this kind of information, handwriting is also a skill that individualizes people Moreover, devices like PDAs, pocket PCs, tablet PCs, or 3G mobile phones might offer handwriting capabilities, due to the fact that handwriting is considered as being more natural for humans and also to the possibility of size reduction by eliminating the keyboard. From this point of view, signature is a social and legal acceptable biometrics personal authentication method.

When a person signs his name, he writes not only the characters, but also his identity, which is implied in the dynamic writing process and the static signature. Computer based online and offline signature verification approaches have been developed to extract the identity. Compared to the static handwriting image of offline approach, online one uses those dynamics during writing and has relatively higher classification

* The work was funded by the Natural Science Foundation of China with grant No. 60375027, No. 60475005.

D. Zhang and A.K. Jain (Eds.): ICB 2006, LNCS 3832, pp. 516 – 522, 2005.

rate. In online signature verification systems, different approaches can be considered to extract signature information; they can be divided into: i) function-based approaches, in which signal processing methodology is applied to the dynamically acquired time sequences (i.e., velocity, acceleration, force, or pressure), and ii) feature-based approaches, in which statistical parameters are derived from the acquired information. One can also specify different levels of classification, so it is possible to use and combine shape-based global static (i.e., aspect ratio, center of mass, or horizontal span ratio), global dynamic (i.e., total signature time, time down ratio, or average speed) or local (stroke direction, curvature or slope tangent) parameters.

The signature is the trajectory of the writing pen's contact movement on the writing surface driven by writing forces. So writing forces are one of the most important information of writing dynamics and many researches have been done on them. Crane and Ostrem developed a three-dimension force-sensitive pen to get the writing forces [1]. With input device of the SmartPen, Martens and Claesen devised an online signature verification system based on three-dimension forces [2]. Tanabe studied signature verification based on the pressure with digital pen device [3]. Sakamoto did research on signature verification incorporating pen position, pen pressure and pen inclination with WACOM Tablet [4]. Shimizu developed an electrical pen using two-dimensional optical angle sensor to get writing forces [5].

Although all kinds of writing pen devices are used to get the writing forces, they can't get the forces accurately, because a writing pen may be rotated during writing which would change the measure coordination. And WACOM Tablets can only get the writing pressure. A new writing tablet, named F_Tablet is used here to capture the three-dimension writing forces. And an improved DTW algorithm is used to verify those signatures. Compared to the general DTW algorithm, this one deals with the varying consistency of signature point, signing duration and weights of writing forces in different direction. The F_Tablet signature capturing device and the signature database are introduced in the next chapter. Chapter 3 discusses the improved DTW algorithm and iterative experiment. The experimental results are given in chapter 4 and conclusions are made in the last chapter.

2 Signature Acquisition

2.1 The F_Tablet

This F-Tablet is capable of capturing three perpendicular forces of the pen-tip to the contacting plane and two dimension torques directly because of the core part of a multi-dimension force/torque sensor. With the specially designed structure, the static trajectory of the pen-tip and other dynamic signals such as velocities, accelerations and writing angles can also be calculated indirectly [6]. The input tablet of 70×70 mm^2 is on the up-left side of the F_Tablet, as shown in Fig.1. The device is connected to computer via USB interface with a maximum sample rate of 120Hz. And there is no special requirement on the writing pen. Fig.2 displays the coordinate of the F_Tablet. The coordinate is fixed during design and it won't change no matter how the writing pen is rotated. Here the device is used to get the three-dimension writing forces, F_x, F_y and F_z.

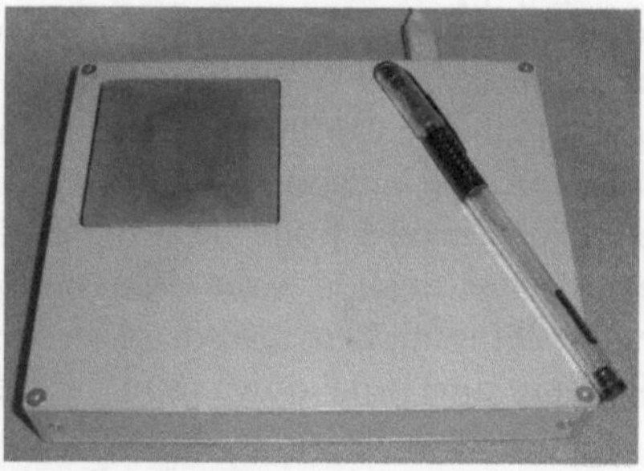

Fig. 1. Photo of the F_Tablet

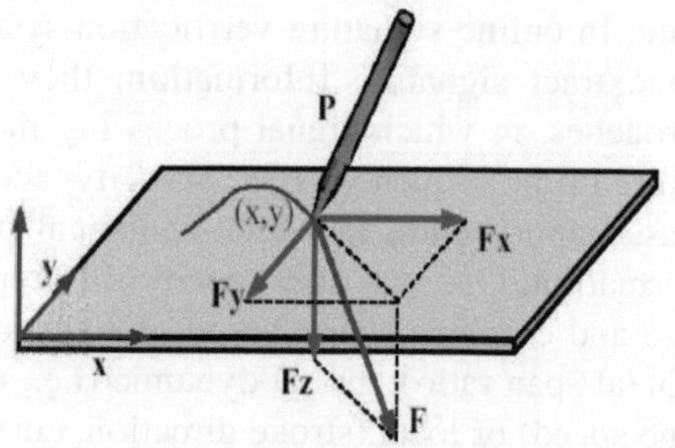

Fig. 2. Coordinate of the F_Tablet

2.2 Signature Database Construction

The database is constructed with 30 persons involved over a one-month period. Each subject donated 40 signatures with 10 ones every week. At the same time, each subject is told to practice and imitate other subject's signature as simple forgeries after the static signature. And 10 persons are recruited to make skilled forgeries. Before the skilled forgeries were collected, each subject can view the signing process of the signature to be imitated with a special program and practice for a period of time what ever long he wants. And a signature database with 1200 genuine signatures, 600 simple forgeries and 300 skilled forgeries is constructed.

2.3 Signature Preprocessing

To improve the classification result, signatures are preprocessed before calculation. The preprocessing methods taken here are filtering, direction adjustment and dehooking.

1) *Filtering.* To remove the noise in the signature data, the Gaussian filter is applied to filter the three dimension forces respectively.
2) *Direction Adjustment.* The posing of the writer or the position of the F Tablet may change when the signatures are collected in several batches, which results in the inconsistency of the signature direction. So force direction adjustment is introduced to adjust the force direction in X and Y direction.
3) *Dehooking.* As the writing surface of the F Tablet is a little smoother than the general paper, so a jerk may occur when a pen collide with the tablet, which will cause the wrong judgement of the pen-down or pen-up status. So dehooking is taken to remove the jerk.

3 Improved DTW Algorithm

Compared to the general DTW algorithm, this improved one takes the varying consistency of different signature point, the signing duration and weights of writing forces in different direction into account. First, weighted multiple signature templates are generated with general DTW algorithm. Then weights of different signature points and stroke signing duration are used to calculate distance. The weights of writing forces in different and classification threshold are decided with iterative experiment.

3.1 Distance Calculation

The difference between the general DTW algorithm and this one lies on the distance calculation. This one takes weight of writing forces in different direction, weight of different signature point and signing duration into account. Euclidean distance is used to measure the difference of the aligned point pair, and it can be expressed as (1):

$$d(T_i, S_j) = \left[w_x (F^{(i)}_x - F^{(j)}_x)^2 + w_y (F^{(i)}_y - F^{(j)}_y)^2 + w_z (F^{(i)}_z - F^{(j)}_z)^2 \right]^{1/2} \quad (1)$$

where w_x, w_y and w_z are weights of F_x, F_y and F_z respectively. Distance between corresponding strokes can be expressed as (3):

$$D_k(T_i, S_j) = \min \begin{cases} D_k(T_{i-2}, S_{j-1}) + \frac{1}{2}\left[w_{i-1} d(T_{i-1}, S_j) + w_i d(T_i, S_j) \right] \\ D_k(T_{i-1}, S_{j-1}) + w_i d(T_i, S_j) \\ D_k(T_{i-1}, S_{j-2}) + \frac{1}{2}\left[w_i d(T_i, S_{j-1}) + w_i d(T_i, S_j) \right] \end{cases} \quad (2)$$

$$D_k^{'}(T, S) = D_k(T, S) \frac{t_s}{t_t} \quad (3)$$

where $D_k(T, S)$ is the k th stroke distance between the sample and weighted template and $D_k^{'}(T, S)$ is the distance which take signing duration into account; t_s is the stroke point number of the sample before being resampled and t_t is that of the template. Only the pen-down points are used here, but the pen-up points can also be regarded as that of the pen-down stroke followed. Then, the distance between the sample and the template can be expressed as (4):

$$D(T, S) = \sum_k D_k^{'}(T, S) \quad (4)$$

3.2 Threshold Setting

Because the consistencies of signing of different subjects are sure to be different, so the threshold is set to reflect both personal and global signing characteristics. The threshold D_i^{th} of subject i is expressed as:

$$D_i^{th} = \mu_i + f \sigma_i \quad (5)$$

where μ_i and σ_i are the average value and standard deviation of the distances between registered signatures and templates; while f is the global coefficient to make sure that all subjects have an optimistic classification performance.

3.3 Iterative Experiment

Iterative experiment is introduced to decide the global threshold coefficient f and the ratio between w_x, w_y and w_z. As F_x and F_y are forces parallel to the writing tablet, they have same weight in distance calculation. So in the experiment, the ratio between w_x, w_y and w_z is $w:w:(3-2*w)$. The goal of the experiment is to get the values of w and f where the signature classification result is the best. First initial values are set for them, then parameters w and f are alternatively changed to do experiments on the signature database to find the other one where FAR equals FRR, until the parameter difference between two experiments is small enough.

4 Experimental Results

4.1 The Trajectory and Forces of Signature

The first purpose of our experiment presented here is to verify the F-Tablet™ the forces consistency in writing the same character. The second is to identify the relationship between trajectory and forces. Test software is developed to display the trajectory and draw the three forces at the same time. The time-interval width of the stroke can be obtained from the pressure between the pen-tip and the input tablet or Fz. And because the coordinates of the trajectory are calculated from Fx, Fy, Fz, Mx and My, so the shape of the displayed character implies all the forces and torques of the writing.

Fig.4 shows the shape of the Chinese character and the three forces during writing. From the shape of the character, we can see that not only the trajectory of the pen, but also the writing style of the writer is recorded. And from the force curve and the static shape we can even get the stroke order. The right part of the image represents the three forces to time series. The value of Fz is negative and defined by the multi-force sensor reference frame. How the pen has gone can be figured out combine Fx with Fy. First, the pen wrote the top point. Then it went from left to right. Then it goes from top right to down left. And then wrote the last stroke.

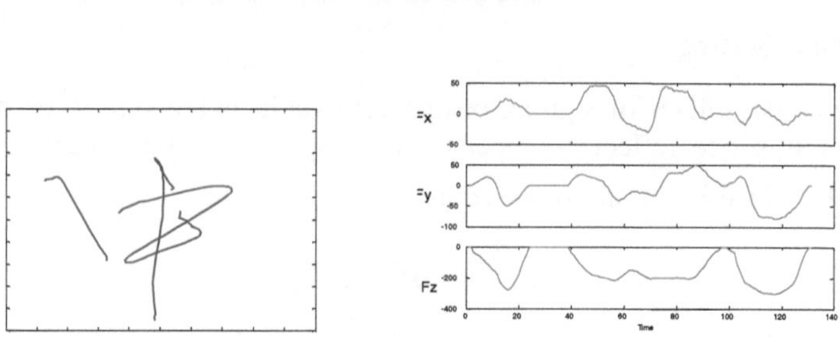

Fig. 3. The trajectory and forces of handwriting (by Wu)

4.2 The Performance of Verification

Experiments are carried out on the constructed signature database. And before experiment, preprocessing such as filtering, dehooking and rotation are applied to the raw signature data. The results of the improved DTW algorithm and the general one without signature point weight setting and signing duration are displayed in Fig.5 and Fig.6 respectively. From the figures, we can see that a great improvement in classification rate has been made. This shows that signature point weight and signing duration are important factors in signature verification. With the iterative experiment, we get the optimum equal error classification rate of zero where the global threshold coefficient f is 2.50 and the ratio between w_x, w_y and w_z is $7:7:1$. But this doesn't imply that F_x and F_y are much more important than F_z, because the amplitude of F_z is much bigger than those of F_x and F_y.

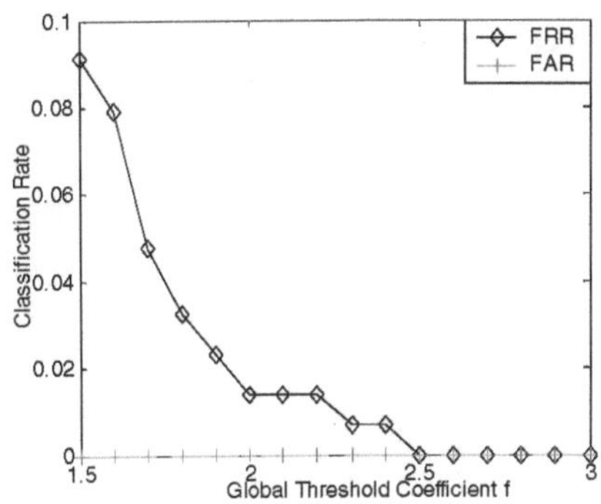

Fig. 4. Performance of the improved DTW

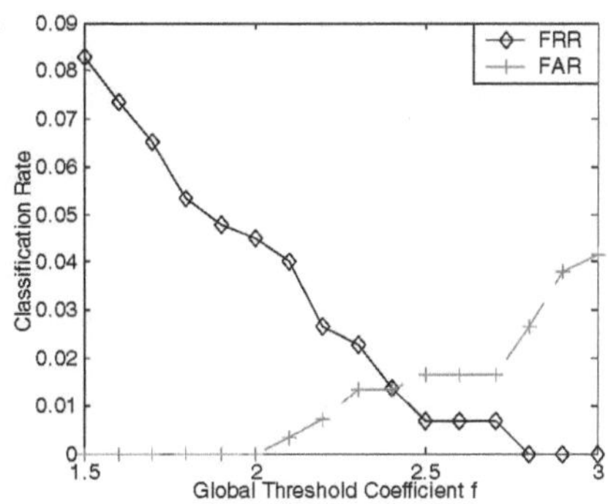

Fig. 5. Performance of the general DTW

5 Conclusion

An improved DTW algorithm is put forward to very online signature based on writing forces. The F_Tablet is used here to capture the three perpendicular writing forces. Compared to the general DTW algorithm, this one deals with the varying consistency of signature point, signing duration and the different weights of writing forces in different direction. And iterative experiment is introduced to decide weights for writing forces in different direction and the classification threshold. With this algorithm, the equal error classification rate of zero is realized base on the constructed signature database.

Although this improved DTW algorithm has better performance, it does have its deficiency.This algorithm consumes more computation time and memory space and these problems are left for the future work.

References

[1] Crane H D, Ostrem J S. Automatic Signature Verification Using a Three-Axis Force-Sensitive Pen. IEEE Transactions on Systems, Man, and Cybernetics, 1983, 3: 329-33

[2] R. Martens and L. Claesen, "Incorporating local consistency information into the online signature verification process". International Journal on Document Analysis and Recognition, vol.1, pp 110-115, 1998.

[3] K. Tanabe, M. Yoshihara, S. Kameya et al, "Automatic signature verification based on the dynamic feature of pressure". Proceedings of Sixth International Conference on Document Analysis and Recognition, vol.1, pp1045 – 1049, 2001.

[4] D. Sakamoto, H. Morita, T. Ohishi, Y. Komiya and T. Matsumoto, "On-line signature verification algorithm incorporating pen position, pen pressure and pen inclination trajectories". IEEE International Conference on Acoustics, Speech, and Signal Processing, vol.2, pp993 – 996, 2001.

[5] H. Shimizu, S. Kiyono, T. Motoki and W. Gao, "An electrical pen for signature verification using a two-dimensional optical angle sensor", Sensors and Actuators, Vol.111, pp216-221, 2004.

[6] Ping F, Zhong C W, Ming M et al. A Novel Tablet for On-Line Handwriting Signal Capture. Proceedings of the 5th World Congress on Intelligent Control and Automation, vol6, pp3714-3717, 2004;

[7] Sheng C L, Xiao Q D and Yan C, "On line signature verification based on weighted dynamic programming matching", Journal of Tsinghua University, vol39.9, pp61-64, 1999.

Improving the Binding of Electronic Signatures to the Signer by Biometric Authentication

Olaf Henniger, Björn Schneider, Bruno Struif, and Ulrich Waldmann

Fraunhofer Institute for Secure Information Technology,
Rheinstr 75, 64295 Darmstadt, Germany
{henniger, struif, waldmann}@sit.fraunhofer.de

Abstract. Due to the fact that the biometric characteristics of a person are bound to that person, biometric methods deployed for signer authentication have the potential of improving the binding of electronic signatures to persons. If there is evidence that a biometric method was used for signer authentication, and if the level of security of this method is sufficiently high, then the receiver of a signed document can trust that the signature creation was indeed initiated by the legitimate holder of the private signature key. To achieve this goal, an approach to provide evidence of the use of biometric signer authentication has been developed. The approach has been implemented in a prototype electronic signature creation system with fingerprint verification.

1 Motivation

Legal regulations permit biometric methods to be deployed for signer authentication also in products for "qualified" electronic signatures (which have the same legal effects as handwritten signatures on paper), provided that the strength of function of the biometric methods and their resistance against penetration attacks are certified to be sufficiently high. Biometric methods are considered more user-friendly than knowledge-based authentication methods because they free the users from the burden of recalling a PIN or password from memory. Moreover, biometric methods can also increase the binding of electronic signatures to persons since the biometric characteristics of a person are bound to that person and cannot easily be presented by others. Knowledge-based signer authentication on the other hand brings along the risk that the PIN or the password are presented by unauthorized persons.

Since biometric characteristics are not always available (e.g., a fingerprint cannot be presented if the finger is injured), biometric user authentication mechanisms must always be accompanied by knowledge-based fallback mechanisms. In order to increase the binding of electronic signatures to persons by deploying biometric methods, the receivers of signed documents need to be informed in a secure way of the signer authentication method (biometric or knowledge-based) that was used at signature creation time. The signer should neither be able to deny it if a biometric method was used, nor to pretend it if a biometric method was not used. If satisfactory evidence is provided that a biometric method was used for signer authentication, and if the strength of function of the biometric method and its resistance against penetration

D. Zhang and A.K. Jain (Eds.): ICB 2006, LNCS 3832, pp. 523 – 530, 2005.

attacks is sufficiently high, the receiver of a signed document can have high confidence that the signature creation was indeed initiated by the legitimate holder of the private signature key.

This paper presents a solution for providing evidence of the deployed signer authentication method. The solution complies with legal regulations on electronic signatures [1–3] and with commonly used formats for electronically signed documents [4]. It does not impede verifying the electronic signatures with the usual programs.

2 System Architecture

In order to prevent the fraudulent use of a smart card with electronic signature creation function (signature card), the user must be authenticated before the signature creation function can be used. User authentication requires the user to present a secret PIN or biometric characteristics. The comparison of the verification data presented by the user with the stored reference data takes place within the smart card (on-card matching).

The authenticity and integrity of the biometric verification data handed over at the card interface must be protected to ensure that these data are captured anew and not fed in by way of bypass or replay attacks (where an impostor, after having stolen or at least temporarily taken possession of a smart card, sends recorded or otherwise acquired biometric data of the legitimate cardholder to the card, evading the regular data capture equipment. The protection of the authenticity and integrity of the biometric verification data is achieved by mutual authentication of both the signature card and the card terminal, the establishment of cryptographic keys, and subsequent application of cryptographic algorithms to the biometric verification data (secure messaging via a trusted channel) [5, 6]. For this purpose, a security module is integrated into the card terminal [7]. To allow flexible handling of this component, the security module is a smart card in plug-in format. Its functionality could also be completely integrated into a tamper-resistant card terminal. However, the advantage of a plug-in card is that it can be easily replaced, e.g. if a public-key certificate is to be renewed.

For establishing the trusted channel between the SMC and the signature card, a hybrid method is used consisting of both an asymmetric and a symmetric cryptographic algorithm. To make the signature creation system useable for multiple signature cards and to solve the cryptographic key distribution problem, the asymmetric cryptographic algorithm (RSA) is used for establishing the trusted channel. The faster symmetric cryptographic algorithm (Triple DES) is then used to allow a fast calculation and verification of secure-messaging objects. Upon successful completion of the card-to-card authentication both cards have available the symmetric session key for cryptographic checksum calculation and the initial value of the send sequence counter, which is used as initial vector for the calculation of cryptographic checksums. Cryptographic checksums are calculated as retail message authentication code (Retail MAC).

Figure 1 shows an example of a system architecture of a signature creation system consisting of a PC with the signature creation application, a card terminal, and a signature card. A fingerprint sensor, the fingerprint feature extraction component as well as the security module card (SMC) are integrated into the tamper-resistant card terminal to prevent bypass and replay attacks at their interfaces. Furthermore, the card terminal contains a smart-card interaction component that controls the security protocol running between the SMC and the signature card.

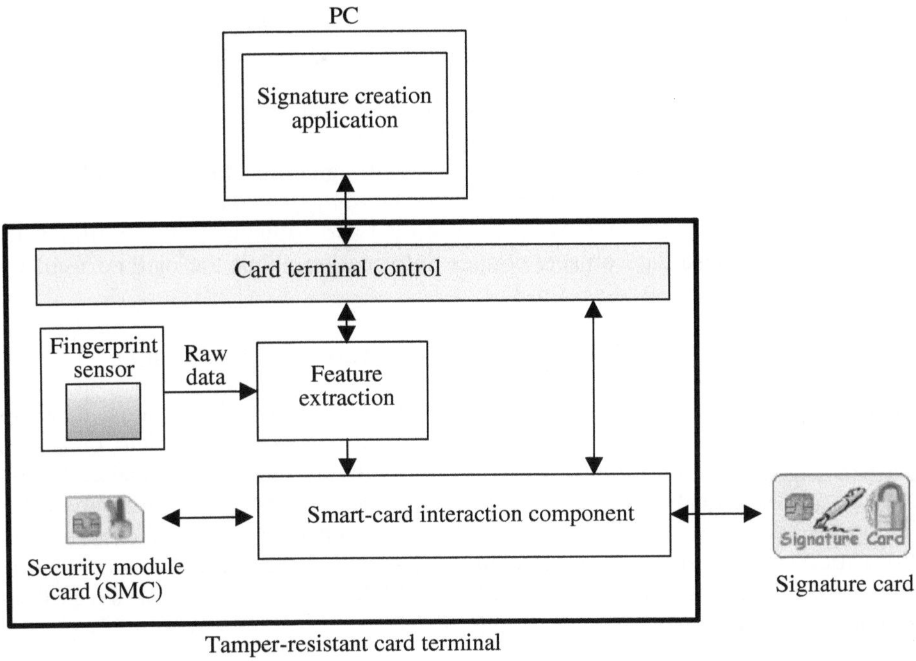

Fig. 1. System architecture of a signature creation system with smart-card interfaces

The signature card is based on a STARCOS SPK 2.4 card of Giesecke & Devrient with a signature creation application certified according to ITSEC E4/high. It has been extended to support minutiae-based fingerprint on-card matching in addition to PIN verification and to provide evidence of the used signer authentication method. The SMC has been implemented on a Java card platform. A prototype of a "Trusted Signature Terminal" [8] serves as a tamper-resistant card terminal.

3 Providing Evidence of the Used Signer Authentication Method

3.1 Different Security Environments for Different Authentication Methods

Each signer authentication method runs in its own security environment on the signature card: The PIN authentication method runs in security environment SE#1, and the fingerprint authentication method in security environment SE#2. The current security environment can be changed by sending a MSE (Manage Security Environment) RESTORE command to the signature card. As soon as the security environment is changed, the local security status, e.g. "Signer authentication successful", is lost.

The applicable signer authentication method depends on the currently selected security environment: It is not possible to carry out biometric authentication in the security environment SE#1 set up for PIN authentication, and it is not possible to carry out PIN authentication in the security environment SE#2 set up for fingerprint authentication.

3.2 Notification of the Used Signer Authentication Method

Only in security environment SE#2, i.e. only after biometric signer authentication, the signature card responds to the command PSO (Perform Security Operation) COMPUTE DS (Digital Signature) with a special signature block that includes, in addition to the signature of the document to be signed, the data objects Control Reference Template (CRT) for Authentication [5] and Biometric Information Template (BIT) [9]. These data objects contain information about the method used for signer authentication.

3.3 Signing the Notification of the Used Signer Authentication Method

The smart-card interaction component in the tamper-resistant card terminal forwards the signature card's PSO COMPUTE DS response within the data field of a PSO VERIFY CC command to the SMC, as it does with any other secure-messaging response from the signature card. Since the SMC receives the information about the method used for signer authentication via the trusted channel, the SMC can trust this information and take on the task of signing the special signature block to confirm it. To achieve this, the functionality of the PSO VERIFY CC command on the SMC is extended as follows:

1. Verify the cryptographic checksum given in the data field.
2. If a signature block with supplementary information about the used signer authentication method is present in the data field, then
   - create a supplementary signature over the signature block by applying the SMC's private key for card authentication PrK.SMC.AUT,
   - store the signature block together with the supplementary signature in a log file.

By signing the supplementary information together with the document signature, the supplementary information is bound to the corresponding signed document. Using the SMC's private key for card authentication PrK.SMC.AUT fulfils the security requirements, as the holder of the signature card cannot control the use of this key. The solution can be considered fraud-resistant because the creation of the supplementary signature is solely under the control of the SMC and cannot be induced from outside. The application range of PrK.SMC.AUT is usually restricted to the INTERNAL AUTHENTICATE command. Its extension to the creation of the supplementary signature must be confirmed by a certification authority responsible for certifying the security of the SMC.

The supplementary signature created by the SMC is appended to the signature block received from the signature card and stored together with it in a log file on the SMC. Afterwards, the smart-card interaction component reads the log file from the

SMC. The log file is selected by a SELECT command and read using READ BINARY commands under the security status "Signature card authentication successful". Then, the signed signature block can be forwarded together with the signed document and the X.509 certificates of both the cardholder's public key for electronic signatures PuK.CH.DS and the SMC's public key for card authentication PuK.SMC.AUT, which are needed for verifying the document signature and the supplementary signature, respectively.

3.4 Format of the Signed Document

The signed document is formatted as a PKCS#7 message [4] of type "signedData" (see Figure 2). This message consists of the signed document ("contentInfo"), the X.509 certificate including the public key of the cardholder for electronic signatures PuK.CH.DS ("certificates"), and the "signerInfo". The "signerInfo" contains

- "authenticatedAttributes": the hash value of the document and the signature creation time,
- "encryptedDigest": the document signature covering the "authenticated-Attributes",
- "unauthenticatedAttributes": optional, informative data that is not signed by the cardholder.

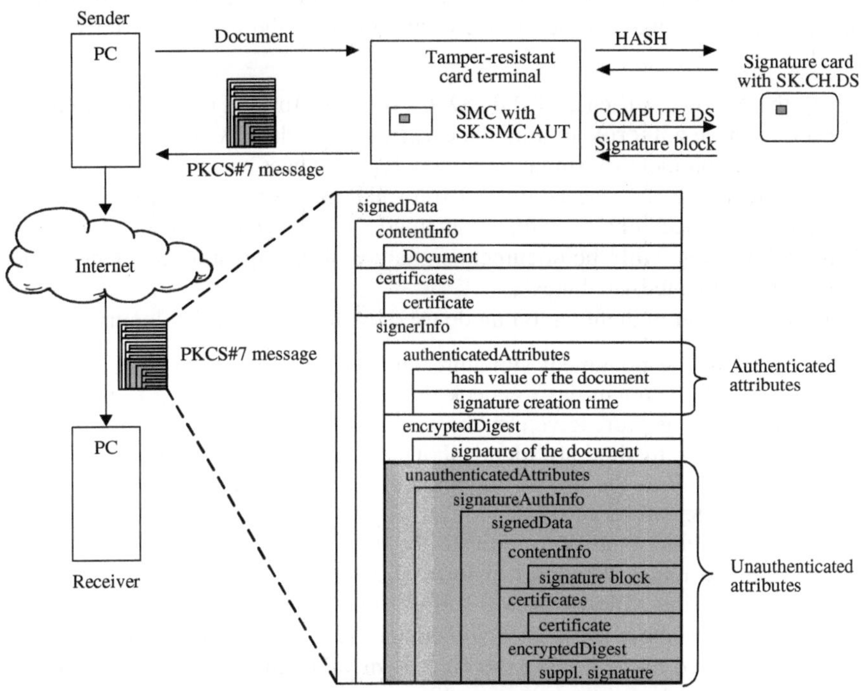

Fig. 2. Informing the document receiver of the signer authentication method

The "unauthenticatedAttributes" include the signature block and the supplementary signature as verifiable evidence that the signer was authenticated by means of a biometric method. The smart-card interaction component formats the notification of the biometric signer authentication mode ("signatureAuthInfo") like a PKCS#7 message of type "signedData", consisting of the signature block including the document signature and the supplementary information about the used signer authentication method ("contentInfo"), the X.509 certificate with the SMC's public key for card authentication PuK.SMC.AUT ("certificates"), and the supplementary signature created using the SMC's private key for card authentication ("encryptedDigest"). This self-contained and signed PKCS#7 message is integrated into the "unauthenticated - Attributes" block of the original document's PKCS#7 message.

4 Examining the Evidence of the Signer Authentication Method

Any program that is capable of interpreting PKCS#7 messages, e.g. the e-mail program Outlook under Windows, and that has got the corresponding certificates can verify an electronic signature created as described in the previous sections. However, most signature verification programs offer only a very restricted view of the individual attribute values, actually nothing with regard to the additional attributes. In particular, the programs cannot display the special unauthenticated attributes to indicate the used signer authentication mode. Thus, for interpreting the "unauthenticated-Attributes" an additional program module that provides the following functionality is required:

1. verify the X.509 certificate of PuK.SMC.AUT (for this purpose, the public key of the certification authority signing the certificate, i.e. the CA certificate, is needed),
2. verify the supplementary signature of "encryptedDigest",
3. in order to prove that the additional information truly belongs to the document, compare the document signature contained in "encryptedDigest" of the overall PKCS#7 message with the document signature of the signature block contained in the "unauthenticatedAttributes",
4. display the signer authentication mode used by the signer of the document.

After verifying the document signature and the supplementary signature, the receiver of a signed document has to check that the signature attached to the document and the document signature given in the supplementary information about the used signer authentication method are identical. This way, the authenticity and integrity of the document as well as the use of the biometric signer authentication method at signature creation time can be verified.

The additional functionality to indicate the used signer authentication method has been implemented under Windows in form of a plug-in extending the Explorer program. For e-mail files (file extension ".eml") that contain a PKCS#7 message and therein information about the signer authentication method, this additional information can be retrieved by activating the context menu (right mouse click on the document) and choosing the new menu item "Signer's authentication info...". One of two possible notification windows is opened showing information about the signer authentication method used at document signature creation time (see Figure 3). In case a

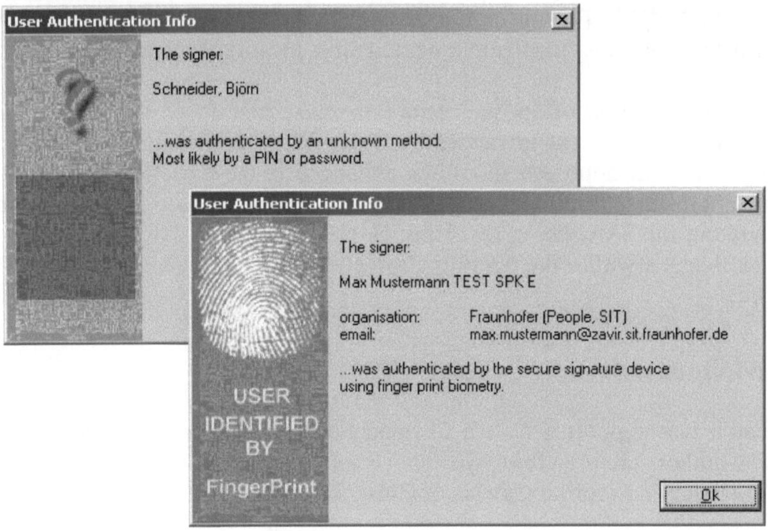

Fig. 3. Pop-up windows "User Authentication Info"

PKCS#7 message is received without supplementary information about the used signer authentication method, knowledge-based signer authentication is assumed by default. In addition to the signer authentication mode some standard information about the signer's certificate is displayed.

5 Summary and Outlook

Since biometric characteristics are bound to a certain person, the binding of electronic signatures to persons can be improved by the deployment of biometric methods for signer authentication. If it is satisfactorily shown that a biometric method was used for signer authentication, and this method's strength of function and resistance against penetration attacks is sufficiently high, then the receiver of a signed document can have high confidence that the electronic signature creation was indeed initiated by the legitimate holder of the signature key.

In this paper, a solution has been presented, which fulfills these requirements in compliance with standardized signature formats. Furthermore, the solution does not interfere with the regular signature verification. Based on the STARCOS SPK 2.4 signature card of Giesecke & Devrient, a prototype of a signature card with fingerprint on-card matching as alternative to PIN verification has been developed. The response to signature creation commands indicates whether or not the biometric signer authentication method was used. In order to protect the authenticity of this information, the signature block is signed again. This supplementary signature is created by the SMC, which is integrated into the tamper-resistant terminal. In order that the notification of the signer authentication mode can be attributed to a certain document, the supplementary signature covers also the corresponding document signature. The SMC has been implemented based on a Java card. The functionality of the card terminal has been implemented in a prototype of a "Trusted Signature Terminal" [8].

The outlined solution offers the chance to make electronic signatures applicable also for high-value business processes by improving the binding of electronic signatures to persons. The approach should be enhanced in the direction that the signature card by itself creates a fraud-resistant information about the used signer authentication method without the SMC being involved. However, such an extension would require substantial changes within the operating system of already existing and certified signature cards.

Acknowledgements

This research was supported by the German Federal Ministry of Education and Research. The authors are grateful to the other members of the project team, in particular to Gisela Meister and Florian Gawlas of Giesecke & Devrient, for fruitful discussions.

References

[1] Directive 1999/93/EC of the European Parliament and of the Council of 13 December 1999 on a Community Framework for Electronic Signatures
[2] German Signature Act, Fed. Law Gaz. 2001, Part I no. 22, May 2001
[3] German Signature Ordinance, Fed. Law Gaz. 2001, Part I no. 59, Nov. 2001
[4] Public-Key Cryptography Standards (PKCS) #7, Cryptographic Message Syntax Standard v1.5, RSA Laboratories, Bedford, Maine, USA, Nov. 1993
[5] Information technology – Identification cards – Integrated circuit cards – Part 4: Organization, security and commands for interchange. Internat. Standard ISO/IEC 7816-4, 2005
[6] Application Interface for Smart Cards used as Secure Signature Creation Devices, Part 1 – Basic requirements, Version 1 Release 10, CWA 14890-1, March 2004
[7] D. Scheuermann, U. Waldmann: Protected Transmission of Biometric User Authentication Data for On-card Matching. In *Proc. of the ACM Symposium on Applied Computing*, Nikosia, Cyprus, March 2004
[8] O. Henniger, B. Struif, K. Franke, R. Ulrich: Trusted Signature Terminal – A trustworthy signature creation environment. In P. Horster (ed.): *Proc. of the D-A-CH Security Workshop*, Erfurt, Germany, 2003. – In German
[9] Information Technology – Identification Cards – Integrated Circuit Cards – Part 11: Personal Verification through Biometric Methods. Internat. Standard ISO/IEC 7816-11, 2004

A Comparative Study of Feature and Score Normalization for Speaker Verification

Rong Zheng, Shuwu Zhang, and Bo Xu

Institute of Automation, Chinese Academy of Sciences, Beijing, China
{rzheng, swzhang, xubo}@hitic.ia.ac.cn

Abstract. In speaker verification, it is necessary to reduce the influence of different environmental conditions. In this paper, two stages of normalization techniques, feature normalization and score normalization, are examined for decreasing the mismatch between training and testing acoustic conditions. At the first stage, cepstral mean and variance normalization (CMVN) is modified to normalize the cepstral coefficients with the similar segmental parameter statistics. Next, due to score variability between verification trials, Test-dependent zero-score normalization (TZnorm) and Zero-dependent test-score normalization (ZTnorm) are comparatively presented to transform the output scores entirely and make the speaker-independent decision threshold more robust under adverse conditions. Experiments on NIST2002 SRE corpus show that the normalizations with CMVN in feature stage and ZTnorm in score stage achieved 20.3% relative reduction of EER and 18.1% relative reduction of the minimal DCF compared to the baseline system using CMN and zero normalization.

1 Introduction

For speaker verification tasks, it is necessary to extract information from the speech signal that is speaker specific and robust to noise and various channels. Under adverse conditions, speech features based on cepstral analysis are corrupted. If there is a big mismatch between training and testing speech, the performance of speaker verification system would be deteriorated.

In recent years, some robust feature normalization methods for reducing noise and/or channel effects were proposed, such as CMN, Cepstral Mean and Variance Normalization (CMVN) [1,2], RASTA [3], and feature warping [4]. CMN is proposed to compensate for linear channel variations. When additive noise exists, a natural extension of CMN is CMVN, which normalizes the distribution of cepstral features over some specific window length by subtracting the mean and scaling the standard deviation. RASTA is a kind of modulation spectrum analysis that aims to reduce the effects of convolutional noise in the communication channel. Feature warping algorithm maps the distribution of a cepstral feature stream to a standard normal distribution over a specified time interval.

In output score domain, score normalization has been introduced to deal with score variability and to make speaker-independent threshold more robust and effective. In previous work, the use of score normalization has significantly improved the performance of speaker recognition systems, such as cohort normalization [5], Znorm (Zero normalization) [6], Hnorm (Handset normalization) [7], Tnorm (Test normali-

D. Zhang and A.K. Jain (Eds.): ICB 2006, LNCS 3832, pp. 531–538, 2005.

zation) [8]. Score normalization is the transformation of the distribution of speaker verification scores to enhance the robustness of decision threshold. Hnorm is derived from Znorm. Both Znorm and Hnorm compute scores from a set of impostor speech segments to normalize a speaker model. In Tnorm, each test segment is scored against a set of impostor models at test time. Tnorm parameters are estimated to normalize the score of the test segment against claimed model. Znorm and Tnorm are the mostly used forms of score normalization.

To enhance the robustness of GMM-based text-independent speaker verification systems under adverse conditions, we compare and evaluate a number of feature and score normalization approaches. By appending different types of delta cepstral coefficients, CMVN is improved to be more robust against the effects of linear channel and slowly varying additive noise. After using more robust features, Test-dependent zero-score normalization (TZnorm) and Zero-dependent test-score normalization (ZTnorm) are comparatively presented for score normalization to cope with score variability caused by different speaker models and various test segments.

The remainder of this paper is organized as follows. Section 2 describes the feature normalization. The score normalization is introduced in section 3. The experimental results based on NIST2002 SRE corpus are discussed in section 4. Finally, we will draw a conclusion in section 5.

2 Feature Normalization

In telephony speaker recognition applications, channel variability is the most significant factor decreasing the recognition performance. In order to fit a speaker model to a given speaker, not to background noise, silence and noisy frames are usually removed.

Alternatively, robust speech recognition techniques have been introduced to reduce the effect of linear channel and slowly variable additive noise. Feature warping and CMVN algorithms were proposed to normalize distribution parameters of cepstral coefficients over specified time interval using sliding windows. In feature warping method [4], mapped features of short-time distribution of acoustic features are quickly found by looking up in a standard normal cumulative distribution function table. It has been proven to outperform CMN. A natural extension of CMN, CMVN, improves normalization of the mean and variance parameters. CMVN yields a better compensation of the mismatch caused by additive noise [4,9].

The use of time derivative cepstral parameters improved recognition performance in previous studies. In original CMVN algorithm, it was reported to only append corresponding delta coefficients after static coefficients have been normalized [1]. If we use NMFCC to represent MFCC coefficients after CMVN performed, then static and delta cepstral coefficients through original CMVN can be represented as NMFCC+$\triangle$NMFCC. All of the feature normalizations mentioned above assumes that the mean and/or the variance, even all the moments of the probability distribution of the cepstral coefficients [10], are irrelevant information for speaker recognition. Since the separation of relevant and irrelevant information contained in the speech is not clear, it is hard to theoretically judge the proper normalization strategies for delta coefficients. We further evaluate two types of additional derivation of delta coefficients compared with the original CMVN method. In figure 1 (a), delta coefficients derived from NMFCC are further processed by CMVN, which is denoted as NMFCC+N△NMFCC. In figure 1 (b), CMVN is applied to static and delta coefficients, respectively. The cepstral coefficients are expressed as NMFCC+N△MFCC.

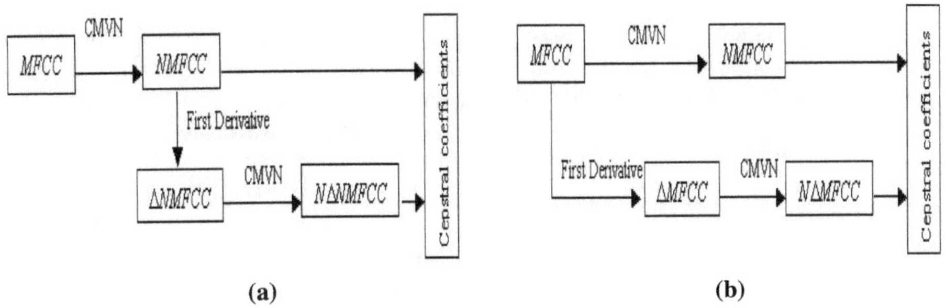

Fig. 1. (a) Delta coefficients derived from NMFCC are further processed by CMVN. (b) CMVN is applied to static and delta coefficients, respectively.

3 Score Normalization

Decision-making process in speaker verification based on GMM-UBM (Gasussian Mixture Models-Universal Background Model) is to compare the likelihood ratio obtained from the claimed speaker model and the UBM model with a decision threshold [7]. Due to the score variability between verification trials, the tuning of decision thresholds is an important and troublesome problem. Score variability mainly consists of two different sources. The first one is the different quality of speaker modeling caused by enrollment data varying. The second one is the possible mismatches and environment changes among test utterances.

Znorm [6] calculates the scores of the target speaker model k against a set of impostor speech utterances. Then the mean μ_k^Z and standard deviation σ_k^Z of these scores are estimated to normalize the target speaker score $S(x,k)$ computed from each test segment x against this target model.

$$S_{znorm}(x,k) = \frac{S(x,k) - \mu_k^Z}{\sigma_k^Z}. \tag{1}$$

Tnorm [8] parameters are estimated from scores of each test segment x against a set of impostor speaker models at test time. Then the mean μ_x^T and standard deviation σ_x^T of the impostor scores are used to adjust the target speaker score $S_{tnorm}(x,k)$.

$$S_{tnorm}(x,k) = \frac{S(x,k) - \mu_x^{\ T}}{\sigma_x^{\ T}}. \tag{2}$$

Znorm is mostly utilized to scale various output scores caused by different speaker models, and Tnorm is to transform output scores caused by various test utterances. These two normalization methods have proven to be effective. Some comparative experiments are shown in [9].

In order to enhance the robustness of decision threshold and normalize the uncertainty of score variability between trials entirely, two kinds of combination mode, Test-dependent zero-score normalization (TZnorm) (Figure 2) and Zero-dependent test-score normalization (ZTnorm) (Figure 3), are presented as follows.

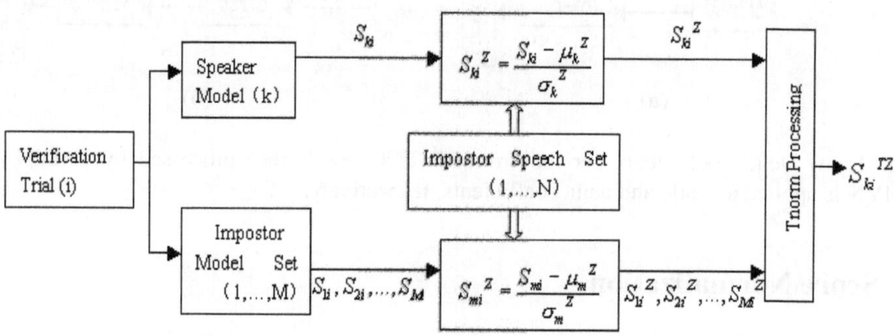

Fig. 2. Schematic diagram of Test-dependent zero-score normalization

Figure 2 shows a block diagram of Test-dependent zero-score normalization (TZnorm). In the application of TZnorm, we first compute the log-likelihood ratio scores for verification trial (i) against a target speaker model (k) and against impostor model set (1,2,…,M). We get the score of target speaker S_{ki} and a set of impostor scores $S_{1i}, S_{2i}, ..., S_{Mi}$, respectively. Then, S_{ki} and $S_{1i}, S_{2i}, ..., S_{Mi}$ perform zero normalization processing separately using impostor speech set (1,2,…N). We assume $S_{1i}^{\ Z}, S_{2i}^{\ Z}, ..., S_{Mi}^{\ Z}$ have a Gaussian distribution and estimate the mean $\mu_i^{\ Z}$ and standard deviation $\sigma_i^{\ Z}$. Finally, TZnorm normalized output score is obtained as follows.

$$S_{ki}^{\ TZ} = \frac{S_{ki}^{\ Z} - \mu_i^{\ Z}}{\sigma_i^{\ Z}}. \tag{3}$$

Figure 3 shows a schematic diagram of Zero-dependent test-score normalization (ZTnorm). When considering ZTnorm, we first compute the log-likelihood ratio scores for speaker model (k) with verification trial (i) and impostor speech set (1,2,…,N). S_{ki} and $S_{k1}, S_{k2}, ..., S_{kN}$ are obtained respectively. Then, S_{ki} and

$S_{k1}, S_{k2}, ..., S_{kN}$ perform test normalization processing using impostor model set (1,2,…,M) separately. Supposing $S_{k1}^{T}, S_{k2}^{T}, ..., S_{kN}^{T}$ has a Gaussian distribution, the mean μ_k^{T} and standard deviation σ_k^{T} are estimated. The output score of ZTnorm is derived from the following equation 4.

$$S_{ki}^{ZT} = \frac{S_{ki}^{T} - \mu_k^{T}}{\sigma_k^{T}}. \tag{4}$$

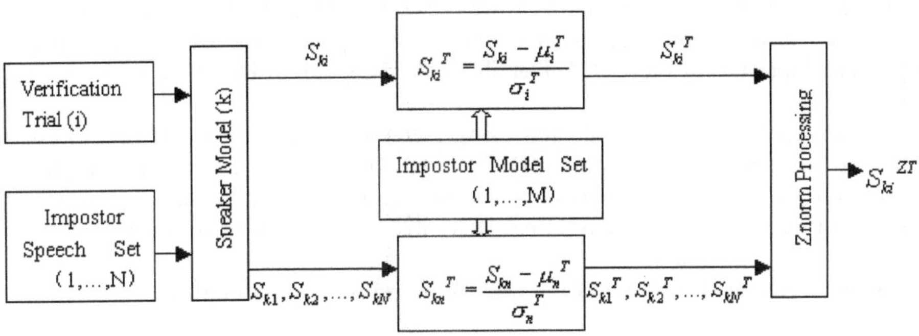

Fig. 3. Schematic diagram of Zero-dependent test-score normalization

4 Experimental Results

The described speaker verification system has been developed and evaluated using data from NIST2001 and NIST2002 speaker recognition evaluations (SRE) [11]. In this section, we will describe the corpus, the baseline system and the performance of speaker verification based on feature normalization and score normalization, respectively.

4.1 Corpus

The NIST2002 corpus includes cellular data extracted from the Switchboard Cellular part 2. This corpus consists of 139 male and 191 female speakers with 2 minutes of training speech from a single cellular phone call. There are 3570 test segments (1442 males and 2128 females) which range from few seconds to a minute (with a primary focus on segments with 15s~45s). All of the verification utterances have to be scored against gender-matching impostors and true speaker. A detailed description of the evaluation corpus can be found in [11].

NIST2001 SRE is used to train universal background model and to collect the needed speech for detection score normalization. This corpus includes 60 development speakers (38 males and 22 females), 174 test speakers (74 males and 100 females) and a total of 2038 verification speech segments (850 males and 1188 females). To avoid bimodal distribution, the impostor data for score normalization is of the same gender as the target speaker [6].

4.2 Baseline System

Speech utterances are divided into 24ms frames, shifted by 12ms, ignoring about 10%~15% low energy frames which do not contain much speaker information. 16 Mel-Frequency Cepstral Coefficients (MFCCs) and 16 delta coefficients are calculated. CMN is applied to mitigate linear channel effects. Znorm is performed to transform output scores. Two gender-dependent Universal Background Models are trained on the 60-speaker development set of the NIST2001 SRE. Each gaussian mixture in the UBM has a diagonal covariance matrix. All speaker models are obtained by Bayesian adaptation of the UBM. Details of the adapted GMM-UBM system can be found in [7].

Performance measure is evaluated using Detection Error Trade-off (DET) curves [12] and Detection Cost Function (DCF) [11]. The DCF is defined to weight two types of errors, that is, miss detections and false alarms. For all results, we report the Equal error rate (EER) and the minimal DCF obtained in a posteriori way.

4.3 Comparisons of Feature Normalization and Score Normalization

For CMVN processing, we have evaluated three types of combinations of static and delta coefficients mentioned in section 2. For short,

(1) CMVN1: NMFCC+△NMFCC; (2) CMVN2: NMFCC+N△NMFCC; (3) CMVN3: NMFCC+N△MFCC.

In our experiments, feature warping is also applied to the cepstral coefficients to obtain corresponding normalized feature parameters. We only report the best experimental results. The length of sliding windows is the same as in [4].

Figure 4 (a) shows the DET curves of different types of feature normalization for speaker verification: CMN, Feature warping, and three kinds of cepstral coefficients derived from CMVN. Table 1 gives the corresponding EER and the minimal DCF. It has shown that CMVN3+Znorm brings 12.9% relative reduction of EER and 4.6% relative reduction of the minimal DCF compared to the baseline system. So this type of robust cepstral parameters is used in the following comparative experiments on score normalization. It is also notable that three kinds of feature normalization strategies for delta coefficients caused different recognition results. CMVN3 strategy produced modest improvement.

Speaker verification performance using various score normalizations based on CMVN3 features is shown in Figure 4 (b). Corresponding EER and the minimal DCF are reported in Table 2. The results showed that recognition performance is significantly improved by employing the combined modes of score normalization. CMVN3+ZTnorm yields the best performance, which achieves 20.3% relative reduction of EER and 18.1% relative reduction of the minimal DCF compared to the baseline system using CMN and zero normalization.

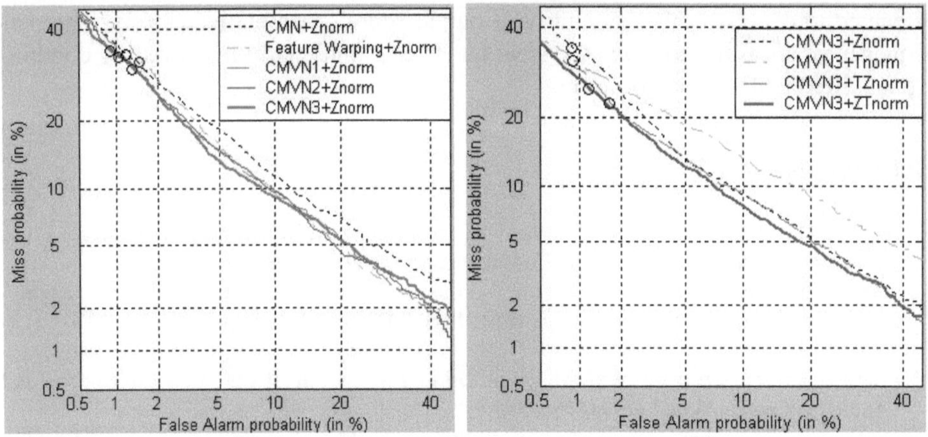

(a) **(b)**

Fig. 4. (a) Speaker verification performance for different types of feature normalization. The minimal DCF operating point is indicated with a circle. (b) Speaker verification performance for different types of score normalization.

Table 1. EER and minimal DCF for different types of feature normalization

Feature normalization	EER (%)	DCF
CMN+Znorm (Baseline)	10.8	0.0457
FeatureWarping+Znorm	9.7	0.0472
CMVN1+Znorm	9.9	0.0439
CMVN2 +Znorm	9.7	0.0437
CMVN3 +Znorm	9.4	0.0436

Table 2. EER and minimal DCF for different types of score normalization

Feature normalization	EER (%)	DCF
CMN+Znorm (Baseline)	10.8	0.0457
CMVN3+Znorm	9.4	0.0436
CMVN3+Tnorm	12.0	0.0410
CMVN3 +TZnorm	9.4	0.0391
CMVN3 +ZTnorm	8.6	0.0374

5 Conclusions

In this paper, we have compared two different transformation methods, feature normalization and score normalization respectively, for speaker verification tasks over cellular data. Experimental results showed that fine-tuning derivation of delta coefficients when normalizing local distributions of cepstral parameters improves the recognition performance. TZnorm and ZTnorm detection score normalization have been presented to cope with the variability of output score distributions entirely. The performance of speaker verification system has been significantly improved by combining modified CMVN with ZTnorm.

References

1. Viikki, O., Laurila, K.: Cepstral domain segmental feauture vector normalization for noise robust speech recognition. Speech Communication, Vol.25 (1998) 133-147
2. Segura, J.C., Benítez, C. et al.: Cepstral domain segmental nonlinear feature transformations for robust speech recognition. IEEE Signal Processing Letters, Vol.11 (2004) 517-520
3. Hermansky, H., Morgan, N.: RASTA processing of speech. IEEE Trans. on Speech and Audio Processing, Vol.2 (1994) 578-589
4. Pelecanos, J., Sridharan, S.: Feature warping for robust speaker verification. Proc. Speaker Odyssey Conf. (2001) 213-218
5. Rosenberg, A.E., Delong, J. et al.: The use of cohort normalized scores for speaker verification Proc. ICSLP Vol.2 (1992) 599-602
6. Reynolds, D.A.: Comparison of background normalization methods for text-independent speaker verification Proc. EuroSpeech (1997) 963-966
7. Reynolds, D.A., Quatieri, T.F., Dunn, R.B.: Speaker verification using adapted gaussian mixture models Digital Signal Processing, Vol.10 (2000) 19-41
8. Auckenthaler, R., Carey, M., Lloyd-Thomas, H.: Score normalization for text-independent speaker verification systems Digital Signal Processing, Vol.10 (2000) 42-54
9. Barras, C., Gauvain, J.: Feature and score normalization for speaker verification of cellular data Proc. ICASSP Vol.2 (2003) 49-52
10. Molau, S., Pitz, M., Ney, H.: Histogram based normalization in the acoustic feature space Proc. ASRU (2001) 21-24
11. NIST SRE plans. http://www.nist.gov/speech/tests/spk/index.htm
12. Martin, A., Doddington, G. et al.: The DET curve in assessment of detection task performance Proc. EuroSpeech (1997) 1895-1898

Dynamic Bayesian Networks for Audio-Visual Speaker Recognition

Dongdong Li, Yingchun Yang, and Zhaohui Wu

Department of Computer Science and Technology,
Zhejiang University, Hangzhou 310027, P.R. China
{lidd, yyc, wzh}@cs.zju.edu.cn

Abstract. Audio-Visual speaker recognition promises higher performance than any single modal biometric systems. This paper further improves the novel approach based on Dynamic Bayesian Networks (DBNs) to bimodal speaker recognition. In the present paper, we investigate five different topologies of feature-level fusion framework using DBNs. We demonstrate that the performance of multimodal systems can be further improved by modeling the correlation of between the speech features and the face features appropriately. The experiment conducted on a multi-modal database of 54 users indicates promising results, with an absolute improvement of about 7.44% in the best case and 3.13% in the worst case compared with single modal speaker recognition system.

1 Introduction

Dynamic Bayesian Networks (DBNs) [1] are knowledge representation schemes that can characterize probability relationships among temporal data and make exact or approximate inferences. Some prior knowledge (e.g. gender, noise) can be described by DBNs in a convenient way [2]. A sea of previous research revealed the power of DBNs in fusing visual and audio sensors cues with contextual information and expert knowledge both for speaker detection and other similar applications [3, 4]. It is assumed that DBN is an instrumental tool for information fusion.

D. Li et al. [5] introduced the DBNs to the audio- visual speaker recognition with a specific topology of feature-level fusion. In this paper, we further discuss the topologies by investigate the correlation between multi- features derived from different modalities. Five topologies are explored to model the speech data and face data. This paper is organized as follows: we give a brief description of the architecture of the audio-visual speaker system in Section 2. In Section 3, we give a detailed illustration of the five types of topologies for the feature-level fusion using Dynamic Bayesian Network. The data set is presented in Section 4 with the experimental setup. And the comparisons between the audio-visual speaker based on the proposed topologies and single modal biometrics system with speech features or face features will be presented in this section as well. Section 5 serves as a conclusion.

2 Biometrics Fusion Architecture

The task of identification is to determine if the speaker is a specific one in the group of enrolled users given his utterance. The speech sequences and face image sequences

D. Zhang and A.K. Jain (Eds.): ICB 2006, LNCS 3832, pp. 539 – 545, 2005.

are processed by feature extractors. The speech features and the face features are integrated to make a DBN model in this framework. Input speech and face data are matched with DBN models. The final decision is made as the highest scoring decision procedure applied to the DBN matcher module outputs.

In the voiceprint feature extraction, the hamming window is 32 mm and the frame shift is 16mm. The silence and unvoiced segments are discarded based on an energy threshold. The feature vectors are composed by 16 MFCC and their delta coefficients. The face feature extraction method is based on standard Principal Component Analysis (PCA) [6]. The 32 largest eigenvectors are taken from the list of eigenvectors, and forming a matrix with these eigenvectors in the columns to form the feature vector.

3 DBN Based Feature-Level Fusion

Dynamic Bayesian Networks are a special case of singly connected Bayesian networks specifically aimed at time series modeling. Consider in a DBN, every observation V representing the speech features is conditionally dependent on a state variable X. If an N slice data, $V = \{v_1, v_2, ..., v_N\}$, corresponds to a series of states, $X = \{x_1, x_2, ..., x_N\}$, then the conditional probability can be represented as $P(v_n \mid x_n)$. We then attempted to incorporate additional face features, $F = \{f_1, f_2, ..., f_N\}$, into the speaker recognition process.

3.1 Five Topologies

We hypothesize that the face features have a certain relationship with the speech features and the hidden state variables. There are three situations for hidden variables: 1) only have relationship with the speech variables; 2) only have relationship with the face nodes; 3) have correlation with both of them. Similarly, there are two relations between speech variables and face ones, interrelated and irrespective. As a consequence, six (3*2) cases are generated according to the relationship among face features, speech features and hidden state variables. Here we propose five types of topologies that can be used for fusion within the framework of DBN. The case of different state affects speech feature and face information respectively which separates these two kinds of data thoroughly is out of consideration.

In each topology, we conform to standardized measurements: shading nodes are observed; clear nodes are hidden. $X_t^i, t = 1,2,...,T, i = 1,2,..N$ are the hidden nodes with discrete values, N is the number of hidden nodes in one time slice. The observed nodes V_t , $t = 1,2,...,T$, represent the speech features. The observed nodes $F_t, t = 1,2,...,T$, represent the face features. Here T is the length of time slices. V_t and F_t satisfy Gaussian distributions.

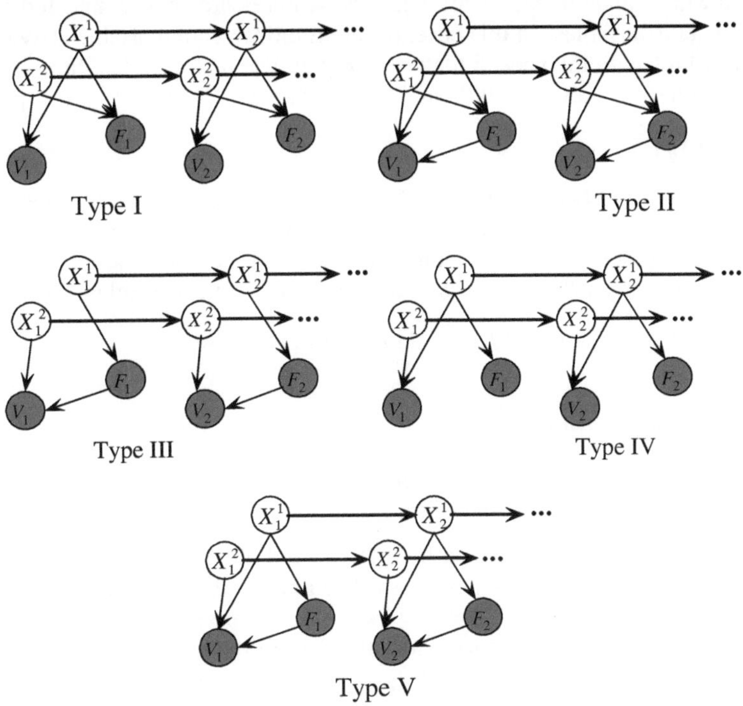

Fig. 1. Five topologies explored for audio-visual speaker recognition. Type I): Identical States and Independent; Type II): Identical States and Dependent; Type III): Different States and Dependent; Type IV): Mixed States and Independent; Type V): Mixed States and Dependent.

The five types of fusion topologies are depicted as follows:

I. Identical States and Independent: The face features are only connected to hidden state variables. Put it another way, both the face features and the speech features are activated by the change of the same hidden state variables. But they themselves have no relation, see Figure 1. Then the joint probability is $\prod_{i=1}^{2} P(v_n \mid x_n^i) \prod_{i=1}^{2} P(f_n \mid x_n^i)$. The topology of this type is the same as in [5].

II. Identical States and Dependent: The face features are connected to speech features and hidden state variables. In other words, observations of the speech features are affected by not only the hidden state variables but also the face features. Then the joint probability is $P(v_n \mid f_n) \prod_{i=1}^{2} P(v_n \mid x_n^i) \prod_{i=1}^{2} P(f_n \mid x_n^i)$.

III. Different States and Dependent: The face features are only connected to speech features. That is to say, the speech features and the face features are controlled by different hidden state variables respectively. However, the face features still has some effect on the speech features. Then the joint probability is $P(v_n \mid f_n) P(v_n \mid x_n^1) P(f_n \mid x_n^2)$.

IV. Mixed States and Independent: The face features are only connected to part of hidden state variables. In this case, different hidden state variables have different relationships with the speech features and the face features. The face features are not affected by all the hidden state variables. Then the joint probability is $P(f_n \mid x_n^1)\prod_{i=1}^{2} P(v_n \mid x_n^i)$.

V. Mixed States and Dependent: The face features are connected to speech features and part of hidden state variables. That is, observations of the speech features are controlled by both the hidden state variables and the face features. But the face features are only affected by part of the hidden state variables. Then the joint probability is $P(v_n \mid f_n)P(f_n \mid x_n^1)\prod_{i=1}^{2} P(v_n \mid x_n^i)$.

3.2 Training and Testing

As in the case of Dynamic Bayesian Networks applied to speaker identification, one may be interested in the following tasks:

Training: Each speaker is modelized by a DBN, and all the models are trained independently in our speaker identification task. We assume the structure of the DBNs be known for simplicity so only the parameters of a DBN are need to be estimated given a sequence of observations so as to model the data in the most appropriate way. The log-likelihood of the training set $C = \{V_1,...V_M, F_1,..., F_M\}$ can be calculated as:

$$\begin{aligned} L &= \log \prod_{m=1}^{M} \Pr(Y_m \mid G) \\ &= \sum_{i=1}^{N}\sum_{m=1}^{M} \log P(X_i \mid parent(X_i), V_m, F_m) \end{aligned} \tag{1}$$

Here G is a DBN model with N variables. These marginal posterior probability terms are computed in the inference engine. The computed marginal posterior probability can be used for the expected counts in expectation-maximization (EM) training for learning the mean μ, and the covariance Σ in the case of conditional linear Gaussian distributions, tailored to our needs of speaker identification.

Testing: The testing procedure of speaker identification is concerned with determining the right person whose features best match the features of the person to identify from a closed-set, given a set of observation. The speaker $\hat{i}$ whose DBN model M_i maximizes the posterior probability $p(M_i \mid C)$ is the identified one.

According to the Bayesian rule,

$$p(M_i \mid C) = p(C \mid M_i) * p(M_i) / p(C) \tag{2}$$

The values of $p(M_i)/p(C)$ for all speaker models M_i are treated to be equal as no prior knowledge about the probability could be retrieved. For simplicity, the decision rule can be formulized as:

$$\hat{i} = \arg_i \max p(C \mid M_i), \quad i = 1,2,...N, \tag{3}$$

The posterior probabilities $p(C \mid M_i)$ can be achieved by computing the joint probability using (1).

4 Experiment and Discussion

4.1 Database

We have built a collected multi-modal corpus of 54 people (17 females and 37 males) to evaluate our system [5]. Each visitor is subjected to two recordings: a speech shot and face recording.

The speech content is varied and enormous, including Mandarin, Dialect and English in terms of language types, and prompted texts and free talks in terms of speech types. The corpus contains 54 subjects with 216 images and 2916 sentences.

The utterance set is divided into seven sessions such as personal information, mandarin digits, dialect digits, English digits, province phrase, paragraph and free talk. The image set consist of frontal images and side profile images with 4 shots for every visitor, 2 frontal ones, 2 side ones. Recording is made in an office with a low level of acoustic noise and sufficient lighting. In this way, we obtain a corpus of 54 subjects, with 4 face images and 54 utterances from per subject.

4.2 Setup

We also compare the feature-level DBN fusion method with the feature concatenation method and the single modal recognition system using acoustic features or facial features.

In the baseline strategy, the speaker recognition expert and face recognition expert are evaluated. The approach to recognize the speaker identity is based on the use of the MFCC as the parameters and the Dynamic Bayesian Networks (DBNs) for the classification task. All the settings are the same as described in [7]. BNT toolkit [8] is used as the interface in our source code. The face identification system uses the Eigenface method as the face matcher. Images are compared by means of their corresponding feature vectors extracted as described in Chapter 2.

In the feature concatenation method, the speech features and the face features are extracted using the same approaches as mentioned in baseline strategy. The 32-dimensional face features are then appended to the 32-dimensional speech features directly resulting in a 64-dimensional feature vector. The restructured features are modeled using DBNs, with the same parameters in speaker recognition expert.

The proposed topologies are evaluated as the third setup. The speech features and the face features are the same as the above two setups. The speaker models are trained and tested as presented in Chapter 3.2.

4.3 Results and Discussion

In order to ascertain whether or not the method is robust with different speech contents and different speech types, we make experiments on some subsets of our multi-modal data corpus: Mandarin, Dialect, English, Phrase, and Free talk.

Generally speaking, the speech features and the face features have no direct causal relationship, but they bear some inherent relation, for these two features are produced by one and the same person. We foresee that the case of "Mixed States and Independent" would outperform other topologies with qualitative analyses. The results are listed in table 1.

Table 1. Experimental results with different speech types and contents of test sets. I stands for the identification rate. Man stands for the Mandarin type. Eng stands for the English type. Dia stands for the dialect type. Phr stands for the phrasal type. FTlk stands for free talk.

Fusion	I for each speech content and type (%)					
Method	Man	Dia	Eng	Phr	FTlk	Average
Voice Only	84.63	85.55	91.11	87.78	87.78	87.37
Face Only	85.18					
concatenation	87.59	88.15	91.85	89.07	89.81	89.29
Type I	90.21	91.11	94.81	92.03	92.96	92.22
Type II	89.63	92.03	93.15	91.85	91.66	91.66
Type III	89.68	89.15	92.33	90.21	91.11	90.50
Type IV	93.33	93.70	96.67	95.18	95.18	94.81
Type V	91.11	92.03	94.07	93.33	92.96	92.70

Conclusions can be drawn from our experiments as follows:

- The bimodal speaker authentication system has improved the identification rate by 7.44% and 9.63% compared with the speaker recognition and face recognition system respectively in the best cases. And the simple concatenation method which only enhances the performance by little degree is still far from satisfactory. That's why researchers are still pursuing superior features fusion methods.
- The bimodal speaker identification system based on the feature-level fusion using DBN outperforms the simple concatenation method by 5.52% in the best situation and 1.21% in the worst one. Indications are that it is a promising way of using feature-level DBN fusion in multi-modal problems.
- Among these five types of fusion, "Mixed States and Independent" works the best, which corresponds with our hypothesis.

5 Conclusions

This paper presents a feature-level fusion approach using Dynamic Bayesian Network for audio-visual speaker recognition. Five types of topologies are explored in the framework of bimodal speaker identification based on the correlations between the speech features and the face features. Encouraging experiment findings from multi-modal corpus including different speech types and contents reveal that the multi-biometric system can be further refined by the DBN-based feature-fusion approach. Further experiments will focus on the automate topology learning of DBN for multi-modals fusion.

Acknowledgements

This work is supported by National Natural Science Foundation of P.R.China (60273059), Zhejiang Provincial Natural Science Foundation (M603229) and National Doctoral Subject Foundation (20020335025).

References

1. Kevin Murphy.: Dynamic Bayesian Networks: Representation, Inference and Learning. Ph.D. thesis, U.C. Berkeley (2002)
2. Dongdong Li, Yingchun Yang, Zhaohui Wu, Wenyao Liu,: Add prior knowledge to speaker recognition. Multisensor, Multisource Information Fusion: Architectures, Algorithms, and Applications 2005, part of the SPIE Defense and Security Symposium 2005. Vol. 5813 (2005) 192-200
3. Ara V Nefian, Lu Hong Liang, Xiao Xing Liu, Xiaobo Pi and Kevin Murphy.: Dynamic Bayesian networks for audio-visual speech recognition. EURASIP, Journal of Applied Signal Processing, vol. 2002, no 11 (2002) 1274-1288
4. V. Pavlovic, A. Garg, J. Rehg, and T. S. Huang,: Multimodal speaker detection using error feedback dynamic Bayesian networks. Computer Vision and Pattern Recognition vol.2. (2000) 34-41.
5. Dongdong Li, LiFeng Sang, Yingchun Yang and Zhaohui Wu,: Bimodal Speaker Identification Using Dynamic Bayesian Network, 5th Chinese Conf. on Biometric Recognition, Lecture Notes in Computer Science, Vol. 3338. (2004) 577-585
6. Y. Wang, T. Tan and A. K. Jain,: Combining Face and Iris Biometrics for Identity Verification. Proc. of 4th Int'l Conf. on Audio- and Video-Based Biometric Person Authentication (AVBPA), Guildford, UK (2003) 805-813
7. Lifeng Sang, Zhaohui Wu, Yingchun Yang, Wanfeng Zhang,: Automatic Speaker Recognition Using Dynamic Bayesian Network. IEEE ICASSP. Vol.1 (2003) 188-191
8. Kevin Murphy. The Bayes Net Toolbox for Matlab. Computing Science and Statistics, vol 33 (2001)

Identity Verification Through Palm Vein and Crease Texture

Kar-Ann Toh[1], How-Lung Eng[1], Yuen-Siong Choo[2],
Yoon-Leon Cha[2], Wei-Yun Yau[1], and Kay-Soon Low[2]

[1] Institute for Infocomm Research,
21 Heng Mui Keng Terrace, Singapore 119613
katoh@ieee.org, {hleng, wyyau}@i2r.a-star.edu.sg
[2] School of Electrical & Electronic Engineering,
Nanyang Technological University, Singapore
ekslow@ntu.edu.sg

Abstract. In this paper, an identity verification framework which combines pattern information from the palm-vein and the palm-crease texture is proposed. Main feature of this system is the use of a low cost Near-Infra-Red (NIR) camera instead of the more expensive infra-red thermal camera for palm image capture. Our preliminary experiments show that useful information from palm-vein and palm-crease texture can be effectively extracted for identity verification using a simple setup to contain the camera.

Keywords: Biometrics, Multimodal Biometrics, Palm-vein Recognition, Palm-print Recognition and Pattern Classification.

1 Introduction

The palm dermatoglyphics (palmprints) [1, 2, 3, 4, 5] and hand vascular network (veins from back of hand) [6, 7] have increasingly gained research attention recently. Main reason for this growth, perhaps, can be attributed to the non-intrusive nature and the good inter class differentiability offered by these biometrics. The non-intrusive property includes whether it is linked to criminal records. Palm dermatoglyphics and hand vascular network are advantages over fingerprint and face biometrics in this aspect. The good inter class differentiability to distinguish individual identities are seen from more and more empirical evidences over recent years from the above cited works.

In this work, a low-cost NIR camera, rather than a considerably more expensive Infra-Red (IR) or specially designed thermal camera, is used for capturing the palm image. Since the NIR camera is operating within the visible electromagnetic spectrum, the images captured from this device contain certain amount of palm-crease texture information on top of the thermal vein patterns. We shall explore in this paper, a biometric system whereby feature information are extracted from these two spectrums of palm images for multimodal biometric decision fusion. Main contributions of this work include: (i) definition of palm-vein and

D. Zhang and A.K. Jain (Eds.): ICB 2006, LNCS 3832, pp. 546–553, 2005.

palm-crease biometric features for NIR palm images which are acquired from a single low cost camera system, and (ii) exploration of fusing these palm-vein and palm-crease texture modalities for identity verification.

2 Proposed Framework

2.1 System Overview

A low cost monochrome NIR CCD camera (JAI CV-M50 IR [8]) was used to capture the frontal palm images. Since the NIR camera was not sensitive enough to detect the IR radiation emitted by the human body (3000 - 14000 nm), an IR light source was used to irradiate the palm [6]. The camera was mounted on a customized rig with the IR source. Each user was asked to rest his/her hand on a rigid platform with palm facing a hollow cutout where the camera was positioned within. Apart from an alignment point for the placement of middle finger, no additional alignment pads were used.

With appropriate processing to extract information from different spectral zones of the NIR CCD camera, a multi-spectrum palm signature could be available. In this work, we focused only on the visible zone and the NIR zone (600 nm to 1000nm) of the NIR CCD for palm biometric identity authentication. The images were digitized into 768×576 pixels (spatial resolution offered by the CCD) with a gray-scale resolution of 8-bit per pixel.

2.2 A Bimodal Framework for Identity Verification

Based on the NIR CCD palm images obtained from the above system, we propose in this paper the following pattern features for identity verification.

Palm-vein feature points. To facilitate fast point-based matching, we propose to represent the palm-vein network structure by a set of points extracted from sub-sampling. A reference grid frame with appropriate size is superimposed on the extracted palm-vein structure, and a palm-vein point is defined as interception point between a vein line and the grid. Although for simplicity reason in this preliminary work, only the location information is included, we shall use the orientation and width information for accuracy enhancement in future.

Palm-crease texture. The palm-crease texture provides another source of information for identity verification. Two main approaches are available for this information extraction, namely local crease-line detection and global texture pattern analysis. In this feasibility study, we adopt the global texture pattern approach by means of elementary wavelet analysis.

Decision fusion. The matching outputs from both the above biometrics are fused at measurement level for final decision to determine whether the query is a genuine-user or an imposter.

3 Palm Patterns Recognition

In this section, the processing steps to extract the palm-vein and palm-crease features are briefly introduced since much of the adopted techniques are based on elementary image processing tools. We shall then show in the experiment section that such elementary modalities can be effectively combined to yield a reasonably accurate bimodal system for identity verification.

3.1 Region of Interest

For both palm-vein and palm-crease texture, the comparison (matching) of two identities is based on an image region common to all for computational efficiency reason. We call this the Region of Interest (ROI) [3] and it will be applied to both palm-vein and palm-crease texture biometrics for matching purpose.

3.2 Verification Using Palm-Vein

Image pre-processing. High pass filtering: Here, we are only interested in the fairly faint vein line patterns obtained from the NIR spectrum. By carefully analyzing the grey scale palm images, Gaussian high pass filtering is performed with the cut-off frequency determined empirically. The Gaussian filter is first applied to the original image (The intensities of the output image is scaled). The filtered image is then passed through a frequency equalization stage to enhance the desired narrow intensity range. Fig. 1(b) shows the outcome of this processing step.

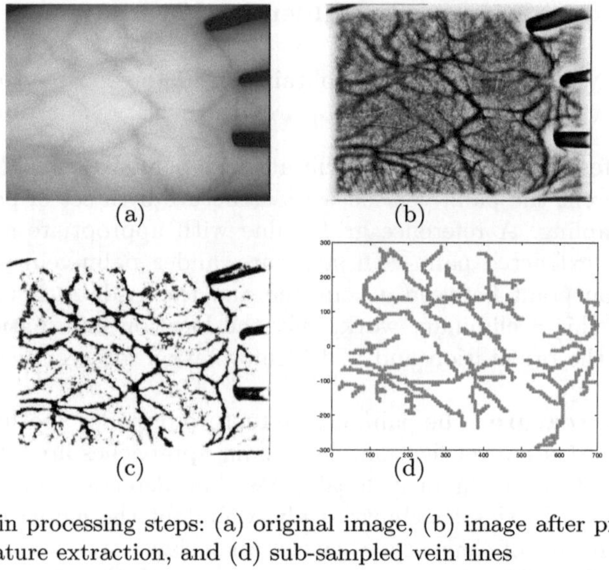

Fig. 1. Palm-vein processing steps: (a) original image, (b) image after pre-processing, (c) vein-lines feature extraction, and (d) sub-sampled vein lines

Vein-lines feature points extraction. The extraction of vein lines feature begins with a morphological gradient operation, which consists of dilation and erosion of an image, to enhance the edges of the vein structure. A morphological gradient operated image is added to the above pre-processed image for further removal of winkles and textures other than the vein. Fig. 1(c) shows the vein patterns after this processing step.

The positional (movement) and granular noises are further handled by mapping the vein signatures onto a new grid with large grid size. The separation between the grid lines is set to 6 pixels, which means there are 6 pixels in between 2 adjacent grid lines. This grid resolution was selected to provide an adequate representation of the vein network, but at the same time matches the inherent noise associated with the positional information of the vein structure. Fig. 1(d) shows the results of sub-sampling the signatures. The descriptors of the vein structure from the same user produced almost similar diagrams with minimum or no positional noises. In addition, the mapping procedure is also effective in reducing the size of the data which directly improved computational efficiency.

Matching. A template library using the above extracted features within the ROI is constructed for identity verification. The template for each identity is obtained by averaging 5 feature sets from 5 palm samples of the same person. The Correlation Coefficient or the Pearson's correlation coefficient is adopted for matching since it is widely used in statistical analysis, pattern recognition, and image processing.

3.3 Verification Using Palm-Crease Texture

Image pre-processing. High-pass filtering: Our interest here is the high frequency palm-crease texture obtained from the visible light spectrum. A Butterworth high-pass filtering is performed to exclude other component frequencies. The cutoff frequency was selected based on empirical observation from applying the filter to images in the database. Fig. 2(b) shows the processed palm image with texture features emphasized.

Palm-crease texture extraction. To extract the features from the ROI subimage (Fig. 2(c)), we use an elementary multi-resolution analysis. At each resolution (scale), multi-resolution analysis decomposes an image into several directions (Horizontal, Vertical and Diagonal). Using the Haar wavelet, three level of decomposition are performed on the image block. At each level of decomposition, the horizontal, vertical and diagonal energy are obtained (Fig. 2(d)). Finally the vector is normalized by the total energy obtained from summing up the elements.

Matching. For identity comparison, a template is first constructed from energy samples of multiple palm images within the ROI. A simple averaging of the multiple energy vectors has been adopted for simplicity and time efficiency reasons. Given two energy representations (test and template resulted from training), the matching module adopts a simple CityBlock (Manhattan or L_1-norm) distance measure. It is noted that smaller distance denotes better similarity between the two palms in comparison.

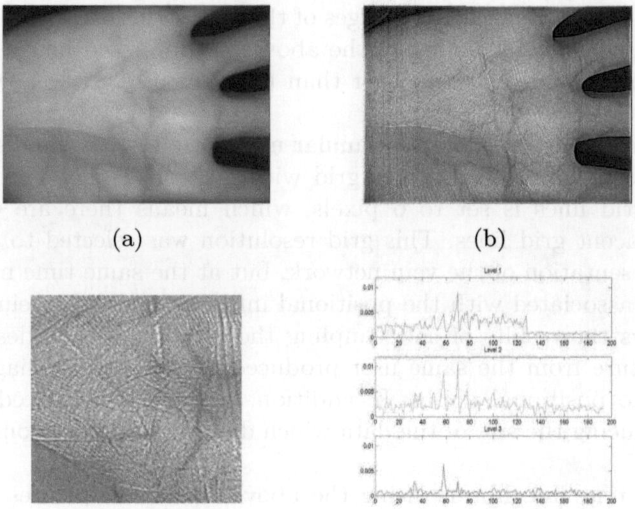

(a) (b) (c) (d)

Fig. 2. Palmprint processing steps: (a) original image, (b) image after high-pass filtering, (c) ROI extraction, and (d) wavelet energy features

4 Experiments

4.1 Database

A total of 1000 NIR images of the palm are collected as the database for our study - 10 images each from the right and left palm, from 50 different individuals. Since the left and right palms are observed to have significantly different vein patterns, we flip the left palm images to be in similar orientation to those from the right palm and make the assumption that left and right palms constitute different identities. We therefore have 100 identities in total with each identity containing 10 palm image samples.

The data is divided into two sets to accommodate for training and test experiments. We have 5 image samples per individual for each of the training set (100×5 images in total) and the test set (100×5 images in total). For each training set and each test set, 1000 ($5 \times 4 \times 100/2$) genuine match scores and 74250 ($100 \times 99 \times 5 \times 3/2$) imposter match scores are generated for each modality (vein and crease) for performance evaluation. The reason for using only 74250 out of the total possible 123750 ($100 \times 99 \times 5 \times 5/2$) is to have a smaller difference between the imposter and the genuine user data sets. We shall swap between the training set and the test set to create a 2-fold experiment. Since training results do not indicate predictivity, only the unseen test results will be reported in the sequel.

4.2 Uni-modal Verifications

The lower dotted lines in Fig. 3-(a) and Fig. 3-(b) show the Receiver Operating Characteristic (ROC) performances for palm-vein and palm-crease-texture biometrics for two-fold experiments. The ROC performance of the palm-vein biometric shows significant superiority over that of the palm-crease-texture biometric for the current experimental setting. Main reason is due to better discriminative features obtained from palm-vein line extraction as compared to those generic wavelet energies extracted from the palm-crease-texture.

4.3 Combining Palm-Vein and Palm-Crease-Texture

In the next set of experiments, the output from both modalities (vein and crease) are fused at measurement level [9] to form a bimodal verification system. Several classifiers with acclaimed performance (good accuracy and/or fast training speed) are experimented. These classifiers include a simple SUM-rule, Support Vector Machines adopting different kernels (SVM-Linear, SVM-Poly, SVM-RBF) [10] and a Reduced Multivariate polynomial (RM) [11].

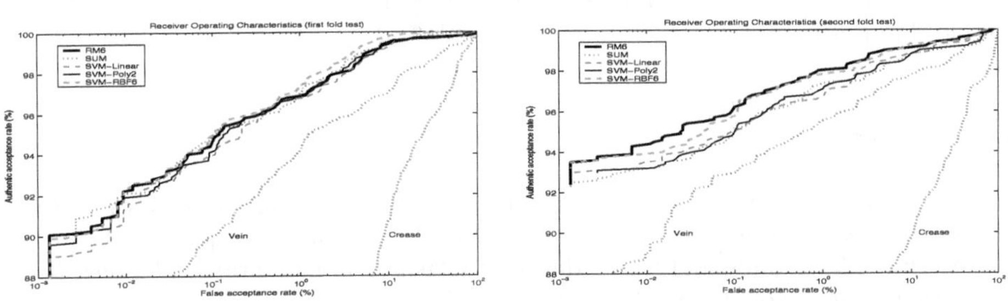

(a) (b)

Fig. 3. ROC plots: (a) First-fold verification test combining palm-vein and palm-crease-texture, (b) Second-fold verification test combining palm-vein and palm-crease-texture

The SUM-rule simply sums up the scores and then divide it by 2. The SVM-Poly was experimented with different polynomial orders ranging from 2 to 8 by validation tests. The best performed order 2 (SVM-Poly2) was presented in Fig. 3-(a) and Fig. 3-(b) respectively for first- and second-fold tests. For SVM-RBF, the scaling parameter Gamma [10] was experimented for values in [0.1,0.5,1,2,3,4,5,6] and the best value was found to be 6 (SVM-RBF6) from validation. The RM was ran from order 2 to order 8 and finally order 6 (RM6) was selected. These selected results for the two-fold tests are shown in Fig. 3-(a) and Fig. 3-(b).

The results from the plots show that significant accuracy improvement have been achieved by fusing the two modalities for all the compared methods. Among the SVMs, the RBF kernel appears to generalize best for both the two-fold experiments. RM6 appears to achieve good accuracy for most operating ranges. Here, we note the simple SUM-rule achieves remarkably good accuracy for the first-fold test while having a much poorer performance in the second-fold as compared to other classifiers. Here we see that SUM-rule could be a simple and effective method for fusion in this application, but main question remains whether the test data fits well with the presumed probabilities for uncorrelated inputs. This finding is congruent with our previous work [12] on decision fusion.

5 Conclusion

We proposed to combine the decision measures from two biometrics which were derived from a single palm image obtained from a low cost near-infra-red CCD camera in this paper. The main pattern features adopted for the palm-vein and the palm-crease texture were respectively a set of sub-sampled vein lines network and a set of directional wavelet energies. The two uni-modals are subsequently combined at measurement level to form a bimodal system. Several well acclaimed classifiers (SUM and SVMs) from the literature were experimented comparing with an in-house developed classifier (RM) for decision fusion. Our empirical experiments show that SVM with RBF kernel and RM generalizes best for the two-fold experiments.

References

1. A. J. Rice, "A quality approach to biometric imaging," in *Proceedings of Image Processing for Biometric Measurement IEE Colloquium*, 1994, pp. 4/1–4/5.
2. N. Duta, A. K. Jain, and K. V. Mardia, "Matching of palmprints," *Pattern Recognition Letters*, vol. 23, no. 4, pp. 477–485, 2002.
3. D. Zhang, W.-K. Kong, J. You, and M. Wong, "Online palmprint identification," *IEEE Trans. Pattern Analysis and Machine Intelligence*, vol. 25, no. 9, pp. 1041–1050, 2003.
4. L. Zhang and D. Zhang, "Characterization of palmprints by wavelet signatures via directional context modeling," *IEEE Trans. Systems, Man and Cybernetics, Part-B*, vol. 34, no. 3, pp. 1335–1347, June 2004.
5. A. Kumar, D. C. M. Wong, H. C. Shen, and A. K. Jain, "Personal verification using palmprint and hand geometry bionetric," in *Proc. 4th International Conference on Audio- and Video-Based Person Authentication (AVBPA)*, Guildford, UK, June 2003, pp. 668–678.
6. J. M. Cross and C. L. Smith, "Thermographic imaging of the subcutaneous vascular network of the back of the hand for biometric identification," in *IEEE 29th Annual International Carnahan Conference on Security Technology*, October 1995, pp. 20–35.
7. C.-L. Lin and K.-C. Fan, "Biometric verification using thermal images of palm-dorsa vein patterns," *IEEE Trans. Circuits and Systems for Video Technology*, vol. 14, no. 2, pp. 199–213, 2004.
8. JAI Camera Solutions, "Near IR industrial CCD camera," in *http://www.jai.com/db_datasheet/cvm50irdb.pdf [on-line]*, 2005, (datasheet).
9. L. Hong and A. Jain, "Integrating faces and fingerprints for person identification," *IEEE Trans. Pattern Analysis and Machine Intelligence*, vol. 20, no. 12, pp. 1295–1307, 1998.

10. J. Ma, Y. Zhao, and S. Ahalt, "OSU SVM classifier matlab toolbox (ver 3.00)," in *http://eewww.eng.ohio-state.edu/~maj/osu_svm/*, 2002, the Ohio State University.
11. K.-A. Toh, Q.-L. Tran, and D. Srinivasan, "Benchmarking a reduced multivariate polynomial pattern classifier," *IEEE Trans. Pattern Analysis and Machine Intelligence*, vol. 26, no. 6, pp. 740–755, 2004.
12. K.-A. Toh and W.-Y. Yau, "Combination of hyperbolic functions for multimodal biometrics data fusion," *IEEE Trans. Systems, Man and Cybernetics, Part-B*, vol. 34, no. 2, pp. 1196–1209, 2004.

Multimodal Facial Gender and Ethnicity Identification

Xiaoguang Lu, Hong Chen, and Anil K. Jain

Michigan State University, East Lansing, MI 48824
{Lvxiaogu, chenhon2, jain}@cse.msu.edu

Abstract. Human faces provide demographic information, such as gender and ethnicity. Different modalities of human faces, e.g., range and intensity, provide different cues for gender and ethnicity identifications. In this paper we exploit the range information of human faces for ethnicity identification using a support vector machine. An integration scheme is also proposed for ethnicity and gender identifications by combining the registered range and intensity images. The experiments are conducted on a database containing 1240 facial scans of 376 subjects. It is demonstrated that the range modality provides competitive discriminative power on ethnicity and gender identifications to the intensity modality. For both gender and ethnicity identifications, the proposed integration scheme outperforms each individual modality.

1 Introduction

Human face contains a variety of information for adaptive social interactions with people. Humans are able to process a face in a variety of ways to categorize it by its identity, along with a number of other demographic characteristics, such as gender, ethnicity, and age. Gender and ethnicity are involved in human face perception and recognition [1–5].

Unlike gender, ethnic categories are loosely defined due to the intermingling of races and the natural variations within races. We reduce the ethnicity classification into a two-category classification problem, *Asian* and *non-Asian*, which was also used in [6]. Anthropometrical statistics showed the ethnic craniofacial morphometric differences [7] and a close relationship between the 3D shape of the human face and ethnicity [8].

A lot of effort has been spent on the gender and ethnicity classification from different modalities. Most of them are focused on a single modality [9–13, 6] Only a few studies have investigated multiple modalities [14].

We address the problem of gender and ethnicity identification using two different facial modalities, range and intensity. With the advances of 3D imaging technology, commercial 3D sensors provide not only the range data, but also registered intensity information [15, 16] (see Fig. 1 for an example of a facial scan). We explore the surface shape (range) of the human face, which captures the craniofacial structure, for determining the ethnicity. Furthermore, since the identification from each individual modality can provide confidence of the assigned class membership for each test sample, the decision accuracy can be enhanced by integrating the confidence from different modalities. Since the precise facial landmark localization is difficult due to the variations of

D. Zhang and A.K. Jain (Eds.): ICB 2006, LNCS 3832, pp. 554–561, 2005.

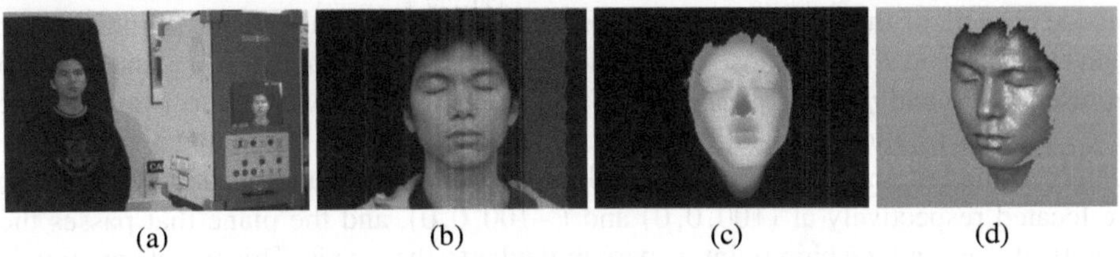

Fig. 1. An example of facial scan captured by Minolta Vivid 910. (a) Data-capture scenario; (b) texture image; (c) range image, with points closer to the sensor displayed in red; (d) 3D visualization.

facial structures, we do not use the anthropometrical measurements. Instead, we explore the appearance-based scheme [17], which has demonstrated its power in image-based face recognition.

2 Methodology

The system architecture is illustrated in Fig. 2. Range images are normalized in 3D space, and intensity images are normalized consequently. Data within a certain region are cropped from the normalized range and intensity images. Two SVMs classify the cropped range data and the intensity data. The classification results are integrated to achieve the final decision.

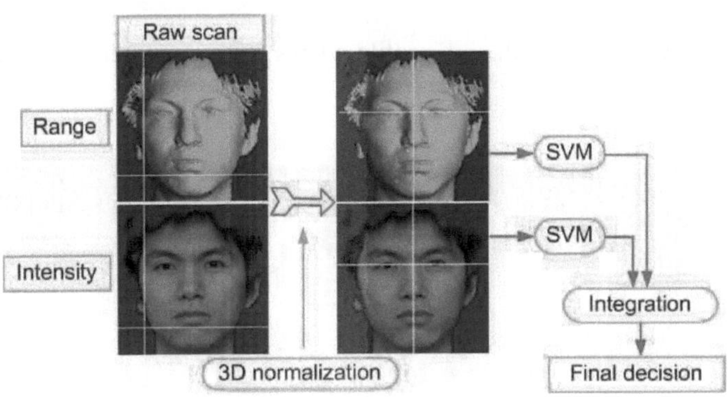

Fig. 2. System Diagram for gender and ethnicity identification

2.1 Normalization

To apply the appearance-based scheme, the raw scans are required to be aligned [18]: the raw scans are translated, scaled, and rotated so that the coordinates of the reference points are aligned.

The scans obtained from the 3D sensor are a set of points $S = \{(x, y, z)\}$. For the purpose of normalization, we manually specify 6 points in the scan: the inside and the outside corners of the left eye, $E_{l,i}$ and $E_{l,o}$, the inside and the outside corners of the right eye, $E_{r,i}$ and $E_{r,o}$, the nose tip N, and the chin point C. We use $E_{l,i,x}$ and $E_{l,i,y}$ to represent the x and y value of $E_{l,i}$, and $E_{r,i,x}$ and $E_{r,i,y}$ to represent the x and y value of $E_{r,i}$. After rotation, translation and scaling, the points are normalized so that the centers of the left and the right eyes (midpoints of the inside and outside eye corners) are located respectively at $(100, 0, 0)$ and $(-100, 0, 0)$, and the plane that passes the centers of eyes and the chin point, is perpendicular to the z-axis. This transformation is defined as:

$$\begin{pmatrix} x' \\ y' \\ z' \end{pmatrix} = s \cdot R \cdot \begin{pmatrix} x \\ y \\ z \end{pmatrix} + \begin{pmatrix} t_1 \\ t_2 \\ t_3 \end{pmatrix}, \tag{1}$$

where

$$\begin{pmatrix} t_1\ t_2\ t_3 \end{pmatrix} = -(\vec{E}_{l,i} + \vec{E}_{l,o} + \vec{E}_{r,i} + \vec{E}_{r,o})/4, \quad s = 400/\|\vec{E}_{l,i} + \vec{E}_{l,o} - \vec{E}_{r,i} - \vec{E}_{r,o}\|,$$

$$\begin{pmatrix} x_0\ y_0\ z_0 \end{pmatrix} = (\overrightarrow{E_{l,i}} - \vec{C}) \times (\overrightarrow{E_{r,i}} - \vec{C}), \qquad R = M_z \cdot M_x \cdot M_y,$$

$$M_x = \begin{pmatrix} 1 & 0 & 0 \\ 0 & \cos\alpha & \sin\alpha \\ 0 & -\sin\alpha & \cos\alpha \end{pmatrix}, \quad M_y = \begin{pmatrix} \cos\beta & 0 & -\sin\beta \\ 0 & 1 & 0 \\ \sin\beta & 0 & \cos\beta \end{pmatrix}, \quad M_z = \begin{pmatrix} \cos\gamma & \sin\gamma & 0 \\ -\sin\gamma & \cos\gamma & 0 \\ 0 & 0 & 1 \end{pmatrix},$$

$$\alpha = -\arctan(y_0/\sqrt{x_0^2 + z_0^2}), \quad \beta = \arctan(x_0/z_0), \qquad \gamma = \arctan(\frac{E_{l,i,y} - E_{r,i,y}}{E_{l,i,x} - E_{r,i,x}}),$$

Figure 3 shows the frontal and profile views of a face scan before and after normalization.

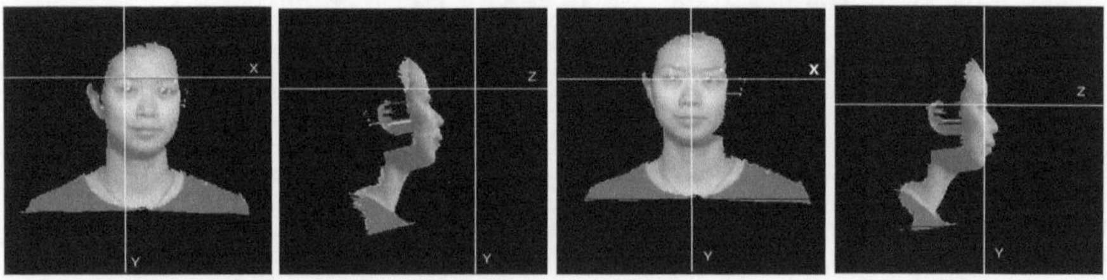

(a) (b) (c) (d)

Fig. 3. (a) Frontal view before normalization. (b) Profile view before normalization. (c) Frontal view after normalization. (d) Profile view after normalization.

2.2 Feature Vector Construction

To avoid the effect of hairstyle and other facial accessories, a close facial scan cropping scheme is applied. Given a normalized 3D face data set C, x and y coordinates of a rectangular area R to be cropped, and the numbers of rows and columns of the grid in the rectangle R, m and n, we crop the face areas and construct feature vectors as follows:

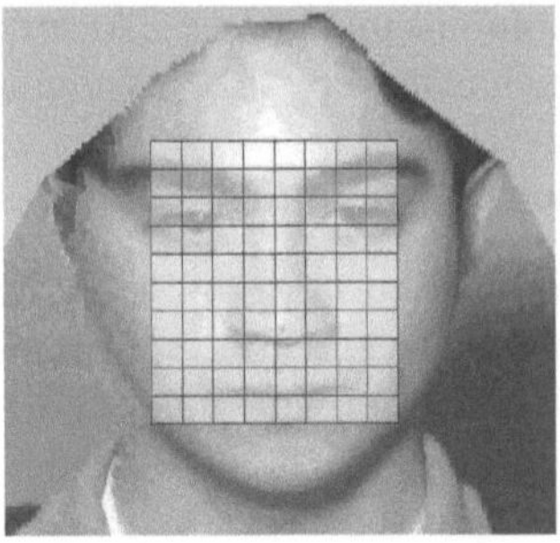

Fig. 4. Cropping face areas for construction of feature vectors. A 10×8 grid is overlaid on the facial scan for demonstration.

(1) Build a grid G. The grid G is in a plane parallel to the x-y plane. It has m rows and n columns. The borders of G are set to be the rectangle R. A grid G is shown in Fig. 4.

(2) Build the $m \times n$ projection matrices XM, YM, ZM. The elements $XM(i,j)$, $YM(i,j)$ and $ZM(i,j)$, $i = 1, \cdots, m$, $j = 1, \cdots, n$, correspond to the grid node $G(i,j)$. Denote the set of points inside $G(i,j)$ as C', where $C' = \{(x,y,z)|(x,y,z) \in C$, and x, y are inside $G(i,j)\}$. If C' is empty, the corresponding element is labeled as a hole (see Fig. 5). Otherwise, the value of each grid is computed as follows:

$$XM(i,j) = \frac{1}{|C'|} \sum_{\text{for all } (x,y,z) \in C'} x,$$

$$YM(i,j) = \frac{1}{|C'|} \sum_{\text{for all } (x,y,z) \in C'} y,$$

$$ZM(i,j) = \frac{1}{|C'|} \sum_{\text{for all } (x,y,z) \in C'} z,$$

where $|C'|$ is the number of elements in C'.

(3) Interpolation. After the 3D rotation, the occluded points in the original scan cause holes in the normalized scan. The holes in XM, YM, and ZM are filled by interpolating the nearest neighbors as shown in Fig. 5.

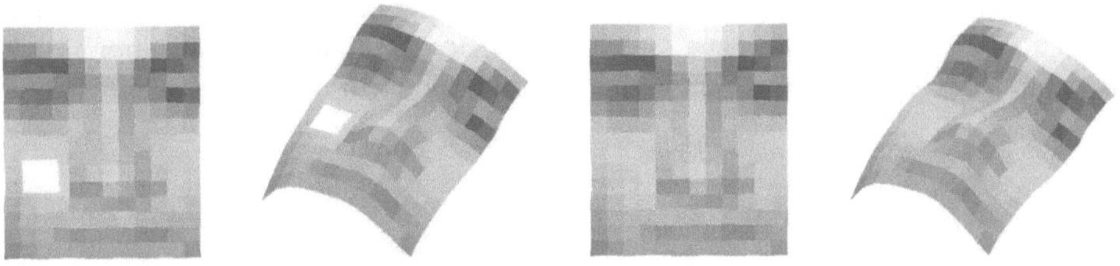

(a) (b) (c) (d)

Fig. 5. (a,b) Examples of the holes (shown as white patches) after 3D normalization. (c,d) The holes are filled by interpolation.

(4) Vector formation. The columns in matrices ZM are concatenated to generate the vector V of length $m \times n$, which is used by the classifiers for identification.

2.3 Identification and Fusion of Modalities

The gender and ethnicity identification using individual modalities are formulated as a two-class classification problem. In the appearance-based scheme, Support Vector Machines are a type of classifiers that provide high gender classification accuracy [13]. We use SVMs in our experiments for both ethnicity and gender classifications. Instead of matching scores, the posterior probabilities are extracted from the SVMs [19].

The combination strategy we used in our experiments is the sum rule [20] conducted at the decision level, which has more generality, when classifiers have physically different types of features.

For gender classification, the fusion process is formulated as:

$$p(male|s) = (p(male|s_{range}) + p(male|s_{intensity}))/2, \quad (2)$$

$$p(female|s) = (p(female|s_{range}) + p(female|s_{intensity}))/2, \quad (3)$$

where s is the subject to be classified, s_{range} and $s_{intensity}$ are respectively the range and the intensity maps of the subject, $p(male|s_{range})$ and $p(female|s_{range})$ are the posterior probabilities provided by the SVM that uses range data for gender classification, and $p(male|s_{intensity})$ and $p(female|s_{intensity})$ are the posterior probabilities provided by the SVM that uses intensity data for gender classification. The final decision is made by comparing $p(male|s)$ and $p(female|s)$. The same fusion scheme is applied to the ethnicity identification.

3 Experiments and Discussion

A mixture of two frontal 3D face databases is used for evaluating the proposed schemes. Representative facial scans are given in Fig.6. One database is from University of Notre Dame (UND) [21], composed of 944 scans from 276 subjects. The other is collected at Michigan State University (MSU), containing 296 scans of 100 subjects. The demographic information of the entire mixed database is summarized in Table 1. The samples of both databases were collected using the Minolta Vivid series 3D scanner [15].

For ethnicity identification, a 10-fold cross-validation is conducted. Each time we use 9 folds as the training set and the remaining fold as the test set. Scans from the same subject are grouped into the same set to ensure that the ethnicity classification results are not biased by the similarity between the testing and the training data in terms of the

Table 1. Number of subjects and scans (given in parenthesis) of the combination of UND and MSU databases in each category

	Non-Asian	Asian	Subtotal
Female	106 (255)	33 (110)	139 (465)
Male	176 (563)	61 (212)	237 (775)
Subtotal	282 (918)	94 (322)	376 (1240)

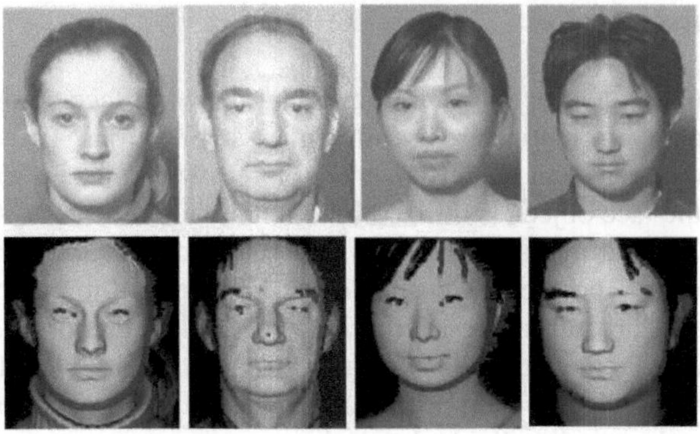

Fig. 6. Scan examples in the database. Intensity images (top) and the corresponding range images (bottom). From left to right, they are non-Asian female, non-Asian male, Asian female, Asian male.

identity. The mean and the standard deviation of the matching error rates from these 10 experiments are reported. The same scheme is applied for gender identification.

The ethnicity and gender identification performance is provided in Tables 2 and 3. For both ethnicity and gender identifications, the experimental results show that 3D (range) information provides competitive results to the 2D (intensity) modality. It is demonstrated that the integration of range and intensity outperforms each individual modality.

Table 2. Ethnicity identification performance. The average and standard deviation of the error rates using 10-fold cross-validation are reported.

	Non-Asian	Asian	Overall
Range	$2.7\% \pm 0.028$	$6.7\% \pm 0.052$	$3.8\% \pm 0.024$
Intensity	$2.1\% \pm 0.027$	$5.9\% \pm 0.051$	$3.2\% \pm 0.029$
Range + Intensity	$0.7\% \pm 0.010$	$5.5\% \pm 0.039$	$2.0\% \pm 0.016$

Table 3. Gender identification performance. The average and standard deviation of the error rates using 10-fold cross-validation are reported.

	Female	Male	Overall
Range	$24.5\% \pm 0.101$	$9.0\% \pm 0.030$	$14.6\% \pm 0.044$
Intensity	$19.2\% \pm 0.123$	$11.3\% \pm 0.066$	$14.0\% \pm 0.047$
Range + Intensity	$17.0\% \pm 0.093$	$4.4\% \pm 0.032$	$9.0\% \pm 0.030$

3D sensors in the current market are not as mature as 2D sensors. Typical problems with range images include missing data near dark regions (e.g., eye regions) and spikes at the region with high reflectivity. The interpolation and smoothing results are the approximations. These problems would deteriorate the gender and ethnicity identification performance using range images even though there exist methods for recovering some data in such areas [21].

4 Conclusions

Gender and ethnicity identifications are important topics in face recognition. The extracted demographic information are useful in many applications. Two different modalities of human faces, range and intensity, are explored. The range information, containing 3D shape of the face object, is utilized for ethnicity identification. A fusion scheme is developed by integrating the range and intensity to identify the gender and ethnicity from facial scans. The proposed scheme can be extended to combine other facial modalities, such as thermal images. Experimental results demonstrate that the range modality provides effective capability for gender and ethnicity identifications. It also shows that the proposed combination strategy obtain better classification accuracy than the classifiers based on each individual modality.

References

1. Malpass, R., Kravitz, J.: Recognition for faces of own and other race. J. Perc. Soc. Psychol. **13** (1969) 330–334
2. Brigham, J., Barkowitz, P.: Do 'they all look alike?' the effect of race, sex, experience and attitudes on the ability to recognize faces. J. Appl. Soc. Psychol **8** (1978) 306–318
3. O'Toole, A., Peterson, A., Deffenbacher, K.: An other-race effect for classifying faces by sex. Perception **25** (1996) 669–676
4. Golby, A., Gabrieli, J., Chiao, J., Eberhardt, J.: Differential responses in the fusiform region to same-race and other-race faces. Nature Neuroscience **4** (2001) 845–850
5. Jain, A.K., Nandakumar, K., Lu, X., Park, U.: Integrating faces, fingerprints, and soft biometric traits for user recognition. In: Proceedings of Biometric Authentication Workshop, in conjunction with ECCV2004, LNCS 3087, Prague (2004) 259–269
6. Shakhnarovich, G., Viola, P.A., Moghaddam, B.: A unified learning framework for real time face detection and classification. In: Proc. IEEE International Conference on Automatic Face and Gesture Recognition. (2002) 14–21
7. Farkas, L.: Anthropometry of the Head and Face. 2nd edn. Raven Press (1994)
8. Enlow, D.: Facial Growth. 3rd edn. W.H. Saunders (1990)
9. Glolomb, B., Lawrence, D., Sejnowski, T.: Sexnet: A neural network identifies sex from human faces. In: Advances in Neural Information Processing Systems (NIPS). Volume 3. (1990) 572–577
10. O'Toole, A., Vetter, T., Troje, N.F., Bulthoff, H.H.: Sex classification is better with three-dimensional structure than with image intensity information. Perception **26** (1997) 75–84
11. Gutta, S., Huang, J., Phillips, P., Wechsler, H.: Mixture of experts for classification of gender, ethnic origin, and pose of human faces. IEEE Trans. Neural Networks **11** (2000) 948–960
12. Davis, J., Gao, H.: Gender recognition from walking movements using adaptive three-mode PCA. In: Proc. IEEE Workshop on Articulated and Nonrigid Motion, Washington DC (2001) 9–16
13. Moghaddam, B., Yang, M.: Learning gender with support faces. IEEE Trans. Pattern Analysis and Machine Intelligence **24** (2002) 707–711
14. Walavalkar, L., Yeasin, M., Narasimhamurthy, A., Sharma, R.: Support vector learning for gender classification using audio and visual cues. International Journal of Pattern Recognition and Artificial Intelligence **17** (2003) 417–439
15. Minolta Vivid 910 non-contact 3D laser scanner. (<http://www.minoltausa.com/vivid/>)
16. Cyberware Inc. (<http://www.cyberware.com>)

17. Li, S., (Eds.), A.J.: Handbook of Face Recognition. Springer (2005)
18. Shan, S., Chang, Y., Gao, W., Cao, B.: Curse of mis-alignment in face recognition: Problem and a novel mis-alignment learning solution. In: Proc. IEEE International Conference on Automatic Face and Gesture Recognition, Korea (2004) 314–320
19. Platt, J.: Probabilistic outputs for support vector machines and comparison to regularized likelihood methods. In Smola, A., Bartlett, P., Schoelkopf, B., Schuurmans, D., eds.: Advances in large Margin Classifiers, MIT Press, Cambridge, MA. (2000)
20. Kittler, J., Hatef, M., Duin, R., Matas, J.: On combining classifiers. IEEE Trans. Pattern Analysis and Machine Intelligence **20** (1998) 226–239
21. Chang, K.I., Bowyer, K.W., Flynn, P.J.: Multi-modal 2D and 3D biometrics for face recognition. In: Proc. IEEE Workshop on Analysis and Modeling of Faces and Gestures, France (2003)

Continuous Verification Using Multimodal Biometrics*

Sheng Zhang, Rajkumar Janakiraman, Terence Sim, and Sandeep Kumar

School of Computing, National University of Singapore,
3 Science Drive 2, Singapore 117543
{zhangshe, janakira, tsim, skumar}@comp.nus.edu.sg

Abstract. In this paper we describe a system that continually verifies the presence/participation of a logged-in user. This is done by integrating multimodal passive biometrics in a Bayesian framework that combines both temporal and modality information holistically, rather than sequentially. This allows our system to output the probability that the user is still present even when there is no observation.

Our implementation of the continuous verification system is distributed and extensible, so it is easy to plug in additional asynchronous modalities, even when they are remotely generated. Based on real data resulting from our implementation, we find the results to be promising.

1 Introduction

For most computer systems, once the identity of the user has been verified at login, the system resources are typically made available to the user until the user exits the system. This may be appropriate for low-security environments but can lead to session "hijacking" (akin to hijacking [1]) in which an attacker targets a post-authenticated session. In high risk environments or where the cost of unauthorized use of a computer is high, continuous verification, if it can be realized efficiently is important to reduce this window of vulnerability. By this we mean that biometric verification is not merely used to authenticate a session on startup, but that it is used in a loop throughout the session to continuously authenticate the presence/particapation of the user. Examples where continuous verification is desirable include the usage of computers for airline cockpit controls, in defense establishments, and in other processing that affects the security and safety of human lives. In such situations, the desirable default action might be to render the computer system ineffective when the authorized user is not the one controlling it.

One way to realize (an approximation of) continuous verification is to use passive but accurate biometric verification. However, a single biometric may be inadequate for passive verification either because of noise in data samples or because of unavailability of a sample at a given time. For example, face verification cannot work when frontal face detection fails because the user presents

* This work was funded by the National University of Singapore, project no. R-252-146-112.

D. Zhang and A.K. Jain (Eds.): ICB 2006, LNCS 3832, pp. 562–570, 2005.

a non-frontal pose. To overcome this limitation, researchers have proposed the use of multiple biometrics, and have demonstrated increased accuracy of verification with a concomitant decrease in vulnerability to impersonation [4]. Use of multiple biometrics has led to the investigation of integrating different types of inputs (modalities) with different characteristics. Kittler et al. [2] experiment with six fusion methods for face and voice biometrics, using the sum, product, minimum, median, and maximum rules. In our work, we follow a similar approach: we combine face and fingerprint to do continuous verification.

For a continuous verification system, three criteria are important with regard to biometrics fusion:

1. The different reliability of the various modalities must be accounted for. That is, any fusion method must factor in the reliability of each modality.
2. Older observations must be discounted, to reflect the increasing uncertainty of the continued presence of the legitimate user.
3. Any fusion method should be able to handle lack of observations in one or more modalities, which arises from a normal usage pattern, i.e., when the user looks away from the camera.

Thus the usual fusion methods of sum, product etc. cannot be directly used because they do not satisfy the above criteria.

The key to continuous verification is the integration of biometric observations across both modality and time. Up to now, the task of integrating data across both modality and time has not been addressed satisfactorily. In this paper, we propose a Holistic Fusion method that combines face and fingerprint across modalities and time simultaneously and in a way that satisfies the above three criteria. This is realized by using the Hidden Markov Model (HMM). We experimentally compare our fusion method with a few alternatives – Time-first, Modality-first, and Naive Integration – and show that our method is superior.

2 Theory

The goal of verification is to determine whether the person with the claimed identity is who he claims to be. Two situations can occur: either the verifier *accepts* the claim as genuine, or the verifier *rejects* it (and decides that the user is an imposter).

In our case, the verification uses two types (modalities) of observations: fingerprint and face images. The challenge is to integrate these observations across modality and over time. To do this, we devised the integration scheme shown in Figure 1. Currently we implement a face verifier and a fingerprint verifier, other modalities are possible in the future. Each verifier computes a score from its input biometric data (fingerprint or face), which is then integrated

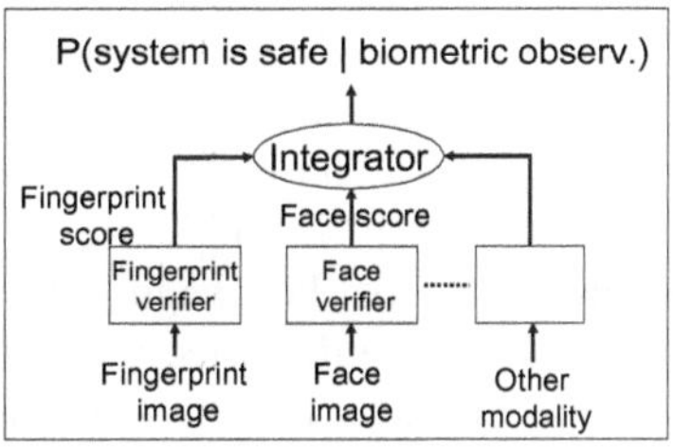

Fig. 1. Integration scheme

(fused) by the Integrator. The output from the Integrator is then used by the operating system kernel to delay or freeze user processes. For implementation details, please refer to [3].

2.1 Fingerprint Verifier

We acquire fingerprint images using the SecureGen^{TM} mouse, which incorporates a fingerprint scanner ergonomically where the thumb would normally be placed. This makes the mouse a passive (non-intrusive) biometric sensor, ideally suited for continuous verification. The mouse comes with a SDK that matches fingerprints, i.e., given two images, it computes a similarity score between 0 (very dissimilar) and 199 (identical). Unfortunately, the matching algorithm is proprietary and is not disclosed by the vendor. Nevertheless, it is enough to get good results using the score generated by the proprietary algorithm.

First, we collect 1000 training fingerprint images from each of four users. For each user, we compute two probability density functions (pdf) - the intra-class and inter-class pdfs (represented by histograms). If we denote the similarity score by s, the intra-class set by Ω_U, and the inter-class set by Ω_I, then these pdfs are $P(s \mid \Omega_U)$ and $P(s \mid \Omega_I)$. The pdfs are similar to those in Figure 2 (which are for faces), but have smaller overlap, indicating that fingerprint verification is reliable (high verification accuracy).

Given a new fingerprint image and a claimed identity, the image is matched against the claimed identity's template (captured at registration time) to produce a score s. From this we compute $P(s \mid \Omega_U)$ and $P(s \mid \Omega_I)$. These values are then used by the Integrator to arrive at the overall decision. See Section 2.3 for more details.

2.2 Face Verifier

Our Face Verifier is also based on intra- and inter-class pdfs, except that the score s is now an image distance, rather than a measure of similarity. To train the Face Verifier, we first capture 500 images of each user under varying head poses, using a Canon VCC4 video camera and the Viola-Jones face detector [6]. The images are resized to 28×35 pixels. For each user, the training images are divided into the intra-class and inter-class sets. For each set, we calculate the pairwise image distance using the L_p norm (described below). This is similar to the ARENA method [5]. These distances are now treated as scores s, and the pdfs $P(s \mid \Omega_U)$ and $P(s \mid \Omega_I)$ estimated as before.

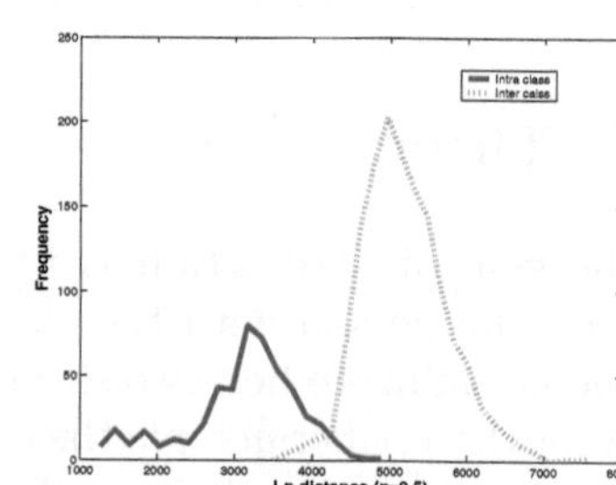

Fig. 2. Face intra-class and inter-class pdfs for a typical user

The L_p norm is defined as $L_p(\mathbf{a}) \equiv (\sum |a_i|^p)^{\frac{1}{p}}$, where the sum is taken over all pixels of image $\mathbf{a}$. Thus the distance between images $\mathbf{u}$ and $\mathbf{v}$ is $L_p(\mathbf{u}-\mathbf{v})$. As in ARENA, we found that $p = 0.5$ works better than $p = 2$ (Euclidean). Given a new face image and a claimed identity, we compute the smallest L_p distance between the image and the intra-class set of the claimed identity. This distance is then used as a score s to compute $P(s \mid \Omega_U)$ and $P(s \mid \Omega_I)$, which in turn are used by the Integrator.

2.3 Holistic Fusion

The heart of our technique is in the integration of biometric observations across modalities and over time. This is done using HMM, which is a sequence of states x_t that "emit" observations z_t, for time $t = 1, 2, \ldots$ Each state can assume one of two values: $x_t \in \{Safe, Attacked\}$. *Safe* means that the logged-in user is still present at the computer console, while *Attacked* means that an imposter has taken over control. It is also possible for the user to be absent from the console, but for a high security environment, this is considered to be the same as *Attacked.* Each observation z_t is either a face or fingerprint image, or equivalently, its corresponding score (See Sections 2.1, 2.2). Note that the states are hidden (unobservable), and the goal is to infer the state from the observations.

The result of the fusion is the calculation of P_{safe}, the probability that the system is still in the *Safe* state. This value can then be compared to a pre-defined threshold T_{safe} set by the security administrator, below which appropriate action may be taken. A key feature of our method is that we can compute P_{safe} at any point in time, whether or not there are biometric observations. In the absence of observations, we decay P_{safe}, reflecting the increasing uncertainty that the system is still *Safe.*

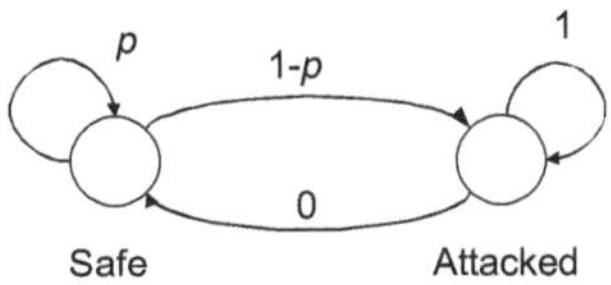

Fig. 3. State transition model

Let $\mathcal{Z}_t = \{z_1, \ldots, z_t\}$ denote the history of observations up to time t. From a Bayesian perspective, we want to determine the state x_t that maximizes the posterior probability $P(x_t \mid \mathcal{Z}_t)$. Our decision is the greater of $P(x_t = Safe \mid \mathcal{Z}_t)$ and $P(x_t = Attacked \mid \mathcal{Z}_t)$. Equivalently, we seek to determine if $P(x_t = Safe \mid \mathcal{Z}_t) > 0.5$, since the probabilities must sum to 1. We may rewrite:

$$P(x_t \mid \mathcal{Z}_t) \propto P(z_t \mid x_t, \mathcal{Z}_{t-1}) \cdot P(x_t \mid \mathcal{Z}_{t-1}) \tag{1}$$

$$P(x_t \mid \mathcal{Z}_{t-1}) = \sum_{x_{t-1}} P(x_t \mid x_{t-1}, \mathcal{Z}_{t-1}) \cdot P(x_{t-1} \mid \mathcal{Z}_{t-1}) \tag{2}$$

This is a recursive formulation that leads to efficient computation[1]. The base case is of course $P(x_0 = Safe) = 1$, because we know that the system is *Safe* immediately upon successful login. Observe that the state variable x_t has the effect of summarizing all previous observations. Because of our Markov assumptions, we note that $P(z_t \mid x_t, \mathcal{Z}_{t-1}) = P(z_t \mid x_t)$, and $P(x_t \mid x_{t-1}, \mathcal{Z}_{t-1}) = P(x_t \mid x_{t-1})$.

However, $P(z_t \mid x_t)$ is simply the intra-class pdf (when $x_t = Safe$) or the inter-class pdf (when $x_t = Attacked$). As for $P(x_t \mid x_{t-1})$, this is described by the state transition model shown in Figure 3. In the *Safe* state, the probability

[1] At time t, if there exists a biometric observation, we use Equation 1 to compute P_{safe}, otherwise Equation 2.

of staying put is p, while the probability of transitioning to *Attacked* is $(1-p)$. Once in the *Attacked* state, however, the system remains in that state and never transitions back to *Safe*.

The value of p is governed by domain knowledge - if there is no observation for a long period of time, we would like p to be small, indicating that we are less certain that the user is still safe (and thus more likely to have been attacked). To achieve this effect, we define $p = e^{k\Delta t}$, where Δt is the time interval between the current time and the last observation, and k is a free parameter that controls the *rate of decay*, which the security administrator can define. For instance, if the security administrator decides that p should drop to 0.5 in 30 seconds, then $k = -(\log 2)/30$.

In general, any decay function may be used to specify p, with a suitable rate of decay. We chose an exponential function for its simplicity: a value of $k = 0$ means that the user is never attacked ($p = 1$), while a very large value of k indicates that attacks are very likely.

3 Discussion

We compare our method with other alternatives: Temporal-first, Modality-first and Naive Integration.

3.1 Temporal-First and Modality-First Integration

Figure 4 shows how observations from different modalities present themselves over time. Observations from a single modality are shown horizontally, while observations across time are shown vertically. Note that at time t_3, only fingerprint is observed and also for ease of understanding, we show observations a and d as aligned vertically. In practice we allow a and d to occur within a small window of time apart.

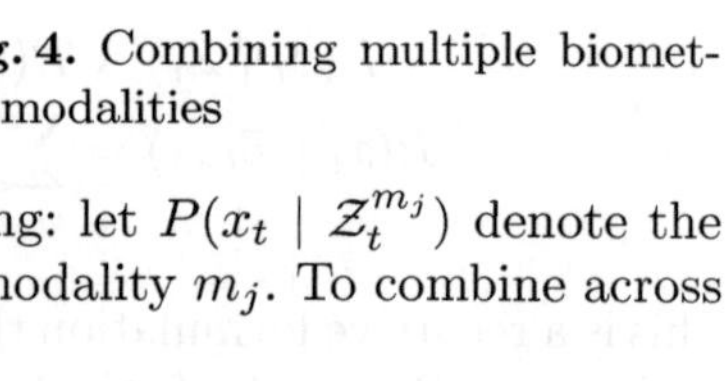

Fig. 4. Combining multiple biometric modalities

One common method of fusion is the following: let $P(x_t \mid \mathcal{Z}_t^{m_j})$ denote the posterior probability of being safe at time t for modality m_j. To combine across time, we compute the weighted sum:

$$P(x_t \mid \mathcal{Z}_t^{m_j}) = \frac{1}{N} \sum p(x_{t_i} \mid z_{t_i}^{m_j}) \cdot e^{k\Delta t_i} \tag{3}$$

Where Δt is the time difference between the current time and observation time, N is the number of observations. This decays older observations by the weight $e^{k\Delta t}$ such that it satisfies Criterion 2 for continuous verification.

To combine over modalities, we may again use a weighted sum:

$$P(x_{t_i} \mid z_{t_i}) = w^{m_1} \cdot P(x_t \mid z_{t_i}^{m_1}) + w^{m_2} \cdot P(x_t \mid z_{t_i}^{m_2}) \tag{4}$$

Note that here the two weights are w^{m_1} and w^{m_2}. They should be chosen to reflect the reliability of each modality, in order to satisfy Criterion 1. We will use the area under the ROC curve to represent the reliability.

Thus, Temporal-first implies the application of Equation 3 followed by Equation 4. Similarly, Modality-first changes this construction by applying Equation 4 first, then Equation 3. Note that if there is only a single modality (i.e., time t_3 in Figure 4), we just use the modality (no weight applied) as the combined result. Likewise if there is only one observation across time, then we just decay the observation by $e^{k\Delta t}$. In practice, for computational efficiency, we combine observations that occur within a recent history H from the current time, since observations that are too old have negligible weights.

3.2 Naive Integration

Since fingerprint is more reliable than face and also more reliable than the two combined (See Section 4.1), the idea of naive integration is to use the most reliable modality available at any time instant. More precisely,

1. At any time t, if a fingerprint observation exists, then $P(x_t \mid \mathcal{Z}_t) = P(x_t \mid z_t^{m_2})$ ($m_2 = fingerprint$) whether or not face observation exists.
2. Otherwise if there exists only face observation, then $P(x_t \mid \mathcal{Z}_t) = P(x_t \mid z_t^{m_1})$ ($m_1 = face$), since now face is the most reliable biometric that is available.
3. Else if no biometric observation is available, then we just decay the probability $P(x_t \mid \mathcal{Z}_t) = P(x_{t-1} \mid z_{t-1}) \cdot e^{k\Delta t}$. Where $P(x_{t-1} \mid z_{t-1})$ is calculated from Step (1) or (2), depending on the last biometric observation (fingerprint or face). Here Δt is the time interval between the current time and the latest observation time.

It is clear that Naive Integration satisfies the three criteria in Section 1.

4 Experiments

All the experiments were conducted on real users using an Intel Pentium 2.4 Ghz Workstation with 512MB RAM. The captured images are 384×288, 24-bit deep taken using a Euresys Picolo capture card with a Canon VCC4 camera.

Ideally all the biometric data are acquired at fixed times. But in reality the observations greatly depends on how the user presents himself to the Biometric system. Following are the possible cases where there could be no observation. (1) User is not using the mouse or not placing his thumb on the fingerprint scanner. (2) User is not presenting a frontal face to the camera.

4.1 ROC Curve Analysis

For assessing the Receiver Operator Characteristic (ROC) our system, we run 6 sets of experiments for each user under the different combinations of legitimate user versus imposter for face and fingerprint modalities.

The area under ROC curve is the reliability measure. From the fused probabilities of the above experiments, we compute ROC's for face verifier, fingerprint verifier and both combined. The ROC areas for fingerprint-only, combined-modality, and face-only verifiers are 0.9995, 0.989, and 0.970, respectively. Thus verification using fingerprint alone is the best, followed by combining the two modalities. Face verification alone is the least reliable.

However, for continuous verification, combining multimodal biometrics is preferred over using just a single modality. The lack of observations from a single modality can be compensated by using a second modality. Also it is more difficult for an imposter to impersonate multiple biometrics.

4.2 Comparing the Fusion Methods

We run four experiments to evaluate how the system behaves when one or both of the biometrics are impersonated. In these we take turns to impersonate each modality one at a time. Because each user presents his biometric in a different way, we cannot average the curves from different users. Figure 5(a) 5(b) 5(c) show five plots each in the following order: individual probabilities, Holistic Fusion, Naive Integration, Modality-first, Temporal-first Integration. In these experiments, $\Delta t = 1.5s$ is used for modality integration, $H = 30s$ for temporal integration and $k = -log(2)/30$ for the decay function. There can be no observation at some time periods. In these situations in order to maintain the system integrity we choose to lock the system. The user has to re-login to regain access. These four setups can be classified into three cases.

Legitimate user using the system. Figure 5(a) shows the biometric observation for 15 minutes. The individual probabilities P_{safe} (5(a)-1) are not consistently high, it occurs in a sporadic manner. This means that any value for the threshold T_{safe} will result in significant False Accept (FAR) and False Reject (FRR) rates. In continuous verification, a False Accept is a security breach, while a False Reject inconveniences the legitimate user, because he must re-authenticate himself. Ideally P_{safe} should not fluctuate, but be equal to 1 as long as observations are available. Of the four fusion methods, Holistic Fusion comes closest to this ideal (5(a)-2). It computes a P_{safe} value close to 1, except for the periods when there are no observations from both modalities (around 300s and 600s). At such times P_{safe} decreases gradually according to the decay function. By comparison, the P_{safe} computed by Naive Integration (5(a)-3) fluctuates wildly, because only a single modality is used any at time. Again, this means no T_{safe} value will make both FRR and FAR small. As for Modality-first (5(a)-4) and Temporal-first (5(a)-5) Integration, the plots are similar. The P_{safe} values are not close to 1. Moreover in the absence of observations P_{safe} drops abruptly to zero resulting in sudden lock outs. From these plots, it is clear that Holistic Fusion is superior to the other fusion methods.

Imposter taking over the system. Figure 5(b) shows the observations when an imposter takes over the system at some time instant (around 38s). The probabilities of individual biometrics (5(b)-1) as well as P_{safe} for all integration methods drop to near zero after the attack. The goal here is to detect the attack as soon as possible so that damage to the system is minimized. Both Holistic Fusion (5(b)-2) and Naive Integration (5(b)-3) detect this situation sooner than the other two methods. However, P_{safe} for Naive Integration does not remain consistently low; it fluctuates widely. This implies that FAR > 0 for most values of T_{safe}. For Modality-first (5(b)-4) and Temporal-first (5(b)-5) Integration, the system takes longer to detect the imposter (when $T_{safe} = 0.5$). Choosing a larger value for T_{safe} can reduce the time to detection, but at the expense of a higher FRR. The best method is Holistic Fusion, which detects the imposter quickly (within 5s in our experiments), and whose P_{safe} remains low after the attack.

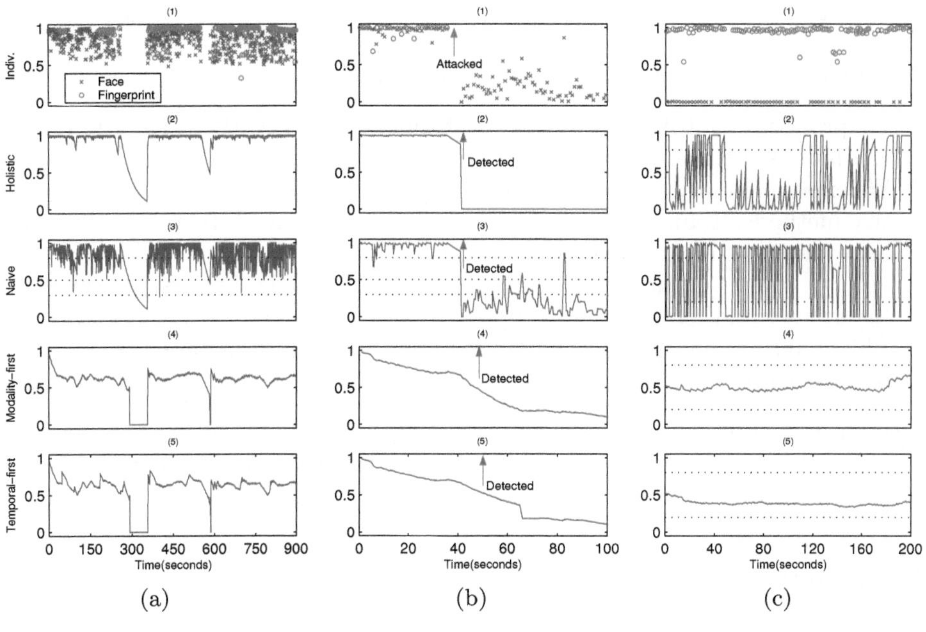

Fig. 5. (a) Legitimate user using the system for 15 minutes. (b) Imposter taking over the system. (c) Partial impersonation: Genuine fingerprint + Fake face. Experiments conducted with Fake fingerprint + Genuine face produced similar results as (c).

Imposter successful in faking one of the biometric (Partial impersonation). Figure 5(c)-1 depicts a situation where the imposter has successfully faked the fingerprint but not face. The individual probabilities contradict each other, and results in wildly fluctuating plots in both Holistic Fusion (5(c)-2) and Naive Integration (5(c)-3). This gives us a way to detect partial impersonation: We may just take two thresholds, one high and one low (say: 0.8 and 0.2) and simply count the number of times within a fixed time interval that P_{safe} jumps between these thresholds. However, comparing Figures 5(c)-3 and 5(a)-3, we see that Naive Integration cannot distinguish between partial impersonation and the legitimate user. Fluctuating P_{safe} values seem to be an inherent property of

Naive Integration. The plots for Modality-first (5(c)-4) and Temporal-first (5(c)-5) Integration are relatively flat, and are in fact similar to those in Figure 5(a) (except when there are completely no biometric observations). Again, this means these two methods cannot distinguish between partial impersonation from legitimate usage. Only Holistic Fusion provides a way to detect partial impersonation that is different from detecting the real user.

What happens if an imposter is careful not to present any observation (neither face nor fingerprint)? In this case, P_{safe} decreases to zero due to the decay function. This is also the situation if the legitimate user has left the console without logging off. In either case, system integrity is ensured.

5 Conclusion

In summary, our work has the following key features:

1. We propose a Holistic Fusion approach that satisfies all the three criteria for continuous verification.
2. We experimentally show that our Holistic Fusion is superior to other alternative methods: Temporal-first, Modality-first and Naive Integration. It is the only method that (a) achieves a low FAR and FRR, (b) detects an attack quickly after it occurs, and (c) is able to detect partial impersonation.
3. In our system, there is only one free parameter k that governs the decay rate. This is intuitively specified by the security administrator based on security requirements.

In the near future, we plan to incorporate keyboard dynamics as another biometric modality. We also plan to make face verification more robust by using incremental training.

References

1. Laurent Joncheray. A Simple Active Attack Against TCP. *Proceedings of the 5th USENIX Security Symposium*, pages 7–19, 1995.
2. J. Kittler, M. Hatef, R. P. W. Duin, and J. Matas. On combining classifiers. *IEEE Transactions on Pattern Analysis and Machine Intelligence*, 20(3):226–239, Mar. 1998.
3. Sandeep Kumar, T. Sim, Rajkumar Janakiraman, and S. Zhang. Using continous biometrics verification to protect interactive login sessions. *To appear in the 21st Annual Computer Security Applications Conference*, 2005.
4. A. Ross and A. K. Jain. Information fusion in biometrics. *Pattern Recognition Letters*, 24(13):2115–2125, 2003.
5. T. Sim, R. Sukthankar, M. Mullin, and S. Baluja. Memory-based Face Recognition for Visitor Identification. In *IEEE International Conference on Automatic Face and Gesture Recognition*, 2000.
6. Paul Viola and Michael Jones. Robust real-time object detection. *International Journal of Computer Vision*, 2002.

Fusion of Face and Iris Features for Multimodal Biometrics

Ching-Han Chen and Chia Te Chu

Institute of Electrical Engineering, I-Shou University, 1, Section 1,
Hsueh-Cheng Rd., Ta-Hsu Hsiang, Kaohsiung County, Taiwan 840, R.O.C
pierre@isu.edu.tw, cld123@giga.net.tw

Abstract. The recognition accuracy of a single biometric authentication system is often much reduced due to the environment, user mode and physiological defects. In this paper, we combine face and iris features for developing a multimode biometric approach, which is able to diminish the drawback of single biometric approach as well as to improve the performance of authentication system. We combine a face database ORL and iris database CASIA to construct a multimodal biometric experimental database with which we validate the proposed approach and evaluate the multimodal biometrics performance. The experimental results reveal the multimodal biometrics verification is much more reliable and precise than single biometric approach.

Keywords: Multimodal biometrics, face, iris, wavelet probabilistic neural network.

1 Introduction

With increasing need for reliable authentication schemes, the demand for high reliable automatic person authentication system is very obvious. Traditional automatic personal identification technologies, which use methods such as Personal Identification Number (PIN), ID card, key, etc., to verify the identity of a person, are no longer considered reliable enough to satisfy the security requirement of person authentication system. Hence biometrics-based person authentication system is gaining more and more attention. Biometrics recognition is the process of automatically differentiating people on the basis of individuality information from their physical or behavioral characteristics like fingerprint, iris, face, voice, and etc.

The biometric recognition can be further divided into two modes: identification and verification. The identification mode is designed for identifying an authorized user when he wants to access a biometric recognition system. The system then attempts to find out whom the biometric feature belongs to, by comparing the query sample with a database of enrolled samples in the hope of finding a match. This is known as a one-to-many comparison. On the other side, the verification mode is a one-to-one comparison in which the recognition system tries to verify an individual's identity. In

D. Zhang and A.K. Jain (Eds.): ICB 2006, LNCS 3832, pp. 571 – 580, 2005.

this case, a query sample is captured and compared with the previously enrolled sample. If the two samples match, the biometric system confirms that the applicant is the one that he claims to be. In the paper, we will only focus on the issue of biometric verification. As we know, the identification problem is a typically binary classification problem, i.e. accept or reject.

If people use only a single biometric authentication system, the results obtained are not always good enough. This is due to the fact that the precision of single biometric system is easily affected by the reliability of the sensor used. Besides, the single biometric system has still some domain-specific limitation. For example, accuracy of face recognition is affected by illumination, pose and facial expression and the voiceprint is affected by environment noise. According to the report of the US Congress [1], approximately 2 percent of the population does not have a legible fingerprint, that is, cannot be enrolled into a fingerprint biometric system.

Many multimodal biometrics methods and strategies have been proposed [2-8]. In these works, the fusion of the various biometric features is used to make the unique recognition decision. Aiming at the same issue, we integrate two biometric recognition systems, such as face and iris. The purpose is to improve overall error rate by utilizing as much information as possible from each biometric modality.

We previously proposed a series of single biometric approach which including the face recognition [9-11], speaker recognition [12], and iris recognition [13] based on wavelet transform feature extraction. The proposed algorithms are very efficient and suitable for real-time system. The probabilistic neural network is adopted as a common classifier in these methods.

In the multimodal biometric system, we select the face and iris features for constructing a high reliable biometric system, because the face recognition is friendly and non-invasive whereas iris recognition is the most accurate biometrics to date among all biometrics systems [14].

2 Face and Iris Feature Extraction

2.1 Face Feature Extraction

In [9-11], we proposed an efficient face feature extraction method. A 2-D face image is transformed into 1-D energy profile signal. The face images as Fig. 1 (a) taken at the Olivetti Research Laboratory (ORL) in Cambridge University, U.K. [9].

Let G be a face image of size 112x92

$$\mathrm{G} = \begin{bmatrix} g_{1x1} & \cdots & g_{1x92} \\ \vdots & \ddots & \vdots \\ g_{112x1} & \cdots & g_{112x92} \end{bmatrix} \tag{1}$$

According to the symmetric property of the face, the horizontal signal can be accumulated as 1-D energy signal as Fig. 1(b).

$$\mathrm{S} = \begin{bmatrix} s_1 \\ \vdots \\ s_{112} \end{bmatrix} \tag{2}$$

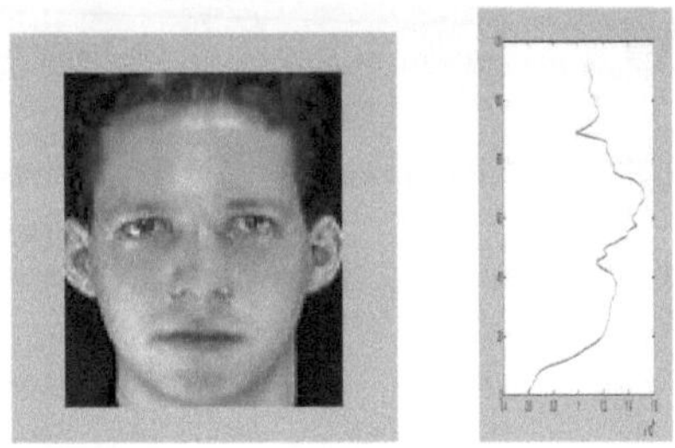

Fig. 1. (a) Facial image (b) 1-D energy signal

2.2 Iris Feature Extraction

In [12], we previously proposed a method for low complexity iris feature extraction. Firstly, in order to reduce the computational complexity, we use 2-D wavelet transform to obtain a lower resolution image and localize the pupil position, as shown in Fig.2(a) and (b). By the center of pupil and the radius of pupil, we can acquire the iris circular rings, as shown in Fig.2(c). The more iris circular rings are acquired, the more information is abundant. Secondly, we segment the iris image into three parts and two parts and adopt Sobel transform to enhance iris texture in each part as a feature vector.

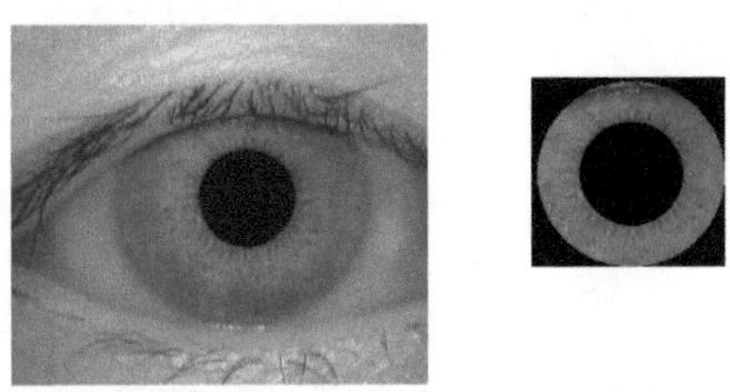

(a) (b)

Fig. 2. Iris location

We extract consecutive circular rings. These circular rings then are stretched horizontally and accumulated, and construct a rectangular-type iris block image, shown as in Fig. 3 (a).

The iris image is divided into three parts, see Fig. 3 (b). The segmented iris image is normalized, see Fig. 3 (d). Subsequently, the Sobel vertical mask

$$\begin{bmatrix} -1 & 0 & 1 \\ -2 & 0 & 2 \\ -1 & 0 & 1 \end{bmatrix} \tag{3}$$

is adopted to enhance iris texture in each segmented part, see Fig. 3(d). The purpose of Sobel operator is for enhancing the high frequency signal.

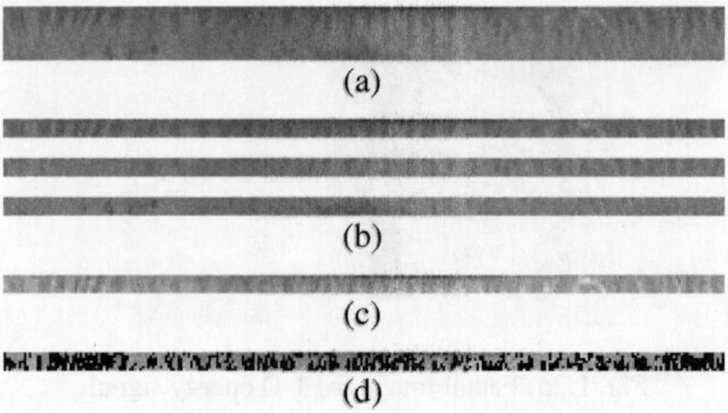

Fig. 3. Iris feature Extraction: (a) stretched iris block image;(b) iris image divided into three parts; (c) normalized iris image; (d) iris image after Sobel transform

The vertical projection is finally used to convert the block image to 1-D energy profile signal. The projected signal is compact and energy-concentrated. We adopt vertical projection to obtain 1-D energy profile signal and to reduce system complexity. In order to concentrate the energy, every row is accumulated as energy signal.

Let G be a segmented iris image of size *mxn*, m is the number of iris circular ring, and n is pixels of each iris circular ring.

$$G = \begin{bmatrix} g_{1xn} & \cdots & g_{1xn} \\ \vdots & \ddots & \vdots \\ g_{mx1} & \cdots & g_{mxn} \end{bmatrix} \quad (4)$$

After vertical projection, the 1-D energy signal Y is obtained.

$$S = \begin{bmatrix} s_1 & \cdots & s_n \end{bmatrix} \quad (5)$$

The m is much smaller than the n. Thus, the information of iris texture after vertical projection is more than the information after horizontal projection.

3 Wavelet Probabilistic Neural Network (WPNN) Classifier

The WPNN classifier has been proposed in [11] which is applied for face recognition. Fig. 4 presents the architecture of a four-layer WPNN, which consists of feature layer, wavelet layer, Gaussian layer and decision layer. In feature layer, $X_1,\ldots,X_N$ are regarded as sets of feature vectors or input data, and N is the dimension of data sets. The wavelet layer is a linear combination of several multidimensional wavelets. Each wavelet neuron is equivalent to a multidimensional wavelet, and the wavelet in the following form

$$\phi_{a,b}(x) = \sqrt{a}\phi \ \frac{x-b}{a} \qquad a,b \in R \quad (6)$$

is a family of function generated from one single function $\phi(x)$ by the scaling and translation, which is localized in both the time space and the frequency space. The $\phi(x)$ is called a mother wavelet and the parameters a and b are named respectively scaling factor and translation factor.

In Gaussian layer, the probability density function of each Gaussian neuron is the following form

$$f_i(X) = \frac{1}{(2\pi)^{\frac{2}{p}}\sigma^p}(\frac{1}{n_i})\sum_{i=1}^{n_a}\exp(\frac{-(X-S_j^i)^2}{2\sigma^2}) \tag{7}$$

where X is the feature vector, p the dimension of training set, n the dimension of input data, j the jth data set, S_j^i the training set and σ the smoothing factor of Gaussian function.

The scaling factor, the translation factor and the smoothing factor are randomly initialized at the beginning and will be trained by PSO algorithm. Once the training is accomplished, the architecture of WPNN and the parameters are fixed for further verification.

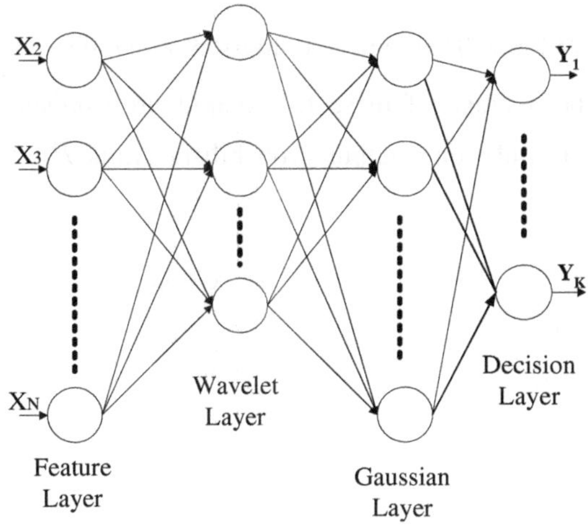

Fig. 4. Wavelet Probabilistic Neural Network

3.1 Learning Algorithm

The Particle Swarm Optimization(PSO) is used for training single neuron to optimize WPNN model. PSO is a new bio-inspired optimization method developed by Kenney and Eberhart [15]. The basic algorithm involves the start from a population of distributed individuals, named particles, which tend to move toward the best solution in the search space. The particles will remember the individual best solution encountered and the swarm population's best solution. At each iteration, every particle adjusts its velocity vector, based on its momentum and the influence of both its individual best solution and the swarm population's best solution.

At time unit t, the position of *i*th particle x_i, $i = 1,2,\ldots,M$, (M is the number of particles) moves by adding a velocity vector v_i. v_i is the function of the best position p_i found by that particle, and of the best position g found so far among all particles of the swarm. The movement can be formulated as:

$$v_i = w(t)v_i(t-1) + c_1u_1(p - x_i(t-1)) + c_2u_2(g - x_i(t-1)) \tag{7}$$

$$x_i(t-1) = x_i(t) + v_i(t) \tag{8}$$

Where $w(t)$ is the inertia weight, c the acceleration constants, and $\mu \in (0,1)$ the uniformly distributed random numbers.

We encode the wavelet neuron by scaling factor and translation factor of wavelet neuron, and Gaussian neuron by smoothing factor. PSO, in offline mode, searches the best set of factors in the three dimensional space.

3.2 Decision Rule

In decision layer of WPNN, There are five inferred probabilistic values $P_{1,} P_2, \ldots, P_5$ for iris features. The average of these five output probabilistic values is P_I . For face features, there is only one output probabilistic value P_f . We take the linear

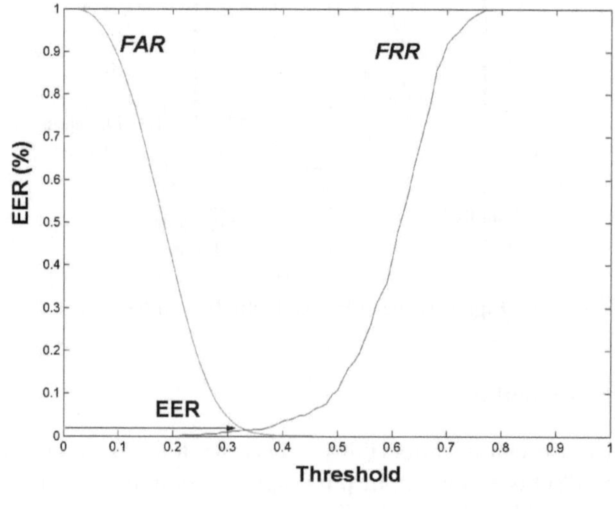

Fig. 5. The curve of FAR-FRR

combination of these two inference probability, that is the resulting output P_{av} is the average of P_I and P_f. The fig 5. shows the false rejection ratios (FRR) and false accept ratios (FAR) obtained by adjusting the threshold of P_{av}. The horizontal axis shows the decision threshold of P_{av} for the multimodal biometrics recognition, and the vertical axis shows each false rate. When the output of P_{av} for an unregistered sample was lower than the decision threshold, false accepted occurred. The FAR was calculated by counting the trails of false acceptance. On the other hand, when the output of P_{av} for a resisted sample was higher than the decision threshold, the registered sample was wrong rejected. The FRR was calculated by counting the trials of false rejection.

4 Experiment and Results

As the adopted database contains face and iris databases, the face database is from two databases: ORL face database, IIS face database, and the iris database is from CASIA iris database. For the verification experiments, the experiments are divided into two sets. The first set contains 40 subjects from ORL face database and CASIA iris database. The second set contains 100 subjects from IIS face database and CASIA iris database. They are well-known public domain face and iris databases. The ORL database contains 40 subjects and 400 images. The IIS database contains 100 subjects and 3000 images. The CASIA iris database contains 108 subjects and 756 images. The multimodal biometrics recognition system is evaluated in the two sets.

4.1 Evaluated in ORL Face Database and CASIA Iris Database

In first set, the multimodal biometric recognition system is evaluated on the ORL face database and CASIA iris database. The face images are sampled from 40 subjects, each subject having 10 images with varying lighting, facial expressions (open / closed eyes, smiling / nonsmiling), facial details (glasses / no glasses) and head pose (tilting and rotation up to 20 degrees). The size of each image is 112x92. For each subject, the five images are randomly sampled as train sample and the remaining five images as test samples. The CASIA iris database contains 756 iris images acquired of 108 subjects (7 images per subject). 40 subjects are randomly selected from CASIA iris database. For each subject, the three images are randomly sampled as train sample and the remaining four images as test samples. Each subject in CASIA iris database is randomly paired with each subject in ORL face database. Such procedures are carried out 100 times. Fig. 7 shows the ROC curve on the three modalities and the experimental results are reported in Table 1.

From the results shown in Fig. 7 and Table 1, they show the recognition performance of different modalities. Fig 7 shows the ROC curves of the three modalities: face, iris, integrated face and iris. From the curves, we can see that the multimodal biometrics recognition system achieves better performance than the single face or iris modalities. In Table 1, the multimodal biometrics recognition system has a corresponding best EER of 0.00% much better than the other two modalities. That is, the multimodal biometrics system is more reliable than single biometrics system.

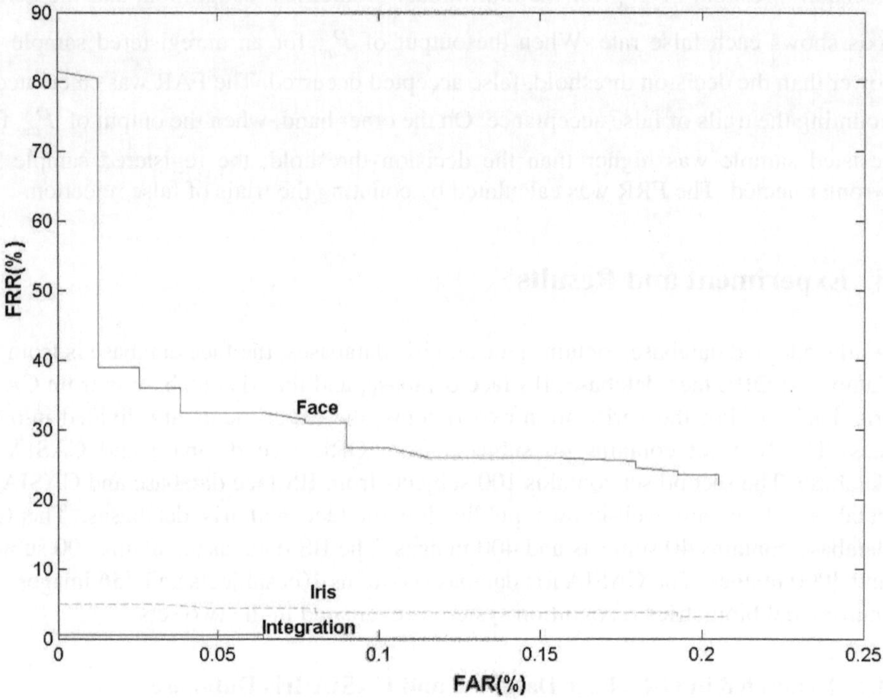

Fig. 7. The ROC curve in different modalities

Table 1. Recognition performance of comparison with different modalities

	ORL Face	Iris	Integration
Best EER	1.76%	0.02%	0.00%
Average EER	3.83%	1.25%	0.33%

4.2 Evaluated in IIS Face Database and CASIA Iris Database

In second set, the multimodal biometric recognition system is evaluated on the IIS face database and CASIA iris database. The face images are sampled from 100 subjects, each subject having 30 images with varying viewpoints and expressions. The size of each image is 175x155. For each subject, the six images are randomly sampled as train sample and the remaining twenty-four images as test samples. The algorithm is evaluated in the IIS face database. In CASIA iris database, there are 100 subjects randomly selected f. For each subject, the three images are randomly sampled as train sample and the remaining four images as test samples. Each subject in CASIA iris database is randomly paired with each subject in IIS face database. Such procedures are carried out 100 times and the experimental results are reported in Table 2.

Table 2. Recognition performance of comparison with different modalities

	IIS Face	Iris	Integration
Best EER	3.59%	0.7%	0.01%
Average EER	4.77%	1.87%	0.64%

Looking at the results shown in Table 2, we can find the multimodal biometric system achieves the best recognition performance among three modalities. The results show further the multimodal biometric modality is better than single biometrics modality. The best EER is 0.01% and the average EER is 0.64 in the multimodal biometric modality.

5 Conclusions

A multimodal biometric system integrating face and iris features is proposed. Firstly, the features of face and iris are separately extracted, and feed into WPNN classifier to make the multimodal decision. We combine a face database ORL and iris database CASIA to construct a multimodal biometric experimental database with which we validate the proposed approach and evaluate the multimodal biometrics performance. The experimental results reveal the multimodal biometrics verification is much more reliable and precise than single biometric approach.

References

[1] Robert Snelick, Mike Indovina, James Yen, Alan Mink, "Multimodal Biometrics: Issues in Design and Testing", *ICMI'03,* pp.68-72, British Columbia, Canada, Nov 5-7. 2003.

[2] A. Ross and J. Z. Qian, "Information fusion in biometrics," 'in *Proc. 3rd International Conference on Audio- and Video-Based Biometric Person Authentication, Halmstad, Sweden,* pp. 354-359, June 2001.

[3] A. Ross, A. Jain, "Information fusion in biometrics", *Pattern Recognition Letters*, vol.24, pp.2115–2125 , 2003.

[4] Andrew L. Rukhin, Igor Malioutov, "Fusion of Biometric Algorithm in the Recognition Problem," *Pattern Recogition Letters*, pp. 299-314, 2001.

[5] R.W. Frischholz and U.Dieckmann,"Bioid: A Multimodal Biometric Identification System", *IEEE Computer*, vol. 33, pp.64–68, Feb 2000.

[6] C.Sanderson and K.K. Paliwal,"Information Fusion and Person Verification Using Speech & Face Information", *IDIAP, Martigny, Research Report*, pp.02-33, 2002.

[7] Vassilios Chatzis, Adrian G..Bors, and Ioannis Pitas, "Multimodal Decision-Level Fusion for Person Authentication", *IEEE Trans. Systems. Man Cybernetics.,* vol. 29, no. 6, pp.674-680, April. 1999.
[8] Y.Wang, T.Tan and A.K. Jain, "Combining Face and Iris Biometrics for Identity Verification", *Proc. Of 4th Int'l Conf. on Audio- and Video-Based Biometric Person Authentication (AVBPA).,* pp.805-813, Guildford, UK, June 9-11. 2003.
[9] Chia-Te CHU,Ching-Han CHEN, "The Application of Face Authentication System for Internet Security Using Object-Oriented Technology", *Journal of Internet Technology, Special Issue on Object-Oriented Technology and Applications on Internet.* (accepted for Oct. 2005 publication)
[10] Ching-Han CHEN, Chia-Te CHU, " Combining Multiple Features for High Performance Face Recognition System ", *2004 International Computer Symposium(ICS2004),* pp.387-392,Taipei, Dec 2004.
[11] Ching-Han CHEN, Chia-Te CHU, "Real-Timee Face Recognition Using Wavelet Probabilistic Neural Network", *Journal of Imaging Science and Technology.* (accepted)
[12] Ching-Han CHEN, Chia-Te CHU, " High Efficiency Iris Feature Extraction Based on 1-D Wavelet Transform ", *2005 Design automation and test in Europe (DATE2005),* Munich,Germany , March, 2005.
[13] Ching-Han CHEN, Chia-Te CHU, " An High Efficiency Feature Extraction Based on Wavelet Transform for Speaker Recognition ", *2004 International Computer Symposium(ICS2004),* pp. 93-98,Taipei, Dec 2004.
[14] J. Daugman, "How Iris Recognition Works", *Proceedings of 2002 International Conference on Image Processing,* Vol. 1, 2002.
[15] J. Kennedy, R. Eberhart, "Particle Swarm Optimization," *Proc. of IEEE international Conference on Neural Networks (ICNN),* vol.IV, pp.1942-1948, Perth, Australia, 1995.

The Role of Statistical Models in Biometric Authentication

Sinjini Mitra[1], Marios Savvides[2], and Anthony Brockwell[1]

[1] Department of Statistics,
Carnegie Mellon University, Pittsburgh, PA 15213
{smitra, abrock}@stat.cmu.edu
[2] Electrical and Computer Engineering Department,
Carnegie Mellon University, Pittsburgh, PA 15213
msavvid@cs.cmu.edu

Abstract. The current paper demonstrates the role of statistical models in authentication tasks – both in system development and in performance evaluation. We first introduce a model-based face authentication system based on the Fourier domain phase using Gaussian Mixture Models (GMM) which yields verification error rates as low as 0.3% on a face database of 65 individuals with extreme illumination variations. We then present a statistical framework for predicting authentication error rates for future populations in a rigorous way. This is in contrast to most evaluation protocols used today that are based on observational studies and valid only for the databases at hand. Applications establish that our model-based approach has better predictive performance than an existing state-of-the-art authentication technique.

1 Introduction

There are two broad approaches to devising authentication systems: (1) feature-based, and (2) model-based. For facial biometrics, feature-based methods use individualized facial characteristics such as distance between eyes, nose, mouth, and their shapes and sizes as the matching criteria. The model-based systems, on the other hand, use a statistical model to represent the pattern of some facial features (often, the ones mentioned above), and then some characteristics of the fitted model, such as, parameters or likelihood, are used as the matching criteria. Feature-based methods are simple and less rigorous usually than the model-based ones and often involve parameters that need to be chosen by extensive experimentation for a particular database.

Although the importance of models is well-understood and has been exploited quite extensively in image re-construction and segmentation, its use it devising face authentication systems has been relatively limited. Some common face models include Gaussian ([1]), Markov ([2], [3]). One class of flexible statistical models is the *Gaussian Mixture Models* (GMM; [4]), which represents complex distributions through an appropriate choice of its components to capture accurately the local areas of support of the true distribution. Apart from statistical

D. Zhang and A.K. Jain (Eds.): ICB 2006, LNCS 3832, pp. 581–588, 2005.

applications, GMMs have also been used in computer vision for modeling the shape and texture of face images ([5]).

Most existing face recognition systems are based on spatial image intensities. Recently much research effort has focused on the frequency domain as well, whose useful properties have been successfully exploited in many signal processing applications ([6]). The frequency domain representation of an image consists of two components, the *magnitude* and *phase*. In 2D images particularly, the phase captures more of the image intelligibility than magnitude and hence is very significant for performing image reconstruction ([7]). [8] showed that correlation filters built in the frequency domain can be used for efficient face verification. Recently, [9] proposed correlation filters based only on phase which performed as well as the original filters, and [10] demonstrated that performing PCA in the frequency domain using only the phase spectrum outperforms spatial domain PCA and also has illumination tolerance. However, no face models have been developed, as per the authors' knowledge, in the frequency domain.

Another important component of biometric authentication is performance evaluation. Most face authentication systems are tested on small to moderately-sized databases, which is not adequate to address bigger questions about the expected performance on large-scale databases to which the system has not been previously exposed to. For example, say a certain system yields a false alarm rate of 1%; this implies that a database of size 1,000,000 will produce 10,000 false alarms and this is quite undesirable in practice. It is known that there are about 500 million border crossings per year in the United States (one-way only), so this system will result in a lot of innocent travelers being unnecessarily harassed and extra overhead (personnel, time) required to attend to them. Large-scale evaluation protocols used today are based on observational studies (like FRVT, 2002, 2004) but from a statistician's perspective, these are at most empirical in nature – there is no statistical basis (e.g.,modeling) and scope for valid inference. Our goal in this paper is to propose a framework for performing such large-scale inference based on statistical models which has the potential to be more reliable.

The paper is organized as follows. Section 2 gives a brief description of the database used and Section 3 presents our GMM-based authentication scheme. The statistical framework for performance evaluation is introduced in Section 4 and its application to our model-based scheme and comparison with an existing non model-based method appears in Section 5. Section 6 contains a discussion.

2 Data

We use a subset of the "CMU-PIE Database" ([11]) which contains frontal images of 65 people under 21 different illumination conditions from balanced to shadows. Images belonging to one person from the database appears in Figure 1.

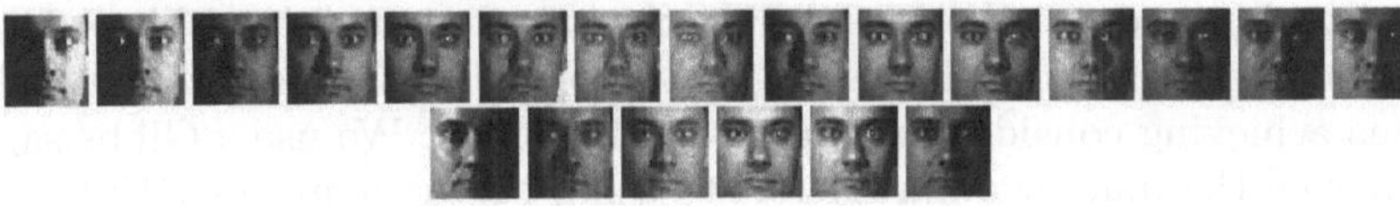

Fig. 1. Sample images of a person from the CMU-PIE database

3 Model-Based System: GMM Based on Phase

Mixture models provide a semiparametric framework for modeling unknown distributional shapes and can handle situations where a single parametric family is unable to provide a satisfactory model for local variations in the observed data.

Let $(\mathbf{Y_1}, \ldots, \mathbf{Y_n})$ be a random sample of size n where $\mathbf{Y_j}$ is a p-dimensional random vector with probability distribution $f(\mathbf{y_j})$ on $\mathcal{R}^p$, and let $\boldsymbol{\theta}$ denote a vector of the model parameters to be estimated. A g-component mixture model can be written in parametric form as:

$$f(\mathbf{y_j}; \boldsymbol{\Psi}) = \sum_{i=1}^{g} \pi_i f_i(\mathbf{y_j}, \boldsymbol{\theta_i}), \tag{1}$$

where $\boldsymbol{\Psi} = (\pi_1, \ldots, \pi_g, \boldsymbol{\theta_1}, \ldots, \boldsymbol{\theta_g})^T$ contains the unknown parameters. Here, $\boldsymbol{\theta_i}$ represents the model parameters for the ith mixture component and $\boldsymbol{\pi} = (\pi_1, \ldots, \pi_g)^T$ is the vector of the mixing proportions with $\sum_{i=1}^{g} \pi_i = 1$. In case of Gaussian mixture models, the mixture components are multivariate Gaussian given by:

$$\begin{aligned} f(\mathbf{y_j}; \boldsymbol{\theta_i}) &= \phi(\mathbf{y_j}; \boldsymbol{\mu_i}, \boldsymbol{\Sigma_i}) \\ &= (2\pi)^{-1} |\boldsymbol{\Sigma_i}|^{-\frac{1}{2}} \exp\{-\frac{1}{2}(\mathbf{y_j} - \boldsymbol{\mu_i})^T \boldsymbol{\Sigma_i}^{-1} (\mathbf{y_j} - \boldsymbol{\mu_i})\} \end{aligned} \tag{2}$$

so that $\boldsymbol{\theta_i} = (\boldsymbol{\mu_i}, \boldsymbol{\Sigma_i}),\ i = 1, \ldots, g$ and the mixture model has the form:

$$f(\mathbf{y_j}; \boldsymbol{\Psi}) = \sum_{i=1}^{g} \pi_i \phi(\mathbf{y_j}; \boldsymbol{\mu_i}, \boldsymbol{\Sigma_i}). \tag{3}$$

Despite the significance of phase in face identification tasks, modeling the phase angle poses several difficulties such as, the "wrapping around" property (it lies between $-\pi$ and π) and its sensitivity to distortions (such as illuminations). This prompted us to use the real and imaginary parts of the frequencies in the "phase-only" spectrum as an alternative representation of the phase. This is a simple yet effective way of modeling phase since it does not suffer from the difficulties associated with direct phase modeling.

Let $R_{s,t}^{k,j}$ and $I_{s,t}^{k,j}$ respectively denote the real and the imaginary parts at frequency (s, t) of the phase spectrum of the jth image from the kth person, $s, t = 1, 2, \ldots$, $k = 1, \ldots, 65$, $j = 1, \ldots, 21$. We model $(R_{s,t}^{k,j}, I_{s,t}^{k,j})$, $j = 1, \ldots, 21$ as a mixture of bivariate Gaussians whose density is given by Eqn.(3) for each

(s,t) and each person k. We model only the low frequencies within a 50×50 grid around the spectrum origin since they capture all the image identifiability ([6]), thus achieving considerable dimension reduction. We use a Gibbs sampler for estimating the unknown parameters $\boldsymbol{\Psi}$ using posterior means ([12]).

3.1 Classification and Verification

Classification of a new test image is done with the help of a MAP (maximum *a posteriori*) estimate based on the posterior likelihood of the data. For a new observation $Y = (R^j, I^j)$ extracted from the phase spectrum of a new image, it is assigned to class C if

$$C = \arg\max_k f(k|Y), \tag{4}$$

where $f(k|Y) \propto g(Y|k)p(k)$ with $g(Y|k) = \Pi_s \Pi_t f_k^{s,t}(\mathbf{y_j}; \boldsymbol{\Psi})$ assuming independence among frequencies and uniform priors over all the people and where $f_k^{s,t}(\mathbf{y_j}; \boldsymbol{\Psi})$ denotes the likelihood of Y for person k at frequency (s,t).

Table 1 shows the classification results for our database using different number of training images and $g = 2$. The training set in each case is randomly selected and the rest are used for testing. This selection of the training set is repeated 20 times (in order to remove selection bias) and the final errors are obtained by averaging over those from the 20 iterations. The results are fairly good,

Table 1. Identification error rates for GMM-based system ($g = 2$)

# of Training images	# of test images	Error Rates	Std. dev. over 20 repetitions
15	6	1.25%	0.69%
10	11	2.25%	1.12%
6	15	9.67%	2.89%

which demonstrates that our model is able to capture the illumination variation suitably. However, an adequate number of training images is required for the efficient estimation of the parameters; in our case, 10 is the optimal number of training images required. The associated standard errors in each case also prove the consistency of the results. Increasing the number of mixture components ($g = 3$ and $g = 4$) do not improve results significantly; hence a 2-component GMM represents the best parsimonious model in this case.

Verification is performed by imposing a threshold on the log-likelihood of the test images. Satisfactory results are achieved with the optimal model ($g = 2$ and 10 training images) that yields an Equal Error Rate (EER) of 0.3% at a threshold value of -1700 (ROC curve not included for space constraints).

4 Large-Scale Performance: Random Effects Model

Traditional performance evaluation tools like linear regression models, ANOVA ([13]) are *fixed effects models* and the inference from them cannot be generalized beyond the people in the database at hand. This difficulty can be obviated by the use of *random effects models* ([12]) which provide a flexible framework for extending inference to a bigger population by assuming that the particular subset of subjects in the current database is a random sample from a bigger population. The regression framework allows the inclusion of any number of potential covariates representing image properties (distortions, clarity) and system design parameters which may influence the system performance, and quantitatively determine which of these factors significantly affect performance in a general population and to what extent. They are thus crucial for large-scale inference and also take into account the heterogeneity across individuals in their regression coefficients with the help of a probability distribution. Some particular questions of interest in this context are:

- What are the effects of certain image properties and system parameters on the score distribution in the population?
- What is the predicted score distribution for authentics and impostors and error rates for an unknown large population?

4.1 The Model Framework

Let Y_{ij} denote the outcome for the jth observation on the ith subject in the database, while $x_{ij}^{(m)}$ denotes the corresponding value for covariate m. We adopt the following hierarchical model:

$$Y_{ij} \overset{ind.}{\sim} N(\alpha_i + \sum_{m=1}^{M} \beta_i^m x_{ij}^{(m)}, \sigma^2), \quad i = 1, \ldots, k, \quad j = 1, \ldots, n_i, \tag{5}$$

where M is the total number of covariates in the study. We assume that the slope-intercept vectors for each individual are drawn from a common population:

$$\boldsymbol{\theta_i} \equiv (\alpha_i, \beta_i^1, \ldots, \beta_i^M)^T \sim MVN(\boldsymbol{\theta_0} \equiv (\alpha_0, \beta_0^1, \ldots, \beta_0^M)^T, \Sigma), \ i = 1, \ldots, k \tag{6}$$

We then select conjugate hyperpriors for the other parameters as follows:

$$\sigma^2 \sim IG(a, b), \quad \boldsymbol{\theta_0} \sim N(\boldsymbol{\eta}, C), \quad \Sigma^{-1} \sim Wishart((\rho R)^{-1}, \rho), \tag{7}$$

where R is a matrix and $\rho \geq 2$ is a scalar "degrees of freedom" parameter. The hyperparameters in the model $a, b, \boldsymbol{\eta}, C, \rho, R$ are assumed known. The unknown parameters are then estimated by using a Gibbs sampler by simulating from the respective full conditionals, assumed independence, which are:

$$\boldsymbol{\theta_i} | \mathbf{y}, \boldsymbol{\theta_0}, \Sigma^{-1}, \sigma^2 \sim N(D_i(\frac{1}{\sigma^2} X_i^T \mathbf{y_i} + \Sigma^{-1}\boldsymbol{\theta_0}), D_i), \quad i = 1, \ldots, k, \quad \text{where}$$

$$D_i^{-1} = \frac{1}{\sigma^2} X_i^T X_i + \Sigma^{-1}, \quad \mathbf{y_i}, X_i = \text{variables for person i.}$$

$$\boldsymbol{\theta_0} | \mathbf{y}, \boldsymbol{\theta_i}, \Sigma^{-1}, \sigma^2 \sim N(V(k\Sigma^{-1}\bar{\boldsymbol{\theta}} + C^{-1}\boldsymbol{\eta}), V), \text{ where } V = (k\Sigma^{-1} + C^{-1})^{-1}, \ \bar{\boldsymbol{\theta}} = \frac{1}{k}\sum_{i=1}^{k} \boldsymbol{\theta_i}.$$

$$\Sigma^{-1}|\mathbf{y},\boldsymbol{\theta_i},\boldsymbol{\theta_0},\sigma^2 \sim Wishart\left(\left[\sum_{i=1}^{k}(\boldsymbol{\theta_i}-\boldsymbol{\theta_0})(\boldsymbol{\theta_i}-\boldsymbol{\theta_0})^T+\rho R\right]^{-1}, k+\rho\right).$$

$$\sigma^2|\mathbf{y},\boldsymbol{\theta_i},\boldsymbol{\theta_0},\Sigma^{-1} \sim IG\left(\frac{n}{2}+a,\frac{1}{2}\sum_{i=1}^{k}(\mathbf{y_i}-X_i\boldsymbol{\theta_i})^T(\mathbf{y_i}-X_i\boldsymbol{\theta_i})+b\right),\ n=\sum_{i=1}^{k}n_{i\cdot}\quad(8)$$

Inference from this model is based on the marginal posteriors for the population parameters $\boldsymbol{\theta_0}=(\alpha_0,\beta_0^1,\ldots,\beta_0^M)$ and posterior predictive distributions $p(y_{ij}|\mathbf{y})$. This is obtained with Gaussian kernel density estimates for new data generated using Equation 5 with the post-convergence values of the parameters.

5 Application

We apply the random effects methodology to the GMM-based system and to an existing system called the *Minimum Energy Correlation Filter* (MACE; [8]) using their authentication results on the PIE database. The response variable in each case is the authentication score – log-likelihood for GMM and log(PSR) for MACE (PSR = Peak-to-Sidelobe Ratio; [8]). We use a covariate denoting authenticity for both systems, but add one representing the illumination effect in MACE (since GMM modeled this). We again use a Gibbs sampler for parameter estimation and posterior means as the estimates. Table 2 shows the estimates of the population parameter $\boldsymbol{\theta_0}$ which quantitatively show the effect of the covariates on the scores. We see that authenticity has a significant effect on authentication score for both systems while illumination has no effect on PSR.

The posterior predictive distributions of the authentication scores for the two systems are shown in Figure 2. We generated 1000 values for each system and 500 each for the authentics and the impostors. As can be seen, there exists a clear separation among the predicted values of authentic and impostor people; in fact, the distributions of the score statistics appear to be a mixture of two distinct distributions. The amount of overlap in the tails of the authentic and impostor distributions indicates the chances of false alarm and this clearly shows that the GMM method has a much reduced risk than the MACE system.

We next estimate the predicted FAR and FRR using closed-form expressions based on Gaussian kernels as functions of the authentication threshold τ. Let T denote the PSR values, and $f_A(\cdot)$ and $g_I(\cdot)$ respectively be the posterior

Table 2. Estimates of $\boldsymbol{\theta_0}$ and 95% predictive intervals for the two systems

System	Parameter	Estimate	Lower 95% CI	Upper 95% CI
MACE	α	1.9737	0.7504	3.1971
	β_0	1.4634	1.2874	1.6395
	γ_0	-0.0184	-0.1965	0.1581
GMM-based	α_0	-1694.4	-1698.0	-1690.8
	β_0	81.7	76.7	86.7

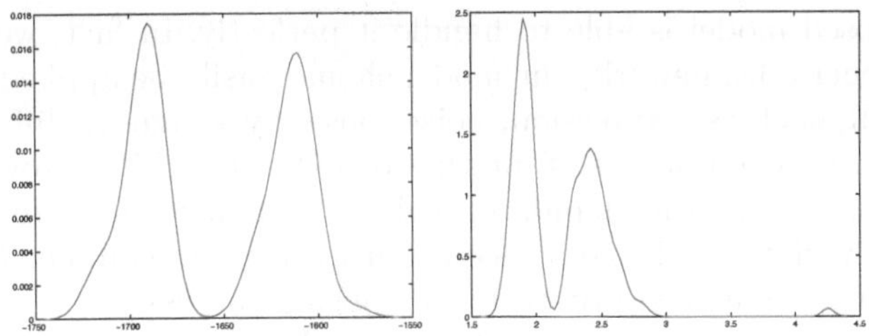

(a) GMM (log-likelihood) (b) MACE (log(PSR))

Fig. 2. Predictive posterior distribution of scores for the two systems

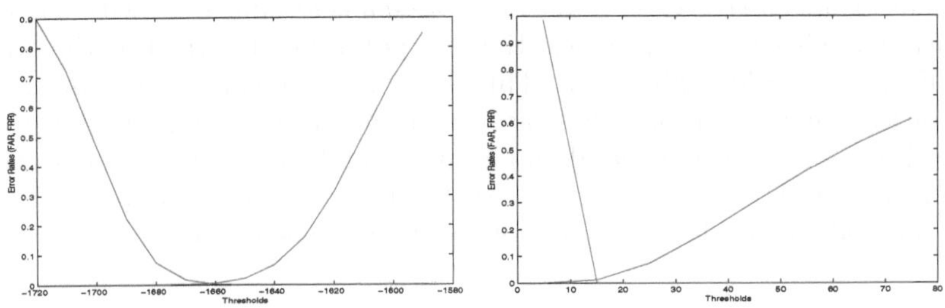

(a) GMM (EER = 0.8%) (b) MACE (EER = 1.5%)

Fig. 3. Predicted false alarm rates for the two systems. The decreasing curve is the FAR, and the increasing one is the FRR.

predictive distributions of $\log(T)$ for the authentics and the impostors. If τ is the given threshold for authentication, the FAR and FRR are defined as:

$$FRR = P(T \leq \tau|\text{Authentic}) = P(\log(T) \leq \log(\tau)|\text{Authentic}) = \int_{-\infty}^{\tau} f_A(x)dx$$
$$FAR = P(T > \tau|\text{Impostor}) = P(\log(T) > \log(\tau)|\text{Impostor}) = \int_{\tau}^{\infty} g_I(y)d(y9)$$

Now if f_A and g_I be Gaussian with means (μ_A, ν_I) and variances (σ_A^2, η_I^2), these can be written in a in terms of Φ (distribution function of standard normal) as:

$$FRR = \Phi\left(\frac{\log(\tau) - \mu_A}{\sigma_A}\right), \qquad FAR = 1 - \Phi\left(\frac{\log(\tau) - \nu_I}{\eta_I}\right). \tag{8}$$

The resulting FAR and FRR for the two systems are shown in Figure 3. The predicted EERs are 0.8% for the GMM system at a threshold log-likelihood value of -1650 and 1.5% for MACE at a threshold PSR value of 15.

6 Discussion

This paper presented a face authentication scheme based on phase and GMM. Although the importance of phase is well-known, this fact had not been utilized in building model-based classification techniques. This is partially because modeling phase variations is a challenging task and our results show convincingly that the proposed model is able to handle it perfectly. In fact, we believe that owing to its general framework, our model should easily be applied to other distortions as well, such as, expression, noise, pose, by assigning different types of images to different components of mixture distributions. This proves the practical utility of this method for handling real life databases that are often subject to extraneous variations. In conclusion, harnessing the combined potential of GMM and phase has indeed proved to be a grand success.

We then proposed a novel statistical framework based on random effects model to predict the performance of a biometric system on unknown large databases. We applied this to the MACE system and our GMM based system, and established that the latter has a superior performance in terms of predictive performance. Development of such a rigorous evaluation protocol is feasible only with the help of statistical models which helps assess the true potential of authentication systems in handling real-world applications. This is the first of its kind and hence replaces the empirical and naive approaches based on observational studies that were being used until now. It is fairly general and easily extends to other biometrics as well. In conclusion, both our techniques have established the significant role played by statistical modeling tools in the technology of biometric authentication.

References

1. Turk, M.A., Pentland, A.P.: Face recognition using eigenfaces. In Proceedings of CVPR (1991)
2. Yuille, A. (1991): Deformable templates for face recognition. Journal of Cognitive Neuroscience **3** (1991)
3. Liu, C., Zhu, S.C., Shum, H.Y.: Learning inhomogeneous gibbs model of faces by minimax entropy. In Proceedings of ICCV (2001) 281–287
4. McLachlan, G., Peel, D.: Finite Mixture Models. John Wiley and Sons (2000)
5. Zhu, S., Wu, Y., Mumford, D.: Minimax entropy principle and its application to texture modeling. Neural Computation **9** (1997)
6. Oppenheim, A.V., Schafer, R.W.: Discrete-time Signal Processing. Prentice Hall, NJ (1989)
7. Hayes, M.H.: The reconstruction of a multidimensional sequence from the phase or magnitude of its fourier transform. ASSP **30** (1982) 140–154
8. Savvides, M., Vijaya Kumar, B.V.K., Khosla, P.: Face verification using correlation filters. In: 3rd IEEE Automatic Identification Advanced Technologies, Tarrytown, NY (2002) 56–61
9. Savvides, M., Kumar, B.V.K.: Eigenphases vs.eigenfaces. In Proceedings of ICPR (2004)
10. Savvides, M., Kumar, B.V.K., Khosla, P.K.: Corefaces - robust shift invariant PCA based correlation filter for illumination tolerant face recognition. CVPR (2004)
11. Sim, T., Baker, S., Bsat, M.: The CMU pose, illumination, and expression (PIE) database. In: Proceedings of the 5th International Conference on Automatic Face and Gesture Recognition. (2002)
12. Gelfand, A.E., Hills, S.E., Racine-Poon, A., Smith,A.F.M.: Illustration of bayesian inference in normal data models using gibbs sampling. Journal of the American Statistical Association **85** (1990) 972–985
13. Weisberg, S.: Applied Linear Regression. Wiley (1985)

Technology Evaluations on the TH-FACE Recognition System

Congcong Li, Guangda Su, Kai Meng, and Jun Zhou

The State Key Laboratory of Intelligent Technology and System,
Electronic Engineering Department, Tsinghua University, Beijing 100084, China
lcc@mails.tsinghua.edu.cn

Abstract. For biometric person authentication, evaluations on a biometric system are very essential parts of the entire process. This paper presents the technology evaluations on the TH-FACE recognition system. The main objectives of the evaluations are to 1) test the performance of the TH-FACE recognition system objectively; 2) provide a method to design and organize a database for evaluations; 3) identify the advantage and weakness for the TH-FACE recognition system. Particular description of the test database used in the evaluations is given in this paper. The database contains different subsets which are sorted by different poses, illuminations, ages, accessory, etc. Results and analysis on the entire performances of the TH-FACE recognition system would be also presented.

1 Introduction

Nowadays Biometric authentication has become one of the most active research areas in the world, as part of which, face recognition technology has made a large development. Therefore, How to evaluate the performance of the technology level among different systems or algorithms and how to test the adjustability to different conditions such as various poses, illuminations, ages, etc. have become urgent and difficult problems in front of researchers on this field. Moreover, evaluations can help customers to understand, to adopt and to identify a new technology [9].

There are already some famous evaluations on face recognition on the world, for example, the FERET [2] and the FRVT [1]. FRVT 2002 computed performance statistics on an extremely large data set; however, the images it used have not been distributed to the public. So FERET, which has publicly distributed its database containing fewer images than those in FRVT, has now been the standard testing set of international face recognition community.

Despite its success in the evaluations of face recognition algorithms, the FERET database has limitations in the relatively simple and unsystematically controlled variations of face images for research purposes [4]. Considering these limitations and aiming at the TH-FACE recognition system which primarily points to the oriental, we design the TH test database for the evaluations. The advantage of using such a database in the evaluations is that the results would more access the performance of the very system in real applications. Meanwhile, the technology evaluations on the

D. Zhang and A.K. Jain (Eds.): ICB 2006, LNCS 3832, pp. 589 – 597, 2005.

TH-FACE recognition system cover different aspects on the system, including not only the recognition algorithms, but also the preprocessing methods.

The rest of the paper is organized as follows: Section 2 addresses the design of the evaluation including the contents of the TH test database and the subsets composing according to our design principles; Section 3 introduces briefly the TH-FACE recognition system. Section 4 presents the results and analysis of the evaluations on the TH-FACE recognition system. Finally, the conclusion is given in section 5.

2 Design of the Evaluation

In this section, we would talk about the composition of the database used for evaluation and the design principles of the evaluations.

2.1 Composition of the Database

The TH test database contains 19,289 images of 750 individuals (394 males and 356 females) with controlled Pose, Expression, Glasses, Background and Lighting variations. All the images in the TH test database can be divided into two main subsets: the frontal subset and the pose subset.

- In the frontal subset, subjects in all the images are looking right into the camera that captures the images. Among the frontal images, each of the 750 subjects has a frontal image with normal expression, no accessory, standard lighting and plaint background. Every subject has another two images which were respectively captured one year and two year apart from the time when the normal frontal images mentioned above were captured. Some subjects may have images captured further before. There is also one image with complicated background outside the room. 554 subjects have images wearing glasses. All of the 750 subjects have images with expressions of a little smile.
- In the pose subset, every subject has 13 (9+4) images with different poses. Among the 13 images of every subject, 9 images present that the subject is looking to the left, the central and the right, with a yaw angle from -40° to +40°, each about 10°apart, supposed the central is 0° and the counter-clockwise is positive. The 4 remaining images and the frontal image of the 9 images mentioned just now, cover different pitching angles from -20° to +20°, each about 10°apart, supposed the central is 0° and the counter-clockwise is positive.

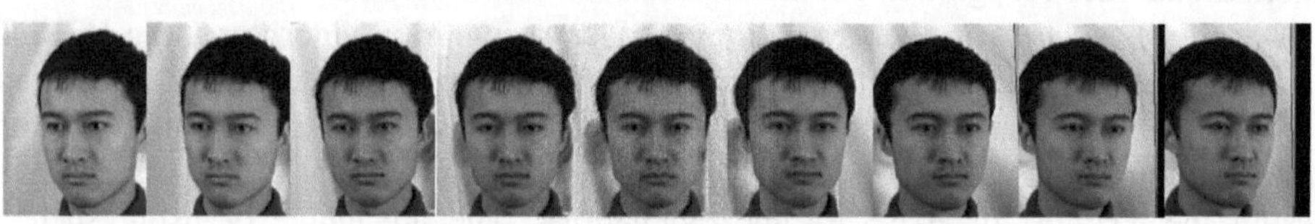

Fig. 1. Examples of left-right poses

The content of the TH test database is summarized in Table 1.

Table 1. The contents of the TH test database

Subsets		**Subject #**	**Image #**
Frontal	Normal	750	750
	Aging	750	2,235
	Glass	554	554
	Background	750	750
	Expression	750	750
	Lighting	750	4,500
Pose	Yaw angles	750	6,750
	Pitching angles	750	3,000
	Total:	19,289	

2.2 Design Principles of the Evaluations

A design principle describes how evaluations are designed. The precepts in these evaluations are:

1. The exploitation of the TH-FACE recognition system and the design of the evaluations are independently carried out;
2. Test data are not transparent at all to the TH-FACE recognition system before the evaluations.
3. Datasets used in the evaluations should reflect multiple performances of the system against different conditions.

Points 1 and 2 ensure the system is evaluated on their ability to generalize performance to new sets of faces, not the abilities of the system to be tuned to a particular set of faces [1]. Point 3 is up to the 'three bears' problem presented by Phillips, which sets guiding principles for designing an evaluation of the right level of difficulty. The goal in designing an evaluation is to have variation among the scores. There are two sorts of variation. One is variation among algorithms for each experiment, and the other type is variation among the experiments in an evaluation. Because at this time the evaluation was only taken on the TH-FACE recognition system, we emphasize on the latter.

According to these principles, we composed the two types of datasets below from the TH test database to implement the evaluations. All the images here can't be used for training in the system before. The datasets are summarized in Table 2.

- Gallery Set. A gallery set is a collection of images of known individuals against which testing images are matched. In the evaluation, the gallery set contains 750 images of 750 subjects (each subject has one image under normal condition). Actually, the gallery set consists of all the normal images mentioned in Table 1.
- Probe Set. A probe set is a collection of probe images of unknown individuals to be recognized. In the evaluation, 18 probe sets are composed from the TH-FACE

database. Among them, 5 probe sets correspond to the 5 subsets in the frontal subset: aging, glass wearing, background, smile expression and lighting as described in Table 1. The other 13 probe sets correspond to the images with different poses.

Table 2. The datasets composed according the evaluation design principles

Datasets			Image #		
Gallery set			750		
Datasets		**Image #**	**Datasets**		**Image #**
Probe sets (frontal)	Aging	2,235	Probe sets (different yew angles)	-40°	750
	Glass	554		-30°	750
	Background	750		-20°	750
	Expression	750		-10°	750
	Lighting	4,500		0°	750
Probe sets (different pitching angles)	-20°	750		+10°	750
	-10°	750		+20°	750
	+10°	750		+30°	750
	+20°	750		+40°	750

3 The TH-FACE Recognition System

3.1 MMP-PCA Face Recognition Method

The baseline algorithm for the TH-FACE recognition system is called multimodal part face recognition method based on principal component analysis (MMP-PCA). Various facial parts are combined in this MMP-PCA method.

The algorithm firstly detaches face parts. According to the face structure, a human face is divided into five parts: bare face, eyebrow, eye, nose and mouth. Next principal component analysis (PCA) is performed on these facial parts to calculate the eigenvector of each facial part. The projection eigenvector of known human faces' facial parts are then stored in the database. In the face recognition procedure, the algorithm first calculates the projection eigenvalue of the human face, and then calculates its similitude degree with the projection eigenvalues stored in the database, after that sorts the faces in the database according to the similarity degrees from large to small. Display the photo and personal information of the person being searched according to this order.

By choosing all the facial parts or arbitrary several facial parts, the algorithm can be adjusted to gain a relative optimal recognition rate according to different situations.

3.2 Preprocessing Method

In the TH-FACE recognition system, the preprocessing of the face images includes several important steps: geometric normalization, illumination normalization and the process of removing the glasses. The details of these steps are described as follows:

In the geometric normalization step, the TH-FACE rsystem automatically positions not only the eyes but also the chin point. Then each face image would be scaled and rotated so that the eyes are positioned in line and the distance between the chin point and the center of the eyes equals a predefined length. After that, the face image is cropped to a certain size which is 360×480 (pixels). Evaluation is taken to compare this method with the traditional method, which positions only eyes and makes the distance between the eyes equal to a predefined length.

In illumination normalization step, multi-linear algebra is applied to obtain a representation of face image, so separates the illumination factor from face images. Then illumination in different regions of the image can be compensated to a relative lighting balance so that decrease the bad effects caused by illumination. More details of the multi-linear algebra for illumination normalization can be referred to [11].

The TH-FACE recognition system uses a method to remove glasses on the face by combining the PCA reconstruction and compensating the face region hidden by the glasses with repeated iterativeness. The details of this method can be referred to [10]. The performance on using the glass-removal preprocessing method is shown in Fig. 2.

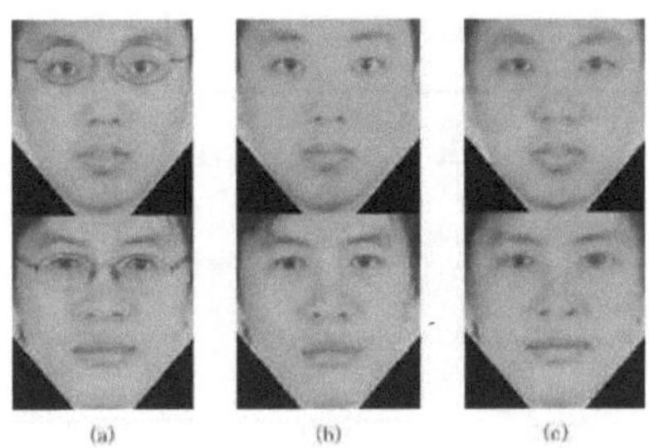

Fig. 2. Results of removing glasses in the TH-FACE recognition system

Images with glasses in the first line and those without glasses in the second line are both real images captured by camera. Image in the third line are synthesized from the first line of images by removing the glasses with the method mentioned above.

4 Evaluation Results

In this section, we carried out a set of experiments to evaluate the identification performances of the TH-FACE recognition system based on the datasets mentioned above. Effects caused by the preprocessing methods are also considered.

4.1 Identification Rates from Different Probe Sets

The face recognition system is evaluated on the 5 frontal probe sets and 13 pose probe sets as described in section 2.2. The statistic results of the identification rates in all these experiments are listed below. This performance statistics is described as follows: A probe has rank k if the correct match is the kth largest similarity score. The identification rate at rank k is the fraction of probes that have rank k or higher [1].

- **Results on the frontal datasets**

From Fig.3 we can draw some conclusions: The TH-FACE recognition system generally has an excellent identification performance on frontal images with all the identification rates in the frontal sets above 70% for Rank 1 and above 80% for Rank 10. Especially, this system solves the glass wearing problem very well. In addition, the expression of a little smile does not impact the system too much, partly because of the MMP-PCA algorithm introduced in section 3.1, where mouse has the smallest projection eigenvalue. However, the system still needs to improve its adjustability to the changing of the lighting.

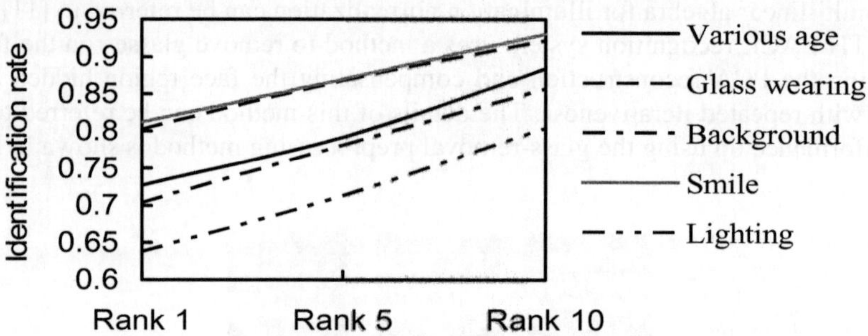

Fig. 3. The identification rates of the TH-FACE recognition system on different frontal datasets

- **Pose datasets**

Fig. 4 and Fig. 5 show the statistic results on different pose probe sets respectively. The angles in Fig. 4 are yaw angles which describe how left or right the subjects are looking to the camera while the pitching angle keeps 0°. In Fig. 5 the angles are pitching angles which describe how up or down the subjects are looking to the camera. The yaw angle keeps 0° at that time.

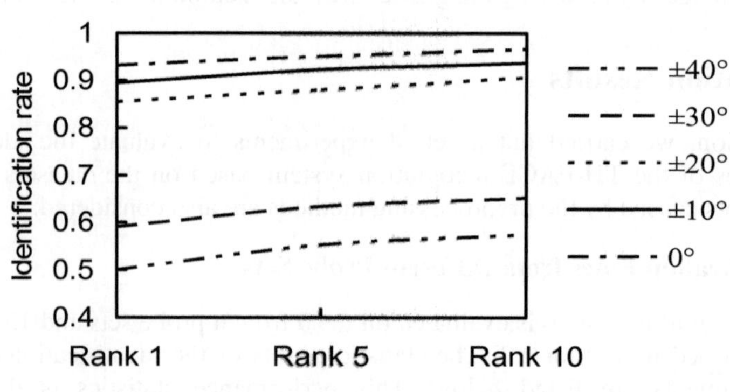

Fig. 4. The identification rates on different left-right pose probe sets

From the results above, we can see that when the viewing perspective ranging from -20° to +20° (with reference to the vertical axis) the performance keeps relatively steady and high. While the angle increases more, the performance decreases rapidly.

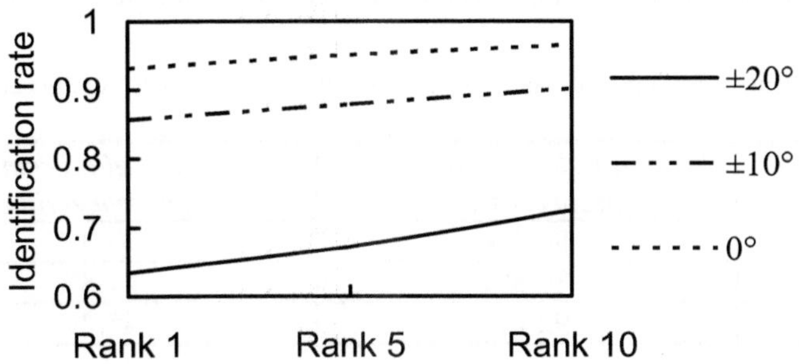

Fig. 5. The identification rates on different up-down pose probe sets

Similarly as the results on the left-right pose datasets, the results in Fig. 5 reflect that the identification works well while the pitching angle ranging from -10° to +10° while the performance decreases rapidly when the angle enlarges.

4.2 The Performance Difference Caused by Preprocessing

In this section, we examine the identification performance influenced by the preprocessing procedures.

In section 4.1, we can see that the TH-FACE system displays nice performance on the left-right pose datasets, which may be partly due to the geometric normalization method it chooses. So Evaluation is taken here to check the identification rate difference (choosing rate at Rank 1) under the two different geometric normalization methods mentioned in section 3.2. The result below shows the eyes-chin geometric normalization does positive effect on the identification rate indeed.

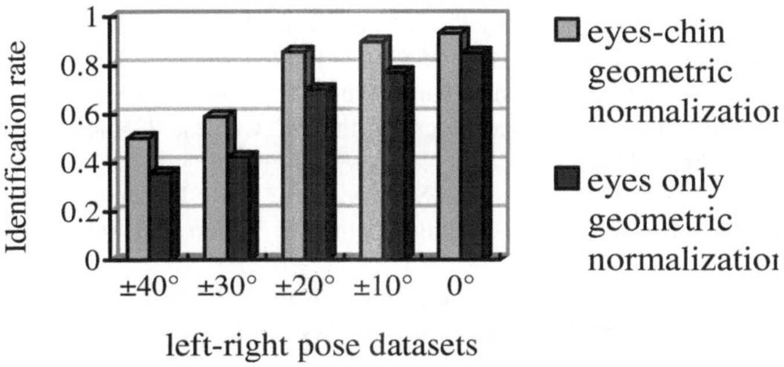

Fig. 6. Identification performance according to different Geometric normalization

We also carry out another evaluation on the effect of the glass removal preprocessing based on the glass wearing probe set. See results in Table 3. From the table above, we can easily understand the great positive effect of the glass removal preprocessing to the face identification performance while the system has to match an image with glasses among a gallery where all subjects are not wearing glasses.

Table 2. Identification performances with and without the glass removal preprocessing

	With glass removal preprocessing	***Without glass removal preprocessing***
Rank *1*	80.3%	37.3%
Rank *5*	86.8%	41.8%
Rank *10*	92.2%	45.2%

5 Conclusion

This paper presents the technology evaluations taken on the TH-FACE recognition system. The evaluations are based on the TH test database, containing 19,289 images of 750 individuals with controlled Pose, Expression, Glasses wearing, Background and Lighting variations. The division of database successfully achieves the object of finding out the performances of the system under different conditions. So this paper sets an example of technology evaluations on the TH test database and provides latest evaluation results on a new system with nice performance to the research community.

References

1. P. J. Phillips, P. Grother, R. J. Micheals, D. M. Blackburn, E. Tabassi, M. Bone: FRVT 2002 Evaluation Report, Technical Report.
Website: http://www.frvt.org/FRVT2002/documents.htm. March 2003.
2. P. J. Phillips, H. Wechsler, J. Huang, and P. Rauss: The FERET Database and Evaluation Procedure for Face Recognition Algorithms. Image and Vision Computing Journal, Vol. 16, No. 5 (1998) 295-306
3. Guangda Su, Cuiping Zhang, Rong Ding, Cheng Du: MMP-PCA face recognition method. Electronics Letters, Volume 38, Issue 25 (2002) 1654 -1656
4. Bo Gao, Shiguang Shan, Xiaohua Zhang, Wen Gao: Baseline Evaluations on the CAS-PEAL-R1 Face Database. Proceedings of the 5th Chinese Conference on Biometric Recognition. (2004) 370-378
5. P. J. Phillips, H. Moon, P. Rauss, and S. Rizvi.: The FERET evaluation methodology for face-recognition algorithms. Proceedings Computer Vision and Pattern Recognition 97 (1997) 137-143
6. Mansfield, T., G. Kelly, D. Chandler, and J. Kane.: Biometric Product Testing Final Report. Technical Report, Website:
http://www.cseg.gov.uk/technology/biometrics/index.htm. 2001.
7. D.M. Blackburn.: Evaluating Technology Properly - Three Easy Steps to Success. Corrections Today, Vol. 63 (1) (2001).

8. Phillips, P. J., A. Martin, C. L. Wilson, and M. Przybocki.: An introduction to evaluating biometric systems. Computer, Vol. 33 (2000) 56-63.
9. P.J. Grother, R.J. Micheals and P. J. Phillips.: Face Recognition Vendor Test 2002 Performance Metrics. Proceedings of the 4th International Conference on Audio Visual Based Person Authentication, 2003
10. Cheng Du, Guangda Su.: Eyeglasses Removal from Facial Images. Pattern recognition letters. Accepted
11. Yuequan Luo, Guangda Su. A Fast Method of Lighting Estimate Using Multi-linear Algebra. Proceedings of 5th Chinese Conference on Biometric Recognition (2004) 205-211

Study on Synthetic Face Database for Performance Evaluation

Kazuhiko Sumi, Chang Liu, and Takashi Matsuyama

Graduate School of Informatics,
Kyoto University, Kyoto 606–8501, Japan
sumi@vision.kueee.kyoto-u.ac.jp
http://vision.kuee.kyoto-u.ac.jp/

Abstract. We have analyzed the vulnerability and threat of the biometric evaluation database and proposed the method to generate a synthetic database from a real database. Our method is characterized by finding nearest neighbor triples or pairs in the feature space of biometric samples, and by crossing over those triples and pairs to generate synthetic samples. The advantages of our method is that we can keep the statistical distribution of the original database, thus, the evaluation result is expected to be the same as original real database. The proposed database, which does not have privacy problem, can be circulated freely among biometric vendors and testers. We have implemented this idea on a face image database using active appearance model. The synthesized image database has the same distance distribution with the original database, which suggests it will deriver the same accuracy with the original one.

1 Introduction

Evaluation of biometric authentication systems, especially accuracy evaluation, requires a large-scale biometric database[1]. As biometric authentication systems become practical, number of volunteers required for evaluation becoming large[2]. Once, the individual data is leaked, there are several scenarios of database abuse and possible social threat. We analyze such social threat and propose a synthetic database as an alternative solution for privacy protection. In the field of fingerprint, an image synthesis tool SFINGE[3] has been developed. It has been applied in the public benchmarking such as FVC2004[4], and proven to have correlation with a real database. However, initial conditions, such as ridge orientation map and locations of fiducial points, should be given a priori. Moreover, this method cannot be applied to other biometrics, whose development process are not modeled well.

In this paper, we propose a method to generate synthetic biometric samples from real biometric examples. We try to maintain the same recognition difficulty as the original database, in order to use the synthetic database for evaluation purpose. Our idea is to find closest triples and pairs in the original database and to cross between those triples and pairs for generation of synthetic samples. As a case study, we apply this idea on a face databese.

D. Zhang and A.K. Jain (Eds.): ICB 2006, LNCS 3832, pp. 598–604, 2005.

2 Threat Analysis of Biometric Evaluation Database

Various types of vulnerability have been alerted in biometric authentication systems. It is much easier to steal personal data from evaluation database than to steal them from templates, because evaluation database has raw images and it is not secured in a safe place.

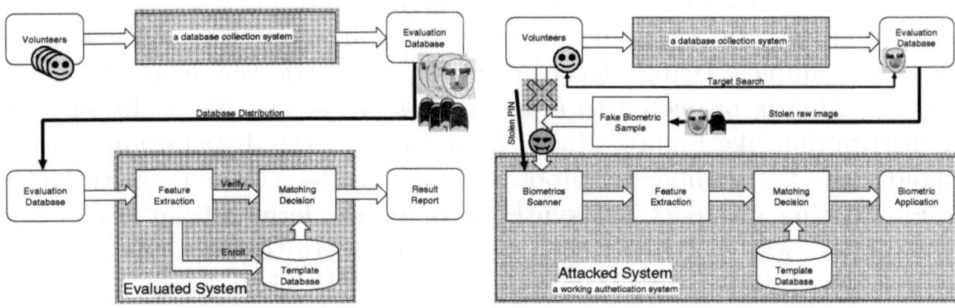

(a) System (b) Vulnerability

Fig. 1. Schematic diagram of database collection and evaluation of a biometric authentication system and its vulnerability

Figure. 1(a) shows the schematic diagram of database collection and evaluation procedure. In this figure, database which has volunteer's individual biometric data in a raw format is transferred from database developper to an evaluater.

The first scenario is to produce a fake biometric example from stolen database. A fake biometric sample, such as a fake fingerprint, an iris and a fake facemask, can be produced from the raw image. This fake example can be used to attack a biometric authentication system under operation shown in Figure 1(b) Suppose the attacker steal the template database DB_B of biometric evaluation system B. If the attacker knows the PIN N_j of the person P_j, $(P_j \in DB_B)$, a fake biometric, which produces the same impression as T_j, is produced from T_j, then it can be used to attack a 1-to-1 authentication system A and obtain the access permit of the owner j. Even if the PIN is not known, but it is certain to be enrolled in the specific system A, the fake example can be used to obtain access permit to the system, if the system allows 1-to-N authentication scheme.

To prevent those privacy invasion, protecting the database is desirable. However, hiding raw image is impossible, because the evaluation usually includes feature extraction algorithm as well as matching and classifying algorithms. The input of the algorithm must be a raw image. So, we propose a synthetic biometric database in the next section.

3 Requirement of Synthetic Biometric Database

A synthetic biometric database consists of virtual individual biometric examples is one of the solution. However, to satisfy accurate evaluation of biometric authentication systems, the database should have the following characteristics:

1. (precision requirement) The evaluation results derived from a synthetic biometric database should be equal to the one from the real database.
2. (universality requirement) The precision requirement should be satisfied for all of authentication algorithms to be evaluated.
3. (privacy requirement) Each biometric data in the synthetic database should not represent any real person.

The precision requirement can be resolved in the following way.

Suppose group A is a real database corrected from existing individuals consist of M_A examples. Using algorithm Θ, an biometric raw example $a_i, (a_i \in A, 1 \leq i \leq M_A)$ is projected to $\theta(a_i)$ in a future space. If we obtain a similarity distribution like Figure. 2(a)A, it means that the distribution of h_{FA} at threshold T_h is the number of impostor samples closer than T_h in the feature space Θ. Another group B is a synthetic database derived from A consists of M_B examples. ($M_B = M_A$ in this case) Using algorithm θ, a biometric raw example $b_i, (b_i \in B, 1 \leq i \leq B_A)$ is projected to $\theta(b_i)$ in a future space. If we like to have the same false rejection rate (FRR) and false accept rate (FAR) at threshold T_h, the number of pairs, which are closer than T_h in the feature space Θ should be the same with the case of A. This suggests that we should be careful not to change the distance of samples, whose distance is less than T_h, but we don't have to be careful about the distance of samples, whose distance is larger than T_h. Figure.2(b)shows an example of such a deformation. Suppose P are the biometric samples. For a arbitrary index i, select samples closer than the threshold T_h in the feature space Θ. In this figure, they are P_{i1}, P_{i2}, and P_{i3}. If we generate synthetic examples Q_{i1}, Q_{i2}, and Q_{i3}, and the distance between Q_{i1} and Q_{i2}, Q_{i1} and Q_{i3}, and Q_{i2} and Q_{i3} are equal to the original distance between P_{i1} and P_{i2}, P_{i1} and P_{i3}, and P_{i2} and P_{i3}, respectively, the synthetic samples satisfy with the three requirements explained in this section.

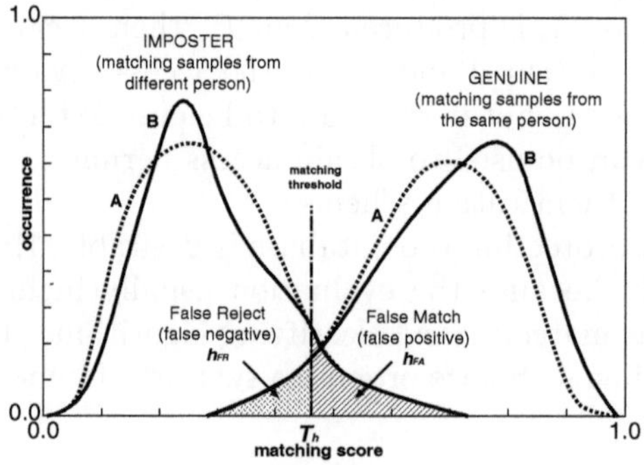

(a) Distribution

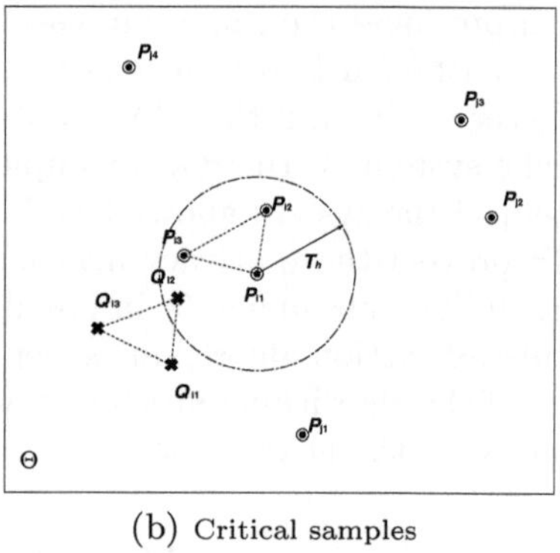

(b) Critical samples

Fig. 2. Similarity distributions of synthetic database B and real database A (a), and relationships of critical samples in a real database and the corresponding synthetic database (b)

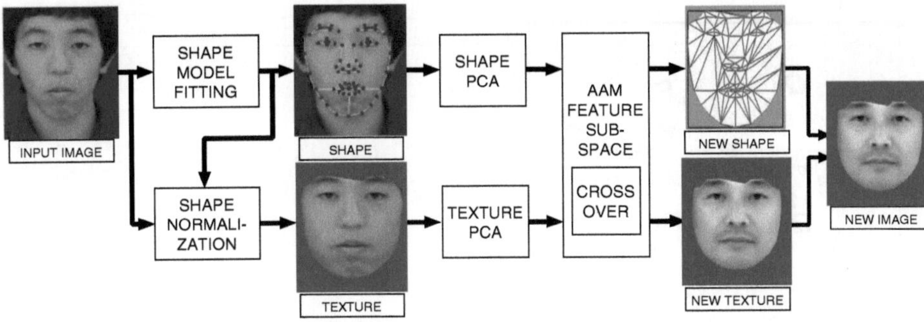

Fig. 3. Schematic diagram of face image deformation based on facial parts regions

In the above deformation, we should consider isolated samples which have only one neighbor or no neighbors within the threshold T_h. In case of doubles, we rotate the pair of samples around its center. In case of standalone, we move the sample along a certain displacement, which has a fixed length and a random direction.

4 A Case Study Using Active Appearance Model

According to the idea in Section. 3, we have synthesized a face database from a real face database. The real faces are from HOIP face database contains 300 subjects of various age (from 20 to 60) and gender (150 males and 150 females), in a illumination controlled environment.

In this study, we deform the faces in PCA subspaces represented by active appearance model[5], and then reconstruct face images. Deformation is performed in the PCA subspaces of active appearance model (AAM), which consists of shape subspace and texture subspace. In the subspace, all the samples are grouped into triples, pairs, and singles according to the distance to the nearest neighbors. Then cross-over operation is performed to generate new samples. Finally, those synthetic samples are back projected and images of synthetic samples are generated.

Regards to the details of deformation, triples are detected in each PCA subspace Θ of AAM feature space. Then the center of the triples P_{i1}, P_{i2}, and P_{i3} are calculates as C. The synthetic face samples Q_{i1}, Q_{i2}, and Q_{i3} are placed at the symmetrical position of P_{i1}, P_{i2}, and P_{i3}, respectively. The relationships of P_{i1}, P_{i2}, P_{i3}, Q_{i1}, Q_{i2}, Q_{i3}, and C are shown in Figure. 4(a). In case of not finding a triple around the focused sample, a pair is detected instead. If there are no sample in a given distance T_h, the sample is regarded as a singlar sample. Random displacement is given for such a singlar sample.

Examples of synthetic faces are shown in Figure. 5. In the figure, upper row is the real faces and the lower row is the synthesized images. As we can see, the pair (upper and lower) of images is apparently different person but has similar impression. The distances between three samples are same within the PCA feature space.

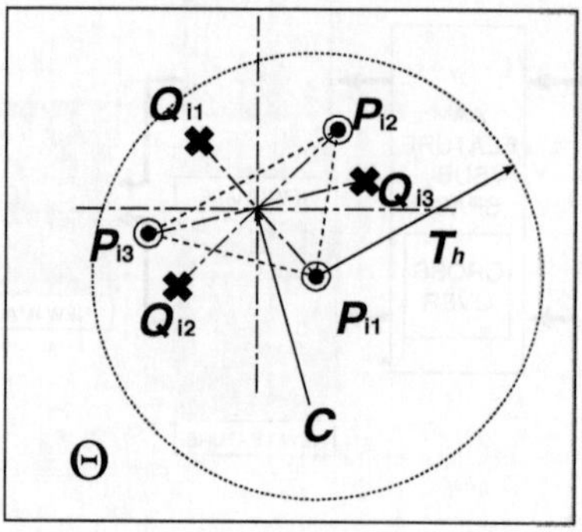

(a) Closest triples

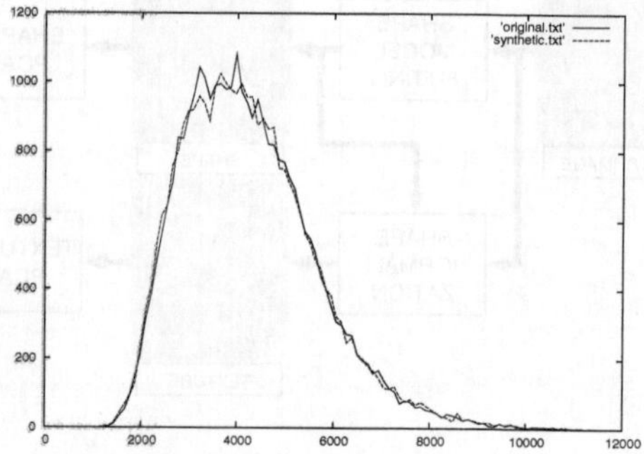

(b) Distance distribution

Fig. 4. Selection of the closest triples and its deformation (a) and the distribution of distance between arbitrary two samples in the original database and in the synthetic database (b)

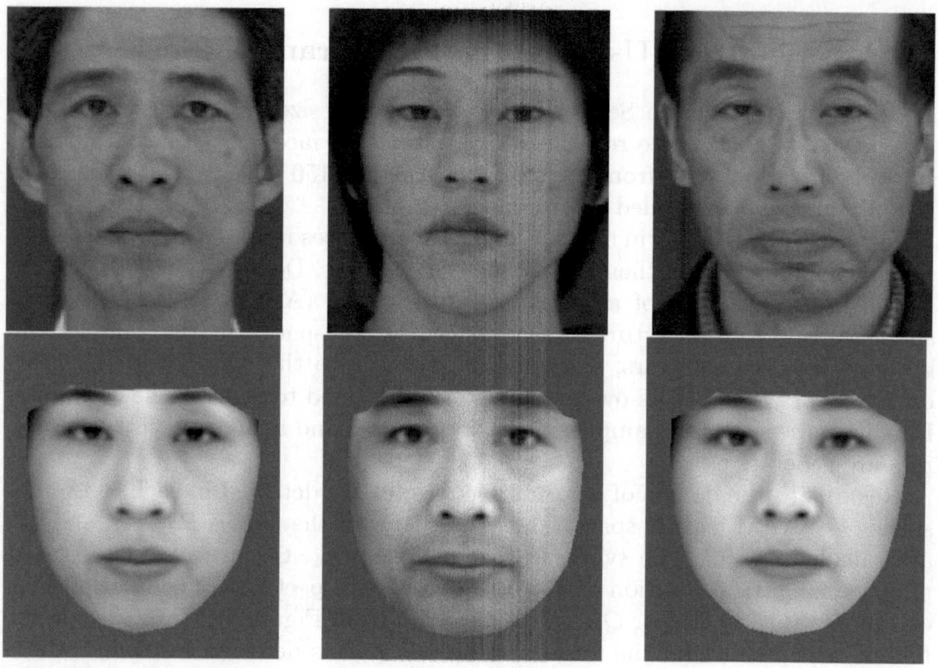

Fig. 5. The real face image (upper) and the synthetic face image (lower) using our proposed method

To confirm the precision requirement described in Section 3, we compared the distribution of distance between arbitrary samples both in original database and in synthetic database. Figure.fig-deformation2(b) shows the distribution of distance. The original database and the synthetic database shows the quite similar distribution. It suggests that evaluation using these two database will derive the same accuracy results.

5 Discussion and Future Direction

At this moment, we have not completed the way to satisfy the universality requirement yet. The synthetic database, whose distances are same with the original database measured in PCA sub-space of AAM, may not be equal-distant in other feature space, such as simple eigenfaces and banch graph matching. Never the less, AAM, which employs both geometric (shape) and photometric (texture) features, is the most promissing approach for 2D image.

Another discussion is how to deal with intra-personal variations. Some of the face recognition algorithms require multiple images with different appearances for enrollment. Also, we need intra-personal variations to evaluate false non-match rate of an algorithm. So, we have to synthesize multiple appearances for a synthesized person. The method to generate multiple appearances for a person depends on the variation of the original images. If the variation of the original images are arbitrary, we have to use the common intra-personal variation space, which is introduced by Moghaddam[6]. First, we will build the intra-personal variation space using the original images, then apply it to the synthetic image and generate multiple views. If the variation of the original images are taken systematically and changes are parameterized, we can use the changed images and apply same deformation to the images.

6 Summary

In this paper, we have analyzed the vulnerability and threat of the biometric evaluation database and proposed a new method to generate synthetic database based on real database. Our method is characterized by finding nearest neighbor triples or pairs in the feature space of biometric samples, and by crossing over those triples and pairs to generate synthetic samples. The advantages of our method is that we can keep the statistical distribution of the original database, thus, the evaluation result is expected to be the same as original real database. The proposed database, which does not have privacy problem, can be circulated freely among biometric vendors and testers. We have implemented this idea on a face image database using active appearance model. The proposed database, which does not have privacy problem, can be circulated freely among biometric vendors and testers. We hope that this technique will accelerate the development practical biometric authentication systems.

Acknowledgments

This research is supported in part by the Informatics Research Center for Development of Knowledge Society Infrastructure, 21st. Century COE program and by contracts 13224051 and 14380161 of the Ministry of Education, Culture, Sports, Science and Technology, Japan. This research is also supported in part by the research contracts with Japan Automatic Identification Systems Association.

References

1. Wilson, C.L.: Large scale usa patriot act biometric testing. In: Proc. International Meeting of Biometrics Expert. (2004) http://www.biometricscatalog.-org/document_area/view_document.asp?pk={5E0CA69A-B4AC-4FE9-9624-6ED3450E9CCF}.
2. Wayman, J.: Technical Testing and Evaluation of Biometric Identification Devices, in, A. Jain, etal(ed): Biometrics: Personal Identification in a Networked Society. Kluwer Academic Press, Higham, MA, USA (1999)
3. Cappelli, R., Erol, A., Maio, D., Maltoni, D.: Synthetic fingerprint-image generation. In: Proc. International Conference on Pattern Recognition. (2000)
4. Maio, D., Maltoni, D., Cappelli, R., Wayman, J., Jain, A.K.: Fvc2004: Third fingerprint verificatin competition. In: Proc. International Conference on Biometric Authentication. (2000) 1–7
5. Cootes, T., Walker, K., Taylor, C.: View-based active appearance models. In: AFGR00. (2000) 227–232
6. Moghaddam, B., Jebara, T., Pentland, A.S.: Baysian face recognition. PR **33** (2000) 1171–1782

Gait Recognition Based on Fusion of Multi-view Gait Sequences

Yuan Wang[1], Shiqi Yu[1], Yunhong Wang[2], and Tieniu Tan[1]

[1] National Laboratory of Pattern Recognition, Institute of Automation, Chinese Academy of Sciences, P.O. Box 2728, Beijing 100080, China
[2] School of Computer Science and Engineering, Beihang University
{ywang, sqyu, wangyh, tnt}@nlpr.ia.ac.cn

Abstract. In recent years, many gait recognition algorithms have been developed, but most of them depend on a specific view angle. In this paper,we present a new gait recognition scheme based on multi-view gait sequence fusion. An experimental comparison of the fusion of gait sequences at different views is reported. Our experiments show the fusion of gait sequences at different views can consistently achieve better results. The Dempster-Shafer fusion method is found to give a great improvement. On the other hand, we also find that fusion of gait sequences with an angle difference greater than or equal to 90° can achieve better improvement than fusion of those with an acute angle difference.

1 Introduction

Gait has recently received an increasing interest from researchers. Gait is an attractive biometric feature for human identification at a distance, which is non-contact, non-invasive and easily acquired at a distance in contrast with other biometrics, so it has been considered as the most suitable biometric for human identification in visual surveillance.

Over the past years many gait recognition algorithms [1, 2, 3, 4, 5] have been proposed, but most of them are dependent on only one view, normally side view and have low recognition rate due to the influence of clothing, background, light, walker's mental state, etc. How to develop a robust and accurate gait recognition system has become an important direction. The purpose of this paper is therefore to present a new gait recognition scheme based on the fusion of multi-view gait sequences which can improve the performance of gait recognition system greatly and can be used in practice conveniently. Because in many surveillance environments, multiple cameras at different view angles are used, this makes it possible to get gait sequences from multi-view directions. Unlike most previous studies that focus on extracting good features, this paper is trying to construct a multi-view gait recognition system which is more robust and accurate.

The remainder of this paper is organized as follows. Section 2 briefly introduces the Key Fourier Descriptor (KFD) method for gait recognition and fusion rules for multi-view gait sequence fusion. Section 3 presents the CASIA multi-view gait database. Then the main scheme of the fusion system is presented in Section 4. Section 5 introduces our experimental results and analysis. Finally, this paper is concluded in Section 6.

D. Zhang and A.K. Jain (Eds.): ICB 2006, LNCS 3832, pp. 605–611, 2005.

2 KFD Gait Recognition Method and Fusion Rule

2.1 KFD Gait Recognition Method

Given a fixed camera, the human silhouette can be extracted by background subtraction and thresholding. We take advantage of the method given in [2] to segment human silhouettes from image sequences. Since the extracted silhouette sizes are not unique, the height of them is normalized to a fixed size.

To extract KFDs feature, all the contours and the gait cycle are normalized to have the same number (N) of samples and the same number (T) of frames, respectively. All Fourier descriptors $g(i)$ can be obtained by discrete Fourier transform. The KFDs are defined as in [5]:

$$G = \left[\frac{|g(2T)|}{|g(T)|}, \frac{|g(3T)|}{|g(T)|}, \ldots, \frac{|g((N-1)T)|}{|g(T)|} \right] \tag{1}$$

where g is Fouries descriptors.

To measure the similarity between two gait sequences a and b, we use the metric shown in Equation (2) as the similarity measure:

$$D(a,b) = \frac{1}{M} \sum_{m=1}^{M} |G_a(m) - G_b(m)| \tag{2}$$

where G_a and G_b are, respectively, feature vectors of sequences a and b, and M is the feature vector length.

2.2 Overview of Fusion Rule

We used 4 traditional fusion methods in our experiments. For the verification mode in biometrics authentication system, the incomer will be compared with the template of the person he claims. We treat the outputs of each authenticaiton system as a feature vector $X = [x_1, x_2, \cdots, x_N]$, where N is the number of subsystems and x_i is the output of each subsystem. Then we can use any known classifier to determine the separation bound between imposter and client. In our paper, we use the following 4 kinds of fusion rules:

1. Sum Rule

$$x = \sum_{i=1}^{N} x_i \tag{3}$$

2. Weighted Sum Rule

$$x = \sum_{i=1}^{N} w_i * x_i \tag{4}$$

Here, w_i is computed by the EER of each fusion system: $w_i = \frac{ERR_i^{-1}}{\sum_{j=1}^{N} ERR_j^{-1}}$

3. Product Rule

$$x = \prod_{i=1}^{N} x_i \tag{5}$$

4. Dempster-Shafer(D-S) rule
   In this frame of the evidence theory, the best representation of support is a belief function rather than a Bayesian mass distribution. The theory embraces the familiar idea of assigning numbers between 0 and 1 to indicate the degree of support but, instead of focusing on how these numbers are determined, it concerns the combination of degrees of belief. Here, we use the algorithm proposed in [6].

The decision of fusion system can be made based on x computed by these methods.

3 CASIA Multi-view Gait Database

In our experiments, we used the CASIA multi-view gait database, which contains gait sequences of 124 subjects (94 males, 30 females) taken from 11 cameras at 11 different views. All the subjects were asked to walk naturally on the concrete ground along a straight line in an indoor environment. The videos were captured by 11 cameras from different view directions The view angle θ between the view direction and the walking direction took on the values of $0°, 18°, 36°, \cdots$, and $180°$, as delineated in Fig. 1. Each subject walked along the straight line 10 times (6 for normal walking, 2 for walking with a bag, and 2 for walking with a coat), and 11 video sequences are captured each time. Thus, 110 sequences were recoded for each subject, and the database contains a total of $110 \times 124 = 13640$ video sequences. All the video sequences have the same resolution of 320×240 pixels. Some sample frames are shown in Fig. 2.

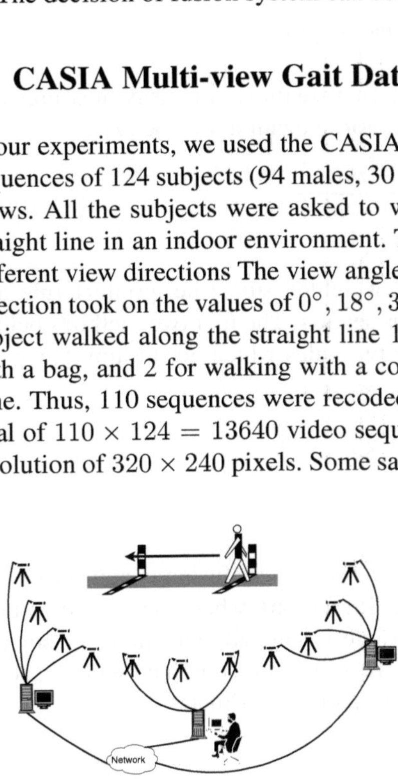

Fig. 1. The schematic diagram of gait data collection system

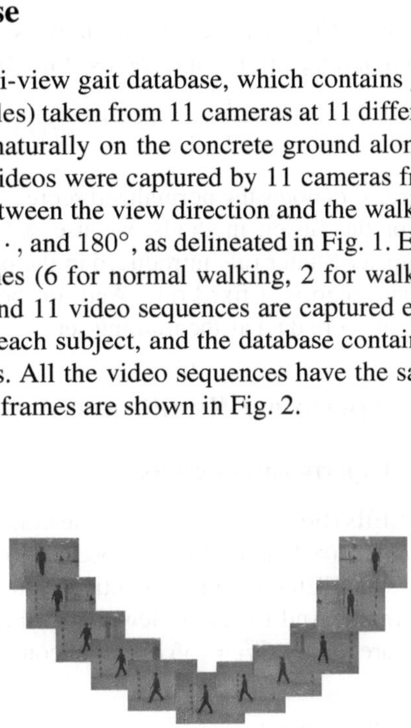

Fig. 2. Sample frames from 11 view directions

In the database those factors affecting gait recognition, such as view direction (11 views), clothing (with or without coat), and carrying condition (with or without bag) are included. Here only view direction is studied, though other factors are interesting to study too.

4 Fusion Scheme

Based on the KFD gait recognition method, we can get the similarity measures between the two gait sequences and the template as shown in Fig. 3. And before combining the

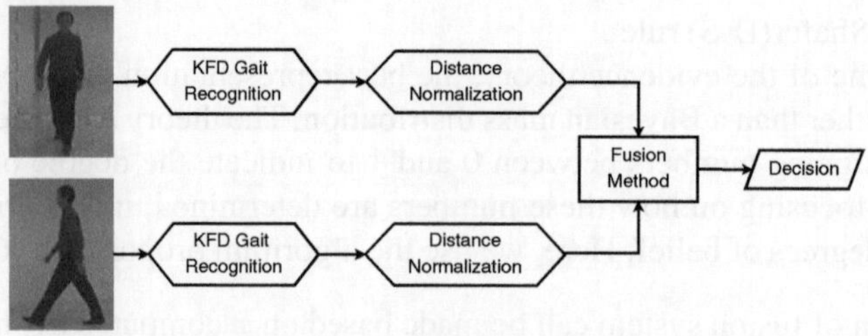

Fig. 3. The scheme of fusion system

two similarity measures by fusion, we should normalize those similarity measures to a common range $[0, 1]$. Here we use the Min-Max normalization method [7]:

$$s' = f(s) = \frac{s - min}{max - min} \tag{6}$$

where, s' denotes the normalized score.

For the fusion methods, we use sum rule, weighted sum rule, product rule and Dempster-Shafer rule introduced in the previous section. The first two methods belong to the category of fixed rule. And for the other two rules which need training, we use 10% of score data as the training set.

5 Experiment Results and Analysis

5.1 Experimental Results

The EERs (Equal Error Rates) of the multi-view gait sequences fusion system are shown in Table 1 (only partial results because of length limitation), where the second column shows the EERs of gait recognition system of Angle 1 using KFD method, and so does Column 4. And for each view angle, the other 10 views are combined with it and so there are totally $C_{11}^2 = 55$ different combinations in our experiments.

5.2 Discussions

Based on the above results, we can draw some conclusions. When using sum rule as the fusion method, the average EER of 55 fusion experiments is 9.08%, and for product rule and weighted sum rule, it becomes 8.56% and 8.85%. D-S rule gives the lowest average EER along the four methods which is only 3.81%. Within the gait recognition system of 11 views, there are only 27% of them whose average EERs are less than 10%. But for the 55 fusion systems using sum rule, 75% of them give EERs less than 10% and if using D-S rule, 85% of fusion systems' EERs are less than 5%. On the other hand, within the 55 fusion experiments, 7 experiments fail to give improvement comparing the best single system when using sum rule and 6 experiments fail when using product rule. While for the trained rules, the number of failures becomes 2 and 0 corresponding to weighted sum rule and D-S rule respectively. So we can draw a conclusion that the trained rules are better than fixed rules from the view of whether fusion system can give improvement.

Table 1. The EERs of Multi-view Gait Sequence Fusion System

Angle1	EER	Angle2	EER	Sum	Product	W-Sum	D-S
0°	13.98%	18°	15.31%	12.00%	11.97%	11.43%	5.86%
0°	13.98%	54°	11.34%	10.28%	8.48%	10.08%	4.70%
0°	13.98%	90°	8.05%	7.38%	7.89%	6.66%	2.69%
0°	13.98%	126°	11.80%	9.54%	8.63%	9.25%	4.11%
0°	13.98%	162°	10.44%	9.02%	8.81%	9.04%	3.90%
18°	15.31%	36°	12.16%	12.70%	12.17%	12.66%	5.21%
18°	15.31%	72°	8.08%	8.94%	8.20%	7.46%	3.21%
18°	15.31%	108°	9.96%	8.79%	7.80%	8.95%	3.12%
18°	15.31%	144°	11.50%	11.08%	10.12%	10.39%	4.67%
18°	15.31%	180°	13.75%	10.94%	9.84%	11.38%	4.46%
36°	12.16%	54°	11.34%	10.73%	9.33%	10.78%	4.65%
36°	12.16%	90°	8.05%	8.16%	8.48%	7.81%	3.00%
36°	12.16%	126°	11.80%	9.36%	8.74%	9.37%	3.98%
36°	12.16%	162°	10.44%	9.30%	8.69%	9.30%	3.56%
54°	11.34%	72°	8.08%	7.96%	8.88%	7.45%	3.40%
54°	11.34%	108°	9.96%	8.97%	8.27%	8.53%	3.58%
54°	11.34%	144°	11.50%	9.02%	8.35%	9.13%	3.76%
54°	11.34%	180°	13.75%	9.52%	8.41%	9.82%	4.06%
72°	8.08%	90°	8.05%	6.61%	6.59%	6.62%	2.46%
72°	8.08%	126°	11.80%	8.53%	8.36%	8.05%	2.99%
72°	8.08%	162°	10.44%	6.41%	6.85%	6.49%	2.47%
90°	8.05%	108°	9.96%	7.42%	7.08%	7.10%	3.43%
90°	8.05%	144°	11.50%	7.39%	6.83%	6.88%	2.59%
90°	8.05%	180°	13.75%	7.54%	6.98%	7.45%	2.68%
108°	9.96%	126°	11.80%	8.71%	7.84%	8.60%	5.06%
108°	9.96%	162°	10.44%	7.48%	7.45%	7.33%	3.55%
126°	11.80%	144°	11.50%	10.66%	9.87%	10.66%	5.23%
126°	11.80%	180°	13.75%	10.30%	10.03%	10.24%	4.84%
144°	11.50%	162°	10.44%	9.40%	8.90%	9.37%	4.53%
144°	11.50%	180°	13.75%	9.96%	9.19%	9.78%	4.68%
162°	10.44%	180°	13.75%	9.48%	9.31%	9.21%	4.46%

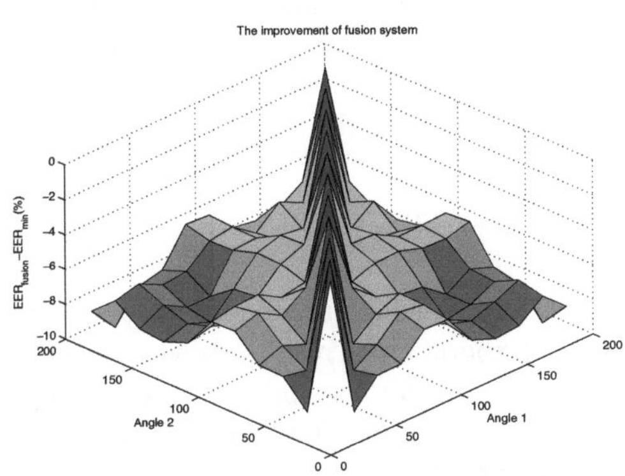

Fig. 4. The Improvement of Combination System Using D-S Rule

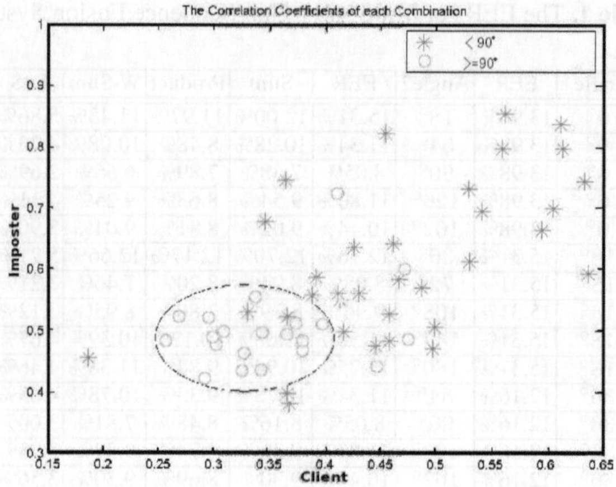

Fig. 5. The Correlation Coefficients of Each Combination

According to Table 1, we can also find that the view differences of those systems without improvements are all less than 90°.

It means that fusion of gait sequences with an angle difference greater than or equal to 90° can achieve better improvement than fusion of those with an acute angle difference. This conclusion is also illustrated in Fig. 4, where the z axis shows the difference between the EERs of fusion system and the lowest EERs of the single systems (the lower of the surface, the greater improvement). It is very clear that the difference values between the two EERs are much larger in the acute angle zone. This is mainly because that the information contained in the two gait sequences with an acute angel difference is more correlated than those with larger angle difference. In other words, much more information is contained in case of an obtuse angle (including a right angle) than an acute angle. In Fig. 5, we computed the correlation coefficients of the output score of two single view gait recognition systems. The x axis denotes the correlation coefficients of the score of client data and y axis denotes that of imposter data. We can find that the correlation coefficients of the gait sequences with an acute angle differences are scattered in the whole area and for those with larger angle differences, the coefficients are gathered into a small ellipse as shown in the figure.

6 Conclusion and Future Work

In this paper, we have presented a new gait recognition scheme based on the fusion of multi-view gait sequences. Experimental results show that the proposed system can help improve the performance of gait recognition system. Specifically, when using Dempster-Shafer fusion rule, the combination EERs mostly drop to around 5%, which achieves a great improvement when compared to the EERs of single view gait recognition.

Multi-view gait recognition is a new direction in this field and there are many open questions to address. For multi-view gait sequence taken from one subject, how to extract some common features existing in all views is an interesting direction. Our current work is focused on score level fusion and further work should include the feature level fusion of multi-view gait sequences. In addition the database should include some outdoor data which is more similar to the data in practical application.

Acknowledgement

This work is partly supported by National Natural Science Foundation of China (Grant No. 60332010 and 60335010) and the National Basic Research Program of China (Grant No. 2004CB318100).

References

1. Kale, A., Sundaresan, A., Rajagopalan, A.N., Cuntoor, N.P., roy Chowdhury, A.K., Krüger, V., Chellappa, R.: Identification of humans using gait. IEEE Transactions on Image Processing **13** (2004) 1163–1173
2. Wang, L., Tan, T., Ning, H., Hu, W.: Silhouette analysis-based gait recognition for human identification. IEEE Transactions on Pattern Analysis and Machine Intelligence **25** (2003) 1505–1518
3. Wang, L., Ning, H., Tan, T., Hu, W.: Fusion of static and dynamic body biometrics for gait recognition. IEEE Transactions on Circuits and Systems for Video Technology **14** (2004) 149–158
4. Yam, C.Y., Nixon, M.S., Carter, J.N.: On the relationship of human walking and running: automatic person identification by gait. In: Proc. of International Conference on Pattern Recognition, Quebec,Canada (2002) 287–290
5. Yu, S., Wang, L., Hu, W., Tan, T.: Gait analysis for human identification in frequency domain. In: Proc. of the 3rd International Conference on Image and Graphics, Hong Kong, China (2004) 282–285
6. Y, S., T, K.: Media-integrated biometric person recognition based on the dempster-shafer theory. In: 16th International Conference on Pattern Recognition. (2002) 381–384
7. Jain, A.K., Nandakumar, K., Ross, A.: Score normalization in multimodal biometric systems. to appear in Pattern Recognition (2005)

A New Representation for Human Gait Recognition: Motion Silhouettes Image (MSI)

Toby H.W. Lam and Raymond S.T. Lee

Department of Computing, The Hong Kong Polytechnic University,
Hung Hom, Kowloon, Hong Kong
{cshwlam, csstlee}@comp.polyu.edu.hk

Abstract. Recently, gait recognition for human identification has received substantial attention from biometrics researchers. Compared with other biometrics, it is more difficult to disguise. In addition, gait can be captured in a distance by using low-resolution capturing devices. In this paper, we proposed a new representation for human gait recognition which is called Motion Silhouettes Image (MSI). MSI is a grey-level image which embeds the critical spatio-temporal information. Experiments showed that MSI has a high discriminative power for gait recognition. The recognition rate is around 87% in SOTON dataset by using MSI for recognition. The recognition rate is quite promising. In addition, MSI can also reduce the storage size of the dataset. After using MSI, the storage size of SOTON has reduced to 4.2MB.

1 Introduction

In this paper, we proposed a new representation for gait recognition, Motion Silhouettes Image (MSI). The idea of MSI was inspired by the Motion History Image (MHI) which is developed by Bobick and Davis [2]. Bobick and Davis used MHI for motion recognition and applied Hu moments for dimension reduction. Experiments had been done in a test set which contains 18 aerboics exercises and the recognition is above 80%. Our proposed MSI is similar to MHI. However, MSI is simpler and easier to implement. MSI is a gray-level image which embeds the spatial and temporal information. Experiments showed that MSI has a high discriminative power. Besides, it greatly reduced the computational cost and the storage size. In our proposed algorithm, we applied Principal Component Analysis (PCA) on MSIs for reducing the dimensionality of the input space and optimizing the class separability of different MSIs. We use the SOTON dataset [3] to demonstrate the efficacy of the proposed algorithm.

The rest of this paper is organized as follows. We show the related research about gait recognition in Section 2. In Section 3, we reveal and explain the detail about Motion Silhouettes Image (MSI) and the proposed recognition algorithm. The experimental results are shown in Section 4. Conclusion appears in Section 5.

2 Related Work

Murase and Sakai [4] proposed a parametric eigenspace representation for moving object recognition. Eigenspace representation is formerly used in face recognition [5].

D. Zhang and A.K. Jain (Eds.): ICB 2006, LNCS 3832, pp. 612 – 618, 2005.

Murase and Sakai applied eigenspace representation in gait recognition and lip reading. In their proposed algorithm, the extracted silhouettes were projected to the eigenspace by using Principle Components Analysis (PCA). The sequence of movement forms a trajectory in the eigenspace which called parametric eigenspace representation. During recognition, the input image sequence of movement was preprocessed to form a sequence of binary silhouette and these formed a trajectory in the eigenspace. The best match was the one which gets the smallest distance between the input trajectory and the reference sequence. Huang, Harris and Nixon [6] applied the similar approach for gait recognition. Instead of using PCA, they used Linear Discriminating Analysis (LDA), or namely canonical analysis, for transformation. Wang and Tan proposed another new representation for gait recognition [7]. Wang and Tan generated a distance signal by unwrapping the human silhouette. The time-varying distance signals were applied to eigenspace transformation based on PCA. The performance of our proposed algorithm is compared with the performance of Wang's algorithm. For more details, please refer to Section 4.

3 Recognition Algorithm

The proposed gait recognition algorithm could be divided into four steps: (i) Image sequences pre-processing, (ii) Motion Silhouettes Image (MSI) generation, (iii) Principle components generation and (iv) Classification. Fig. 1 shows a flow diagram of our proposed algorithm.

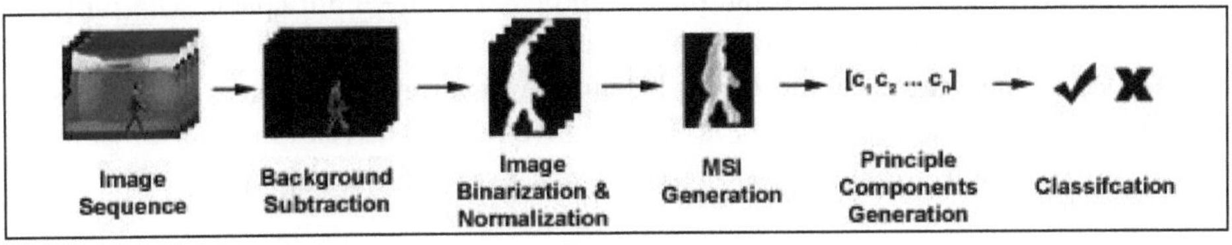

Fig. 1. Flow diagram of gait recognition by using MSI

3.1 Preprocessing

In our proposed gait recognition algorithm, silhouettes are the basis of the feature for recognition. By using background subtraction and thresholding, the silhouettes are extracted from the image sequences [5]. To further eliminate the scaling effect, the silhouettes are extracted according to the size of bounding box and resize to a standard size (128 x 88 pixels). Fig. 2 shows some examples of normalized silhouettes.

Fig. 2. Normalized silhouettes

3.2 Motion Silhouettes Image

Motion Silhouettes Image (MSI) is a gray-level image where the pixel intensity is a function of the temporal history of motion of that pixel. The intensity of the MSI represents motion information. MSI embeds the critical spatial and temporal information and it could be formulated by (eqn. 1) . Fig. 3 shows some examples of MSI.

$$MSI(x, y, t) = \begin{cases} 255 & \text{if } I(x, y, t) = 1 \\ \max(0, MSI(x, y, t-1) - 1) & \text{otherwise} \end{cases} \quad (1)$$

where I is the silhouette image, t is the current time, x and y are the horizontal and vertical coordinates of the image respectively.

Frame 1 Frame 15 Frame 30 MSI

Fig. 3. Examples of Motion Silhouette Image

3.3 Training and Transformation

PCA is used for capturing the principle components of the input space. The purpose of using PCA is reducing the feature space to a subspace which maximizing the variance of classes. Suppose there are C classes for training, each class $c \in C$ has N_c of q-dimensional MSI $m_{c,i}$,where i is the instance label. The total number of training samples is $N_{total} = N_1 + N_2 + \ldots + N_c$. The average MSI of all samples $\mu \in \Re^q$ is defined as

$$\mu = \frac{1}{N_{total}} \sum_{c \in C} \sum_{j=1}^{N_c} m_{c,j} \quad (2)$$

and the covariance matrix Σ_{MSI} is defined as

$$\sum_{MSI} = \frac{1}{N_{total}} \sum_{c \in C} \sum_{j=1}^{N_c} (m_{c,j} - \mu)(m_{c,j} - \mu)^T \quad (3)$$

A transformation matrix $W_{pca} = [w_1, w_2, \ldots w_p]$ is obtained where $w_1, w_2, \ldots w_p$ are the eigenvectors of the samples covariance matrix Σ_{MSI} corresponding to p (p<q) largest eigenvalues. These eigenvectors are orthonormal to each other and span the eigenspace. The original MSI $m_{c,i}$ is projected to points $p_{c,i}$ in p-dimensional eigenspace by using the transformation matrix W_{pca}.

$$p_{c,i} = W_{pca}^{\ T} m_{c,i} = [w_1, w_2, \ldots, w_p]^T m_{c,i} \quad (4)$$

where $c \in C$ and $i = 1, 2, \ldots, N_c$

3.4 Recognition

Suppose an input image sequence be r(t), where t = 1, 2...T. MSI m_r is generated by using eqn. (1). The MSI is projected into the eigenspace by

$$z_r = [w_1, w_2, ..., w_p]^T m_r \tag{5}$$

The Euclidean distance E between the projected testing feature vector and the projected training feature vector is calculated (see eqn. 6).

$$E(z_r, z_{c,i}) = \sqrt{(z_r - z_{c,i})^T (z_r - z_{c,i})} \tag{6}$$

where z_r is the projected feature vector of testing MSI m_r and $z_{c,i}$ is the projected feature vector of training MSI $m_{c,i}$. The calculated Euclidean distance E is used for classification by using the nearest neighbor (NN) classifier. The image sequence r(t) is classified as class c if the Euclidean distance E is the minimum among other training samples.

4 Experiments and Results

We used SOTON dataset for evaluating the proposed algorithm. SONON dataset was developed by University of University of Southampton [3]. SOTON dataset contains 115 subjects and total 2,128 walking sequences. The walking sequences were filmed in a controlled laboratory environment and captured with a digital video cam at 25 frames per second. The sequence could be divided into two types: (I) subject walking from right to left and (II) subject walking from left to right. For our purposes, the silhouettes is extracted and normalized to a standard size (128 x 88 pixels). We adopted the scheme of FERET [8] for the evaluation and measure the recognition rate by cumulative match score. All experiments are implemented by Matlab and run in a PC computer with P4 2.26GHz and 512MB memory. Fig. 4 shows some walking sequences in SOTON dataset.

In our proposed algorithm, we used MSI as the basis for recognition. As we mentioned before, SOTON dataset contains two types of walking sequences: people walking from left to right and people walking from right to left. We did not separate these sequences for the experiments. In the testing and training subsets, they included sequences from different walking direction. As a result, it is more difficult for the recognition. The experiment could reveal the discriminative power of the proposed representation in gait recognition.

Fig. 4. Examples of walking sequence in SOTON dataset: (a) walking from right to left (b) walking from left to right

4.1 Recognition with Different Number of Training and Testing Samples

To evaluate the recognition effects with different number of training and testing samples, we conduct three tests: (A) 90% of the image sequences in each class is used for training and other 10% is used for testing;(B) 50% of the image sequences in each class is used for training and other 50% is used for testing; (C) 75% of the image sequences in each class is use for training and other 15% is used for testing. The recognition procedure is mentioned in section 3.2 and the result is shown in Table 1. In this experiment, 95% of the accumulated variance of eigenvalues was used. The best identification rate was 87.15% when using (C) 90% train, 10% test. Fig. 5a shows the ranking order statistics for three different tests. The best identification rate was 87.15% by using (A) 90% train and 10%. The experiment revealed that the overall recognition rate by using (A) 90% train and 10% train is higher than using (B) 75% train and 15% test and (C) 50% train and 50% test (see Fig. 5a).

Table 1. Recognition rates for different number of training and testing samples in SOTON

	Top 1	Top 5	Top 10
(A) 90% train 10% test	87.15%	95.18%	95.98%
(B) 75% train 15% test	84.85%	93.76%	95.37%
(C) 50% train 50% test	82.10%	92.31%	94.94%

4.2 Recognition with Different Eigenvalues

In this experiment, we used seven different percentages of accumulated variance of eigenvalues for evaluating the influence of the eigenvalues. The result is tabulated in Table 2. The result showed that the recognition rate is directly proportional to the number of eigenvectors (see Fig. 5b). Again, the overall recognition rate by using (A) 90% train and 10% test is better than the others (B & C).

Table 2. Recognition rates with different eigenvalues

	Percentages of accumulated variance of eigenvalues						
	65%	**70%**	**75%**	**80%**	**85%**	**90%**	**95%**
(A) 90% train, 10% test	(i) 77.11% (ii) 87.55% (iii) 93.98%	(i) 81.53% (ii) 90.36% (iii) 94.38%	(i) 83.13% (ii) 92.37% (iii) 95.98%	(i) 85.54% (ii) 93.57% (iii) 95.98%	(i) 87.15% (ii) 95.58% (iii) 95.98%	(i) 86.75% (ii) 94.78% (iii) 95.98%	(i) 87.15% (ii) 95.18% (iii) 95.98%
(B) 75% train, 15% test	(i) 74.51% (ii) 88.77% (iii) 92.51%	(i) 79.32% (ii) 90.37% (iii) 93.94%	(i) 81.64% (ii) 91.27% (iii) 94.83%	(i) 83.42% (ii) 92.51% (iii) 94.83%	(i) 85.03% (ii) 93.76% (iii) 95.01%	(i) 84.14% (ii) 93.58% (iii) 95.19%	(i) 84.85% (ii) 93.76% (iii) 95.37%
(C) 50% train, 50% test	(i) 71.51% (ii) 87.82% (iii) 91.75%	(i) 76.76% (ii) 89.32% (iii) 92.88%	(i) 79.19% (ii) 90.91% (iii) 94.38%	(i) 80.79% (ii) 91.47% (iii) 94.56%	(i) 81.82% (ii) 92.22% (iii) 94.75%	(i) 81.91% (ii) 92.41% (iii) 94.56%	(i) 82.10% (ii) 92.31% (iii) 94.94%

(i) – Top 1, (ii) – Top 5, (iii) – Top 10

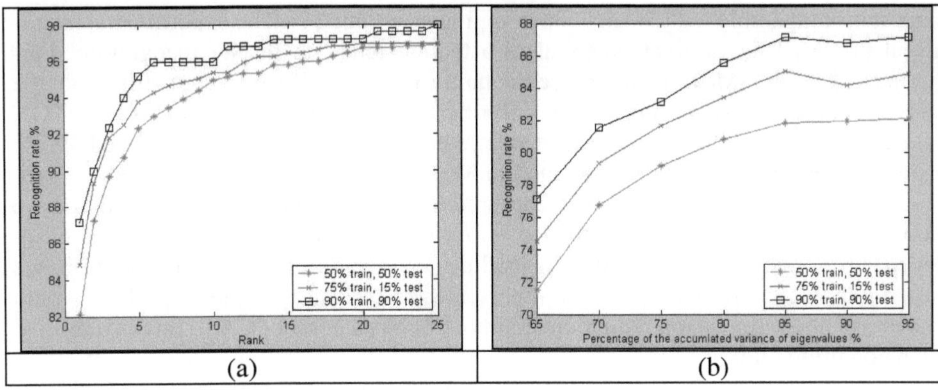

Fig. 5. (a) Identification performance in terms of rank order statistics. (b) The recognition performance under different eigenvalues.

4.3 MSI vs. Unwrapping

We implemented Wang's unwrapping transformation algorithm by using Matlab [7]. The performance of our proposed algorithm is compared with the performance of the Wang's unwrapping algorithm. We also used PCA for dimension reduction before classification. We chose 95% of the accumulated variance of eigenvalues in this experiment. Table 3 shows the recognition rate of the unwrapping algorithm and our proposed algorithm. The recognition rate of our proposed algorithm is comparable with unwrapping algorithm.

Table 3. Recognition rates by using proposed algorithm and unwrapping algorithm in SOTON database (PCA)

Method	**MSI**			**Unwrapping**		
Rank	**Top 1**	**Top 5**	**Top 10**	**Top 1**	**Top 5**	**Top 10**
a. 50% train 50% test	87.15%	95.18%	95.98%	93.16%	97.56%	98.31%
b. 75% train 25% test	84.85%	93.76%	95.37%	94.47%	97.33%	98.04%
c. 90% train 10% test	82.10%	92.31%	94.94%	94.38%	97.59%	98.39%

Remark: 95% of the accumulated variance of eigenvalues is used in this experiment

5 Conclusion and Future Work

In this paper, we proposed a new representation for gait recognition. We used Motion Silhouettes Image (MSI) as the feature template for the recognition. The experiments showed that MSI has a high discriminative ability. The recognition performance is quite promising. The recognition rate is over 80 percent in SOTON dataset by using

MSI. It is comparable with Wang's unwrapping algorithm [7]. The experimental results reveal that MSI has a great potential as a feature template in gait recognition. Furthermore, by using MSI, it can reduce the data storage significantly, but still retain the critical spatial-temporal information for recognition. The original size of SOTON dataset is around 1.92GB. The dataset size reduced to 4.2MB which is reduced 99.7% of the original dataset after transformed to MSIs. In this paper, we evaluated the proposed algorithm in the dataset which captured in the indoor environment. In future, we would like to evaluate the performance of our proposed algorithm under different independent variables situation such as loading, surface and footwear. In addition, we would like to explore other potential representation for gait recognition and we would like to use other dimension reduction technique such as kernel PCA to further improve the recognition rate.

Acknowledgements

We are grateful to the partial support from the CERG grant for the iMASS Project (B-Q569) and the CRG Project (A-PF74) from the Hong Kong Polytechnic University.

References

1. M.P. Murray, A.B. Drought and R.C. Kory, "Walking patterns of normal men," Journal of Bone and Joint Surgery, Vol. 46 – A, No. 2, pp. 335-60
2. Bobick, A.F.; Davis, J.W., "The recognition of human movement using temporal templates", IEEE Trans. on PAMI, Vol. 23, No. 3, pp. 257 – 267, March 2001
3. J. D. Shutler, M. G. Grant, M. S. Nixon, and J. N. Carter "On a Large Sequence-Based Human Gait Database", Proc. 4th International Conference on Recent Advances in Soft Computing, Nottingham (UK), pp 66-71, 2002
4. H. Murase, R. Sakai,"Moving object recognition in eigenspace representation: gait analysis and lip reading, " Pattern Recognition Letters, Vol. 17, pp. 155-62, 1996
5. M. Turk and A. Pentland, "Face Recognition using Eigenfaces," in Proceedings of the Computer Vision and Pattern Recognition, 1991
6. P.S. Huang, C.J. Harris and M.S. Nixon, "Human Gait Recognition in Canonical Space Using Temporal Templates," IEE Proceedings - Vision, Image and Signal Processing, 146(2), pp. 93-100, 1999
7. L. Wang and T. Tan, "Silhouette Analysis-Based Gait Recognition for Human Identification," IEEE Trans on PAMI, Vol. 25 (12), 1505-1518, 2003
8. J. Phillips, H. Moon, S. Rizvi, and P. Rause, "The FERET Evaluation Methodology for Face Recognition Algorithms," IEEE Trans. PAMI, Vol. 22, No.8, pp. 1090-1104, Oct. 2000

Reconstruction of 3D Human Body Pose for Gait Recognition

Hee-Deok Yang and Seong-Whan Lee*

Department of Computer Science and Engineering,
Korea University, Anam-dong, Seongbuk-gu,
Seoul 136-713, Korea
{hdyang, swlee}@image.korea.ac.kr

Abstract. In this paper, we propose a novel method to reconstruct 3D human body pose for gait recognition from monocular image sequences based on top-down learning. Human body pose is represented by a linear combination of prototypes of 2D silhouette images and their corresponding 3D body models in terms of the position of a predetermined set of joints. With a 2D silhouette image, we can estimate optimal coefficients for a linear combination of prototypes of the 2D silhouette images by solving least square minimization. The 3D body model of the input silhouette image is obtained by applying the estimated coefficients to the corresponding 3D body model of prototypes. In the learning stage, the proposed method is hierarchically constructed by classifying the training data into several clusters recursively. Also, in the reconstructing stage, the proposed method hierarchically reconstructs 3D human body pose with a silhouette image. The experimental results show that our method can be efficient and effective to reconstruct 3D human body pose for gait recognition.

1 Introduction

There has been a growing interest in improving automatic person identification. Established methods of automatic person identification range from fingerprint to face, iris and gait recognition. Gait recognition is an attractive biometric for passive surveillance system. Unlike other biometrics, gait recognition can be measured at a distance.

Most of gait recognition systems consist of two main stages: a feature extraction stage and a recognition stage and features for gait recognition are extracted from silhouette image [2, 6, 9, 10]. Collins et al. [4] used silhouette corresponding to certain poses such as the double-support and mid-stance poses. Classification of person is achieved by comparing their silhouette sequence. He and Debrunner [6] computed a quantized vector of Hu moments from the subject's silhouette image and used them for recognition by using an HMM(Hidden Markov Model). In Murray's research [11], ankle rotation and spatial displacements were shown to have individual consistency in repeated. To extract features for gait recognition, a number of researchers have been developed for estimating and reconstructing 2D or 3D body pose [1, 3, 7, 12].

* To whom all correspondence should be addressed.

D. Zhang and A.K. Jain (Eds.): ICB 2006, LNCS 3832, pp. 619–625, 2005.

In this paper, we propose a solution to the problem of reconstructing 3D human body pose for gait recognition using a hierarchical linear learning method. The proposed method is related to machine learning models [2, 6] that use a hierarchical method to reduce the complexity of the learning problem by splitting it into several simpler ones.

2 Top-Down Learning

2.1 3D Human Model

The 3D human modeling includes human body representation and kinematics. For human body representing, we build a 3D human model consists of body segments, joints, and a perfect coordination among them. It has 17 body parts with 17 joints and 37 DOF(Degree of Freedom) and has 5 additional joints which are end-effectors used to calculate angle of each body segment in inverse kinematics [1]. Our 3D human model and the corresponding tree structure are shown in Fig. 1.

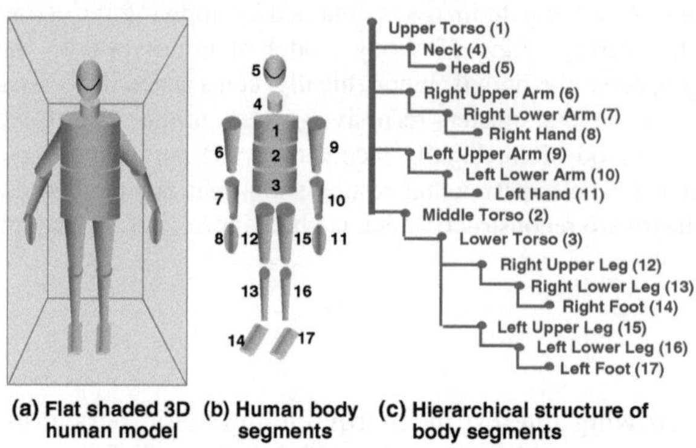

Fig. 1. The proposed 3D human model

2.2 3D Gesture Representation

In order to reconstruct 3D human body pose from continuous silhouette images, we used a learning based approach. If we have sufficiently large amount of pairs of a silhouette image and its 3D body model as prototypes of 3D body models, we can reconstruct an input 2D silhouette image by a linear combination of prototypes of 2D silhouette images. Then we can obtain its reconstructed 3D body model by applying the estimated coefficients to the corresponding 3D body model of prototypes as shown in Fig. 2. Our goal is to find an optimal parameter set α which best reconstructs a given silhouette image. To make various prototypes of 2D silhouette images and theirs 3D body models, we generate data using the 3D human model described in Sec. 2. 1.

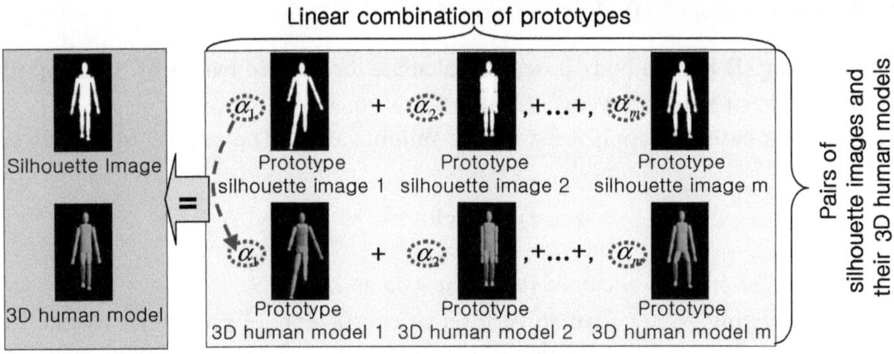

Fig. 2. Basic idea of the proposed method

The silhouette image is represented by a vector $s=(s'_1,\ldots,s'_n)^T$, where n is the number of pixels in the image and s' is the intensity value of a pixel in the silhouette image. The 3D body model is represented by a vector $p=((x_1,y_1,z_1),\ldots,(x_q,y_q,z_q))^T$, where x, y and z are the position of body joint in the 3D world and q is the number of joints in 3D human model. Eq. (1) explains training data.

$$S=(s_1,\ldots,s_m),\ P=(p_1,\ldots,p_m) \tag{1}$$

where m is the number of prototypes.

A 2D silhouette image is represented by a linear combination of a number of prototypes of 2D silhouette images $\tilde{S}$ and its 3D body model $\tilde{P}$ represented by estimated coefficients to the corresponding 3D body model of prototypes by such as:

$$\tilde{S}=\sum_{i=1}^{m}\alpha_i s_i,\ \tilde{P}=\sum_{i=1}^{m}\alpha_i p_i \tag{2}$$

2.3 Hierarchical Statistical Model

In order to reduce search area, we construct our algorithm hierarchically. Given a set of silhouette images and their 3D body models for training, we classify them into several clusters. A set of cluster is built in which each has similar shape in 2D silhouette image space. Then, for each of the cluster, we divide it into several sub-clusters recursively. To divide training data into sub-clusters, we apply K-means algorithm.

The lower level is the mean value of the data in higher-cluster. In our model, the value of the first level is the mean value of each cluster in the second level, the values of the second and third level are the mean value of each cluster in the higher level respectively, and all leaf nodes are about 100,000 and each cluster in the third level has about 10~60 leaf nodes. Each cluster in the hierarchical model is presented by one gait representation described by Eq. (2).

2.4 Reconstruction of 3D Human Body Pose

To reconstruct 3D human body pose, we calculate the inverse matrix of S in Eq. (2). The inverse S^{-1} of a matrix S exists only if S is square. However, a matrix S is not square. In this case, we apply least square minimization. The pseudo inverse is defined such as:

$$S^{+} = (S^{T}S)^{-1}S^{T} \tag{3}$$

And, the solution $\alpha S = \widetilde{S}$ can be rewritten such as $\alpha = S^{+}\widetilde{S}$.

After calculating $\alpha S = \widetilde{S}$, we solve a set of coefficients of prototypes by Eq. (3). The silhouette image and the position of each segment of 3D human model calculated such as:

$$\widetilde{S}_i = \sum_{k=1}^{m} \alpha_k s_k \,,\quad \widetilde{P}_i = \sum_{k=1}^{m} \alpha_k p_k \tag{4}$$

3 Gait Recognition

For recognizing gait, we compute 5 angles as shown in Fig. 3. The trajectory of the above features used to recognize gait of subject.

$$F = (\theta_1, \theta_2, \ldots, \theta_c) \tag{5}$$

where c is the number of angle used as features.

Given s subjects and each subject represents feature sequences of subject's gait. Let $D_{i,j}$ be the jth F in class I and N_i the number of F in the ith class. We can easily obtain the mean m_d of full data set and the global covariance matrix such as:

$$m_d = \frac{1}{N_t} \sum_{i=1}^{s} \sum_{j=1}^{N_t} D_{i,j} \tag{6}$$

By applying PCA for training data, we can obtain eigenvalues $\lambda_1, \lambda_2, \ldots, \lambda_k$ and the associated eigenvectors $e_1, e_2, \ldots, e_k$. We construct transform matrix $E = [e_1, e_2, \ldots, e_k]$ to project an $D_{i,j}$ into a point $P_{i,j}$ in the eigensapce. A gait sequence will be represented as a trajectory in the eigenspace. The projection average C_i of each training sequence in the eigenspace is given by

$$P_{i,j} = [e_1 \; e_2 \; \ldots \; e_k]^T D_{i,j} \tag{7}$$

$$C_i = \frac{1}{N_t} \sum_{j=1}^{N_t} P_{i,j} \tag{8}$$

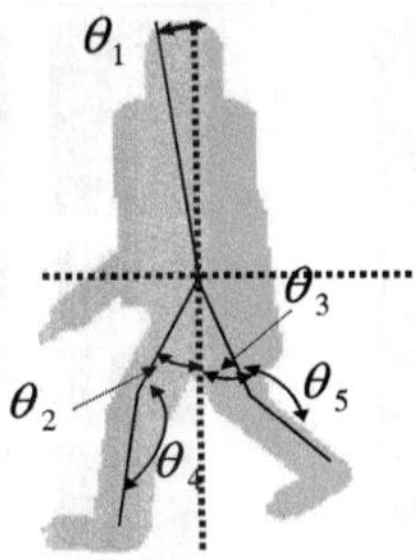

Fig. 3. The space domain features

4 Experimental Results and Analysis

4.1 Experimental Results

For training the proposed method, we generated approximately 100,000 pairs of silhouette images and their 3D human models. We use a perspective projection transform to achieve silhouette and depth images.

For testing the performance of our gait recognition method, we use two data sets. One is Georgia Tech Database [14] and the other is CMU Motion of Body(BoMo) database [5]. For testing the reconstruction performance of 3D body pose, we used the KU Gesture Database [8, 15].

4.2 Experimental Results

Fig. 4. shows the reconstructed results from the Georgia Tech. Fig. 4(a) and Fig. 4(b) are source image and silhouette image respectively. Fig. 4(c) represents silhouette image, left side, front, right side view of reconstructed 3D human body respectively. Fig. 5. shows the reconstructed results from the CMU Mobo database.

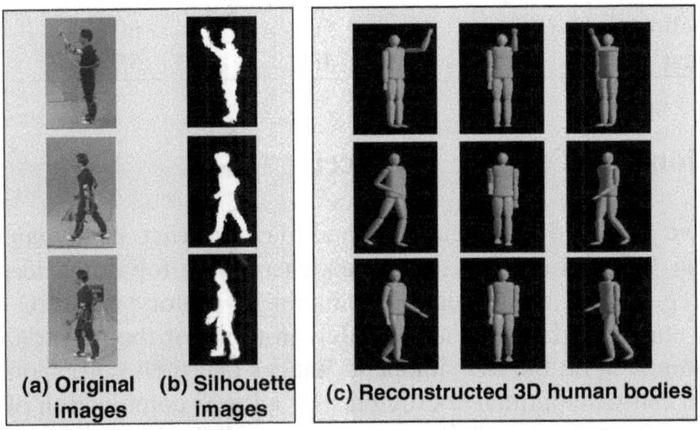

Fig. 4. Examples of the reconstructed 3D human body pose with the Georgia Tech database

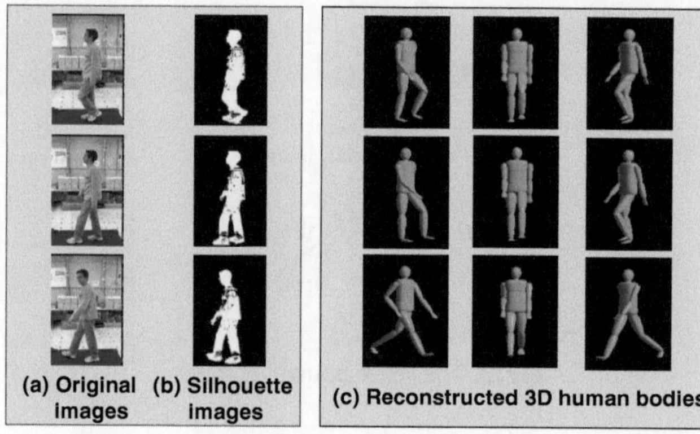

Fig. 5. Examples of the reconstructed 3D human body pose with the CMU MoBo database

For recognizing subject, we select the normalized Euclidean distance measure. The accumulated distance between the test sequence $P(t)$ and projection averages $C(i)$ obtained such as:

$$d(i) = \sum_{t=1}^{T} \left\| \frac{P(t)}{\|P(t)\|} - \frac{C(i)}{\|C(i)\|} \right\|^2 \tag{9}$$

The classification result is accomplished by choosing the minimum of $d(i)$. The sequences are divided into two training and testing sets, by the ratio 4:1. The result of four experiments is shown in Table 1.

Table 1. Performance for CMU MoBo database in terms of PI at rank 1, 2, 5, 10

Train vs Test	PI (%)(at rank)			
	1	2	5	10
Slow vs Slow	60	65	80	100
Fast vs Fast	55	60	75	90
Incline vs Incline	50	50	60	85
Slow vs Fast	30	40	55	75

5 Conclusion and Further Research

In this paper, we proposed an efficient method to reconstruct 3D human body pose for gait recognition from monocular image sequence using top-down learning. Human body pose is represented by a linear combination of prototypes of 2D silhouette images and their corresponding 3D body models in terms of the position of a predetermined set of joints. With the 2D silhouette images and their corresponding 3D body models, we can estimate optimal coefficients for a linear combination of prototypes of the 2D silhouette images and their corresponding 3D body models by solving least square minimization.

The performance of the presented method shows that reconstructing 3D human body pose from visual features obtained from a single image is possible and the extracted feature by model 3D human body model is effective for gait recognition.

Acknowledgements

This research was supported by the Intelligent Robotics Development Program, one of the 21st Century Frontier R&D Programs funded by the Ministry of Commerce, Industry and Energy of Korea.

References

1. Benadbdlkader et al.: Gait Recognition using Image Self-Similarity. EURASIO Journal on Applied Signal Processing, Vol. 4, (2004) 1-14
2. Bissacoo, A. et al.: Recognition of Human Gaits. Proc. of the IEEE Computer Society Conference on Computer Vision and Pattern Recognition, Hawaii, USA, (Dec 2001) 52-57
3. Bowden, R., Mitchell, T. A., Sarhadi, M.: Non-linear Statistical Models for 3D Reconstruction of Human Pose and Motion from Monocular Image Sequences. Image and Vision Computing, Vol. 18, No. 9, (2000) 729-737
4. Collins, R., Gross, R., Shi, J.: Silhouette-based Human Identification from Body Shape and Gait. Proc. of the IEEE International Conference on Automatic Face and Gesture Recognition, Washington D.C., USA, (May 2002) 351-356
5. Gross, R., Shi, J.: The CMU Motion of Body(MOBO) Database. Technical report CMU-RI-TR-01-18, Robotics Institute, Carnegie Mellon University, (June 2001)
6. He, Q., Debrunner, C.: Individual Recognition from Periodic Activity using Hidden Markov Models. Proc. of the IEEE Workshop on Human Motion, Austin Texas, USA, (Dec. 2000) 47-527
7. Heap, T., Hogg, D.: Improving Specificity in PDMs using a Hierarchical Approach. Proc. of 8th British Machine Vision Conference, Colchester, UK, (Sep. 1997) 590-599
8. Hwang, B.-W., Kim, S. Lee, S.-W. "Full-Body Gesture Database for Analyzing Daily Human Gestures," Proc. of 1st Int'l Conf. on Intelligent Computing, Hefei, China, (Aug. 2005) 611-620.
9. Kale, A. et al.: Identification of Humans using Gait. IEEE Trans. on Image Processing, Vol. 13, (2004) 1163-1173
10. Little, J., Boyd, J.: Recognizing People by Their Gait: the Shape of Motion. Videre, Vol. 1, No. 2, (1998) 1-32
11. Murray, M.: Gait as a Total Pattern of Movement. American Journal of Physical Medicine, Vol. 46, No. 1, (1967) 290-332
12. Rosales, R., Sclaroff, S.: Specialized Mapping and the Estimation of Human Body Pose from a Single Image. Proc. of IEEE Workshop on Human Motion, Texas, USA (Dec. 2000) 19-24
13. Yam, C. et al.: Automated Person Recognition by Walking and Running via Model-based Approaches. Pattern Recognition, Vol. 37, No. 5, (2004) 1057-1072
14. The Gait Recognition Database at Georgia Tech, http://www.cc.gatech.edu/cpl/ projects hid/images.html.
15. The KU Gesture Database, http://gesturedb.korea.ac.kr/.

Artificial Rhythms and Cues for Keystroke Dynamics Based Authentication

Sungzoon Cho and Seongseob Hwang

Department of Industrial Engineering, Seoul National University,
San 56-1, Shillim-dong, Kwanak-gu, Seoul 151-744, Korea
{zoon, hss9414}@snu.ac.kr
http://dmlab.snu.ac.kr

Abstract. Biometrics based user authentication involves collecting user's patterns and then using them to determine if a new pattern is similar enough. The quality of the user's patterns is as important as the quality of the classifier. But, the issue has been ignored in the literature since the popular biometrics are mostly trait based such as finger prints and iris so that its pattern quality depends on the quality of the input device involved. However, the quality of the user's patterns of behavior based biometric such as keystroke dynamics can be improved artificially by increasing the peculiarity of the typing style. In this paper, we propose several ways to improve the quality. But, first we define the quality of patterns in terms of two factors: uniqueness and consistency. Finally, the results of a preliminary experiment are presented that support the utility of the proposed methods.

1 Introduction

User authentication based on keystroke dynamics has been around for several decades with many papers, patents and products available [1, 2, 3, 4, 5, 6, 7]. There are three steps involved. First, a user registers or enrolls his/her timing vector patterns. Second, a classifier is built using the timing vector patterns. Third, whenever a new timing vector pattern is presented, it is either accepted or rejected based on the classification made by the classifier. Advantages include a low cost, usability and ease for remote access control. A relatively lower accuracy was reported, however, since being a behavior based biometric, keystroke dynamics tends to be less consistent. Recently, however, fairly high accuracies have been achieved through a combination of rather complex models such as neural network, support vector machine, and genetic algorithm [6, 7]. But, when only a small number of patterns are available, it is difficult to achieve practically acceptable error rates.

Most research efforts related to biometrics based authentication have focused on improving classifier accuracy. In this paper, however, we focus on a different aspect of the problem, i.e. improving the quality of the timing vector patterns. Since keystroke dynamics is a biometric based on user's behavior, patterns can be made "better" by conscious efforts of the user. The quality of the patterns of trait based biometrics such as fingerprint, face, iris and palm prints depends less on the user, and more

D. Zhang and A.K. Jain (Eds.): ICB 2006, LNCS 3832, pp. 626 – 632, 2005.

on the input device involved. Thus, improving the quality involves increasing the total cost of the system. The quality of keystroke dynamics patterns can be defined in terms of two factors: uniqueness and consistency. Uniqueness is concerned with how different the impostor's patterns are to those enrolled in the registration stage. Uniqueness tends to depend on the peculiarity of the typing style. Consistency is concerned with how similar the user's patterns are to those enrolled in the registration stage. Consistency depends on the typing skill and the concentration level of the user. A combination of a high consistency and a high uniqueness will lead to a better discriminability or the ability to make better classification of user's patterns and impostors' patterns. In this paper, we propose several ways to increase uniqueness and consistency of keystroke dynamics. For uniqueness, we propose artificially designed rhythms to be used. They include pause, musical rhythm, staccato, legato, and slow tempo. For consistency, we propose timing cues to be used. They include auditory, visual, and audiovisual cues. A preliminary experiment shows that they improve the quality of patterns, thus discriminability.

This paper is structured as follows. The next section presents measures of uniqueness, consistency, and discriminability. Then, several ways to enhance the uniqueness as well as the consistency are proposed along with empirical evidences to support their utility. Finally, a summary and a list of future work are presented.

2 Uniqueness, Consistency, and Discriminability

Uniqueness is concerned with how different the valid user's keystroke dynamics is from those of potential impostors. A simple measure of uniqueness can be defined as the average distance of the impostors' typing patterns from the prototype or centroid of the user's typing patterns registered in the enrollment step. Let $\{\vec{x_i} \mid i=1,\cdots,N_x\}$, $\{\vec{y_j} \mid j=1,\cdots,N_y\}$ and $\{\vec{z_k} \mid k=1,\cdots,N_z\}$ denote the valid user's training (enrollment) pattern set, the valid user's test pattern set and the impostors' pattern set, respectively. Given the prototype pattern $\vec{m}=\sum_i \vec{x_i}/N_x$, Uniqueness is defined as

$$\text{Uniqueness} = \sum_{k=1}^{N_z} |\vec{z_k}-\vec{m}|/N_z - \sum_{i=1}^{N_x} |\vec{x_i}-\vec{m}|/N_x . \tag{1}$$

Consistency is concerned with how similar the valid user's future keystroke dynamics is to the current keystroke dynamics. A simple measure of inconsistency, the opposite concept, can be defined as the average distance of the valid user's own future typing patterns to the prototype or centroid of the user's typing patterns registered in the enrollment step. Inconsistency is defined as

$$\text{Inconsistency} = \sum_{j=1}^{N_y} |\vec{y_j}-\vec{m}|/N_y - \sum_{i=1}^{N_x} |\vec{x_i}-\vec{m}|/N_x . \tag{2}$$

Of course, neither measure can be actually calculated because neither impostor patterns nor future user patterns are available. We can measure these here from the data sets that include user training data, user future test data, and impostor patterns.

As a discriminability measure, we propose to use the difference between the smallest distance from the impostor's pattern to the prototype and the largest distance from the user's future pattern to the prototype.[1] Now, discriminability is defined as

$$\text{Discriminability} = \min_k |\vec{z_k} - \vec{m}| - \max_j |\vec{y_j} - \vec{m}|. \tag{3}$$

If the former (minimum impostor distance to prototype) is smaller than the latter (maximum user distance to prototype), we obtain a negative discriminability value. Now, one can achieve a perfect discrimination with the use of a proper threshold. We will show that the proposed ways to increase uniqueness and consistency result in a better discrimination.

3 Strategies to Increase Typing Uniqueness

In this section, we propose four different ways to increase typing uniqueness. First is inserting any number of pauses. For instance, "pa__ss__word" shows an artificial rhythm containing two pauses of two beats long each. A user types p, and a in a natural rhythm, and then inserts a pause of two beats long. Typing s and s in a natural rhythm is followed by another pause of two beats long. Finally, w, o, r, and d are typed in a natural rhythm. Second is typing a password according to a rhythm from certain tune, chant, or rooting. In the experiment, the user employed the rooting rhythm used and popularized by Korean supporters during World Cup 2002 Korea-Japan. The musical rhythm has an advantage. It is easy for a user to remember, thus results in more consistent patterns. A potential disadvantage might be its applicability, particularly to those users with a less rhythmical sense. Third is typing a password with a minimum duration time of each character. So called "staccato" was adopted from a bowing style for string instruments characterized by "being cut short crisply and detached." Staccato typing results in patterns that are very short in duration lengths and that are very uniform in interval lengths. Fourth is the opposite of staccato. In so called "legato" style, one tries to keep each character key down as long as possible, i.e. to maximize duration time of each character. Fifth is typing a password in a slow tempo. Tempo is manifest in the length of interval.

In order to check if these methods are useful, we conducted a simple preliminary test involving one password and one user. Password "password" was chosen. A user typed it in a natural rhythm for 20 times. Then, each of the above mentioned strategies was employed for typing. Two kinds of pauses were tried. For short pauses, "pa__ss__word" explained above was used. For long pauses, "p___assword____" was used which contains two long pauses of three beats and four beats each. Musical rhythm patterns were obtained with the soccer rooting chant mentioned above. Two

---

[1] We have in mind a simple classifier that is based on the distance to the prototype. A typing pattern is accepted if its distance to the prototype is small and rejected if it is large. A proper threshold can be identified.

kinds of staccato were tried: single character staccato and double (two consecutive) character staccato. The double staccato patterns were collected with 'p' and 'a' typed together as fast as possible, followed by the pair of 's' and 's' typed together as fast as possible. The user did not pay any attention to the interval between the two pairs. The pairs of 'w' and 'o' and 'r' and 'd' were typed in the same manner. The legato patterns were collected as explained above. Also, the slow tempo patterns were obtained as explained above. Note that one password can be converted into many different typing pattern sets, each of which corresponds to a uniqueness enhancing method.

Figure 1 shows the values of inconsistency, uniqueness, and discriminability of seven passwords. Their uniqueness values all increased from at least 200% (short pauses) to 500% (slow tempo). Note that the uniqueness values of Long Pauses and Slow Tempo are 1,300 msec and 1,540 msec, respectively so that the corresponding bars in the figure were chopped to fit 1,000. Inconsistency values did not increase much with exceptions of Long Pauses and Slow Tempo. What really matters is the fact that a negative discriminability value of Natural was turned into positive values by employing artificial rhythms. Now all the pattern sets can be perfectly discriminated with a proper threshold.

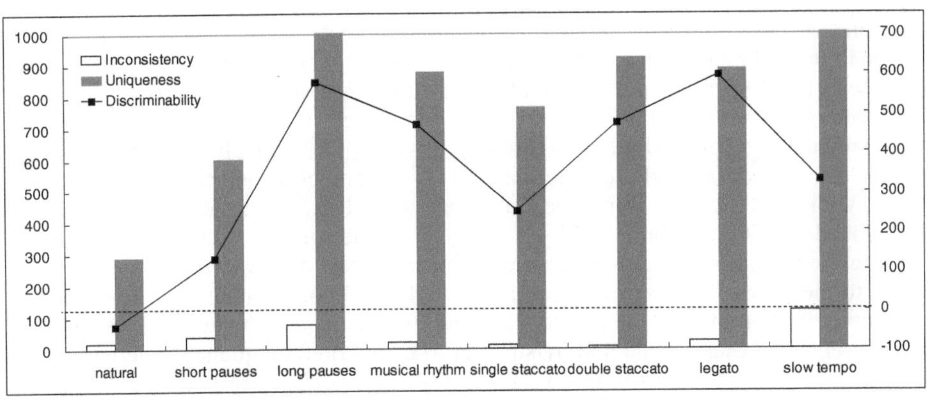

Fig. 1. Uniqueness, Inconsistency (*left scale*) and Discriminability (*right scale*) of typing pattern set obtained using natural as well as artificial rhythms. Uniqueness values of Long pauses and Slow tempo are 1,300 and 1,540, respected, but shown here chopped to fit to 1,000.

Figure 2 shows a more detailed picture of what really happened. Each figure contains cumulative distributions of the distances from the training prototype of the user's training patterns (Tr20), the user's test patterns (Test) and the impostors' patterns (Impostor). Discriminability is related to the distance between the solid curve in the middle (Test) and the thick solid curve to the right (Impostor). The farther, the better. When the figure of natural (a) is compared with that of long pauses (b), one thing is clear: the impostor curve is shifted right away from the user test curve. This separation of test patterns and impostor patterns would make a perfect discrimination possible. For (b), its test pattern curve is also pushed right, which was caused by a decrease of consistency in typing.

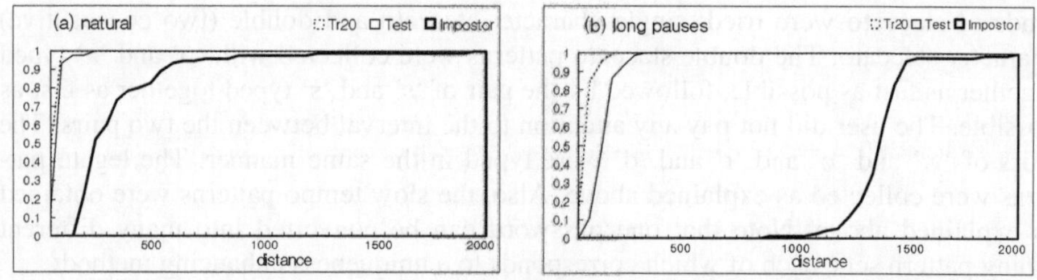

Fig. 2. Cumulative distributions of Distances from Training Prototypes of Tr20 (dotted), Test (solid) and Impostor (thick solid) when the employed method is (a) natural, (b) long pauses

4 Strategies to Increase Typing Consistency

In this section, we propose ways to increase typing consistency. First, let us take a look at the slow tempo patterns that had a high inconsistency value in Figure 1. The patterns of slow tempo were collected again in the presence of an auditory cue ticking every 750 msec. The inconsistency value was reduced to 8 from 121. The uniqueness value was slightly reduced to 1436 from 1540. Thus, the discriminability value was increased to 728 from 330. In short, consistency was improved almost 15 times while discriminability was improved more than twice with a simple auditory cue. See Figure 3 for comparison. User's typing patterns during and after enrollment are quite similar now.

Encouraged by the improvement, we set out to test the effectiveness of various cues with a long pause rhythm of "pass____word____," which contains two long pauses of four beats each. Five different users typed and collected patterns of the long pauses with the following three cues: auditory, visual and audiovisual cue. First, the same speed of 160 per minute ticking sound from a metronome was used for the auditory cue. Second, a video clip of a hammer hitting a nail on a wooden block was presented to the users on the same screen at a speed of 160 per minute. Third, a synchronized combination of both the auditory cue and the visual cue the audiovisual cue was also presented to users as the audiovisual cue. The average uniqueness, inconsistency, and discriminability values over five users for various cues are shown in Figure 4. Note that uniqueness of visual cue is 2,086 so it was chopped at 2,000. Overall, the use of cues clearly decreased inconsistency and increased discriminability.

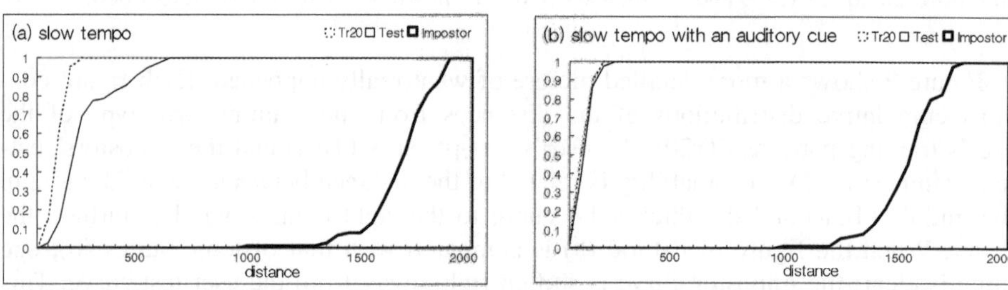

Fig. 3. Cumulative distributions of Distances from Training Prototypes of Tr20 (dotted), Test (solid) and Impostor (thick solid) when the employed method was (a) slow tempo (b) slow tempo with an auditory cue

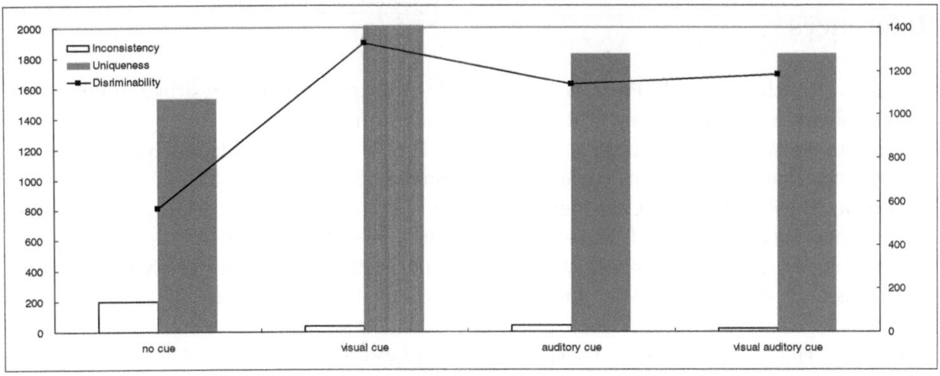

Fig. 4. Uniqueness, Inconsistency (*left scale*) and Discriminability (*right scale*) of typing pattern set obtained with various cues used

5 Conclusion

This paper proposes several ways to improve consistency and uniqueness in keystroke dynamics. First, for uniqueness improvement, artificial rhythms were suggested. A preliminary test involving one user and five such rhythms was performed. Typing patterns obtained with the proposed artificial rhythms used were found to be significantly more unique than those patterns obtained with a natural rhythm used. The improvement of uniqueness led to better discriminability. Second, for consistency improvement, the use of visual, audio and audiovisual cues was suggested. A preliminary test involving five users and three such cues was performed. Typing patterns obtained with those cues played were found to have decreased inconsistency in all cases. Some cues were found to be more useful to some users.

The ideas and results presented in this work are preliminary. There are many issues to solve and experiments to do. First, much more users and passwords need to be involved in the experiment. Second, artificial rhythms other than those presented here need to be identified. Questions that need to be answered include how easy it is to remember, how long it is to remember, and how consistent typing patterns are in the future. Third, those cues introduced here are the most basic ones. What are the useful cues other than those? Which cues are useful to which users? Is it better to allow a user to choose whatever he/she likes from a menu of various cues? Fourth, it will be interesting to see how the tempo of the cue affects the consistency. Various tempos need to be investigated. Fifth, how this idea of using artificial rhythms and helpful cues translates into situations where an input device other than a keyboard is involved? Examples include cellular phone pads, ATM pads, and some digital door lock pads. Finally, other measures such as ROC should be employed.

Acknowledgements

This work was supported by grant No. R01-2005-000-103900-0 from the Basic Research Program of the Korea Science and Engineering Foundation. The authors would like to thank Mr. Sung Hoon Park for developing modules that collect keystroke data.

References

1. Gaines, R., Lisowski, W., Press, S., Shapiro, N.: Authentication by keystroke timing: some preliminary results. Rand Report R-256-NSF. Rand Corporation (1980)
2. Leggett, J., Williams, G., Usnick, M., Longnecker, M.: Dynamic identity verification via keystroke characteristics. Int. J. Man-Machine Studies. 35 (1991) 859-870
3. Brown, M., Rogers, S.J.: User identification via keystroke characteristics of typed names using neural networks. Int. J. Man-Machine Studies. 39 (1993) 999-1014
4. Obaidat, M., Sadoun, S.: Verification of computer users using keystroke dynamics. IEEE Transactions on Systems, Man and Cybernetics, Part B:P Cybernetics. 27(2) (1997) 261-269
5. Cho, S., Han, C., Han, D., Kim, H.: Web-based keystroke dynamics identity verification using neural network. J. Organizational computing and electronic commerce. 10(4) (2000) 295-307
6. Cho, S., Han, D: Apparatus for Authenticating an Individual Based on a Typing Pattern by Using a Neural Network System, Patent No. 6,151,593, Nov. 21, 2000, US Patent and Trademark Office, Washington DC 20231, 2000. (2000)
7. Yu, E., Cho, S.: Keystroke dynamics identity verification - its problems and practical solutions. Computers and Security. 23(5) (2004) 428-440

Retraining a Novelty Detector with Impostor Patterns for Keystroke Dynamics-Based Authentication

Hyoung-joo Lee and Sungzoon Cho*

Department of Industrial Engineering, Seoul National University,
San 56-1, Shillim-dong, Kwanak-gu, 151-744, Seoul, Korea
{impatton, zoon}@snu.ac.kr

Abstract. In keystroke dynamics-based authentication, novelty detection methods have been used since only the valid user's patterns are available when a classifier is built. After a while, however, impostors' keystroke patterns become also available from failed login attempts. We propose to retrain the novelty detector with the impostor patterns to enhance the performance. In this paper the support vector data description (SVDD) and the one-class learning vector quantization (1-LVQ) are retrained with the impostor patterns. Experiments on 21 keystroke pattern datasets show that the performance improves after retraining and that the one-class learning vector quantization outperforms other widely used novelty detectors.

1 Introduction

While passwords are the most popular in identity verification, they become vulnerable when they have been leaked out or stolen. To make up for the weakness, keystroke dynamics-based authentication has been motivated by the observation that a user's keystroke patterns are repeatable and distinct from those of other users [1]. It can be combined with passwords in almost a user-transparent fashion. Even if an impostor has obtained the password, the account can be protected from the intrusion through the keystroke-based authentication as illustrated in Figure 1. Other biometric methods have been also proposed for complementing or replacing the password method, e.g. fingerprint, iris, voice, etc. [2]. However, these methods need very expensive devices [3] and, more importantly, users may be reluctant to provide their biometric information. On the other hand, the keystroke-based method needs no additional device and keystroke data can be collected relatively easily.

Every time a user types his or her password, a keystroke pattern is defined. The times that each key is stroked and then released are measured by millisecond. A "duration" denotes a time interval during which a key is pressed and an "interval" a latency between a key and the next key. Then, a password of

* Corresponding author.

D. Zhang and A.K. Jain (Eds.): ICB 2006, LNCS 3832, pp. 633–639, 2005.

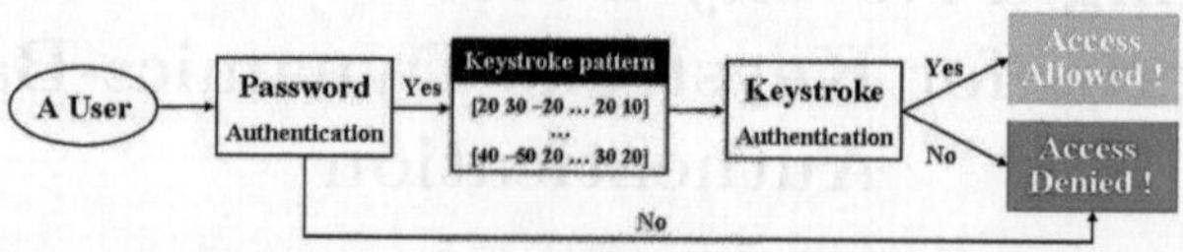

Fig. 1. The framework of a keystroke dynamics-based authentication: If a user types an incorrect password, the access will be immediately denied. Even if the correct password is presented, the keystroke pattern should be more or less accordant to the registered patterns to be allowed.

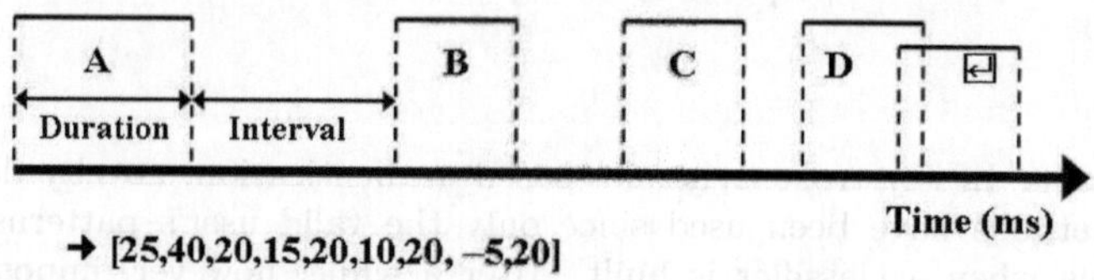

Fig. 2. Transforming a keystroke pattern into a timing vector when a user inputs a string 'ABCD': The duration and interval times are measured by millisecond

m characters can be transformed into a $(2m + 1)$-dimensional timing vector. Figure 2 illustrates how a string 'ABCD' can be represented as a 9D timing vector. The negative value means that the user released the 'D' after stroking the enter key.

After a number of the user's keystroke patterns are collected, a classifier is constructed based on them. One can employ a statistical model [3], [4], [5] or a neural network model [6]. Many of previous researches have used discrimination-based learning techniques, for which one needs to collect patterns from not only the valid user but also impostors. However, when building a classifier in the beginning, impostor patterns are impossible and undesirable to collect. This limitation can be overcome by the novelty detection framework [7], [8]. In novelty detection, the valid user's patterns are denoted as normal and all other individuals' patterns as novel. Then, a model learns characteristics of normal patterns and detects novel patterns that are different from the normal ones. In a geometric sense, the novelty detector defines closed boundaries around the normal patterns in input space [9].

Most of novelty detectors also have limitations that they are unsupervised learners assuming that novel patterns do not exist. So they utilize only normal patterns during training even though in some cases there exist, if few, novel patterns. For instance, intrusions may be attempted and somehow detected. Or even valid users may type their own passwords much differently from their typical keystroke patterns, so those patterns become impostor patterns. Although the impostor patterns are not sufficient to train a classifier, they can help a novelty detector generate more accurate and tighter boundaries. After impostors' keystroke patterns become available from those failed login attempts, the novelty detector which was trained only with the user's patterns can be retrained with the impostor patterns. A few methods have been proposed to exploit novel patterns [10], [11]. It has been experimentally shown that one can achieve a higher accuracy by utilizing them. Recently, the authors proposed a method to take advantage of the novel patterns [11]. We propose to utilize, if available, the novel patterns for retraining. In particular, so-called one class learning vector quantization (1-LVQ) method [11] is recommended.

This paper is organized as follows. The next section introduces two novelty detectors which can utilize impostor patterns, the support vector data description (SVDD) [10] and the 1-LVQ. In Section 3, the SVDD and the 1-LVQ are applied to keystroke pattern datasets and compared with other novelty detectors. Finally, in Section 4 conclusions and discussion are given.

2 Retraining a Novelty Detector with Impostor Patterns

A valid user's timing vector pattern set constitutes training data $\mathbf{U} = \{(\mathbf{x}_i, y_i)|i = 1, 2, \cdots, N_{\mathbf{U}}\}$, where $\mathbf{x}_i \in \mathbb{R}^d$ is a keystroke pattern represented as a timing vector and $y_i = +1$ is its class label. Then, a novelty detector can be trained with this training dataset. When a person tries to access a system, his keystroke pattern is measured and input to the novelty detector. If the detector recognizes the pattern as normal, he will be allowed access. If the detector rejects the pattern as novel, on the other hand, he will be denied access.

In the beginning, only the user's keystroke patterns are available. However, as time passes, impostor patterns may appear in some cases. First, some impostors may attempt access and be somehow detected and denied access. Second, even the valid user may type inconsistently. When he is rejected as an impostor, that pattern is regarded as an impostor pattern. In these cases, a set of impostor patterns can be denoted as $\mathbf{I} = \{(\mathbf{x}_i, y_i)\}, y_i = -1$. Then, a combined training dataset $\mathbf{X} = \mathbf{U} \cup \mathbf{I}$ can be formed. Usually, the number of the valid user's patterns is much greater than that of impostor patterns, i.e. $|\mathbf{U}| \gg |\mathbf{I}|$, making it impossible to train a binary classifier. Instead, a novelty detector is trained with the user's patterns and then later retrained when the impostor patterns become available. Among novelty detectors that can be retrained with the impostor patterns, the SVDD [10] and the 1-LVQ [11] are considered in this paper.

2.1 Support Vector Data Description (SVDD)

An SVDD tries to define a hypersphere with a minimal volume so that it surrounds as many normal patterns and as few novel patterns as possible [10]. The radius and the center of the hypersphere is denoted respectively as R and $\mathbf{a}$. They can be found by, like a support vector machine (SVM), the standard quadratic programming techniques. In the authentication process, an unknown keystroke pattern $\mathbf{z}$ is accepted as the genuine user's if $\|\mathbf{z} - \mathbf{a}\|^2 \leq R^2$, or rejected as an impostor's otherwise. Using Mercer kernels allows to define boundaries more flexible than just a hypersphere. If a radial basis function (RBF) kernel is employed, the SVDD provides a solution in essentially the same form as the Parzen window or the one-class support vector machine (1-SVM).

2.2 One-Class Learning Vector Quantization (1-LVQ)

The 1-LVQ algorithm is a modified form of the original LVQ [11]. Just as an LVQ, a 1-LVQ is initialized by updating codebooks using a conventional SOM. When updating initial codebooks, only the normal patterns are used. The SOM generates a set of codebooks $\mathbf{W} = \{\mathbf{w}_k | k = 1, 2, \ldots, K\}$, $K \ll N$ to represent the normal data. When codebook update is done, the codebook $\mathbf{m}(\mathbf{x})$ of an input pattern $\mathbf{x}$ and the Voronoi region $\mathbf{S}_k$ of each codebook $\mathbf{w}_k$ are defined.

When the training set includes the novel patterns, a learning rule different from the conventional one can be obtained as follows,

$$\mathbf{w}_k \leftarrow \begin{cases} \mathbf{w}_k, & \text{if } \mathbf{x}_i \notin \mathbf{S}_k, \\ \mathbf{w}_k + \eta(\mathbf{x}_i - \mathbf{w}_k), & \text{if } \mathbf{x}_i \in \mathbf{U}_k, \\ \mathbf{w}_k - \eta(\mathbf{x}_i - \mathbf{w}_k), & \text{if } \mathbf{x}_i \in \mathbf{I}_k, \end{cases} \tag{1}$$

where $\mathbf{U}_k = \mathbf{U} \cap \mathbf{S}_k$ and $\mathbf{I}_k = \mathbf{I} \cap \mathbf{S}_k$. According to this rule, the normal patterns "pull" their codebooks while the novel patterns "push away" theirs.

Since the 1-LVQ, unlike the LVQ, assigns all the codebooks to the normal data, thresholds should be explicitly determined. While some codebooks lie inside dense lumps of input patterns, others lie in regions where patterns are sparsely scattered. For that reason, it is desirable to set different thresholds for different codebooks. For each Voronoi region, a hypersphere with a center at $\mathbf{w}_k$ and a minimal radius can be obtained, so that it surrounds as many normal patterns and as few novel patterns as possible. Now, classification is done as follows. Given a keystroke pattern $\mathbf{z}$, its codebook $\mathbf{m}(\mathbf{z}) = \mathbf{w}_q$ is found. Then, $\mathbf{z}$ is accepted if $\|\mathbf{z} - \mathbf{w}_q\|^2 \leq (r_q^*)^2$, or rejected otherwise.

3 Experimental Results

A program was developed to measure keystroke patterns. The data were collected via the keyboard connected to a workstation from 21 valid users, whose passwords are listed in the first column of Table 1. Many of them would not be understandable, since they are written in Korean alphabet, e.g. '`rhkdwo`', '`dhfpql.`', and '`tjddmswjd`'. They are simply shown in the corresponding English characters, with regard to their positions on the keyboard. The 21 users typed their own passwords with lengths ranging from six to ten characters, generating the normal class of data, and to simulate potential intrusion attempts, 15 "impostors", who had practiced the passwords, typed the 21 users' passwords. In all, 21 datasets were constructed for 21 passwords. For each user's password, 76 to 388 normal patterns were collected for training and 75 for test and 75 novel patterns were also collected. Since we assumed that the training set should be highly imbalanced, 50 user's patterns and 5 impostor patterns were randomly sampled for training. The 75 normal patterns and the rest of 70 novel patterns constituted the test set. A total of 30 different training and test sets were randomly sampled for each password to reduce a sampling bias.

We applied a total of six novelty detectors including the SVDD and the 1-LVQ. The other models are the Gaussian (Gauss) and the Parzen (Parzen) density estimator, the auto-associative neural network (AANN) and the 1-SVM, all of which cannot utilize impostor patterns even when they are available. The keystroke authentication has two types of error, i.e. false acceptance rate (FAR) and false rejection rate (FRR) [12]. Since one type of error can be reduced at the expense of the other, choosing an appropriate trade-off point should be critical to authentication accuracy. In order to avoid the biases introduced by possibly arbitrary parameters and threshold selection, we compared these models in terms of the ingetrated error [10] which was obtained from an ROC curve by integrating the FARs over the FRR from 0 to 50%.

We are interested in the effects of utilizing impostor patterns. For each password, the SVDD and the 1-LVQ were trained with two types of training dataset, one with both the user's and impostors' keystroke patterns and the other with only the user's. In other words, we compared the novelty detectors retrained with impostor patterns and the detectors trained only with the user's patterns. The average integrated errors for 21 passwords are listed in Table 1. For 16 out of 21 passwords, the 1-LVQ trained with the both classes performed better and its integrated errors were statistically lower with a significant level of 10%. For three passwords, '`autumnman`', '`dusru427`', and '`yuhwa1kk`', both models could achieve the minimum error, i.e. 0%. The 1-LVQ trained only with the user's patterns has never produced lower error rates. To sum up, when utilizing impostor patterns, the 1-LVQ produced errors as much as 67% lower and, on average, 27% lower. The SVDD trained with the both classes gave lower, though at best marginally, errors only for 4 passwords. Table 1 shows that by utilizing impostor patterns, the 1-LVQ improved much more than the SVDD did.

The SVDD and the 1-LVQ were compared with other four novelty detectors. It was assumed that the SVDD and the 1-LVQ were retrained with the impostor patterns while the other four models were trained only with 50 user's patterns. It may sound unfair, but in practice there is nothing that the four models can do with the impostor patterns anyway. The integrated errors of six models for 21 passwords are listed in Table 1. The 1-LVQ turns out to be the best, resulting in the lowest errors for 13 out of 21 passwords. Among them, the errors for 11 passwords were statistically lower than those of other models with a significance level of 10%. On average, the 1-LVQ produced 26% lower errors than the second best model, the 1-SVM. The 1-SVM has not produced the lowest error for any of the passwords, but ranked second in most cases. For four passwords, '`autumnman`', '`dltjdgml`', '`dusru427`', and '`yuhwa1kk`', most models achieved a perfect classification. The users of these passwords probably have very unique keystroke patterns. The errors of the SVDD were slightly, but not significantly, higher than those of the 1-SVM, another support vector-based method. The SVDD were as accurate as the other models except the 1-LVQ. The Gauss produced the lowest errors for two passwords and, on average, was comparable to the Parzen, suggesting that keystroke patterns probably are distributed in a unimodal manner and that just a hyperellipsoid can be a decent solution. The

Table 1. The average integrated errors (%) for six novelty detectors: The columns denoted as 'Both' and 'Normal' for the 1-LVQ and the SVDD respectively indicate models trained with the both classes of patterns and with only the user's patterns. The bold faced figures indicate the lowest errors for the corresponding password. An asterisk indicates that the marked model is better than the other models with a significance level of 10%.

Passwords	1-LVQ		SVDD		Gauss	Parzen	AANN	1-SVM
	Both	Normal	Both	Normal				
90200jdg	1.88	2.15	1.97	1.97	**1.58***	2.66	3.43	1.91
ahrfus88	**0.33***	0.46	0.50	0.50	0.73	0.59	0.77	0.48
anehwksu	0.28	0.38	0.33	0.33	**0.16***	0.43	1.46	0.30
autumnman	0.00	0.00	0.00	0.00	0.00	0.00	0.40	0.00
beaupowe	0.02	0.02	0.01	0.01	0.01	0.02	0.81	0.01
c.s.93/ksy	**0.14***	0.19	0.19	0.19	0.82	0.22	0.23	0.19
dhfpql.	**0.71***	0.89	0.86	0.87	0.85	1.12	1.99	0.80
dirdhfmw	**0.37***	0.83	0.96	0.96	1.30	1.37	2.13	0.87
dlfjs wp	**0.42***	0.45	0.45	0.45	0.49	0.50	2.26	0.45
dltjdgml	0.00	0.00	0.00	0.00	0.02	0.00	0.17	0.00
drizzle	**0.06**	0.10	0.09	0.09	0.26	0.18	0.74	0.08
dusru427	0.00	0.00	0.00	0.00	0.00	0.00	0.02	0.00
i love 3	**0.86**	1.04	1.13	1.14	0.87	1.22	2.68	1.05
love wjd	**0.85***	1.39	1.70	1.72	2.24	2.14	3.38	1.59
loveis.	**0.32***	0.44	0.42	0.42	0.93	0.50	0.98	0.41
manseiii	**0.59***	1.02	1.19	1.20	2.45	1.46	1.94	1.08
rhkdwo	**0.77***	1.32	1.45	1.45	1.58	2.23	2.66	1.38
rla sua	0.01	0.03	0.01	0.01	0.06	0.06	0.38	0.01
tjddmswjd	**0.24***	0.38	0.46	0.46	1.23	0.96	2.31	0.40
tmdwnsl1	**1.13***	1.18	1.22	1.22	1.36	1.32	2.07	1.20
yuhwa1kk	0.00	0.00	0.00	0.00	0.00	0.00	0.00	0.00
TotalAvg	**0.43**	0.59	0.62	0.62	0.81	0.81	1.47	0.58

AANN might have difficulties in training, since a lot of training patterns are necessary to train an AANN in a high-dimensional space.

4 Conclusions and Discussion

We applied two novelty detectors to keystroke dynamics authentication. Even though impostor patterns are not available in the beginning, they become available later from failed login attempts. We proposed to retrain the novelty detectors using them. Experiments on 21 keystroke pattern datasets have demonstrated that the 1-LVQ can take advantage of the impostor patterns, though the improvements of the SVDD were no more than marginal. Compared with other widely used novelty detectors, the 1-LVQ has shown its competence as an authenticator, resulting in significantly lower integrated errors.

A few limitations and future directions should be addressed. First, while we have not considered the issue of parameter selection, in practice a particular set of parameters should be specified in advance. It is tricky to select proper parameters for both the SVDD and the 1-LVQ. Second, while we have arbitrarily sampled 5 impostor patterns, it demands an investigation on how many impostor patterns are needed for the SVDD or the 1-LVQ. Third, we have included all the features for training. However, some features are more important than others while some features may be useless or even be harmful. So, a good feture selection scheme can improve the accuracy. Fourth, one might want to know in advance whether retraining with impostor patterns will be useful. Certain quality measures may be employed to determine the utility of retraining.

Acknowledgement

This work was supported by grant No. R01-2005-000-103900-0 from the Basic Research Program of the Korea Science and Engineering Foundation.

References

1. Gaines, R., Lisowski, W., Press, S., Shapiro, N.: Authentication by keystroke timing: some preliminary results. Rand Report R-256-NSF. Rand Corporation. (1980)
2. Jain, A.K., Bolle, R., Pankanti, S.: Biometrics: Personal Identification in Networked Society. Kluwer, Norwell (1999)
3. Monrose, F., Rubin, A.D.: Keystroke Dynamics as a Biometric for Authentication. Future Generation Computer System 16(4) (2000) 351-359
4. Araújo, L.C.F., Sucupira Jr., L.H.R., Lizárraga, M.G., Ling, L.L., Yabu-Uti, J.B.T.: User Authentication through Typing Biometrics Features. IEEE Transactions on Signal Processing 52(2) (2005) 851-855
5. Bleha, S., Slivinsky, C., Jussein, B.: Computer-access Security Systems using Keystroke Dynamics. IEEE Transactions on Pattern Analysis and Machine Intelligence 12(12) (1990) 1217-1222
6. Obaidat, M.S.. Sadoun, B.: Verification of Computer Users using Keystroke Dynamics. IEEE Transactions on Systems, Man, and Cybernetics, Part B: Cybernetics 27(2) (1997) 261-269
7. Cho, S., Han, C., Han, D., Kim, H.: Web Based Keystroke Dynamics Identity Verification using Neural Networks. Journal of Organizational Computing and Electronic Commerce 10(4) (2000) 295-307
8. Yu, E., Cho, S.: Keystroke Dynamics Identity Verification - Its Problems and Practical Solutions. Computer and Security 23(5) (2004) 428-440
9. Schölkopf, B., Platt, J.C., Shawe-Taylor, J., Smola, A.J., Williamson, R.C.: Estimating the Support of a High-dimensional Distribution. Neural Computation 13 (2001) 1443-1471
10. Tax, D.M.J., Duin, R.P.W.: Support Vector Data Description. Machine Learning 54 (2004) 45-66
11. Lee, H., Cho, S.: SOM-based Novelty Detection Using Novel Data. In: Proceedings of Sixth International Conference on Intelligent Data Engineering and Automated Learning, Lecture Notes in Computer Science 3578 (2005) 359-366
12. Golfarelli, M., Maio, D., Maltoni, D.: On the Error-Reject Trade-off in Biometric Verification Systems. IEEE Transactions on Pattern Analysis and Machine Intelligence 19(7) (1997) 786-796

Biometric Access Control Through Numerical Keyboards Based on Keystroke Dynamics

Ricardo N. Rodrigues, Glauco F.G. Yared, Carlos R. do N. Costa, João B.T. Yabu-Uti, Fábio Violaro, and Lee Luan Ling

Laboratory of Pattern Recognition and Computer Networks,
Department of Communications, School of Electrical and Computer Engineering,
State University of Campinas, Albert Einstein Av., 400,
PO Box 6101, Postal Code 13083-852, Campinas, SP, Brazil
ricardonagel@gmail.com,
{glauco, ccosta, yabuuti, fabio, lee}@decom.fee.unicamp.br

Abstract. This paper presents a new approach for biometric authentication based on keystroke dynamics through numerical keyboards. The input signal is generated in real time when the user enters with target string. Five features were extracted from this input signal (ASCII key code and four keystroke latencies) and four experiments using samples for genuine and impostor users were performed using two pattern classification technics. The best results were achieved by the *HMM* (EER=3.6%). This new approach brings security improvements to the process of user authentication, as well as it allows to include biometric authentication in mobile devices, such as cell phones.

1 Introduction

Access control to computational systems has been becoming more important nowadays, and the most well known and usual mechanism to guarantee the security of the information systems is through user authentication by a password. However, this type of security mechanism is fragile. A neglegent user compromise the security mechanism when one uses fragile passwords like the birth date, phone numbers, etc. On the other hand, the cost and the simplicity of this classic security mechanism justify its use, and in some situations it remains as the principal mechanism, supplemented with other security strategies. The intention of this work is to improve the process of password authentication using biometric features.

The biometric technology employed in this work is typing biometrics, also known as keystroke dynamics. The typing biometrics is an authentication process of analyzing the user's typing rhythm in a terminal at a keyboard during the identification. Authentication based on keystroke dynamics is a continuous or static process. The static manner analyzes the input keyboard at a particular moment, for example when the user types his password, while the continuous approach analyzes all inputs in the keyboard during a user session [1].

D. Zhang and A.K. Jain (Eds.): ICB 2006, LNCS 3832, pp. 640–646, 2005.

The personal authenticational system proposed in this paper only captures the keystroke dynamics from numerical keyboards (numerical passwords). This numerical password based approach was firstly introduced in [11]. The use of only a numerical keyboard causes more complex problems than using a full computer keyboard since only one hand is used to enter the password; as consequences, less information is avaible for authentication.

The Numerical Typing Dynamic biometrics can be incorporated into mobile phones, Automated Teller Machine (ATM) systems and to control the access to restricted areas. The methodology adopted in this work has low processing cost, is non-intrusive and statically authenticate users (only consider the input when the user types his password).

2 Related Works

Biometric authentication by keystroke dynamics has been an active research area since 1990 [1]-[7]. In this section, we briefly provide an review of the previous.

- **Target string.** It is the string that will be provided by the user and monitored by the system. In [2] four target strings were used during the authentication (username, password, first name and last name). However, in some works the password is the only target. Another important aspect about the target string is the string size. In [3] the authors concluded that the numbers of errors in classification process increases as the length of target string decreases.
- **Amount of samples for obtaining the template.** Samples are collected during the user enrollment phase. Some or total of these form the system training set classifier. The number of samples varies largely in literature, varying from three to thirty samples per user [4]. In [1] the authors have observed that the minimum number of samples that does not compromise the system performance is around six samples per user.
- **Feature extraction.** Two of more often observed features during the typing are the period of time in which the key remains pressed and the keystroke latency that corresponds to the time interval to move between successive keys [4].
- **Adaptation mechanism.** Biometric features are subject to small changes over the time. Most of previous works in the literature seldomly mention this important aspect. To solve this problem, a suitable adaptation mechanism or a re-enrollment procedure can be implemented to keep the users templates updated. In [6] whenever a new positive authentication occurs, the users template are updated. The database updating consist of including the new sample, discarding the oldest one and re-training the user's reference feature model.
- **Classification.** In [1]-[3] the authors used statistical Classifiers in their experiments, such as k-means classifiers, Bayes decision rules, etc. In [4], [5] and [7], artificial neural networks have been used to identify the user.

3 Methodology

In this work, we limit our investigation over numerical passwords collected from numerical keyboards. This keyboard type can be found in cell phones, ATM machines and most other access control systems. Each password is composed of a sequence of eight numerical characters, which is robust and easily memorized. For classifier a Hidden Markov Models (*HMM*) is implemented and tested.

3.1 Feature Extraction and Test Database

Ten samples, each with eight numerical characters were collected from each user in each session of 4 sessions, totalizing in 40 samples per user. Twenty people have been invited to contribute with their password samples for the experiment.

Let n denote the password sample size (n characters) can contain some keystroke features. The keystroke feature vector of sample w of user account a can be expressed as $k_{a,w} = (k_1(a,w), k_2(a,w), \ldots, k_n(a,w))$. Each element $k_i(a,w)$, where $i \leq n$, represents one of the following features [4]:

- The time interval when a key remains pressed (Down-Up or DU). This is represented by the expression $DU_{a,w} = \{DU_1(a,w), \ldots, DU_n(a,w)\}$, where $DU_i(a,w) = T_{i.up}(a,w)$ - $T_{i.down}(a,w)$. $T_{i.up}(a,w)$ is the instant where the key i is released and $T_{i.down}(a,w)$ is the time instant when key i is pressed;
- The time interval until the next key is pressed (Up-Down or UD). This is represented by the expression $UD_{a,w} = \{UD_1(a,w), \ldots, UD_{n-1}(a,w)\}$, where $UD_i(a,w) = T_{i+1.down}(a,w)$ - $T_{i.up}(a,w)$;
- The time interval between two consecutive pressed keys (Down-Down or DD). This is represented by $DD_{a,w} = \{DD_1(a,w), \ldots, DD_{n-1}(a,w)\}$, where $DD_i(a,w) = T_{i+1.down}(a,w)$ - $T_{i.down}(a,w)$;
- The interval between two consecutive released keys (Up-Up or UU). This is represented by $UU_{a,w} = \{UU_1(a,w), \ldots, UU_{n-1}(a,w)\}$, where $UU_i(a,w) = T_{i+1.up}(a,w)$ - $T_{i.up}(a,w)$.

For illustration purpose, figure 1 shows the feature extraction for a user typing a given target string.

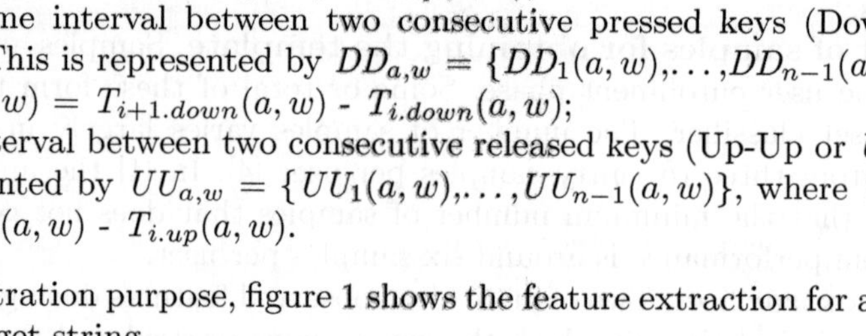

Fig. 1. Representation of the features observed during the typing of a given target string. DU is the time when a key remains pressed, UD is the time interval until the next key is be pressed, DD is the time interval between two consecutive pressed keys and UU is the time interval between two consecutive released keys.

3.2 Statistical Classifier

In [8] the authors suggest that the user template be composed of the mean and standard deviation of sample feature vector acquired during the enrollment. Every time a user needs to authenticate his password, the system calculates the distance of the target string to the template. If the distance is larger than the threshold, the user is authenticated. The template implemented is derived from sample feature vector K={DD, UD, DU, UU} according to equations (1) and (2)

$$\mu_{K_i(a)} = \frac{1}{N}\sum_{j=1}^{N} K_i(a,j), \tag{1}$$

$$\sigma_{K_i(a)} = \frac{1}{N-1}\sum_{j=1}^{N} |K_i(a,j) - \mu_{K_i(a)}|. \tag{2}$$

In authentication, through the ASCII code provided by the system determines the intending user and the retrieves the template from the corresponding account for authentication. The 1 to 1 comparison consists of computing the distance between the template and the input sample feature vector through the equation (3)

$$D_K(a,w) = \frac{1}{n}\sum_{i=1}^{n} \frac{K_i(a,w) - \mu_{K_i(a)}}{\sigma_{K_i(a)}}, \tag{3}$$

where n is the number of latency feature in K, K={DD, UD, DU, UU}. If $D_K(a,w) \leq \tau_k(a)$, for all K, the user is considered authentic. $\tau_k(a)$ is a empirically defined threshold for user account a. In [8] an data updating mechanism for the model is considered similarly with [6]. The authors reported error rates are below 2%.

3.3 Classifier Using *HMM*

HMM based systems have been widely used in pattern recognition [9]. This is due to the necessity of constructing models that analyze pattern's temporal variability. Moreover, the use of Gaussian mixtures allows ones to model complex distributions, and consequently build complex decision boundaries for classification processes.

The probabilistic model used to represent each user's password is modeled by a continuous *HMM* with 15 states and *left-to-right* topology. Each state is associated with the time when the user presses the key (down-up or DU) or with the time interval between two consecutive pressed keys (up-down or UD).

Let $A_{i(i+1)}$ denote the transition probability from state i to state $i+1$. Each new input feature value $K_i(a,w)$, i varying from 1 to 15 state is modeled by 6 Gaussian distribution. The input feature value $K_i(a,w)$ makes the *HMM* system advance from state i to state $i+1$.

The basic idea of *HMM* model as to associate each latency measure (observed feature) to a state of the model. Therefore, a 8 digits password will resulting in 15 latencies. For each typed password, we estimate the likelihood probability $P(O_{a,w}|\lambda_a)$ of the corresponding *HMM* model through the Viterbi algorithm [9], where

$$O_{a,w} = \{DU_1(a,w), UD_1(a,w), DU_2(a,w), UD_2(a,w), \ldots, \\ DU_{n-1}(a,w), UD_{n-1}(a,w), DU_n(a,w)\},$$

and λ_a is the set of feature value associated with user a, used for the model estimation.

If the likelihood probability estimate is superior to a threshold value, say a $\tau(a)$, the user is declared authentic; otherwise, he is considered an impostor. In this work, the threshold is obtained from the training data as follows:

$$\tau(a) = \mu_{P_a} - 3\sigma_{P_a} \tag{4}$$

where

$$\mu_{P_{(a)}} = \sum_{w=1}^{N} \frac{P(O_{a,w}|\lambda_a)}{N} \tag{5}$$

and

$$\sigma_{P_{(a)}} = \frac{1}{N-1} \sum_{w=1}^{N} |\mu_{P_{(a)}} - P(O_{a,w}|\lambda_a)| \tag{6}$$

For the system training the Baum-Welch algorithm implemented in Hidden Markov Toolkit (*HTK* [10]) is used. Although the system model was initially trained by a set of fixed N samples for each user a, the system has been trained again by most recent N true samples every time a new sample is positive authenticated.

4 Experiments and Results

The experiments were performed on a *Pentium IV* microcomputer platform, using only numeric keyboard part. Twenty users of both sex aging from 20 to 60 years old with different levels of familiarity with the numerical keyboard have participated in the experiment. The target strings are composed of eight numbers freely chosen by the users. Two kinds of data were collected: *Authentic database* were each user was undergone 4 sessions. In each session 10 sample was collected. This results in 800 samples in total. *Faked database* were for each true password, 30 samples are collected from ones other than the true user. This results in 600 faked samples in total.

For the statistical classifier design and system evaluations, three sets of training samples were used for model dimensioning:

I. Ten samples (N=10) to construct the model with K={DD,UD,DU,UU};
II. Twenty samples (N=20) to construct the model with K={DD,UD,DU,UU};
III. Thirty samples (N=30) to construct the model with K={DD,UD,DU,UU}.

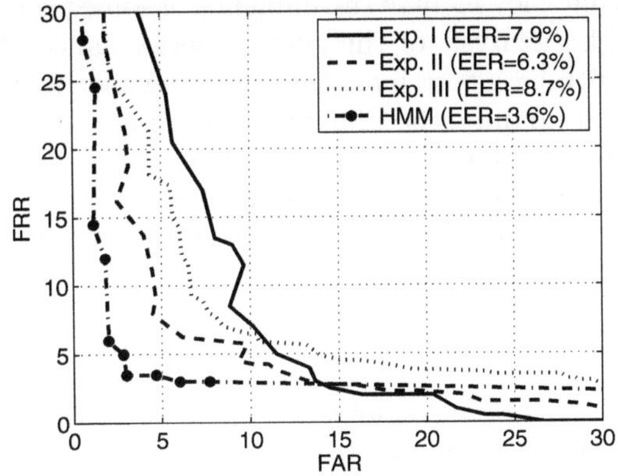

Fig. 2. The *ROC* performance of the experiments with the statistical classifiers (trained by N=10 samples, N=20 samples and N=30 samples) and HMM classifier trained by N=30 samples

In addition to the statistical classifiers, an *HMM* classifier was implemented, based on the using K={UD, DU} feature pattern. For model training, thirty samples (N=30) were used.

To represent False-acceptance and false-rejection rates (FAR and FRR), we used *ROC* (*Receiver Operating Characteristics*) curve. Each point of the *ROC* curve represent a specific operating condition of the biometric system, wich is a function of decision thresholds.

Figure 2 shows the *ROC* performance of the statistical classifiers design with different amounts of training samples and the HMM trained by 30 samples. Notice that the *EER* for the statistical classifier is considerably higher that supported in [8] (*EER* of 1.6%). However, this result is not surprising due to the fact that the experiments were made in an alphanumerical keyboard, in each user provided with his passwords with target string much bigger than 8 characters, what resulted in more representative latencies than in numerical keyboards. The performance of the *HMM* classifier (with N=30) wich outperforms all the 3 statistical classifiers in experiments. The method of updating the model allowed a more efficient form for modeling the typing dynamics of users.

5 Conclusions and Future Works

This work presented a novel methodology for biometric authentication based on the latency features extracted from the keystroke dynamics and *HMM* modelling approach. The biometric systems has the goal of improving the process of access control to restricted areas or to increase the security of banking transactions. The experimental results reveal the potentially of hidden Markov models, resulting in an *EER* of 3.6%. This rate is competitive when is compared to that obtained by the statistical classifier considered in [8].

We also observed that they include the influence of some practical aspects need to be considered in the analysis of the system performance. They include the familiarity of users with target strings, the updating mechanism, the precision of data acquisition and, mainly, the number of training samples.

For the future work, we intend to make our database even more robust in forms of population size and number of data acquisition sessions. Also other topologies of the *HMM* with different model structures should be investigated.

References

1. F. Monrose and A. D. Rubin, *Keystroke Dynamics as a Biometric for Authentication*, Future Generation Computer Systems, Vol. 16, no. 4, pp. 351-359, March 1999.
2. R. Joyce and G. Gupta, *Identity authentication based on keystroke latencies*, commun. ACM, vol. 33, no. 2, pp. 168-176, 1990.
3. d. Bleha and M. Obaidat, *Dimensionality reduction and feature extraction applications in identifying computer users*, IEEE Trans. Syst., Man, Cybern., Vol. 21, no. 2, pp. 452-456, Mar.-Apr. 1991.
4. D. T Lin, *Computer-access authentication with neural network based keystroke identity verification*, in Proc. Int. Conf. Neural Networks, vol. 1, 1997, pp. 174-178.
5. M. S. Obaidat and B. Sadoun, *Verification of computer user using keystroke dynamics*, IEEE Trans. Syst., Man, Cybern., vol. 27, no. 2, pp. 261-269, Mar.-Apr. 1997.
6. F. Monrose, M. K. Reiter, and S. Wetzel, *Password hardening based on keystroke dynamics*, in Proc. 6th ACM Conf. Computer Security, Singapore, Nov. 1999.
7. F. W. M. H. Wong, A. S. M. Supian, A. F. Ismail, L. W. Kin, and O. C. Soon, *Enhanced user authentication through typing biometrics with artificial neural network and k-nearest neighbor algorithm* in Conf. Rec. 35th Asilomar Conf. Signals, Syst., comput., Vol. 2, 2001, pp. 911-915.
8. L. C. F. Araújo, L. H. R. Sucupira Jr., M. G. Lizárraga, L. L. Ling, and J. B. T. Yabu-uti, *User authentication through typing biometrics features*, IEEE Trans. on Signal Processing, vol. 53, No. 2, Feb. 2005.
9. R. O. Duda, P. E. Hart and D. G. Stork, *Pattern Classification* Wiley-Interscience Publication, 2nd Edition, Oct. 2000.
10. Cambridge University Engineering Departament, *The HTK Book*, Cambridge University, 2002.
11. T. Ord and S. M. Furnell, *User authentication for keypad-based devices using keystroke analysis*, Proc. Second International Network Conference (INC 2000), Plymouth, UK, pp. 263-272.

Keystroke Biometric System Using Wavelets

Woojin Chang

Department of Industrial Engineering, Seoul National University,
San 56-1, Sillim-dong, Gwanak-gu, Seoul 151-742, Korea
changw@snu.ac.kr

Abstract. We developed the keystroke biometric system (KBS) using the statistical features of the discrete wavelet transformed keystroke pattern in the frequency domain in addition to those of the original keystroke pattern in the time domain. Only 20 keystroke patterns of user's password typing, where the length of password is no more than 10, are used for building a KBS. The features in the time domain and those in the frequency domain are separately scored by the rules that we developed, and arbitrary given keystroke patterns are classified on the basis of total scores. The results show that our KBS is competitive in comparison with others due to its cheap computational cost, cheap usability cost, and the practically acceptable classification accuracy.

Keywords: Keystroke Dynamics, Keystroke Biometric System, Keystroke Authentication, Discrete Wavelet Transform.

1 Introduction

Keystroke biometric system (KBS) authenticates the legitimate user by his or her keystroke dynamics. KBS classifies an arbitrary keystroke typing pattern as either the legitimate user's or imposter's, and has two types errors for false acceptance and false rejection. The false acceptance rate (FAR) is the percentage of imposters' keystroke typing patterns identified as the genuine user's, and the false rejection rate (FRR) is the percentage of legitimate user's keystroke typing patterns identified as the imposters'. From the nature of two errors, FAR can be reduced at the cost of FRR, and vice versa.

One possible way to achieve high classification accuracy, cheap computational cost and cheap usability cost together is to use the statistical features of the transformed keystroke patterns in the frequency domain in addition to those of the original keystroke patterns in the time domain. The novelty of our research is that the statistical features of keystroke dynamics are directly detected and measured not only in the time domain, but also in the frequency domain by simple and basic statistical method that we developed. Furthermore, we use only 20 keystroke patterns of user's password typing to build a KBS.

2 Keystroke Timing Vector and Wavelet Transformation

A keystroke pattern is expressed as a keystroke timing vector (KTV) consisting of the sequences of duration and interval time pairs measured at the accuracy of

D. Zhang and A.K. Jain (Eds.): ICB 2006, LNCS 3832, pp. 647–653, 2005.

milliseconds(ms). In the timing vector, each keystroke duration time is followed by the interval time which is calculated by subtracting the key-hit time from the previous key-release time. Thus, the interval time can be a negative value when a key is stroked before a previous key is released. This research includes the "Enter" keystroke as the last element of KTV. In this set-up, typing a string of n characters results in a KTV of length $2n+1$, which consists of $n+1$ keystroke duration times including the "Enter" key and n keystroke interval times.

Discrete wavelet transformation (DWT) is applied to a KTV in the time domain, and then the corresponding keystroke wavelet coefficient vector (KWV) is produced in the frequency domain. Since DWT separates a KTV into multiresolution components, the latent features in KTV can be well observed and extracted in the KWV. Since the adequate data size for DWT should be 2 to the power of any natural number, the dummy data points need to be padded to the unfit KTV. The resulting modified vector $\mathbf{y}$ for DWT is as follows,

$$\mathbf{y} = (y_1, \ldots, y_{2^m}) = (\underbrace{0, \ldots, 0}_{\lceil \frac{2^m - N}{2} \rceil}, v_1, \ldots, v_N, \underbrace{0, \ldots, 0}_{\lfloor \frac{2^m - N}{2} \rfloor}).$$

For a KTV of length N, $\mathbf{v} = (v_1, \ldots, v_N)$, the smallest 2^m larger than N ($2^{m-1} < N \leq 2^m$) is the adequate vector length for DWT. In this research, two zero vectors of length $\lceil \frac{2^m-N}{2} \rceil$ (where $\lceil x \rceil$ is the smallest integer larger than or equal to x) and $\lfloor \frac{2^m-N}{2} \rfloor$ (where $\lfloor x \rfloor$ is the largest integer less than or equal to x) need to be put into the front side and the back side of the KTV respectively. The piecewise constant function f on $[0, 1)$ generated by $\mathbf{y}$ can be represented as the corresponding wavelet decomposition as $f(x) = \sum_{k=0}^{2^m-1} y_k \cdot \mathbf{1}\{k2^{-m} \leq x < (k+1)2^{-m}\} = c_{00}\phi(x) + \sum_{j=1}^{m-1} \sum_{k=0}^{2^j-1} d_{jk}\psi_{jk}(x)$, where $\mathbf{1}\{A\}$ is 1 if the condition 'A' is satisfied and 0 otherwise, $\phi(x) = \mathbf{1}\{0 \leq x < 1\}$ is the scaling function, and $\psi_{jk}(x) = 2^{\frac{j}{2}}\psi(2^j x - k)$ is a dilation and translation of Haar wavelet, $\psi(x) = \mathbf{1}\{0 \leq x < \frac{1}{2}\} - \mathbf{1}\{\frac{1}{2} \leq x < 1\}$. Note the set $\{\psi_{jk}, j \in \mathbb{Z}, k \in \mathbb{Z}\}$ defines an orthogonal basis of $L^2(\mathbb{R})$ [1].

The wavelet coefficients, c_{00} and d_{jk}'s are obtained from y_k's in the following way. When we let $\mathbf{c}_m = (c_{m0}, \ldots, c_{m2^m-1})$ denote $\mathbf{y}$, and define $\mathbf{c}_j = (c_{j0}, \ldots, c_{j2^j-1})$ and $\mathbf{d}_j = (d_{j0}, \ldots, d_{j2^j-1})$ for $j = 1, \ldots, m-1$, $\mathbf{c}_j$ and $\mathbf{d}_j$ are calculated as follows. $c_{m-1\ell} = c_{m2\ell}/\sqrt{2} + c_{m2\ell+1}/\sqrt{2}$, $d_{m-1\ell} = c_{m2\ell}/\sqrt{2} - c_{m2\ell+1}/\sqrt{2}$, where $\ell = 0, \ldots, 2^{m-1}-1$. For each $j = 1, \ldots, m-1$, $c_{j-1\ell} = c_{j2\ell}/\sqrt{2} + c_{j2\ell+1}/\sqrt{2}$, $d_{j-1\ell} = d_{j2\ell}/\sqrt{2} - d_{j2\ell+1}/\sqrt{2}$, where $\ell = 0, \ldots, 2^{j-1}-1$. Then the wavelet coefficients, $\mathbf{w} = (\mathbf{d}_{m-1}, \mathbf{d}_{m-2}, \ldots, \mathbf{d}_1, d_{00}, c_{00})$ can be obtained through cascade algorithm whose computational complexity is $O(2^m)$. $\mathbf{w}$ is the keystroke wavelet coefficient vector (KWV) corresponding to $\mathbf{v}$.

Let $\#_0(\cdot)$ denote the number of 0's in a vector. It can be shown that $\#_0(\mathbf{w}) = \sum_{j=1}^{m-2} \left\lfloor \lceil \frac{2^m-N}{2} \rceil \cdot 2^{-j} \right\rfloor + \sum_{j=1}^{m-2} \left\lfloor \lfloor \frac{2^m-N}{2} \rfloor \cdot 2^{-j} \right\rfloor < \#_0(\mathbf{y}) = 2^m - N$. This shows that the number of nonzero elements in $\mathbf{w}$ is larger than the number in $\mathbf{y}$. This implies that through DWT the information contained in N elements of $\mathbf{v}$ can be diffused into more than N elements of $\mathbf{w}$ in the frequency domain. By eliminating the zero elements from $\mathbf{w}$ and retaining nonzero elements of $\mathbf{w}$, the reduced

KWV, $\mathbf{u}$ of length $(2^m - \#_0(\mathbf{w}))$ is constructed. The overall DWT process in this section can be illustrated as $\mathbf{v} \rightarrow \mathbf{y} \rightarrow \mathbf{w} \rightarrow \mathbf{u}$, and $\dim(\mathbf{v}) < \dim(\mathbf{u})$, where $\dim(\cdot)$ is the length of a vector dimension.

3 Authentication of Keystroke Dynamics

In this paper, we measure how much different the given KTV is from the average user's pattern, and determine whether it belongs to user's or imposter's. Thus, the first step of KBS design should be the construction of user's keystroke dynamics reference consisting of the statistical characteristics of the elements of user's KTV and KWV.

Since the elements having a consistent pattern compose user's keystroke dynamics features, the KTV or KWV element having high consistency need to be selected for user's keystroke dynamics reference. This selection process results in the dimension reduction effect of KTV and KWV.

Let $\boldsymbol{\nu}$ and $\boldsymbol{\omega}$ denote user's sample KTV and the corresponding KWV, respectively. The consistency level at each element of user's KTV and KWV can be estimated from $\boldsymbol{\nu}$'s and $\boldsymbol{\omega}$'s, respectively. Assume that M KTVs, $\boldsymbol{\nu}_i = (\nu_{i1}, \ldots, \nu_{iN})$ $i = 1, \ldots, M$, are collected from the user. The corresponding KWV, $\boldsymbol{\omega}_i = (\omega_{i1}, \ldots, \omega_{i2^m})$ is obtained from $\boldsymbol{\nu}_i$, and the zero elements in $\boldsymbol{\omega}_i$ deleted, so that $\boldsymbol{\omega}_i$ is reduced to become $\boldsymbol{\mu}_i = (\mu_{i1}, \ldots, \mu_{i2^m - \#_0(\boldsymbol{\omega}_i)})$. The consistency at the jth element of $\boldsymbol{\nu}$ and $\boldsymbol{\mu}$ are represented as p_j and q_j respectively, as follows. After letting $p_1 = q_1 = 1$, for $j = 2, \ldots, N$,

$$p_j = 1/M \cdot \max\Big(\sum_{i=1}^{M} \mathbf{1}\{\text{sgn}(\nu_{ij} - \nu_{ij-1}) = 1\}, \sum_{i=1}^{M} \mathbf{1}\{\text{sgn}(\nu_{ij} - \nu_{ij-1}) = -1\}\Big),$$

and for $j = 2, \ldots, 2^m - \#_0(\boldsymbol{\omega})$,

$$q_j = 1/M \cdot \max\Big(\sum_{i=1}^{M} \mathbf{1}\{\text{sgn}(\mu_{ij} - \mu_{ij-1}) = 1\}, \sum_{i=1}^{M} \mathbf{1}\{\text{sgn}(\mu_{ij} - \mu_{ij-1}) = -1\}\Big),$$

where $\text{sgn}(x) = \mathbf{1}\{x > 0\} - \mathbf{1}\{x < 0\}$.

To construct the keystroke dynamics reference comprising the vector elements with high consistency level only, the following thresholding rule is applied to $\boldsymbol{\nu}_i$ and $\boldsymbol{\mu}_i$ $(i = 1, \ldots, M)$.

$$\tilde{\nu}_{ij} = \nu_{ij} \cdot \mathbf{1}\{p_j > T\} \; j = 1, \ldots, N$$
$$\tilde{\mu}_{ij} = \mu_{ij} \cdot \mathbf{1}\{q_j > T\} \; j = 1, \ldots, 2^m - \#o(\boldsymbol{\omega}_i)$$

where T $(0 \leq T < 1)$ is the threshold. By eliminating the zero elements from $\tilde{\boldsymbol{\nu}}_i = (\tilde{\nu}_{i1}, \ldots, \tilde{\nu}_{iN})$ and $\tilde{\boldsymbol{\mu}}_i = (\tilde{\mu}_{i1}, \ldots, \tilde{\mu}_{i2^m - \#o(\boldsymbol{\omega}_i)})$ and retaining nonzero elements, the reduced vector $\boldsymbol{\nu}_i^* = (\nu_{i1}^*, \ldots, \nu_{iR}^*)$ and $\boldsymbol{\mu}_i^* = (\mu_{i1}^*, \ldots, \mu_{iS}^*)$ are constructed. In comparisons with $\boldsymbol{\nu}_i$ and $\boldsymbol{\mu}_i$, $\boldsymbol{\nu}_i^*$ and $\boldsymbol{\mu}_i^*$ are reduced in size and their overall patterns become more consistent.

After these processes, we obtain user's keystroke dynamics reference consisting of the sample averages, $\bar{\boldsymbol{\nu}}^* = (\bar{\nu}_1^*, \ldots, \bar{\nu}_R^*)$ and $\bar{\boldsymbol{\mu}}^* = (\bar{\mu}_1^*, \ldots, \bar{\mu}_R^*)$, and the corresponding sample standard deviations, $\mathbf{s}_{\nu^*} = (s_{\nu_1^*}, \ldots, s_{\nu_R^*})$ and $\mathbf{s}_{\mu^*} = (s_{\mu_1^*}, \ldots, s_{\mu_S^*})$. By using these sample statistics $\bar{\boldsymbol{\nu}}^*$, $\bar{\boldsymbol{\mu}}^*$, $\mathbf{s}_{\nu^*}$, and $\mathbf{s}_{\mu^*}$ in user's keystroke dynamics reference, the difference between any given keystroke pattern and user's average keystroke pattern is measured.

Let $\mathbf{v}$ denote an arbitrary KTV, which is either the user's or the imposter's, and $\mathbf{w}$ denote the corresponding KWV. To classify $\mathbf{v}$ on the basis of the statistical features of keystroke dynamics defined in the framework of user's keystroke dynamics reference, $\mathbf{v}$ and $\mathbf{w}$ are processed in the same way as $\boldsymbol{\nu}$ and $\boldsymbol{\mu}$. By eliminating zero elements and retaining nonzero elements in $\mathbf{w}$, the reduced KWV, $\mathbf{u}$ is obtained. In $\mathbf{v}$ and $\mathbf{u}$, the jth element is kept if the corresponding jth element in $\boldsymbol{\nu}$ and $\boldsymbol{\mu}$ is used for user's keystroke dynamics reference, and eliminated otherwise. Then, $\mathbf{v}^* = (v_1^*, \ldots, v_R^*)$ and $\mathbf{u}^* = (u_1^*, \ldots, u_S^*)$, which correspond to $\boldsymbol{\nu}^*$ and $\boldsymbol{\mu}^*$ are constructed. Note that $\dim(\mathbf{v}^*) = R \leq \dim(\mathbf{v}) = N$ and $\dim(\mathbf{u}^*) = S \leq \dim(\mathbf{u}) = 2^m - \#o(\mathbf{w})$.

Using the elements of $\mathbf{v}^*$ and $\mathbf{u}^*$, the features of $\mathbf{v}$ are defined in the following three ways: size of vector element (v_j^* for $j = 1, \ldots, R$, and u_j^* for $j = 1, \ldots, S$), sign of vector element ($\text{sgn}(v_j^*)$ for $j = 1, \ldots, R$, and $\text{sgn}(u_j^*)$ for $j = 1, \ldots, S$), and sign of the difference between two values of vector elements ($\text{sgn}(v_j^* - v_\ell^*)$ for $j = 1, \ldots, R-1$, $\ell = j+1, \ldots, R$, $\text{sgn}(u_j^* - u_\ell^*)$ for $j = 1, \ldots, S-1$, $\ell = j+1, \ldots, S$).

An arbitrary keystroke pattern is scored by the appropriate rule incorporating the statistical keystroke dynamics features. For the convenient descriptions of the rules, the early defined $\boldsymbol{\nu}^*$, $\boldsymbol{\mu}^*$, $\mathbf{v}^*$, and $\mathbf{u}^*$ are used from this point on. Before the description of each rule, the followings need to be mentioned. First, α_n's and β_n's are the constant values giving weight to the scoring sources of v_j^*'s in the time domain and to those of u_j^*'s in the frequency domain, respectively. Second, the scoring sources of v_j^* and u_j^* are multiplied by p_j and q_j respectively, to have weight.

- **Rule 1:** Measure the sizes of v_j^* and u_j^*, and give penalty for v_j^* and u_j^* with abnormal size.

$$\mathsf{score}_1^\alpha(\mathbf{v}^*) = \sum_{j=1}^{R} \left(\alpha_1 p_j \mathbf{1}\{\bar{\nu}_j^* \neq 0\} \frac{|v_j^* - \bar{\nu}_j^*|}{s_{\nu_j^*}} + \sum_{k=2}^{5} \alpha_k p_j \mathbf{1}\{|v_j^* - \bar{\nu}_j^*| > k s_{\nu_j^*}\} \right)$$

$$\mathsf{score}_1^\beta(\mathbf{u}^*) = \sum_{j=1}^{S} \left(\beta_1 q_j \mathbf{1}\{\bar{\mu}_j^* \neq 0\} \frac{|u_j^* - \bar{\mu}_j^*|}{s_{\mu_j^*}} + \sum_{k=2}^{5} \beta_k q_j \mathbf{1}\{|u_j^* - \bar{\mu}_j^*| > k s_{\mu_j^*}\} \right)$$

Note $\bar{\nu}_j^*$, $\bar{\mu}_j^*$ are the sample means, and $s_{\nu_j^*}$, $s_{\mu_j^*}$ are the sample standard deviations.

- **Rule 2:** Give penalty when $\text{sgn}(v_j^*) \neq \text{sgn}(\bar{\nu}_j^*)$ or $\text{sgn}(u_j^*) \neq \text{sgn}(\bar{\mu}_j^*)$.

$$\mathsf{score}_2^\alpha(\mathbf{v}^*) = \sum_{j=1}^{R} \alpha_6 p_j |\text{sgn}(v_j^*) - \text{sgn}(\bar{\nu}_j^*)|, \quad \mathsf{score}_2^\beta(\mathbf{u}^*) = \sum_{j=1}^{S} \beta_6 q_j |\text{sgn}(u_j^*) - \text{sgn}(\bar{\mu}_j^*)|$$

- **Rule 3:** Give penalty when $\mathrm{sgn}(v_j^* - v_{j+1}^*) \neq \mathrm{sgn}(\bar{\nu}_j^* - \bar{\nu}_{j+1}^*)$ or $\mathrm{sgn}(u_j^* - u_{j+1}^*) \neq \mathrm{sgn}(\bar{\mu}_j^* - \bar{\mu}_{j+1}^*)$.

$$\mathsf{score}_3^\alpha(\mathbf{v}^*) = \sum_{j=1}^{R-1} \alpha_7 p_{j+1} \mathbf{1}\{\mathrm{sgn}(v_j^* - v_{j+1}^*) \neq \mathrm{sgn}(\bar{\nu}_j^* - \bar{\nu}_{j+1}^*)\} \frac{|v_{j+1}^* - \bar{\nu}_{j+1}^*|}{s_{\nu_{j+1}^*}}$$

$$\mathsf{score}_3^\beta(\mathbf{u}^*) = \sum_{j=1}^{S-1} \beta_7 q_{j+1} \mathbf{1}\{\mathrm{sgn}(u_j^* - u_{j+1}^*) \neq \mathrm{sgn}(\bar{\mu}_j^* - \bar{\mu}_{j+1}^*)\} \frac{|u_{j+1}^* - \bar{\mu}_{j+1}^*|}{s_{\mu_{j+1}^*}}$$

- **Rule 4:** Give penalty when $\mathrm{sgn}(v_j^* - v_k^*) \neq \overline{\mathrm{sgn}(\nu_j^* - \nu_k^*)} = \sum_{i=1}^M \mathrm{sgn}(\nu_{ij}^* - \nu_{ik}^*)/M$ and $\mathrm{sgn}(u_j^* - u_k^*) \neq \overline{\mathrm{sgn}(\mu_j^* - \mu_k^*)} = \sum_{i=1}^M \mathrm{sgn}(\mu_{ij}^* - \mu_{ik}^*)/M$.

$$\mathsf{score}_4^\alpha(\mathbf{v}^*) = \sum_{j=1}^{R-1} \sum_{k=j+1}^{R} \alpha_8 p_j p_k |\mathrm{sgn}(v_j^* - v_k^*) - \overline{\mathrm{sgn}(\nu_j^* - \nu_k^*)}|$$

$$\mathsf{score}_4^\beta(\mathbf{u}^*) = \sum_{j=1}^{S-1} \sum_{k=j+1}^{S} \beta_8 q_j q_k |\mathrm{sgn}(u_j^* - u_k^*) - \overline{\mathrm{sgn}(\mu_j^* - \mu_k^*)}|$$

Combining the above four rules, the total score for $\mathbf{v}$ is made as $\mathsf{score}(\mathbf{v}) = \sum_{n=1}^4 \left(\mathsf{score}_n^\alpha(\mathbf{v}^*) + \mathsf{score}_n^\beta(\mathbf{u}^*)\right)$ We also define $\mathsf{score}^\alpha(\mathbf{v}^*) = \sum_{n=1}^4 \mathsf{score}_n^\alpha(\mathbf{v}^*)$ as the scoring function for $\mathbf{v}^*$ in the time domain, and $\mathsf{score}^\beta(\mathbf{u}^*) = \sum_{n=1}^4 \mathsf{score}_n^\beta(\mathbf{u}^*)$ as the scoring function for $\mathbf{u}^*$ in the frequency domain. To discriminate imposter's keystroke patterns from user's, the distribution of user's keystroke dynamics scores is obtained by the calculation of $\mathsf{score}(\boldsymbol{\nu}_i)$ for $i = 1, \ldots, M$. The truncated sample score mean ($\overline{\mathsf{score}}_{95}$) and the truncated sample standard deviation ($s_{95}(\mathsf{score})$) are calculated after excluding highest 5% of $\mathsf{score}(\boldsymbol{\nu}_i)$s. An arbitrary $\mathbf{v}$ is classified as the user's if $\mathsf{score}(\mathbf{v}) \leq \overline{\mathsf{score}}_{95} + t \cdot s_{95}(\mathsf{score})$, and $\mathbf{v}$ is classified as imposter's otherwise. Note that t value relates to user's security setting, and the small t can result in low FAR and high FRR, and the large t can result in high FAR and low FRR. When $\mathsf{score}^\alpha(\mathbf{v}^*)$ or $\mathsf{score}^\beta(\mathbf{u}^*)$ is used only as the scoring function for $\mathbf{v}$, the same classification rule as in $\mathsf{score}(\mathbf{v})$ is applied.

4 Experimental Results

For the evaluation of the KBS using wavelets, the data set from Yu and Cho [3] were used in this paper. In this research, to construct user's keystroke dynamics reference, 20 keystroke patterns were randomly selected from user's training data set. Thus, only 20 keystroke patterns ($M = 20$) were provided to build a KBS. For the evaluation of the KBS, two test sets of 75 keystroke patterns, one from the user and the other from the imposters were used. Thresholding with $T = 0.8$ was applied to the corresponding KTVs and KWVs.

Table 1 shows the test results for the keystroke dynamics of 21 passwords typing, the dimensions of $\mathbf{v}$, $\mathbf{v}^*$, $\mathbf{u}^*$, and the FARs and FRRs of the classifications using $\mathsf{score}^{\alpha}(\mathbf{v}^*)$, $\mathsf{score}^{\alpha}(\mathbf{u}^*)$, and $\mathsf{score}(\mathbf{v})$. $\mathsf{score}^{\alpha}(\mathbf{v}^*)$, $\mathsf{score}^{\beta}(\mathbf{u}^*)$, and $\mathsf{score}(\mathbf{v})$ are evaluated in terms of accuracy by using the data mentioned above. The values of α_n and β_n for $n = 1, \ldots, 8$ were determined using 20 user's keystroke patterns ($\boldsymbol{\nu}_i$, $i = 1, \ldots, 20$) in the heuristic way that prioritizes the scoring sources of v_j^*'s and u_j^*'s and makes the user's keystroke dynamics scores, $\mathsf{score}(\boldsymbol{\nu}_i)$ $i = 1, \ldots, 20$ around 100.

Table 1. The test results for keystroke dynamics of 21 passwords typing, the dimensions of $\mathbf{v}$, $\mathbf{v}^*$, $\mathbf{u}^*$, and the FARs and FRRs of the classifications using $\mathsf{score}^{\alpha}(\mathbf{v}^*)$, $\mathsf{score}^{\alpha}(\mathbf{u}^*)$, and $\mathsf{score}(\mathbf{v})$. Note 'c.s.93/ksy 8' contains special characters.

User's				$\mathsf{score}^{\alpha}(\mathbf{v}^*)$		$\mathsf{score}^{\beta}(\mathbf{u}^*)$		$\mathsf{score}(\mathbf{v})$	
Password	dim($\mathbf{v}$)	dim($\mathbf{v}^*$)	dim($\mathbf{u}^*$)	FAR	FRR	FAR	FRR	FAR	FRR
loveis.	15	12	9	0	9.33	4.00	5.33	0	5.33
i love 3	17	14	12	6.67	1.33	6.67	0	5.33	0
90200jdg	17	7	11	2.67	16.00	1.33	16.00	0	16.00
autumnman	19	15	16	0	4.00	0	8.00	0	4.00
tjddmswjd	19	16	17	0	14.67	0	16.00	0	12.00
dhfpql.	15	11	7	24.00	4.00	0	10.67	2.67	5.33
love wjd	17	10	15	10.67	9.33	2.67	12.00	2.67	10.67
ahrfus8	17	11	15	21.33	4.00	4.00	5.33	0	5.33
dusru427	17	17	14	0	9.33	0	1.33	0	2.67
manseii	17	11	11	6.67	12.00	4.00	10.67	5.33	12.00
drizzle	15	11	10	0	2.67	2.67	1.33	0	2.67
beaupowe	17	10	11	2.67	2.67	0	2.67	0	1.33
tmdwnsl1	17	15	15	2.67	4.00	0	13.33	0	10.67
yuhwa1kk	17	15	17	0	0	0	0	0	0
anehwksu	17	16	12	10.67	1.33	1.33	6.67	1.33	2.673
rhkdwo	13	13	8	0	6.67	20.00	4.00	2.67	5.33
rla sua	17	15	16	0	1.33	0	2.67	0	1.33
dlfjs wp	17	15	18	1.33	2.67	1.33	1.33	0	2.67
dltjdgml	17	16	16	0	0	0	2.67	0	1.33
dirdhfmw	17	11	17	6.67	2.67	4.00	5.33	2.67	6.67
c.s.93/ksy 8	21	18	21	1.33	4.00	0	6.67	0	4.00
Minimum				0	0	0	0	0	0
Maximum				24.00	16.00	20.00	16.00	5.33	16.00
Average				4.64	5.40	2.48	6.29	1.08	5.33

Note: For $\mathsf{score}^{\alpha}(\mathbf{v}^*)$ calculation, we used $\alpha_1 = \alpha_2 = 1, \alpha_3 = \alpha_4 = \alpha_5 = 10, \alpha_6 = 1, \alpha_7 = 2.5, \alpha_8 = 1.9$ and $\beta_k = 0$ for $k = 1, \ldots, 8$. For $\mathsf{score}^{\alpha}(\mathbf{u}^*)$ calculation, we used $\alpha_k = 0$ for $k = 1, \ldots, 8$ and $\beta_1 = \beta_2 = 1, \beta_3 = \beta_4 = \beta_5 = 10, \beta_6 = 1, \beta_7 = 2.5, \beta_8 = 1.9$. For $\mathsf{score}(\mathbf{v})$ calculation, we used $\alpha_1 = \alpha_2 = \beta_1 = \beta_2 = 1, \alpha_3 = \alpha_4 = \alpha_5 = \beta_3 = \beta_4 = \beta_5 = 10, \alpha_6 = \beta_6 = 1, \alpha_7 = \beta_7 = 2.5, \alpha_8 = \beta_8 = 1.9$. We used $t = \frac{\overline{\mathsf{score}_{95}}}{s_{95}(\mathsf{score})} \cdot \mathbf{1}\{\frac{\overline{\mathsf{score}_{95}}}{s_{95}(\mathsf{score})} < 4.45\} + 4.45 \cdot \mathbf{1}\{\frac{\overline{\mathsf{score}_{95}}}{s_{95}(\mathsf{score})} \geq 4.45\}$ empirically.

From the table, it can be said that $\mathsf{score}(\mathbf{v})$ performs best overall, and $\mathsf{score}^{\beta}(\mathbf{u}^*)$ does better than $\mathsf{score}^{\alpha}(\mathbf{v}^*)$ in that FAR has priority over FRR when the difference between average FRRs is small. This implies that the distinct features of keystroke dynamics tend to be better expressed by using wavelet transformed keystroke patterns in the frequency domain than original keystroke patterns in the time domain, and the classification incorporating both the statistical features in the time and frequency domains are more effective than the classification incorporating those either in the time domain or in the frequency domain only.

5 Conclusions

The nonzero FARs in table 1 indicate the need for the improvement of classification accuracy. However, in the practical view, the KBS using $\mathsf{score}(\mathbf{v})$ is quite competitive due to the following reasons. First, the computational cost is very cheap since the complexity of algorithm required for the KBS model building and testing is $O\big(\text{the largest among } 2^m, R^2, S^2\big)$ where 2^m is the smallest integer larger than or equal to $N = \dim(\mathbf{v}) = \dim(\boldsymbol{\nu}^*)$, $R = \dim(\boldsymbol{\nu}_i^*) = \dim(\mathbf{v}^*) \leq N$, and $S = \dim(\boldsymbol{\mu}_i^*) = \dim(\mathbf{u}^*) \leq N$. Second, the usability cost is very cheap since the KBS is built by only 20 user's keystroke patterns, $\boldsymbol{\nu}_i$, $i = 1, \ldots, 20$, whose size N ranges from 13 to 21. Third, the practically acceptable classification accuracy is obtained (average FAR = 1.08%, average FRR = 5.33%) at the low cost of usability and computational complexity.

Acknowledgement

We would like to thank Professor Sungzoon Cho at Seoul National University for sharing his data on keystroke dynamics. This work was supported by grant No.R01-2005-000-103900-0 from the Basic Research Program of the Korea Science and Engineering Foundation.

References

1. Vidakovic, B.: Statistical Modeling by Wavelets. Wiley (1999)
2. Peacock, A., Ke, X., Wilkerson, M.: Typing Patterns: A Key to User Identification. IEEE Securiy & Privacy **2** (2004) 40–47
3. Yu, E., Cho, S.: Keystroke dynamics identy verification–its problems and practical solutions. Computers & Security **23** (2004) 428–440
4. Sheng, Y., Phoha, V.V., Rovnyak, S.M.: A Parallel Decision Tree-Based Method for User Authentication Based on Keystroke Patterns. IEEE Trans. Sytems, Man and Cybernetics, Part B. **35** (2005) 826–833

GA SVM Wrapper Ensemble for Keystroke Dynamics Authentication

Ki-seok Sung and Sungzoon Cho*

Department of Industrial Engineering, Seoul National University,
San 56-1, Shillim-dong, Kwanak-gu, Seoul 151-744, Korea
{zoro81, zoon}@snu.ac.kr
http://dmlab.snu.ac.kr

Abstract. User authentication based on keystroke dynamics is concerned with accepting or rejecting someone based on the way the person types. A timing vector is composed of the keystroke duration times interleaved with the keystroke interval times. Which times or features to use in a classifier is a classic feature selection problem. Genetic algorithm based wrapper approach does not only solve the problem, but also provides a population of "fit" classifiers which can be used in ensemble. In this paper, we propose to add uniqueness term in the fitness function of genetic algorithm. Preliminary experiments show that the proposed approach performed better than two phase ensemble selection approach and prediction based diversity term approach.

1 Introduction

Keystroke dynamics based authentication (KDA) is concerned with accepting or rejecting someone based on the way that person types. In typing a phrase or a string of characters, the keystroke dynamics or its timing pattern can be measured and used for identity verification. More specifically, a timing vector consists of the keystroke duration times interleaved with the keystroke interval times. The times can be measured in a scale of milliseconds (ms). When a key is stroked before a previous key is released, a negative interval results. When a password of n characters is typed, a (2n +1) dimensional timing vector results, which consists of n keystroke duration times and (n+1) keystroke interval times, with the return key included (see Figure 1).

Feature selection, a major step in pattern classification, determines the minimum number of essential features to be used in building a classifier. There have been some works investigating which elements are useful in KDA, but it seems there is not a clear winner [1]. There are two different feature selection approaches, filter and wrapper approach [2]. In wrapper approach, a subset of features is tentatively selected and fed to a classifier. And this process repeats until a good subset is found (see Figure 2). Combinatorial optimization in the search process is often performed by genetic algorithm, thus it is called GA based wrapper [3]. GA wrapper results in not just one subset of features, but a set of subsets of features (see Figure 3). Repetitive application of genetic operators such as crossover and mutation transforms a randomly generated

* Corresponding author.

D. Zhang and A.K. Jain (Eds.): ICB 2006, LNCS 3832, pp. 654 – 660, 2005.

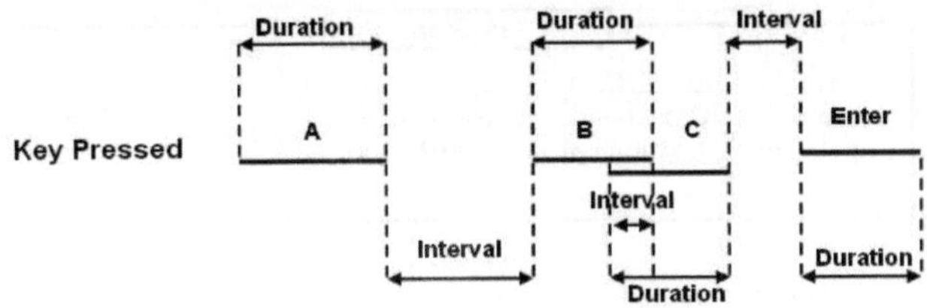

Fig. 1. Timing vector of Password "ABC"

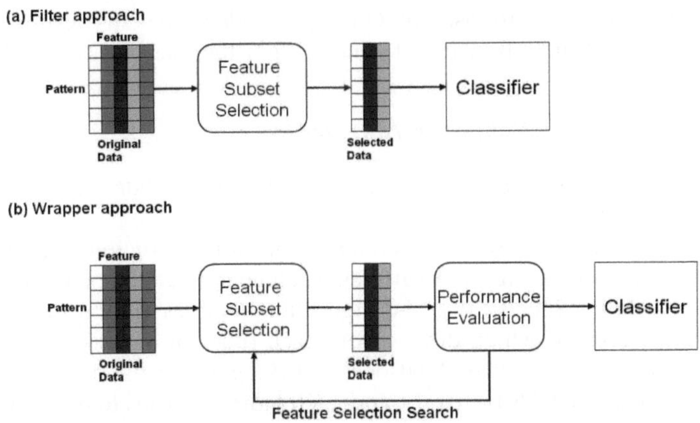

Fig. 2. (a) filter and (b) wrapper approach in feature selection

population of classifiers into a population of highly fit classifiers. In KDA, given a set of timing vectors of D dimension, feature selection tries to find "reduced" yet "optimal" timing vectors of d dimension where d < D. By optimal, we mean achieving the minimum error or highest accuracy of the classifier which employs the reduced set of features. In GA based wrapper approach, a candidate is represented by a D bit binary string. The value of an element is 0 or 1 when the corresponding feature is absent or present, respectively. Started with a randomly generated population of D bit chromosomes, GA process repeats application of evolutionary operations to the population. In the end, fit chromosomes are expected to emerge. The classifiers that correspond to the fit chromosomes are identified and used in the ensemble.

Ensemble is a set of classifiers trained differently: by different data sets, by different features, or by different models [4]. After individual classifiers are trained, they are combined by either majority voting or averaging to output a single value. The performance of an ensemble classifier has been found to be quite high in practice in a variety of applications. Bagging and Boosting are two of the most popular methods [5, 6, 7]. Individual classifiers participating in an ensemble have to be accurate as well as diverse in order to result in a accurate ensemble. It is only natural to combine

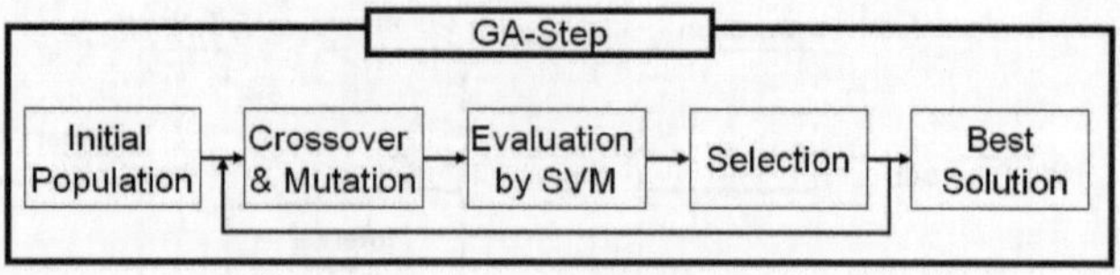

Fig. 3. GA wrapper based feature subset selection

GA wrapper and ensemble since the former generates a population of accurate classifiers. Of course, it has be made sure that they are diverse. So called Genetic Ensemble Feature Selection (GEFS) proposed by Opitz [8], adds a diversity term in the fitness function of GA. The fitness function of genetic algorithm has two terms, the accuracy and diversity:

$$\text{Fitness}(x)=A(x)+\lambda D(x). \tag{1}$$

where A denotes accuracy and D denotes diversity with lambda a constant weighing between the two terms.

The accuracy measures how well each neural network predicted of each validation pattern. The diversity measures how different each neural network's prediction is from that of the ensemble. Specifically, the algorithm involves finding a population of neural networks, each of which differs from each other in terms of the predictions. The GEFS performed better than AdaBoost and Bagging for the data sets tested. But major disadvantage of GEFS is that the approach indirectly tries to diversify the population through the difference in prediction. A more direct approach would consider the difference in the features actually employed in each neural network.

Recently, a similar but more elaborate approach has been proposed for KDA by Yu and Cho [9]. Other differences include use of SVM as base classifier for quick training and a different fitness function for GA.

$$\text{Fitness}(x)=\alpha A(x)+\beta\frac{1}{\text{LrnT}(x)}+\gamma\frac{1}{\text{DimRat}(x)}. \tag{2}$$

where A refers to false rejection rate, LrnT training time, and DimRat dimensionality reduction ratio. If the dimensionality of full feature set was 15, and the dimensionality of currently selected feature subset is 6, for instance, then DimRat(x) = 6/15 = 40%. Since the fitness function clearly does not force diversity, the post processing step was required. Major disadvantage of this approach is that the post processing step involves a time consuming heuristic procedure.

Here in this paper, we propose one step approach similar to that of GEFS, yet with a more direct diversity term in the fitness function and SVM as base classifier and similar to that of Yu and Cho, yet with a diversity term and no more post processing step. In particular, so called "uniqueness" term is used in a fitness function, measuring how unique each classifier is from others in terms of the features used.

This paper is structured as follows. The next section presents the proposed approach. Then, experimental settings and results follow. Finally, a conclusion and future work is discussed.

2 Proposed Method

Contrary to the ordinary GA, GA wrapper has to find not only good strings but also diverse strings. In order to enforce diversity, the fitness function needs a diversity term as in GEFS. What we propose to use here is "uniqueness" term, which measures for each chromosome how different it is from other chromosomes. Since more unique chromosomes are preferred, uniqueness is simply added to accuracy just like the diversity term in GEFS.

Before defining uniqueness, let us define S-distance between the two chromosomes. The S-distance between two chromosomes i and j, $S(d_{ij})$ is defined as follows:

$$S(d_{ij}) = \begin{cases} (\frac{d_{ij}}{C})^2, \text{ if } d_{ij} < C; \\ 1, \qquad\qquad \text{otherwise}. \end{cases} \tag{3}$$

where d_{ij} denotes the Hamming distance between two chromosomes and C a constant. Inspired by sharing function proposed in [11], S-distance is upper bounded at 1.

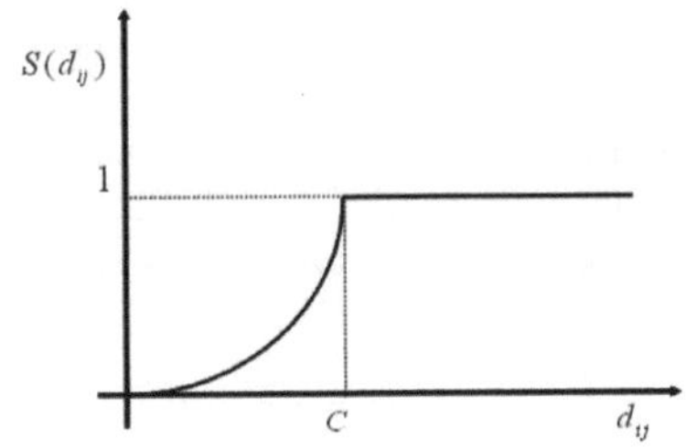

Fig. 4. $S(d_{ij})$ against d_{ij}

Now the uniqueness of xth chromosome is defined as an arithmetic average of S-distances to all other chromosomes.

$$U(\mathrm{x}) = \frac{\sum_{\mathrm{x} \neq j} S(d_{\mathrm{x}j})}{n-1}. \tag{4}$$

Finally, the fitness of chromosome x is defined as a simple sum of accuracy and uniquness:

$$\text{Fitness(x)=A(x)+U(x)}. \tag{5}$$

Of course, accuracy A here represents 1 – false-rejection-rate (1-FRR) since only the user's patterns are available in training.

The proposed approach differs from that of Opitz [8] in that diversity is not measured by the indirect approach,_difference in the predictions, but by the direct approach, difference in the actual features selected and. The proposed approach differs from that of Yu and Cho [9] in that diversity is introduced in wrapper GA step through the use of uniqueness term so that the subsequent post processing is not necessary and it makes term qualitatively simple.

3 Experimental Setting

The proposed method was applied to 21 sets of password typing patterns used in other research [9, 10]. Even though the original data sets contain hundreds of user's typing patterns, only 50 patterns were used in order to improve the reality of the experiments. Generally, it is hard to expect a user to type a password hundreds of times in enrollment. Out of 50 patterns for each password, 35 of them were used for training while 15 of them were used for validation, in particular to measure FRR in the fitness function of GA wrapper.

It has to be noted that one timing vector set was found to be very poor in its consistency. Figure 5 compares the mean timing vectors of training and test patterns. For "90200jdg" on the left, they are quite different. In particular, note the first, second, sixth and eighth interval values. They are all negative for the test while they are all positive for the training. It is obvious that the user was originally not quite familiar to the password, but later on, after hundreds of typing "practice," he became familiar to it. There is no way to discriminate user and impostor based on the user's past typing patterns if they changed over the time. Thus, we removed this particular password, "90200jdg," in the experiment.

In order to understand the performance of the proposed approach, we also implemented related approaches: the work of Opitz and that of Yu and Cho. Even though Yu and Cho also used the same data set, they used a randomly selected 50 patterns. So we performed experiment again with the different 50 patterns.

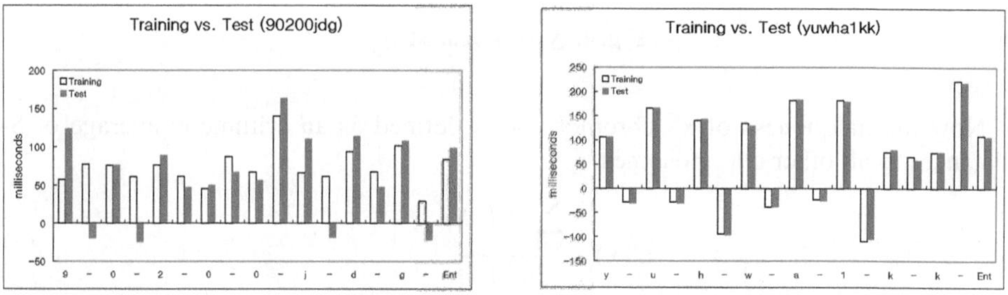

Fig. 5. Comparison of training and test timing vectors of two passwords "90200jdg" and "yuhwa1kk"

A population of 100 chromosomes was run 50 generations with cross over rate of 0.2 and mutation rate of 0.01. The SVM employed Gaussian kernel. The values of its parameters γ, cost, υ were determined in an empirical way. Of course, these values were shared also by all three approaches. The C value for the proposed approach was set to 30% of the original dimension. Early stopping criterion and classifier HD percentage used by Yu and Cho approach was set to 0.2 and 30%, respectively.

4 Results

Table 1 shows the performance of three approaches for 20 password timing vector sets. Since GA is stochastic in nature, five GA runs were made for each. The every entry in the table is an average from the five runs. There are 75 user's test patterns and 75 impostor patterns. They were used to calculate the accuracy, false acceptance rate and false negative rate. The number of ensemble denotes the number of classifiers in ensemble. The proposed approach and Opitz approach has a same fixed number, but Yu Cho approach has various numbers since it is the post processing phase that determines the exact number of classifiers in ensemble. On average, the proposed approach results in the best numbers, closely followed by that of Opitz and Yu and Cho in that order, although the difference may not be statistically significant. The FAR is much smaller than the FRR in the proposed approach, which is quite desirable considering that FAR is much more costly than FRR. By comparing best performing approach of each password, the proposed approach was best by coming first in nine passwords.

Table 1. Performance of three approaches

Password	Models											
	Sung-Cho Fitness = A(x) + U(x)				Yu-Cho Fitness = 10A(x) + 1/(100 * LrnT(x)) + 1/DimRat(x)				Opitz Fitness = A(x)+D(x)			
	Ensemble Accuracy	FAR	FRR	Num of Ensemble	Ensemble Accuracy	FAR	FRR	Num of Ensemble	Ensemble Accuracy	FAR	FRR	Num of Ensemble
ahrfus88	89.60	5.86	14.93	31	80.00	36.26	3.73	10.20	89.60	4.80	16.00	31
anehwksu	90.53	1.60	17.33	31	86.26	12.80	14.66	13.40	90.53	3.46	15.46	31
autumnman	93.60	0.00	12.80	31	92.00	10.93	5.06	11.40	92.93	0.53	13.60	31
beaupowe	86.00	17.33	10.66	31	78.00	38.66	5.33	9.20	85.33	21.86	7.46	31
c.s.93/ksy	93.20	1.33	12.26	31	92.66	6.66	8.00	23.60	93.60	1.33	11.46	31
dhfpql.	94.40	0.00	11.20	31	95.73	1.33	7.20	11.20	96.00	0.00	8.00	31
dirdhfmw	96.93	0.00	6.13	31	98.13	0.80	2.93	11.20	96.26	0.00	7.46	31
dlfjs wp	85.46	0.00	29.06	31	93.06	1.86	12.00	12.40	85.60	0.00	28.80	31
dltjdgml	90.93	0.00	18.13	31	95.73	1.60	6.93	10.60	91.86	0.00	16.26	31
drizzle	92.13	6.66	9.06	31	87.46	21.06	4.00	11.60	91.33	6.13	11.20	31
dusru427	90.13	0.00	19.73	31	93.06	1.33	12.53	15.40	90.53	0.00	18.93	31
i love 3	94.93	1.06	9.06	31	91.06	10.66	7.20	8.60	95.06	1.06	8.80	31
love wjd	88.80	14.13	8.26	31	84.40	27.20	4.00	11.80	86.13	11.20	16.53	31
loveis.	92.13	8.00	7.73	31	89.06	20.00	1.86	12.40	91.06	7.20	10.66	31
manseiii	83.06	18.40	15.46	31	74.00	46.13	5.86	13.00	81.33	24.53	12.80	31
rhkdwo	93.06	0.53	13.33	31	93.60	4.53	8.26	7.80	92.53	0.80	14.13	31
rla sua	97.20	1.86	3.73	31	89.73	16.80	3.73	10.80	95.86	3.46	4.80	31
tjddmswjd	90.93	0.26	17.86	31	91.20	2.40	15.20	14.40	90.13	0.00	19.73	31
tmdwnsl1	90.26	0.00	19.46	31	93.60	1.60	11.20	11.00	91.20	0.00	17.60	31
yuhwa1kk	97.06	0.00	5.86	31	97.33	0.00	5.33	11.80	96.53	0.00	6.93	31
Min	83.06	0.00	3.73	31	74.00	0.00	1.86	7.8	81.33	0.00	4.80	31
Max	97.20	18.40	29.06	31	98.13	46.13	15.20	23.6	96.53	24.53	28.80	31
Average	91.52	3.85	13.10	31	89.80	13.13	7.25	12.09	91.17	4.32	13.33	31

5 Conclusion and Future Work

In this paper, we proposed a GA based wrapper approach to be applied to keystroke dynamics based authentication. Compared to the previous work by Yu and Cho, we proposed to introduce diversity of the population by adding a term in fitness function that measures the uniqueness of a chromosome. This renders a rather complicated post processing unnecessary. Compared to the work by Opitz, we used one class SVM as base classifier and forced diversity through the uniqueness of each chromosome. A preliminary experiment involving 20 passwords shows that the proposed approach performed best. It is our contribution that a simpler approach produced a slightly better or similar performance.

There are limitations to the approach. First, the SVM used as a base classifier does not involve a threshold thus a balance between FAR and FRR cannot be controlled. We can indirectly control FRR in training, by using training parameters, like γ and cost. Second, fitness is computed as a sum of accuracy and diversity. A multi-objective optimization technique can be used instead. Third, removing outliers from user's training patterns might help achieve better performance.

Acknowledgements

This work was supported by grant No.R01-2005-000-103900-0 from the Basic Research Program of the Korea Science and Engineering Foundation.

References

1. Araujo, L., Sucupira, L., Lizarraga, M., Ling, L., and Yabu-Uti, J.: User Authentication through Typing Biometrics Features. IEEE Transactions on Signal Processing. 53(2), (2005) 851-855
2. Liu, H., Motoda, H. : Feature Selection for Knowledge Discovery and Data Mining. Kluwer Academic Publishers (1998)
3. Yang, J., Honavar, V. : Feature Subset Selection using a Genetic Algorithm. in Feature Selection for Knowledge Discovery and Data Mining. Liu, H. and Motoda, H. (eds.), Kluwer Academic Publishers, (1998) 117-136
4. Dietterich, T. G. : Ensemble methods in machine learning. First International Workshop on Multiple Classifier Systems, (2000) 1-15.
5. Breiman, L. : Bagging predictors. Machine Learning. 24(2), (1996) 123-140
6. Freund, Y., Schapire, R.E. : Experiments with a new boosting algorithm. Proceedings of the 13th International Conference on Machine Learning. Morgan Kaufmann, (1996) 148–156.
7. Sullivan, J., Langford, J., Caruana, R., Blum, A. : Featureboost: A meta-learning algorithm that improves model robustness. Proceedings of the Seventeenth International Conference on Machine Learning. (2000)
8. Opitz, D. : Feature selection for ensembles. AAAI/IAAI, (1999) 379-384.
9. Yu, E., Cho, S.: Keystroke dynamics identity verification - its problems and practical solutions. Computers and Security. 23(5) (2004) 428-440
10. Cho, S., Han, C., Han, D., Kim, H.: Web-based keystroke dynamics identity verification using neural network. J. Organizational computing and electronic commerce. 10(4) (2000) 295-307
11. Srinivas, N., Deb, K. : Multiobjective Optimization Using Nondominated Sorting in Genetic Algorithms, Evolutionary Computation, 2(3) (1994) 221-248

Enhancing Login Security Through the Use of Keystroke Input Dynamics

Kenneth Revett[1], Sérgio Tenreiro de Magalhães[2], and Henrique M.D. Santos[2]

[1] University of Westminster,
Harrow School of Computer Science, London, UK HA1 3TP
revettk@westminster.ac.uk
[2] Universidade do Minho,
Department of Information Systems, Campus de Azurem,
4800-058 Guimaraes, Portugal
{psmagalhaes, hsantos}@dsi.uminho.pt

Abstract. Security is a critical component of most computer systems – especially those used in E-commerce activities over the Internet. Global access to information makes security a critical design issue in these systems. Deployment of sophisticated hardware based authentication systems is prohibitive in all but the most sensitive installations. What is required is a reliable, hardware independent and efficient security system. In this paper, we propose an extension to a keystroke dynamics based security system. We provide evidence that completely software based systems based on keystroke input dynamics can be as effective as expensive and cumbersome hardware based systems. Our system is behavioral based that captures the typing patterns of a user and uses that information, in addition to standard login/password security to provide a system that is user-friendly and very effective at detecting imposters.

1 Introduction

With the increasing number of E-commerce based organizations adopting a stronger consumer-orientated philosophy, web-based services (E-commerce) must become more user-centric. As billions of dollars worth of business transactions occur on a daily basis, E-commerce based enterprises must ensure that users of their systems are satisfied with the security features in place. As a starting point, users must have confidence that their personal details are secure. Access to the user's personal details is usually restricted through the use of a login ID/password protection scheme. If this scheme is breached, then a user's details are generally open for inspection and possible misuse. Hardware (physiological) based systems are not yet feasible over the Internet because of cost factors and in addition, the question as to their ability to reduce intruder detection has not yet been answered equivocally. Our system is based on what has now become known as "keystroke dynamics" with the addition of keyboard partitioning [1,2]. We also consider in this study the affect of typing speed and the use of a rhythm when a user enters their login details. Keystroke dynamics

D. Zhang and A.K. Jain (Eds.): ICB 2006, LNCS 3832, pp. 661 – 667, 2005.

was first introduced in the early 1980s as a method for identifying the individuality of a given sequence of characters entered through a traditional computer keyboard. Researchers focused on the keystroke pattern, in terms of keyboard duration and keyboard latency [2,10]. Evidence from preliminary studies indicated that when two individuals entered the same login details, their typing patterns would be sufficiently unique as to provide a characteristic signature that could be used to differentiate one from the another. If one of the signatures could be definitively associated with a proper user, then any differences in typing patterns associated with that particular login ID/password must be the result of a fraudulent attempt to use those details. Thus, the notion of a software based biometric security enhancement system was born. Indeed, there are commercial systems such as BioPassword that have made use of this basic premise.

A critical issue with respect to enhancement of login based security systems is the criteria for success. There are two basic errors associated with biometric applications with respect to verification: false rejection (FRR -type I error) and false acceptance (FAR - type II error). One wishes to develop a system that minimises type II errors without increasing type I errors. In this paper, we employ the Crossover Error Rate (CER) as our measure of the balance between false acceptance ratio (FAR) and the false rejection ratio (FRR), as depicted in Figure 1. Striking the balance between sensitivity and specificity is a difficult balancing act. Traditional approaches have employed either machine-learning or deterministic algorithms. Among the solutions based on machine learning, the work presented by Chen [3] achieved a CER less than 1% and a 0% FAR. Ord and Furnell [4] also tested this technology, with a 14 people group, to study the viability of applying this technology on PINs (Personal Identification Numbers) typed on a numeric-pad. Although the results were initially promising, it was found that the results did not scale up well and the authors indicated that this technology was not feasible for community based applicability. Deterministic algorithms have been applied to keystroke dynamics since the late 70's. In 1980 Gaines [5] presented the results of a study of the typing patterns of seven professional typists. The typists were asked to enter a specified text (3 paragraphs) repeatedly over a period of several months. The authors collected data in the form of keystroke latencies from which they constructed digraphs were constructed and analysed statistically. Unfortunately, no real conclusion could be drawn from this study regarding the uniqueness of each typist's style – most likely resulting from the small sample size and/or inadequate data sample. The method used to establish a keystroke pattern was a breakthrough, which introduced the concept of a digraph, the time spent to type the same two letters (digraph), when together in the text. Since then, many algorithms based on Algebra and on probability and statistics have been presented. Joyce & Gupta presented in 1990 [6] an algorithm to calculate a metric that represents the distance between acquired keystroke latency times over time, thus introducing a dynamic approach. In 1997 Monrose and Rubin employed an Euclidean Distance and probabilistic method based on the assumption that the latency times for one-digraph exhibits a Normal Distribution [7]. Later, in 2000, the same authors presented an algorithm for identification, based on the similarity models of Bayes [8], and in 2001 they presented an algorithm that employed polynomials and vector spaces to generate complex passwords from a simple one, using keystroke patterns [9].

In our research, we examine various typing characteristics that might provide subtle but consistent signatures that we can use for keystroke verification purposes. Our initial study was designed to provide a baseline case for the CER from a group of informed users that were asked to participate in this study. Once we established a baseline CER< we then wished to determine if there were factors related to typing styles that could alter the CER. We selected two basic factors: length of the passphrase and typing speed. In the next section we describe in detail the algorithms deployed in this study, followed by a Results section, and lastly a brief discussion of this work.

2 Implementation

Our primary goal is to produce a software-based system that is capable of performing automated user ID/password verification. We employ the following steps when a new user is added to the system (or is required to change their login details):

1. The login ID/password or simply the new password is entered a certain number of times (enrollment).
2. A profile summarising the keystroke dynamics of the input is generated and stored for access to the verification component.
3. A verification procedure is invoked which compares stored biometric attributes to those associated with a given login ID/password entry after the enrollment process.

The enrollment process, made by the user once on the first use of the service, consists on typing the users usual password, or passphrase, twelve times. If the user mistyped the passphrase, they were prompted to continue entering until all twelve entries were entered. During the enrollment procedure, statistics were calculated and stored for the verification process. Specifically, our algorithm calculates and stores the average, median, standard deviation, and the coefficient of variation for the latency times for each digraph (13 in all) and the total time spent entering each passphrase. Our enrollment phase was based on a series of 14 character passphrases entered into our system by a group of 8 volunteers, all of whom were fully aware of the purpose of this study and all reasonably computer literate. Each volunteer was requested to input a passphrase a minimum of twelve times in order to generate the statistics required for the verification phase. In addition, each volunteer served as their own control for FRR rates by entering their respective passphrases for an additional period of four weeks after the start of the study (yielding an average of 10,000 entries for FRR determination). The stored data table for the enrollment statistics was updated over time, with the oldest entry replaced by the most recent enrollment episode.

For our verification stage, we recruited a group of 43 volunteers (34 through the internet version of this software) and 9 users via a laptop running our software. All participants in the verification phase of this study (including the volunteer group) were required to enter at least 16 entries per user. For the volunteers (enrollment and verification), this provided use with the means to calculate the FRR (the first 12 entries were for enrollment and the rest for verification) and also for FAR on passphrases entered by other volunteers. All verification participants (43) only participated in determination of the FAR of the system. In total, we had over 187,000 login attempts in the baseline determination phase, with less than 0.01% successful attacks.

To allow a comparison of our FAR/FRR values with existing published results [6], we used a threshold of 60% for the time latencies for a positive match between a verification request and stored data for that passphrase. When a verification entry was input into our system, we used the following measure to determine if the digraph latency time was appropriate for a given passphrase. For each pair of keystrokes (digraphs) the algorithm will measure the time latency, defined as *TLP*, and compare it with the one stored.

$$Lowest(Average; median)*\left(0{,}95-\frac{SDesviation}{Average}\right)\leq TLP$$

$$TLP\leq Higher(Average; median)*\left(1{,}05+\frac{SDesviation}{Average}\right)$$

Equation 1. Crtieria for acceptance of a given input for the digraph latency

The comparison result will be a hit if and only if this criterion has been met. A total of 13 digraphs exist for each 14-character passphrase, and the results for each digraph are stored in a temporary Boolean array. A '1' is placed in the table if the LTP is within the specified boundary conditions and is the first occurrence of a '1' in the passphrase (always true for the first correctly entered character). Subsequent correctly input keystrokes would result in a '1' being replaced by a '1.5' for that digraph entry in the array. If the keystrokes did not result in a hit, then a '0' is entered for that digraph position in the array. Then the elements in the array for a particular passphrase are added together. If the sum is greater than a given threshold, then the entry is considered valid, otherwise it is invalid. For instance, if the threshold is set on 70%, users will only be authenticated to the system if the value A obtained from a given attempt is over 70% of the highest possible value, which is given by: $(number\_of\_characters-1)*1.5+1$.

Finally, if and only if the login attempt is accepted, the oldest values stored for the latencies are substituted by the corresponding values collected in this successful attempt. This last procedure will allow the data stored to evolve with the user. This allows the system to evolve over time, as the user's familiarity with their passphrase improves with time and practise, so will the statistics. The system administrator can change the sensitivity of the system at will. For instance, to maintain a 60% threshold, all users must generate a score given by (number_of_characters –1)*1.5 +1 that is over 60% of the maximum score. For a 14-character passphrase, this would yield a score of 12.3, which would be set to 12, since our threshold is a multiple of 1/2. Thus any score greater than 12 would be considered a legitimate entry into the system. By varying this threshold, we can extract an estimate of the FAR/FRR as a function of the sensitivity threshold. What we wish to produce is a system that yields a very low FAR without incurring a large FRR in the process. A reasonable criterion is when the FAR/FRR intersect – the Crossover Error Rate What we wish to do is reduce the CER to the lowest possible value without placing an undue burden on the user community. We have explored two basic techniques in a previous work [11], focusing on keyboard partitioning and the typing speed. We present the results of an extended study on these factors, which we present in the following Results section.

3 Results

This algorithm presented a CER of 5,58% and it can achieve, at the lowest thresholds, a FRR of near zero that maximizes the comfort of the user. At the higher demanding thresholds the algorithm presents a near zero FAR, maximizing the security. The results of our baseline experiment, with a single 14-character passphrase are presented in Figure 1 below. The results presented in Figure 1 can be summarized by the CER – which was 5.58%. It is important to notice that the results obtained in this experiment are the worst case scenario, when a passphrase breech has occurred. If the passphrase was not disclosed, then we could extrapolate the FAR (considering a brute force attack) by:

$$FAR_{Brute\_force} = (1/(Number\_of\_possible\_passphrases))* FAR_{Know\_passphrase} \quad (2)$$

Equation 2 states that if the passphrase were not known, then the FAR would be equivalent to the FAR when the passphrase was known, multiplied by the probability of guessing the passphrase. With a 14-character passphrase, the success rate of a brute force is near astronomical.

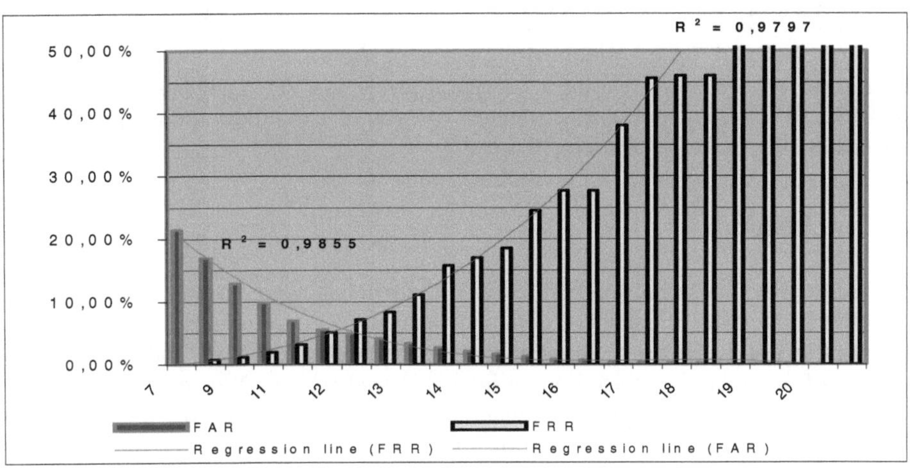

Fig. 1. False Acceptance Rates and False Rejection Rates for the range of possible thresholds for a 14 character passphrase. The x-axis is the threshold according to equation 1 and the y-axis is the resulting FAR/FRR. The data was generated from over 10,000 entries of the same passphrase.

3.1 Additional Experiments

We wanted to determine whether we could improve on our based CER of 5.58%. We investigated the length of the passphrase to see if it has an influence on the CER value. Our previous work [2] along with the work presented in this paper so far utilised long passphrases (14 characters). Generally, most IDs/passwords, PINs etc. are much shorter – on average between 4-8 characters in length. We therefore investigated a series of 7 character passphrases selected randomly by a computer programme. We enlisted a group of 10 volunteers to participate in this study. The results indicate that the FAR/FRR was reduced to approximately 2% (see Figure 2). We also incorporated keyboard gridding into our performance criteria by weighting characters that are in contiguous keyboard partition more heavily than those that were within the same partition or in non-contiguous partitions (by a factor of 2). The results indicate that the CER could be reduced to less then 0.01% using a combination of 14-character passphrase and keyboard partitioning.(data not shown).

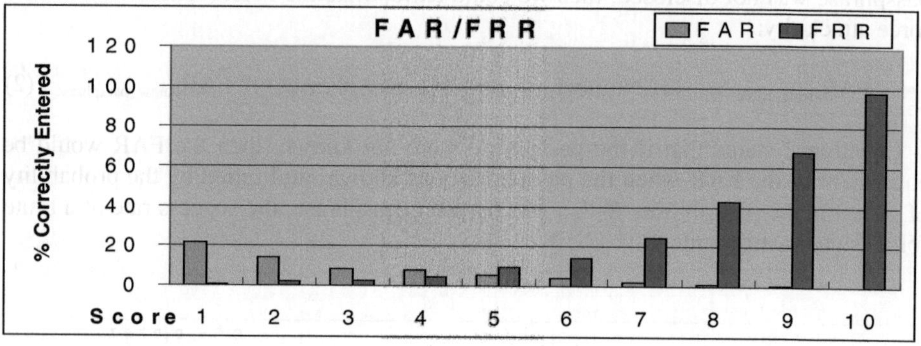

Fig. 2. FAR/FRR for the study using a 7 character passphrase. The CER (not on this display) was 4.1%. These results were obtained through 10 volunteers, entering a specific passphrase of 6 characters for a total of 1,000 trials (10 users).

4 Conclusions

This study provides supporting evidence to the role software based security systems can bring to the issue of enhanced computer security. Our system, based on keystroke dynamics, is not overly burdensome to the user, very cost-effective, and very efficient in terms of the overhead placed on an internet based server. We achieve a very low FAR/FRR (each less than 5%), compatible with those produced by very expensive hardware based systems. In addition, we have begun investigating additional strategies that can be combined with keystroke hardening, such as keyboard partitioning. Partitioning provides an added layer of security, but requires users to limit their selection of login IDs and passwords.

Our system incorporates the evolving typing styles of individuals. This is an important property of any software based biometric system. Users may experience through personal development, variations in their typing styles and/or speed. For instance, when a user is forced to change their password, they will take time to adjust to it, which will certainly have an impact on their typing signature. Any system that fails to take this into account will yield an undue burden on the user if it is not capable of dynamically adjusting the required acceptance thresholds.

References

[1] Yan, J., Blackwell, A.F., Anderson, R. & Grant, A. , 2004, Password memorability and security: Empirical results, *IEEE Security and Privacy* **2**(5), 25-31.

[2] Magalhães, S. T. and Santos, H. D., 2005, An improved statistical keystroke dynamics algorithm, *Proceedings of the IADIS MCCSIS 2005.*

[3] Chen, Z., 2000. *Java Card Technology for Smart Cards.* Addison Wesley, U.S.A.

[4] Ord, T. and Furnell, S. M., 2000. User authentication for keypad-based devices using keystroke analysis. *Proceedings of the Second International Network Conference – INC 2000.* Plymouth, U.K.

[5] Gaines, R. et al, 1980. Authentication by keystroke timing: Some preliminary results. Rand Report R-256-NSF. Rand

[6] Joyce, R. and Gupta, G., 1990. Identity authorization based on keystroke latencies. *Communications of the ACM.* Vol. 33(2), pp 168-176.

[7] Monrose, F. et al, 2001. Password Hardening based on Keystroke Dynamics. *International Journal of Information Security.*

[8] Monrose, F. and Rubin, A. D., 1997. Authentication via Keystroke Dynamics. *Proceedings of the Fourth ACM Conference on Computer and Communication Security.* Zurich, Switzerland.

[9] Monrose, F. and Rubin, A. D., 2000. Keystroke Dynamics as a Biometric for Authentication. *Future Generation Computing Systems (FGCS) Journal: Security on the Web.*

[10] Alen Peacock, Xian Ke, Matthew Wilkerson. "Typing Patterns: A Key to User Identification, IEEE. *Security and Privacy*, vol. 02, no. 5, pp. 40-47, September-October, 2004

[11] Revett, K. and Khan, A., 2005, Enhancing login security using keystroke hardening and keyboard gridding, *Proceedings of the IADIS MCCSIS 2005.*

A Study of Identical Twins' Palmprints for Personal Authentication

Adams Kong[1,2], David Zhang[2], and Guangming Lu[3]

[1] Pattern Analysis and Machine Intelligence Lab, University of Waterloo,
200 University Avenue West, Waterloo, Ontario N2L 3G1, Canada
adamskong@ieee.org

[2] Biometric Research Centre, Department of Computing,
The Hong Kong Polytechnic University, Kowloon, Hong Kong
csdzhang@comp.polyu.edu.hk

[3] Biocomputing Research Lab, School of Computer Science and Engineering,
Harbin Institute of Technology, Harbin, China
csglu@comp.polyu.edu.hk

Abstract. Biometric recognition based on human characteristics for personal identification has attracted great attention. The performance of biometric systems highly depends on the distinctive information in the biometrics. However, identical twins having the closest genetics-based relationship are expected to have maximum similarity between their biometrics. Classifying identical twins is a challenging problem for some automatic biometric systems. In this paper, we summarize the exiting experimental results about identical twins' biometrics including face, iris, fingerprint and voice. Then, we systemically examine identical twins' palmprints. The experimental results show that we can employ low-resolution palmprint images to distinguish identical twins.

1 Introduction

Biometric systems measuring our biological and behavioral features for personal authentication have inherent advantages over traditional knowledge-based approach such as password and over token-based approach such as physical key. Various biometric systems such as face, iris, retina, fingerprint and signature, were proposed, implemented and deployed in the last thirty years [1]. Biometric systems use the distinctive information in our biometric traits to identify different people. Nevertheless, not all biometrics have sufficient information to classify identical twins having the same genetic expression.

There are two types of twins, monozygotic and dizygotic twins. Dizygotic twins result from different fertilized eggs. Consequently, they have different Deoxyribo Nucleic Acid (DNA). Monozygotic twins, also called identical twins are the result of a single fertilized egg splitting into two individual cells and finally developing into two persons. Thus, identical twins have the same DNA. The frequency of identical twins is about 0.4% across different populations [2]. Some people believe that this is the limit of face recognition systems [18].

D. Zhang and A.K. Jain (Eds.): ICB 2006, LNCS 3832, pp. 668 – 674, 2005.

1.1 From DNA to Biometrics

DNA contains all the genetic information required to generate an organ of a species. The mapping from the genetic information to an organ is very complicated. First of all, the genetic information in DNA molecule is copied to RNA (Ribo Nucleic Acid) molecule. Next, the information in RNA is used to generate amino acids and the amino acids are converted into functioning proteins. The functioning proteins are assembled to be an organ. In this process, genetic information is not the only one factor affecting the organ. It can be influenced by various other factors. As a result, identical twins who share the same genetic expression have many different biometrics including fingerprint, iris and retina [3, 15, 17]. In fact, some biometrics such as faces continually change after we are born. The changes depend on environmental conditions such as living style, diet and climate. They make identical twins more different when they age. Fig. 1 shows two pairs of identical twins at different ages. The older twins in Fig. 1(b) are easier to be distinguished.

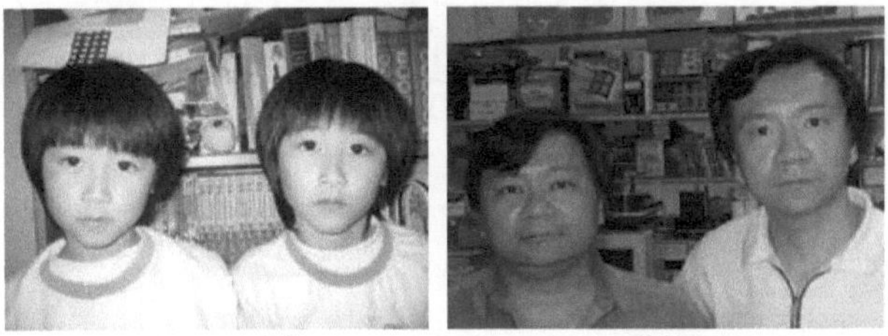

(a) (b)

Fig. 1. Two pairs of identical twins at different ages

1.2 Motivations

Identifying identical twins is important for all biometric systems. The systems that cannot handle identical twins have a serious security hole. According to our best knowledge, so far no paper summarizes the testing results of identical twins. In addition, no one investigates the similarity between low-resolution identical twins' palmprints.

The rest of this paper is organized as follows. Section 2 summarizes the testing reports from different sources. Section 3 gives the experimental results of identical twins' palmprints. Section 4 discusses the experimental results and the summary. Finally, Section 5 offers some concluding remarks.

2 Summary of the Existing Reports

In this paper, we discuss only the biological features including retina, iris, face, voice and fingerprint that are directly affected by genetic factors. Fig. 2 illustrates identical twins' retinas, irises, fingerprints and palmprints. These images are collected from different pairs of twins. The iris and palmprint images are collected using our self-designed devices [20] and the retina images are obtained from [6] with permission to reprint. The fingerprint images are collected using a standard optical fingerprint scanner. Fig. 2 shows that the retinas, irises and palmprints can easily be distinguished by human vision. For the fingerprints, we have to pay more attention at the minutiae points, commonly utilized in fingerprint systems. Based on the positions and directions of the minutiae points, the twins' fingerprints can be distinguished without any problem.

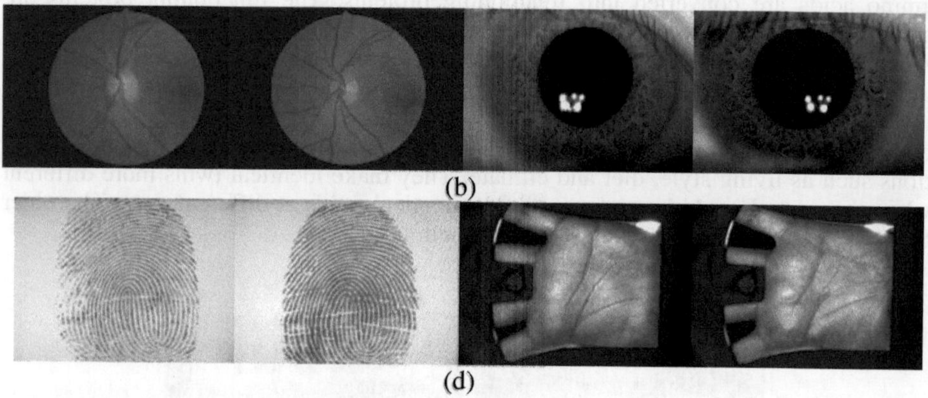

Fig. 2. Different features from identical twins', (a) retinas, (b) irises, (c) fingerprints and (d) palmprints

In many cases, biometrics are proposed by medical doctors or ophthalmologists [15-16] but almost all the biometric identification systems are designed by engineers. The features discovered by doctors or ophthalmologists and the features applied to authentication systems may not be the same. The iris is a typical example [6, 17]. Ophthalmologists distinguish irises based on the structural features including moles, freckles, nevi and crypts while current iris recognition systems use binary sequences to represent the textural features. Therefore, the experimental results or observation given by doctors or ophthalmologists about identical twins may not be applicable to automatic biometric systems. In other words, it is essential to test automatic biometric systems on identical twins. Table 1 summarizes the testing results including, iris, face, palmprint and voice. We also give the sizes of testing databases and age ranges of the testing samples in Table 1. The database size refers to the number of different biometrics, not the number of twin pairs. The testing results are represented by the symbols "+" and "−". The symbol "+" denotes that the tested methods can distinguish identical twins, just as normal persons. The symbol "−" denotes that the tested method cannot correctly distinguish them.

All the results in Table 1 are positive, except voice recognition. Some of the results are not significant since their testing databases are too small. Based on [7, 9] and experimental results in Section 3, we ensure that iris, palmprint and fingerprint can be used to separate identical twins. However, testing on large databases is required to verify the results of 3D face, 2D face and fusion of lip motion and voice [10-11, 13, 14]. It is generally believed that faces cannot be used for separating identical twins. Experts in National Institute of Standards and Technology (USA) said "although identical twins might have slight facial differences, we cannot expect a face biometric system to recognize those differences."[18]. Interestingly, the results in Table 1 contradict our general beliefs.

In addition to fingerprint, palmprint and iris, retina and thermogram are considered as distinctive features for identical twins [9]. So far, we have not obtained any testing report about them.

Table 1. Summary of the existing twin tests

Biometric	Results	Age Ranges	Database Size	Refer-ence
Iris	+	*	648#	[7]
3D face	+	*	Several	[10-11]
2D face	+	*	20	[13]
Fingerprint	+	*	188	[9]
Palmprint	+	6-45	106	Section 3
Voice	–	*	32	[12]
Lip motion and speech	+	18-26	4	[14]

* The age ranges are not available.

In this test, 648 right/left iris pairs from 324 persons are tested since our left and right irises are generated from the same DNA.

3 Study of Twins' Palmprints

According to our best knowledge, no one studies identical twins' palmprints for automatic personal authentication. In this experiment, we utilize the orientation fields of palmprints as feature vectors to represent low-resolution palmprint images and use angular distance to compare the feature vectors. Readers can refer to [8] for the computational detail of this method. Shorter angular distance represents more similarity between two palmprint images. This method is a modification of our previous work [20]. To compare with the palmprints from general persons and identical twins, we prepare two databases for this study. The details of the databases are given in the following sub-sections.

3.1 Twin and General Palmprint Databases

The twin database contains 1028 images collected from 53 pairs of identical twins' palms. We collect the images from their left and right palms. Around 10 images are collected from each palm. All the images are collected by our self-designed palmprint capture device [20]. The image size is 384×284.

To produce a reliable genuine distribution, we prepare a palmprint database containing 7,752 images from the right and left palms of 193 individuals. This database is called general palmprint database. The images in this database are collected on two separate occasions, two months apart. On each occasion, the subject was asked to provide about 10 images, each of the left palm and the right palm. The average interval between the first and second collections was 69 days. More information about this database can be referred to [20].

3.2 Experimental Results

To study the similarity between identical twins' palmprints and to obtain twin imposter distribution, we match a palmprint in the twin database with his/her identical twin sibling's palmprints (twin match). We also match every palmprint in the general database with other palmprints in the general database to obtain genuine and imposter (general match) distributions of normal persons. In addition, we match different person's left palmprints and match different person's right palmprints to obtain a side imposter distribution (side match). Total number of genuine matchings, general imposter matchings, side imposter matchings and twin imposter matchings are 74,068, 29,968,808, 14,945,448 and 4,900, respectively. The genuine distribution and imposter distributions of general match, twin match and side match are given in Fig. 3(a). The genuine distribution along with the three imposter distributions in Fig. 3(a) is used to generate the Receiver Operating Characteristics (ROC) curves given in Fig. 3(b). Fig. 3(b) shows that we can use low-resolution palmprint images to distinguish identical twins but identical twins' palms have some inherent correlation, which is not due to side matching.

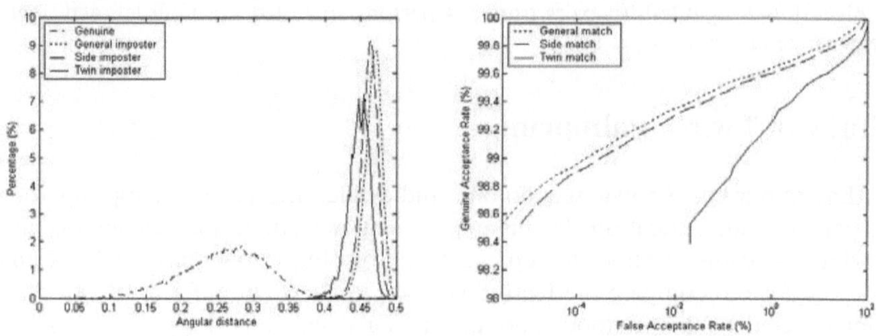

(a) (b)

Fig. 3. Verification results. (a) Twin imposter, side imposter, general imposter and genuine distributions and (b) ROC curves for corresponding distributions.

4 Discussion

According to the summary and the experimental results in Section 3, we have confidence to say that, iris, fingerprint and palmprint are three effective biometrics to distinguish identical twins. The subjective comparisons of these three biometrics are given in Table 2. The comments of fingerprint and iris are obtained from [1]. We also agree the comments about palmprint in [1], except collectability. The palmprints discussed in this paper are collected by a CCD camera-based palmprint scanner. Thus, the collectability of palmprint should be similar to that of hand geometry (High). According to Table 2, none of them is perfect. Each of them has strengths and weaknesses. Our low-resolution palmprint recognition method has combined the advantages of hand geometry and fingerprints, with high collectability and high

Table 2. Comparison of palmprint, fingerprint and iris

	Palmprint [8, 20]	Fingerprint [1]	Iris [1]
Universality	Middle	Middle	High
Distinctiveness	High	High	High
Permanence	High	High	High
Collectability	High*	Middle	Middle
Performance	High	High	High
Acceptability	Middle	Middle	Low
Circumvention	Middle	Middle	Low

* The authors' comments are different from [1].

performance. In addition, low-resolution palmprints do not have the problem of latent prints, which can be used to make artificial fingerprints to fool current commercial fingerprint systems [4].

5 Conclusion

In this paper, we have summarized the testing reports about examining biometric systems on identical twins. Although identical twins have the same DNA, their biometric traits including iris palmprints and fingerprint are different. Currently, biometric systems can effectively classify identical twins' irises and fingerprints. The existing reports about face recognition for identical twins give some encouraging results. They show that face is possible to be a foolproof way to tell the differences between identical twins. However, the methods should be tested on larger twin databases. Since their testing databases are too small, their results may not be reliable. In addition to the summary, the experimental results show that identical twins' palmprints are distinguishable but they have some inherent correlation.

References

1. A.K. Jain, A. Ross and S. Prabhakar, "An introduction to biometric recognition", *IEEE Trans. CSVT*, vol. 14, no. 1, pp. 4-20, 2004.
2. J.J. Nora, F.C. Fraser, J. Bear, C.R. Greenberg, D. Patterson, D. Warburton, "Twins and theirs use in genetics," in Medical Genetics: Principles and Practice 4th ed, Philadelphia: Lea & Febiger, 1994.
3. E.P. Richards, "Phenotype vs. genotype: why identical twins have different fingerprints" available at http://www.forensic-evidence.com/site/ID/ID_Twins.html
4. D. Cyranoski, "Detectors licked by gummy fingers" *Nature*, vol. 416, pp 676, 2002.
5. http://www.deleanvision.com/
6. Retinal technologies, http://www.retinaltech.com/technology.html
7. J. Daugman and C. Downing, "Epigenetic randomness, complexity and singularity of human iris patterns," *Proceedings of the Royal Society, B*, vol. 268, pp. 1737-1740, 2001.
8. A.W.K. Kong and D. Zhang, "Competitive coding scheme for palmprint verification," in Proc. *ICPR*, vol. 1, pp. 520-523, 2004.

9. A.K. Jain, S. Prabhakar and S. Pankanti, "On the similarity of identical twin fingerprint," *Pattern Recognition*, vol. 35, no. 11 pp. 2653-2663, 2002.
10. R. Kimmel, *Numerical Geometry of Image*, Springer, New York, 2003
11. D. Voth, "Face recognition technology," *IEEE Magazine on Intelligent Systems*, vol. 18, no. 3, pp. 4-7, 2003
12. "Large scale evaluation of automatic speaker verification technology: dialogues spotlight technology report," The Centre for Communication Interface Research at The University of Edinburgh, May 2000, Available at http://www.nuance.com/assets/pdf/ccirexecsum.pdf.
13. K. Kodate, R. Inaba, E. Watanabe and T. Kamiya, "Facial recognition by a compact parallel optical correlator," *Measurement Science and Technology*, vol. 13, pp. 1756-1766, 2002.
14. C.C. Chibelushi, F. Deravi and J.S.D. Mason, "Adaptive classifier integration for robust pattern recognition," *IEEE Trans. on SMC, Part B*, vol. 29, no. 6, pp. 902-907, 1999
15. C. Simon and I. Goldstein, "A new scientific method of identification," *New York state journal of medicine*, vol. 35, no. 18, pp. 901-906, 1935.
16. P. Tower, "The fundus oculi in monozygotic twins: report of six pairs of identical twins," *Archives of ophthalmology*, vol. 54, pp. 225-239, 1955.
17. L. Flom and A. Safir, U.S. Patent No. 4641349, U.S. Government Printing Office, Washington, DC, 1987.
18. P.J. Phillips, A. Martin, C.L Wilson, M. Przybocki, "An introduction to evaluating biometric systems," *Computer*, vol. 33, no. 2, pp. 56-63, 2000.
19. Veinid, "http://www.veinid.com/product/faq.html"
20. D. Zhang, W.K. Kong, J. You and M. Wong, "On-line palmprint identification," *IEEE Trans. PAMI*, vol. 25, no. 9, pp. 1041-1050, 2003.

A Novel Hybrid Crypto-Biometric Authentication Scheme for ATM Based Banking Applications

Fengling Han[1], Jiankun Hu[1], Xinhuo Yu[2], Yong Feng[2], and Jie Zhou[3]

[1] School of Computer Science and Information Technology,
Royal Melbourne Institute of Technology, Melbourne VIC 3001, Australia
{fengling, jiankun}@cs.rmit.edu.au
[2] School of Electrical and Computer Engineering,
Royal Melbourne Institute of Technology, Melbourne VIC 3001, Australia
{feng.yong, x.yu}@ems.rmit.edu.au
[3] Department of Automation, Tsinghua University, Beijing 100084, China
jzhou@tsinghua.edu.cn

Abstract. This paper studies the smartcard based fingerprint encrytion/authentication scheme for ATM banking systems. In this scheme, the system authenticates each user by both his/her possession (smartcard) and biometrics (fingerprint). A smartcard is used for the first layer of authentication. Based on the successful pass of the first layer authentication, a subsequent process of the biometric fingerprint authentication proceeds. The proposed scheme is fast and secure. Computer simulations and statistical analyze are presented.

1 Introduction

With rapidly increasing number of break-in reports on traditional PIN and password security systems, there is a high demand for greater security for access to sensitive/personal data. These days, biometric technologies are typically used to analyze human characteristics for security purposes [1]. Biometrics based authentication is a potential candidate to replace password-based authentication [2]. In conjunction with smartcard, biometrics can provide strong security. Various types of biometric systems are being used for real-time identification. Among all the biometrics, fingerprint-based identification is one of the most mature and proven technique [3].

Smartcard based fingerprint authentication has been actively studied [4-6]. A fingerprint based remote user authentication scheme by storing public elements on a smartcard was proposed, each user can access to his own smartcard by verifying himself using his fingerprint [4]. In [5] and [6], the on-card-matching using fingerprint information was proposed. However, these schemes require high resource on the smartcard and the smartcard runs a risk of physical attack. Together with the development of biometric authentication, incorporate biometric into cryptosystems has also been addressed [2]. However, instability of fingerprint minutiae matching hinders its direct use as encryption/decryption key. With the widely studied of automatic personal identification, a representation scheme which combines global and local information in a fingerprint was proposed [3, 7], this scheme is suitable for matching as well as storage on a smartcard.

D. Zhang and A.K. Jain (Eds.): ICB 2006, LNCS 3832, pp. 675–681, 2005.

Biometric authentication is image based. For remote biometric authentication, the images need to be encrypted before transmitted. Chaotic map used in image encryption has been demonstrated [8-10]. The permutation of pixels, the substitution of gray level values, and the diffusion of the discretized map can encrypt an image effectively.

In this paper, a biometric authentication protocol is proposed. Based on the modified Needham-Schroeder PK protocol [11], strong smartcard public key system for the first layer of authentication and then fingerprint authentication for the remaining parts are used. The primary application of our scheme is ATM based banking systems due to its popularity and trusted physical terminal that has 24 hours camera surveillance.

The rest of the paper is organized as follows: Section 2 provides the description of the new hybrid crypto-biometric authentication protocol. Generation of encryption key is studied in Section 3. Evaluation of the encryption scheme is conducted in Section 4. Conclusions are presented in Section 5.

2 Hybrid Crypto-Biometric Authentication Protocol (HCBA)

Generally, there are two basic fingerprint authentication schemes, namely the local and the centralized matching. In the central matching scheme, fingerprint image captured at the terminal is sent to the central server via the network, then is matched against the minutiae template stored in the central server.

There are three phases in HCBA: registration, login and authentication. In the registration phase, the fingerprints of a principal A are enrolled and the derived fingerprint templates are stored in the central server. The public elements and some private information are stored on smartcard. The login phase is performed at an ATM terminal equipped with a smartcard reader and a fingerprint sensor. The hybrid smartcard and ATM based fingerprint authentication protocol is shown in Fig.1.

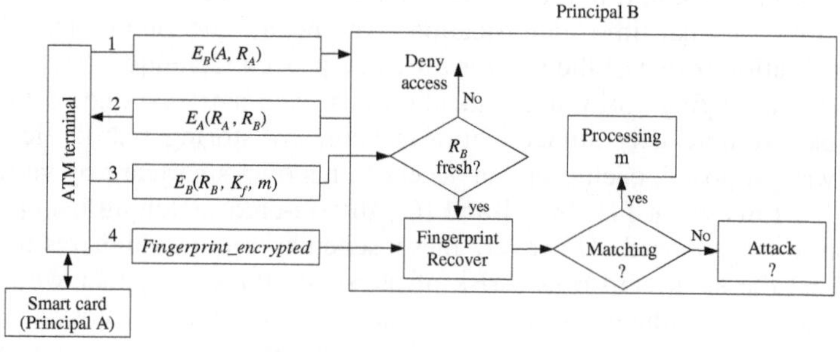

Fig. 1. Diagram of the new hybrid chaotic-biometric authentication protocol (HCBA)

The smartcard releases its ID and private key after being input at the terminal. The first layer of mutual authentication is done via messages 1 and 2 as following:

1. Alice sends message 1 $\boldsymbol{E}_\mathrm{B}$ (**A**, $\boldsymbol{R}_\mathrm{A}$) to identify herself **A** together with a random number (nonce) $\boldsymbol{R}_\mathrm{A}$, by using the principal **B** (bank)'s public key.
2. Message 1 can only be read by principal **B** with its private key. Then **B** generates its own random number (nonce) $\boldsymbol{R}_\mathrm{B}$ and sends it together with $\boldsymbol{R}_\mathrm{A}$ in message 2 $\boldsymbol{E}_\mathrm{A}$ ($\boldsymbol{R}_\mathrm{A}$, $\boldsymbol{R}_\mathrm{B}$) encrypted with Alice's public key.

When Alice sees $\boldsymbol{R}_\mathrm{A}$ inside the message 2, she is sure **B** is responding and it is fresh for she sent $\boldsymbol{R}_\mathrm{A}$ milliseconds ago and only **B** can open the message 1 with **B**'s private key. Conventional public key cryptographic protocols (modified Needham-Schroeder PK protocol [11]) can be used to exchange further challenge-response messages.

Fingerprint is integrated to complete the process of mutual authentication which is illustrated via messages 3, 4 and diagrams within the bank server as shown in Fig.1. In this process, Alice needs to provide her fingerprint, then the terminal will encrypt it. The encryption key $\boldsymbol{K}_f$ can be generated from the raw fingerprint image, and is transmitted to the central server via secure channel (such as RSA cryptography).

When **B** finds $\boldsymbol{R}_\mathrm{B}$ in message 3, it knows that the message 3 must come from Alice's smartcard and also fresh. Message 4 is the encrypted fingerprint of Alice.

After being verified that the smartcard belongs to the claimed user Alice, the ***En***(***FP***) in message 4 is recovered. At this stage, the bank **B** can still not be sure the fingerprint is from Alice. The recovered fingerprint is then matched against Alice's fingerprint template. If the minutiae matching are successful, then **B** will process the message *m*. Till now, the authentication phase is finished.

2 Improved Pixels Permutation and Key Generation

One complete encryption process consists of (1) One permutation with simultaneous gray level mixing, (2) One diffusion step during which information is spread over the image. The detail procedures are referred to [10]. The image encryption technique is based on [10], which assigns a pixel to another pixel in a bijective manner. The improvement of this proposed scheme is the permutation and the key generation.

3.1 Improved Permutation of Pixels

An image is defined on a lattice of finitely many pixels. A sequence of i integer, $n_1, \ldots, n_i$ such that $\sum n_i = N$ ($i \leq N$) is employed as the encryption key for the permutation of pixels. The image is again divided into vertical rectangles $N \times n_i$, as shown in Fig.2(a). Inside each column, the pixels are divided into N/n_i boxes, each box containing exactly N pixels. Take an example of 8×8 image shown in Fig.2(b), it is divided into 2 column (n_1=3, , n_2=5). The pixels permutation is shown in Fig.2(c), the key is (3, 5).

The key is an arbitrary combination of integers, which add up to the pixels number N in a row. One can choose whatever digits in the key arbitrarily.

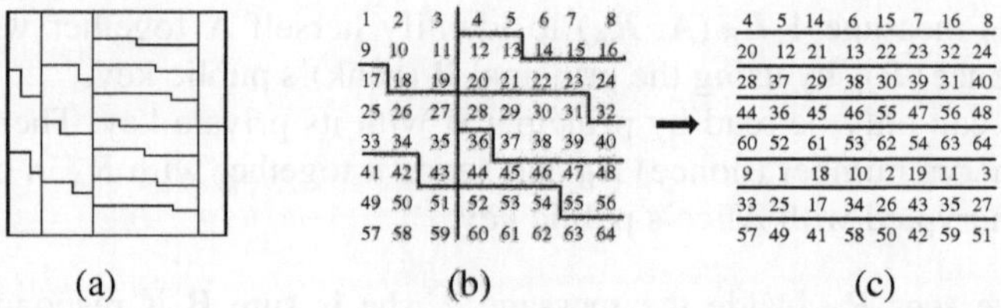

Fig. 2. Permutation of pixels. (a) $N \times 4$ blocks; (b) A 8×8 block; (c) After permutation.

If the raw fingerprint image is a $P \times Q$ rectangular, it can be reformed into a square $N \times N$ image first, where N is the integer makes $(N \times N - P \times Q)$ minimum.

3.2 Key Generation

Encryption keys are vital to the security of the cipher, which can be derived in the following three ways:

- From the randomly chosen values of pixels and their coordinates in raw image.

  Randomly choose 5-10 points in the raw fingerprint image. The vertical and horizontal position of pixels, as well as the gray level values of each point is served as key. Mod operations are conducted. The key consists of the remainders and a supplementary digit that makes the sum of key equals to N. For example, in a 300×300 gray level fingerprint image, there are five points picked up, their coordinates and pixels values are: (16,17,250); (68,105,185); (155,134,169); (216,194,184); (268,271,216). After conducting mod(40) and mod(10) operations for the coordinates and the gray level values, respectively. The result is: (16,17,0); (28,25,5); (35,14,9); (16,34,4); (28,31,6). The sum of above five groups numbers is S_8=268. At last, a supplementary digit $N - S_m$ =300-268=32 is the last digit of the key. The encryption key is: {16, 17, 0, 28, 25, 5, 35, 14, 9, 16, 34, 4, 28, 31, 6, 32}.
- From the stable global features (overall pattern) of fingerprint image.

  Some global features such as core and delta are highly stable points in a fingerprint [13], which have the potential to be served as cryptography key. Some byproduct information in the processing of fingerprint image can be used as the encryption key. For example, the Gabor filter bank parameters are: concentric bands is 7, the number of sectors considered in each band is 16, each band is 20 pixels wide; there are 12 ridge between core and delta, the charges of the core and delta point are 4.8138e-001 and 9.3928e-001, and the period at a domain is 16. Gabor filter with 50 cycles per image width. Then the key could be: {7, 16, 20, 12, 4, 8, 13, 8, 9, 39, 28, 27, 1, 16, 50, 42}. The last digit is the supplementary digit to make the sum of key equals to N.
- From the pseudo random number generator based on chaotic map.

  One can also use the pseudo random number generator introduced in [10] to produce the key.

The users can choose how to generate keys in their scheme. To encrypt a fingerprint image, three to six rounds of iterations can hide the image perfectly; each iteration is suggested to use different key, and different way to generate the keys.

4 Simulation and Evaluation

In this section, the proposed encryption scheme is tested. Simulation results and its evaluation are presented.

4.1 Simulations

The gray level fingerprint image is shown Fig.3(a). The first 3D permutation is performed with the key {16, 17, 0, 28, 25, 5, 35, 14, 9, 16, 34, 4, 28, 31, 6, 32}. After first round 3D permutation, the encrypted fingerprint image is shown in Fig.3(b). The second round permutation is performed with the key {7, 16, 20, 12, 4, 8, 13, 8, 9, 39, 28, 27, 1, 16, 50, 42}. After that, the image is shown in Fig.3(c). The third round permutation is finished with a key {1, 23, 8, 19, 32, 3, 25, 12, 75, 31, 4, 10, 14, 5, 25, 13}. After this, the image is shown in Fig.3(d), which is random looking.

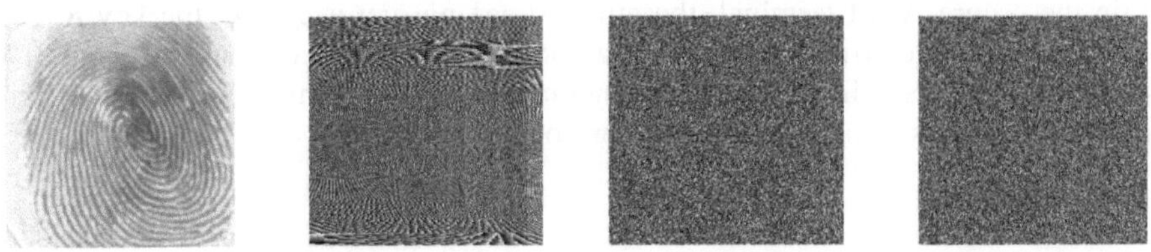

(a) (b) (c) (d)

Fig. 3. Fingerprint and the encrypted image. (a) Original image; (b) One round of iteration; (c) Two rounds of iterations; (d) Three rounds of iterations.

4.2 Statistical and Strength Analysis

- Statistical analysis.

The histogram of original fingerprint image is shown in Fig.4(a). After 2D chaotic mapping, the pixels in fingerprint image can be permuted, but as the encrypted fingerprint image has the same gray level distribution, they have the same histogram as that in Fig.4(a). As introduced in Section 3, 3D chaotic map can change the gray level of the image greatly. After one round and three rounds 3D substitution, the histograms are shown in Fig.4(b) and (c) respectively, which is uniform, and has much better statistic character, so the fingerprint image can be well hidden.

- Cryptographic strength analysis.

In [10], the known plaintext and ciphertext only type of attack were studied: the cipher technique is secure with respect to a known plaintext type of attack. With the diffusion mechanism, the encryption technique is safe to ciphertext type of attack. As the scheme proposed here use different keys in different rounds of iterations, and the length is not constrained, it can be chosen according to the designer's requirement, there is a much large key space than that Fridrich claimed.

- Compared with Data Encryption Standard (DES).

The computational efficiency of the proposed fingerprint encryption scheme is compared with DES. The computation time use DES to encrypt the fingerprint image in Fig.4(a) is 24185ms in 33MHz 386 computer. To encrypt this fingerprint image with the proposed scheme in this paper, three rounds of iterations with 16 digits key in each iteration costs 5325ms with the same computer. Around one-fifth time of the DES did.

- Key transmission and decryption.

The security strength of messages 1, 2, and 3 in Fig.1 relies the asymmetric cryptography, such as RSA scheme which is widely employed. Even in the worst case that the attacker has Alice's smartcard, he/she can successfully proceed the whole authentication process in terms of exchanging messages 1 through 4 in Fig.1, the attack will fail at the final fingerprint matching phase conducted in the bank sever **B** as the attacker does not have Alice's fingerprint. If the attacker has Alice's smartcard and legitimate messages from Alice's last session, there seems a risk of breaking the security system. However as the encryption/decryption as well as key generation are within the secure ATM terminal, the attacker can not get access to the key $\boldsymbol{K_f}$ to recover the legitimate Alice's fingerprint as only the bank **B** can open message 3. We also propose to use different keys generated with different methods in different rounds of iterations. This will make the protocol more secure.

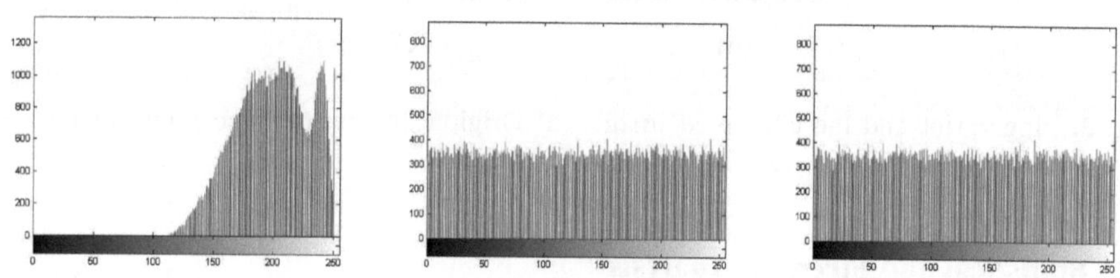

(a) (b) (c)

Fig. 4. Histograms of fingerprint image and the encrypted image. (a) Original fingerprint image; (b) One round of 3D iteration; (c) Three rounds of 3D iterations.

5 Conclusions

A smartcard based ATM fingerprint authentication scheme has been proposed. The possession (smartcard) together with the claimed user's biometrics (fingerprint) is required in a transaction. The smartcard is used for the first layer of mutual authentication when a user requests a transaction. Biometric authentication is the second layer. The fingerprint image is encrypted via 3D chaotic map as soon as it is captured, and then is transmitted to the central server via symmetric algorithm. The encryption keys are extracted from the random pixels distribution in a raw image of fingerprint, some stable global features of fingerprint and/or from pseudo random number generator. Different rounds of iterations use different keys.

Some parts of the private key are transmitted to central server via asymmetric algorithm. The stable features of the fingerprint image need not to be transmitted; it can be extracted from the templates at the central server directly.

After decryption, the minutia matching is performed at the central server. The successful minutia matching at last verifies the claimed user.

Future work will focus on the study of stable features (as part of encryption key) of fingerprint image, which may help to set up a fingerprint matching dictionary so that to narrow down the workload of fingerprint matching in a large database.

Acknowledgments

The work is financially supported by Australia Research Council linkage project LP0455324. The authors would like to thanks Associate professor Serdar Boztas for his valuable discussion on keys establishment protocol.

References

1. Soutar, C., Roberge, D., Stoianov, A., Gilory, R., Kumar, B.V.: Biometric encryption, www.bioscrypt.com.
2. Uludag, U., Pankanti, S., Prabhakar, S., Jain, A.K.: Biometric cryptosystems: Issue and challenges, *Proceedings of the IEEE*, 92 (2004) 948-960
3. Jain, A.K., Prabhakar, S., Hong, L., Pankanti, S.:Filterbank-based fingerprint matching, *IEEE Trans. on Image Processing*, 9 (2000) 846-859
4. Lee, J.K., Ryu S.R., Yoo, K.Y.: Fingerprint-based remote user authentication scheme using smart cards, *Electronics Lett.*, 38 (2002) 554-555
5. Clancy, T.C., Kiyavash, N., Lin, D.J.: Secure smartcard-based fingerprint authentication, *ACM workshop on Biometric Methods and Applications*, Berkeley, California, Nov. (2003)
6. Waldmann, U., Scheuermann D., Eckert, C.: Protect transmission of biometric user authentication data for oncard-matching, *ACM symp. on Applied Computing*, Nicosia, Cyprus, March (2004)
7. Jain, A.K., Prabhakar S., Hong, L.: A multichannel approach to fingerprint classification, *IEEE Trans. on Pattern Anal. Machine Intell.*, 21 (1999) 348-359
8. Kocarev, L. Jakimoski, G., Stojanovski T., Parlitz, U.: From chaotic maps to encryption schemes, *Proc. IEEE Sym. Circuits and Syst.*, 514-517, Monterey, California, June (1998)
9. Chen, G., Mao, Y., Chui, C.: A symmetric encryption scheme based on 3D chaotic cat map, *Chaos, Solitons & Fractals*, 21 (2004) 749-761
10. Fridrich, J.: Symmetric Ciphers Based on two-dimensional chaotic maps, *Int. J. Bifurcation and Chaos*, 8 (1998) 1259-1284
11. Menezes, A., Oorschot, P., Vanston, S.A.: Handbook of Applied Cryptography. CRC Press, (1996)
12. Uludag, U., Ross, A., Jain, A.K.: Biometric template selection and update: a case study in fingerprints, *Pattern Recognit.*, 37 (2004) 1533-1542
13. Ratha, N.K, Karu, K. Chen, S., Jain, A.K.: A real-time matching system for large fingerprint databases, *IEEE Trans. on Pattern Anal. Machine Intell.*, 18 (1996) 799-813
14. Zhou, J., Gu, J.: A model-based method for the computation of fingerprints' orientation field, *IEEE Trans. on Image Processing*, 13 (2004) 821-835

An Uncorrelated Fisherface Approach for Face and Palmprint Recognition

Xiao-Yuan Jing[1], Chen Lu[1], and David Zhang[2]

[1] Shenzhen Graduate School of Harbin, Institute of Technology,
Shenzhen 518055, China
jingxy_2000@yahoo.com, cssandylu@hitsz.edu.cn
[2] Dept. of Computing, Hong Kong Polytechnic University, Kowloon, Hong Kong
csdzhang@comp.polyu.edu.hk

Abstract. The Fisherface method is a most representative method of the linear discrimination analysis (LDA) technique. However, there persist in the Fisherface method at least two areas of weakness. The first weakness is that it cannot make the achieved discrimination vectors completely satisfy the statistical uncorrelation while costing a minimum of computing time. The second weakness is that not all the discrimination vectors are useful in pattern classification. In this paper, we propose an uncorrelated Fisherface approach (UFA) to improve the Fisherface method in these two areas. Experimental results on different image databases demonstrate that UFA outperforms the Fisherface method and the uncorrelated optimal discrimination vectors (UODV) method.

1 Introduction

The linear discrimination analysis (LDA) technique is an important and well-developed area of image recognition and to date many linear discrimination methods have been put forward. The Fisherface method is a most representative method of LDA [1]. However, there persist in the Fisherface method at least two areas of weakness. The first weakness is that it cannot make the achieved discrimination vectors completely satisfy the statistical uncorrelation while costing a minimum of computing time. Statistical uncorrelation is a favorable property useful in pattern classification [2-3]. The uncorrelated optimal discrimination vectors (UODV) method requires that the achieved discrimination vector satisfy both the Fisher criterion and the statistical uncorrelation [2]. However, it uses considerable computing time to calculate every discrimination vector satisfying the constraint of uncorrelation, when the number of vectors is large. The second weakness is that not all the discrimination vectors are useful in pattern classification. In other words, vectors with the larger Fisher discrimination values should be chosen, since they possess more between-class than within-class scatter information. In this paper, we propose an uncorrelated Fisherface approach (UFA) that improves the Fisherface method in the foregoing two areas. The rest of this paper is organized as follows. In Section 2, we describe the UFA. In Section 3 we provide the sufficient experimental results on different image databases and offer our conclusions.

D. Zhang and A.K. Jain (Eds.): ICB 2006, LNCS 3832, pp. 682 – 687, 2005.

2 Description of UFA

In this section, we first present two improvements in the Fisherface method, and then propose the UFA which synthesizes the suggested improvements.

(i) Improvement of the statistical uncorrelation of discrimination vectors:

Lemma 1 [2]. Suppose that the between-class scatter matrix and the total scatter matrix are S_b and S_t, the discrimination vectors obtained from UODV are $(\varphi_1, \varphi_2, \cdots, \varphi_r)$, where r is the rank of $S_t^{-1}S_b$. The nonzero eigenvalues of $S_t^{-1}S_b$ are represented in descending order as $\lambda_1 \geq \lambda_2 \geq \cdots \geq \lambda_r > 0$, and the kth eigenvector ϕ_k of $S_t^{-1}S_b$ corresponds to λ_k $(1 \leq k \leq r)$. If $(\lambda_1, \lambda_2, \cdots, \lambda_r)$ are mutually unequal, that is $\lambda_1 > \lambda_2 > \cdots > \lambda_r > 0$, then φ_k can be represented by ϕ_k.

Lemma 1 shows that when the non-zero Fisher discrimination values are mutually unequal, the discrimination vectors generated from the Fisherface method can satisfy the statistical uncorrelation. That is, in this situation, the Fisherface method and UODV obtain identical discrimination vectors with non-zero discrimination values. Therefore, Lemma 1 reveals the essential relationship between these two methods. Although UODV satisfies the statistical uncorrelation completely, it requires more computational time than the Fisherface method. Furthermore, it is not necessary to use UODV if the non-zero Fisher discrimination values are mutually unequal, because the Fisherface method can take the place of UODV. In the application of the Fisherface method, we find that only a small number of the Fisher values are equal respectively, and the others are unequal mutually. How, then, can computational time be reduced, while simultaneously guaranteeing the statistical uncorrelation for the discrimination approach? Here, we propose an improvement on the Fisherface method. Use the assumption in Lemma 1, our measure is: (a) use the Fisherface method to obtain the discrimination vectors $(\phi_1, \phi_2, \cdots, \phi_r)$. If the corresponding Fisher values $(\lambda_1, \lambda_2, \cdots, \lambda_r)$ are unequal mutually, end; else go to the next step. (b) For $2 \leq k \leq r$, if $\lambda_k \neq \lambda_{k-1}$, then keep ϕ_k, else replace ϕ_k by φ_k from UODV. Obviously, the proposal not only satisfies the statistical uncorrelation, it reduces the computing time. This will be further demonstrated by our experiments.

(ii) Improvement of the selection of discrimination vectors:

Suppose that the within-class scatter matrix is S_w, the discrimination vectors are expressed by $(\phi_1, \phi_2, \cdots, \phi_r)$, where r is the number of discrimination vectors. For ϕ_i $(i = 1, \cdots, r)$, we have $\phi_i^T S_t \phi_i = \phi_i^T S_b \phi_i + \phi_i^T S_w \phi_i$. The discriminative value $F(\phi_i)$ is defined as $F(\phi_i) = \dfrac{\phi_i^T S_b \phi_i}{\phi_i^T S_t \phi_i}$. If $\phi_i^T S_b \phi_i > \phi_i^T S_w \phi_i$, then $F(\phi_i) > 0.5$. In this situation, according to the Fisher criterion, there is more between-class separable information than within-class scatter information. So, we choose those discrimination vectors whose Fisher discrimination values are more than 0.5, and discard the others. This improvement allows efficient linear discrimination information to be kept and non-useful information to be discarded. Such a selection of the effective discrimination vectors is important to the recognition effect, where the number of vectors is large. The experiment will demonstrate the importance of this.

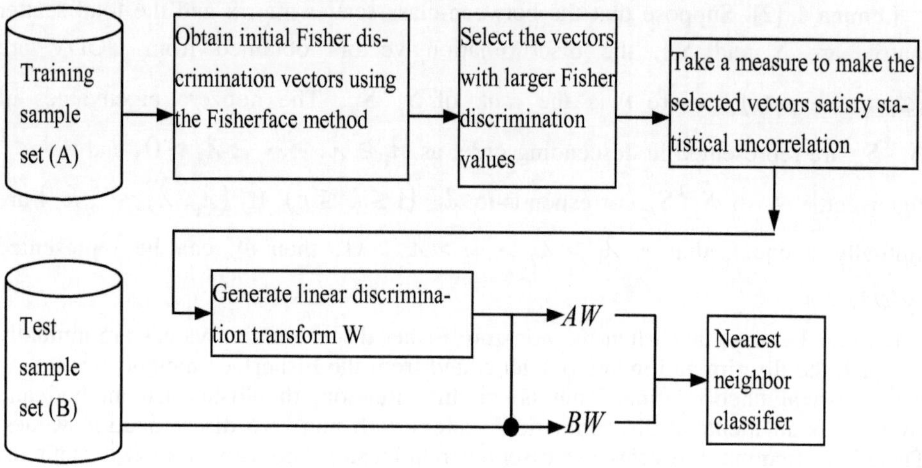

Fig. 1. Recognition procedure of UFA

The UFA can be described in the following three steps:

Step 1. From the discrimination vectors that are obtained, select those whose Fisher discrimination values are more than 0.5.

Step 2. Using the proposed measure to make the selected vectors satisfy statistical uncorrelation, obtain the discrimination vectors $(\phi_1, \phi_2, \cdots, \phi_r)$. The linear discrimination transform W is then defined as: $W = [\phi_1 \ \phi_2 \ \cdots \ \phi_r]$, where ϕ_i is the ith column of W.

Step 3. For each sample x in X, extract the linear discriminative feature y where $y = Wx$. This obtains a new sample set Y with the linear transformed features corresponding to X. Use the nearest neighbor classifier to classify Y. Here, the distance between two arbitrary samples, y_1 and y_2, is defined by $d(y_1, y_2) = \| y_1 - y_2 \|_2$, where $\| \ \|_2$ denotes the Euclidean distance. Figure 1 shows a flowchart of UFA.

3 Experimental Results and Conclusions

In this section, we compare the experimental results of UFA, the Fisherface method [1] and UODV [2], using different image data. The experiments are implemented on a

Pentium 1.4G computer with 256MB RAM and programmed using the MATLAB language. In the following experiments, the first two samples of every class in each database are used as training samples and the remainder as test samples. Generally, it is more difficult to classify patterns when there are fewer training samples. The experiments take up that challenge and seek to verify the effectiveness of the proposed approach using fewer training samples.

(i) Experiment on ORL face database:

The ORL database (http://www.cam-orl.co.uk) contains images varied in facial expressions, facial details, facial poses, and in scale. The database contains 400 facial images: 10 images of 40 individuals. The size of each image is 92×112 with 256 gray levels per pixel. Each image is compressed to 46×56. We use 80 (=2*40) training samples and 320 (=8*40) test samples. Table 1 shows the Fisher discriminative values that are obtained from UFA on the ORL database, ranged from 0 to 1. We find that only 2 values equal to 1.0 in the total 39 discriminative values and 9 values are less than 0.5. It means that 2 discrimination vectors are statistically correlated and 9 vectors with less discriminative values should be discarded in the UFA. Table 2 shows a comparison of the classification performance of UFA and other methods on this database. The improvements in UFA's recognition rates over Fisherface and UODV are 3.12% and 2.81%, respectively. UFA is much faster than UODV and its training time is slightly less than that of Fisherface. It is 50.29% faster than UODV and 1.47% faster than Fisherface. Compared with Fisherface and UODV (which use the same number of discriminative features), UFA reduces the feature dimension by 23.08%.

Table 1. An illustration of Fisher discriminative values $F(\phi_i)$ obtained using UFA

Databases	Fisher discriminative values
ORL	Number of discrimination vectors: 39
	1.0000 1.0000 0.9997 0.9981 0.9973 0.9962 0.9950 0.9932 0.9917 0.9885 0.9855 0.9845 0.9806 0.9736 0.9663 0.9616 0.9555 0.9411 0.9356 0.9151 0.9033 0.8884 0.8517 0.8249 0.8003 0.7353 0.7081 0.6930 0.6493 0.5515 0.4088 0.3226 0.2821 0.2046 0.0493 0.0268 0.0238 0.0081 0.0027
Palmprint	Number of discrimination vectors: 189
	1.0000 1.0000 1.0000 1.0000 1.0000 0.9999 0.9999 0.9999 0.9998 0.9998 0.9998 0.9997 0.9997 0.9996 0.9996 0.9995 0.9995 0.9994 0.9993 0.9993 0.9992 0.9991 0.9990 0.9989 0.9987 0.9986 0.9985 0.9983 0.9983 0.9982 0.9982 0.9979 0.9976 0.9976 0.9974 0.9971 0.9970 0.9968 0.9967 0.9965 0.9962 0.9960 0.9959 0.9952 0.9948 0.9947 0.9945 0.9943 0.9941 0.9937 0.9932 0.9930 0.9928 0.9922 0.9917 0.9912 0.9910 0.9908 0.9903 0.9900 0.9897 0.9892 0.9888 0.9883 0.9878 0.9870 0.9869 0.9862 0.9858 0.9846 0.9843 0.9836 0.9833 0.9825 0.9822 0.9816 0.9800 0.9795 0.9792 0.9787 0.9783 0.9767 0.9759 0.9752 0.9743 0.9731 0.9723 0.9718 0.9703 0.9701 0.9686 0.9679 0.9656 0.9646 0.9635 0.9621 0.9613 0.9605 0.9591 0.9557 0.9551 0.9535 0.9521 0.9507 0.9486 0.9481 0.9439 0.9436 0.9390 0.9384 0.9371 0.9331 0.9318 0.9313 0.9273 0.9225 0.9194 0.9186 0.9147 0.9118 0.9112 0.9088 0.9069 0.9050 0.9036 0.8889 0.8845 0.8821 0.8771 0.8747 0.8709 0.8659 0.8607 0.8507 0.8488 0.8424 0.8340 0.8280 0.8220 0.8157 0.8070 0.8007 0.7959 0.7825 0.7751 0.7639 0.7626 0.7434 0.7378 0.7284 0.7060 0.6944 0.6613 0.6462 0.6372 0.6193 0.6121 0.5663 0.5436 0.5061 0.4753 0.4668 0.4343 0.3730 0.3652 0.3024 0.2900 0.2273 0.2014 0.1955 0.1758 0.1541 0.1270 0.1159 0.0858 0.0741 0.0683 0.0591 0.0485 0.0329 0.0243 0.0205 0.0184 0.0107 0.0090 0.0049 0.0026 0.0004 0.0001

(ii) Experiment on palmprint database:

Palmprint recognition has become an important complement to personal identification [4]. Our palmprint database (http://www4.comp.polyu.edu.hk/~biometrics/) contains a total of 3,040 images from 190 different palms, each 16 images with size 64×64. The major differences between them are the illumination, position and pose. We use 380 (=2*190) training samples and 2,660 (=14*190) test samples. Table 1 also shows the Fisher discriminative values that are obtained from UFA on the palmprint database, ranged from 0 to 1. We find that 25 values have their respective equivalents in the total 189 discriminative values and 29 values are less than 0.5. It means that 25 discrimination vectors are statistically correlated and 29 vectors with less discriminative values should be discarded in the UFA. Table 2 also shows a comparison of the classification performance of UFA and other methods on this database. The improvements in UFA's recognition rates over Fisherface and UODV are 10.12% and 3.09%, respectively. UFA is much faster than UODV and its training time is slightly more than that of Fisherface. It is 43.31% faster than UODV and 8.71% slower than Fisherface. Compared with Fisherface and UODV (which use the same number of discriminative features), UFA reduces the feature dimension by 15.34%.

Table 2. Classification performance of all methods on the ORL and palmprint database

Classification performance	Different databases	Discrimination methods		
		UFA	Fisherface	UODV
Recognition rates (%)	ORL	84.06	80.94	81.25
	Palmprint	91.47	81.35	88.38
Training time (second)	ORL	14.07	14.28	31.58
	Palmprint	39.17	36.03	69.1
Extracted feature dimension	ORL	30	39	39
	Palmprint	160	189	189

This paper presents an uncorrelated Fisherface approach for image recognition. UFA makes the achieved discrimination vectors satisfy the statistical uncorrelation using less computing time and improves the selection of discrimination vectors. We verify UFA on different image databases. Compared to the Fisherface method and UODV, UFA improves the recognition rates up to 10.12% and 3.09%, respectively. The training time of UFA is similar to that of Fisherface and UFA is at least 43.31% faster than UODV. In addition, UFA reduces the feature dimension by up to 23.08% than Fisherface and UODV. Consequently, we conclude that UFA is an effective linear discrimination approach.

Acknowledgment

The work described in this paper was fully supported by the National Natural Science Foundation of China (NSFC) under Project No. 60402018.

References

[1] P.N. Belhumeur, J.P. Hespanha, D.J. Kriegman: Eigenfaces vs. fisherface: recognition using class specific linear projection, IEEE Trans. Pattern Anal. and Machine Intell., 19(7) (1997) 711–720.

[2] X.Y. Jing, D. Zhang, Z. Jin: UODV improved algorithm and generalized theory, Pattern Recognition, 36(11) (2003) 2593–2602.

[3] Z. Jin, J. Yang, Z. Hu, Z. Lou: Face recognition based on the uncorrelated discrimination transformation, Pattern Recognition, 34(7) (2001) 1405–1416.

[4] D. Zhang, W.K. Kong, J. You and M. Wong: On-line palmprint identification, IEEE Trans. Pattern Anal. and Machine Intell., 25(9) (2003) 1041–1050.

Fast and Accurate Segmentation of Dental X-Ray Records

Xin Li, Ayman Abaza, Diaa Eldin Nassar, and Hany Ammar

Lane Dept. of Computer Science and Electrical Engineering,
West Virginia University, Morgantown, WV 26506-6109
{xinl, ayabaza, dmnassar, ammar}@csee.wvu.edu

Abstract. Identification of deceased individuals based on dental characteristics is receiving increased attention. Dental radiographic films of an individual are usually composed into a digital image record. In order to achieve high level of automation in postmortem identification, it is necessary to decompose dental image records into their constituent radiographic films, which are in turn segmented to localize dental regions of interest. In this paper we offer an automatic hierarchical treatment to the problem of cropping dental image records into films. Our approach is heavily based on concepts of mathematical morphology and shape analysis. Among the many challenges we face are non-standard assortments of films into records, variability in record digitization as well as randomness of record background both in intensity and texture. We show by experimental evidence that our approach achieves high accuracy and timeliness.

1 Introduction

Law enforcement agencies have exploited biometrics for decades as key forensic identification tools. Dental features, resist early decay of body tissues as well as withstand severe conditions usually encountered in mass disasters, which make them the best candidates for PM identification [1] [2]. Recent works on developing a research prototype of an automated dental identification system (ADIS) reveal a couple of challenging image segmentation problems [4]. First, the digitized dental X-ray record of a person often consists of multiple films, as shown in Fig. 1(a), which we recognize as a global segmentation problem of cropping a composite digitized dental record into its constituent films. Second, each cropped film contains multiple teeth, as shown in Fig. 1 (b), which we recognize as a local segmentation problem of isolating each tooth in order to facilitate the extraction of features (e.g., crown contour and root contour) for identification use. The latter problem was studied in [5] [6]. Though performing the film cropping task may seem trivial for a human observer, it is desirable to automate this process and to integrate it with the framework ADIS. In this paper, we focus on the global segmentation (cropping) problem of dental X-ray records and seek a solution to achieve a good tradeoff between accuracy and complexity. On one hand, we want segmentation results to be as accurate as possible since inaccuracy in cropping of

D. Zhang and A.K. Jain (Eds.): ICB 2006, LNCS 3832, pp. 688–696, 2005.

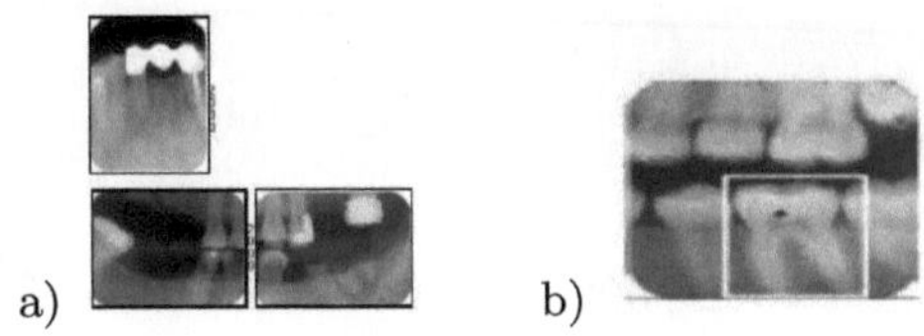

Fig. 1. a) Global segmentation (cropping), b) local segmentation (teeth isolation) [5]

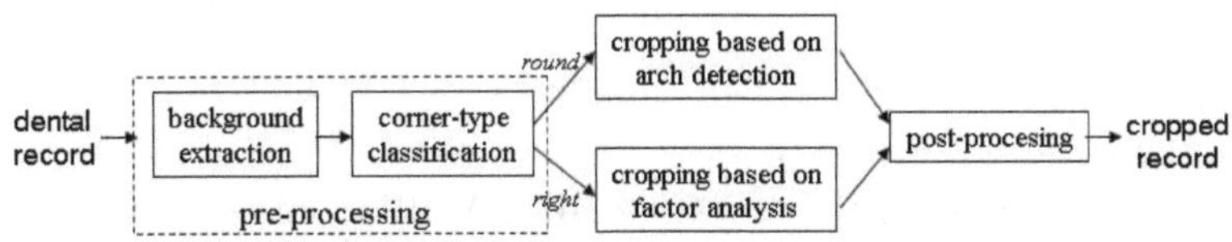

Fig. 2. The three-stages approach for dental record cropping

dental records is likely to hinder the performance of subsequent processing steps and accordingly the overall performance of the entire identification system. On the other hand, we want the computational cost to be reasonably low, especially with the large volume of records that need to be processed. Fast and accurate cropping of dental X-ray records is a nontrivial challenge due to the heterogeneity of dental records. Traditionally, dental X-ray records are digitized by different facilities where human intervention is inevitable during the digitization process. Therefore, characteristics of dental X-ray records vary not only from database to database but also from image to image [3]. We propose a three-stage approach for cropping as depicted in Fig. 2: First a preprocessing stage whereby we extract the background layer of the image record, extract connected components and classify them as either round-corner or right-corner connected components. The second stage performs arch detection and dimension analysis, realization of this stage differs according to the outcome of the preprocessing stage. The third stage is a postprocessing stage that performs topological assessment of the cropping results in order to eliminate spurious objects. In section 2 we introduce the notations and terminology that we will use through the remainder of the paper. In sections 3, 4, and 5 we elaborate on the preprocessing stage, the arch detection and dimension analysis stage, and the postprocessing stage respectively. In section 6 we present experimental results and a discussion on these results. Finally in section 7, we conclude the paper and sketch our plans for future work.

2 Notations and Terminology

In this section, we introduce some notations and definitions for later use. The dental X-ray record is assumed to be a gray-scale image denoted by $X(i,j) \in [0,255]$ where $(i,j) \in \Omega = [1,H] \times [1,W]$ (H, W are the height and width of the image). The dimension of individual dental films is denoted by h, w.

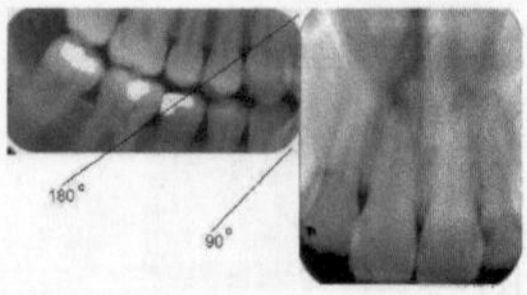

a)

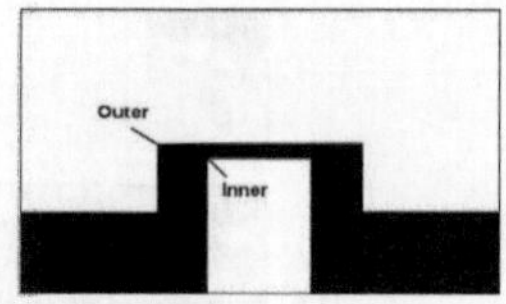

b)

Fig. 3. a)90° V-corner and 180° V-corner; b)Inner L-corner and Outer L-corner

Level set and its size. Level set L_k is a binary image defined by $L_k(i,j) = \begin{cases} 1 & X(i,j)=k \\ 0 & else \end{cases}$. The size of a binary image L_k, denoted by $|L_k|$, is simply the total number of ones in the image.
Connected Film Set and Boundary film. Multiple films that are not completely separated form a connected film set. Within a connected film set, a film is called boundary film if removing it does not affect the connectivity of remaining in the set.
Morphological area-open and area-close operators. Area open operator is the extension of morphological opening - it consists of three consecutive filters, namely erosion, small objects removal and dilation. Area close operator is the extension of morphological closing - it consists of three consecutive filters, namely dilation, small objects removal and erosion.
90° V-corner and 180° V-corner. 90° V-corner refers to the corner formed by a straight line and an arc segment; 180° V-corner refers to the corner formed by two adjacent arc segments (refer to Fig. 3).
Inner L-corner and Outer L-corner. Inner L-corner refers to a right-corner with one quadrangle of white and three quadrangles of black; outer L-corner refers to a right-corner with one quadrangle of black and three quadrangles of white (refer to Fig. 3).

Note: Inner and outer L-corners can be easily detected by using morphological hit-or-miss operators (as shown in Section 3).

3 Preprocessing

3.1 Background Extraction

Although the background typically consists of uniform color such as white, gray or black, intensity alone is not sufficient to distinguish the background from dental films. For example, the empty (no tooth) areas in a film often appear as dark regions and could become confused with a black background. Similarly, dental filling often appear as bright regions in a film and could cause problem when the background is white as well. A more robust approach of extracting the background color is to count on *geometric* clues such as the shape of dental films. Since any dental film can be bounded by a rectangle, the boundary of background largely consists of vertical and horizontal lines. Suppose the histogram of input image $X(i,j)$ and the largest three peaks are n_1, n_2, n_3. We consider their corresponding level sets $\partial L_k, k = n_1 - n_3$ and apply morphological filtering to extract the boundary of those three sets ∂L_k. For dental films whose boundary is largely rectangular, the fitting ratio of ∂L_k by vertical or horizontal lines reflects its likelihood of being the true background. Specifically, we propose to extract vertical and horizontal lines from ∂L_k by direct run-length counting and define the fitting ratio by

$$r_k = \frac{|R_k|}{\partial L_k}, k = n_1 - n_3. \tag{1}$$

where R_k is the binary image recording the extracted vertical and horizontal lines. The set with the largest fitting ratio among the three level sets is declared to be the background L_b. As soon as background is detected, we do not need intensity information but only the geometry of L_b for further processing (refer to Fig. 4).

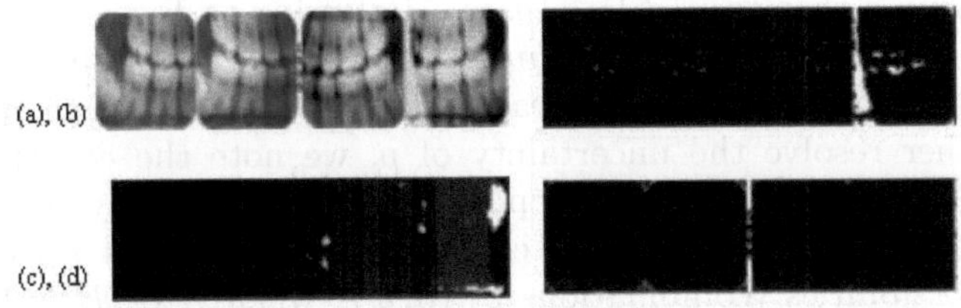

Fig. 4. Background extraction example, a) original dental record, b) level set L1, fitting ratio r1=0.09, c) level set L2, fitting ratio r2=0.05, d) level set L3, fitting ratio r3=0.36

3.2 Arc Detection

The complement of detected background ($\bar{L}_b$) consists of non-cropped dental films as well as various noises. The noise could locate in the background (e.g., textual information such as the year) or within dental films (e.g., dental fillings that have similar color to the background). To eliminate those noise, we propose to apply morphological area-open operator to $\bar{L}_b$ and L_b sequentially. Suppose the Euler number of filtered $\bar{L}_b$ is N, then we can label the N connected components in $\bar{L}_b$ by integers $1 - N$. For each connected component (a binary map), we need to classify its corner type since a record could contain the mixture of round-corner and right-corner films. The striking feature of a round-corner film is the arc segments around the four corners. In the continuous space, those arc segments are essentially 90°-turning curves (they link a vertical line to a horizontal one). In the discrete space, we propose to use Hit-or-Miss operator to detect corner pixels first and then morphological area-close operator to locate arc segments. The area-close operator is suitable here because it connects the adjacent corner pixels around a round corner to makes them stand out as an arc segment. By contrast, the corner pixels around a right corner will be suppressed by the area-close operator (refer to Fig. 5).

3.3 Dimension Analysis

For round-corner films, detected arc segments will be shown sufficient for cropping purpose in the next section. However, the location uncertainty of right-corner film boundary is more difficult to resolve because the inner boundary could disappear in the case of parallel superposition. If all right-corner films are placed in such a way to form a seamless rectangle, it is impossible to separate them apart using geometric information in L_b alone. Fortunately, such seamless concatenation does not exist in the database we have, which indicates it is a rare event in practice. Instead, we propose a purely geometric approach for estimating the dimension (h, w) of right-corner films. Our dimension analysis techniques are based on the heuristic observation that concatenation of two right-corner films could only give rise to inner L-corners. Therefore, the distance between any two *outer* L-corners must be $d_i = m_i h + n_i w, i = 1 - k_1$ where (m_i, n_i) are a pair of nonnegative integers. Moreover, since the borders of right-corner films would not all align to each other in the case of non-parallel concatenation, it is reasonable to assume that $min\{d_i\} \in \{h, w\}$. Referring to Fig. 3, if we mark the two outer corners corresponding to $min\{d_i\}$ by A, B, then the closest corner to A, B must form a rectangular area of phw (p is an unknown positive integer). To further resolve the uncertainty of p, we note the constraint on film aspect ratio (i.e., $h \in [\frac{2}{3}w, \frac{3}{2}w]$)as supported by an exploratory experiment that we elaborate on in section 6). Such constraint often reduces p to be at most two viable possibilities (combinations of (h, w)) when A or B is linked to an inner L-corner. If we denote the distance between any two *inner* L-corners by $d'_i = m'_i h + n'_i w, i = 1 - k_2 > k_1$, then the weighting coefficients m'_i, n'_i could be arbitrary integers. There exist efficient Euclidean algorithm for solving such diophantine equations. By comparing the solutions (m'_i, n'_i) given by different combinations of (h, w), we pick out the one whose solution is closest to integers to be the most likely dimension.

4 Cropping Techniques

Preprocessing classifies each connected component in L_b to be either round-corner or right-corner. In this section, we present tailored cropping techniques for round-corner and right-corner components respectively. For round-corner component, we demonstrate that two types of V-corners associated with arc segments are sufficient for cropping. While for right-corner component, we recursively crop out films one-by-one based on the estimated film dimension (h, w).

4.1 Round-Corner Component

When multiple round-corner films are placed side by side, they form two types of V-corners as defined above. For 90° V-corner, its straight edge indicates where the cropping should occur. For 180° V-corner, we note that it is symmetric with respect to the target cropping line (refer to Fig. 3). Therefore, the cropping of round-corner films can be fully based on locating and classifying the two types of V-corners. A V-corner is characterized by the intersection of two segments where the curvature experiences a sharp change. Such geometric singularity can be identified by local analysis of digital arc segment. Specifically, we define "curvature index" at a location (i, j) to be the maximum length of consecutive white pixels in $\bar{L}_b$ as we traverse its eight nearest neighbors in a clockwise order. The position is declared to be a V-corner if its curvature index is above 5 and it is close to one of the detected arc segments in corner-type classification. Further classification of V-corner into 90° and 180° can be done based on symmetry analysis. For a 90° V-corner, neither horizontal nor vertical line passing the V-corner divides the corner symmetrically. While for 180° corner, there exists a symmetric axis which tells the place to cut. We note that unlike generic symmetry detection problems, the direction of symmetric axis is known to be either horizontal or vertical. Therefore, symmetry analysis can be conveniently carried out by correlation or differential techniques.

4.2 Right-Corner Component

The cropping of right-corner films is based on the following intuitive observation with the boundary films. Due to the special location of boundary films, they can be properly cropped out with a higher confidence than the rest. Moreover, cropping out boundary films could make other non-boundary films become boundary ones and therefore the whole process of cropping boundary films can be recursively performed until only one film is left. Formally, we propose to characterize the boundary films under a graph theoretic framework. Each film is viewed as a vertex; an edge between two vertexes is induced if the corresponding two films are adjacent to each other. Fig. 6 shows an example of connected film set and its graph representation. It is easy to see that boundary films correspond to the vertexes with degree of one (removing them do not affect the connectivity of the remaining graph). Unless the connected film set forms a loop, it is always possible to reduce the graph by successively removing unit-degree vertexes without affecting its connectivity. To implement recursive cropping, we require reliable detection of boundary films. It follows from the definition of boundary films that they must satisfy: 1) any boundary film contains a pair of outer L-corners; 2) the distance between the L-corner pair is either h or w. Therefore, detected outer L-corners in dimension analysis give useful clue for locating boundary films. We note that since the area of connected film set A_{cfs} and the film dimension are both known, the number of films contained in the set it approximately known ($n = \lfloor \frac{A_{cfs}}{hw} \rfloor$). The iteration number of recursive cropping is given by $n - 1$.

5 Postprocessing

Cropping techniques in the previous section mainly target separating connected films. There are other factors that can not be handled by cropping and may

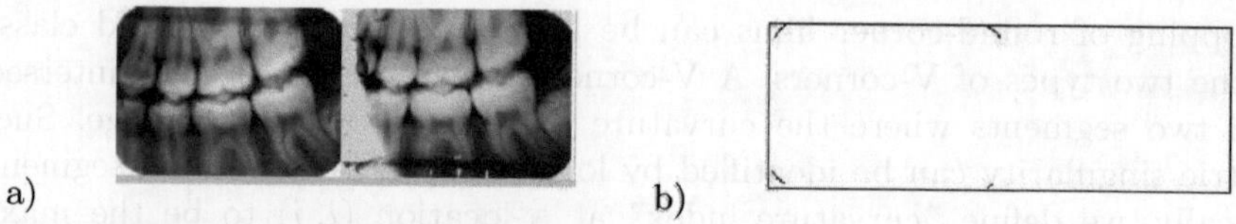

Fig. 5. Arc detection for corner-type classification. a) a dental record with both round-corner and right-corner films; b) arc detection result.

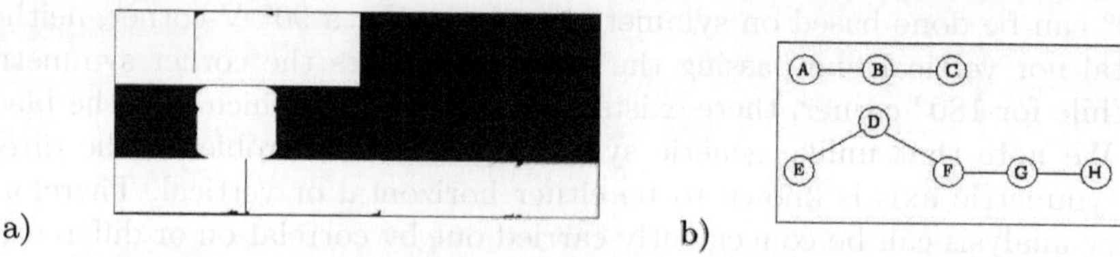

Fig. 6. a)An example of connected film set, b)its graph representation

affect final segmentation results. For example, some dental film is contaminated before digitization such that a portion of film becomes indistinguishable from the background (Fig. 5). The accuracy of cropping itself could also degrade due to errors in dimension analysis. For example, in the case of right-corner films, there might be some leftovers after cropping out $\lfloor \frac{A_{cfs}}{hw} \rfloor$ films. Consequently, it is desirable to have a post-processing stage to consummate the segmentation results. One of the prior information about dental films is that they are all convex sets, regardless of the corner-type. Such knowledge implies that the hole or cracks of any segmented component be filled in by finding its convex hull. Therefore, the first step in post-processing is to enforce the convexity of all connected components after cropping. Secondly, we propose to check the size and shape of each convex component. If the size of a component is too small or its shape significantly deviates from rectangle, we detect its outer L-corners and check if they correspond to one border of the film. If yes, we conclude that the film was contaminated and derive its boundary using dimension information. Otherwise, we decide it is likely to be a non-film object and put it back to the background layer.

6 Experiments and Results

In this section we report three types of experiments pertaining to our film cropping approach along with their outcomes: (i) An exploratory experiment to study the constraints on dental films dimensions ratio: Since there are 5 standard film sizes [7], ideally the minimum sides ratio, which we define as the minimum of the aspect ratio and its reciprocal i.e. ($\gamma \equiv min[\frac{w}{h}, \frac{h}{w}]$), would assume only 5 discrete values (.5, .64, .66, .75, .77). However the manual procedure followed for mounting films onto records may result in some variation in the observed values of . In this experiment we used a random sample of 500 manually cropped periapical and bitewing films to study the distribution of γ. We observed the following: $0.49 \leq \gamma \leq 0.91$ for the entire sample; $\gamma < 0.5$ for $\sim 1\%$; $\gamma > 0.9$ for $\sim 0.2\%$; $(0.6 \leq \gamma \leq 0.8)$ in $\sim 94\%$ of the sample films; and $\sim 6\%$ of films have $\gamma < 0.6$ or $\gamma > 0.8$ (almost equally distributed). (ii) A performance assessment experiment: We evaluate both the yield and timeliness aspects of our film cropping approach using a randomly selected test sample of 100 dental records (images) from the CJIS ADIS database [3], the total film count in the test set is 722. We verified that the test sample has variability in background and contains films with both corner types (48 round-corner, and 52 right-corner records). We marked the cropped segments using the following convention: a perfect segment contains exactly one film, an under segmented region contains several whole films 7(b), and an erroneous segment is that contains part of the film, or region from the background texture, or both. In Fig 7(a), we summarize the yield analysis: $\sim 73.7\%$ of the films were perfectly cropped, $\sim 23.8\%$ were under segmented, and only $\sim 2.5\%$ developed into erroneous segments. Further cropping-yield analysis show that in the right corner records perfect segmentation rate is $\sim 70.1\%$, under segmentation rate is $\sim 28.2\%$, and erroneous segmentation rate is $\sim 1.7\%$. While in round corner records these rates are $\sim 76.9\%$, $\sim 19.9\%$, and $\sim 3.2\%$ respectively. We measured the record cropping time of our algorithm using an uncompiled MATLAB implementation running on a 2.0 GHz 512 MB RAM Intel Pentium IV PC platform. The average cropping time is 30 kpix/sec and it varies depending on the number of films in the record and the amount of separation between films. (iii) An exploratory experiment to examine potential future yield-boosting opportunities: Some film geometric properties, like the minimum sides ratio , may provide clues to judge cropped segments. We found that by checking the rule $\gamma > .49$, we could mark under segments, containing $\sim 14.8\%$ of the films, as γ-violating. Furthermore, by observing that most records comprise films of about the same area (except for panoramic films), we could also mark under segments, containing $\sim 8.4\%$ of the films, as area-violating. In the future we may exploit these properties for defining additional postprocessing rules whose violations call for further subsequent processing and hence boost the yield.

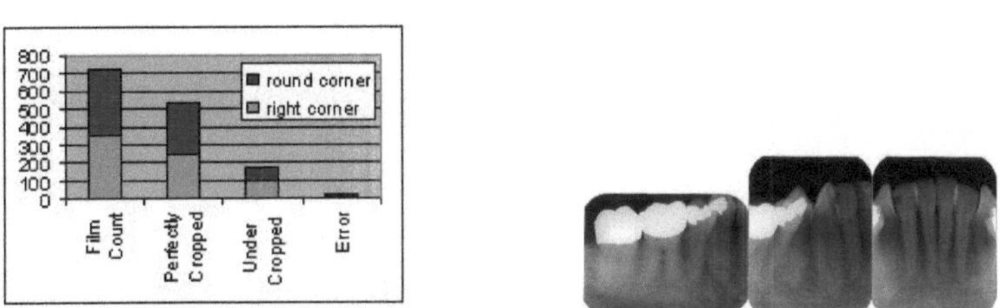

a) b)

Fig. 7. a)Expiremental results, b)example of undersegmented

7 Conclusions and Future Work

In this paper, we presented a global segmentation technique of cropping the dental films from the dental X-ray records. We started by using the rectangular film property and separating the background that have various colors and textures. Then we classify the connected components according to the corners being right or round. Cutting the round corner components depending on whether they are 90° or 180° V-corners; and cutting the right corner ones by viewing the boundary films under a graph theoretic framework, where it is always possible to reduce the graph by successively removing unit vertexes without affecting its connectivity. In the future we will exploit more geometric properties of films to develop additional postprocessing rules, which will identify segments that require further processing by a complementary, more computationally expensive cropping.

References

1. American Society of Forensic Odontology, Forensic Odontology News, vol (16), No. (2) - Summer 1997
2. The Canadian Dental Association, Communique. May/June 1997.
3. CJIS Division - ADIS, Digitized Radiographic Images (Database), August 2002.
4. G. Fahmy, D. Nassar, E. Haj-Said, H. Chen, O. Nomir, J. Zhou, R. Howell, H. H. Ammar, M. Abdel-Mottaleb and A. K. Jain, "Towards an automated dental identification system (ADIS)", *Proc. ICBA (International Conference on Biometric Authentication), pp. 789-796, Hong Kong*, July 2004.
5. A. K. Jain and H. Chen, "Matching of Dental X-ray Images for Human Identification", *Pattern Recognition, vol. 37, no. 7, pp. 1519-1532*, July 2004
6. E. Haj-Said, D. Nassar, G. Fahmy, and H. Ammar, "Dental X-ray Image Segmentation", *in Proc. SPIE Biometric Technology for Human Ident., vol. 5404, pp. 409-417*, August 2004.
7. S. White, and M. Pharoah, Oral Radiology Principles and Interpretation. Mosby, Inc. Fourth Edition 2000.

Acoustic Ear Recognition

Ton H.M. Akkermans, Tom A.M. Kevenaar, and Daniel W.E. Schobben

Philips Research, Prof. Holstlaan 4, 5656 AA Eindhoven, The Netherlands
{ton.h.akkermans, tom.kevenaar, daniel.schobben}@philips.com

Abstract. We investigate how the acoustic properties of the pinna – i.e. the outer flap of the ear- and the ear canal can be used as a biometric. The acoustic properties can be measured relatively easy with an inexpensive sensor and feature vectors can be derived with little effort. Classification results for three platforms are given (headphone, earphone, mobile phone) using noise as an input signal. Furthermore, preliminary results are given for the mobile phone platform where we use music as an input signal. We achieve equal error rates in the order of 1%-5%, depending on the platform that is used to do the measurement.

1 Introduction

Well-known biometric methods for identity verification are based on modalities such as fingerprints, irises, faces, or speech to distinguish individuals. In some situations, however, these well-known modalities cannot be used due to the price and/or form factor of the required sensor or the required effort to derive feature vectors from measurements. Therefore we investigated if the acoustic properties of the pinna - i.e., the outer flap of the ear - and the ear canal can be used as a biometric. The acoustic properties can be measured relatively simple and economically and we found that the acoustic properties differ substantially between individuals. Therefore ear recognition is a possible candidate to replace pin codes in devices such as mobile phones or to automatically personalize headphones or other audio equipment. An additional advantage of ear recognition is that, unlike real fingerprints that are left behind on glasses or desks, "ear-fingerprints" are not left behind and can also not be captured as easily as an image of a face. In this respect acoustic ear recognition may lead to a more secure biometric.

The shape of the outer ear, such as the folds of the pinna, the length and shape of the ear canal are very different between humans as can be observed when visually comparing the ears of two individuals. These differences are even more pronounced for acoustic measurements of the transfer function of the pinna and ear canal using a loudspeaker close to the ear and a microphone close to, or in, the ear canal as shown in Figure 1. Such transfer functions can be seen as a kind of "fingerprint" of the ear canal and/or pinna. The spectrum of an acoustic transfer function can be used almost directly as the feature vector for a given individual. Using the acoustic properties of the ear as a biometric has first been published in [1] but there has been no public data on performance and application of the technology.

D. Zhang and A.K. Jain (Eds.): ICB 2006, LNCS 3832, pp. 697–705, 2005.

2 Acoustic Properties of the Ear Canal

It is well known that the optical properties of human ears can be used in biometric identification [2,3,4]. In [5], the authors investigate the relationship between optical and acoustic properties of the ear. In [6], acoustic ear biometrics have been used to develop and evaluate a recently developed biometric template protection system [7,8,9]. In [1], the Sandia Corporation US claimed the first US patent on acoustic ear recognition. In the current paper we focus mainly on the acoustic properties of the ear and its potential to be used as biometric modality.

The ear canal is a resonant system, which together with the pinna provides rich features. In a coarse approximation it is a one-dimensional system that resonates at one quarter of the acoustic wavelength. The resonance will typically be around 2500Hz but it will vary from person to person. Typical resonance frequencies are corresponding to typical lengths and shapes of pinna and ear canal.

The length of the ear canal and the curvatures of the pinna have dimensions that vary from millimeters to a few centimeters. To be able to detect these shapes and curvatures the acoustic probing waves should have proper wavelengths. Restricting ourselves to low cost loudspeakers and microphones we can easily generate and measure sound waves from 100Hz up to 15 kHz. Assuming that we can resolve structures in the order of 1/10 of the wavelength, the minimum resolving power becomes about 2mm, which seems appropriate to capture most distinguishing features.

3 Set-Up

The principle of the measurement set-up is shown in Figure 1. A loudspeaker close to the ear canal generates an excitation signal while a microphone measures the reflected echo responses. In general the excitation can be any acoustic signal like noise or music that has a fairly flat frequency spectrum. Alternatively the excitation signal may be preprocessed in such a way that those frequencies are emphasized that allow for a good discrimination between individuals.

In our current set-up we measure the transfer function of the ear by sending a noise signal into the pinna and outer ear. Figure 2 shows a possible method for determining

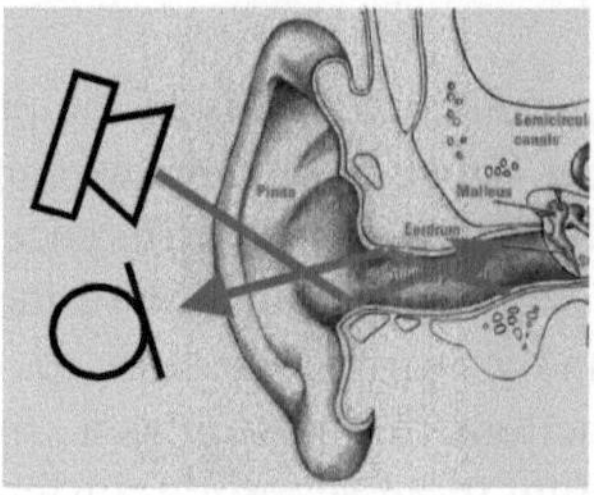

Fig. 1. An acoustic probe wave is send into the ear canal while a microphone receives the response

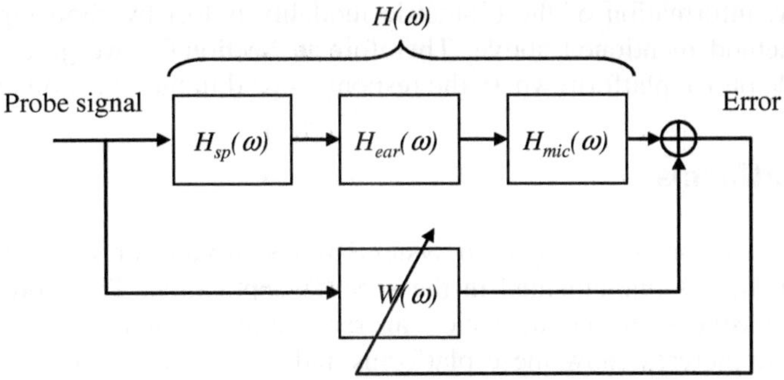

Fig. 2. Measuring the transfer function

this transfer function. The excitation signal is fed into the transfer function $H(\omega)$ that should be identified. The finite impulse response filter $W(\omega)$ is adaptively optimized using a steepest descent adaptive filter that minimizes the error signal which is the difference between the microphone signal and the output of the adaptive filter $W(\omega)$. Both the system $H(\omega)$ to be identified and its estimate $W(\omega)$ consist of the cascade of the transfer functions of the loudspeaker, pinna and ear-canal, and microphone. An alternative approach for determining the transfer function is to directly divide, in the frequency domain, the response signal coming from the microphone by the input signal [10]. Although both approaches gives similar results for noisy signals, the approach depicted in Figure 2 is more flexible when non-stationary input signals such as music are used as a probe signal (see also section 6.4).

4 Acquiring a Feature Vector

The estimate of the transfer function is a complex entity. Although it is expected that delays and phase shifts still contain significant discriminating information about an individual, they may also lead to larger intra-class variations, i.e. variations amongst various measured transfer functions for the same subject due to unwanted phase shift introduced by the measurement system. In order to eliminate these phase shifts we extracted the amplitude of the ear transfer function frequency spectrum as the biometric feature vector. As an example, Figure 3 shows transfer functions for three individuals.

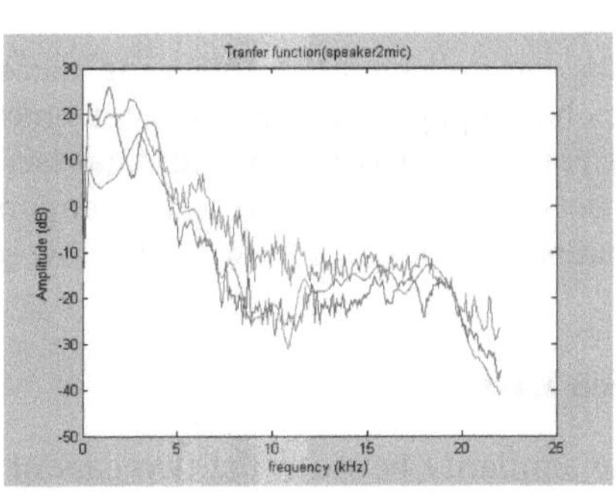

Fig. 3. Amplitude of the frequency response of the ear transfer function

Obviously, information of the biometric modality is lost by choosing the feature extraction method mentioned above. Therefore in Section 6.4 we give some results for the mobile phone platform where the response signal in the time domain is used.

5 Test Platforms

Often the performance of recognition systems relies strongly on the way the feature extraction method is implemented in the specific application. Therefore we investigated the robustness of the acoustic ear recognition system based on different platforms. The pictures show these platforms and the position of their microphones marked by the arrow.

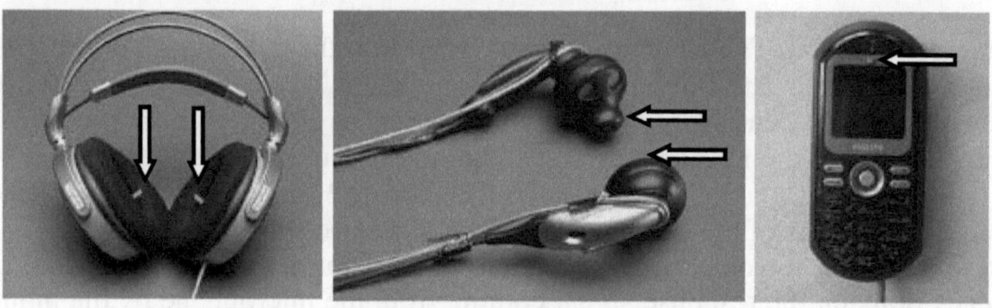

Fig. 4. The three platforms: headphones, earphones and mobile phone all with extra microphones indicated by arrows

The headphones in Figure 4 (Philips SBC HP 890) have 1 microphone per side that is mounted underneath the cloth that covers the loudspeaker. A tube is mounted onto each microphone that allows for measuring the sound pressure at the entrance of the ear canal. The earphones in Figure 4 have 1 microphone per ear-piece which is mounted underneath the original factory fit rubber cover.

The mobile phone of Figure 4 has 1 microphone next to the speaker whereas the other two platforms of Figure 4 each have 2 sensing microphones (1 microphone per ear) resulting in feature vector lengths of 256 and 512 components, respectively.

6 Results

In order to derive results, we collected the following measurements. For both the headphone and earphones based platform, 8 ear transfer functions were measured for each of 31 subjects and collected in two separate databases. For the mobile phone platform we enrolled 17 persons with 8 measurements per person that were stored in a third database. In the remainder of this section we will show some results obtained using these databases.

6.1 Correlation Between Ears

In order to determine the similarity between the two ears of an individual we determined the average correlation between the measurements of the two ears. We define the correlation as

$$C = \frac{x^T y}{\|x\| \|y\|} \tag{1}$$

where x and y are two feature vectors taken relative to the mean of the whole population. The average correlation C_j between the left and right ear of an individual j is taken as the average over the correlations between every possible combination of a measurement in the headphone database of the left and the right ear of this individual. The overall correlation between the left and right ear of the whole population is then defined as the average over the C_j's of all individuals. The reason for using the headphone database was that it shows the lowest intra-class variability per ear and is therefore most suitable to determine the biometric difference between left and right ear. In order to minimize the loss of information, the time responses rather than frequency responses were used and they were manually compensated for undesirable time delays. It turns out that the correlation between measurements of one ear of one individual is on average 90%. Comparing left and right ear gives an average correlation of roughly 80%. In conclusion we can say that using both ears only gives marginally better discrimination capabilities since the acoustic left and right ear responses are quite similar and differs 10% in terms of correlation.

6.2 Recognition Performance

To test the performance of the acoustic ear recognition system the FAR (False Acceptance Rate) and FRR (False Rejection Rate) have been investigated using the impostor and genuine distributions using the correlation measure (1). The probing noise signal contained frequencies in the range 1.5kHz-22kHz. Figure 5 shows the Receiver Operating Characteristics (ROC) of the unprocessed frequency response data. We observe that the headphones and earphones give roughly the same performance resulting in an equal error rate of respectively 7% and 6 %.

As a second experiment, Fischer Linear Discriminant Analysis (LDA) was applied to the three ear databases to selects the most discriminating components among the subjects. In order to determine the eigenvalues and eigenvectors, the generalized eigenvalue problem

$$S_b q = \lambda S_w q \tag{2}$$

was solved for q and λ where S_b and S_w are the estimated between-class and within-class covariance matrices, respectively. We used a regulation parameter to avoid singularity problems in S_w.

Figure 6 again shows the ROC performance but now a Fisher LDA transformation is applied to the frequency impulse responses. It can be seen from Figure 6 that the performance improves significantly, especially for the headphones and earphones platform. Furthermore a slight increase in FRR will significantly reduce the FAR leading to a high security level. The mobile phone performance is worse due to two reasons. Firstly, the between-class variation of mobile phones is much larger due to uncontrolled position and pressing of the mobile phone against the pinna. This is also observed when we consider the 'signal-to-noise ratio' of the feature vector

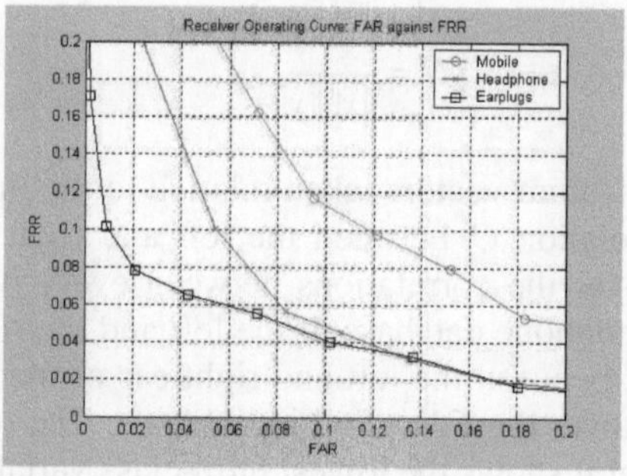

Fig. 5. Receiver operating curves without Fisher LDA transformation

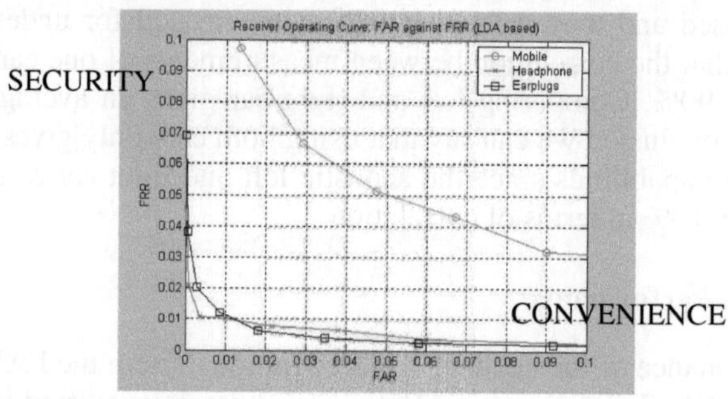

Fig. 6. Receiver Operating Curves using Fisher LDA a transformation

components after LDA. The average over all users and all components for the headphone and earphone database is in the order of 40 while for the mobile phone is it in the order of 16. A second reason is that, although the correlation between the two ears of one individual is very high, measuring two ears rather than one still gives slightly better discrimination between individuals.

6.3 Relevant Frequency Ranges

We also investigated how the applied frequency range used in the excitation signal influences the classification performance. Table 1 gives an overview of the Equal Error probability as function of the applied frequency range of the acoustic probe signal using the Fisher LDA transformation.

Table 1. Ear recognition performance (EER) as a function of the frequency range of the excitation signal

Freq.Range (Hz)	Headphones	Earphones	Mobile Phone
1.5k-22k	0.8	1	5.5
1.5k-10k	0.8	1.4	6.5
10k-22k	2.5	2.5	10
16k-22k	8	6	18

Although these figures depend quite heavily on the individual loudspeaker and microphone performances (especially in mobile phones the loudspeaker transfer at frequencies above 10 kHz deteriorate significantly), it can be seen that a wider frequency range gives better classification results. It is further interesting to notice that the frequency range 16kHz-22kHz still leads to reasonable classification results indicating that ultrasonic characterisation might be an option.

6.4 Experiments with Music and Time Domain Signals

In order to enhance user convenience we performed experiments where the excitation signal is a music signal rather than a noise signal. In our case we used a music signal in MP3 format which has the advantage that it has inaudible noise components in its spectrum due to the underlying Human Auditory System model used to compress music signals. These noise components improve the estimate of the transfer function. The initial experiments used a database of 12 persons with 10 measurements per person. The output signal from the microphone in the frequency domain rather than the transfer function $H(\omega)$ was used directly as a feature vector. Consequently, a user should always be probed with the same piece of music.

In Figure 7 two ROCs are given, one for a noise input and one for a music input where the curve referring to a noise input signal is copied from Figure 6. It can be seen that both systems give similar classification results.

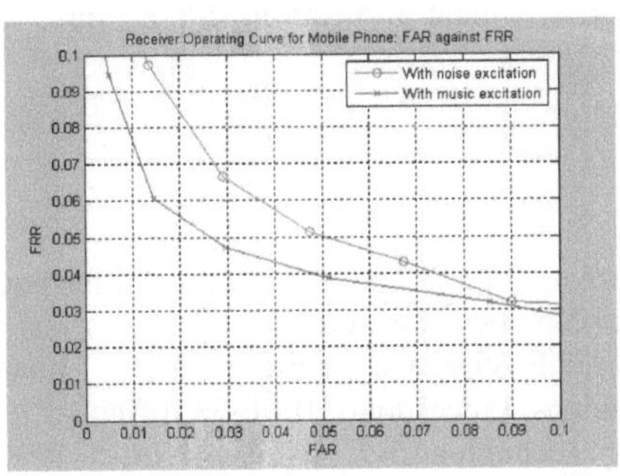

Fig. 7. The Receiver Operating Curves for a noise and music input signal for a mobile phone

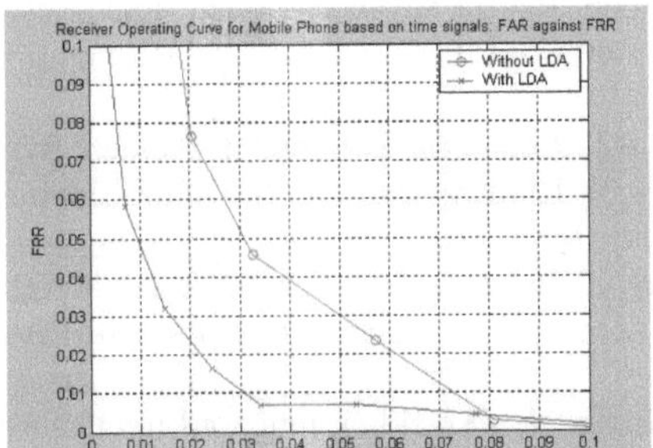

Fig. 8. The Receiver Operating Curves for a mobile phone based on time signals

As mentioned above, discarding the phase information in the feature vectors might deteriorate classification results but is practically necessary to handle random phase shifts in the measurement system. In order to estimate the influence of discarding the phase, we used the time-domain signal coming from the microphone as a feature vector where we manually compensated for the system delay. The results are given in Figure 8 where, compared to Figure 6, we see an improvement in classification results. In practical systems a pilot tone can be inserted to handle random system delays.

7 Conclusions

This paper describes a novel biometric system based on the acoustic properties of the human ear. Three practical platforms were developed including a mobile phone, headphones and earphones where using noise as a probing signal. The amplitude of the frequency spectrum of the ear transfer function has been found to provide stable and rich features. False acceptance and rejection rates have been derived from measurements taken from various subjects. Applying a Fisher LDA transform greatly improve the performance. In order to enhance user convenience we also used music as a probing signal which resulted in comparable ROCs. Finally we used a time signal rather than the amplitude of the transfer function as a feature vector resulting in improved classification results. Further research consists of deriving the transfer function for an arbitrary piece of music and retaining the phase information in the measurement signal.

References

1. Sandia Corporation, patent US 5,787,187, "Systems and methods for biometric identification using the acoustic properties of the ear canal."
2. B. Moreno, A. Sanchez, J.F. Velez,On the use of outer ear images for personal identification in security applications, Proceedings. IEEE 33rd Annual 1999 International Carnahan Conference on Security Technology,5-7, pp469 – 476, Oct. 1999.
3. M. Burge, W. Burger, Ear biometrics in computer vision, Proc. 15th International Conference on Pattern Recognition, Vol. 2, 3-7, pp822 – 826, Sept 2000.
4. K.H. Pun and Y.S. Moon, Recent advances in ear biometrics, Proc. Sixth IEEE International Conference on Automatic Face and Gesture Recognition, 17-19, 164 – 169, May 2004.
5. Y. Tao, A.I. Tew, and S.J. Porter, The Differential Pressure Synthesis Method for Estimating Acoustic Pressures on Human Heads, 112th Audio Engineering Society Convention, 10–13 Munich, Germany, May 2002.
6. P. Tuyls, E. Verbitskiy, T. Ignatenko, D. Schobben and T.Akkermans. Privacy Protected Biometric Templates:Acoustic Ear Identification. Proc.SPIE,Vol.5404, pp176-182, April, 2004.
7. J-P. Linnartz and P. Tuyls, New shielding functions to enhance privacy and prevent misuse of biometric templates, Proc.4th Int. Conf. on Audio- and Video-Based Biometric Person Authentication (AVBPA 2003), Springer LNCS 2688, pp 393-402, 2003.

8. P. Tuyls and J. Goseling, Capacity and Examples of Template Protection in Biometric Authentication systems, Biometric Authentication Workshop (BioAW 2004), LNCS 30087, pp158-170, Prague, 2004.
9. P. Tuyls, A. Akkermans, T. Kevenaar, G-J Schrijen, A. Bazen and R. Veldhuis. Practical Biometric Authentication with Template Protection. Proc. 5th Int. Conf. on Audio- and Video-Based Biometric Person Authentication (AVBPA2005), Springer LNCS 3546, pp436-446, 2005.
10. A.H.M. Akkermans, T.A.M. Kevenaar, D.W.E. Schobben. Acoustic Ear Recognition for Person Identification. Accepted for the IEEE AutoID Workshop, October 17-18, 2005 Buffalo, New York, USA.

Classification of Bluffing Behavior and Affective Attitude from Prefrontal Surface Encephalogram During On-Line Game

Myung Hwan Yun, Joo Hwan Lee, Hyoung-joo Lee, and Sungzoon Cho

Department of Industrial Engineering,
Seoul National University, Seoul, 151-742 South Korea
{mhy, leejh337, impatton, zoon}@snu.ac.kr

Abstract. The purpose of this research was to detect the pattern of player's emotional change during on-line game. By defining data processing technique and analysis method for bio-physiological activity and player's bluffing behavior, the classification of affective attitudes during on-line game was attempted. Bluffing behavior displayed during the game was classified into two dimensions of emotional axis based on prefrontal surface electroencephalographic data. Classified bluffing attitudes were: (1) pleasantness/unpleasantness; and (2) honesty/bluffing. A multilayer-perception neural network was used to classify the player state into four attitude categories. Resulting classifier showed moderate performance with 67.03% pleasantness/unpleasantness classification, and 77.51% for honesty/bluffing. The classifier model developed in this study was integrated to on-line game as a form of 'emoticon' which displays facial expression of opposing player's emotional state.

1 Introduction

Although bio-electrical signal was known since late 1840s, forms of BCI (Brain Computer Interface), which facilitates human-system interaction using various bio-signals, were introduced only in 1970s. Later on, through various researches (Hesham, 2003; Yuko, 2002; Pfurtscheller et al., 1996), basic interface functions such as controlling mouse cursors by EEG signal are being reported. On the other hand, there were substantial researches attempting to discriminate emotional state of human beings in real life situation such as fatigue, drowsiness, and general stress level using bio-physiological signals (Eoh et. al., 2005). With continuous improvement of signal measurement and digital processing technology, BCI is rapidly becoming an important option for biometric interaction. While it is far from being a realistic authentication tool, BCI is continuously expanding its application area. Brain is the primary center for the regulation and control of bodily activities, re-ceiving and interpreting bio signals, and transmitting information (Andreassi, 1995). By attaching electrodes on the scalp, the electrical activity of the brain can be recorded. Details of the EEG and its processing are out of the scope of this paper. There are numerous sources of information related to this area (Gevins et al., 1998; Wilson et al., 1999). A polygraph is an instrument that records changes in physiological processes such as heart rate,

D. Zhang and A.K. Jain (Eds.): ICB 2006, LNCS 3832, pp. 706–712, 2005.

blood pressure, respiration and other bio-signals. The polygraph test is frequently used in practical situation. The underlying assumption of the polygraph test is that when people deviate from their normal state, they produce measurable changes in their physiological signal. A baseline for testing these physiological characteristics is established by asking various questions whose answers the investigator already knows. Deviation from the baseline for truthfulness is taken as sign of lying or bluffing (Luan K., 2005; Vance, 2001). There are three basic approaches to the logic behind affective state classification with polygraph test (shown the table1) and they are used as the underlying concept for developing the affective attitude classification model of this study.

Table 1. Three approaches to affective attitudes classification

Approaches	Method
The Control Question Test (CQT).	This test compares the physiological response to relevant questions about the crime with the response to questions relating to prior misdeeds. "This test is often used to determine whether certain criminal suspects should be classified as uninvolved in the crime"
The Directed Lie (or bluffing) Test (DLT)	This test tries to detect lying by comparing physiological responses when the subject is told to deliberately bluffing to responses when they tell or act the truth.
The Guilty Knowledge Test (GKT)	This test compares physiological responses to multiple-choice type questions about the crime, one choice of which contains information only the investigators and the criminal would know about.

2 Experiment

Fifteen students (10 males, 5 females, mean age, 25.8, S.D. ± 3.51) who have fair amount of experience in on-line game participated for data collection. The experimental equipment used for signal measurement and analysis was developed specifically for this study. The equipment consists of a hair band with EEG sensors, heart rate sensors and signal transmission unit. For the EEG measurement, four prefrontal channels (two forehead channels, one Reference, and one Ground) were used. The band pass filter was butter-worth with 4-46 Hz range with the gain of 4,700. For the heart rate measurement, a direct reflex sensor (exposing the skin to direct-current infra-red beam and extracting blood pulse signal from the IR reflexion data, band-pass range, 2 to 10 Hz, gain 6,000) was attached inside the EEG hair band. Figure 1 illustrates the experimental settings. Based on the polygraph paradigm, game situations and player reactions were classified as pleasant (advantageous) and unpleasant (disadvantageous) situations.

Player reaction was classified to two separate state; aggressive and conservative. Table 2 also shows the scheme of affective attitude classification used in this study; honesty and bluffing. Figure 2 is the task analyses chart which is used to branch and bound the possibility of various outcomes from the attitudes classification structure in Table2.

Fig. 1. Experimental set up (players, EEG/ECG input devices (hair band), signal transmission unit, and heart rate monitors, attached to subject's chest area)

Table 2. Classification of player bluffing and affective attitudes

	Pleasantness (Advantageous situation)	Unpleasantness (Disadvantageous situation)
Aggressive betting	Honesty	Bluffing
Conservative betting	Bluffing	Honesty

Four type of EEG band, alpha, beta, theta, and delta were filtered by FFT analysis. Changes in affective attitudes was detected through an EEG signal coming from right frontal region, Fp2 (based on 10-20 International System, Waldstein et al, 2000, Kimbrell et al., 1999). After the measurement equipments were attached, a session of the on-line game was conducted with fifteen matches (10 minutes of session were used for the measurement of baseline EEG). After 10 minutes rest, 30 sets of on-line game were conducted (including 10 minutes rest between each 15 sets). Saved video files were later shown to each subject for the post-hoc evaluation. First, input variable of EEG signal was selected. The raw data collected through experiment was divided at 0.33 second interval and classified into four frequency bands (α, β, δ, θ). The following is the input vector that was created by calculating the RMS value of each wave.

$$\text{RMS value: } \chi_i = \sqrt{1/|P_i| \sum_i P_i^2} \quad (i = \alpha, \beta, \delta, \theta) \tag{1}$$

$$\text{Input vector: } X = [x_\alpha, x_\beta, x_\delta, x_\theta]^T$$

Output variable was evaluated as the follows; Output variable1 (y_1): pleasantness (advantage) or unpleasantness (disadvantage).

$$y_1 \begin{cases} = 1\text{: When the participants feel advantageous} \\ = 0\text{: When the participants feel disadvantageous} \end{cases} \tag{2}$$

Output variable2 (y_2): honesty or bluffing

$$y_2 \begin{cases} = 1\text{: When the participants play bluffing} \\ = 0\text{: When the participants play honestly} \end{cases} \quad (3)$$

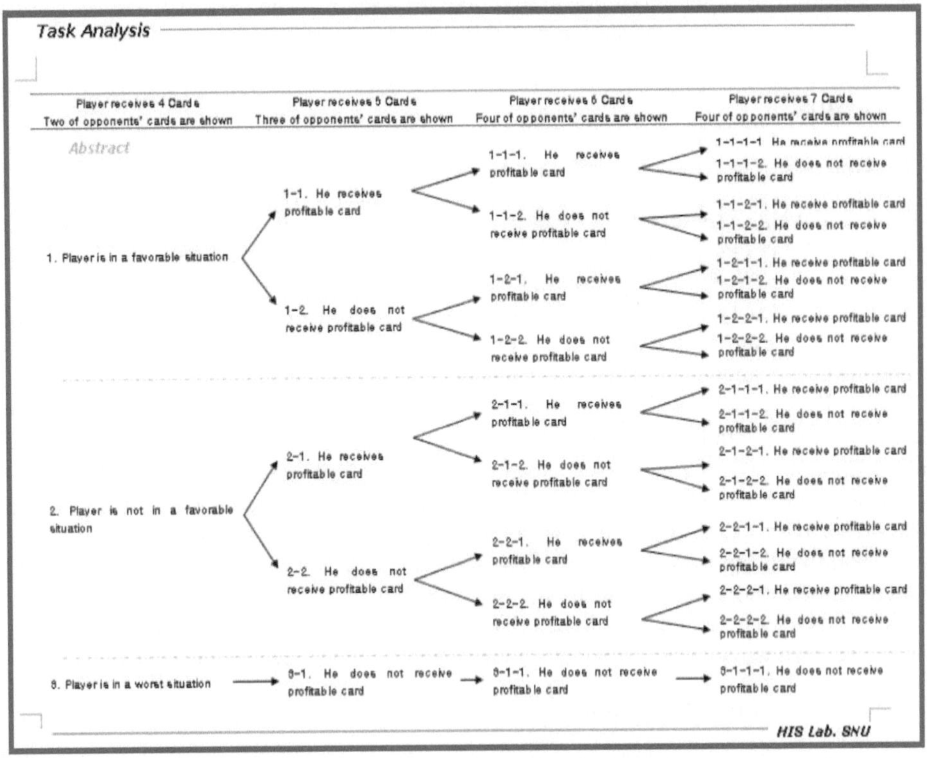

Fig. 2. Task structures and game plan used for attitudes classification (on-line poker game)

Table 3. Data distribution classified by output variables

		Output variable 1 (y1)		Total
		Pleasantness (1)	Unpleasantness (0)	
Output variable 2 (y2)	Bluffing(1)	135 (2.48%)	1,159 (21.31%)	1,293 (23.80%)
	Honesty(0)	3,192 (58.75%)	948 (17.45%)	4,140 (76.20%)
Total		3,327 (61.24%)	2,106 (38.76%)	5,433

The EEG data was collected through fifteen subjects and the data for subject evaluation was also collected. In general, change of emotion was relatively more noticeable in the latter half of the trial than in the early stage in the EEG data. Therefore, only EEG data at the time of the sixth and seventh card (latter half) was selected and the EEG pattern was developed from selecting the data when the subject evaluation score was over 9 point (very highly confident). A total 5,433 set of data was created from four input variable and two output variables. The data distribution classified by output variable is as shown in Table 3.

3 Results

3.1 Analysis of Various Classification Model

Model Analysis, which compares each candidate model by accuracy, generates the optimal EEG analysis model. For selecting the model, 1,800 data among whole data were used. The criterion of comparing each model was classification accuracy. Linear Discriminant Analysis (LDA) spent less training time than the other data mining models. So, it can be suitable for an on-line game in real time. But LDA produced higher training and test errors than others. RBF kernel SVM was used for model analysis (Burges, 1998). For RBF kernel SVM, 2 parameters should be specified: the kernel width, 'r' and the penalty for misclassification, 'cos*t*'. The highest classification accuracy was obtained at $r = 1$ and cos*t* = 30. ANN (Artificial Neural Network) was also used for model analysis. A back propagation multi-layer perceptron (1 hidden layer) using the Levenberg-Marquadt algorithm was employed. Hidden layer activation functions were as follows. The highest classification accuracy was obtained with 10 hidden nodes. The results are summarized in Table 4. Bagging Neural Network (BNN) was carried out. When 10 networks with 10 hidden nodes were combined, the highest classification performance was obtained (Breiman, 1996).

3.2 Selection of the Best Model

Among the 5,433 data, 1,800 data were extracted for candidate model selection. To select the suitable model, criteria such as; (1) possibility of real time analysis; (2) high classification accuracy; and (3) easiness to develop the program was considered. Table 5 is a summary of experimental results. ANN turns out to be the most suitable model. LDA spent the shortest training time and was very easy to apply but its accuracy was too low. SVM was more accurate than LDA but is impossible for real time analysis and spent relatively longer time for training. ANN achieved the highest accuracy, spent relatively shorter training time and was easy to apply. Eventually, ANN was selected for final EEG analysis model.

3.3 EEG Index Model

Using the basis of the ANN model, the final EEG index was developed. EEG index is as follows:

Discrimination of pleasantness/unpleasantness: $$Score1 = round(y1 \times 4 + 1) \quad (5)$$

Discrimination of honesty/bluffing: $$Score2 = round(y2 \times 99 + 1) \quad (6)$$

Table 4. The analysis results using ANN

(Unit: %)

	Hidden nodes						
	3	5	10	15	20	25	30
Output variable 1	70.10	72.30	72.95	64.73	72.30	72.79	72.62
Output variable 2	70.64	67.69	67.58	71.77	68.64	67.48	67.62
Combination	55.34	54.75	56.15	5.019	56.11	55.51	55.65

Table 5. Experimental results and features of each model

(Unit: %)

	LDA	SVM	ANN	BNN
Output variable 1	50.60	62.38	67.03	62.23
Output variable 2	54.72	75.83	77.51	76.40
Combination	36.61	55.79	58.92	58.37
Accuracy	too bad	good	good	good
Training time	very good	good	good	normal
Easiness of application	very good	normal	good	bad

In pleasantness/unpleasantness index, index score range was set to 1 to 5. Higher score means higher level of pleasantness displayed by the gamer. In honesty/bluffing classification index, index score range was set to 1 to 100 such that higher score means higher probability of bluffing behavior. Eventually, the two kind of index developed from the model was programmed as a form of 'emoticon' so that facial expression of opposing gamer can be displayed together with the cards being played in the game. Using the blood pulse rate obtained from the IR sensor, conversion to Heart Rate (HR) was conducted. The variation of HR during the game was used for detecting the difference between pleasantness and unpleasantness. As the result, when HR was increased, pleasantness/unpleasantness score increased. Since the relationship between pleasantness/unpleasantness index and HR was not statistically significant, HR was used only as the weight value of pleasantness/unpleasantness index.

4 Conclusion

The purpose of this research was to classify the affective attitudes during on-line game using a EEG-based data processing technology. This study also suggested an index approach to quantify the player's behavior through the psychophysiological dimension of pleasantness/unpleasantness and honesty/bluffing. Since the approach used a real-time, continuously updating strategy, the classification scheme will be improved continuously as the game progresses. Although resulting classifier model showed moderate performance with 67.03% pleasantness/unpleasantness classification, and 77.51% for honesty/bluffing classification, it is higher than expected level considering that the model will be used in real-time, continuously updated situation. Together with the classification model, an on-line game with EEG measurement was also developed and implemented in this study. The EEG of game players were measured, transferred, and displayed to the other player during an on-line game in the form of an 'emoticon' displaying a various facial expression according to pleasantness/unpleasantness and honesty/bluffing scores calculated from the classifier.

References

1. Andreassi, J.L.: Psychophysiology: Human Behavior & Physiological Response, 3rd ed., Lawrence Erlbaum Associates, New Jersey (1995)
2. Bishop, C.M.: Neural Network for Pattern Recognition. Oxford University Press (1995)
3. Breiman L.: Bagging predictors, Machine Learning, 24(2) (1996) 123-140
4. Burges, C.J.C.: A tutorial on support vector machines for pattern recognition, Data Mining and Knowledge Discovery, 2 (1998) 121-167
5. Eoh, H.J., Chung, M.K. and Kim, S. H.: Electroencephalographic study of drowsiness in simulated driving with sleep deprivation, International Journal of Industrial Ergonomics, 35(4) (2005) 307-320
6. Gevins, A., Smith, M.E., Leong, H., Mcevoy, L., Whitfield, S., Du, R., and Rush, G.: Monitoring working memory load during computer-based tasks with EEG pattern recognition methods, Human Factors, 40(1) (1998) 79-91
7. Hesham Sheikh, Dennis J. McFarland, William A. Sarnacki and Jonathan R. Wolpaw: Electroencephalographic (EEG)-based communication- EEG control versus system performance in humans, Neuroscience Letters, 345(2) (2003), 89-92
8. Kimbrell T.A., Mark S. George, Priti I. Parekh, Terence A. Ketter, Daniel M. Podell, Aimee L. Danielson, Jennifer D. Repella, Brenda E. Benson, Mark W. Willis, Peter Herscovitch and Robert M. Post: Regional brain activity during transient self-induced anxiety and anger in healthy adults, Biological Psychiatry, 46(4) (1999) 454-465
9. Luan, K.: Neural correlates of telling lies: A functional magnetic resonance imaging study at 4 Tesla, Academic Radiology, 12(2) (2005) 164-172
10. Pfurtscheller G., G.R. Miller, C. Guger: Direct control of a robot by electrical signals from the brain, proceeding EMBEC '99, Part 2, (1999) 1354-1355
11. Vance V.: A quantitative review of the guilty knowledge test, Journal of applied psychology, 86(4) (2001) 674-683
12. Waldstein, S.R., Kop W.J., Schmidt, L.A.: Frontal electro cortical and cardiovascular reactivity during happiness and anger, Biological Psychology, 55(1) (2000) 3-23
13. Wilson, G.F., Swain, C.R., and Ullsperger, P.: EEG Power Changes during a Multiple Level Memory Retention Task, International Journal of Psychophysiology, 32 (1999) 107-118
14. Yuko Ishiwaka, Hiroshi Yokoi and Yukinori Kakazu: EEG on-line analysis for autonomous adaptive interface, International Congress Series, 1232 (2002) 271-275
15. Yun, M.H.: Development of an adaptive computer game interface based on biophysiological signal processing technique, Ministry of Science and Technology, South Korea (2000) (unpublished research report, in Korean)

A Novel Strategy for Designing Efficient Multiple Classifier

Rohit Singh[1], Sandeep Samal[2], and Tapobrata Lahiri[3,*]

[1] Wipro Technologies, K-312, 5th Block, Koramangala, Bangalore – 560095, India
rohit.singh@wipro.com
[2] Tata Consultancy Services, Bangalore
sandeep.sam@tcs.com
[3] Indian Institute of Information Technology, Allahabad – 211012, India
tlahiri@iiita.ac.in

Abstract. In this paper we have shown that systematic incorporation of decision from various classifiers following a simple decision decomposition rule, gives better decision in comparison to the existing multiple classifier systems. In our method each classifier were graded according to their effectiveness of providing more accurate results. This approach first utilizes the best classifier. If this classifier classifies the test sample into more than one class or fails to classify the test data then the feature next to the best is summoned to finish up the remaining part of the classification. The continuation of this process, along with the judicious selection of classifiers, yields better efficiency in identifying a single class for the test data. The results obtained after the experiments on a set of fingerprint images shows the effectiveness of our proposed classifier.

1 Introduction

Personal Identification systems based on fingerprints or facial images, diagnosis of diseases by analyzing the histopathological images, etc. are some applications where accuracy cannot be compromised with, as it may be a case of identifying an authorized person for access to critical or highly restrictive places, or it might be the case of saving the life of a patient through proper diagnosis. More often it is seen that a single classifier struggles to give a high accuracy and reliability level that some critical applications demands. As a result of this, a multiple classifier can be a viable solution for the accuracy and reliability constraints. Work has been going in this field from last decade. From the point of view of analysis, the classification scenarios can be of two types. In the first scenario, all the classifiers use the *same representation* of the input pattern. Here each classifier produces an estimate of the same aposteriori class probability. In the second scenario each classifier uses its *own representation* of the input pattern. They can be either sequential or pipelined [1], [7], or hierarchical [8], [9]. Other studies done in the gradual reduction of the set of possible classes are shown in [3], [4], [6]. The combination of ensembles of neural networks (based on

* Corresponding author.

D. Zhang and A.K. Jain (Eds.): ICB 2006, LNCS 3832, pp. 713 – 720, 2005.

different initializations) has been studied in the neural network literature [10], [11]. In another approach, a theoretical framework based on Bayesian decision rule has also been used [2]. But this approach has serious limitation. The assumptions made are unrealistic for most applications [13]. The first assumption is that joint probability distributions of measurements extracted by classifier are conditionally statistically independent. The second assumption is that posterior class probability does not deviate much from a priory probability.

The approach proposed by us, takes the result of the best classifier at any instant. The classifiers considered here use their *own representation* of the input pattern. This approach makes use of predicted class to classify a test image. The basic assumption is that the decision can be decomposed into multi-classifier space that can be thought as analogous to the decomposition of any general type of data into multidimensional space. Classifier that accurately classifies the maximum number of test data is chosen as the topmost level classifier regarding their usability. In this way the classifiers are organized into levels. The test data to be classified goes through these levels from top to bottom.

2 Underlying Principle – Decision Decomposition

The methodology proposed is based on ensemble of diverse classifiers working with fractal, ridge and wavelet features extracted from a fingerprint image database. For better comprehension of the underlying principle let us draw an analogy between data and decision.

Decomposition of any data into more than one dimension provides more detail and discriminatory information about this data. Let us take the help of one example of 2 dimensional Gel Electrophoresis (2D Gel) to isolate different proteins according to two-dimensional separation method. In figure 1(a) the separation of proteins has been shown as different separated bands according to their charge increasing from top to down. If we assume that each band represents only one protein, it may be a mistake. This is because two proteins of different masses may have same charge. Further horizontal separation of the proteins from each band according to their mass

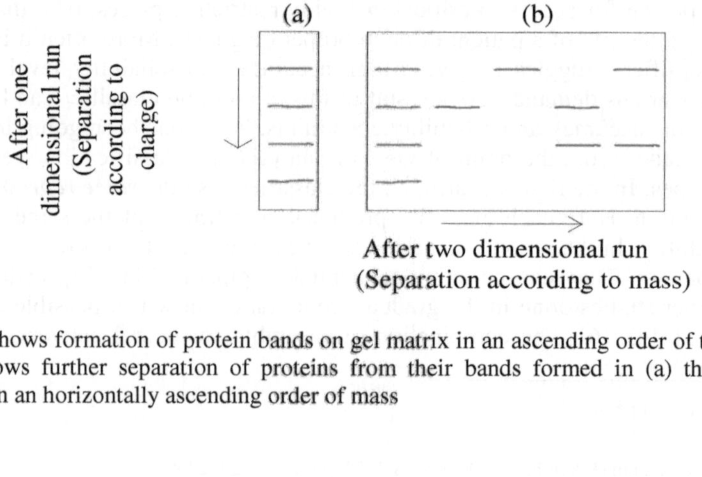

Fig. 1. (a) shows formation of protein bands on gel matrix in an ascending order of their charge and (b) shows further separation of proteins from their bands formed in (a) through their separation in an horizontally ascending order of mass

increasing from left to right, shows that this assumption holds true for the fourth band only. Other bands are further decomposed to give the idea of, more than one protein within each band as shown in figure 1(a).

The above figure gives the idea that more is the attributes (or dimensions) better is the discrimination among classes. Drawing analogy from this experiment we can say that one classifier may initially give some cluster or assembly of classes that can be further separated with the use of other classifier. This approach forms the basis of our work.

Suppose that there are N different classes

$$\{C_i\}^{N}_{i=1}. \quad (1)$$

each of which is represented by a set of data

$$\{d_j\}^{M}_{j=1}. \quad (2)$$

Also suppose that we have a test data d_t whose class is unknown. The proposed algorithm can be generalized as follows:

Step1: First we design P number of classifiers. In this design different classifiers are characterized by different type of logical features extracted from the same data (say image) while their classification rule (distance metric, nature of class boundary) is kept as same.

Step2: Next the efficiency of each classifier is tested. The classifiers' levels are fixed according to the descending order of their efficiency value.

Step3: We start our hunt for appropriate class against the test data using Classifier of level 1, say, CL1. If the test data falls within the overlapping boundary of more than one classes, go to step 4, else if, it falls within a single non-overlapping class boundary, it is assigned as the appropriate class against the test data.

Step4: If the number of overlapping classes defining the test data are

$$\{C_k\}^{K}_{k=1}. \quad (3)$$

we take the help of Classifier of level 2, say, CL2 and focus our attention on K number of overlapping classes only. If, with CL2 test data falls within the overlapping boundary of more than one classes again (say L number of classes where $L \le K$) repeat step 4 with next level of classifier and so on.

In our methodology we have applied the above Multiple Classifier System by Decision Decomposition (MCSDD). However, as this algorithm is very strict regarding rejection of wrongly classified data and also because of the problem of comparison of this rule with the existing Multiple Classifier Systems (MCS) we have also applied a modified and somewhat flexible rule to compare it with existing MCS.

3 Implementation

3.1 Database and Software Used

The experiment was carried out in Matlab 7.0 environment with 168 fingerprint images, i.e. 8 images taken from each of the 21 persons (i.e., classes). The images were downloaded from Biometric System Lab., University of Bologna – ITALY (http://www.csr.unibo.it/research/biolab/). Out of 8 images per classes, 6 were used as training data while the rest 2 were kept for testing purpose.

3.2 Designing Classifiers

We designed three classifiers based on three different input feature sets that are multi-fractal parameter, third level decomposed wavelet coefficients using 'haar' wavelet and ridge features.

3.3 Extraction of Multi-fractal Parameter

The multi-fractal parameter extraction from different intensity plane was carried out in a number of steps. Firstly the RGB image is converted to gray scale one. From the gray-scale image (say, I), *n* numbers of binary images $\{B_i\}^n_{i=1}$ are obtained by splitting the intensity level of the image at (n + 1) equally spaced different intensity intervals applying the following rule.

$$B_i = 1 \text{ if } g_i \leq \mathbf{I} \leq g_{i+1}. \tag{4}$$

where g_i is the i-th gray level. Otherwise,

$$B_i = 0. \tag{5}$$

Thus the pixels of the i-th binary image B_i having gray value 1 are marked as occupied for the corresponding intensity interval. The well-known Box-counting algorithm [6] was then applied to find the fractal dimension D_i for B_i. In practice we have chosen the value of n equals to 8 and the uppermost and lowermost gray levels g_U and g_L respectively as:

$$g_U = 1.4 \times g_m \text{ and } g_L = 0.6 \times g_m. \tag{6}$$

where g_m is the mean gray value of **I**.

3.4 Extraction of Wavelet Coefficients

The wavelet coefficients are obtained from the converted gray image. Two dimensional discrete wavelet transformations is applied using *dwt2* function of MATLAB with *'haar'* wavelet. It gives the output as approximation and detail coefficients. The approximation coefficient of third level has been considered as our second feature.

3.5 Extraction of Ridge Feature

For extraction of ridge features following steps were executed [12]:

Step1: We determine a reference point and region of interest for the fingerprint image.

Step2: Tessellate the region of interest around the reference point.

Step3: The region of interest is then filtered in eight different directions using a bank of Gabor filters.

Step4: We then compute the average absolute deviation from the mean (AAD) of gray value in individual sectors in filtered images, which defines the feature vector.

4 Classification Rule

After the feature parameters are obtained they were assigned to three classifiers as mentioned in the previous section. A clustering algorithm is used for each level of classifiers starting from level 1 up to the level that gives single class decision subjected to some modifications and flexibility as discussed in the latter section. The general steps are as follows:

Step1: Choose CL1

Step2: Find class boundary of each class. For this, first find the average feature value of say i-th class and keep it as class center, C_i of that class. Also find the distance of the maximally distant feature point from C_i, r_i and keep it as the class radius assuming each class as sphere.

Step3: Before presenting a test data to the classifier, extract its feature parameters and consider it as the feature point, T.

Step4: Find the Euclidian distances between T and all the class centers, $\{d_i\}^{21}_{i=1}$ for 21 classes.

Step5: Count the number of classes J for which the following expression holds true

$$d_i \leq r_i$$

If
the classifier level is not the end level (i.e., the 3rd level in our case),
Then,
if J = zero or J ≥ 2, chose the next level classifier.
else if J = 1, stop classification and assign the corresponding class to the test data.
Else if
the classifier level is the 3rd level,
Then,
if J = 1, assign the corresponding class to the test data.
else reject the test data for classification

Step6: Repeat step 2 to 5 for next level classifiers till the final decision about the test data (whether to be accepted to a particular class or fully rejected) is obtained.

However, the above rule is very stringent and also it requires a large amount of data per class for getting an accurate class boundary. Hence, we have incorporated the following modifications in our algorithm.

If, at the last level of classifier (refer step 6 above),
J = zero or J ≥ 2, Go To step 1
Assign the test data to the j-th class for which,
d_j = minimum of $\{d_i\}^{21}_{i=1}$

Figure 2 (a), (b) and (c) shows the three classification criteria discussed above

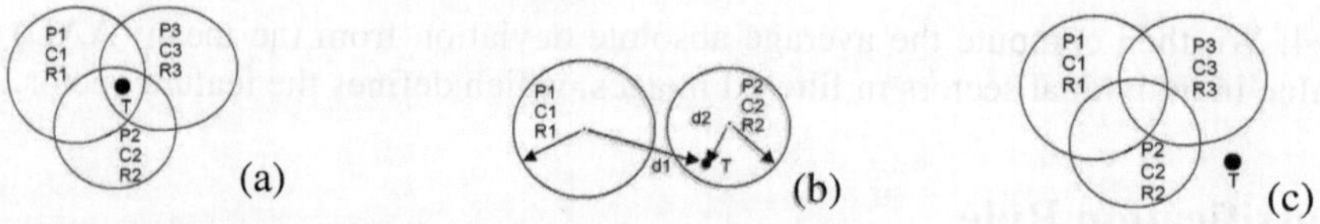

Fig. 2. (a): Test feature point lying in more than one classes (P, C and R are the respective person, class center and class radius). **(b):** Test feature point lying in one class and satisfying radius test i.e $d_{T2} < R_2$. **(c):** Test feature point lying outside all the classes (the case of rejection).

5 Benchmarking

For benchmarking purpose we compare the result of our approach to that of Kittler's sum rule, which is considered to be best among all combination schemes. For this we have to find out the efficiency of the Kittler's sum rule [2].

Kitler's sum rule: assign test feature point, θ to class w_j if,

$$(1-R)P(w_j) + \sum_{i=1}^{R} P(w_j \mid x_i) = \max_{k=1}^{m} \left[(1-R)P(w_k) + \sum_{i=1}^{R} P(w_k \mid x_i) \right] \quad (7)$$

where, R is number of features, m is the number of classes, x_i is the i-th classifier based on the feature, x_i, $P(w_j)$ = prior probability of the class w_j for raw data, and $P(w_k|x_i)$ = aposteriori probability of the class w_k given the feature, x_i

6 Result and Discussions

Table 1 shows that, when the wavelets, ridges and multi-fractal classifiers were used independent of each other; the recognition accuracy produced was very low at 30.95%, 21.43 and 19.05% respectively. But on combining the three classifiers through our above-described method, the recognition accuracy increased to 80.95% while the recognition accuracy of the Kittler's sum rule also increased to 57%. This shows that the performance of a pattern recognition system can be improved significantly by multiple classifier systems proposed by us. It also shows that the judicious selection and combination of classifier can increase the efficiency of the recognition system by many folds.

Table 1. shows some representative results of applying the proposed methodology to a set of test images

Number of Queries	Efficiency of Wavelet based Classifier (in %)	Efficiency of Ridge based Classifier (in %)	Efficiency of multi-fractal based Classifier (in %)	Efficiency of proposed multiple classifier (in %)	Efficiency of Kittler's Sum Rule (in %)
42	30.95	21.43	19.05	80.95	57.14

7 Conclusions

The problem of combining classifiers, which use different representations of the patterns to be classified was studied. We have developed a decision decomposition based framework for utilizing decision obtained from multiple classifiers. Our proposed MCSDD showed that passing the input pattern through several classifier levels does analogous discriminatory function based on multidimensional data. The multiple classifier system designed here does not degrade the quality of any classifier and rather fully utilize the quality of each individual classifier.

Our result shows that MCSDD has far better efficiency than the existing multiple classifier systems. Incorporation of more features and large database is expected to enhance the classification efficiency further.

Acknowledgement

We gratefully acknowledge the financial support received by the corresponding author for continuing this work in the form of Grant-in-aid from ILTP Cooperation between India (DST) and Russia (RAS) for the Indo-Russian collaborative project.

References

1. Pudil, P.; Novovicova, J.; Blaha, S. and Kittler, J., "Multistage Pattern Recognition with Reject Option," Proc. 11th IAPR Int'l Conf. Pattern Recognition, Conf. B: Pattern Recognition Methodology and Systems, vol. 2, 1992, pp. 92-95.
2. Kittler, J.; Hatef, M.; Duin, R.P.W.; Matas, J.; "On combining classifiers.", IEEE Transactions on Pattern Analysis and Machine Intelligence, Volume: 20 , Issue: 3, March 1998 ,pp:226 – 239.
3. Denisov, D.A. and Dudkin, A.K.; "Model-Based Chromosome Recognition via Hypotheses Construction/Verification," Pattern Recognition Letters, vol. 15, no. 3, 1994, pp. 299- 307.
4. Fairhurst, M.C. and Abdel Wahab, H.M.S., "An Interactive Two-Level Architecture for a Memory Network Pattern Classifier," Pattern Recognition Letters, vol. 11, no. 8, 1990, pp. 537-540.
5. Feder, "Fractals", Plenum Press, New York, 1988.
6. Kimura, F. and Shridhar, M., "Handwritten Numerical Recognition Based on Multiple Algorithms," Pattern Recognition, vol. 24, no. 10, 1991, pp. 969-983.
7. El-Shishini, H.; Abdel-Mottaleb, M.S.; El-Raey, M. and Shoukry, A."A Multistage Algorithm for Fast Classification of Patterns," Pattern Recognition Letters, vol. 10, no. 4, 1989, pp. 211-215.
8. Kurzynski, M.W., "On the Identity of Optimal Strategies for Multistage Classifiers," Pattern Recognition Letters, vol. 10, no. 1, 1989, pp. 39-46.
9. Zhou, J.Y. and Pavlidis, T. "Discrimination of Characters by a Multi-Stage Recognition Process," Pattern Recognition, vol. 27, no. 11, 1994, pp. 1,539-1,549.
10. Hashem, S. and Schmeiser, B. "Improving Model Accuracy Using Optimal Linear Combinations of Trained Neural Networks," IEEE Trans. Neural

11. Networks, vol. 6, no. 3, 1995, pp. 792-794.
12. Lahiri T. and Samal S., "A novel technique for making multiple classifier based decision", Proc. WSEAS International Conference on MATHEMATICAL BIOLOGY and ECOLOGY, Corfu, Greece, August 17-19, 2004
13. Jain, A.K, Prabhakar, S. Hong, L. and Pankanti, S., 2000 "Filterbank-Based Fingerprint Matching" IEEE Transaction on Image Processing, Vol. 9, No. 5, May 2000

Hand Geometry Based Recognition with a MLP Classifier

Marcos Faundez-Zanuy[1], Miguel A. Ferrer-Ballester[2], Carlos M. Travieso-González[2], and Virginia Espinosa-Duro[1]

[1] Escola Universitària Politècnica de Mataró (UPC), Barcelona, Spain
{faundez, espinosa}@eupmt.es
http://www.eupmt.es/veu

[2] Dpto. de Señales y Comunicaciones, Universidad de Las Palmas de Gran Canaria, Campus de Tafira, E-35017, Las Palmas de Gran Canaria, Spain
{mferrer, ctravieso}@dsc.ulpgc.es
http://www.gpds.ulpgc.es

Abstract. This paper presents a biometric recognition system based on hand geometry. We describe a database specially collected for research purposes, which consists of 50 people and 10 different acquisitions of the right hand. This database can be freely downloaded. In addition, we describe a feature extraction procedure and we obtain experimental results using different classification strategies based on Multi Layer Perceptrons (MLP). We have evaluated identification rates and Detection Cost Function (DCF) values for verification applications. Experimental results reveal up to 100% identification and 0% DCF.

1 Introduction

In recent years, hand geometry has become a very popular biometric access control, which has captured almost a quarter of the physical access control market [1]. Even if the fingerprint [2], [3] is the most popular access system, the study of other biometric systems is interesting, because the vulnerability of a biometric system [4] can be improved using some kind of data fusion [5] between different biometric traits. This is a key point in order to popularize biometric systems [6], in addition to privacy issues [7].

Although some commercial systems rely on a three-dimensional profile of the hand, in this paper we study a system based on two dimensional profiles. Even though three dimensional devices provide more information than two dimensional ones, they require a more expensive and voluminous hardware.

A two-dimensional profile of a hand can be get using a simple document scanner, which can be purchased for less than 100 USD. Another possibility is the use of a digital camera, whose cost is being dramatically reduced in the last years.

In our system, we have decided to use a conventional scanner instead of a digital photo camera, because it is easier to operate, and cheaper. This paper can be summarized in three main parts: section two describes a database which has been specially acquired for this work. In section three, we describe the pre-processing and feature extraction. Section four provides experimental results on identification and verification rates using neural net classifiers. Finally, conclusions are summarized.

D. Zhang and A.K. Jain (Eds.): ICB 2006, LNCS 3832, pp. 721 – 727, 2005.

2 Database Description

Our database consists of 10 different acquisitions of 50 people, acquiring the right hand of each user. We have used a conventional document scanner, where the user can place the hand palm freely over the scanning surface; we do not use pegs, templates or any other annoying method for the users to capture their hands [8]. The images have been acquired with a typical desk-scanner using 8 bits per pixel (256 gray levels), and a resolution of 150dpi. To facilitate later computation, every scanned image has been scaled by a factor of 20%.

3 Feature Extraction

This stage can be split into the following steps: binarization, contour extraction, work out the geometric measurements, and finally the features are stored in a new reduced database (see figure 1).

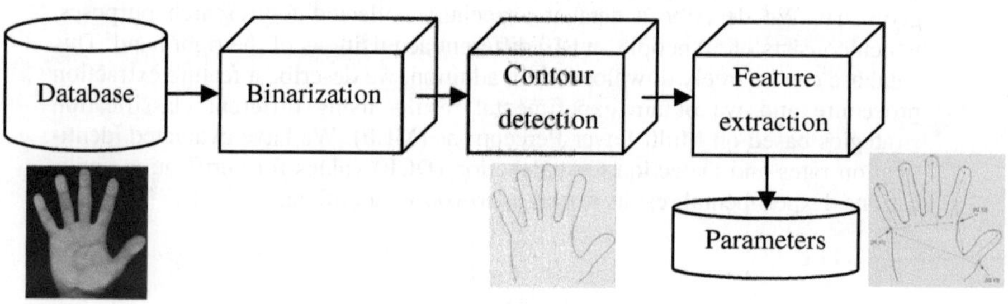

Fig. 1. Steps in the feature extraction process

Step 1. The goal of this step is the conversion from a 8 bit per pixel image to a monochrome image (1 bit per pixel). As the contrast between the image and the background is quite high, it reduces the complexity of the binarization process. After several experiments changing the threshold and evaluating the results with different images extracted from the database, we reach the conclusion that with a selected threshold of 0.25 the results were adequate for our purposes. We discard to use other binarization algorithm such as the suggested by Lloyd, Ridler-Calvar and Otsu [9] because the results are similar and the computational burden is higher.

Step 2. The goal is to find the limits between the hand and the background and obtain a numerical sequence describing the hand-palm shape. Contour following is a procedure by which we run through the hand silhouette by following the image's edge. We have implemented an algorithm, which is a modification of the method created by Sonka, Hlavac and Boyle [10].

Step3. Several intermediate steps have been performed to detect the main points of the hands from the image database. The method for the geometric hand-palm features extraction is quite straightforward. From the hand image, we locate the following main points: finger tips, valleys between the fingers and three more points that are necessary to define the hand geometry precisely. Finally, using all the main points previously computed, the geometric measurements are obtained. We take the eight following distances: Length of the 5 fingers, distances between points (X1, Y1) and (X2,Y2), points (X2,Y2) and the valley between the thumb and first finger and the points (X3,Y3) and (X1,Y1). Figure 1 shows the final results along with the geometric measurements taken into account.

4 Experimental Results

Biometric systems can be operated in two ways:

- Identification: In this approach no identity is claimed from the person. The automatic system must determine who is trying to access.
- Verification: In this approach the goal of the system is to determine whether the person is who he/she claims to be. This implies that the user must provide an identity and the system just accepts or rejects the users according to a successful or unsuccessful verification. Sometimes this operation mode is named authentication or detection.

For identification, if we have a population of N different people, and a labeled test set, we just need to count the number of identities correctly assigned.

Verification systems can be evaluated using the False Acceptance Rate (FAR, those situations where an impostor is accepted) and the False Rejection Rate (FRR, those situations where a user is incorrectly rejected), also known in detection theory as False Alarm and Miss, respectively. There is trade-off between both errors, which has to be usually established by adjusting a decision threshold. The performance can be plotted in a ROC (Receiver Operator Characteristic) or in a DET (Detection error trade-off) plot [11]. DET plot uses a logarithmic scale that expands the extreme parts of the curve, which are the parts that give the most information about the system performance.

In order to summarize the performance of a given system with a single number, we have used the minimum value of the Detection Cost Function (DCF). This parameter is defined as [11]:

$$DCF = C_{miss} \times P_{miss} \times P_{true} + C_{fa} \times P_{fa} \times P_{false} \qquad (1)$$

where C_{miss} is the cost of a miss (rejection), C_{fa} is the cost of a false alarm (acceptance), P_{true} is the a priori probability of the target, and $P_{false} = 1 - P_{true}$. We have used $C_{miss} = C_{fa} = 1$.

Multi-Layer Perceptron classifier trained in a discriminative mode. We have trained a Multi-Layer Perceptron (MLP) [12] as discriminative classifier in the following fashion: when the input data belongs to a genuine person, the output (target of the NNET) is fixed to 1. When the input is an impostor person, the output is fixed to –1. We have used a MLP with 40 neurons in the hidden layer, trained with gradient descent algorithm with momentum and weight/bias learning function. We have trained the neural network for 2500 and 10000 epochs using regularization. We also apply a multi-start algorithm and we provide the mean, standard deviation, and best obtained result for 50 random different initializations. The input signal has been fitted to a [–1, 1] range in each component.

Error correction codes. Error-control coding techniques [13] detect and possibly correct errors that occur when messages are transmitted in a digital communication system. To accomplish this, the encoder transmits not only the information symbols, but also one or more redundant symbols. The decoder uses the redundant symbols to detect and possibly correct whatever errors occurred during transmission.

Block coding is a special case of error-control coding. Block coding techniques map a fixed number of message symbols to a fixed number of code symbols. A block coder treats each block of data independently and is a memoryless device. The information to be encoded consists of a sequence of message symbols and the code that is produced consists of a sequence of codewords. Each block of k message symbols is encoded into a codeword that consists of n symbols; in this context, k is called the message length, n is called the codeword length, and the code is called an [n, k] code.

A message for an [n, k] BCH (Bose-Chaudhuri-Hocquenghem) code must be a k-column binary Galois array. The code that corresponds to that message is an n-column binary Galois array. Each row of these Galois arrays represents one word.

BCH codes use special values of n and k:

- n, the codeword length, is an integer of the form 2m–1 for some integer m > 2.
- k, the message length, is a positive integer less than n.

However, only some positive integers less than n are valid choices for k. This code can correct all combinations of t or fewer errors, and the minimum distance between codes is:

$$d_{\min} \geq 2t+1 \tag{2}$$

Table 2 shows some examples of suitable values for BCH codes,

Table 1. Examples of values for BCH codes

n	7	5			31				
k	4	11	7	5	26	21	16	11	6
t	1	1	2	3	1	2	3	5	7

Multi-class learning problems via error-correction output codes. Multi-class learning problems involve finding a definition for an unknown function $f(\vec{x})$ whose range is a discrete set containing $k > 2$ values (i.e. k classes), and $\vec{x}$ is the set of measurements that we want to classify. We must solve the problem of learning a k-ary classification function $f: \Re^n \rightarrow \{1,\cdots,k\}$ from examples of the form $\{\vec{x}_i, f(\vec{x}_i)\}$. The standard neural network approach to this problem is to construct a 3-layer feed-forward network with k output units, where each output unit designates one of the k classes. During training, the output units are clamped to 0.0, except for the unit corresponding to the desired class, which is clamped at 1.0. During classification, a new $\vec{x}$ value is assigned to the class whose output unit has the highest activation. This approach is called [14], [15], [16] the *one-per-class* approach, since one binary output function is learnt for each class.

Experimental Results. We use a Multi-layer perceptron with 10 inputs, and h hidden neurons, both of them with *tansig* nonlinear transfer function. This function is symmetrical around the origin. Thus, we modify the output codes replacing each "0" by "–1". In addition, we normalize the input vectors $\vec{x}$ for zero mean and maximum modulus equal to 1. The computation of Mean Square Error (MSE), and Mean Absolute Difference (MAD) between the obtained output and each of the codewords provides a distance measure.

We have converted this measure into a similarity measure computing (1 – distance). We will summarize the Multi-Layer Perceptron number of neurons in each layer using the following nomenclature: *inputs*× *hidden*× *output*. In our experiments, the number of inputs is fixed to 10, and the other parameters can vary according to the selected strategy.

We have evaluated the following strategies (each one has been tested with 5 and 3 hands for training, and the remaining ones for testing):

- One-per-class: 1 MLP 10×40×50 (table 2)
- Natural binary code: 1 MLP 10×40×6 (table 3)
- Error Correction Output Code (ECOC) using BCH (15, 7) (table 4) and BCH (31, 6) (table 5).
- Error Correction Output Code (ECOC) using random generation (table 6).

Table 2. 1 MLP 10×40×50 (one-per-class)

	Train=5 hands, test=5 hands						Train=3 hands, test=7 hands					
	Identif. rate (%)			Min(DCF) (%)			Identif. rate (%)			Min(DCF) (%)		
Epoch	mean	σ	max	mean	σ	min	mean	σ	max	mean	σ	min
2500	98.30	0.39	98.80	0.69	0.2	0.34	97.71	0.70	99.14	0.86	0.2	0.50
10000	98.23	0.47	99.2	0.67	0.16	0.37	97.79	0.64	98.57	0.86	0.18	0.53

Table 3. 1 MLP 10×40×6 (Natural binary code)

		Train=5 hands, test=5 hands						Train=3 hands, test=7 hands					
		Identif. rate (%)			Min(DCF) (%)			Identif. rate (%)			Min(DCF) (%)		
	Ep-och	mean	σ	max	mean	σ	min	mean	σ	max	mean	σ	min
MAD	2500	95.97	1.25	98.4	3.94	0.52	2.74	92.43	1.49	96.57	5.70	0.45	4.79
	10000	96.42	1	98.4	3.80	0.49	2.77	92.66	1.17	95.14	5.58	0.39	4.81
MSE	2500	95.97	1.25	98.4	0.88	0.3	0.38	92.43	1.49	96.57	2.60	0.41	1.87
	1000	96.42	1	98.4	0.83	0.3	0.29	92.66	1.17	95.14	2.53	0.42	1.60

Table 4. 1 MLP 10×40×50 (ECOC BCH (31, 6))

		Train=5 hands, test=5 hands						Train=3 hands, test=7 hands					
		Identif. rate (%)			Min(DCF) (%)			Identif. rate (%)			Min(DCF) (%)		
	Epoch	mean	σ	max	mean	σ	min	mean	σ	max	mean	σ	min
MAD	2500	99.58	0.15	100	0.04	0.05	0	98.54	0.60	99.71	0.49	0.21	0.12
	10000	99.62	0.21	100	0.03	0.04	0	98.50	0.60	99.43	0.45	0.18	0.11
MSE	2500	99.58	0.15	100	0.03	0.04	0	98.59	0.57	99.71	0.46	0.20	0.06
	10000	99.62	0.22	100	0.02	0.03	0	98.53	0.59	99.43	0.43	0.17	0.11

Table 5. 1 MLP 10×40×14 (ECOC BCH (15, 7))

		Train=5 hands, test=5 hands						Train=3 hands, test=7 hands					
		Identif. rate (%)			Min(DCF) (%)			Identif. rate (%)			Min(DCF) (%)		
	Epoch	mean	σ	max	mean	σ	min	mean	σ	max	mean	σ	min
MAD	2500	99.58	0.15	100	0.04	0.05	0	98.06	0.58	99.43	0.94	0.26	0.49
MAD	10000	99.62	0.21	100	0.03	0.04	0	98.30	0.58	99.43	0.85	0.26	0.44
MSE	2500	99.58	0.15	100	0.03	0.04	0	98.07	0.58	99.14	0.47	0.18	0.18
MSE	10000	99.61	0.22	100	0.02	0.03	0	98.35	0.61	99.43	0.39	0.19	0.06

Table 6. 1 MLP 10×40×50 (random ECOC generation)

		Train=5 hands, test=5 hands						Train=3 hands, test=7 hands					
		Identif. rate (%)			Min(DCF) (%)			Identif. rate (%)			Min(DCF) (%)		
	Epoch	mean	σ	max	mean	σ	min	mean	σ	max	mean	σ	min
MAD	2500	99.50	0.23	100	0.26	0.12	0.01	98.09	0.78	99.71	1.22	0.38	0.41
MAD	10000	99.58	0.23	100	0.23	0.12	0	98.30	0.70	99.71	1.10	0.31	0.60
MSE	2500	99.50	0.21	100	0.13	0.01	0.004	98.14	0.78	99.71	0.85	0.33	0.22
MSE	10000	99.58	0.23	100	0.09	0.09	0	98.32	0.67	99.71	0.71	0.32	0.09

5 Conclusions

Taking into account the experimental results, we observe the following conclusions:

- Comparing tables 3 and 4, we observe better performance using the one-per-class approach. We think that is due to the larger number of weights when using the first strategy, which lets to obtain a better classifier. Additionally, we can interpret that the larger hamming distance of one-per-class approach lets to improve the results.
- ECOC lets more flexibility with the MLP architecture, because it has a wide range of possibilities for the number of outputs, given a set of users. In addition, experimental results outperform the one-per-class approach. Comparing tables 5 and 6 we see similar performance. Thus, we prefer BCH (15, 7) because it is simpler.
- Although it is supposed that random generation for ECOC should outperform BCH codes, our experimental results reveal better performance when using the latest ones.
- Our results offers better efficacy than other works with similar database size [17-18].

Acknowledgement

This work has been partially funded by FEDER and MCYT TIC2003-08382-C05-02.

References

1. Jain A. K., Bolle R., Pankanti S., "Introduction to biometrics" in Biometrics Personal identification in networked society. Kluwer Academic Publishers 1999
2. Faundez-Zanuy, M., "Door-opening system using a low-cost fingerprint scanner and a PC" IEEE Aerospace and Electronic Systems Magazine. Vol. 19 nº 8, pp.23-26, August 2004.
3. Faundez-Zanuy M., Fabregas J. "Testing report of a fingerprint-based door-opening system". IEEE Aerospace and Electronic Systems Magazine. Vol.20 nº 6, pp 18-20, June 2005
4. Faundez-Zanuy, M., "On the vulnerability of biometric security systems". IEEE Aerospace and Electronic Systems Magazine. Vol.19 nº 6, pp.3-8, June 2004.
5. Faundez-Zanuy M., "Data fusion in biometrics" IEEE Aerospace and Electronic Systems Magazine. Vol.20 nº 1, pp.34-38, January 2005.
6. Faundez-Zanuy, M., "Biometric recognition: why not massively adopted yet?" IEEE Aerospace and Electronic Systems Magazine. Vol.20 nº 9, pp.1-4 September 2005
7. Faundez-Zanuy, M., "Privacy issues on biometric systems". IEEE Aerospace and Electronic Systems Magazine. Vol.20 nº 2, pp13-15. February 2005
8. Carlos M Travieso-González, J. B. Alonso, S. David, Miguel A. Ferrer-Ballester, "Optimization of a biometric system identification by hand geometry" Complex systems intelligence and modern technological applications, Cherbourg, France, pp. 581-586, 19-22, September 2004.
9. L. O'Gorman and R. Kasturi, Document Image Analysis, IEEE Computer Society Press, 1995.
10. Milan Sonka, Vaclav Hlavac, Roger Boyle, Image Processing, Analysis and Machine Vision. 2nd edition, 30 September 1998.
11. Martin A. et alt. "The DET curve in assessment of detection performance", V. 4, pp.1895-1898, European speech Processing Conference Eurospeech 1997
12. Haykin S., "Neural nets. A comprehensive foundation", 2on edition. Ed. Prentice Hall 1999
13. Wicker, Stephen B., Error Control Systems for Digital Communication and Storage, Upper Saddle River, N.J., Prentice Hall, 1995.
14. Dietterich T. G., Bakiri G., "Error-correcting output codes: A general method for improving multiclass inductive learning programs". Proceedings of the Ninth National Conference on Artificial Intelligence (AAAI-91). Anaheim, CA: AAAI Press.
15. Dietterich T., "Do Hidden Units Implement Error-Correcting Codes?" Technical report 1991
16. Kuncheva, L. I. "Combining pattern classifiers". Ed. John Wiley & Sons 2004.
17. YingLiang Ma; Pollick, F.; Hewitt, W.T, "Using B-spline curves for hand recognition" Proceedings of the 17th International Conference on Pattern Recognition, Vol. 3, pp. 274 – 277, Aug. 2004.
18. R. Sanchez-Reillo, C. Sanchez-Avila, and A. Gonzalez-Marcos, "Biometric Identification Through Hand Geometry Measurements", IEEE Transactions on Pattern Analysis an Machine Intelligence, 22(10), pp 1168-1171, 2000.

A False Rejection Oriented Threat Model for the Design of Biometric Authentication Systems

Ileana Buhan, Asker Bazen, Pieter Hartel, and Raymond Veldhuis

University of Twente, Faculty of Electrical Engineering,
PO 217, 7500AE Enschede, The Netherlands

Abstract. For applications like Terrorist Watch Lists and Smart Guns, a false rejection is more critical than a false acceptance. In this paper a new threat model focusing on false rejections is presented, and the "standard" architecture of a biometric system is extended by adding components like crypto, audit logging, power, and environment to increase the analytic power of the threat model. Our threat model gives new insight into false rejection attacks, emphasizing the role of an external attacker. The threat model is intended to be used during the design of a system.

1 Introduction

Biometric authentication systems are used to identify people, or to verify the claimed identity of registered users when entering a protected perimeter. Typical application domains include air-and seaports, banks, military installations, etc. For most of these systems the main threat is an authorized user gaining access to the system. This is called a *false acceptance* threat. Currently, new applications that have a completely different threat model are emerging. For example, *Terrorist Watch List* applications and *Smart Guns* applications are characterized by the fact that a false rejection could lead to life threatening situations. Terrorist watch list applications currently use facial recognition or fingerprint recognition [1]. Watch lists are mainly used in ports to identify terrorists. For this application, the main threat is a *false rejection* which means that a potential terrorist on the list is not recognized. A *false acceptance* results in a convenience problem, since legitimate subjects are denied access and their identity needs to be examined more carefully to get access.

Smart guns are weapons that will fire only when operated by the rightful owner. Such guns are intended to reduce casualties among police officers whose guns are taken during a struggle. The most promising biometric for this application is grip pattern recognition [15]. Again, a *false rejection* is the most serious threat as this would result in a police officer not being able to use the weapon when necessary. For a police officer to trust his gun the *false reject rate* must be below 10^{-4}, which is the accepted failure rate for police weapons in use.

We propose 3W trees (Who, hoW, What) for identifying false rejection threats to biometric security systems. Analysis based on a 3W tree leads to concrete questions regarding the security of the system. Questions raised by other methods (e.g. attack trees) do not lead to the same level of specific questions. A similar approach is taken

D. Zhang and A.K. Jain (Eds.): ICB 2006, LNCS 3832, pp. 728–736, 2005.

by de Cock et al. in [3], when modeling threats for security tokens in web applications. Our method is more concrete than other methods because we make explicit assumptions about the generic architecture of the system, thus exposing all main components in the architecture that are vulnerable to attack. Our method is not less general than other methods because other architectural assumptions can be plugged in easily. Our method is intended to be used as a design aid.

Section 2 is an overview of points of vulnerability in biometric authentication systems. The extended architecture of a biometric authentication system is presented in Section 3. Section 4 describes 3W trees the method proposed for identifying *false rejection* attacks and in Section 5 we apply this 3W tree to the *Terrorist Watch List* and to the *Smart Gun*. The last section concludes and suggests further work.

2 Related Work

Like all security systems, biometric systems are vulnerable to attacks [7, 12]. One specific attack consists of presenting fake inputs such as false fingerprints [4] to a biometric system. To analyze such threats systematically various threat models have been developed. We discuss the most important models: the Biometric Device Protection Profile (BDPP) [6], the Department of Defense & Federal Biometric System Protection Profile for Medium Robustness Environments (DoDPP) [8], the U.S. Government Biometric Verification Mode Protection Profile for Medium Robustness Environments (USGovPP) [10] and Information Technology-Security techniques -A Framework for Evaluation and Testing of Biometric Technology (ITSstand) [5]. In the sequel we refer to these three protection profiles and the ITSstand simply as "the standards".

In many ways, the standards are similar. In particular, they do not make a clear distinction between a *false rejection* and a *false acceptance* attack. A total of 48 distinct threats are identified of which only 3 are *false rejection* threats. These are: (1) cutting the power to the system, (2) flooding hardware components with noise and (3) exposing the device to environmental parameters that are outside its operating range. In addition, there are 12 "catch all" threats with both *false rejection* and *false acceptance* threats.

It is difficult to compare threats amongst the four standards. For example, *BDPP* contains one T.TAMPER threat while *ITSstand* contains three tamper related threats: one for hardware tampering another for software or firmware tampering and one for channel tampering . In *ITSstand* tampering and bypassing is mentioned when describing the same threat while BDPP explicitly mentions the T.BYPASS threat.

ITSstand is the most complete in identifying *false rejection* threats, it identifies the largest number (8) of such rejections (See [5] [threats 8.4, 10.2, 11.2, 13.1, 13.3, 14.1, 14.3, 15.1]). However, only threat 13.3 is a clear false rejection. All the others are "catch all" threats. There are three tamper related threats: one related to hardware tampering (13.1), one related to software tampering (14.1) and one for channel tampering (15.1). These threats are general, not specifying the exact point in the system that is vulnerable, or the circumstances that make the system vulnerable to attack. The method of attack is also not clear, all that is said is that hardware can be tampered with, bypassed or deactivated. These threats lack the exact how and where. The key idea of our 3W tree is that it provides the missing how and where to the analyst.

Attack trees offer a related method of analyzing attacks [14]. The root of the tree is identified with the goal of compromising a system. The goals of the children of a node could be the compromise of a sub-system or a contribution thereof, and so on recursively. The main disadvantage of attack trees is that they provide only the choice between and-/or-nodes. This does only provides a low level way of breaking up a goal up into sub-goals. The general recommendation is to think hard, which does not provide much guidance.

Bolle et al. [13] identifies 9 threats that plague biometric systems. Their opinion is that many questions about how to make biometric authentication work without creating additional security loopholes remain unanswered and that little work is being done presently in this area. Our paper contributes to filling this gap.

3 Biometric Authentication Generic System Architecture

Ratha et al. [12] provide a systematic analysis of different points of attack in a biometric authentication system. Their analysis is based on a generic architecture of a biometric system, as illustrated in Fig. 1.

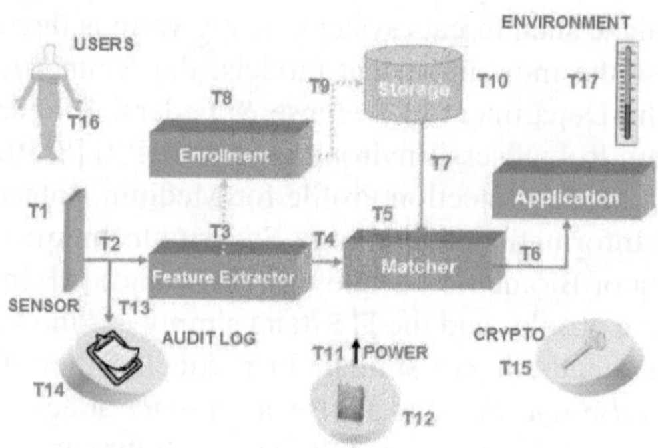

Fig. 1. General view of a Biometric Authentication System showing 17 points of attack

Each of the components as well as the connecting channels are potential targets of attack. Comparing these targets of attack to the threats identified in the standards we discovered some threats that do not have a corresponding target of attack in the architecture. For example in the architecture nothing is mentioned about the power that makes the electric equipment work. Cutting the power to the system will make the system fail. Therefore, we extend the generic biometric architecture to include the following components also shown in *figure* 1:

(a) *Cryptography,* for ensuring the authenticity and integrity of data stored and transmitted on channels. The standards identify threats related to cryptography as follows: T.CRYPT_ATTK in *DoDPP*, T.CRYPT_ATTACK and T.CRYPTO_COMPROMISE in *USGovPP*.

(b) *Audit,* important actions need to be recorded for later analysis. In the case of the *Smart Gun* application it is particularly important to have a record of which user fired the gun at what time. The auditing process itself can be subject to an attack for example T.AUDIT_COMPROMISE, *DoDPP*.

(c) *Power,* is a major concern especially when the biometric device is portable. For example, replacing the power source might restart the application causing the biometric system to enter an unknown or unstable state. This attack is related to threat T.POWER in *BDPP*, *DoDPP*, *ITSstand*, and T.UNKOWNSTATE in *USGovPP*.
(d) *Environment and users,* this is general but we also include in this category: operating parameters such as temperature, humidity, etc. Threats related to users identified in the standards are T.BADUSER, T.BADADMIN, T.BADOPER in *BDPP* and *DoDPP* (T.BADOPER is not present in that document), USGovPP does not contain T.BADUSER and T.BADOPER but it contains two threats related to a bad administrator, namely T.ADMIN_ERROR and T.ADMIN_ROGUE and in *ITSstand* they are labeled as: 8.1, 8.2, 8.3 and 8.4. Other threats are T.FAILSECURE, T.DEGRADE presented in *DoDPP*.

This concludes the extension of the architecture of Ratha et al. [13], by adding 7 components that could influence the performance and security of a biometric system.

4 3W Trees

The attack classifications from the standards are too coarse. For example threat T.UNDETECT in *BDPP* says: *An undetected attack against the TOE security functions is mounted by an attacker, which eventually succeeds in either allowing illegal access to the portal, or denying access to authorized users.* Nothing is said about the type of attack except that it is undetected and that the result can be either a false acceptance or a false rejection. To solve this problem we propose a more detailed analysis using 3W trees to give concrete insights in potential attacks, without burdening the analyst with irrelevant detail.

Three relevant grounds of distinctions are identified in the general security taxonomies in the literature, namely the *who*, the *how* and the *what.* We use each of these grounds of distinction at different levels of the 3W tree (*Fig.* 2).

The first level of the 3W tree is a classical *who* taxonomy from the attacker's position relative to the system [9]. Attackers are divided in three classes. *Class I* attackers or *external* attackers, lack knowledge about the system and have moderately sophisticated

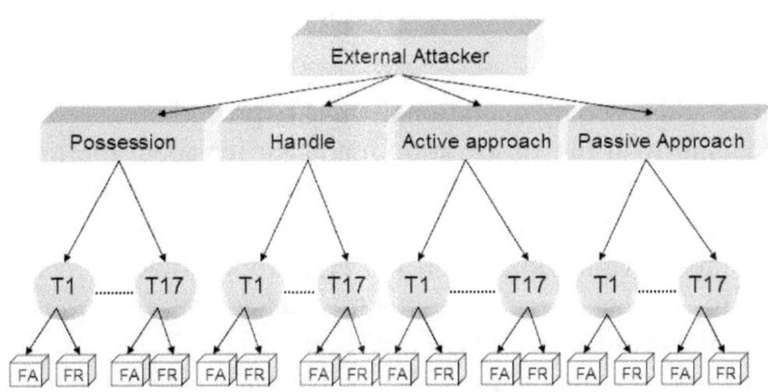

Fig. 2. 3W tree of attacks on biometric systems. T1-T17 are points of attack shown in Fig. 1.

equipment. *Class II* attackers or *internal* attackers are knowledgeable insiders, which are highly educated and have access to most parts of the system. *Class III* attackers are funded organization with ample resources that are able to assemble teams and design sophisticated attacks. It is widely acknowledged that there is no protection against class III attackers. The general opinion is that a system is considered secure if it can withstand class I and class II attackers.

As a second level the 3W tree we use the Rae and Wildman taxonomy for secure devices [11]. This is a *how* taxonomy:

- *passive approach*, the attacker may be in the proximity of the device, but cannot touch the device;
- *active approach*, the attacker can interfere with the device (e.g. over a network) and transmit data to the device from either an insecure or a secure domain.
- *handles* the device physically, but cannot break tamper evident seals on the device;
- *possesses* the device i.e. can open the device and break tamper evident seals with impunity;

The classes presented are related to one another. *Possessing* the device means that the attacker can *handle* the device and of course may *approach* the device. This relationship can be formalized as :

$$\textit{passive approach} \subset \textit{active approach} \subset \textit{handle} \subset \textit{possession}$$

The third level of the 3W tree , the *what*, deals with the threats our system might be subject to. For a description of the first 10 attacks T1-T10 we refer the reader to the Bolle et al. [13]. In addition to threats T1-T10 of Bolle et al. [13] we identify threats T11-T17:

T11. The channel that links the power source to the system is destroyed.
T12. The power source of the system is tampered with.
T13. An attacker may prevent future audit records from being recorded by attacking the channel that transports the audit information.
T14. Audit records may be deleted or modified, thus masking an intruder action.
T15. Security functions may be defeated through cryptanalysis on encrypted data, i.e. compromise of the cryptographic mechanisms.
T16. Users, regardless of the role that they play in the system, can compromise the security functions.
T17. The environment (temperature, humidity, lighting, etc.) and extensive usage can degrade the security function of the system

In our opinion, threats T1-T13 should be addressed by security mechanisms and threats T14-T17 should be addressed by operational security procedures.

Finally, in keeping with our observation made earlier about the increasing importance of studying *false rejections* we add as a fourth layer the distinction between *false acceptance* and *false rejection*. What makes our layered taxonomy biometric specific is that: (1) the points of vulnerability T1-T17 refer to a biometric system and (2) we consider two specific effects of each attack: a *false acceptance* or a *false rejection*.

This concludes the presentation of the 3W tree for identifying attacks on a general biometric authentication system in the design phase, which allows us to classify known attacks and to identify the possibility of new attacks in a systematic manner. This is the subject of the next section.

5 External Attack Scenarios

A scenario is a path in the 3W tree of *figure* 2. A scenario is named as xiy where:

- $x \in \{PA, AA, HA, PO\}$, PA stands for *passive approach*, AA stands for *active approach*, HA stands for *handle* and PO for *possession*.
- $i \in \{1..17\}$ indicates threat Ti.
- $y \in \{A, R\}$, where A means an attack leading to a *false acceptance* attack and R means an attack leading to a *false rejection* attack.

Each path in the tree corresponds to a threat that has to be evaluated. For example, scenario PO1A identifies the following: in the possession situation (denoted by the letters PO), threat $T1$ (presenting a fake biometric/tampering with the sensor) to obtain a false acceptance (A). To describe and evaluate scenarios we use the following attributes:

I *Scenario:* name of the evaluated scenario.
I *Tactics:* describe a possibility to realize this attack.
I *Name:* the name of the attack in the literature or a link to a paper that describes this attack (if known).
II *Damage:* the estimated consequence of the attack for the device. The possibilities are: *minor*, *moderate*, *major*. An attack with *minor* consequences will temporarily damage the device. A *moderate* consequence attack will temporarily damage the device but it needs specialized personnel to repair it. An attack with *major* consequence will completely ruin the device, and the whole or parts of it need to be replaced.
II *Knowledge:* lists the knowledge that an intruder must have to launch the attack. The categories are: *common sense, high school education, expert.*
II *Occurrence:* an educated guess of the probability that such an attack occurs. The estimators are: *low* (unlikely to have such an attack), *medium* (it might happen), *high* (likely to happen).
III *Countermeasures:* some notes on how this attack might be prevented, or how at least to diminish its consequence.

Below we present two examples, showing that analysis based on the 3W tree leads to asking relevant questions about threats on biometric authentication systems. In the Technical Report version of this paper all $4 \times 17 = 68$ threats are analyzed [2]. From 68 possible threats, 13 are considered serious threats. From these 13 threats, 6 have are likely to occur and 12 have major consequences for the integrity of the device.

Example 1: Smart Gun Significant numbers of police weapons are lost or stolen. Each year several police officers die or are injured because their own weapons are used against them. The Smart Gun application is designed for a police force, which would like to render a weapon inoperative when it is captured by the assailant of a police officer. The requirements include that a gun should recognize all members of a police patrol, and that wearing gloves should not affect the operation. The PO4R attack, shown in Table 1 is a tamper attack. All standards mention tamper attacks but do not detail the point in the system where the tampering might occur. However, a tamper attack is relatively easy to perform and the consequences are high: the gun is not working. By

Table 1. *PO4R Scenario in the Smart Gun application*

I. Scenario	Can an attacker in the *possession* situation attack the communication channel between the feature extractor and the matcher in order to produce a *false rejection*?
I. Tactics	Physically breaking the channel is the most obvious choice. To destroy wires/connections inside the electronic device we have the following possibilities: exposing the object to extreme values of pressure, temperature etc. and at some point the mechanical connections will break.
I. Name	Physical tampering.
II. Damage	High. If the template extractor is out of order the gun will not work correctly.
II. Knowledge	Expert. The attacker must know how to open the gun and which device is the template extractor and then reassemble the gun.
II. Occurrence	Medium. The result of such an attack is a gun that is not working properly in the hands of the rightful user. If he wants to harm the user there are other ways in which he has more control over what is happening. (i.e pulling a knife)
III. Counter measures	A seal on the gun handle seems to be most appropriate. The seal must ensure that even if the attacker can open the gun, resealing the device would be easily detectable. It should be possible to discover the details of such an attack from an audit log.

pointing out the specific points of attack, our analysis, suggests that a seal is needed on the gun handle where the electronics are located. A tamper evident seal would indicate the police officer whether the integrity of the weapon has been violated.

Example 2: Terrorist Watch Lists are used to detect terrorists while traveling. Applications like this are usually installed at airports, seaports, main railway stations etc. Peo-

Table 2. *AA1R Scenario in Terrorist Watch List Application*

I. Scenario	Can an *active attacker* produce a false rejection by tampering with the input device (video camera)?
I. Tactics	An active attacker can interfere with the camera using mirrors to reflect sun light on the camera, affecting the quality of the image. The similarity between the newly acquired sample and stored biometric sample might then be below the threshold.
I. Name	Unknown.
II. Damage	Minor. The personnel in charge of supervising the cameras will eventually notice that something is wrong.
II. Knowledge	Common sense. Children play with watches projecting light on surfaces to annoy their teachers.
II. Occurrence	High. It is easy to perform such an attack from a safe distance. No special tools are required.
III. Counter-measures	To ensure that light beams cannot be projected on the camera. This can be done by carefully positioning the camera, detecting changes in lighting conditions,etc..

ple who want to travel are checked against a central database with potentially dangerous persons. There are at least two ways to do the matching: using the name (which can easily be forged) or a biometric feature like face or fingerprint. We consider the case where the terrorist watch list is implemented using face recognition. The intended use is as follows: a camera is placed at a passport control point and before issuing the stamp the person is asked to look at the camera using a neutral expression. The officer in charge will check if the individual is acting as asked. We show that attacking the camera following an *active approach* is feasible, see table 2. We could not find any mention of this attack in the literature. Again, our 3W tree helps to ask the right question during the analysis.

6 Conclusions

Existing biometric protection profiles and standards by and large define the same set of attacks. However, their focus is mainly on *false acceptance* attacks. Attacks that result in a false acceptance or false rejection are often put in the same class. Threats that could only lead to a *false rejection* are largely ignored.

In new applications like *Terrorist Watch Lists* or *Smart Guns*, *false rejection* attacks are more important than *false acceptance* attacks. We propose 3W trees as a flexible tool to highlight *false rejection* or *false acceptance* attacks depending on the type of application. Our threat model gives new insight into false rejection attacks emphasizing the role of an external attacker.

The advantage of the 3W tree is that (1) its fosters a systematic approach to threat analysis, (2) allows asking concrete questions, and (3) does not burden the analysis with irrelevant detail. Analyzing a 3W tree helps us to develop scenarios. For evaluating and describing scenarios we propose a model consisting of: *tactics, name, consequence, estimated knowledge, estimated probability, countermeasure.* In two detailed examples we identify appropriate countermeasures to attacks. For the smart gun example we argue that there must be a seal on the gun handle to protect the electronics inside the gun. For the terrorist watch list we argue that the camera should be positioned in a way that would prevent a light beam to be reflected on the camera. The main advantage of the 3W tree is that relevant threats are identified.

This research is supported by Technology Foundation STW. We thank Jeroen Doumen and Ruud van Munster for their comments on the paper.

References

1. J. M. Bone and D. M. Blackburn. Biometrics for narcoterrorist watch list applications. Technical report, Crane Division, Naval Surface Warfare Center and DoD Counterdrug Technology Development Program Office, July 2003.
2. I. Buhan and P. Hartel. The state of the art in abuse of biometrics. Technical report to appear, Centre for Telematics and Information Technology, Univ. of Twente, The Netherlands, June 2005.
3. D. De Cock, K. Wouters, D. Schellekens, D. Singelee, and B. Preneel. Threat modelling for security tokens in web applications. In D. Chadwick and B. Preneel, editors, *8th IFIP TC-6 TC-11 Conference on Communications and Multimedia Security*, pages 131–144, Lake Windermere, England, Sep 2004. Springer-Verlag, Berlin.

4. T. Van der Putte and J. Keuning. Biometrical fingerprint recognition: Don't get your fingers burned. *Smart Card Research and Advanced Applications, IFIPTC8/W68.8 Fourth Working Conference on Smart Card Reserch and Advanced Applications*, pages 289–303, Sep 2001.
5. Germany DIN-Deutsches Institut Fur Normung E.V., Berlin. Information technology - security techniques - a framework for security evaluation and testing of biometric technology. Technical Report ISO/IEC JTC 1/SC 27 N 3806, DIN - Deutsches Institut fur Normung e.V. Berlin, Germany, 2003.
6. UK Government Biometrics Working Group. Biometric device protection profile (BDPP). Technical Report Draft Issue 0.82, UK Goverment Biometrics Working Group, 2001.
7. A. K. Jain, S. Pankanti, S. Prabhakar, A. Ross, and J.L. Wayman. Biometrics: A grand challenge. *Proceedings of International Conference on Pattern Recognition*, Volume 2:935–942, 2004.
8. A. Kong, A. Griffith, D. Rhude, G. Bacon, and G. Shahs. Department of defense federal biometric system protection profile for medium robustness environments. Technical Report Technical Report Draft Version 0.02, U.S Department of Defense, 2002.
9. P.G. Neuman and D.B. Parker. A summary of computer misuse techniques. *12th National Computer Security Conference, Baltimor, MaryLand*, pages 396–407, 10-13 October 1989.
10. The Biometrics Management Office and National Security Agency. U.s. government biometric verification mode protection profile for medium robustness environments. Technical Report Version 1.0, The Biometrics Management Office and the National Security Agency, 2003.
11. A.J. Rae and L.P. Wildman. A taxonomy of attacks on secure devices. *Australian Information Warfare and IT Security, 20-21 November 2003, Australia*, pages 251–264, 2003.
12. N.K. Ratha, J.H. Connell, and R.M. Bolle. Biometrics break-ins and band-aids. *Pattern Recognition Letters*, 24(13):2105–2113, Sep 2003.
13. R.M.Bolle, J.H. Connel, S. Pankanti, N.K.Ratha, and A.W. Senior. *Guide to Biometrics*. Springer-Verlag, 175, Fifth Avenue, New York ,NY 10010, USA, 2004.
14. B. Schneier. Attack trees: Modeling security threats. *Dr. Dobb's Journal [on-line: www.ddj.com]*, 1999.
15. R.N.J. Veldhuis, A. M. Bazen, J. Kauffman, and P. H. Hartel. Biometric verification based on grip-pattern recognition (invited paper). In E. J. Delp III and P. W. Wong, editors, *IS&T/SPIE 16th Annual Symp. on Electronic Imaging - Security, Steganography, and Watermarking of Multimedia Contents*, volume 5306, pages 634–641, San Jose, California, Jan 2004. SPIE – The Int. Society for Optical Engineering, Washington.

A Bimodal Palmprint Verification System

Tai-Kia Tan[1], Cheng-Leong Ng[1], Kar-Ann Toh[2],
How-Lung Eng[2], Wei-Yun Yau[2], and Dipti Srinivasan[1]

[1] Dept. of Electrical & Computer Engineering,
National University of Singapore, Singapore 117576
dipti@nus.edu.sg
[2] Institute for Infocomm Research,
21 Heng Mui Keng Terrace, Singapore 119613
katoh@ieee.org, {hleng, wyyau}@i2r.a-star.edu.sg

Abstract. Hand-based biometrics such as fingerprint and palmprint had been widely accepted because of their convenience and ease in usage without intruding much to one's privacy such as face. The aim of this work is to develop a new point-based algorithm for palmprint feature extraction and perform reliable verification based on the extracted features. This point-based recognition system is next used as part of a bimodal palmprint recognition system combining with a DCT-based (Discrete Cosine Transform) algorithm for identity verification. The performance of the integrated system is evaluated using physical palmprint images.

Keywords: Biometrics, Palmprint Recognition, Multimodal Biometrics, and Identity Verification.

1 Introduction

Most of the literature reported on palmprint recognition had been based on the use global analysis such as Gabor Filters [1], Discrete Wavelet Transform [2] or global texture energy [3]. Principle lines and wrinkles obtained from edge detectors had also been used directly in some recognition systems [4]. Point based approach for palmprint recognition, however, had not been extensively explored except for [5] where paper palmprints were scanned into computer for processing. Main difference between our method and that in [5] is that we use RGB palm images directly captured from a low cost color CCD camera with VGA resolution whereas in [5] a specially designed handprint box was used together with a 200dpi scanner. A simple and yet efficient point-based system is also defined in this paper for palmprint verification. This point-based recognition system will then be combined with a DCT-based method to form a bimodal verification system.

Main contributions are summarized as follows: (i) proposal of a new point-based method for palmprint verification, (ii) proposal of a bimodal palmprint verification system incorporating the point-based method and a DCT-based method. Some preliminary experiments are reported to show viability of the system.

D. Zhang and A.K. Jain (Eds.): ICB 2006, LNCS 3832, pp. 737–743, 2005.

2 System Overview

A low cost 24-bit colour CCD camera with 768 × 576 resolution was used to capture the frontal palm images. The camera was mounted on a customized rig with fluorescence illumination to optimize the image quality. Each user was asked to rest his/her hand on a rigid platform with palm facing a hollow cutout where the camera was positioned within. Apart from an alignment point for the placement of middle finger, no additional alignment pads were used.

3 Point-Based Verification

Preprocessing. The preprocessing consists of two main parts – image alignment and image enhancement. Finger gaps are used for the purpose of image alignment in this paper. The palmprint is first binarized using a global threshold. The area of each connected component and the position of its centroid (C1) is determined and used as a criterion for identifying the finger gap objects. After identifying the three finger gap objects, pixels in each finger gap object that are to the right of its corresponding centroid are removed from consideration and a new centroid (C2) of the remaining pixels determined. The process is illustrated in Fig. 1(a). The red dot marks the position of the finger gap object. Pixels that are to the right of the green line are removed from consideration and the centroid of the truncated finger gap object is shown in blue.

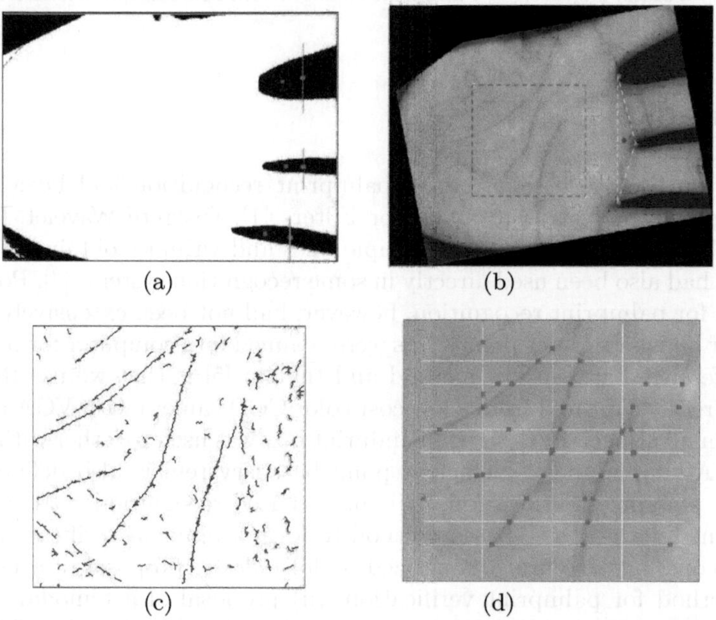

(a) (b) (c) (d)

Fig. 1. (a) Detection of centroids of finger gap objects, (b) Determining ROI from finger webs, (c) ROI after image enhancement, and (d) An example of feature extraction

Boundary tracing is then carried out. A line that passes through the two centroids (C1 and C2) of each finger gap object is determined and the point of interception between this line and the boundary of the palmprint is recorded as the finger gap location.

The palmprint image is rotated until the finger gap between the index finger and middle finger is directly above the finger gap of the ring finger and last finger. The centroid of the triangle formed by the three finger gap locations is then determined and used as a reference point for defining the Region Of Interest (ROI) of the palmprint. Fig. 1(b) illustrates the above process.

Image enhancement is then carried out on the extracted ROI. A Gaussian filter is first applied to smoothen the image and to remove any inherent noise. At each pixel position, the difference in gray value between the current pixel and its neighbours is calculated. That pixel will be stored as part of the feature line if the difference in gray level is larger then a certain threshold value, T.

As some of the feature lines are poorly contrasted from the skin colour for some palmprint images, an adaptive procedure, which varies the threshold T dynamically based on the nature of the palmprint image, is adopted. The basis for determining whether the threshold T needs further adjustment is the number of feature points detected (detection of feature points are presented in next subsection). If too little feature points are detected, it would mean that very few feature lines are detected in the image enhancement procedure. This is a direct consequence of using a threshold that is too high for a particular image. The threshold is therefore lowered and the image enhancement process repeated. The above procedure is iterated until the number of feature points detected is satisfactory.

Line bridging and line thinning is then carried out in order to further improve the quality of the image. Further processing is done to remove lines that do not satisfy the minimum length requirement. Fig. 1(c) shows the ROI after image enhancement.

Feature extraction. Grid lines are superimposed onto the enhanced image. The positions of the points where the gridlines intercept a feature line are recorded as the spatial positions of feature points. The recorded positions are to be used for the computation of error scores. An example of feature extraction is shown in Fig. 1(d).

Orientation information is then computed at each of the detected feature points. The algorithm used for computation of orientation is similar to that proposed by Bazen in [6] for the computation of directional fields for fingerprints. Computation of orientation is done in similar fashion but within a 13 x 13 window in the vicinity of feature point instead.

In addition to orientation, the coherence of the feature lines in the vicinity of the feature points can also be computed [6]. If high coherence values are obtained for the feature lines in the vicinity, it would mean that the gradient operator returns largely invariant values for the pixels in the vicinity. This would in turn imply that the feature lines in the region concerned are more or less in the same direction and that there is strong orientation information. Therefore, feature points taken at regions with higher coherence values are likely to be more reliable. In this paper, the feature points that have the thirty highest coherence values are used for the computation of error score for each palmprint image.

This orientation information, together with the spatial coordinates of the feature point computed earlier, makes up the feature point vector which characterizes the important information needed for the computation of error score in the next stage.

Point-based matching. After the feature points are detected and its spatial coordinates and orientation information computed and recorded, the error scores are computed. As spatial coordinates and orientation information of the feature points are stored as feature vectors, computation of error scores is based on the Euclidean distances between feature vectors from two different palmprints. The following equations are used to compute the error score between each corresponding pair of feature points.

$$\theta_{norm} = k_1(\alpha/\frac{\pi}{2}) \tag{1}$$

$$\delta_{norm} = k_2\left[(x_1 - x_2)^2 + (y_1 - y_2)^2\right] \tag{2}$$

where α is the smaller angle corresponding to the orientation difference of the two feature points. (x_1, y_1) and (x_2, y_2) are the spatial coordinates of the corresponding pair of feature points. k_1 and k_2 are weights attached to orientation and spatial information respectively. The error score contribution by each pair of feature point is given by

$$Error\ Score = \theta_{norm} + \delta_{norm} \tag{3}$$

The total error score for each pair of palmprint images is obtained by summing the error scores of all the thirty pairs of corresponding feature points. A smaller total error score will indicate a better match between the two palmprints.

4 DCT-Based Verification

DCT processing. Ahmed, Natarajan, and Rao (1974) first introduced the discrete cosine transform (DCT) in the early seventies. Ever since, the DCT has grown in popularity and several variants have been proposed. In this work, since we have a square image ROI, a 2D DCT is performed on the ROI which is cropped from the grayscale palm-print image. From the resulting coefficients, only a subset of coefficients is chosen such that it can sufficiently represent the palm. A 64 by 64 window of coefficients are obtained from the original 300 by 300 window of coefficients.

The original 64×64 2D coefficients map is converted into a 1D vector by scanning the DCT matrix in a zigzag fashion which is analogous to that of JPEG/MPEG image coding. This is done so that in the 1D vector, the coefficients are arranged in an order of increasing frequency.

Feature vector coefficient selection. Even after truncating the coefficients to a smaller window of lower frequency coefficient, the performance of the system is not acceptable. This is because in a 64 by 64 window, there are a total of 4096 coefficients. And not all of them are useful for recognition. Some coefficients are more susceptible to noise, while others are coefficients that characterize a palm image in general and are not distinctive between different palms. For example, the d.c. coefficient is very robust to noise, but it is invariant between different palms, hence it has little use in recognition of palms. In addition, the d.c. coefficient corresponds to the illumination of the image which is not desirable, since illumination plays no part in the recognition of palms. As such, some means of selection of coefficients that can be used for recognition has to be employed.

In order to identify which are the coefficients that are distinct between different palms, we can calculate the variance of that particular coefficient across different palms and select those coefficients with high variance among different palms. On the other hand, to identify which are the coefficients that are robust to noise, we can calculate the variance of that particular coefficient among images from the same palm. From the initial 4096 coefficients of the 64x64 window, only 2928 coefficients were selected.

Feature matching. To match a particular input palm, the system compares this palm's feature vector to the feature vectors of a palm from the database. The system compares by calculating the Euclidean distance between the two palms. A match is obtained by minimizing this Euclidean distance.

5 A Bimodal System

The point-based system was combined with the DCT-based syetms to form a bimodal palmprint verification system. Both *parallel* and *serial* integration was attempted, with parallel integration focusing primarily on accuracy and serial integration looking for a compromise between speed and accuracy.

Although the parallel integration method is likely to exhibit greater accuracy, computation is expected to be time consuming as error scores have to be computed for both the point-based algorithm and DCT based algorithm. Serial integration aims to strike a compromise between accuracy and speed by processing the data in two layers. The first layer consists of the DCT-based recognition system. Two predetermined threshold, T1 and T2 are set. After the error score from the DCT-based recognition system have been computed, palmprint images with an error score less than T1 are classified as "genuine users" while those with an error score of more than T2 are classified as "impostors". Only palmprint images that have an error score between T1 and T2 are passed to the second layer, which consists of the point-based recognition system, for further classification.

In this paper, T1 is set at the lowest error score computed for palmprints from impostors using the DCT method while T2 is set at the highest error score computed for palmprints from genuine users using the DCT method. Using these parameters, decisions can be arrived at the first layer for 23.5% of the palmprints.

6 Results

42,230 error scores had been generated from 206 palmprint images taken from 21 different users using the point-based recognition system. 40,410 are the result of false matches while the remaining 1820 are obtained from genuine matches. These error scores are used to determine the accuracy of the point-based recognition system. An Equal Error Rate of 8.455% is achieved for the point-based recognition system. The histogram of error score is shown in Fig. 2(a) and the Receiver Operating Characteristics (ROC) of the point-based recognition system is shown in Fig. 2(b). The graph of error rates against error score is shown in Fig. 2(c) below. The point of intersection in the graph constitutes the equal error rate.

109 out of these 206 palmprint images were used to test the accuracy of the bimodal system. Integration of the two recognition systems produced marked improvement in system accuracy. For the set of 109 palmprints used to test the bimodal recognition system, an Equal Error Rate of 9.985% was achieved for the point-based system while the DCT based system produced an Equal Error Rate of 9.864%. The parallel integrated system is able to achieve an Equal Error Rate of 2.895% while an Equal Error Rate of 5.965% is achieved for serial integration. A comparison of the ROC curves is shown in Fig. 2(d). Detailed tabulation of the error rates for each system is given in the Table 1.

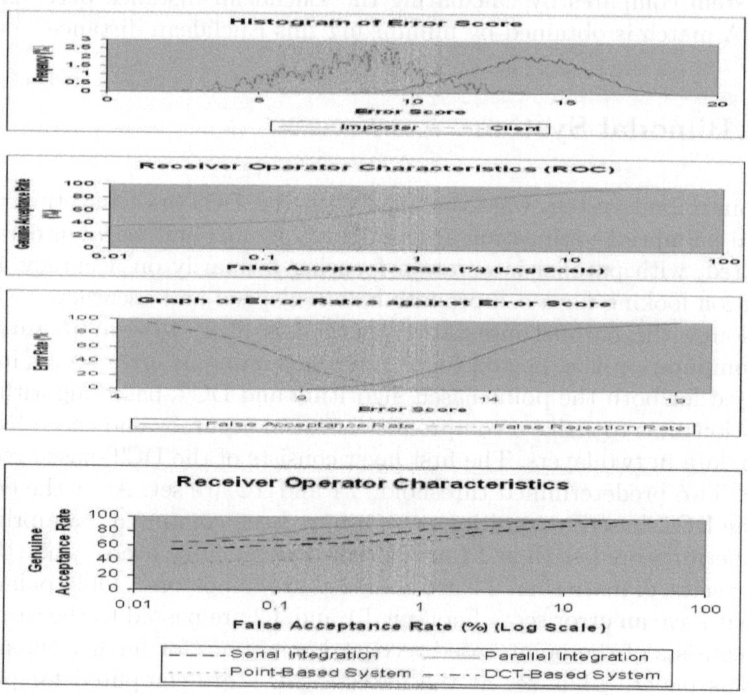

(a) (b) (c) (d)

Fig. 2. (a) Histogram of error scores, (b) ROC of point based verification system, (c) Error rates versus error scores, (d) Comparison of ROC curves for different methods

Table 1. Comparison of Error Rates

	Point-based	DCT-based	Parallel	Serial
EER	9.99%	9.86%	2.90%	5.97%
$FAR_{FRR=0}$	50.48%	49.79%	29.60%	49.80%
$FRR_{FAR=0}$	65.64%	78.98%	34.36%	78.87%

7 Conclusion

In this paper, a point-based method for palmprint recognition was proposed. The accuracy of the system was observed to give fairly good results on a small database. The system was then extended to form part of a bimodal recognition through serial and parallel integration with a DCT-based method. Accuracy of the bimodal systems were determined and compared with the individual systems. It was found that both serial and parallel integration produced improvements in recognition results with the system utilizing parallel integration faring better then the serially integrated system in terms of accuracy but at the expense of higher computation time. Our immediate future work is to test the system on a large database.

References

1. D. Zhang, W.-K. Kong, J. You, and M. Wong, "Online palmprint identification," *IEEE Trans. Pattern Analysis and Machine Intelligence*, vol. 25, no. 9, pp. 1041–1050, 2003.
2. L. Zhang and D. Zhang, "Characterization of palmprints by wavelet signatures via directional context modeling," *IEEE Trans. Systems, Man and Cybernetics, Part-B*, vol. 34, no. 3, pp. 1335–1347, June 2004.
3. J. You, W.-K. Kong, D. Zhang, and K. H. Cheung, "On hierarchical palmprint coding with multiple features for personal identification in large databases," *IEEE Trans. Circuits and Systems for Video Technology*, vol. 14, no. 2, pp. 234–243, 2004.
4. C.-C. Han, H.-L. Cheng, C.-L. Lin, and K.-C. Fan, "Personal authentication using palm-print features," *Pattern Recognition*, vol. 36, pp. 371–381, 2003.
5. N. Duta, A. K. Jain, and K. V. Mardia, "Matching of palmprints," *Pattern Recognition Letters*, vol. 23, no. 4, pp. 477–485, 2002.
6. A. M. Bazen and S. H. Gerez, "Systematic methods for the computation of the directional fields and singular points of fingerprints," *IEEE Trans. Pattern Analysis and Machine Intelligence*, vol. 24, no. 7, pp. 905–919, 2002.

Feature-Level Fusion of Hand Biometrics for Personal Verification Based on Kernel PCA

Qiang Li, Zhengding Qiu, and Dongmei Sun

Institute of Information Science,
Beijing Jiaotong University, Beijing 100044, P.R. China
liqianglq@126.com

Abstract. This paper presents a novel method of feature-level fusion (FLF) based on kernel principle component analyze (KPCA). The proposed method is applied to fusion of hand biometrics include palmprint, hand shape and knuckleprint, and we name the new feature as "handmetric". For different kind of samples, polynomial kernel is employed to generate the kernel matrixes that indicate the relationship among them. While fusing these kernel matrixes by fusion operators and extracting principle components, the handmetric feature space is established and nonlinear feature-level fusion projection could be implemented. The experimental results testify that the method is efficient for feature fusion, and could keep more identity information for verification.

1 Introduction

Fusion of different kind of data for a better decision is a hot topic in many research areas. While in the filed of personal authentication, multimodal biometric technology is becoming an important approach to alleviate the problems intrinsic to stand-alone biometric systems. According to Jain and Ross [1], the information of different biometrics could be fused in three levels: feature extraction level, matching score level and decision level. Though feature-level fusion (FLF) could keep the identity information to its most and expected to perform better than at the other two levels, the study on it is seldom reported. There are mainly two reasons of it [12]. First, the feature spaces of different biometric traits may not compatible. That is, different features may have different dimension and measurement, and their dynamic variation ranges lie in different complicated nonlinear spaces. Second, FLF may lead to the "curse of dimensionality" problem by concatenating several features as one. While solving these problems, we propose a new strategy for FLF based on KPCA.

The choice and number of biometric traits is another issue in multimodal biometric system. In this paper, fusion of hand based biometrics including palmprint [10][11], hand geometry and knuckleprint [3] are investigated. All these three biometrics have the advantage of robust to noise and change of environment, and available in low-resolution images (<50dpi). Furthermore, they could be extracted in the same original hand image and overcome several disadvantages of multimodal biometrics such as inconvenience to use, parameters drifting for different capture devices and higher database requirement.

D. Zhang and A.K. Jain (Eds.): ICB 2006, LNCS 3832, pp. 744 – 750, 2005.

2 Hand Feature Extraction and Fusion Algorithm

2.1 Handmetric: Palmprint, Hand Geometry and Knuckleprint

Palmprint contains the majority of identity information in low-resolution hand images, and could achieve rather high accuracy [10][11]. As one of the most reliable biometric traits, palmprint and have got widely attention in recent years. However, palmprint recognition is still underdevelopment. Besides the problem of device size, illumination and environment variation, there are mainly four difficulties in palmprint recognition: a) ROI (region of interest) is hard to determine precisely; b) change of hand pose brings about variable shadow and nonlinear distortion, which is hard to eliminate; c) the feature lines/points are hard to define exactly; d) palmprint may change dynamically during long period of time.

Fuse with other kind of biometric, palmprint recognition could be more accurate and more ready to business use. By adding hand geometry information on palmprint, Kumar et al. [4] suggests that fusion in decision level using max rule could outperform traditional serial concatenate FLF method and single palmprint system. Ribaric et al. [5] testifies weighted sum rule in a similar manner. Based on these works, we define the outline of the hand and cuticle in the inner surface of the hand as "handmetric". As to low-resolution images, handmetric contains palmprint, hand geometry and knuckleprint with their geometrical correlativity. In this paper, handmetric refers to the FLF of three kind of hand biometrics. Because hand shape is mainly "finger shape", it can be combined with knuckleprint easily [3]. Without loss of generality, we choose the image of middle finger (as shown in Fig.1 (a)) to obtain the whole finger feature. While fusing palmprint (as shown in Fig.1 (b)) with finger feature by FLF, handmetric feature could be obtained for personal verification.

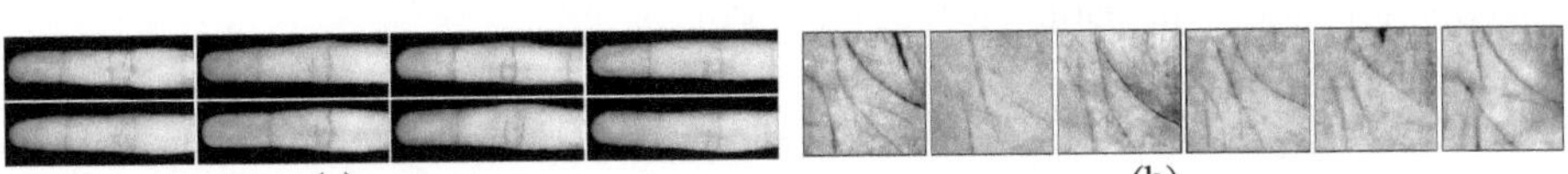

(a) (b)

Fig. 1. Image examples in handmetric database: (a) the middle finger images from different persons, (b) the palmprint images from different persons

2.2 KPCA Algorithm

KPCA is the combination of kernel projection and PCA dimension reduction method [6]. It has been widely used in subspace feature extraction, especially in face recognition task [7]. The main idea of KPCA is: transform the original sample space **X** into high dimensional space **D** by applying a nonlinear mapping Φ to each of the samples. PCA is then performed in **D** to get a feature space **F**. KPCA is the nonlinear extent of PCA, and PCA can be interpreted as a linear version of KPCA.

Given M original training samples $x_1, x_2, \ldots, x_M \in R^N$, the space **D** should be $\{ \Phi(x_i) \mid i=1,\ldots,M\}$. The correlation matrix of **D** is defined as kernel matrix **K**:

$$K_{ij} = (\Phi(x_i) \cdot \Phi(x_j)) \qquad i, j = 1, 2, \cdots, M \tag{1}$$

K can be regarded as generalized similarity measure among the data [2]. Centering the embeddings in **D** is needed for further processing. According to [6], centering could acts on **K** and can be implemented by operator $\mathcal{D}$:

$$\widetilde{K}_{ij} = \mathcal{D}(\mathbf{K}) = [\mathbf{K} - \mathbf{1}_M\mathbf{K} - \mathbf{K}\mathbf{1}_M + \mathbf{1}_M\mathbf{K}\mathbf{1}_M]_{ij} \tag{2}$$

where $[\mathbf{1}_M]_{ij} = 1/M$ for all i,j. Then eigenvalues and eigenvectors of $\widetilde{\mathbf{K}}$ are worked out:

$$\widetilde{\mathbf{K}}\mathbf{A} = M\mathbf{A}\Lambda, \text{ where } \mathbf{A} = [\alpha_1\alpha_2 \ldots \alpha_M], \Lambda = diag\{\lambda_1, \lambda_2, \ldots, \lambda_M\} \tag{3}$$

A is the orthogonal component space of **D**. For dimension reduction, $d(d<<N)$ principle components (that is, α corresponding to larger λ) are extracted to form feature space **F**. For a given testing sample y, the feature f can be calculated by (4).

$$f = \mathbf{F}^T\mathcal{D}([\Phi(x_1)\cdot\Phi(y)\ \Phi(x_2)\cdot\Phi(y)\ \cdots\ \Phi(x_M)\cdot\Phi(y)]^T) \tag{4}$$

It should be noticed that **K** but not Φ is used directly in the algorithm, thus kernel projection is represented by the kernel matrix but not Φ. In another words, nonlinear mapping is implemented by defining a relation measure (1) between all the samples. Three most popular classes of kernel functions are polynomial kernels, Gaussian kernels and sigmoid kernels. The choice of kernel is always done by experiment.

$$K(x, y) = (x \cdot y)^{degree} \tag{5}$$

Fractional power polynomial model [7] is employed in our system, where the power *degree* is set to 0.7 and feature dimension d is set to 60 to maximize the performance to palmprint authentication. To lower the system complexity, finger feature extraction uses the same parameters as palmprint.

2.3 Fusion Algorithm

Given training samples of palmprint $\{p_i\}$ and finger samples $\{s_i\}$ (i=1,…,M), corresponding new handmetric feature $\{h_i\}$ could be acquired using kernel method. First, kernel matrixes of palmprint and finger are:

$$[\mathbf{K}_p]_{ij} = (p_i \cdot p_j)^{0.7}, [\mathbf{K}_s]_{ij} = (s_i \cdot s_j)^{0.7}, i, j = 1, 2, \cdots, M \tag{6}$$

For different biometrics, the same kernel is applied. Therefore, the kernel matrixes have general measurement. According to decision template method proposed by Kuncheva et al. [9], these matrixes could be fused reasonably. We define a fusion operator $\mathcal{B}$ to generalize different fusion rules. The four most popular rules in decision level fusion (sum, product, min and max) are adopted to evaluate the performance of our method. The new kernel matrix for handmetric $\mathbf{K}_h$ using different $\mathcal{B}$ could be:

$$\begin{aligned} &\mathcal{B}_1 : [\mathbf{K}_{h-avg}]_{ij} = \mathcal{D}(([\mathbf{K}_p]_{ij} + [\mathbf{K}_s]_{ij})/2); \quad \mathcal{B}_2 : [\mathbf{K}_{h-prd}]_{ij} = \mathcal{D}([\mathbf{K}_p]_{ij} \cdot [\mathbf{K}_s]_{ij}) \\ &\mathcal{B}_3 : [\mathbf{K}_{h-\min}]_{ij} = \mathcal{D}(\min([\mathbf{K}_p]_{ij}, [\mathbf{K}_s]_{ij})); \quad \mathcal{B}_4 : [\mathbf{K}_{h-\max}]_{ij} = \mathcal{D}(\max([\mathbf{K}_p]_{ij}, [\mathbf{K}_s]_{ij})) \end{aligned} \tag{7}$$

Note that operator $\mathcal{D}$ should act on the hold kernel matrix. We simplified the form (7) for better understanding.

The principle components of $\mathbf{K}_h$ span the handmetric feature space $\mathbf{F}$. The dimension of handmetric is set to 60, which is the same as palmprint and finger features.

After the training stage, handmetric feature of all the samples could be calculated. Suppose p_t, s_t to be a pair of testing palmprint and finger samples, their kernel projections are:

$$k_p = [(p_1 \cdot p_t)^{0.7}\ (p_2 \cdot p_t)^{0.7}\ \cdots\ (p_M \cdot p_t)^{0.7}]^T,\ k_s = [(s_1 \cdot s_t)^{0.7}\ (s_2 \cdot s_t)^{0.7}\ \cdots\ (s_M \cdot s_t)^{0.7}]^T \quad (8)$$

The relation between the testing handmetric to trained handmetric is:

$$k_h = \mathcal{D}\{\mathcal{B}(k_p, k_s)\} \quad (9)$$

And the handmetric f_h would be:

$$f_h = \mathbf{F}^T k_h \quad (10)$$

By using (8)(9)(10), all the training and testing samples are projected in the handmetric space to extract handmetric features for classification.

3 MAP Classifier for Verification

The maximum a posteriori (MAP) classifier is designed for handmetric verification. In the training stage, every handmetric feature would minus all the other features to get the feature differences in the training set. If the difference is got form the same person, then it should be marked as "G (Genuine)" class. Otherwise, it is signed to "I (Impostor)" class. Classifier would build on the G and I classes in training samples.

In the testing stage, suppose a testing handmetric feature f_t has the identity statement Z, according to [8], the difference Δ between f_t and class center of Z is treated as a sample input to the classifier, that is:

$$\Delta = f_t - \overline{f_Z},\quad \overline{f_Z} = \frac{1}{L}(\sum_{i=1}^{L} f_i),\ f_i \in Z \quad (11)$$

The likelihood of Δ to G and I are:

$$P(\Delta / G) = \frac{\exp\{-1/2(\Delta^T \Sigma_G^{-1} \Delta)\}}{(2\pi)^{D/2} |\Sigma_G|^{1/2}}, P(\Delta / I) = \frac{\exp\{-1/2(\Delta^T \Sigma_I^{-1} \Delta)\}}{(2\pi)^{D/2} |\Sigma_I|^{1/2}} \quad (12)$$

Under the assumption that the prior is $P(G)=P(I)=1/2$ [14], we could get the posteriori probability $P(G/\Delta)$:

$$P(G / \Delta) = \frac{P(\Delta / G)P(G)}{P(\Delta / G)P(G) + P(\Delta / I)P(I)} = \frac{P(\Delta / G)}{P(\Delta / G) + P(\Delta / I)} \quad (13)$$

Finally, comparing $P(G/\Delta)$ to similarity threshold T, we could draw the conclusion whether the identity statement is true.

4 Experiments and Discuss

A handmetric verification system is implemented based on our KPCA FLF method. A special device for hand image acquisition is designed. The device is made up of a CCD camera and a platform to fix the hand and illumination. 1,853 right hand images from 98 individuals are captured and stored in the database. For each person, up to 28 images are captured in the period of 6 months for 4 times at most (the average is 2.7 times), 4 samples of each person are taken out to form a training set, while the remaining 1,481 images are taken as testing set. After preprocessing, palmprint database and finger database derived from hand image database. For each biometric (including palmprint, finger and handmetric) in the verification test, a total of 145,138(1481 × 98) comparisons are performed for the testing images, in which 1,481(1481 × 1) are genuine matching. The verification system is programmed using Matlab6.2 under Microsoft Windows XP environment, with Intel P4 2.6G CPU and 256MB memory.

4.1 Comparison the Performance of Palmprint, Finger and Handmetric

The aim of the first test is to verify the performance of handmetric. Whereas we proposed four fusion operators in 2.3, here we choose $\mathcal{B}_2$ in (7) to demonstrate the

Table 1. Error rates for different biometrics

3/4 training samples	FAR(%)	FRR(%)	HTER(%)	EER(%)
Palmprint	0.51/0.45	0.61/0.61	0.56/**0.54**	0.61/**0.55**
Finger	0.70/0.72	1.49/1.28	1.10/**1.00**	1.20/**1.07**
Handmetric	0.23/0.29	0.34/0.20	0.48/**0.25**	0.35/**0.28**

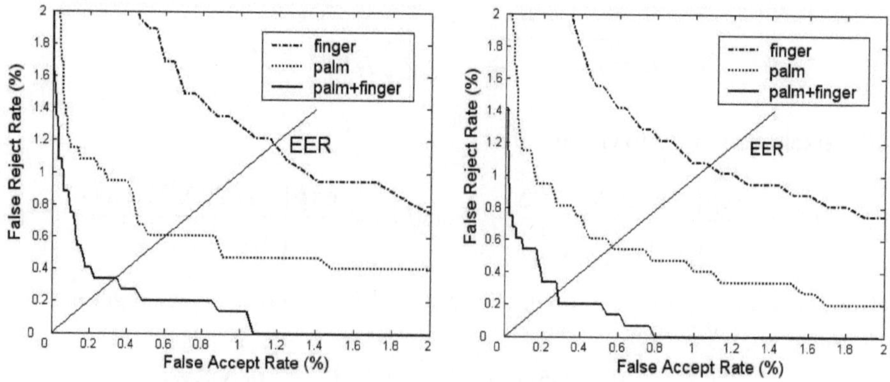

Fig. 2. ROCs of palmprint, finger and handmetric verification: (a) 3 training samples for each person, (b) 4 training samples for each person

effectiveness of fusion algorithm. The test is carried out using 3 and 4 training samples for each person respectively (none of the biometrics works well using 2 samples through the test, the EERs (equal error rates) are higher than 10%) to get a more generalized result.
Experimental results are given in Tab.1 and Fig.2. It is evident that feature extraction using KPCA and the MAP classifier are effective, most of the error rates are under 1%. Furthermore, the improvement of handmetric is outstanding in both test, the EER of it reaches 0.28% using 4 training samples for each person, whereas palmprint is 0.55% and finger is 1.07% respectively in the same case. And, more training samples could make better result. As to handmetric, HTER turns to 0.25% from 0.48% when using one more sample.

4.2 Comparison of Different Fusion Operators

Different fusion operators affect the performance of handmetric greatly. Though the results in 4.1 show that 4 samples case would achieve better result, here we use only 3 samples to make the results of comparison more clearly.

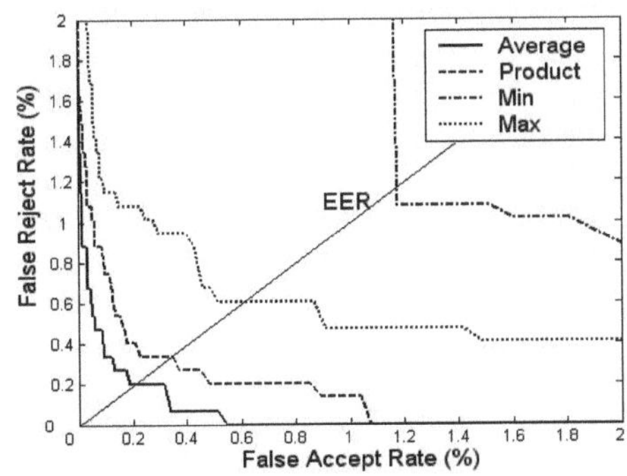

Fig. 3. ROCs of different fusion operators

As shown in Fig.3, the sum operator $\mathcal{B}_1$ performs the best in all the four operators, which is consist with most experimental results of decision level fusion [1][5][12]. Both of HTER (half total error rate) and EER reach 0.20% using only 3 samples for each person, which shows the power of FLF method. However, the max operator $\mathcal{B}_3$ and min operator $\mathcal{B}_4$ couldn't work as well as $\mathcal{B}_1$ and $\mathcal{B}_2$. $\mathcal{B}_3$ could only maintain the accuracy of palmprint with little improvement, while $\mathcal{B}_4$ just gain a little than finger feature. Therefore, fusion cannot ensure the system performance absolutely, and proper fusion algorithm is needed to make it feasible.

Based on the test of 4.2 and 4.3, we use operator $\mathcal{B}_1$ in 4 samples case to buildup the final handmetric verification system. The HTER and EER reach 0.11% and 0.07% respectively, which testifies that our FLF algorithm based on KPCA is effective, and handmetric is reliable and ready to application.

5 Conclusion

A novel FLF method based on KPCA is presented in this paper, and a handmetric verification system is set up based on it. Space incompatible and dimension curse problem of FLF are figured out by proposed method. The experimental results demonstrate the effectiveness of proposed method. The HTER and EER reach 0.11% and 0.07% respectively in the test database include 1,853 samples with 145,138 comparisons.

References

1. Ross, A., Jain A. K.: Information Fusion in Biometrics. Pattern Recognition Letters, Vol. 24, (2003) 2115-2125
2. Lanckriet, G., Deng, M., Cristianini, N., Jordan, M. I.: Kernel-based Data Fusion and Its Application to Protein Function Prediction in Yeast. Proceedings of the Pacific Symposium on Biocomputing, (2004) 300-311
3. Li, Q., Qiu Z., Sun, D.: Personal Identification Using Knuckleprint, Sinobiometric04', Lecture Notes in Computer Science, Vol. 3338. Springer-Verlag, (2004) 680-689
4. Kumar, A., Wong, D., Shen, H. C., Jain, A. K.: Personal Verification using Palmprint and Hand Geometry Biometric. Proceedings of the fourth International Conference on audio- and video-based biometric personal authentication, (2003) 668-678
5. Rabaric, S., Ribaric, D., Pavesic, N.: A Biometric Identification System Based on the Fusion of Hand and Palm Features. Proceedings of The Advent of Biometrics on the Internet, A Cost 275 Workshop. (2002)
6. Scholkopf, B., Smola, A., Muller, K. R.: Nonlinear Component Analysis as a Kernel Eigenvalue Problem. Neural Computation, Vol. 10, (1998) 1299-1319
7. Liu, C-j.: Gabor-based Kernel PCA with Fractional Power Polynomial Models for Face Recognition. IEEE Trans. PAMI, Vol. 26, (2004) 572-581
8. Moghaddam, B.: Principal Manifolds and Probabilistic Subspaces for Visual Recognition. IEEE Trans. PAMI, Vol. 24 (2002) 780-788
9. Kuncheva, L. I., Bezdek, J. C., Duin, R. P. W.: Decision Templates for Multiple Classifier Fusion: An Experimental Comparison. Pattern Recognition, Vol, 34 (2001) 299-314
10. 10.Zhang, D., Kong, W. K., You, J.: Online Palmprint Identification. IEEE Trans. PAMI, Vol 25 (2003) 1041-1050
11. Lu, G., Zhang, D., Wang, K.: Palmprint Recognition using Eigenpalms Features. Patter Recognition Letters, Vol. 24 (2003) 1463-1467
12. Jain, A. K., Ross, A.: Multibiometric Systems. Communication of the ACM, Special Issue on Multimodal Interfaces, Vol. 47 (2004) 34-40

Human Identification System Based on PCA Using Geometric Features of Teeth

Young-Suk Shin[1] and Myung-Su Kim[2]

[1] Department of Information Communication Engineering, Chosun University, #375 Seosuk-dong, Dong-gu, Gwangju, 501-759, South Korea
ysshin@chosun.ac.kr

[2] College of Dentistry, Chosun University, #375 Seosuk-dong, Dong-gu, Gwangju, 501-759, South Korea
msakim@chosun.ac.kr

Abstract. We present a new human identification system based on PCA using geometric features of teeth such as the size and shape of the jaws, size of the teeth and teeth structure. In this paper we try to set forth the foundations of a biometric system for information encrypting of living people using dental features. To create a biometric matching system, a template based on principal component analysis(PCA) is created from dental data collected the plaster figures of teeth which were done at dental hospital, department of oral medicine. Templates of dental images based on PCA representation include the 100 principle components as the features for individual identification. The PCA basis vectors reflects well the features for individual identification in the whole of teeth and the part of teeth. The classification for human identification is generated based on the distance between the whole of teeth and the part of teeth with the nearest neighbor(NN) algorithm. The identification performance in 300 dental image is 97% for the part of teeth missed the right-molar and back teeth, 98.3% for the part of teeth missed the front teeth and 96.6% for the part of teeth missed the left-molar and back-teeth.

1 Introduction

Biometrics refers to the identity authentication of living people using their enduring physical or behavioural characteristics. Biometric identifiers are pieces of information encrypting a representation of a person's unique biological makeup. The most generalized biometric techniques include the recognition of fingerprints, faces, iris, retina, hand geometry, voice, signature and teeth [1,2,3,4,5,6,7].

Jain and Chen[7] utilized dental radiographs for human identification. They studied a biometric system based on the distance between the antemortem and postmortem tooth shape. The dental radiographs have a number of challenges to overcome. For poor quality images where tooth contours are indiscernible, the shape extraction is a difficult problem for dental radiographs. Our algorithm utilizes information about differences in the size and shape of the jaws, size of the teeth and teeth structure.

In this paper, we present a new human identification system based on PCA using geometric features of teeth like the size and shape of the jaws, size of the teeth and

D. Zhang and A.K. Jain (Eds.): ICB 2006, LNCS 3832, pp. 751 – 755, 2005.

teeth structure. First, we collected the plaster figures of teeth from the department of oral medicine in dental hospital. Second, we developed a representation of dental images based on PCA included the 100 principle components as the features for individual identification. Finally, the nearest neighbor algorithm for the classification of individual identification was applied.

2 PCA Representation of Dental Features

2.1 Image Data

Dental data was a database of the plaster figures of teeth which were done at Chosun University dental hospital, department of oral medicine. The data set contained images of 348 individuals of males and females. Each person has two images in a upper jaw and lower jaw. The data set used for research contained 300 gray level images in a upper jaw, each image using 800 by 600 pixels. Examples of the original images are shown in figure 1.

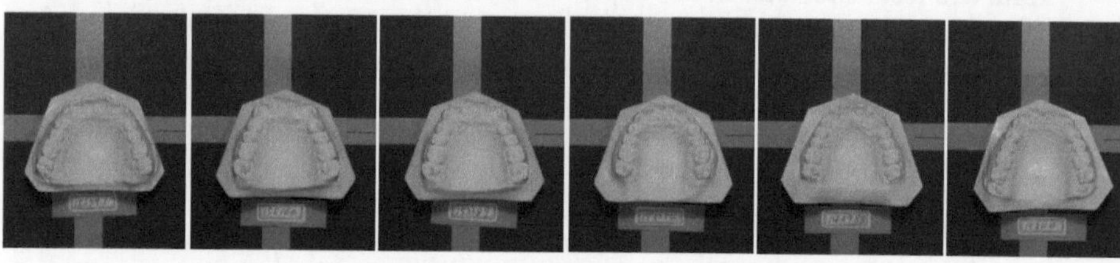

Fig. 1. Examples in a upper jaw from the dental database

2.2 Preprocessing

The dental images were centered with fixed coordinates locations, and then cropped and dug the palatine by semi-automatic method with a teeth template. Finally, The images were scaled to 30x30 pixels. Figure 2(b) shows the image dug the palatine with a teeth template(Figure 2(a)).

The luminance was normalized in two steps. First, a "sphering" step prior to principal component analysis is performed. The rows of the images were concatenated to produce 1×900 dimensional vectors. The row means are subtracted from the dataset, X. Then X is passed through the zero-phase whitening filter, V, which is the inverse square root of the covariance matrix:

(a) (b)

Fig. 2. (a) A teeth template (b) A dental image dug by a teeth template (a)

$$V = E\{XX^T\}^{-\frac{1}{2}} \tag{1}$$
$$W = XV$$

This indicates that the mean is set to zero and the variances are equalized as unit variances. Secondly, we subtract the local mean gray-scale value from the sphered each patch. From this process, W removes much of the variability due to lightening.

2.3 PCA Representation

Some of the most successful algorithms applied PCA representation are "eigen faces[8]" and "holons[9]". These methods are based on learning mechanisms that are sensitive to the correlations in the images. PCA provides a dimensionality-reduced code that separates the correlations in the input. Atick and Redlich[10] have argued for such compact, decorrelated representations as a general coding strategy for the visual system. Redundancy reduction has been discussed in relation to the visual system at several levels. A first-order redundancy is mean luminance. The variance, a second order statistic, is the luminance contrast. PCA is a way of encoding second order dependencies in the input by rotating the axes to corresponding to directions of maximum covariance.

For individual identification based on dental feature, we employed the 100 PCA coefficients, P_n. The principal component representation of the set of images in W in Equation(1) based on P_n is defined as $Y_n = W * P_n$. The approximation of W is obtained as:

$$\overline{W} = Y_n * P_n^T . \tag{2}$$

The columns of $\overline{W}$ contains the representational codes for the training images. The representational code for the test images was found by $\overline{W}_{test} = Y_{test} * P_n^T$ (see figure 3). Best performance for individual identification based on dental feature was obtained using 100 principal components.

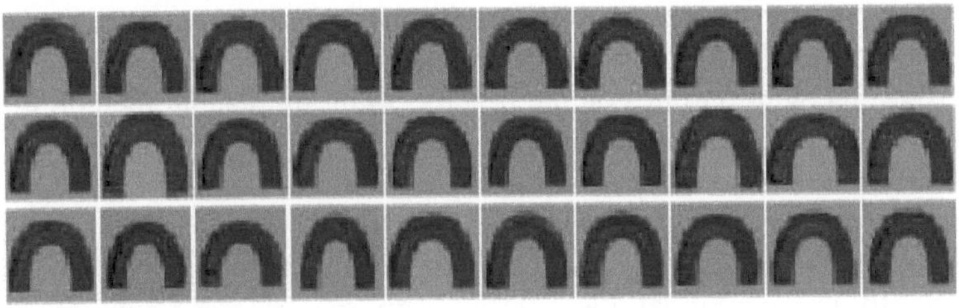

Fig. 3. PCA representation included the 100 principle components

3 Results

The first test verifies with 300 person image of upper jaw trained already. The recognition result was produced by 300 image trained previously showed 99.3%

recognition rates. Since we had 300 person image of upper jaw in the training set, for testing, in total 900 image of 300 person excluded the three parts in the training set was used. The three parts consists of right-molar and back-teeth, left-molar and back-teeth and front-teeth. Each individual has three different images and each part includes 300 person image. That is to say, the algorithms were tested for recognition under three different conditions: right-molar and back-teeth, left-molar and back-teeth, and front-teeth.

Classification for individual identification was applied the nearest neighbor algorithm(NN). The principle of the NN algorithm is that of comparing input image patterns against a number of paradigms and then classifying the input pattern according to the class of the paradigm that gives the closest match.

The recognition performance was 97% for the part of teeth missed the right-molar and back-teeth, 98.3% for the part of teeth missed the front-teeth and 96.6% for the part of teeth missed the left-molar and back-teeth.

4 Summary and Conclusions

This paper propose a new biometric system for information encrypting of living people using dental features. By PCA representation based on geometric features of teeth such as the size and shape of the jaws, size of the teeth and teeth structure, our system can be used to authenticate an individual's identity by comparing a biometric reading from a person with a single stored template. This simulation demonstrates that PCA representation can solve a challenging problem such as a correct object recognition from the part shape in object recognition.

Our system extracts PCA representation included only 100 principle components from image scaled to 30x30 image and individual identification was successfully produced over 96% recognition rates. This results can reflect the fact that the holistic analysis is important for individual identification based on geometric features of teeth such as the size and shape of the jaws, size of the teeth and teeth structure. From the verification test with 300 person image of upper jaw trained already, we showed 99.3% recognition rates. It seems to be a erosion of detail shape of the jaws on the preprocessing caused by semi-automatic method with a teeth template. While we produced 100% recognition rates on the preprocessing by manual method in verification test trained already, the result on the preprocessing by semi-automatic method showed 99.3% recognition rates.

We suggest that in order to recognize the objects from the parts, the patterns within the scenes have to be holistic information reduced redundancy. In the future study we are planning to try individual identification on the preprocessing by automatic method.

Acknowledgements

This study was supported by research funds from Chosun University, 2005.

References

1. Jain, A.K. Pankanti, S. Prabhakar, A., Ross, A.: Recent advances in fingerprint verification, Lecture Notes Comput. Sci. 2091 (2001) 182-190
2. Phillips, P.J.: Support vector machines applied to face recognition, Technical Report, NISTIR 6241, National Institute of Standards and Technology, 1999
3. Daugman, J., Downing, C.: Epigenetic randomness, complexity, and singularity of human iris patterns, Proc. Roy. Soc. 268 (2001) 1737-1740
4. Jain, A. K., Ross, A., Pankanti, S.: A prototype hand geometry-based verification system, Second International Conf. on Audio and Video-based Biometric Person Authentification, Washington DC, USA, March (1999)
5. Furui, S.: Recent advances in speaker recognition, in:J. Bigun, G.Borgeford (Eds.), Audio and Video-based Biometric Person Authentification, Springer, Berlin, (1997)
6. Nalwa, V.S.: Automatic on-line signature verification, Proc. IEEE 85 (2), (1997) 215-239
7. Jain, A. K., Chen, H.: Matching of dental X-ray images for human identification, Pattern Recognition **37**(7), (2004) 1519-1532
8. Turk, M, Pentland, A. : Eigenfaces for recognition. Journal of Cognitive Neuroscience **3**(1) (1991) 71-86
9. Cottrell, G., Metcalfe, J.: Face, gender and emotion recognition using holons. In Touretzky, D., editor, Advances in Neural information processing systems (3) San Maleo, CA. Morgan aufmann (1991) 564-571
10. Atic, J., Redlich, A.: What does the retina know about natural scenes?, Neural Computation (4) (1992) 196-210

An Improved Super-Resolution with Manifold Learning and Histogram Matching

Tak Ming Chan[1] and Junping Zhang[1,2]

[1] Shanghai Key Laboratory of Intelligent Information Processing,
Department of Computer Science and Engineering,
Fudan University, 200433, China
{0272366, jpzhang}@fudan.edu.cn

[2] The Key Laboratory of Complex Systems and Intelligence Science,
Institute of Automation, Chinese Academy of Sciences,
Beijing, 100080, China

Abstract. Biometric Person Authentication such as face, fingerprint, palmprint and signature depends on the quality of image processing. When it needs to be done under a low-resolution image, the accuracy will be impaired. So how to recover the lost information from down-sampled images is important for both authentication and preprocessing. Based on Super-Resolution through Neighbor Embedding algorithm and histogram matching, we propose an improved super-resolution approach to choose more reasonable training images. First, the training image are selected by histogram matching. Second, neighbor embedding algorithm is employed to recover the high-resolution image. Experiments in several images show that our improved super-resolution approach is promising for potential applications such as low-resolution mobile phone or CCTV (Closed Circuit Television) image person authentication.

1 Introduction

The super-resolution problem arises in a number of biometric applications, for example, person authentication from a low-resolution input such an image sent by mobile phones or taken from CCTV. However, a low-resolution image loses detailed information of important features in biometric person authentication such as suspect identification. Therefore, how to recover lost information from a low-resolution image to a high- resolution one is important for building effective image based biometric applications.

Classical recovery methods include interpolation and smoothing approaches [1]. However, images may suffer from block effect and aliasing and lose details such as facial texture and edges. Better methods of super-resolution [2, 3] are developed. Recently, a novel and outstanding method with manifold learning is proposed [4]. In the paper, neighbor embedding with training images is adopted to recover the super-resolution image. One disadvantage of the approach is that the recovery of super-resolution image is easily affected by the training image which needs to be

D. Zhang and A.K. Jain (Eds.): ICB 2006, LNCS 3832, pp. 756–762, 2005.

selected within related contents manually. Meanwhile, the original paper didn't consider how to apply the approach into the preprocessing of biometric person authentication.

Considering those mentioned, we propose an improved approach where the training image is automatically selected based on histogram matching from a set of unlabeled images. Neighbor embedding is then employed. Experiments in several facial images show that the proposed approach has a potential ability to choose the reasonable training image to reconstruct super-resolution images better.

The rest of the paper is outlined as follows. In Section 2 we propose the improved super-resolution with manifold learning and histogram matching. Experimental results are reported in Section 3. In the final section we make a conclusion on this paper.

2 An Improved Super-Resolution with Manifold Learning and Histogram Matching

Our proposed approach is based on super-resolution through manifold learning. For better understanding our work, the original approach will be briefly introduced in the following subsection.

2.1 Super-Resolution Through Manifold Learning Approach

From the manifold learning point of view, data in low-dimensional subspace should have as similar neighborhood relationship as corresponding one in high-dimensional observation space [5,6]. Therefore, one patch in a low-resolution image can be represented by locally linear weighted sum of neighbor patches. And weights can be calculated based on least square criterion of locally linear embedding algorithm [7]. Similarly, the weights combining neighbor patches of some high-resolution image are adopted to reconstruct the unknown low-resolution image. It is the main idea and more details can be seen in [4].

2.2 The Proposed Approach

The disadvantage of super-resolution with manifold learning is that training images need to be manually selected with similar contents. When there are large-scale images, one automated way to select proper training image is desirable. Hence we propose the simple but powerful histogram matching approach to select the training image from a collection of images.

Histogram applies the probability of pixels to represent some statistical properties hidden in an image. The basic formulation in gray level is as follows:

$$h(r_k) = n_k \tag{1}$$

$$p(r_k) = n_k/n \quad n_k < n, r_k = 0, 1, \cdots, L-1 \tag{2}$$

where r_k is the kth gray level, n_k is the number of pixels in the image having gray level r_k, and L is the number of gray level, $p(r_k)$ is an estimate of the

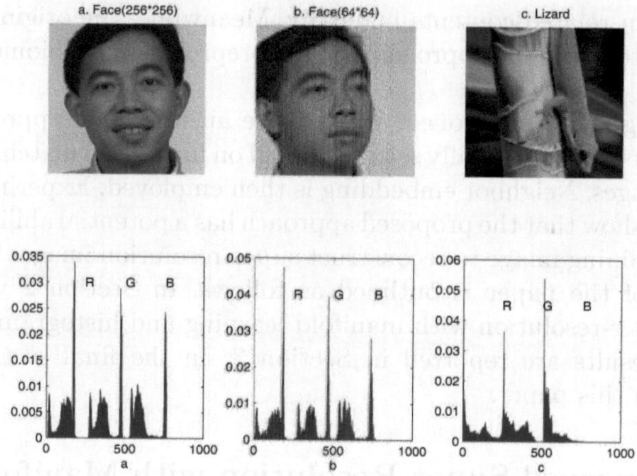

Fig. 1. Color histograms based on Y component of YIQ color space

probability of gray level r_k in an image. While conceptually simple, histogram can partially represent the contents of an image.

From a and b from Figure 1 it is obvious that there are similar normalized histograms between high-resolution and low-resolution, frontal and lateral viewpoint face images. Furthermore, when objects belong to different classes, for example, b and c in Figure 1, the normalized histograms will have remarkable differences. Considering the aforementioned properties, we employ histogram matching for the automated selection of relative training image from a collection of unlabeled images.

In this paper, color histograms are adopted to perform histogram matching. The color space of a image is discretized into n distinct (discretized) colors. A color histogram H is a set of vector $h_1, h_2, \cdots, h_{n_i}$, in which each bucket h_j contains the number of pixels of color j in the image. For a given image I, the color histogram H_I is a compact summary of the image. A database of images can be queried and the most similar image to I, the image I_0 will be returned with the most similar color histogram H_{I_0} . We use the measurement of the sum of squared differences ($L2$-norm), which is formulated as follows:

$$H(I, I')_{L_2} = \|H_I - H_{I'}\|_{L_2} = \sum_{j=1}^{n} (H_I(j) - H_{I'}(j))^2 \tag{3}$$

Then the most similar image to image I would be the one I_0 minimizing distances among I and images from the collection set. The objective criterion is:

$$C(I) = \min_j H(I, I_j) \quad j = 1, 2, \cdots, M \tag{4}$$

Where M denotes the number of training images, $C(I)$ denotes the final training image selected based on Eq. 4. We need a set of images other than only one image. The pseudo-code of the proposed approach is tabulated as in Table 1.

Table 1. The Pseudo-code of The Proposed Algorithm

Input: low resolution image X_t, training set T_r, Neighbor Numbers k,
Patch Size s and Magnification Factor n.

Procedure 1: Histogram matching

1. Compute the normalized histogram H_t of low input X_t.
2. For each image Y_i from T_r, do
   {
   Compute the normalized histogram H_i of Y_i
   Compute the $H(t,i)_{L_2}$ between H_t and H_i
   }
3. Select the image Y_I which has the minimum $H(t,I)_{L_2}$ to be Y_s and blur and downsample it by $1/n$ as X_s.

Procedure 2: Super-Resolution through Neighbor Embedding

1. Cut X_t and also X_s into patches of size s by s with overlapping by one or two pixels.
2. Cut Y_s into patches of size $n \times s$ by $n \times s$ with overlapping by n or $n \times 2$ pixels accordingly
3. For each patch x_t^q from X_t, do
   {
   Find k nearest neighbors among all patches from X_s
   Compute the reconstruction weights to minimize the error of reconstructing x_t^q
   Compute the high-resolution embedding y_t^q using the reconstruction weights combining the patches in Y_s corresponding to the k nearest neighbors in X_s.
   }
4. Enforce local compatibility and smoothness constraints between adjacent patches among all y_t^q and get Y_t.

---

3 Experiments

To evaluate the preprocessing performance of the proposed approach for biometric person authentication, experiments are performed on the pool of images from figure 2. There are 10 eye region images [8] and 6 images of irrelevant topics for testing. Each time we take one out of the set, downsample it to be the low resolution input and leave the rest (15 images) to make up the training set.

The low input size is 70×20 pixels and our goal is to compute the $4X$ magnification. We set the parameters as the paper [4] does, using 5 nearest neighbors, patches size of 3×3 , overlapping 2 pixels, according to its satisfactory performance. With histogram matching we can compute a series of ranking subplots of choices of training images. The ranking k can be apprehended that without those images having ranking higher than k in the training set, the one with ranking k is the choice as training image. Examples of reconstruction of high resolution images ranked by histogram matching are shown in Figure 3 and Figure 4.

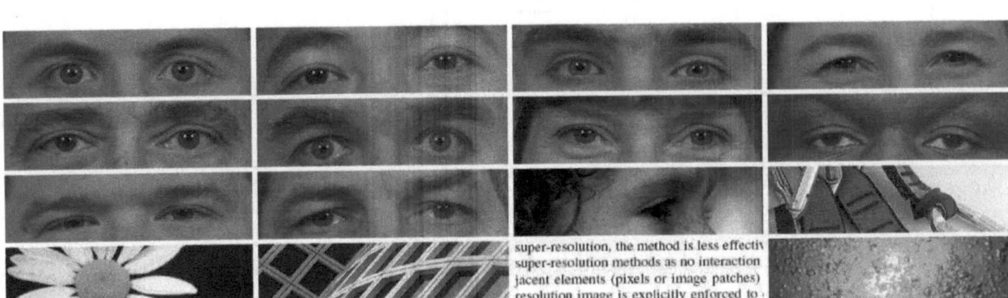

Fig. 2. Training images pool, from left to right, top to bottom: labeled No.1 to No. 16

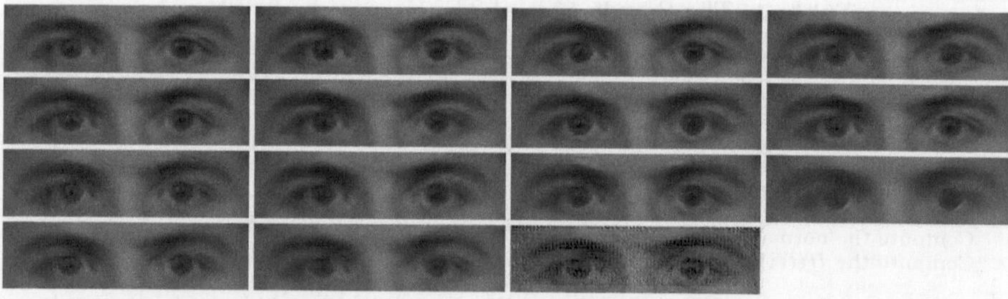

Fig. 3. Results of (YIQ) histogram matching and neighbor embedding. Rankings of the results descending from left to right, top to bottom. The corresponding numbers in training set are: 5, 10, 2, 4, 3, 9, 6, 7, 16, 11, 8, 14, 12, 13 and 15.

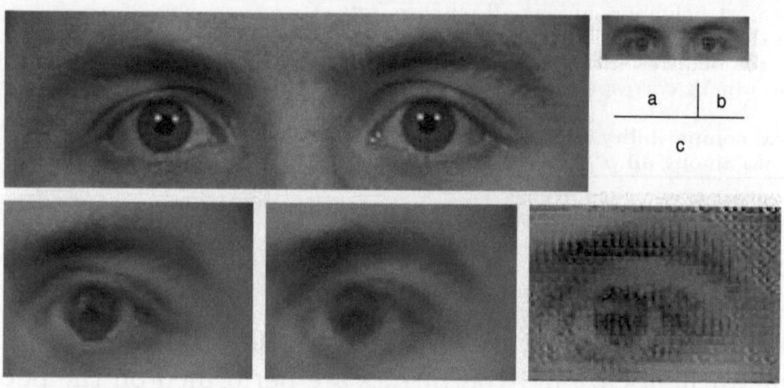

Fig. 4. a: High resolution target (Label No.1 in our pool). b: Low resolution input. c: Parts of results, Training images used: Left: No.5, Middle: No.16, Right: No. 15.

We can see that histogram matching chooses topic related training images prior to those irrelevant images (No.12 to No.16 in training set). Furthermore, it is easy to find that mosaic effect is reinforced as ranking increases in Figure 4.

To quantitative analyze the performance of the reconstruction of super-resolution image, RMS (Root Mean Square) errors are introduced which have the formulation of

$$\mathrm{RMSe} = \left(\sum_{i=1}^{n} \frac{(\hat{y}_i - y_i)^2}{n}\right)^{\frac{1}{2}} \tag{5}$$

Where $\hat{y}_i$ stands for the values of pixel in the ideal target Y and y_i stands for the values of corresponding pixels in output Y_t. And n stands for the number of total pixels in Y.

According to the ranking of histogram matching, average ranked RMS errors and standard deviations can be computed from 15 images of which each is not the low-resolution test image. The ranked RMS errors are illustrated as in Figure 5.

Although the method with histogram matching may not always choose the optimal training image left in the training set, it chooses image good enough and only increases the RMS error by a trivial little comparing to the optimal one.

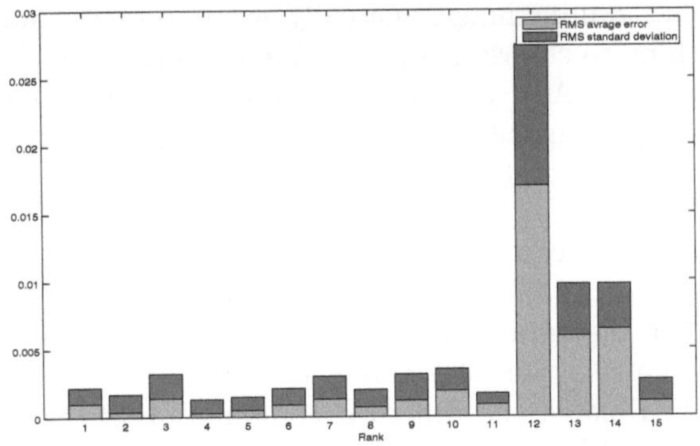

Fig. 5. Average RMS errors and standard deviations of normalized histogram matching-based ranking with 16 test images

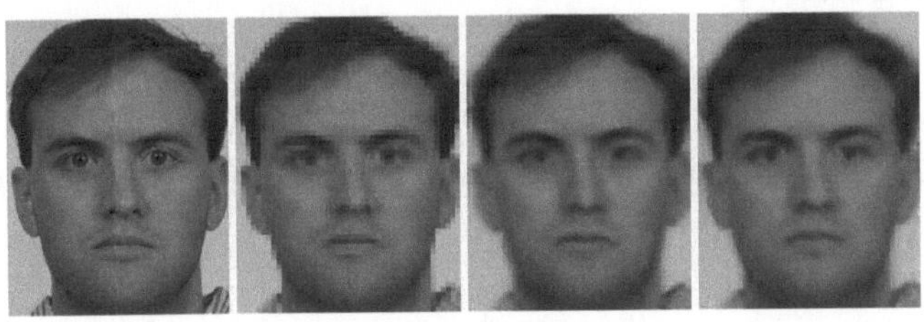

Fig. 6. a: High-resolution target; b: Low-resolution input(Enlarged); c: Histogram matching result (No.4 chosen, RMSe=0.0535); d: Optimal choice result (No. 10 chosen, RMSe=0.0519)

The method of histogram matching is efficient enough to automatically choose training image instead of choosing manually.

At last we show an example of a whole face using YIQ histogram matching and compare the performance with optimal RMS choice. The results are illustrated in c, d in Figure 6. Our training set is identical to our experiment pool while the size of each image is a little smaller to save running time. Notice that our choice of training image based on histogram matching are second best in all RMS errors, i.e., it is the optimal choice if no.10 does not exist in training set.

It is worth noting that the result is not as good in overall details as before. One reason is that we just use parts of eye region, which are not very so similar with the whole face. During our further research of using geometric division to separate facial features and choose training images for each feature under the same principle,in other regions good recovery is obtained as that of eye regions.

4 Conclusion

In this paper, to carry out super-resolution of face images, we improve the novel method of Super-Resolution through Neighbor Embedding. We indicate the problem of the choice of the training image affecting the quality of results. Instead of selecting the training image manually, we propose the automatic method of histogram matching to choose the proper image from the training set and obtain fairly good results. And it is effective and costless to carry out and as a result it explores the capacity of the training set with limited images. The proposed approach has potential application in the preprocessing of biometric person authentication.

Several problems deserve to make further research. First, the performance of neighbor embedding can be further improved and specified for biometric authentication. Second, histogram matching only provides a principle and coarse approximation to the selection of training image, more elaborate methods are under our research. Finally, the practical combination of our proposed super-resolution approach and biometric person authentication systems is desirable.

Acknowledgement

The authors are very grateful to PhD Hong Chang and Professor Dit-Yan Yeung for generous providing source code and invaluable comments. And Portions of the research in this paper use the Gray Level and Color database of the FERET program.

References

1. R. C. Gonzalez and R. E.Woods, Digital Image Processing(Second Edition), Prentice Hall, 2002.
2. William T. Freeman, Thouis R. Jones, and Egon C. Pasztor, "Example-Based Super-Resolution," in *Proceedings of Computer Graphics and Applications, IEEE*, March/April 2002, pp. 56–65.
3. Simon Baker, Takeo Kanade, "Limits on Super-Resolution and How to Break Them," *IEEE Transactions on Pattern Analysis and Machine Intelligence*,vol.24, NO.9, September 2002.
4. H. Chang, D. Y. Yeung, Y. Xiong, "Super-resolution through neighbor embedding," *Proceedings of the IEEE Computer Society Conference on Computer Vision and Pattern Recognition (CVPR)*, vol.1, pp.275-282, Washington, DC, USA, 27 June - 2 July 2004.
5. J. Zhang, S. Z. Li, and Jue Wang, "Manifold Learning and Applications in Recognition,"in *Intelligent Multimedia Processing with Soft Computing*. Yap Peng Tan, Kim Hui Yap, Lipo Wang (Ed.), Springer-Verlag, Heidelberg, 2004.
6. J. Zhang, "Several Problmes in Manifold Learning," *Machine Learning and Applications*, Zhi-Hua Zhou et. al. (Eds.), Tsinghua University Press, 2005.
7. S. T. Roweis and K. S. Lawrance, "Nonlinear Dimensionality Reduction by Locally Linear Embedding," *Science*, 290, pp. 2323-2326, 2000.
8. P. J. Phillips and H. Moon and S. A. Rizvi and P. J. Rauss, "The FERET Evaluation Methodology for Face Recognition Algorithms," *IEEE Trans. Pattern Analysis and Machine Intelligence*, Volume 22, October 2000, pp. 1090-1104.

Invertible Watermarking Algorithm with Detecting Locations of Malicious Manipulation for Biometric Image Authentication

Jaehyuck Lim[2], Hyobin Lee[1], Sangyoun Lee[2], and Jaihie Kim[2]

[1] Biometric Engineering Research Center (BERC),
Graduate Program in Biometrics
[2] Department of Electrical and Electronic Engineering, Yonsei University,
134, Shinchon-dong, Seodaemon-ku, Seoul, 120-749, Korea
Tel:+82-2-2123-5768, Fax:+82-2-362-5563
{jhlim, leehb00, syleee, jhkim}@yonsei.ac.kr

Abstract. In this paper, we present a new method for authentication of biometric images. Our method uses an invertible watermark that can also detect malicious manipulations simultaneously. While virtually all watermarking schemes introduce a small amount of non-invertible distortion in original biometric images, our new method is invertible in the sense that, if the data is deemed authentic, distortion due to authentication can be removed if it becomes necessary to obtain the original biometric image. This technique provides cryptographic strength when verifying image integrity because the probability of making an undetectable modification to the image can be directly related to a secure cryptographic element, such as a hash function. Also, if the biometric image is manipulated, the positions of intentional manipulation can be clearly identified.

1 Introduction

With the present widespread utilization of biometric identification systems, establishing the authentication of biometric images (face, fingerprint, iris etc.) themselves has emerged as an important research topic. Cryptography and digital watermarking techniques are two possible ways of achieving this. While cryptography focuses on methods of making encrypted information meaningless to unauthorized parties [1], digital watermarking techniques can be used to embed proprietary information in host biometric images in order to protect the intellectual property rights and authentication of those images [2]. Encrypted templates are secure since they cannot be utilized or modified without decrypting them with the correct key, which is typically secret. But cryptography does not provide any security once data has been decrypted. Since there is a possibility that decrypted data can be intercepted, cryptography does not address the overall security of biometric images. However, since digital watermarking involves embedding information directly into host biometric images, these images are secure even after decryption. The watermark, which is not related to encryption-decryption operations, provides another line of defense against the illegal utilization of biometric images.

D. Zhang and A.K. Jain (Eds.): ICB 2006, LNCS 3832, pp. 763 – 769, 2005.

One possible drawback of authentication based on watermarking is that the authenticated biometric image will inevitably be distorted by a small amount of noise due to authentication itself [3]. In virtually all previously proposed watermarking schemes, this distortion cannot be completely removed even when the image is deemed authentic. Although the distortion is often quite small, it may be unacceptable in certain kinds of biometric images, for example, iris and fingerprint images. In this paper, we analyze the conditions under which it is possible to "undo" changes introduced by authentication if the biometric image is verified as authentic. We present techniques that make this kind of invertible authentication possible. Then, we embed an invertible message authentication code into the biometric image so that anyone who possesses the authentication key can revert to an exact copy of the original biometric image that existed before authentication occurred. If the biometric image is also manipulated, the proposed method can protect the biometric contents from malicious manipulation and detect modified parts of the biometric data at the same time.

The rest of this paper is organized as follows. In section 2, we explain the proposed invertible watermarking algorithm. Simulation results are given in section 3. Finally, the main contributions of this work and suggestions for future research are summarized in section 4.

2 The Proposed Watermarking Algorithm

Figure 1 shows the overall structure of our proposed invertible watermarking algorithm. The algorithm is able to authenticate biometric images as well as detect manipulated parts. The proposed watermark embedding consists of two stages: invertible biometric image authentication and detection of intentional biometric image attacks. These two stages are explained in more detail in the following subsections.

2.1 Embedding Watermark for the Authentication and Invertibility of the Biometric Image

For the authentication and invertibility of the biometric image, we applied the conventional image watermarking method to the biometric image [3]. Let us assume that the original biometric image is a grayscale image with pixel values from the set $P = \{0, \ldots, 255\}$. We started to calculate the hash value of the biometric image which consists of 128 bits. In our experiments, we used the *MD5* hash function. After that, we divided the image into disjoint groups of n adjacent pixels $(x_1, x_2, \ldots, x_n)$. We also defined the so-called discrimination function f that assigns a real number $f(x_1, x_2, \ldots, x_n)$ to each disjoint group $G = (x_1, x_2, \ldots, x_n)$ as follows.

$$f(G=(x_1,x_2,\ldots,x_n)) = \sum_{i=1}^{n-1} |x_{i+1} - x_i| \tag{1}$$

Then, we defined an invertible operation as F on P. This operation is known as flipping and features a permutation of gray levels that entirely consists of two cycles. Thus, F has the properties of F^2 = Identity or $F(F(x)) = x$ for all $x \in P$. We used the discrimination function f and the flipping operation F to define three pixel groups: regular (R), singular (S), and unusable (U).

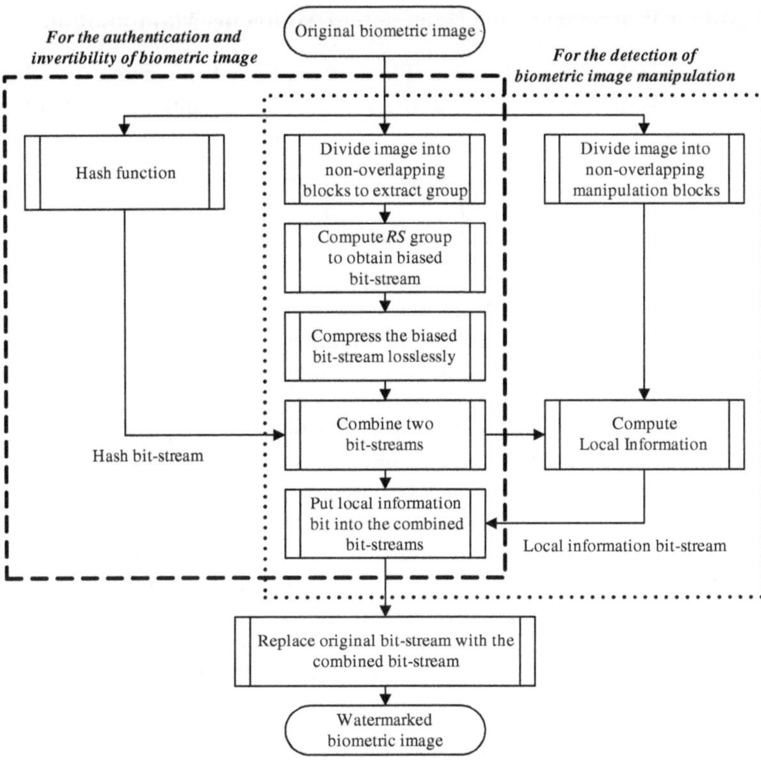

Fig. 1. Flowchart of the proposed algorithm

$$
\begin{array}{lll}
\underline{R}\text{egular groups:} & G \in R \ulcorner \text{ if} & f(F(G)) > f(G) \\
\underline{S}\text{ingular groups:} & G \in S \quad \text{if} & f(F(G)) < f(G) \\
\underline{U}\text{nusable groups:} & G \in U \ulcorner \text{ if} & f(F(G)) = f(G).
\end{array} \qquad (2)
$$

Thus, the R and S groups are flipped into each other under the flipping operation F, while the unusable groups U do not change their status. In a symbolic form, $F(R)=S$, $F(S)=R$, and $F(U)=U$. We then formulated the results using the data embedding method. By assigning a 1 to R and a 0 to S, we embedded one message bit in each R or S group. The RS bit-stream normally has a large number of '0' or '1'. So, to reduce the data, we adopted the context-free lossless compression algorithm for the RS bit-stream in order to obtain the compressed RS bit-stream. So, after lossless compression, the length of the bit-stream was reduced significantly. Namely, the compression created an empty space that was used to store the hash bit-stream, which provided authentication, and the local information bit-stream, which detected malicious manipulation. Details of the local information and embedding process are discussed in the next subsection.

2.2 Embedding Watermarks for Detection of Malicious Manipulation

Details of the algorithm are as follows. First, as explained in section 2.1, the 128-bit hash bit-stream was computed from the original biometric image as shown in Eq. (3), and the biased *RS* bit-stream was also obtained as shown in Eq. (4).

$$\text{Hash bit-stream} = \{ H_k \in \{0,1\} | 1 \le k \le 128 \} \tag{3}$$

$$RS \text{ bit-stream} = \{ RS_{ij} \in \{0,1\} | 1 \le i \le I, 1 \le j \le J \} \tag{4}$$

In Eq. (4), J is the number of non-overlapped manipulation blocks obtained from the original biometric image, and I is the number of *RS* groups in each block. Next, lossless compression was performed on the obtained *RS* bit stream. The compression created an empty space that was used to store the hash bit stream already obtained from the original biometric image. This made it possible to authenticate the watermarked image and to store the local information bit-stream to detect the location of malicious manipulations in the attacked image. When the corresponding bit-stream was put into the space, a user-defined bit-stream shuffling method was used to increase the security level of the proposed algorithm. As a result, the combined bit-stream was obtained by concatenating and shuffling the compressed *RS* bit-stream and the hash bit-stream as shown in Eq. (5). The size of the combined bit-stream is designed to be equal to that of the *RS* group. In the combined bit-stream, the last bit of each manipulation block, CRS_{Ij}, was left empty to store local information. To obtain the combined bit-stream with local information bits, a user-defined sequence with size of J was generated as shown in Eq. (6).

$$\text{Combined bit-stream} = \{ CRS_{ij} \in \{0,1\} | 1 \le i \le I, 1 \le j \le J \} \tag{5}$$

$$\text{User-defined sequence} = \{ P_j \in \{0,1\} | 1 \le j \le J \} \tag{6}$$

J is the number of blocks used to detect the manipulated location. All bits of each manipulation block excluding the last local information bit were summed. Then, the local information bit was assigned to equalize the LSB of the summed value and the corresponding bit of the user-defined sequence as shown in Eq. (7).

$$\text{Local information } L_j = \begin{cases} 1 \ \ if & \left(\left(\sum_{i=1}^{I-1} CRS_{ij} \right) \%2 \right) \ne P_j \\ 0 \ \ if & \left(\left(\sum_{i=1}^{I-1} CRS_{ij} \right) \%2 \right) = P_j \end{cases} \tag{7}$$

$$\text{Watermark embedding} \begin{cases} F(RS_{ij}) & if \ \ CRS_{ij} \ne RS_{ij} \\ skip & if \ \ CRS_{ij} = RS_{ij} \end{cases} \tag{8}$$

Finally, the CRS_{ij} bit stream assigned local information bit was put into the *RS* group as shown in Eq. (8). This means that when both the RS_{ij} and CRS_{ij} were compared, if the two values were the same, RS_{ij} was used as the bit-stream, otherwise the *RS* group mode was flipped according to function F, defined in Section 2.1. When the watermarked image is manipulated, the $R(S)$ group may be converted to the $S(R)$ group, or local information bit may be changed. Based on this information, if the number of 1s of the *R* group in each manipulation block is different from the generated user-defined sequence as shown in Eq. (6), the block is assumed to have been manipulated. Then, the proposed algorithm can only check the LSB of the number of 1s in each block.

2.3 Integrity and Manipulation Verification

Figure 2 shows the overall structure of our watermark detection process. The user extracts the bit-streams from all *R* and *S* groups (*R*=1, *S*=0) by scanning the watermarked biometric images in the same order as they were scanned during embedding. The extracted bit-stream is separated into the embedded original hash bit-stream, the embedded original local information bit-stream, and the embedded compressed *RS* bit-stream. The compressed *RS* bit-stream is decompressed to reveal the original status of all the *R* and *S* groups. The image is then processed once more and the status of all the groups is adjusted as necessary by flipping them back to their original states. Thus, exact copies of the original images are obtained. Then, we calculate the hash values of the restored images and compare the calculated hash values with the embedded hash ones. If the two hash values are identical, the biometric image is authentic. This means that the restored biometric images can be verified to be the original ones. Otherwise, we check the relationship between the number 1s of each manipulation block and of the corresponding local information bit. If they are different, the corresponding blocks are shown to have been manipulated intentionally.

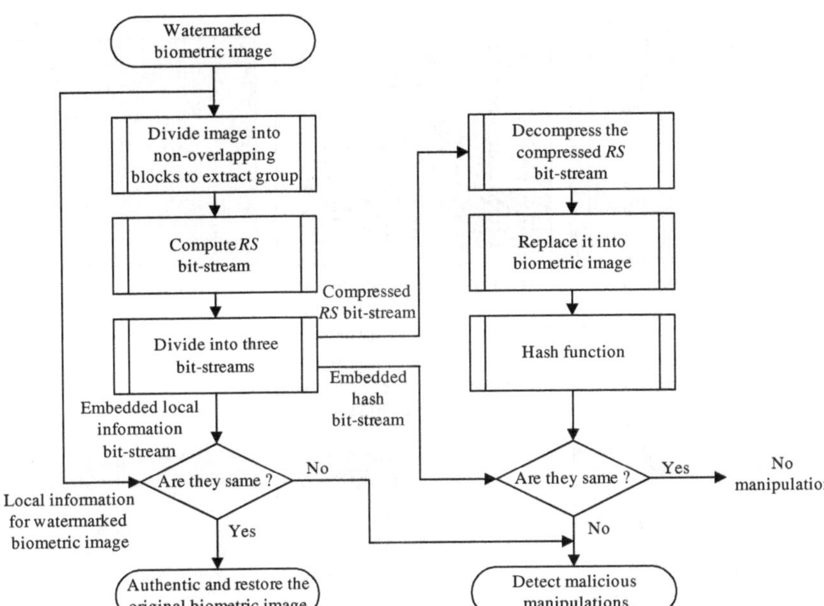

Fig. 2. The watermark detection process

3 Experimental Results

To demonstrate the proposed biometric image authenticator, we used several images in our experiments (face, fingerprint, and iris), as shown in Fig. 3. The corresponding watermarked images are shown in Fig. 4. As can be seen, distortion due to watermarking is invisible and can be removed from the watermarked image if it is deemed authentic. For the detection of malicious manipulation, the watermarked biometric images in Fig. 4 were manipulated partly (in case of face and iris images) and were

(a) Face image (b) Fingerprint image (c) Iris image

Fig. 3. Original biometric images

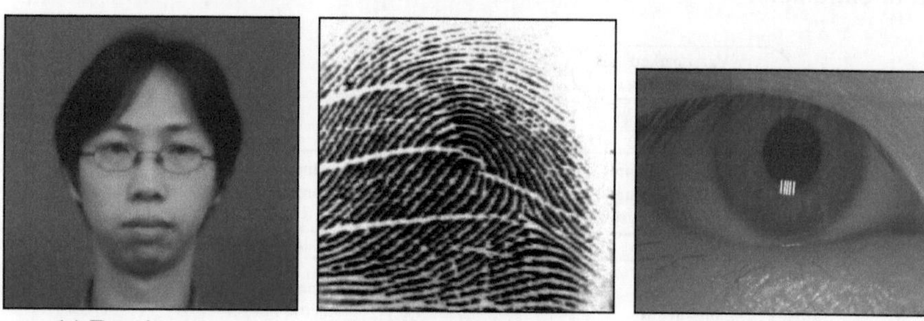

(a) Face image (b) Fingerprint image (c) Iris image

Fig. 4. Watermarked biometric images

(a) Face image (b) Fingerprint image (c) Iris image

Fig. 5. Manipulated biometric images

(a) Face image (b) Fingerprint image (c) Iris image

Fig. 6. Block-based detection of the manipulated biometric images in Fig. 5

modified by content-based manipulations (in case of fingerprint images) as shown in Fig. 5. Applying the proposed algorithm, we were able to identify the block-based locations of the intentional manipulations as shown in Fig. 6.

4 Conclusions

In the paper, we proposed an invertible biometric image watermarking algorithm which can also locate the positions of intentionally manipulated blocks. While virtually all watermarking schemes introduce a small amount of non-invertible distortion in original biometric images, our new proposed method is invertible in the sense that, if the image is deemed authentic, distortion due to authentication can be removed to obtain the original biometric image. Also, if the biometric image is manipulated, the positions of the intentional manipulation can be identified.

Acknowledgements

This work was supported by the Korea Science and Engineering Foundation (KOSEF) through the Biometrics Engineering Research Center (BERC) at Yonsei University.

References

1. O. Khalifa, M. Islam, S. Khan, M. Shebani, "Communications cryptography", RF and Microwave Conference, 2004. RFM 2004. Proceedings, 5-6 Oct. pp. 220-223, 2004.
2. A. Kejariwal, "Watermarking", Potentials, IEEE, vol. 22, Issue 4, pp. 37-40, Oct.-Nov. 2003.
3. J. Fridrich, M. Goljan, R. Du, "Invertible authentication watermark for JPEG images", International Conference on Information Technology: Coding and Computing (ITCC), Las Vegas, Nevada, USA, pp. 223-227, April 2001.

The Identification and Recognition Based on Point for Blood Vessel of Ocular Fundus

Zhiwen Xu[1], Xiaoxin Guo[1], Xiaoying Hu[2], Xu Chen[1], and Zhengxuan Wang[1]

[1] Open Symbol Computation and Knowledge Engineering Laboratory of State Education Department, College of Computer Science and Technology
[2] The First Clinical Hospital,
Jilin University, Changchun City, 130012, Jilin Province, China
xuzhiwen@public.cc.jl.cn

Abstract. Today, iris recognition, fingerprint recognition, face recognition, voice recognition and other biometric technology are experiencing rapid development. This paper addresses a new biometric technology–the identification and recognition based on point of blood vessel skeleton for ocular fundus. The image for green gray scale of ocular fundus is utilized. The cross point of skeleton shape of blood vessel for ocular fundus using contrast-limited adaptive histogram equalization is extracted at first. After filtering treatment and extracting shape, shape curve of blood vessels is obtained. The cross point of shape for curve matching is later carried out by means of cross point matching. The recognition based on shape for blood vessel of ocular fundus has been demonstrated in this paper to possess high Identification and recognition rate, low rejection recognition rate as well as good universality, exclusiveness and stability. With more and more progress made in extracting technology, the recognition for blood vessel of optic fundus is to become an effective biometric technology.

1 Related Work

The shape of blood vessel for ocular fundus is important indicator to diagnose such diseases as hypertension, vascular sclerosis, coronary artery sclerosis and diabetes.The certain achievements have been fulfilled in image processing of blood vessel of ocular fundus. Considering that zero-crossing can neither always correspond to real edge nor always signal edge position precisely, F. Ulupinar[1] makes relevant revision. Canny[2] raises optimal operator suitable for any randomly shaped edge extraction. However, due to utilization of Gaussian filter, it still remains imprecise occasionally. Though arithmetic operator of morphological gradient[3] is easy and quick, it's confined to image with the noise of pepper and salt. The relax method is used to extract linear blood vessels. Since arteries and veins cross over each other, blood vessels have to be segmented into several pieces. Hueckel uses arithmetic operator to apply fitting method of blood vessel to extract edge[4]. It is insensitive to noise and effective in region of intense texture, but it demands tremendous calculation. Taking into consideration the inherent features of blood vessel [5], being that

D. Zhang and A.K. Jain (Eds.): ICB 2006, LNCS 3832, pp. 770–776, 2005.

blood vessel is of linear shape and gradient direction on the left threshold is just opposite to that on the right threshold, Tascini G. puts forward the method of search edge direction [6]. The edge of blood vessel in the second period of hypertension is rather blur and very low contrast so that it can't be traced. In order to improve that, Rangayyan R. M. et al suggests reinforcing linear feature of blood vessel at the frequency field. But the result isn't good due to exudation of ocular fundus. Chauduri S. et al. employs the Gaussian models of twelve different directions to filter fundus blood vessel [7]. Then, those methods aren't nice at enhancing the fundus image of blood vessel for hypertension patients. Otherwise, utilization of Gaussian madel of fixed size makes it fail in dealing with blood vessels with salient variation in diameter and shape highly curved. Substitution of Gaussian models of different sizes is sure to cause excessive computation. The literatures[8,9] make a study on matching for blood vessel of ocular fundus. The literatures [10,11] make a study of algorithm of fingerprint based on point pattern. Based on the researches mentioned above, together with the blood vessel of Ocular Fundus for biometric features, this paper addressed the recognition for blood vessel of Ocular Fundus, using contrast-limited adaptive histogram equalization by employing gradient vector curve to analyze the features of blood vessel of ocular fundus.

2 Feature to Extract Blood Vessel of Ocular Fundus

The diameter of capillary vessels for ocular fundus may vary with passage of time and various diseases, then their direction for distribution remains the same for long time. This encourages vector curve that analyzes distribution of blood vessel orientation to be utilized as biometric features for blood vessel of ocular fundus. The concrete steps extracts feature as the following:

(1) Enhance contrast of gray scale image and utilize contrast limited adaptive histogram equalization. At first, deal with subregions in the image and then converge adjacent small regions with method of bilinear interpolation in order to get rid of artificial edges. Fig.1 is fundus image of gray scale; Fig.2 is gray scale enhanced image; Fig.3 is inversion to enhance result of gray scale.

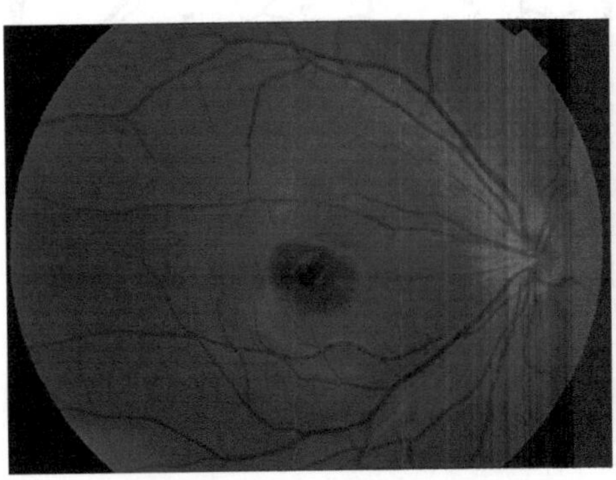

Fig. 1. Fundus image of gray scale

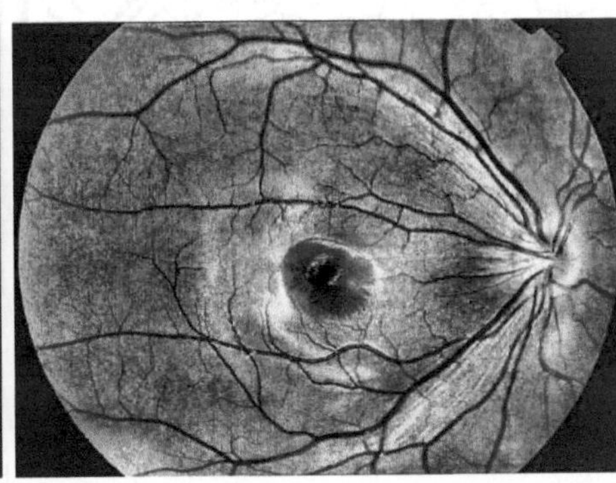

Fig. 2. Gray scale enhanced image

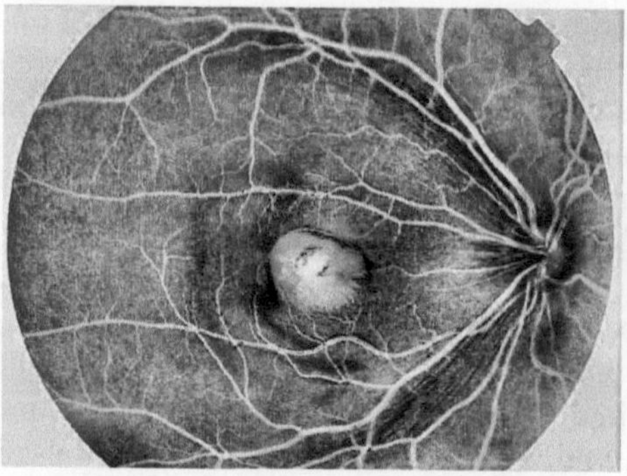

Fig. 3. Inversion to enhance gray scale

Fig. 4. Filtering

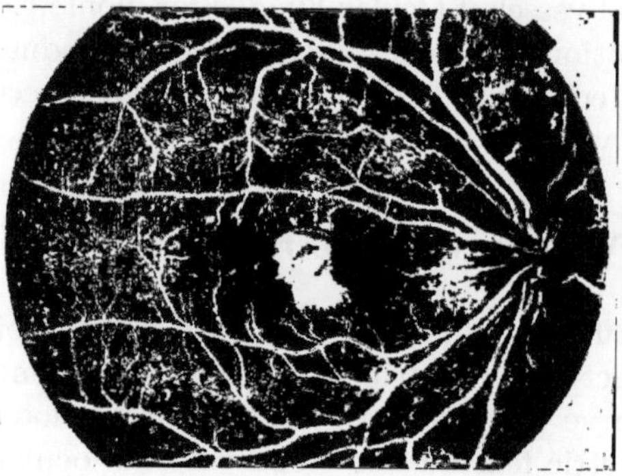

Fig. 5. Binary processing

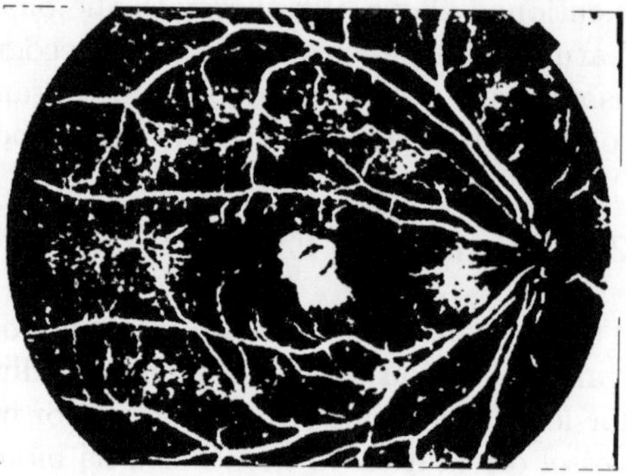

Fig. 6. Median filter

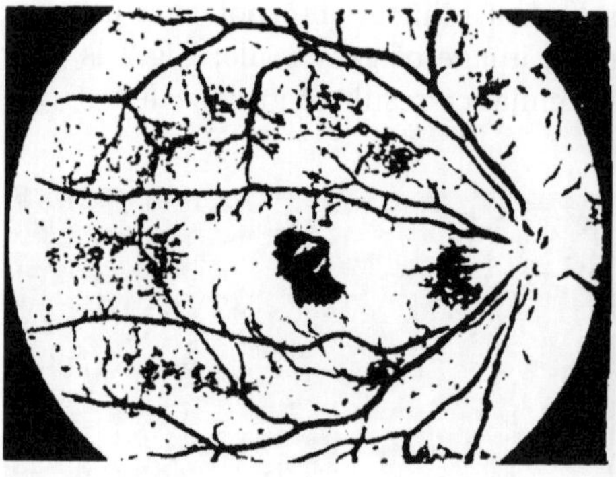

Fig. 7. Fill holes

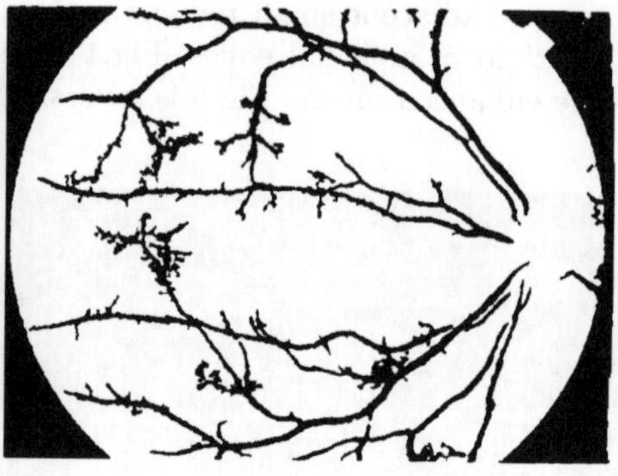

Fig. 8. Erase holes

(2) Filter image by means of un-sharp algorithm(fig.4).

(3) Binary processing. Binary processing deal with all the local maximums over a supposed threshold and input gray scale image as a parameter. In a

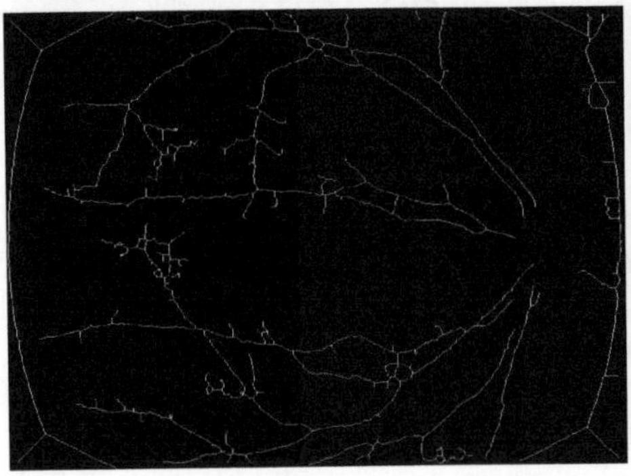

Fig. 9. Extract skeleton

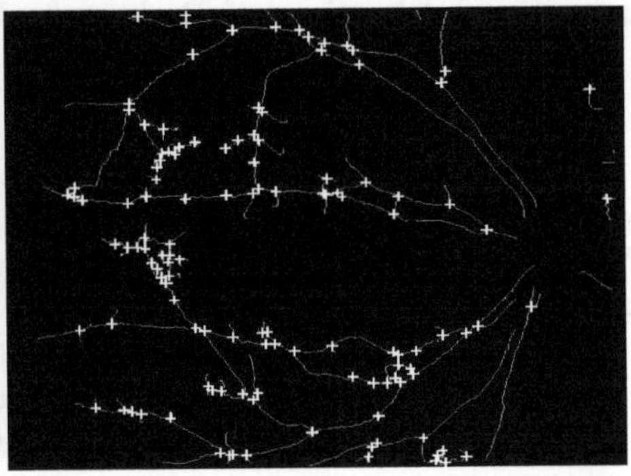

Fig. 10. The cross point of shape of blood vessel

dimorph image, local maximum of output dimorph image is assumed to be 1, and the remainder 0, which are used to search out region whose brightness changes most. (Fig.5). At last, median filter is utilized (Fig.6).

(4) Fill holes. Holes are those dark regions surrounded by whiter setting. Exchange o with 1 in the binary image (Fig.7) and erase holes (Fig.8).

(5) Extract skeleton on the basis of erosion(Fig.9). Cut edge disturbance and find out the embranchment point of shape of blood vessel. The "+" mark the embranchment point and cross point of shape of blood vessel(Fig.10).

3 Matching Algorithm for Blood Vessel of Ocular Fundus

In a bid to transform a given feature point of input image of blood vessel to a corresponding position in the template image of blood vessel, the corresponding transformation factor should be known. The algorithm in this paper deals with image of blood vessel with the same differentiation. In ideal condition, the zoom factor is Z. Assume a given point in the input point set P is $p_i(x_{pi}, y_{pi}, \theta_{pi}, 1)$, which is transformed by the following formula into $p_i(x^T_{pi},\ y^T_{pi},\ \theta^T_{pi},\ Z^T_{pi})$; and assume a given point in the template point set is $q_j(x_{qj}, y_{qj}, \theta_{qj}, z_{qj})$. If $(x^T_{pi}, y^T_{pi}, \theta^T_{pi} = (x_{qj}, y_{qj}, \theta_{qj}, z_{qj})$, the transforming factor is that p_i is similar to q_j on the condition of ($\triangle$x, $\triangle$y, $\triangle\theta$, $\triangle$ z). The images of blood vessel are captured using the same facilities and the same distance. So p_i is simplified as $(x_{pi}, y_{pi}, \theta_{pi})$ and q_j as $(x_{qj}, y_{qj}, \theta_{qj})$. The zoom factor in the course of transformation is ruled out.

$$\begin{vmatrix} x^T_{p_i} \\ Y^T_{p_i} \\ \theta^T_{p_i} \\ 1 \end{vmatrix} = \begin{vmatrix} \cos\triangle\theta & -\sin\triangle\theta & 0 & \triangle x \\ \sin\triangle\theta & cos\triangle\theta & 0 & \triangle y \\ 0 & 0 & 1 & \triangle\theta \\ 0 & 0 & 0 & 1 \end{vmatrix} \begin{vmatrix} x_{p_i} \\ Y_{p_i} \\ \theta_{p_i} \\ 1 \end{vmatrix} \tag{1}$$

Where $\triangle$x and $\triangle$y are translation factors in x direction and y direction respectively, $\triangle\theta$ is rotation factor, and $\triangle$Z=1 is zoom factor. The first three factors need to be determined in order to identify the two images of blood vessel precisely.

3.1 Determination of Reference Point and Calculation of Transformation Factor

In the process of the point of blood vessel matching, the calculation of matching reference point is of a great importance. Clustering method is utilized to get a precise matching reference point and a group of transformation parameters. Though this method can lead to a quite precise matching reference point, it involves excessive computation.Two images of blood vessel can be transformed in accordance with the obtained transforming parameter, which leads to a further examination of similarity between two triangles.

(1) Calculate respectively corresponding side-length to vertex p_i and q_j, $|p_1P_2|$ and $|q_1q_2|$.

(2) If $||p_1P_2|$-$|q_1q_2|| > D_1$, one corresponding side of the two triangles isn't of the same length, then the two triangles aren't congruent. The examination ends. Re-choose vertices p_i and q_j and two nearest feature points (p_1, p_2) and (q_1, q_2). Return to the first step.

(3) Otherwise, calculate respectively the distance form p_i to p_1, q_j to p_1 and from p_2 to q_1, p_2 to q_2:$|p_ip_1|$, $|p_ip_2|$ and $|q_jq_1|$, $|q_jq_2|$. If $||p_ip_1| - |q_jq_1|| \leq D_2$ and $||p_ip_2| - |q_jq_2|| \leq D_2$, or$||p_ip_2| - |q_jq_1|| \leq D_2$ and $||p_ip_1| - |q_jq_2|| \leq D_2$, it's proved the three sides of the two are of similar length and the two triangles are almost congruent. Otherwise Re-choose vertices p_i and q_j and two nearest feature points (p_1, p_2) and (q_1, q_2). Return to the first step.

(4) According to corresponding vertex of the two triangles, calculate orientation disparity between possibly matching feature points, $\triangle\theta_{p_1q_j}$, $\triangle\theta_{p_1q_1}$, $\triangle\theta_{p_2q_2}$. The formula is.

$$\triangle Q_{pq} = \begin{cases} \theta_p - \theta_q, if(\theta_p - \theta_q \geq 0) \\ theta_p - \theta_q + 180, if(\theta_p - \theta_q < 0) \end{cases} \tag{2}$$

If angle disparity between corresponding vertices is similar i.e. $\triangle\theta_{p_1q_j} \approx \triangle\theta_{p_1q_1} \approx \triangle\theta_{p_2q_2}$, the angle between the two feature points (p_i, p_1, p_2) and (q_j, q_1, q_2) is supposed to satisfy a rotation relationship. The formula of this rotation is.

$$\triangle\theta = \frac{1}{3}(\theta_{p_iq_j} + \theta_{p_1q_1} + \theta_{p_2q_2}) \tag{3}$$

Otherwise no matching is formed between the sets. Re-choose vertices p_i and q_j and two nearest feature points (p_1, p_2) and (q_1, q_2). Return to the first step.

(5) Choose (p_i, q_j) as a transforming circle for the rotation and then rotate (q_j, q_1, q_2). The consequent point is . Calculate spatial disparity in x direction and y direction respectively which are $(\triangle x_{p_iq_j}, \triangle y_{p_iq_j})$, $(\triangle x_{p_1q_1}, \triangle y_{p_1q_1})$, $(\triangle x_{p_2q_2}, \triangle y_{p_2q_2})$. The formula is.

$$\triangle x_{pq} = x_p - y_q \tag{4}$$

$$\triangle y_{pq} = y_p - y_q \tag{5}$$

Now if $\triangle x_{p_iq_j}) \approx \triangle x_{p_1q_1}) \approx \triangle x_{p_2q_2})$ and $\triangle y_{p_iq_j}) \approx \triangle y_{p_1q_1}) \approx \triangle y_{p_2q_2})$, the two feature point subsets (p_i, p_1, p_2) and (q_j, q_1, q_2) meet a kind of transformation relationship in x, y direction. The two subset match whose rotation, translation and transformation factor are $(\triangle\theta, \triangle x, \triangle y)$ respectively, and

$$\triangle x = \frac{1}{3}(x_{p_i q_j} + x_{p_1 q_1} + x_{p_2 q_2}) \tag{6}$$

$$\triangle y = \frac{1}{3}(y_{p_i q_j} + y_{p_1 q_1} + y_{p_2 q_2}) \tag{7}$$

According to the obtained reference point $(p_i, q_j)$and transformation factor $(\triangle\theta, \triangle x, \triangle y)$, the skeleton ban of blood vessel be judged whether the same or not.

4 Experimentation

In order to identify that the embranchment point of skeleton shape of blood vessel can serve as biometric feature for fundus image of blood vessel, a TRC-50/50VT fundus camera produced form Japanese Topcon company, and fundus images of blood vessel of 2000 people are collected as experimental database where each people has ten images taken at different time. To obtain False Non Match Race, or False Rejection Rate, a matching algorithm is made between every fundus image of blood vessel T_{ij} and its other sample fundus images of blood vessel F_{ik} $(0 \leq j \leq k \leq 9)$ and then the total matching times should be ((10x9)/2)x 2000=360000. To obtain False Match Rate, or False Acceptance Rate, a matching algorithm is made between the first sample template T_{i0} of every fundus blood vessel in the database and the first image F_{i0} of other fundus blood vessels in the same database. Calculate the ultimate matching result between fundus blood vessel images. The total matching times should be (2000x199)/2=199000.

Extracting features from two fundus images of blood vessel follows several steps. At first, according to gray scale image, fix the brightest region of the window, The region's center of the optical disk acts as the origin. Then starting with the horizontal direction; search out the first point of branch of the three vector curves of blood vessel skeleton. This point of branch is regarded as feature point. And then calculate the matching reference point of the two fundus images of blood vessel. The results of the cross-comparison experiment carried out on the 2000 fundus images of blood vessel are: zero false recognition, 25 false rejection and 0.0125 recognition rejection. In this paper, similarity between different fundus skeleton of blood vessel is measured, where four units are adopted as threshold box: 0.65 of fundus skeletons of blood vessel of different people overlap less than 0.292; 0.886 overlap less than 0.51; 0.973 overlap less than 0.73 and 0.9995 overlap less than 90.

Everyone has contrasting ocular fundus of blood vessel skeleton, which remains constant for long; The diameter of fundus blood vessel does change and capillary vessels do increase but they have no effect on fundus skeleton feature of blood vessel. The feature of embranchment point of blood vessel for ocular fundus is universal, unique and stable but due to certain difficulty as to feature extraction, it hasn't been paid enough attention. At present, with continuous progress in extracting technology, the identification of blood vessel for ocular fundus tends to become an effective method of recognition.

References

1. Ulupinar F, Medioni G: Refining edges detected by LoG operator [J].Comput Vis Graph and Image Process, 51:275 298, 1990.
2. Canny J: Acomputational approach to edge detection [J]. IEEE Trans, 8:679 698, PAMI,1986.
3. Peng J, Rusch Ph: Morphological filters and edge detection application to medical imaging [J]. Annual International Conference of the IEEE Engineering In Medicine and Biology Society, 13 (1): 0251 0252, 1991.
4. Huang C C, Li C C, Fan N, et al: A fast morphological filter for enhancement of angiographic images [J]. Annual International Conference of the IEEE Engineering in Medicine and Biology Society, 13(1): 0229 0230, 1991.
5. Tascini G, Passerini G, Puliti P, et al: Retina vascular network recognition [J]. Proc SPIE, 1898: 322 329, 1993.
6. Chauduri S, Chatterjee S, Katz N, et al: Detection of blood vessels in retinal images using two-dimensional matched filters [J]. IEEE Trans Med Imaging,8: 263 269, 1989.
7. T-L Ji, Sundareshan M K, Roehrig H: Adaptive image contrast enhancement based on human visual properties [J]. IEEE Trans Med Imaging,13:573 586, 1994.
8. Matsopoulos George K, Mouravliansky Nicolaos A, Delibasis Konstantinos K, et al: Automatic retinal image registration scheme using global optimization techniques [J]. IEEE Trans On Information Technology In Biomedicine, 3(1): 47 60, 1999.
9. Maes F, Collignon A, Vandermeulen D, et al: Multi-modality image registration by maximization of mutual information [J]. IEEE Trans Med Img, 16 (2): 187 198,1997.
10. Zhan Xiao-Si, Ning Xin-Bao, Yin Yi-Long, Chen Yun: An Improved Point Pattern Algorithm for Fingerprint Matching. Journal of Nanjing University, Vol.39, No.4,pp491 498, July, 2003.
11. Qi Y, Tian J, Deng X: Genetic algorithm based fingerprint matching algorithm and its application on automated fingerprint identification system. Journal of Software,11(4),pp488 493, 2000.

A Method for Footprint Range Image Segmentation and Description

Yihong Ding, Xijian Ping, Min Hu, and Tao Zhang

Zhengzhou Information Science and Technology Institute,
Zhengzhou, Henan, China, 450002
dingyihong@126.com

Abstract. In this paper, we firstly present a novel footprint range image segmentation method using the principal curvatures and the principal directions. Utilizing the principal curvatures information, we detect the peak areas as the seeds, and apply region growing to locate the edges of each patch. We apply the edge detection technology to the region growth rules, so the boundary localization is precise. To obtain more stable edge information, a multi-scale fusion approach is proposed to integrate the segmentation results calculated at different fitting sizes. After the segmentation, according to the shape characteristics of footprint, we use superquadric and saddle models to describe shape features of each patch. The experiments results on footprint range images show that the segmented patches and the descriptions represent footprint biometric information effectively and set a reliable basis for the further recognition.

1 Introduction

Biometric technologies are automated methods of recognizing a person based on physiological or behavioral characteristics. In the recent years, as the development of the information and automation technologies, biometrics is becoming the foundation of an extensive array of highly secure identification and personal verification solutions. Many features have been widely studied such as face, fingerprints, hand geometry, handwriting, iris, retinal, vein, and voice [1].

Determined by the personal characteristics of stature, skelecton and gait etc, footprint is a kind of singular and stable biometric feature and has been used in criminal detection by artificially recognition for many years [2]. But it is paid less attention in personal identification because the measure of footprint is easy to be interfered by ground conditions and personal behaviors. These interferences bring great difficulties for automatic data collection and analysis.

According to foot's dress, footprint can be classified into bare footprint, shoe footprint, sock footprint etc. Based on the imprint face, it also can be classified into stereo footprint, for example step in the mud or sand, and plane footprint such as step on the floor [2]. The bare footprint and the stereo footprint contains more physiological and behavioral characteristics, because the bare footprint reflects foot impress more directly than shoe and sock footprint and the stereo footprint holds 3-Dimensional range information but the plane footprint shows only 2-Dimensional

D. Zhang and A.K. Jain (Eds.): ICB 2006, LNCS 3832, pp. 777 – 785, 2005.

information. At the same time, the researches on the bare and stereo footprint show more challenges and difficulties. In the past years, some researches have been done in the field of footprint recognition. For instance, Tian et al have proposed a heavy impress slice based method to recognize the 3-Dimensional shoe footprint [3]. Nakajima and his colleagues have developed a personal recognition method by analyzing the gait information from a pair of bare footprints [4]. Kennedy et al have studied on making measurements on the inked barefoot impressions on paper [5].

We take the stereo bare footprint as our research object. The stereo bare footprint is obtained by stepping on the soft impressed face of great plasticity with barefoot. To extract the shape and range information of the footprint, the plaster cast of the footprint is produced by the processes of perfusing, molding and airing. Then a special distance measuring instrument is used to measure the range data of the footprint surface from the plaster cast. In this way, the bare stereo footprint can be described with discrete range data matrix.

In this paper, we propose a novel algorithm to segment the range image of the stereo bear footprint based on principal curvatures and principal directions information. And then a new method is put forward to descript the footprint. Section 2 presents our segmentation algorithm for footprint surface. Section 3 shows the footprint description method using the superquadric and saddle models. And section 4 gives the experiment results and the conclusion.

2 Footprint Range Image Segmentation

The technologies of range image segmentation have been extensively studied and can be roughly classified into three categories: edge based [6][7][8], region based [9][10] and hybrid techniques [11][12].

The segmentation characteristics of commonly used can derive through the principal curvature and the principal direction (for example: the normal vector is two principal directions cross products, the Gaussian curvature is the product of two principal curvatures, the mean curvature is the mean value of two principal curvatures). We may know by the differential geometry theory, two principal curvatures and its the principal direction had reflected completely the curved surface bending strain and the change tendency, and it is important basis of the range image segmentation. Many segmentation algorithms [9][10][11][12] often neglect the direction that has reflected the change tendency of the curved surface. Therefore the principal directions are important to the determination curved surface patch boundary position and the boundary direction.

In this section, we propose a region-growing algorithm based on principal curvatures information and principal directions information, which is important and often neglect by many segmentation algorithm. We apply the edge detection technology to the region growth rules, so the region-based approach can obtain precise boundary localization. According to the practical requirements of footprint analysis, our method stresses on extracting convex local surfaces. Our approach consists of three steps as follows: 1) preprocessing and pre-segmenting, 2) region growing based on the principal curvatures and their directions, 3) fusing segmentation results at different scales to reduce over-segmentation.

2.1 Preprocessing and Pre-segmenting

The range image of the stereo bare footprint in our paper is obtained by measuring the discrete range data of the footprint plaster cast surface with an automated instrument. The collected range data can be digitalized and sent to a computer and saved as a range image. This kind of footprint data collection method has the advantage of cheap cost. But the collected foot range image has a large amount of noise that is caused by the complex ground condition and the quality of manufacturing the plaster cast. We firstly remove the impulsive noise with a 3x3 median filter. Then a Gaussian filter is used to smooth the image.

Limited by the precision of the collection instrument, the minimum practical interval between the adjacent pixels of the range image is 2.5mm. The corresponding image resolution is 128x64 pixels, which cannot meet the requirement of the analysis. So it is necessary to interpolate between the adjacent pixels to improve the data precision. The bicubic spline interpolation is used to improve the image resolution to 512x256 pixels, and the range image is named as $f(x, y)$. Fig 1(a) is the original collected range image. Fig 1(b) is the range image after the processing of noise moving and interpolating.

Suppose the footprint surface is twice differentiable. Because the footprint image data is sampled on a regular grid. The five partial derivatives (fx, f_y, f_{xx}, f_{xy} and f_{yy}) can be estimated respectively by fitting the local patch in a square window with the discrete orthogonal bi-quadratic polynomial [10]. The principal curvatures (k_1 and k_2) and their directions ($\boldsymbol{e}_1$ and $\boldsymbol{e}_2$) can be calculated based on the above five partial derivatives. The mean $H = \frac{1}{2}(k_1 + k_2)$ is called the mean curvature. The product $K = k_1 k_2$ is called the Gaussian curvature.

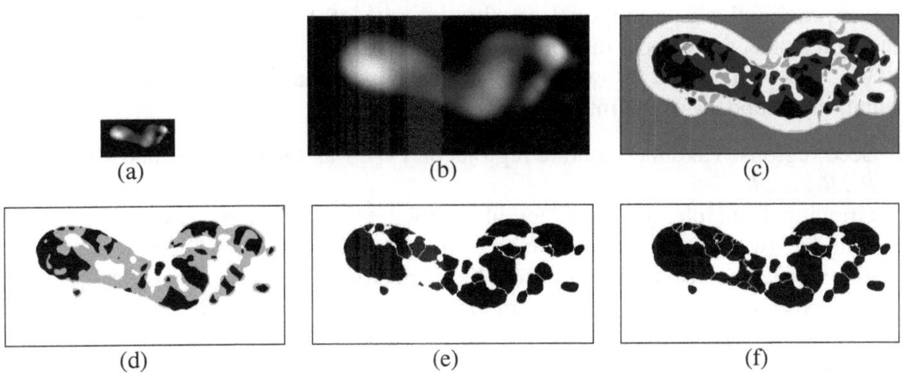

Fig. 1. Some footprint range image (a) range image (b) interpolated range image (c) curvature sign labeling (d) pre-segmentation result (e) result after the first growing (f) iterative growing segmentation result

According to the expert knowledge, the biometric features of bare footprint are mostly expressed by some interesting patches lying on toes, front sole, arch and heel. These interesting patches are generally peaks or ridges with the mean curvatures

$H(i,j) < \varepsilon_H$. Based on the degree of containing biometric characteristics, we classify the regions of sign image into three categories and showed in Fig. 1(d):

$$\begin{aligned} I &= \{(i,j) | H(i,j) < \varepsilon_H \text{ and } K(i,j) > \varepsilon_K\} \\ II &= \{(i,j) | H(i,j) < \varepsilon_H \text{ and } K(i,j) \le \varepsilon_K\} \\ III &= \{(i,j) | H(i,j) \ge \varepsilon_H\} \end{aligned} \tag{1}$$

where ε_H (> 0) and ε_K (> 0) are two preset zero thresholds. Fig. 1(c) shows the curvature sign image of the bear footprint according literature [10].

2.2 Region Growing

In the pre-segmentation result, the positions of the interesting patches which containing most biometric characteristics can be located with Type *I* areas. In this step, we take the regions of Type *I* as seeds to track the boundaries of each interesting patches. Type *II* areas will be disintegrated and absorbed into the interesting patches.

Suppose that footprint range image is a twice-differentiable surface. The principal curvatures and directions are continuous. Let the principal curvatures of pixel (i,j) are $k_1(i,j)$ and $k_2(i,j)$. Without loss of generality, we assume $k_1<k_2$. The definition of Type *I* areas means $k_1+k_2<0$ and $k_1k_2>0$. So it can be deduced that $k_1$and k_2 are both negative for type *I* areas. For Type *II* areas, $k_1+k_2<0$ and $k_1k_2 \le 0$, which means that k_1 is negative and k_2 is positive. Fig. 2(a) shows the zoomed-in local surface of the middle toe of the footprint in Fig. 1. Fig. 2(b) and Fig. 2(c) give respectively the intensities and directions ($\boldsymbol{e}_2$ projected to the zero-range plane) of its principal curvature k_2. We can see that at the boundaries of this interesting patch, the principal directions of k_2 are approximatively orthogonal to the tangent of the boundary curve, and the intensities of k_2 reach local maxima at their directions.

We design a region growing method based on the curvature characteristics to search the boundary of the interesting patches:

1. Seed regions. Take the separate regions of Type *I* as seed regions, and sign them as $R_1, R_2, \ldots, R_n$.
2. Growing Principle. For any neighbor point P of region R_i (i=0, 1, …, n), we examine two adjacent points in the direction of k_2 (no consider the sign of the direction). The two adjacent points lie respectively inside and outside R_i, denoted respectively as P_{in} and P_{out}. We embody P into R_i if it meets all the following three conditions: (i) P is the point of Type *II* areas; (ii) $k_2(P_{in}) < k_2(P)$; (iii) P_{out} is not the point of Type *II*, or if P_{out} is in Type *II* it must be satisfied that $k_2(P) < k_2(P_{out})$.

The above growing conditions lead the boundary of the interesting patches to grow and cease at the local maxima of k_2. But after growing there are still some points in Type *II* not included by any interesting patch (see Fig. 1(e)). For these residual areas, we take the small areas as noise and mark them as background. The residual areas of larger size are taken as the new interesting patches, and the same growing procedure are performed on them by taking the least mean curvature points of each area as the seeds. The growing procedure is iteratively applied until all the points of type *II* areas are classified into the interesting patches or marked as background.

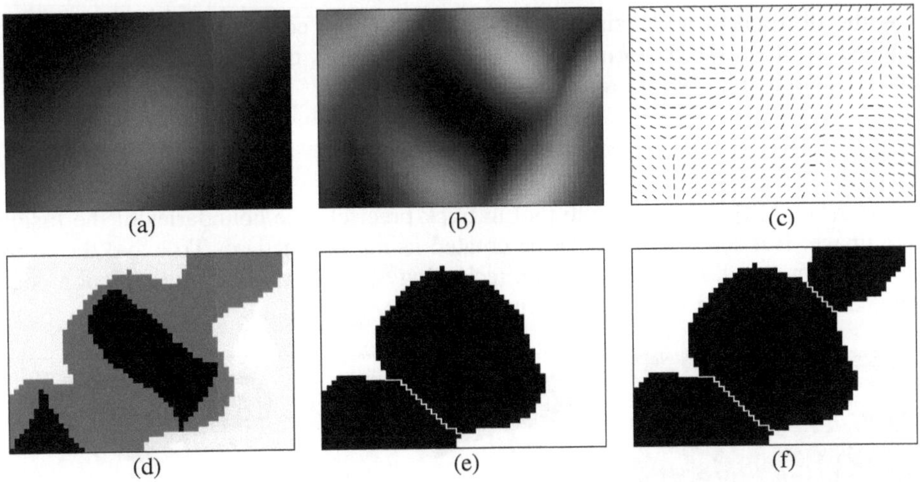

Fig. 2. Local patch round the middle toe (a) range image (b) the bigger principal curvature k2 (c) projection form the principal direction of k2 to the xoy plane (d) pre-segmentation result (e) result after growing once (f) result after iterative growing

In Fig. 1 and Fig. 2, we mark the interesting patches with black areas and the borderlines between two adjacent patches are drawn manually with white curves for clearly demarcated. The segmentation result of Fig. 2 shows that even if there is no peak in the index toe part, the interesting patch of index toe can still be found.

2.3 Multi-scale Fusion

The difficulties of computerized analysis of footprint are mainly caused by the instabilities of ground conditions and physical behavior. To lessen the disturbances of noises and instable factors, in section 2.1 we fit the footprint surface with different discrete orthogonal bi-quadratic polynomials which have different window sizes. Based on these fitting results with different window sizes, we can get different segmentation results. Then we fuse these results to get a more stable result.

Suppose that the footprint surface is fitted with a window of $l \times l$ size. With this fitting window we evaluate the curvatures, then after pre-segmentation and iteratively region growing the segmentation result of this size can be obtained. Let the bigger principal curvature is depicted with k_{l2} and the edge point set of the segmentation result is S_l. We define an edge strength image E_l to represent the edge information of the footprint. For any pixel of E_l, we calculate the value of $g(d(i, j, m, n))k_{l2}(m,n)$ to measure the edge impact of the edge point (m, n) on the pixel (i,j), where $d(i,j,m,n)$ is the Euclidean distance between (i,j) and (m,n) and $g(\cdot)$ is a monotone descending function at the range of $[0,+\infty)$. Then we choose the maximum edge impact from all the edge points as the value of $E_l(i,j)$:

$$E_l(i, j) = \max_{(m,n)\in S_l} g(d(i, j, m, n))k_{l2}(m, n) \tag{2}$$

With t different window sizes l_1, l_2, …, l_m, t different edge strength images can be obtained for the same footprint. A fusion edge image is calculated by averaging the edge strength images of different sizes:

$$E(i,j) = \frac{1}{t}\sum_{s=1}^{t} E_{l_s}(i,j) \tag{3}$$

Watershed [13] is an utilizable tool to track precisely the boundaries of the fusion edge image. But some shivers can be created using this method. To avoid the over-segmentation, geodesic reconstruction technology [14] is applied on the fusion edge image before the watershed processing (see Fig. 3).

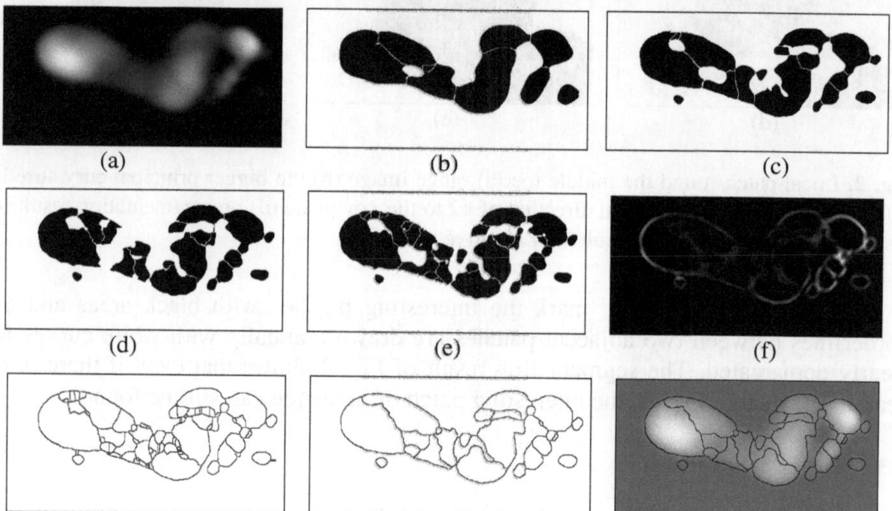

(a) (b) (c) (d) (e) (f) (g) (h) (i)

Fig. 3. Segmentation and fusion results (a) interposed range image (b)(c)(d)(e) segmentation results with 49×49, 25×25, 13×13 and 7×7 window operator (f) fusion image of five segmentation results above (g) watershed segmentation result of the fusion image (h) geodesic reconstruction and watershed segmentation result of the fusion image (i) the rang image covered with segmentation result

3 Footprint Description

As the complexity of the footprint shape and the multiplicity of the segmentation result, we use the man-machine interactive method to select five toes, heavy pressed patches, the foot arch area and the heel region from the segmentation results. It can guarantee the correctness of the characteristic regions, and is of advantage to merge over-segmentation regions.

After the segmentation, an effective and reasonable description to the footprint is another key to obtain the stable characteristic for further recognition. Although the footprint is easy to be interfered by ground conditions and personal behaviors, it contains the skeleton and muscle characteristic of each person. To effectively extract the biometric features of the footprint, we use different parameter surface to represent each regions of the footprint surface. According to the shape characteristic of each patch, we use superquadric fitting for five toes, heavy pressed patches and the heel region, and saddle (a hyperbolic paraboloid) fitting for the foot arch area.

Superquadrics are three-dimensional models suitable for part-level representation of objects, and it can be reconstructed from range images. A superquadric surface is defined by the following implicit equation

$$\left(\left(\frac{x}{a_1}\right)^{2/\varepsilon_2}+\left(\frac{y}{a_2}\right)^{2/\varepsilon_2}\right)^{\varepsilon_2/\varepsilon_1}+\left(\frac{z}{a_3}\right)^{2/\varepsilon_1}=1 \tag{4}$$

where a_1, a_2 and a_3 define the superquadric size, and ε_1 and ε_2 define a smoothly changing family of shapes from rounded to square. Saddles also can be reconstructed from range images. A saddle surface is defined by the following equation

$$(ax)^2-(by)^2=cz \tag{5}$$

According to the shape of the foot arch, we set $c = 1$.

To recover above surfaces in a general position, we should add six parameters: Euler angle (ϕ,θ,ψ) define the orientation in space, and p_x, p_y, p_z define the position in space. The superquadrics and saddles for general position is defined as following implicit equation:

$$F(x,y,z;a_1,a_2,a_3,\varepsilon_1,\varepsilon_2,\phi,\theta,\psi,p_x,p_y,p_z)=0 \tag{6}$$

$$G(x,y,z;a,b,\phi,\theta,\psi,p_x,p_y,p_z)=0 \tag{7}$$

For superquadric and saddle models are nonlinear, the Levenberg-Marquardt method allows one to iteratively find their parameters [15][16][17].

Fig. 4 shows the superquadric and saddle fitting results for the range image of Fig. 3(a). By surface fitting, we describe the foot shape characteristics with several groups of superquadric and saddle parameters, which can be taken as the feature parameters for foot recognition.

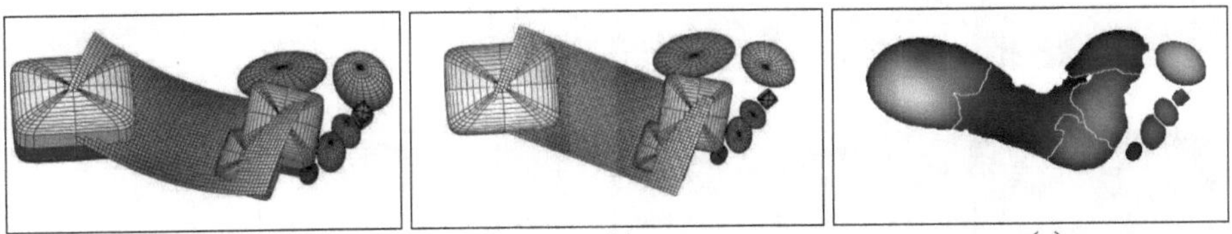

(a) (b) (c)

Fig. 4. Superquadrics and saddles fitting results of Fig. 3(a). (a)(b) the 3-dimensional fitted models of each regions from different viewpoints. (c) a result obtained by replacing the toe and foot arch regions with the fitted data.

4 Experiments and Conclusion

The curvature sign based segmentation method is the most generally used in the segmentation of range surface. So we compare the result of our method with the curvature sign image. We test our method on a number of footprint range images and obtain excellent segmentation results. Fig 5 gives some instances of segmentation results of our method and the corresponding curvature sign images [10]. It can be seen that our method can track the boundaries correctly and the extracted interesting patches comprise significant biometric information which is the foundation of feature analysis and recognition.

The proposed segmentation method for stereo bare footprint has the following merits:

1. The feature parameters used in the segmentation method, including the principle curvatures and directions, contain the entire geometric characteristics of the surface. The characteristics used in the paper are invariable to the translation and rotation.
2. The growing principle can accurately track the boundaries of the interesting patches. For those interesting patches that have no peak, the region can also be detected and segmented. The under-segmentation, such as two or more toes being embodied in the same patch, is on the whole avoided.
3. The applications of multi-scale fusion, geodesic reconstruction and watershed extract the stable edge information efficiently and simultaneously remove most fake edges caused by noises.
4. The superquadric and saddle parameters describe the biometric characteristics and fade the influence of ground conditions and personal behaviors.

The experiment results show that the performance of our segmentation method is excellent. The segmentation results provide a secure platform for the future works on biometric feature extraction and recognition.

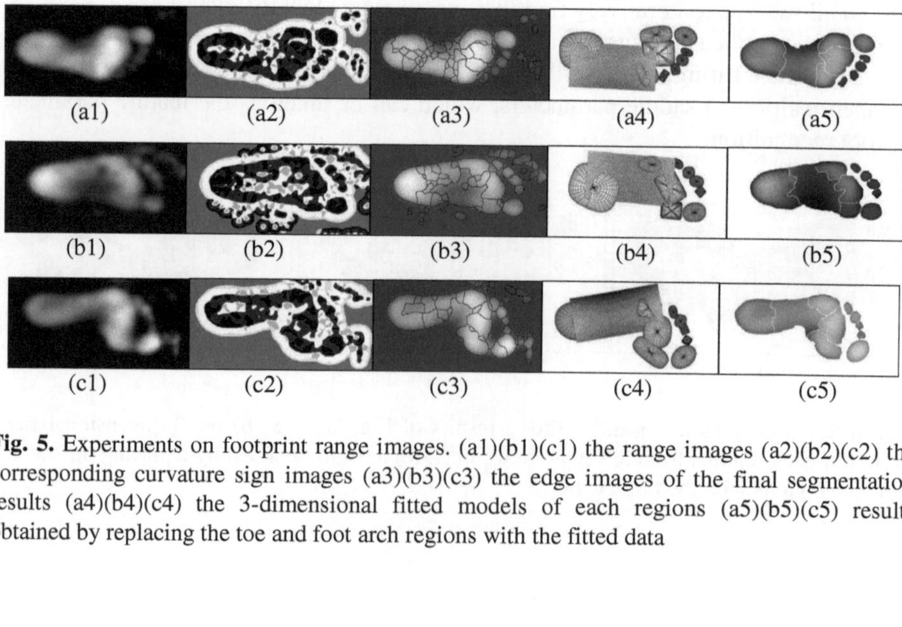

Fig. 5. Experiments on footprint range images. (a1)(b1)(c1) the range images (a2)(b2)(c2) the corresponding curvature sign images (a3)(b3)(c3) the edge images of the final segmentation results (a4)(b4)(c4) the 3-dimensional fitted models of each regions (a5)(b5)(c5) results obtained by replacing the toe and foot arch regions with the fitted data

Acknowledgements

This research is supported by the National Natural Science Foundation of China under Grant No. 60272004.

References

1. Chirillo, John., Blaul, S.: Implementing Biometric Security. John Wiley & Sons. (2003)
2. Wang, Q.J., Han, J.L., Zheng, D.C., Wu Z.L.: The Quantitative Inspection of Footprint and Gait. Chinese People's Public Security University Press (1992) (in Chinese)
3. Tian, Y., Ping X.J., Wang Y.J.: A New Method about 3D Surface Recognition. Proceedings of the 5th Joint Conference on Information Sciences. 1 (2000) A147–A150
4. Nakajima, K., Mizukami, Y., Tanaka, K., Tamura, T.: Footprint-Based Personal Recognition. IEEE Trans. on Biomedical Engineering. 47 (2000) 1534–1537.
5. Kennedy, R.B., Pressman, S., Chen, S., Petersen, P.H., Pressman, A.E.: Statistical Analysis of Barefoot Impressions. Journal of Forensic Sciences. 48 (2003) 55–63
6. Fan, T., Medioni, G., Nevatia, R.: Segmented description of 3-D surfaces. IEEE Journal of Robotics and Automation, , 3 (1987) 527-538
7. Bellon, O.R.P., Silva, L.: New Improvements to Range Image Segmentation by Edge Detection. IEEE Signal Processing Letters. 9 (2002) 43–45
8. Jiang, X.Y., Bunke, H.: Edge Detection in Range Images Based on Scan Line Approximation. Computer Vision and Image Understanding. 73 (1999) 183–199
9. Zhu, S.C., Yuille, A.: Region Competition: Unifying Snakes, Region Growing, and Bayes/ MDL for Multi-band Image Segmentation. IEEE Trans on PAMI. 18 (1996) 884–900
10. Besl, P.J., Jain, R.C.: Segmentation Through Variable Order Surface Fitting. IEEE Trans. on PAMI. 10 (1988) 167–192
11. Koster, K., Spann, M.: MIR: An Approach to Robust Clustering Application to Range Image Segmentation. IEEE Trans. on PAMI. 22 (2000) 430–444
12. Yokoya, N., Levine, D.: Range Image Segmentation Based on Differential Geometry: A Hybrid Approach. IEEE Trans. on PAMI. 11 (1989) 634–649
13. Vincent, L., Soille, P.: Watersheds in Digital Spaces: An Efficient Algorithm Based on Immersion Simulations. IEEE Trans. on PAMI. 13 (1991) 583–598
14. Vincent, L.: Morphological Grayscale Reconstruction in Image Analysis: Applications and Efficient Algorithms. IEEE Trans. on Image Processing. 2 (1993) 176–201
15. Leonardis, A., Jaklic, A., Solina, F.: Superquadrics for Segmenting and Modeling Range Data. IEEE Trans. on PAMI. 19 (1997) 1289-1295
16. Whaite, P., Ferrie, F. P.: From Uncertainty to Visual Exploration. IEEE Trans. on PAMI. 13 (1991) 1038-1049
17. Press, W. H., Teukolsky, S. A., Vetterling, W. T., Flannery, B. P.: Numerical Recipes in C: The Art of Scientific Computing. London: Cambridge University Press, 1992

Human Ear Recognition from Face Profile Images

Mohamed Abdel-Mottaleb* and Jindan Zhou

Department of Electrical & Computer Engineering, University of Miami,
1251 Memorial Dr., Coral Gables, FL 33146

Abstract. In this paper, we present a novel system for ear identification from profile images of the face. The system has two steps. In the first step, the ear is automatically detected from the profile image of the face. In the second step, the ear image is transformed to a force field, then feature points are extracted and the best match is found from a database. We propose a method based on differential geometry to extract ear feature points. We use a transformation of the ear image to make it suitable for extracting the feature points using differential geometry. During recognition, the feature points obtained from a query image are aligned and compared with those in the database using Hausdorff distance. The experimental results show that our method is effective.

1 Introduction

Ear biometric has the advantage of being invariable with respect to changes in face expressions as compared to face recognition [4]. In 1989, Alfred Ianarelli [8], conducted a study on personal identification using ear features. Iannarelli gathered over 10000 ear images and used 12 features to represent each ear. His experimental results showed that indeed ear features vary from person to person in such a manner that they can be used to distinguish between people. His work provided a foundation for the feasibility of using the ear shape for biometrics.

Some approaches have been developed for ear recognition from 2D intensity images. Chang et al. [10] used principal component analysis for ear recognition. Hurley et al. [4][13] applied force field transform to ear images, then, used the positions of the extracted energy wells as features for recognition.. Burge and Burger [9] used Voronoi diagram and curve segments with a novel subgraph matching algorithm for ear authentication. Recently, Chen and Bhanu [11][12] proposed a local surface descriptor and used the ICP (Iterative Closest Point) procedure for matching 3D ear data from range images.

In this paper, we propose a biometric system based on the shape of the ear. The system automatically detects the ear region from color profile images of the face, and then extracts ear features and applies matching for ear recognition. For feature extraction, we apply a force field transform to the detected ear region and extract, from the transformed image, a set of feature points to represent the ear. The recognition is performed by calculating the Hausdorff distance between the set of feature points that represent the query and those that represent every candidate in a database.

---

* Corresponding author.

D. Zhang and A.K. Jain (Eds.): ICB 2006, LNCS 3832, pp. 786 – 792, 2005.

The rest of the paper is organized as follows: Section 2 describes the details of our ear identification system. This includes our method for ear segmentation, feature extraction and matching. Section 3 presents the experimental results of ear identification. The conclusions are given in Section 4.

2 Ear Identification System

2.1 Ear Detection

The goal of the this step is to extract a sub-image that contains only the ear region from the profile image of a person's face. The detection of the ear is achieved in three steps: 1) skin-tone region detection, 2) edge detection within the region of the skin-tone followed by size filtering operation, 3) candidate ear region detection using Hausdorff matching with a simplified model of the ear. Then, only one candidate region, which has the minimum mean square error with an average ear image, is selected as the detected ear region. Figure 1 shows an example with the results of the three-step process of ear detection. The details are given in the following subsections.

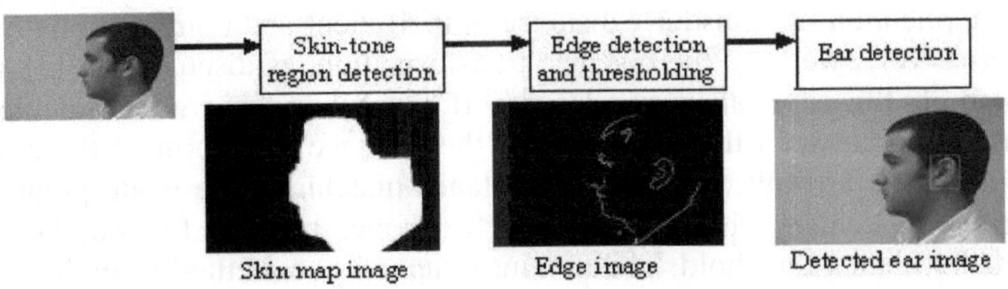

Fig. 1. System diagram for ear detection from profile images

2.1.1 Skin-Tone Region Segmentation

We use Flek's method [1] for skin-tone detection to locate the face region. This skin detection method uses a skin filter which relies on color and texture information.

Since the ear region is not as smooth as other skin regions, sometimes it will not be intact in the output of the skin filter. Therefore, we need to enlarge the output skin region to include the whole ear; this is achieved by applying a morphological dilation operation to the output of the skin filter.

2.1.2 Edge Detection on Skin Region

Canny edge detection is applied only to the region where skin is detected. The result is then filtered using an edge size filter to remove short and isolated edges that appear due to noise. Figure 2 shows an example of the results.

2.1.3 Ear Detection Using Template Matching

In this step, the ear region is detected using template matching. We use a simplified ear template that consists only of an edge that represents the ear's helix, as shown in Figure 3. Then, we use Hausdorff distance [2] to search for the template in the edge

Fig. 2. Edge detecting in the skin-tone region

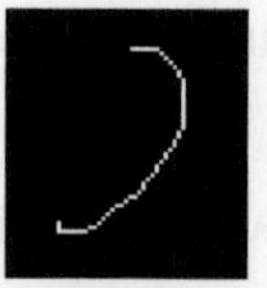

Fig. 3. Left ear's helix edge template

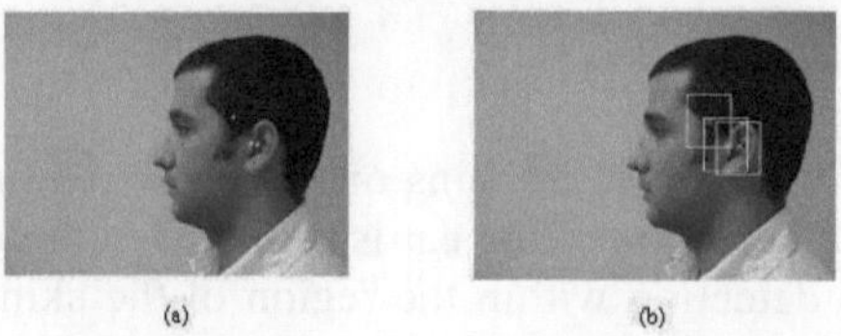

Fig. 4. Template matching and non-max suppression

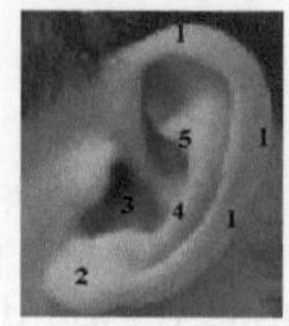

Fig. 5. Anatomy of ear 1. Helix 2. Lobulo 3. Concha 4. Pliegue inferior 5. Pilegue anterior

map of the skin region, while allowing for affine transformation between the template and the data. The details of Hausdorff distance that we use for ear detection are similar to those we use for the recognition step and are given in Section 2.3.

Direct application of Hausdorff distance is inefficient and computationally expensive. Alternatively, we used the distance transformation, as discussed in [3], to solve this problem. In this case calculating Hausdorff distance is achieved by calculating the cross-correlation between the template and the distance transform of the edge map. Figure 4(a) shows a result of Hausdorff distance matching. The white points are the cross-correlation centers where the Hausdorff distance between the template and the area is smaller than a threshold. The non-max suppression method is applied to prune the matching results as shown in Figure 4(b), where only three points remain.

To obtain the correct ear region from the candidate regions, the candidate regions are scaled to the same size as an average ear image, and then we compute MSE (Mean Square Error) between each detected candidate region and the average ear image. The average ear image is defined as the average intensity of a set of training ear images. The detected region is the candidate region that gives the minimum MSE.

2.2 Ear Feature Extraction

Figure 5 shows the anatomy of the ear, the most distinctive features of the human ear are the shapes and locations of these anatomy structures, such as helix rim, inner ear, antihelix, lobule etc. The following sections present our method for feature extraction.

2.2.1 Force Field Transformation for Ear Feature Extraction

We use a differential geometry based method to extract feature points from ear images. In this approach, the 2D intensity image $f(x, y)$ is treated as a surface. The surface is defined as $S = (x, y, z)$, where the z value is represented by the image intensity at pixel (x, y), e.g. $z = f(x, y)$. Under this representation, the most distinctive features of the ear can be considered as surface feature points that lie in ridge-ravine lines. The extraction of ridges and ravines of a surface is based on differential geometry of the surface and need computation of principle curvatures.

Since the calculation of principle curvatures involves second derivatives. To get a robust feature extraction results, the image needs to be smoothed while preserving the important features. We use an energy transformation method called force field transformation, which was used in [4], to preprocess the ear image before feature extraction. In this method, the image is transformed by considering the image to consist of an array of Gaussian attractors, which act as the sources of force field. It smoothes the original grayscale ear image and preserves the important structure of the ear as well.

The force field energy function method transforms the whole ear image into a force field by assuming that each pixel exerts an isotropic force on all the other pixels in the image [4]. The equations below summarize the force field transformation.

$$F_i(r_j) = P(r_i)\frac{r_i - r_j}{|r_i - r_j|^3} \tag{1}$$

$$F(r_j) = \sum_{i \in 0, N-1 | i \neq j} F_i(r_j) = \sum_{i \in 0, N-1 | i \neq j} P(r_i)\frac{r_i - r_j}{|r_i - r_j|^3} \tag{2}$$

Where N is the number of pixels in the image, and $F_i(r_j)$ is the exerted force on a pixel at position r_j by any other pixel at position r_i with pixel intensity $P(r_i)$. The force is proportional to the pixel intensity and inversely proportional to the distance from the pixel. Figure 6 is an example of the force field energy transformation. From the surface representation, it is clear that the transformed image is much smoother than the original image and preserves the image features as well.

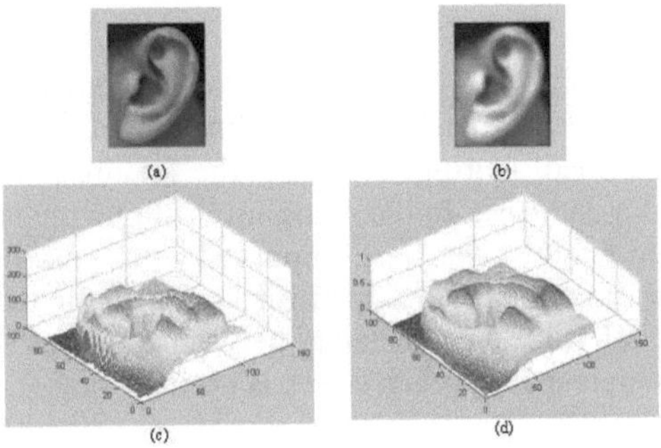

Fig. 6. Force field transformation (a) Original grayscale image and it's surface representation (c). (b) The force field transformation and it's surface representation (d).

2.2.2 Ear Image as Surface for Feature Extraction

Assuming a surface is represented by the function $z = f(x, y)$, the most important curvatures of the surface include the Gaussian curvature K, mean curvature H, principal curvatures k_1 and k_2. The extraction procedure involves calculations of the second derivative as follows, where $\vec{n}$ is the surface normal [5][6]:

$$\vec{n} = \frac{(-f_x, -f_y, 1)}{\sqrt{1 + f_x^2 + f_y^2}} \quad K = \frac{(f_{xx}f_{yy} - f_{xy}^2)}{(1 + f_x^2 + f_y^2)^2} \tag{3}$$

$$H = \frac{f_{xx} + f_{yy} + f_{xx} f_y^2 + f_{yy} f_x^2 - 2 f_x f_y f_{xy}}{2(1 + f_x^2 + f_y^2)^{1.5}}$$

$$k_1 = H + \sqrt{H^2 - k}, \quad k_2 = H - \sqrt{H^2 - k}$$

Our goal is to extract and use the points lying in ridge-ravine line as the feature points. These points correspond to the extreme ridge points on the considered surface. An extreme ridge point is a point where k_1, has a local positive maximum. There are different ways to locate the ridges, here we threshold the k_1 value to find these points.

Figure 7 shows three examples for extracted feature points using the proposed method based on the force field energy image, where the feature points are superimposed on the energy images.

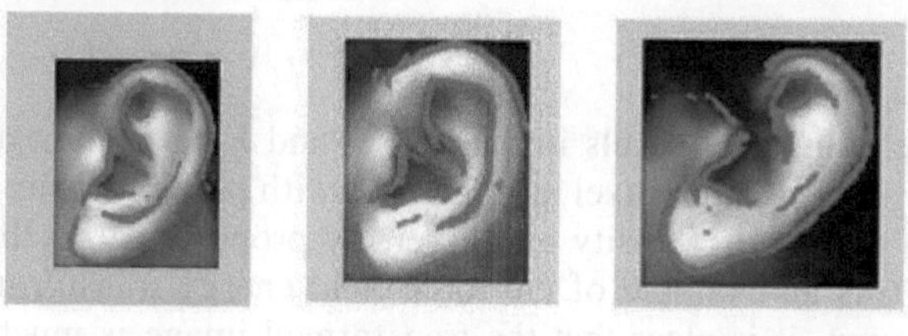

Fig. 7. Extracted feature points superimposed on the energy images

2.3 Recognition

Like most biometric systems, our ear identification system has two stages: off-line stage to prepare and archive the database and online stage in which recognition of an unknown query is performed. During archiving, the feature points extracted from an ear image are stored in a database. In the recognition stage, the test ear image is processed, and the extracted feature points are compared with those in the ear database for a match. We apply bi-directional Hausdorff distance [2] for the matching criterion. The partial bi-directional Hausdorff distance between the two sets of feature points that are transformed with respect to each other, is defined as:

$$H_{LK}(T(P), P') = \max(h_L(P', T(P)), h_K(T(P), P')) \tag{4}$$

Where *P* is the ear's feature points from the test image, *T*(*P*) is the transformation of *P*, which can be represented as:

$$\begin{aligned} T(P) &= A * P + t \\ &= \begin{pmatrix} \cos\theta, & \sin\theta \\ -\sin\theta, & \cos\theta \end{pmatrix} * \begin{pmatrix} s & 0 \\ 0 & s \end{pmatrix} * \begin{pmatrix} x \\ y \end{pmatrix} + \begin{pmatrix} t_x \\ t_y \end{pmatrix} \end{aligned} \tag{5}$$

P' is the set of feature potions of an ear in the database, $h_k(T(P), P')$ is the partial directed distance from *T*(*P*) to *P*', where $1 \leq K \leq q$ (*q* is the number of points of *P*), and $h_L(P', T(P))$ is the partial directed distance from *P*' to *T*(*P*), $1 \leq L \leq q'$ (*q*' is the number of points of *P*'). The definition of partial directed distance, e.g., $h_k(T(P), P')$ denotes the Kth ranked value in the set of distances from *T*(*P*) to *P*'. This process automatically selects the K best matching points in $T(P)$, which means that matching only takes a portion of the feature points into consideration. In practice, to compute the partial directed distance, we specify some fraction f_1, and f_2, where $0<f_1, f_2<1$, and let $K=f_1\times q$, $L=f_2\times q'$. The fractions f_1 and f_2 determine how many missing points between the two sets of feature points can be tolerated.

The matching is performed by finding a rigid transformation, (i.e., rotation, scale and translation), that results in a minimum Hausdorff distance. The database ear image with the minimum Hausdorff distance to the test ear image is considered the best match. The minimum Hausdorff distance is found using Quadratic Programming optimization method [7].

3 Results

For the experiments, we collected a database of profile face images for 103 subjects. For each person, one image was used for training, where the ear region was detected and the extracted ear feature points were stored in the database. During recognition, we used different images from the ones used for the training. The recognition starts by locating the ear region in the test image and extracting the feature points. Then, the feature points are compared with all the sets stored in the database using Hausdorff distance.

The proposed ear recognition method was applied to 58 query images for 29 subjects, two images from each subject. Out of the 58 queries, 51 obtained the correct match as the first match. For the remaining 7 queries, four queries had their correct matches among the top three best matches, while for the other three queries, two obtained their correct matches in the top five best matches and one was ranked the 10th top match.

Our ear identification system was also tested for computational efficiency. The process of ear detection from profile images takes on average 4.6 seconds, i.e., 13 images per minute. The processes of ear recognition takes 3.1 second to calculate the Hausdorff distance between the feature points of a query and the feature points of a database image (this is the average time on the database of the 103 ear images). The result of the computational time is based on a Pentium 4/3GH processor desktop and Matlab platform. The computational time is expected to be much lower if a more efficient language is used.

4 Conclusions

We proposed an ear recognition system for human identification using profile face images. The proposed system includes a new automated method for ear detection. It also includes a new method for extracting ear features by using differential geometry approach applied to a transform of the ear image. The experiments show that the system is effective for ear recognition.

References

1. M. Fleck, D. Forsyth, and C. Bregler, "Finding Naked People," European Conference on Computer Vision, vol. 2, pp. 592-602, 1996.
2. D. P. Huttenlocher, G. A. Klanderman, W. J. Rucklidge, "Comparing images using the Hausdorff distance," IEEE Trans. on PAMI., vol. 15, no. 9, pp. 850-863, 1993.
3. G. Wei, D. Li, I.K. Sethi, "Detection of side-view faces in color images," Proc. Fifth IEEE Workshop on Applications of Computer Vision, pp. 79-84. Dec, 2000.
4. D. J. Hurley, M. S. Nixon and J. N. Carter, "Force Field Energy Functionals for Image Feature Extraction," In Proceedings of Proc. 10th British Machine Vision Conference BMVC99 2, pp. 604-613.
5. R. Huang, T.L. Kumii, "Parallel algorithms for extracting ridges and ravines," proceedings of the First Aizu International Symposium on Parallel Algorithms/Architecture Synthesis. Mar. 1995.
6. A. B. Hamza and H. Krim, "A topological variational model for image singularities," Proc. 2002 IEEE International Conference on Image Processing, 2002.
7. S. P. Han, "A globally convergent method for nonlinear programming," Journal of Optimization Theory and Applications, vol. 22, pp. 297, 1977.
8. A. Ianarelli. "Ear Identification," Forensic Identification Series. Paramount Publishing, California 1989.
9. M. Burge and W. Burger, "Ear Biometrics in Computer Vision," In the 15th International Conference of Pattern Recognition, ICPR 2000, pp. 826-830.
10. K. Chang, K. Bowyer and V. Barnabas, "Comparison and combination of ear and face images in appearance-based biometrics," IEEE Trans. Pattern Analysis and Machine Intelligence. vol. 25, pp.1160–1165, 2003.
11. B. Bhanu and H. Chen, "Human ear recognition in 3D," in Workshop on Multimodal User Authentication, pp. 91–98, 2003.
12. H. Chen and B. Bhanu, "Human Ear Detection from Side Face Range Images," International Conference of Pattern Recognition, ICPR 2004, vol3, p574-577.
13. D. J. Hurley, M. S. Nixon, and J. N. Carter, "Force field feature extraction for ear biometrics," Computer Vision and Image Understanding, 2005.

Author Index

Vol. 3797: S. Maitra, C. E. V. Madhavan, R. Venkatesan (Eds.), Progress in Cryptology - INDOCRYPT 2005. XIV, 417 pages. 2005.

Vol. 3796: N.P. Smart (Ed.), Cryptography and Coding. XI, 461 pages. 2005.

Vol. 3795: H. Zhuge, G.C. Fox (Eds.), Grid and Cooperative Computing - GCC 2005. XXI, 1203 pages. 2005.

Vol. 3794: X. Jia, J. Wu, Y. He (Eds.), Mobile Ad-hoc and Sensor Networks. XX, 1136 pages. 2005.

Vol. 3793: T. Conte, N. Navarro, W.-m.W. Hwu, M. Valero, T. Ungerer (Eds.), High Performance Embedded Architectures and Compilers. XIII, 317 pages. 2005.

Vol. 3792: I. Richardson, P. Abrahamsson, R. Messnarz (Eds.), Software Process Improvement. VIII, 215 pages. 2005.

Vol. 3791: A. Adi, S. Stoutenburg, S. Tabet (Eds.), Rules and Rule Markup Languages for the Semantic Web. X, 225 pages. 2005.

Vol. 3790: G. Alonso (Ed.), Middleware 2005. XIII, 443 pages. 2005.

Vol. 3789: A. Gelbukh, Á. de Albornoz, H. Terashima-Marín (Eds.), MICAI 2005: Advances in Artificial Intelligence. XXVI, 1198 pages. 2005. (Sublibrary LNAI).

Vol. 3788: B. Roy (Ed.), Advances in Cryptology - ASIACRYPT 2005. XIV, 703 pages. 2005.

Vol. 3787: D. Kratsch (Ed.), Graph-Theoretic Concepts in Computer Science. XIV, 470 pages. 2005.

Vol. 3785: K.-K. Lau, R. Banach (Eds.), Formal Methods and Software Engineering. XIV, 496 pages. 2005.

Vol. 3784: J. Tao, T. Tan, R.W. Picard (Eds.), Affective Computing and Intelligent Interaction. XIX, 1008 pages. 2005.

Vol. 3783: S. Qing, W. Mao, J. Lopez, G. Wang (Eds.), Information and Communications Security. XIV, 492 pages. 2005.

Vol. 3782: K.-D. Althoff, A. Dengel, R. Bergmann, M. Nick, T. Roth-Berghofer (Eds.), Professional Knowledge Management. XXIII, 739 pages. 2005. (Sublibrary LNAI).

Vol. 3781: S.Z. Li, Z. Sun, T. Tan, S. Pankanti, G. Chollet, D. Zhang (Eds.), Advances in Biometric Person Authentication. XI, 250 pages. 2005.

Vol. 3780: K. Yi (Ed.), Programming Languages and Systems. XI, 435 pages. 2005.

Vol. 3779: H. Jin, D. Reed, W. Jiang (Eds.), Network and Parallel Computing. XV, 513 pages. 2005.

Vol. 3778: C. Atkinson, C. Bunse, H.-G. Gross, C. Peper (Eds.), Component-Based Software Development for Embedded Systems. VIII, 345 pages. 2005.

Vol. 3777: O.B. Lupanov, O.M. Kasim-Zade, A.V. Chaskin, K. Steinhöfel (Eds.), Stochastic Algorithms: Foundations and Applications. VIII, 239 pages. 2005.

Vol. 3776: S.K. Pal, S. Bandyopadhyay, S. Biswas (Eds.), Pattern Recognition and Machine Intelligence. XXIV, 808 pages. 2005.

Vol. 3775: J. Schönwälder, J. Serrat (Eds.), Ambient Networks. XIII, 281 pages. 2005.

Vol. 3774: G. Bierman, C. Koch (Eds.), Database Programming Languages. X, 295 pages. 2005.

Vol. 3773: A. Sanfeliu, M.L. Cortés (Eds.), Progress in Pattern Recognition, Image Analysis and Applications. XX, 1094 pages. 2005.

Vol. 3772: M. Consens, G. Navarro (Eds.), String Processing and Information Retrieval. XIV, 406 pages. 2005.

Vol. 3771: J.M.T. Romijn, G.P. Smith, J. van de Pol (Eds.), Integrated Formal Methods. XI, 407 pages. 2005.

Vol. 3770: J. Akoka, S.W. Liddle, I.-Y. Song, M. Bertolotto, I. Comyn-Wattiau, W.-J. van den Heuvel, M. Kolp, J. Trujillo, C. Kop, H.C. Mayr (Eds.), Perspectives in Conceptual Modeling. XXII, 476 pages. 2005.

Vol. 3769: D.A. Bader, M. Parashar, V. Sridhar, V.K. Prasanna (Eds.), High Performance Computing – HiPC 2005. XXVIII, 550 pages. 2005.

Vol. 3768: Y.-S. Ho, H.J. Kim (Eds.), Advances in Multimedia Information Processing - PCM 2005, Part II. XXVIII, 1088 pages. 2005.

Vol. 3767: Y.-S. Ho, H.J. Kim (Eds.), Advances in Multimedia Information Processing - PCM 2005, Part I. XXVIII, 1022 pages. 2005.

Vol. 3766: N. Sebe, M.S. Lew, T.S. Huang (Eds.), Computer Vision in Human-Computer Interaction. X, 231 pages. 2005.

Vol. 3765: Y. Liu, T. Jiang, C. Zhang (Eds.), Computer Vision for Biomedical Image Applications. X, 563 pages. 2005.

Vol. 3764: S. Tixeuil, T. Herman (Eds.), Self-Stabilizing Systems. VIII, 229 pages. 2005.

Vol. 3762: R. Meersman, Z. Tari, P. Herrero (Eds.), On the Move to Meaningful Internet Systems 2005: OTM 2005 Workshops. XXXI, 1228 pages. 2005.

Vol. 3761: R. Meersman, Z. Tari (Eds.), On the Move to Meaningful Internet Systems 2005: CoopIS, DOA, and ODBASE, Part II. XXVII, 653 pages. 2005.

Vol. 3760: R. Meersman, Z. Tari (Eds.), On the Move to Meaningful Internet Systems 2005: CoopIS, DOA, and ODBASE, Part I. XXVII, 921 pages. 2005.

Vol. 3759: G. Chen, Y. Pan, M. Guo, J. Lu (Eds.), Parallel and Distributed Processing and Applications - ISPA 2005 Workshops. XIII, 669 pages. 2005.

Vol. 3758: Y. Pan, D.-x. Chen, M. Guo, J. Cao, J.J. Dongarra (Eds.), Parallel and Distributed Processing and Applications. XXIII, 1162 pages. 2005.

Vol. 3757: A. Rangarajan, B. Vemuri, A.L. Yuille (Eds.), Energy Minimization Methods in Computer Vision and Pattern Recognition. XII, 666 pages. 2005.

Vol. 3756: J. Cao, W. Nejdl, M. Xu (Eds.), Advanced Parallel Processing Technologies. XIV, 526 pages. 2005.

Vol. 3754: J. Dalmau Royo, G. Hasegawa (Eds.), Management of Multimedia Networks and Services. XII, 384 pages. 2005.

Vol. 3753: O.F. Olsen, L.M.J. Florack, A. Kuijper (Eds.), Deep Structure, Singularities, and Computer Vision. X, 259 pages. 2005.

Vol. 3752: N. Paragios, O. Faugeras, T. Chan, C. Schnörr (Eds.), Variational, Geometric, and Level Set Methods in Computer Vision. XI, 369 pages. 2005.